Human Molecular Genetics

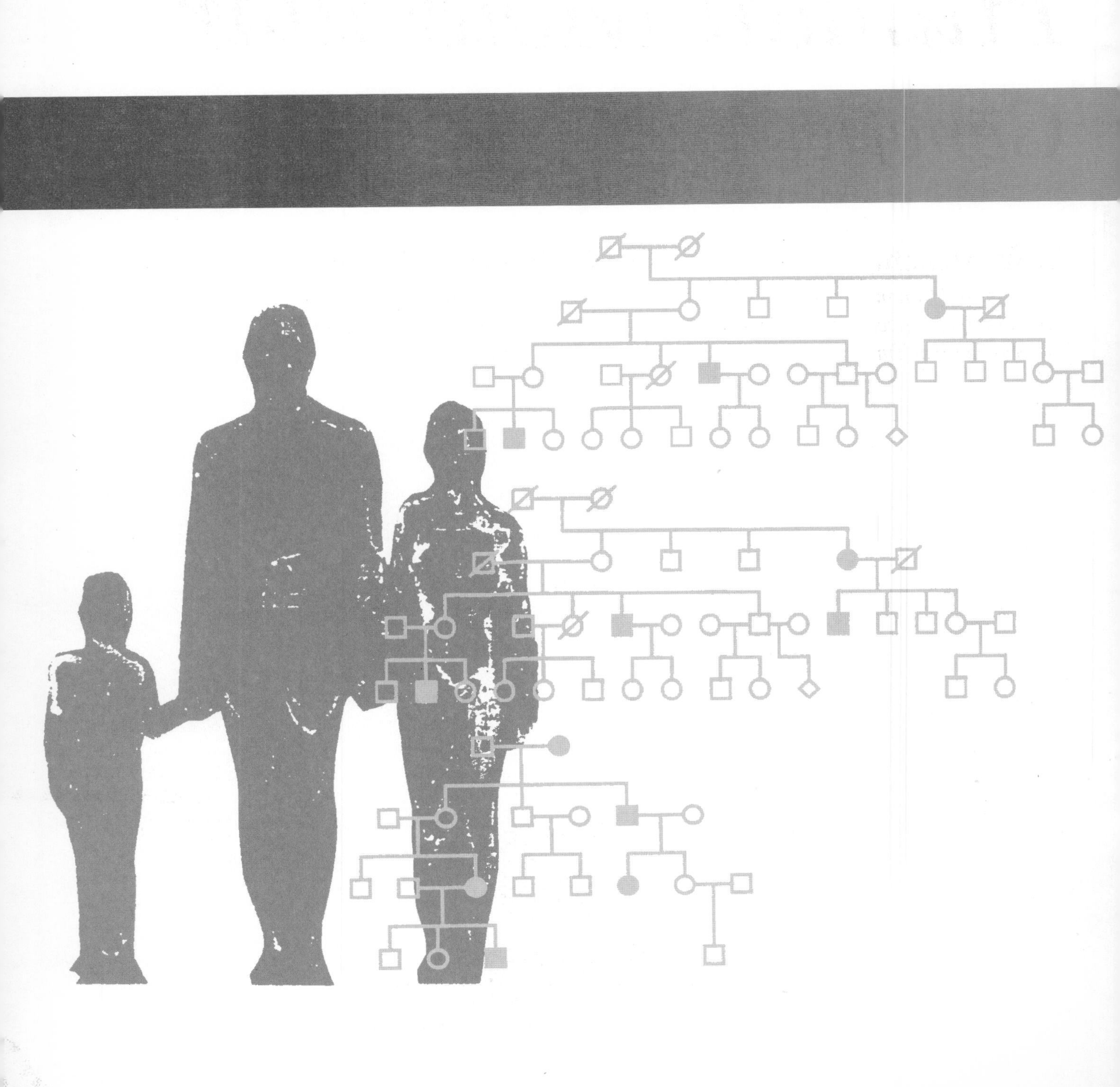

Human Molecular Genetics

Tom Strachan *and* **Andrew P. Read**

Professor of Human Molecular Genetics
University of Newcastle upon Tyne
Newcastle upon Tyne
UK

Professor of Human Genetics
University of Manchester
Manchester
UK

WILEY-LISS

A JOHN WILEY & SONS, INC., PUBLICATION
New York • Chichester • Brisbane • Toronto • Singapore

Published in the United States of America, its dependent territories and Canada by arrangement with BIOS Scientific Publishers Ltd, 9 Newtec Place, Magdalen Road, Oxford OX4 1RE, UK.

First published 1996

A CIP catalog record for this book is available from the British Library.

Library of Congress Cataloging in Publication Data
Strachan, T.
Human molecular genetics / by Tom Strachan and Andrew Read.
p. cm.
Includes index.
ISBN 0-471-13373-6 (pbk. ; alk. paper)
1. Molecular genetics. I. Read, A.P. (Andrew P.) II. Title.
QH442.S775 1996 95–52801
CIP

ISBN 0 471 13373 6

Typeset by AMA Graphics Ltd, Preston, UK.
Printed by Butler and Tanner Ltd, Frome, UK.

Contents

Section B. Fundamentals of DNA cloning and molecular hybridization

Section C. Features of the human genome

12 Genetic mapping 313

13 The Human Genome Project 335

Section E. Human genetic diseases

14 Identifying human disease genes 367

15 Molecular pathology 401

Section F. Dissecting and manipulating genes

Abbreviations

AAV	adeno-associated virus
ADA	adenosine deaminase
AFP	α-fetoprotein
AIDS	acquired immune deficiency syndrome
AMCA	aminomethylcoumarin acetic acid
APRT	adenosine phosphoribosyltransferase
ARMS	amplification refractory mutation system
ARS	autonomously replicating sequence
AS	Angelman syndrome
AT	ataxia telangiectasia
ASO	allele-specific oligonucleotide
BAC	bacterial artificial chromosome
BCG	bacille Calmette-Guérin
BMD	Becker muscular dystrophy
bp	base pair
cAMP	cyclic AMP
CCM	chemical cleavage of mismatches
cDNA	complementary DNA
CDR	complementarity-determining region
CEN	centromere element
CEPH	Centre d' Études du Polymorphisme Humaine
CF	cystic fibrosis
CGH	comparative genome hybridization
CGRP	calcitonin gene-related peptide
CHLC	co-operative human linkage center
cM	centiMorgan
CMGT	chromosome-medicated gene transfer
CMT	Charcot–Marie–Tooth disease
CNS	central nervous system
cR	centiRay
cRNA	complementary RNA
DAPI	4′, 6-diamidino-2-phenylindole
ddNTP	dideoxynucleoside triphosphate
DGGE	denaturing gradient gel electrophoresis
DM	myotonic dystrophy
DMD	Duchenne muscular dystrophy
DNase	deoxyribonuclease
dNTP	deoxynucleoside triphosphate
ddNTP	dideoxynucleoside triphosphate
DOP-PCR	degenerate oligonucleotide-primed polymerase chain reaction
DZ	dizygotic
EBV	Epstein–Barr virus
EMSA	electrophoretic mobility shift assay
ENU	ethylnitrosourea
ER	endoplasmic reticulum
ES cell	embryonic stem cell
EST	expressed sequence tag
EtBr	ethidium bromide
FH	familial hypercholesterolemia
FISH	fluorescence *in situ* hybridization
FITC	fluorescein isothiocyanate
FRAXA	fragile-X syndrome
ftp	file transfer protocol
GAT	gene augmentation therapy
GDB	Genome Data Base
GM-CSF	granulocyte–macrophage colony-stimulating factor
GSS	Gerstmann–Straussler–Scheinker disease
HAT	hypoxanthine–aminopterin–thymidine
HD	Huntington disease
HERV	human endogenous retrovirus
HIV	human immunodeficiency virus
HLA	human leukocyte antigen
HLH	helix–loop–helix
HPFH	hereditary persistence of fetal hemoglobin
HNPCC	hereditary nonpolyposis colon cancer
HPRT	hypoxanthine phosphoribosyltransferase
HRR	haplotype relative risk
HSR	homogeneously staining region
HSV	herpes simplex virus
HTH	helix–turn–helix
HTLV-1	human T-lymphotrophic virus 1
http	hypertext transmission protocol
HUGO	The Human Genome Organization
IBD	identical by descent
IBS	identical by state
IFN	interferon

Ig	immunoglobulin
IGF	insulin-like growth factor
IL	interleukin
IRE	(i) iron-response element, or (ii) interspersed repeat element
IRE-PCR	interspersed repeat element polymerase chain reaction
IRP	island rescue polymerase chain reaction
ISCN	International System for Human Cytogenetic Nomenclature
kb	kilobase
LCR	locus control region
LDL	low density lipoprotein
LHON	Leber's hereditary optic atrophy
LINE	long interspersed nuclear element
LoH	loss of heterozygosity
LTR	long terminal repeat
mAb	monoclonal antibody
MAGP	microfibril-associated glycoprotein
Mb	megabase
MCS	multiple cloning site
MER	medium reiteration frequency
MIM	Mendelian Inheritance in Man
MIN	minisatellite instability
MODY	maturity-onset diabetes of the young
mRNA	messenger RNA
MZ	monozygotic
NF	neurofibromatosis
NIH	National Institutes of Health (USA)
OD	optical density
ODN	oligodeoxynucleotide
OLA	oligonucleotide ligation assay
OMIM	On-line Mendelian Inheritance in Man
ORF	open reading frame
PAC	phage P1-derived artificial chromosome
PAGE	polyacrylamide gel electrophoresis
PAR	pseudoautosomal region
PCR	polymerase chain reaction
PEG	polyethylene glycol
PFD	polyostotic fibrous dysplasia
PFGE	pulsed-field gel electrophoresis
PIC	polymorphism information content
PKU	phenylketonuria
PTT	protein truncation test
Pu	purine
PWS	Prader–Willi syndrome
Py	pyrimidine
RACE	rapid amplification of cDNA ends
RDA	representational difference analysis
rDNA	ribosomal DNA
RE	response element
RER	rough endoplasmic reticulum
RF	replicative form
RFLP	restriction fragment length polymorphism
RH	radiation hybrid
rNTP	ribonucleoside triphosphate
RP	retinitis pigmentosa
rRNA	ribosomal RNA
RSP	restriction site polymorphism
RT	reverse transcriptase
RTLV	retrovirus-like element
RT-PCR	reverse transcriptase polymerase chain reaction
SAR	scaffold attachment region
SCA1	spinocerebellar ataxia type 1
SDA	sex-determining allele
SDS	sodium dodecyl (lauryl) sulfate
SINE	short interspersed nuclear element
snRNA	small nuclear RNA
snRNP	small nuclear ribonucleoprotein
SRP	signal recognition particle
SSCA	single-strand conformation analysis
SSCP	single-strand conformation polymorphism
STRP	short tandem repeat polymorphism
STS	sequence-tagged sites
SV40	simian virus 40
TCR	T-cell receptor
TDT	transmission disequilibrium test
TFO	triplex-forming oligonucleotide
THE	transposable human element
TIGR	The Institute of Genome Research
TIL	tumor-infiltrating lymphocyte
TK	thymidine kinase
T_m	melting temperature
TNF	tumor necrosis factor
TPA	tissue plasminogen activator
TRITC	tetramethylrhodamine isothiocyanate
tRNA	transfer RNA
TS	tumor suppressor
UPD	uniparental disomy
URL	uniform resource locator
UTS	untranslated sequence
UTR	untranslated region
UV	ultraviolet
VNTR	variable number of tandem repeats
VPC	vector-producing cell
WWW	world-wide web
Xgal	5-bromo, 4-chloro-3-indolyl β-D-galactopyranoside
YAC	yeast artificial chromosome

Preface

The idea for this book grew from two earlier efforts, *The Human Genome* (TS; BIOS Scientific Publishers, 1992) and *Medical Genetics, an Illustrated Outline* (APR; Gower Medical Publishing, 1989). In these small books we tried to develop a treatment of human genetics based on understanding the structure and function of the normal human genome. Traditionally, textbooks of human genetics tended to start by considering meiosis and the way diseases segregate in pedigrees, whilst textbooks of molecular genetics rarely emphasized human topics. Until recently this was inevitable because so little was understood about the normal human genome. The Human Genome Project has changed all that, and the present book is an attempt to provide a comprehensive integrated study of human molecular genetics.

It would be hard to overstate the importance for biomedical science of the Human Genome Project. As the first 'big science' project in biology, it forms the focus for a mass collective effort by thousands of researchers world-wide to move our understanding of biology on to a new plane. Human molecular genetics not only forms the cutting edge of biomedical research, but at the same time it has immediate application to the diagnosis of disease and has great potential for treating disease. Thus it is of major interest to all students of biological science and medicine, and to a wide range of biomedical researchers.

Human molecular genetics is a large subject. We have tried to make it more digestible by organizing the text into clearly demarcated sections, using statement headings to define what the reader can find in each section, and identifying important new terms by bold typeface (most of which, apart from basic terms, can be found in the Glossary). The first section (Chapters 1–3) provides introductory-level material on DNA, chromosomes and pedigree patterns but the second section (Chapters 4–6), which describes general principles and applications of cloning and molecular hybridization, contains some advanced examples. The third section (Chapters 7–10) is the one which we believe truly distinguishes this book from the others, and provides a comprehensive guide to the structure, function, evolution and mutational instability of the human genome and human genes. It provides a solid base for relating the subsequent sections on mapping the human genome (Chapters 11–13), studying human genetic diseases (Chapters 14–18), and dissecting and manipulating genes (Chapters 19 and 20).

Currently, the pace of research in this field is extremely rapid, and new information and insights are pouring out of the laboratories at an almost unimaginable rate. At the time of writing, new partial human gene sequences (expressed sequence tags) are being added to the databases at a rate of 8000 per week. To cope with this flood, students need two tools: a good framework of principles and the ability to use modern informatics. One of our aims has been to encourage students to use the wonderful genetic resources available on the Internet (in particular, we have provided accession numbers for the OMIM online database for genetic diseases, rather than listing references).

Revolutionary times are nothing if not exciting. We have tried to convey the feel of fast-moving research, while providing a description in some depth of the techniques and data that are helping us to understand the evolution, nature and function of our genome. This book will have succeeded if readers finish it sharing our excitement and enthusiasm for the continuing voyage of discovery into our DNA. The journey is far from finished.

Tom Strachan and Andrew P. Read

Acknowledgments

We owe thanks to many colleagues who have read drafts and contributed thoughts and material, particularly Mike Jackson (University of Newcastle upon Tyne, UK), Susan Lindsay (University of Newcastle upon Tyne, UK), Riccardo Fodde (University of Leiden, The Netherlands), David Craufurd (University of Manchester, UK), Nalin Thakker (University of Manchester, UK), David Cooper (University of Wales, UK), Pedro Lowenstein (University of Wales, UK), Steve J. Higgins (University of Leeds, UK) and R.G.H. Cotton (The Murdoch Institute, Australia). We thank Margaret Weddle (University of Newcastle upon Tyne, UK) for administrative assistance. We would also like to thank Barbara Czepulkowski and Julie Howard (King's College Hospital, London) for providing the *in situ* hybridization image which is used on the front cover. Jonathan Ray, Lisa Mansell and Rachel Robinson at BIOS Scientific Publishers have done their best to keep us on the rails, and we are very grateful for all their work and patience.

We are indeed grateful for the support, encouragement and patience of our families and we take this opportunity to express our gratitude to Meryl, Alex and James and to Gilly, respectively. Finally, we thank Professor Rodney Harris, CBE who was so important in fostering our early careers in human genetics when we were colleagues at the Department of Medical Genetics, University of Manchester.

DNA structure and function

Chapter 1

DNA, RNA and polypeptides are large polymers defined by a linear sequence of simple repeating units

DNA (deoxyribonucleic acid) and RNA (ribonucleic acid) are information macromolecules which are present in almost all cells. Genetic information is normally stored in DNA molecules which are used to synthesize RNA molecules (using selected DNA sequences as a template) and also to permit the synthesis of the polypeptides of proteins (using selected RNA molecules as templates).

Individual DNA molecules are found in the chromosomes of the nucleus and in the mitochondria of animal cells and the chloroplasts of plant cells. They are large polymers, with a linear backbone composed of residues of a 5-carbon sugar, deoxyribose, which are successively linked by covalent phosphodiester bonds. Covalently attached to carbon atom number 1′ (one prime) of each sugar residue is a nitrogenous base, either a **purine** (adenine, A, or guanine, G) or a **pyrimidine** (cytosine, C, or thymine, T). A sugar with an attached base and phosphate group therefore constitutes the basic repeat unit of a DNA strand, a **nucleotide** (*Figures 1.1* and *1.2*). The composition of RNA molecules is similar to that of DNA molecules, but differs in that they contain ribose sugar residues in place of deoxyribose and uracil (U) instead of thymine (*Figures 1.1* and *1.2*).

Proteins are composed of one or more polypeptide molecules which may be modified by the addition of various carbohydrate side chains or other chemical groups. Each polypeptide is a polymer consisting of a linear sequence of **amino acids**. The latter consist of a positively charged amino group and a negatively charged carboxylic acid (carboxyl) group connected by a central carbon atom to which is attached an identifying side chain. The 20 different amino acids can be grouped into different classes depending on the nature of their side chains (*Figure 1.3*). Classification is based as follows:

(i) **Basic** amino acids carry a side chain with a net positive charge; an amino (NH_2) group or histidine ring in the side chain acquires a H^+ ion at physiological pH.

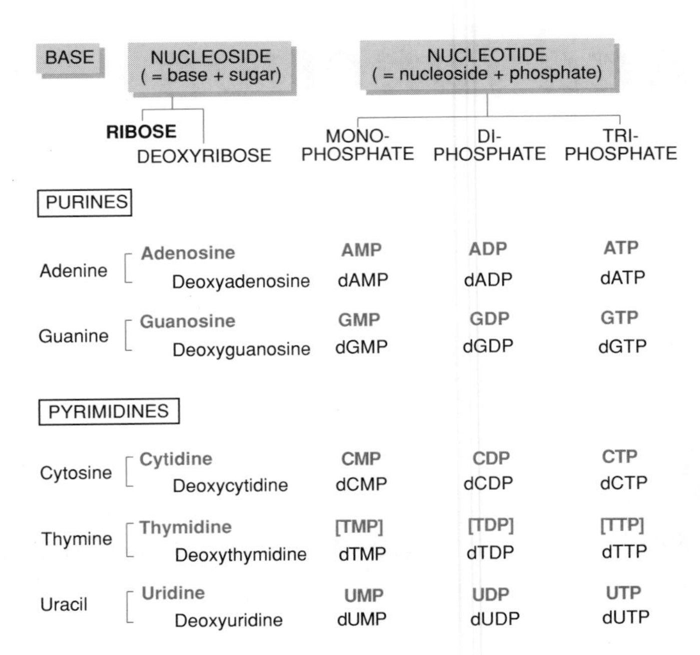

Figure 1.1: Common bases found in nucleic acids and corresponding ribo- and deoxyribonucleosides and nucleotides.

(ii) **Acidic** amino acids carry a side chain with a net negative charge; a carboxyl group (COOH) in the side chain loses a H^+ ion at physiological pH to form COO^-.

(iii) **Uncharged polar** amino acids are electrically neutral but carry side chains with polar chemical groups, which are distinguished by having fractional electronic charges [e.g. hydroxyl ($O^{\delta-}$–$H^{\delta+}$) groups, sulfhydryl ($S^{\delta-}$–$H^{\delta+}$) groups, etc.]. Like the basic and acidic amino acids, polar amino acids may react with other groups bearing electric charges.

(iv) **Nonpolar neutral** amino acids are **hydrophobic** (water-repelling). They often interact with one another and with other hydrophobic groups.

Polypeptides are formed by a condensation reaction between the amino group of one amino acid and the carboxyl group of the next to form a repeating backbone (–NH–CHR–CO–), where the R side chains differ from one amino acid to another (see *Figure 1.21*).

The stability of the nucleic acid and protein polymers is primarily dependent on the strong **covalent** bonds that connect the constituent atoms of their linear backbones. Because of the phosphate charges present in their component nucleotides, both DNA and RNA are negatively charged. Depending on their amino acid composition, proteins may carry a net positive charge (**basic** proteins) or a net negative charge (**acidic** proteins). However, even electrically neutral proteins are often readily soluble, like DNA and RNA molecules, if they contain an appreciable number of charged or neutral polar amino acids. In contrast, membrane-bound proteins are often characterized by a high content of hydrophobic amino acids which are thermodynamically more stable in the hydrophobic environment of a lipid membrane.

In addition to covalent bonds, a number of weak noncovalent interactions, notably hydrogen bonds (see below), occur in these m]acromolecules and help to stabilize their conformations. Typically, however, such noncovalent bonds are weaker than covalent bonds by a factor of more than 10. In addition, unlike covalent bonds whose strength is determined only by the particular atoms involved, the strength of

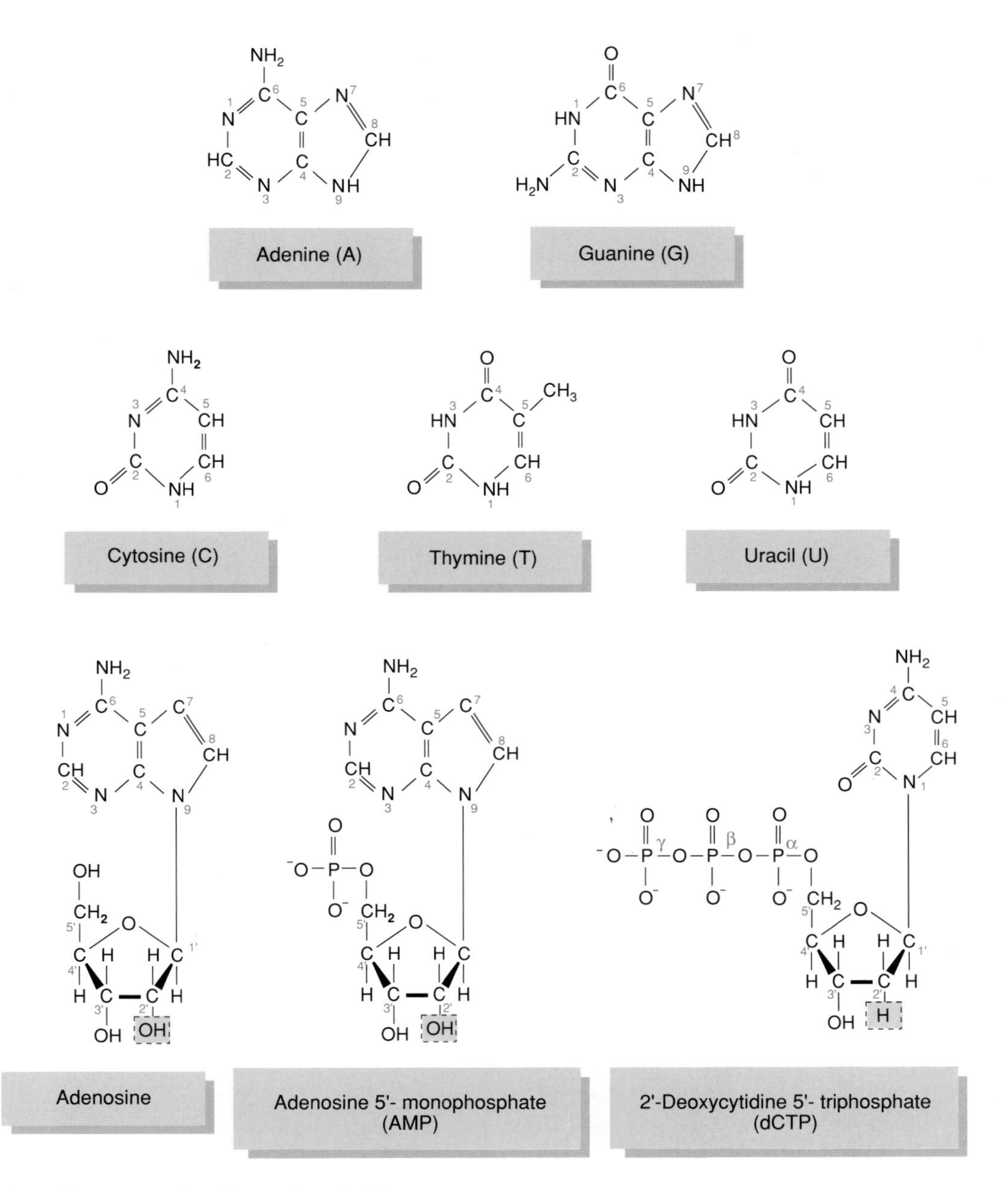

Figure 1.2: Structure of bases, nucleosides and nucleotides.

The bold lines at the bottom of the sugar rings are meant to indicate that the plane of the ring is set at an angle of 90° with respect to that of the corresponding base [i.e. if the plane of a base is represented as lying on the surface of the page, carbon atoms 2′ and 3′ (three **prime**) of the sugar can be viewed as projecting upwards out of the page and the oxygen atom as projecting down below the surface of the page]. *Note* that numbering in deoxyribose and ribose sugars is confined to the five carbon atoms, numbered 1′–5′, but the numbering of the bases includes both carbon and nitrogen atoms which occur within the heterocyclic rings. The highlighted hydroxyl and hydrogen atoms connected to carbon 2′ indicate the essential difference between the ribose and deoxyribose sugar residues. Phosphate groups are denoted sequentially as α, β, γ, etc., according to proximity to the sugar ring.

noncovalent bonds is crucially dependent on their aqueous environment. The structure of water is particularly complex: a rapidly changing network of noncovalent bonding occurs between the individual H_2O molecules. The predominant force in this structure is the **hydrogen bond**, a weak electrostatic bond formed between a partially positive hydrogen atom and a partially negative atom which, in the case of water molecules, is an oxygen atom.

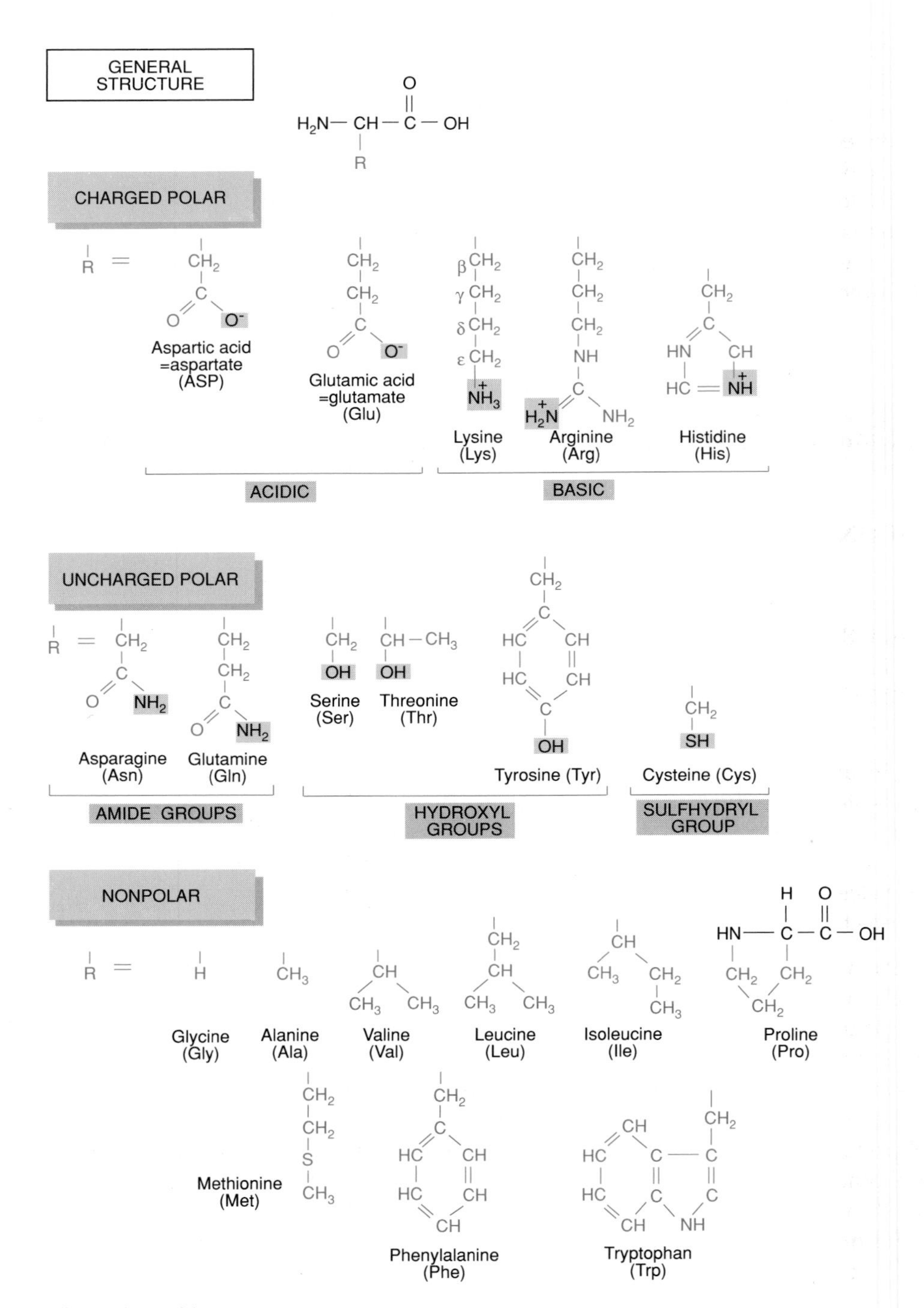

Figure 1.3: Structures of the amino acids.

Amino acids in a subclass (e.g. the acidic amino acids or the basic amino acids) are chemically very similar. Highlighted groups are polar chemical groups. The convention of numbering carbon atoms is to designate the central carbon atom as α and subsequent carbons of linear side chains as β, γ, δ, etc. (see figure for lysine side chain). Although, in general, polar amino acids are hydrophilic and nonpolar amino acids are hydrophobic, glycine (which has a very small side chain) and cysteine (whose sulfhydryl group is not so polar as an hydroxyl group) occupy intermediate positions on the hydrophilic $\rightleftharpoons$ hydrophobic scale. *Note* that proline is unusual in that the side chain connects the nitrogen atom of the NH_2 group as well as the central carbon atom.

The hydrogen bonding potential of water molecules means that molecules with polar groups (including DNA, RNA and proteins) can form multiple interactions with the water molecules, leading to their solubilization. Because many hydrogen bonds can form between polar groups in macromolecules such as nucleic acids and proteins, they can make large contributions to the stability of the structure (**conformation**) of these molecules. However, individual hydrogen bonds are weak and, unlike covalent bonds, can be broken readily under physiological conditions. This permits transient interactions between polar groups which can be of immense importance in the functions of nucleic acids and proteins (see below).

DNA structure and replication

The structure of DNA is an antiparallel double helix

The base composition of DNA from different cellular sources is not random: the amount of adenine equals that of thymine, and the amount of cytosine equals that of guanine. As a result, the base composition of DNA can be specified unambiguously by quoting its % GC (= % G + % C) composition. For example, if a source of DNA is quoted as being 42% GC, the base composition can be inferred to be: G, 21%; C, 21%; A, 29%; T, 29%.

Whereas RNA molecules normally exist as single molecules, the structure of DNA is a double helix in which two DNA molecules (**DNA strands**) are held together by weak hydrogen bonds to form a **DNA duplex**. Hydrogen bonding occurs between laterally opposed bases, **base pairs**, of the two strands of the DNA duplex according to Watson–Crick rules: adenine (A) specifically binds to thymine (T) and cytosine (C) specifically binds to guanine (G) (*Figure 1.4*).

Although DNA can adopt different structures, the most common one in nature is the B conformation. In this conformation, each DNA helix has a pitch (the distance occupied by a single turn of the helix) of 3.4 nm, accommodating 10 nucleotides. As the phosphodiester bonds link carbon atoms number 3′ and number 5′ of successive sugar residues, one end of each DNA strand, the so-called **5′ end**, will have a terminal sugar residue in which carbon atom number 5′ is not linked to a neighboring sugar residue (*Figure 1.5*). The other end is defined as the **3′ end** because of a similar absence of phosphodiester bonding at carbon atom number 3′ of the terminal sugar residue. The two strands of a DNA duplex are said to be **antiparallel** because they always associate (**anneal**) in such a way that the 5′ → 3′ direction of one DNA strand is the opposite to that of its partner (*Figure 1.5*).

Genetic information is encoded by the linear sequence of bases in the DNA strands (the **primary structure**). Consequently, two DNA strands of a DNA duplex are said to have **complementary** sequences (or to exhibit **base complementarity**) and the sequence of bases of one DNA strand can readily be inferred if the DNA sequence of its complementary strand is already known. It is usual, therefore, to describe a DNA sequence by writing the sequence of bases of one strand only, and in the 5′ → 3′ direction. This is the direction of synthesis of new DNA molecules during DNA replication, and also of transcription when RNA molecules are synthesized using DNA as a template (see below). However, when describing the sequence of a DNA region encompassing two neighboring bases (really a dinucleotide) on one DNA strand, it is usual to insert a 'p' to denote a connecting phosphodiester bond [e.g. CpG means that a cytidine is covalently linked to a neighboring guanosine on the

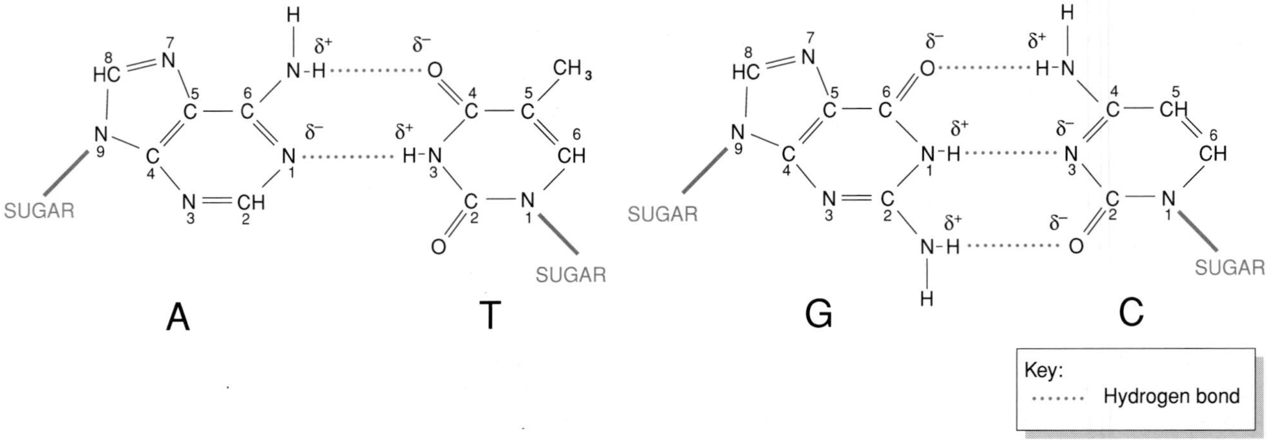

SUGAR SUGAR SUGAR SUGAR

A T G C

Key:
······ Hydrogen bond

Figure 1.4: A–T base pairs have two connecting hydrogen bonds; G–C base pairs have three.

Fractional positive charges on hydrogen atoms and fractional negative charges on oxygen and nitrogen atoms are denoted by δ^+ and δ^-, respectively.

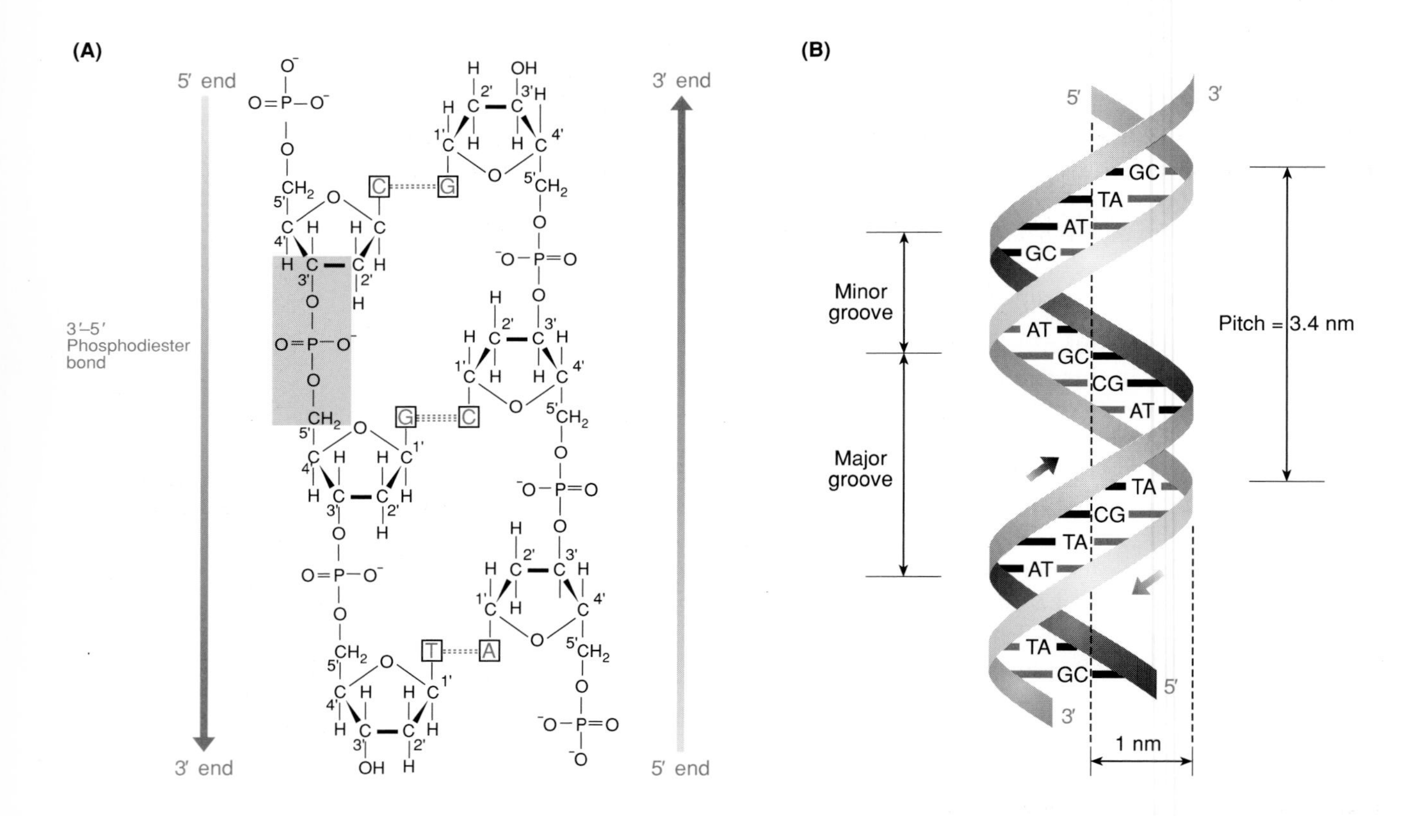

Figure 1.5: The structure of DNA is a double-stranded, antiparallel helix.

(A) Phosphodiester bonding and antiparallel nature of the two DNA strands. The two strands are **antiparallel** because they have opposite directions for linking of 3′ carbon atom to 5′ carbon atom. The structure shown is a double-stranded trinucleotide whose sequence can be represented as: 5′ pCpGpT–OH 3′ (DNA strand on left) / 5′ pApCpG–OH 3′ (DNA strand on right) (where p = phosphodiester bond and –OH = terminal OH group at 3′ end). This is normally abbreviated by deleting the 'p' *and* 'OH' symbols and giving the sequence on one strand only (e.g. the sequence could equally well be represented as 5′ CGT 3′ or 5′ ACG 3′).
(B) The double helical structure of DNA. *Note* that the two strands are wound round each other to form a plectonemic coil. The pitch of each helix represents the distance occupied by a single turn and accommodates 10 nucleotides.

same DNA strand, while a CG base pair means a cytosine on one DNA strand is hydrogen-bonded to a guanine on the complementary strand (see *Figure 1.5*)].

As well as double-stranded DNA helices, RNA–RNA and RNA–DNA double helices can form, given suitable complementary molecules. In addition, hydrogen bonding can occur between bases within a single DNA or RNA molecule. In the latter case, as in DNA, cytidine residues hydrogen-bond with guanosine residues, while adenine residues specifically base-pair with uracil residues. However, guanine residues can also occasionally pair with uracil residues, a form of base pairing which is not particularly stable, but does not significantly disrupt the RNA–RNA helix.

Sequences having closely positioned complementary inverted repeats are prone to forming **hairpin** structures or loops which are stabilized by hydrogen bonding between bases at the neck of the loop (*Figure 1.6*). Such structural constraints, which are additional to those imposed by the primary structure, contribute to the **secondary structure** of the molecule. Certain RNA molecules, such as transfer (t) RNA, show particularly high degrees of secondary structure (*Figure 1.6*).

DNA replication is semi-conservative and synthesis of DNA strands is semi-discontinuous

During the process of DNA synthesis (**DNA replication**), the two DNA strands of each chromosome are unwound by a helicase and each DNA strand directs the synthesis of a complementary DNA strand to generate two daughter DNA duplexes, each of which is identical to the parent molecule (*Figure 1.7*). As each daughter DNA duplex contains one strand from the parent molecule and one newly synthesized

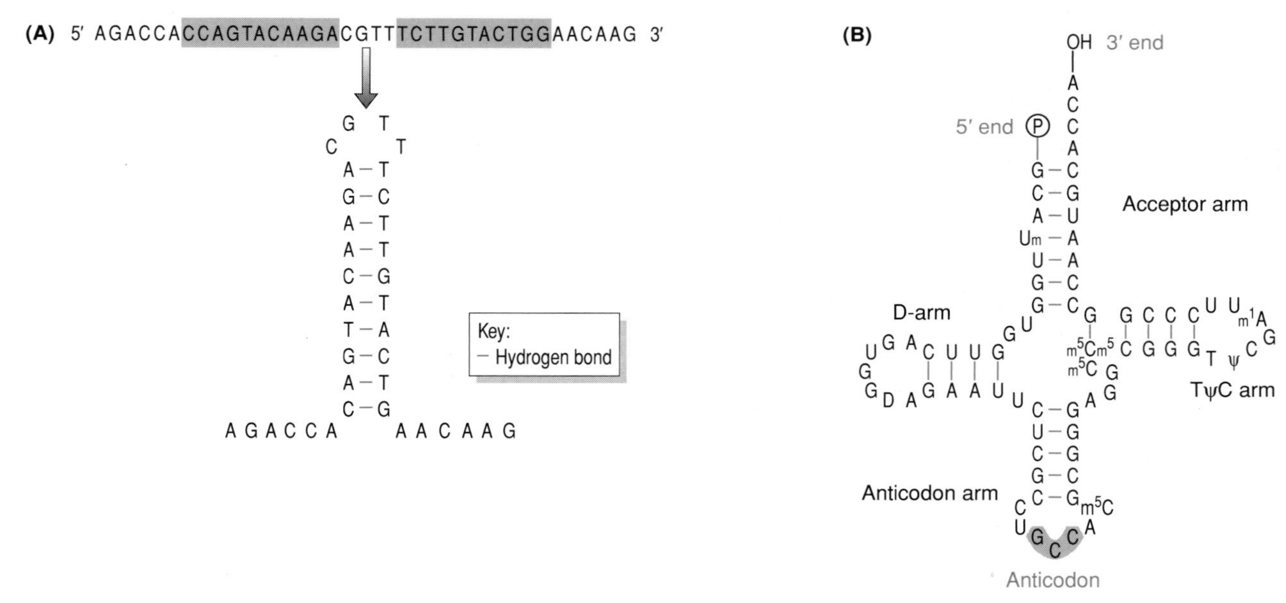

Figure 1.6: Intramolecular hydrogen bonding in DNA and RNA.

(A) Formation of a double-stranded hairpin loop within a single DNA strand. The highlighted sequences in the DNA strand above represent inverted repeat sequences which can hydrogen-bond to form a hairpin structure (below).
(B) Transfer RNA (tRNA) has extensive secondary structure. The example shown is a human $tRNA^{Glu}$ gene. *Note* the minor nucleosides: D, 5, 6-dihydrouridine; ψ, pseudouridine (5-ribosyl uracil); m^5C, 5-methylcytidine; m^1A, 1-methyladenosine. The cloverleaf structure is stabilized by extensive intramolecular hydrogen bonding, with conventional G–C and A–U base pairs, but also the occasional G–U base pair. Four arms are recognized: the **acceptor arm** is the one to which an amino acid can be attached (at the 3′ end); the **TψC arm** is defined by this trinucleotide; the **D arm** is named because it contains dihydrouridine residues; and the **anticodon arm** contains the anticodon trinucleotide in the center of the loop. The secondary structure of tRNAs is virtually invariant: there are always seven base pairs in the stem of the acceptor arm, five in the TψC arm, five in the anticodon arm, and three or four in the D arm.

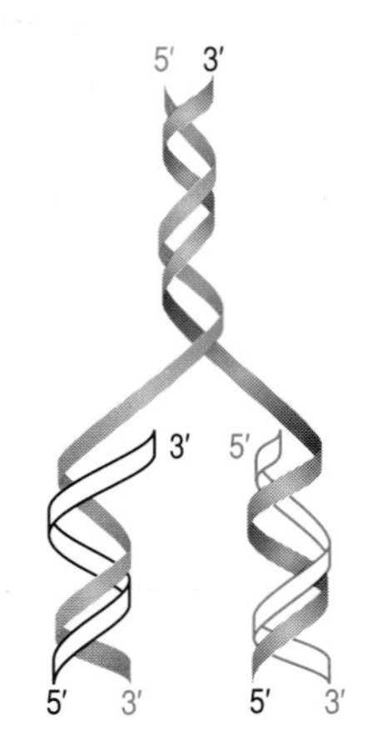

Figure 1.7: DNA replication is semi-conservative.

The parental DNA duplex consists of two complementary, antiparallel DNA strands (dark colors), which unwind and then individually act as templates for the synthesis of new complementary, antiparallel DNA strands (pale colors). Each of the daughter DNA duplexes contains one parental DNA strand and one new DNA strand, forming a DNA duplex which is structurally identical to the parental DNA duplex. *Note* that this figure shows the result of DNA duplication but not the way the process works (for which, see *Figure 1.8*).

DNA strand, the replication process is described as **semi-conservative**. The enzyme DNA polymerase catalyzes the synthesis of new DNA strands using the four deoxynucleoside triphosphates (dATP, dCTP, dGTP, dTTP) as nucleotide precursors.

DNA replication is initiated at specific points, which have been termed **origins of replication.** Starting from such an origin, the initiation of DNA replication results in a **replication fork**, where the parental DNA duplex bifurcates into two daughter DNA duplexes. As the two strands of the parental DNA duplex are antiparallel, but act individually as templates for the synthesis of a complementary antiparallel daughter strand, it follows that the two daughter strands must run in opposite directions [i.e. the direction of chain growth must be $5' \rightarrow 3'$ for one daughter strand, the **leading strand**, but $3' \rightarrow 5'$ for the other daughter strand, the **lagging strand** (*Figure 1.8*)].

The reactions catalyzed by DNA polymerases involve addition, to the free 3′ hydroxyl group of the growing DNA chain, of a dNMP moiety provided by a dNTP precursor [the two distal phosphate residues of the dNTP – that is, the β and γ residues (see *Figure 1.2*) – are cleaved and the resulting pyrophosphate group discarded]. This requirement introduces an asymmetry into the DNA replication process: only the leading strand will have a free 3′ hydroxyl group at the point of bifurcation. This will permit sequential addition of nucleotides and continuous elongation in the same direction in which the replication fork moves. However, synthesis of the lagging strand has to be accomplished as a progressive series of small (typically 100–1000 nucleotides long) fragments, often referred to as **Okazaki fragments**. As only the leading strand is synthesized continuously, the synthesis of DNA strands is said to be **semi-discontinuous**. Each fragment of the lagging strand is synthesized in the $5' \rightarrow 3'$ direction, which will be in the opposite direction to that in which the replication fork moves. Successively synthesized fragments are covalently joined at their ends using the enzyme DNA ligase so as to ensure chain growth in the direction of movement of the replication fork (*Figure 1.8*).

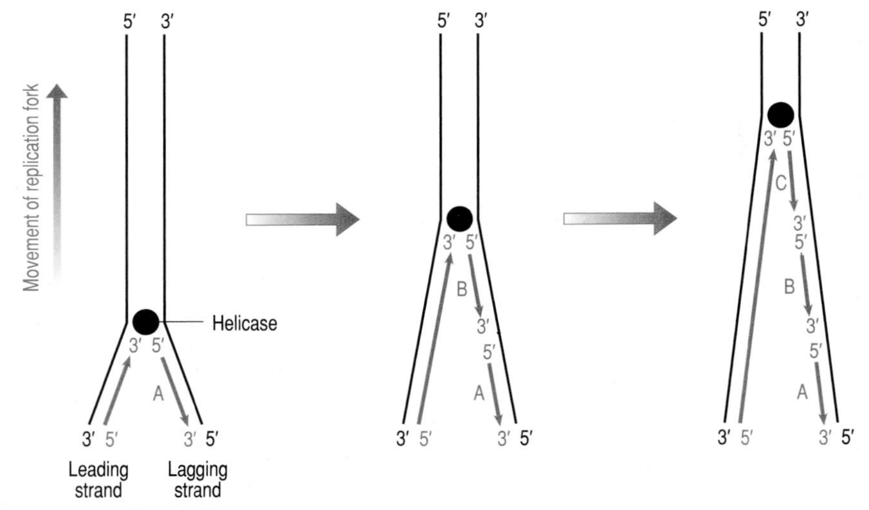

Figure 1.8: Asymmetry of strand synthesis during DNA replication.

Parental strands are shown in black, newly synthesized strands in red. The helicase unwinds the DNA duplex to allow the individual strands to be replicated. As replication proceeds, the $5' \rightarrow 3'$ direction of synthesis of the leading strand is the same as that in which the replication fork is moving, and so synthesis can be continuous. The $5' \rightarrow 3'$ synthesis of the lagging strand, however, is in the direction opposite to that of movement of the replication fork. It needs to be synthesized in pieces (Okazaki fragments), first A, then B, then C which are subsequently sealed by the enzyme DNA ligase to form a continuous DNA strand. *Note* that synthesis of these fragments is initiated using an RNA primer. In eukaryotic cells, the leading and lagging strands are synthesized, respectively, by DNA polymerases δ and α (see *Table 1.1*).

Five classes of mammalian DNA polymerases are known, including a polymerase that is dedicated to replication of the mitochondrial genome (*Table 1.1*). In individual mammalian chromosomes, DNA replication proceeds bidirectionally, to form **replication bubbles** from multiple initiation points. The distance between adjacent replication origins is about 50–300 kb, a distance which may be significant in chromosome structure (see page 36). At different origins, DNA replication is initiated at different times in the S phase of the cell cycle (see Chapter 2) but eventually neighboring replication bubbles will fuse (*Figure 1.9*). DNA replication is time-consuming: human cells in culture require about 8 h to complete the process.

Table 1.1: The five classes of mammalian DNA polymerase

	Class				
	α	β	γ	δ	ε
Location	Nuclear	Nuclear	Mitochondrial	Nuclear	Nuclear
Function	Synthesis and priming of lagging strand	DNA repair	Replicates mitochondrial DNA	Synthesis of leading strand	DNA repair
3′–5′ exonuclease	No	No	Yes	Yes	Yes

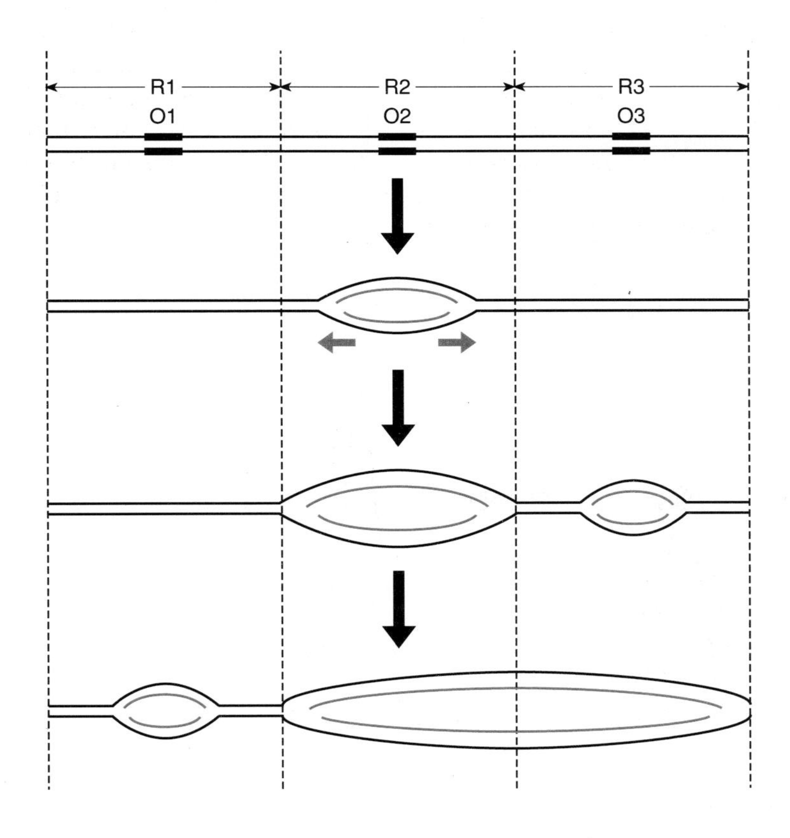

Figure 1.9: The chromosomes of complex organisms have multiple replication origins.

R1, R2 and R3 denote adjacent replication units (**replicons**) located on the same chromosome and have internally located origins of replication O1, O2 and O3, respectively. In this example, bidirectional replication is envisaged to proceed initially from O2, then O3 and finally O1. The bottom panel shows fusion of the replication bubbles initiated from O2 and O3, before the R1 replicon has completed replication.

RNA transcription and gene expression

The flow of genetic information is almost exclusively one way: DNA → RNA → protein

The expression of genetic information in all cells is very largely a one-way system: DNA specifies the synthesis of RNA and RNA specifies the synthesis of polypeptides, which subsequently form proteins. Because of its universality, the DNA → RNA → polypeptide (protein) flow of genetic information has been described as the **central dogma** of molecular biology. The first step, the synthesis of RNA using a DNA-dependent RNA polymerase, is described as **transcription** and occurs in the nucleus of eukaryotic cells and, to a limited extent, in mitochondria and chloroplasts (*Figure 1.10*). The second step, polypeptide synthesis, is described as **translation** and occurs in **ribosomes**, large RNA–protein complexes which are found in the cytoplasm and also in mitochondria and chloroplasts. The RNA molecules which specify polypeptide are known as **messenger RNA (mRNA)**. The expression of genetic information follows a **colinearity principle**: the linear sequence of nucleotides in DNA is decoded to give a linear sequence of nucleotides in RNA which can be decoded in turn to give a linear sequence of amino acids in the polypeptide product.

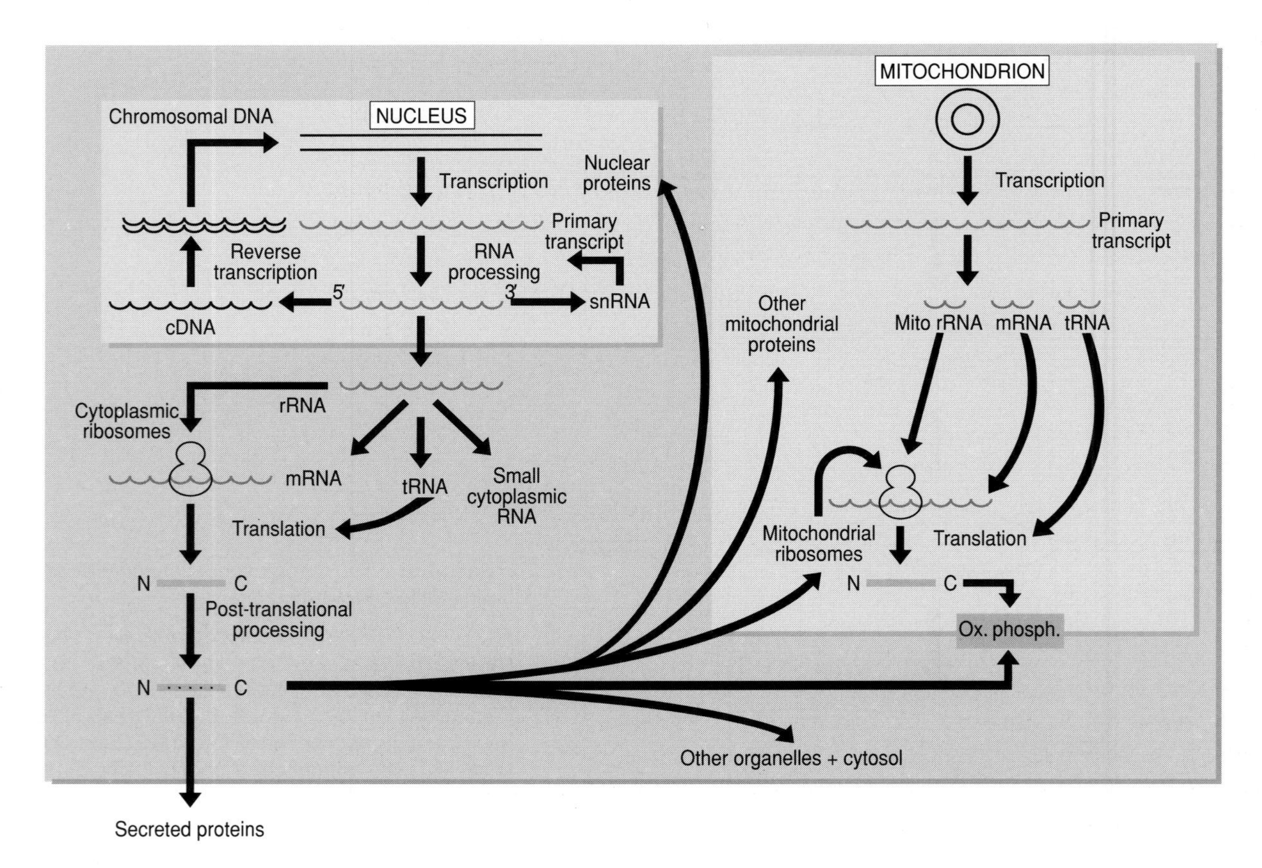

Figure 1.10: Gene expression in an animal cell.

Note that a small proportion of nuclear RNA molecules can be converted naturally to cDNA by virally encoded and cellular reverse transcriptases, and thereafter integrate into chromosomal DNA at diverse locations. *Note* also that the mitochondrion synthesizes its own rRNA and tRNA and a few proteins which are involved in the oxidative phosphorylation (Ox.phosph.) system. However, the proteins of mitochondrial ribosomes and the majority of the proteins in the mitochondrial oxidative phosphorylation system and other mitochondrial proteins are encoded by nuclear genes and translated on cytoplasmic ribosomes, before being imported into mitochondria.

Although DNA is the hereditary material in all present-day cells, it is most likely that early in evolution RNA served that function. RNA molecules can, like DNA, undergo self-replication. However, the 2′ hydroxyl group on the ribose residues of RNA makes the sugar–phosphate bonds comparatively unstable chemically. In DNA the deoxyribose residues carry only hydrogen atoms at the 2′ position and so DNA is much more suited than RNA to being a stable carrier of genetic information. Many different classes of present-day viruses nevertheless have a genome that consists of RNA, not DNA. Retroviruses such as HIV are a subclass of RNA viruses in which the RNA replicates via a DNA intermediate, using a **reverse transcriptase**, an RNA-dependent DNA polymerase. Recently it has become clear that eukaryotic cells, including mammalian cells, contain nonviral chromosomal DNA sequences which encode cellular reverse transcriptases (see page 198 for examples in the human genome). Because some nonviral RNA sequences are known to act as templates for cellular DNA synthesis, the principle of unidirectional flow of genetic information is no longer strictly valid.

Only a small fraction of the DNA in complex organisms is expressed to give a protein or RNA product

Only a small proportion of all the DNA in cells is ever transcribed. According to their needs, different cells transcribe different segments of the DNA (**transcription units**) which are discrete units, spaced irregularly along the DNA sequence. However, the great majority of the cellular DNA is never transcribed in any cell.

Moreover, only a portion of the RNA made by transcription is translated into polypeptide. This is because:

(i) the end product of some transcription units is not a polypeptide but an RNA molecule, as in the case of ribosomal RNAs (rRNAs), tRNAs, and diverse small nuclear (sn) and cytoplasmic RNA molecules;

(ii) the **primary transcript** (initial transcription product) of those transcription units which do encode polypeptides is subject to RNA processing events in which much of the initial RNA sequence is discarded to give a much smaller mRNA (page 17);

(iii) only a central part of the mature mRNA is translated; sections of variable length at each end of the mRNA remain untranslated (page 21).

Although DNA is found in both the nucleus and the mitochondria, the vast majority of the DNA of a cell is located in the chromosomes of the nucleus.

The fraction of **coding DNA** in the genomes of complex eukaryotes is rather small. This is partly a result of the noncoding nature of much of the sequence within genes. Another reason is that a considerable fraction of the genome of complex eukaryotes contains repeated sequences which are nonfunctional or which are not transcribed into RNA. The former include defective copies of functional genes (**pseudogenes** and **gene fragments**), and highly repetitive noncoding DNA, which includes sequences that are thought to contribute to specific chromosomal structures such as centromeres and telomeres (see Chapter 2).

Transcription is the process whereby genetic information in some DNA segments (genes) specifies the synthesis of RNA

RNA synthesis is accomplished using an RNA polymerase enzyme, with DNA as a template and ATP, CTP, GTP and UTP as RNA precursors. The RNA is synthesized as a single strand, with the direction of transcription being 5′ → 3′. Chain elongation occurs by adding the appropriate ribonucleoside monophosphate residue (AMP, CMP, GMP or UMP) to the free 3′ hydroxyl group at the 3′ end of the growing RNA chain. Such nucleotides are derived by splitting a pyrophosphate residue (PP_i) from the appropriate ribonucleoside triphosphate (rNTP) precursors. This means that the nucleotide at the extreme 5′ end (the initiator nucleotide) will differ from all others in the chain by carrying a 5′ triphosphate group.

Normally, only one of the two DNA strands acts as a template for RNA synthesis. Since the growing RNA chain is complementary to this **template strand**, the transcript has the same 5′ → 3′ direction and base sequence (except that U replaces T) as the opposite, nontemplate strand of the double helix. For this reason the nontemplate strand is often called the **sense strand**, and the template strand is often called the **antisense strand** (*Figure 1.11*). In documenting gene sequences it is customary to show only the DNA sequence of the sense strand. Orientation of sequences relative to a gene sequence is commonly dictated by the sense strand and by the direction of transcription (e.g. the 5′ end of a gene refers to sequences at the 5′ end of the sense strand, and sequences **upstream** or **downstream** of a gene refer to sequences which flank the gene at the 5′ or 3′, respectively, ends of the sense strand).

In eukaryotic cells, three different RNA polymerase molecules are required to synthesize the different classes of RNA (*Table 1.2*). The vast majority of cellular genes encode polypeptides and are transcribed by RNA polymerase II. Increasing importance, however, is being paid to genes which encode RNA as their mature product: functional RNA molecules are now known to play a wide variety of roles, and catalytic functions have been ascribed to some of them.

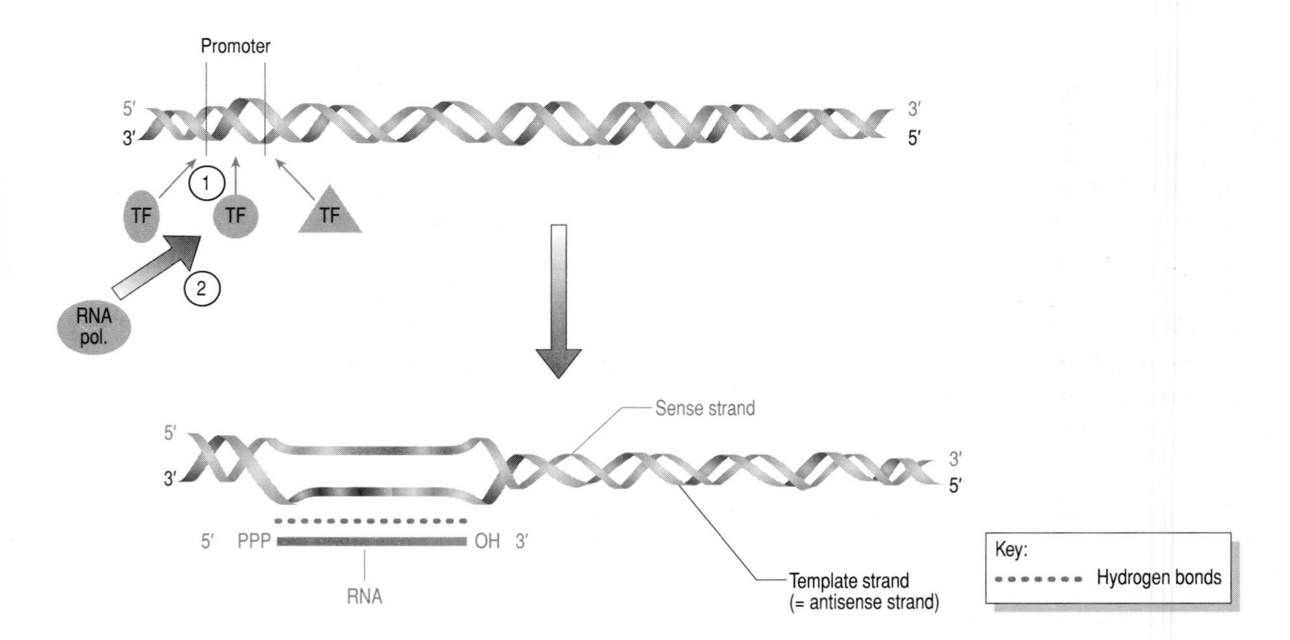

Figure 1.11: RNA is transcribed as a single strand which is complementary in base sequence to one strand (template strand) of a gene.

Various transcription factors (TF) are required to bind to a promoter sequence in the immediate vicinity of a gene ①, in order to subsequently position and guide the RNA polymerase that will transcribe the gene ②. Chain synthesis is initiated with a nucleoside triphosphate and chain elongation occurs by successive addition of nucleoside monophosphate residues provided by rNTPs to the 3′ OH. This means that the 5′ end will have a triphosphate group, which may subsequently undergo modification (e.g. by capping; see page 29) and the 3′ end will have a free hydroxyl group. *Note* that the sequences of the RNA will normally be identical to the sense strand of the gene (except U replaces T) and complementary in sequence to the template strand.

Table 1.2: The three classes of eukaryotic RNA polymerase

Class	Genes transcribed	Comments
I	28S rRNA; 18S rRNA; 5.8S rRNA	Localized in the nucleolus. A single primary transcript (45S rRNA) is cleaved to give the three rRNA classes listed
II	All genes that encode polypeptides; most snRNA genes	Polymerase II transcripts are unique in being subject to capping and polyadenylation
III	5S rRNA; tRNA genes; U6 snRNA; 7SL RNA; 7SK RNA; 7SM RNA	The promoter for some genes transcribed by RNA polymerase III (e.g. 5S rRNA, tRNA, 7SL RNA) is internal to the gene (see *Figure 1.12*) and for others (e.g. 7SK RNA) is located upstream

Transcription of eukaryotic genes requires interaction between *cis*-acting transcription elements and *trans*-acting transcription factors

Eukaryotic RNA polymerases cannot initiate transcription by themselves. Instead, combinations of short sequence elements in the immediate vicinity of a gene act as recognition signals for **transcription factors** to bind to the DNA in order to guide and activate the polymerase. A major group of such short sequence elements is often clustered upstream of the coding sequence of a gene, where they collectively constitute the **promoter**. After a number of general transcription factors bind to the promoter region, an RNA polymerase binds to the transcription factor complex and is activated to initiate the synthesis of RNA from a unique location. The transcription factors are said to be *trans*-acting, because they are synthesized by genes which are remotely located, and they migrate to their sites of action. In contrast, the promoter elements are *cis*-acting; their function is limited to the DNA duplex on which they reside (*Table 1.3*).

In the case of genes which are actively transcribed by RNA polymerase II, either at a specific stage in the cell cycle (e.g. histones) or in specific cell types (e.g. β-globin), the promoter elements always include a **TATA box**, often TATAAA or a variant, at a position about 25 bp upstream (–25) from the transcriptional start site (*Figure 1.12*). Mutation at the TATA element does not prevent initiation of transcription, but does cause the startpoint of transcription to be displaced from the normal position.

The promoters of many other genes, including housekeeping genes (see page 15), lack TATA boxes but often have a **GC box**, containing variants of the consensus sequence GGGCGG. Other common promoter elements include the **CAAT box**, often at about –80, which is usually the strongest determinant of promoter efficiency. Note, however, that the GC and CAAT boxes appear to be able to function in either orientation, although their sequences are asymmetrical (*Figure 1.12*). In

Table 1.3: Examples of *cis*-acting elements recognized by ubiquitous transcription factors

Cis element	DNA sequence is identical to, or a variant of	Associated *trans*-acting factors	Comments
GC box	GGGCGG	Spl	Spl factor is ubiquitous
TATA box	TATAAA	TFIID	TFIIA binds to the TFIID–TATA box complex to stabilize it
CAAT box	CCAAT	Many, e.g. C/EBP, CTF/NFI	Large family of *trans*-acting factors
TRE (TPA response element)	GTGAGTA/CA	AP-1 family, e.g. JUN/FOS	Large family of *trans*-acting factors
CRE (cAMP response element)	GTGACGTA/CAA/G	CREB/ATF family, e.g. ATF-1	Genes activated in response to cAMP

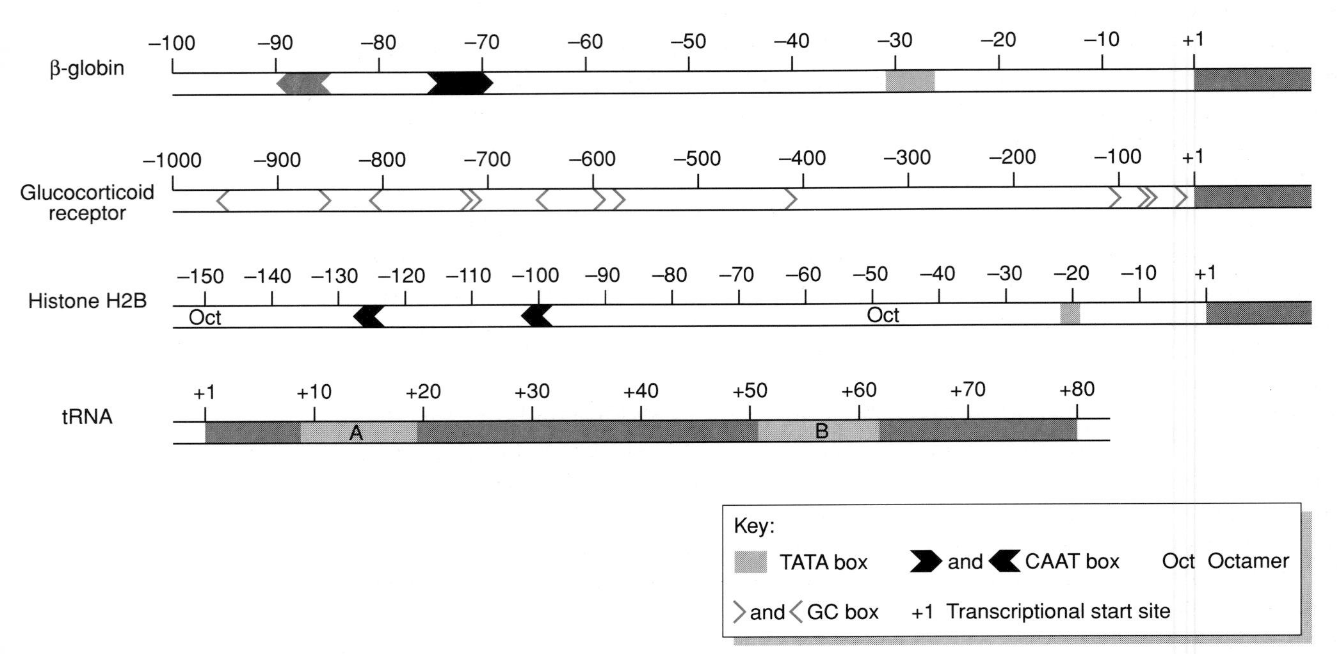

Figure 1.12: Eukaryotic promoters consist of a collection of conserved short sequence elements located at relatively constant distances from the transcription start site.

Alternative orientations for GC and CAAT box elements are indicated by chevron orientation: > = normal orientation; < = reverse orientation. The glucocorticoid receptor gene is unusual in possessing 13 upstream GC boxes (10 in the normal orientation; three in the reverse orientation). The tRNA genes are transcribed by RNA polymerase III and have an internal bipartite promoter comprising element A (usually within the nucleotides numbered +8 to +19 according to the standard tRNA nucleotide numbering system) and element B (usually between nucleotides +52 and +62). Specific transcription factors bind to these elements and then guide RNA polymerase III to start transcribing at +1.

addition to general upstream transcription elements which are recognized by ubiquitous transcription factors, more specific recognition elements are known which are recognized by tissue-restricted transcription factors (see *Table 7.10*).

Enhancers comprise groups of *cis*-acting short sequence elements, which can enhance the transcriptional activity of specific eukaryotic genes. However, unlike promoter elements whose positions relative to the transcriptional initiation site are relatively constant (*Figure 1.12*), enhancers are located a variable, and often considerable, distance from the transcriptional start site, and their function is independent of their orientation. They appear to bind gene regulatory proteins and, subsequently, the DNA between the promoter and enhancer loops out, allowing the proteins bound to the enhancer to interact with the transcription factors bound to the promoter, or with the RNA polymerase. **Silencers** are equivalent regulatory elements which can inhibit the transcriptional activity of specific genes.

Tissue-specific gene expression involves selective activation of specific genes and regions of transcriptionally active chromatin adopt an open conformation

The DNA content of a specific type of eukaryotic cell, a myocyte for example, is virtually identical to that of a lymphocyte, liver cell or any other type of nucleated cell from the same organism. What makes the different cell types different is the pattern of genes which are expressed in the cell. Some cells, particularly brain cells, express a large number of different genes. In many other cell types, a large fraction of the genes is transcriptionally inactive. Clearly, the genes that are expressed are the ones which define the functions of the cell. Some of these functions are common ones which are essential for general cell functions and are specified by so-called **housekeeping genes**. The expression of other genes may be largely restricted to a specific cell type (**tissue-specific gene expression**). Note, however, that even in the case of genes which show considerable tissue specificity in expression, some gene transcripts occur at very low levels in all cell types (**illegitimate** or **ectopic transcription**).

The distinction between transcriptionally active and inactive regions of DNA in a cell is reflected in the structure of the associated chromatin. Transcriptionally inactive chromatin generally adopts a highly condensed conformation and is often associated with regions of the genome which undergo late replication during S phase of the cell cycle. It is associated with tight binding by the histone H1 molecule. By contrast, transcriptionally active DNA adopts a more open conformation and is often replicated early in S phase. It is marked by relatively weak binding by histone H1 molecules and extensive acetylation (see *Table 1.5*) of the four types of nucleosomal histones (see page 36). Additionally, in transcriptionally active chromatin the promoter regions of vertebrate genes are generally characterized by absence of methylated cytosines (see below). Transcription factors can displace nucleosomes, and so the open conformation of transcriptionally active chromatin can be distinguished experimentally because it also affords access to nucleases: at very low concentrations, the enzyme DNase I will digest long regions of nucleosome-free DNA. Although the regulatory regions may contain several sequence-specific binding proteins, the open chromatin structure is marked by the presence of **DNase I-hypersensitive sites** (see pages 190–191).

Unmethylated CpG islands frequently mark the position of actively transcribing genes in vertebrate genomes

Methylated bases, notably 5-methylcytosine, are an important feature of vertebrate genomes, although the DNA of some eukaryotes such as *Drosophila* is not methylated. Methylation of vertebrate DNA is related to a general repression of transcription and is thought to be involved in certain specialized selective gene repression mechanisms (see pages 171–172). Actively transcribed vertebrate genes are marked by the presence of so-called **CpG islands:** short stretches of DNA (often 1–2 kb long), in which the frequency of the dinucleotide CpG (i.e. C connected by a 3′ – 5′ phosphodiester bond to a G) approximates to the frequency expected from the bulk GC content of the DNA. By comparison, the rest of the DNA in the vertebrate genome is markedly deficient in CpGs. Although housekeeping genes have CpG islands which are constitutively methylated (Bird, 1986), genes that are expressed in selected tissues tend to be unmethylated in these tissues but methylated in tissues where they are not expressed. The zygote contains methylated DNA inherited from the contributing sperm and egg cells, but the attached methyl groups are removed very early in embryogenesis (in the morula and early blastula) and the methylation profile is re-established during implantation of the embryo.

Cytosine residues occurring in CpG dinucleotides are targets for methylation at carbon atom 5. Only about 3% of the cytosines in human DNA are methylated, but most that are methylated are found in the CpG dinucleotide, producing 5-methylcytosine, which can spontaneously deaminate to give thymine. 5-Methylcytosine is not a particularly unstable base; instead the deficiency in CpG really results from an ineffective DNA repair process. Nonmethylated cytosine residues are also prone to accidental **deamination** in which the amino group at carbon 4 is replaced by an oxygen atom. This may result in conversion to the base uracil. Since uracil is not normally found in DNA, it is efficiently recognized as being an altered base by the repair enzyme uracil DNA glycosylase, whereupon it is excised and replaced by a

C. However, when 5-methylcytosine is spontaneously deaminated, a thymine residue results, a base normally found in DNA (*Figure 1.13*). As the thymine is now mispaired with a guanine, the base mismatch can be recognized by a specific DNA glycosylase which excises the T and replaces it with a C. However, this DNA repair process is clearly inefficient. As a result, over evolutionary time-scales, deamination of 5-methylcytosine residues has led to CpG being depleted in vertebrate DNA (currently about 20% of the expected frequency) and being slowly replaced by TpG (or CpA on the complementary strand; *Figure 1.13*).

Methylation of a CpG island at a promoter usually prevents expression of the gene; the MeCP-1 and MeCP-2 proteins can bind to methylated CpGs, leading to silencing of expression. Accordingly, high transcriptional activity requires that the promoter and neighboring regions be kept methylation-free. Unmethylated CpG islands have been shown to be present at the 5′ end of constitutively expressed housekeeping genes. However, they have also been found to be associated with many genes which show tissue-specific expression. As a result, the presence of unmethylated CpG islands may be necessary, but is not sufficient, to ensure transcription.

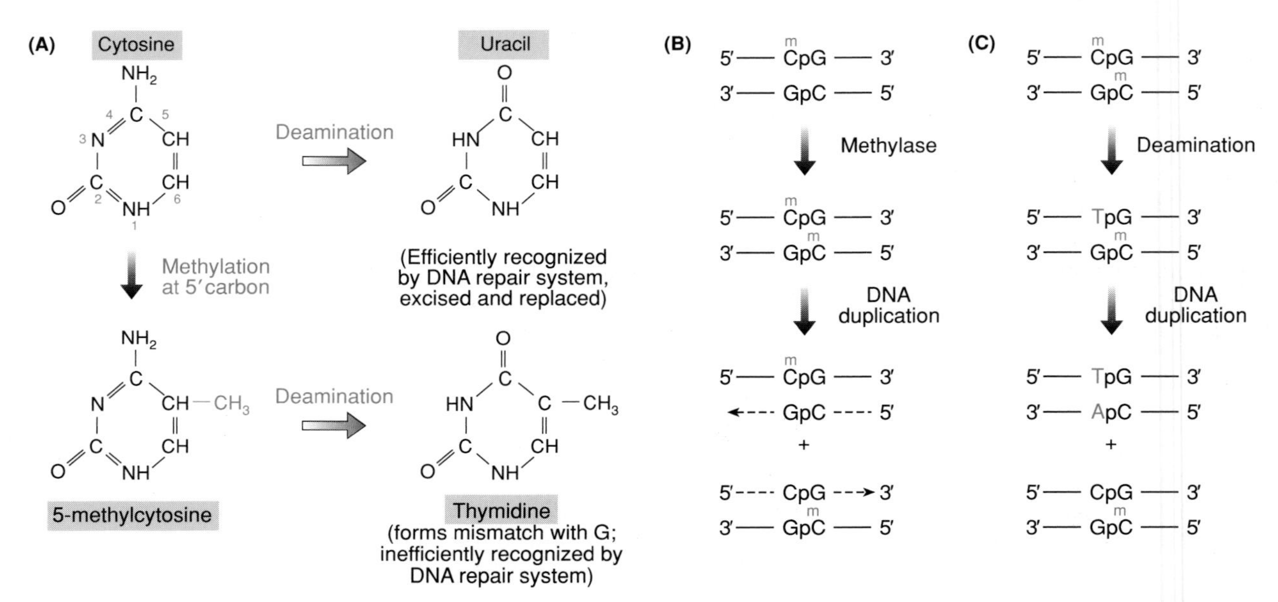

Figure 1.13: The CpG dinucleotide is underrepresented in vertebrate DNA because it is prone to methylation and deaminated 5-methylcytosine is subject to ineffective DNA repair.

(A) The deaminated products of cytosine and its methylated derivative 5-methylcytosine are differentially recognized by DNA repair enzymes.
(B) The methylation pattern of CpGs is perpetuated by a requirement for the specific methylase to recognize a hemi-methylated target sequence. Following methylation of a hemi-methylated target (i.e. methylated on one strand only), the two methylated strands will separate at DNA duplication and act as templates for the synthesis of two unmethylated daughter strands. The resulting daughter duplexes will now provide new hemi-methylated targets for continuing the same pattern of methylation.
(C) Deamination of 5-methylcytosine in the sequence CpG results in conversion of CpG dinucleotides to TpG and CpA dinucleotides.

Post-transcriptional RNA processing

Maturation of the primary RNA transcript of eukaryotic genes often involves removal of unwanted internal segments and rejoining of the remaining segments (**RNA splicing**). Additionally, in the case of RNA polymerase II transcripts, a specialized nucleotide linkage (7-methylguanosine triphosphate) is added to the 5′ end of the primary transcript (**capping**), and adenylate (AMP) residues are sequentially added to the 3′ end of mRNA to form a poly(A) tail (**polyadenylation**).

RNA splicing ensures removal of intronic RNA sequences from the primary transcript and fusion of exonic RNA sequences

The coding sequences of most vertebrate genes, both polypeptide-encoding genes and genes encoding RNA molecules other than mRNA, are split into segments (**exons**) which are separated by noncoding intervening sequences (**introns**). The primary RNA transcript from each such gene contains sequence transcribed from the full length of the gene, encompassing both exons and introns. Subsequently, however, the primary RNA transcript undergoes **RNA splicing,** a series of processing reactions whereby the intronic RNA segments are snipped out and discarded and the exonic RNA segments are joined end-to-end (spliced) to give a shorter RNA product (*Figure 1.14*).

The mechanism of RNA splicing is dependent upon the identity of the nucleotide sequences at the exon/intron boundaries (**splice junctions**). In particular, it is critically dependent on what has been called the **GT–AG rule**: introns virtually always start with GT and end with AG (see *Figure 1.15*).

Although the conserved GT and AG dinucleotides are crucially important for splicing, they are not by themselves sufficient to signal the presence of an intron. Comparisons of documented sequences have revealed that sequences adjacent to the GT and AG dinucleotides show a considerable degree of conservation (*Figure 1.15*). In addition, a third conserved intronic sequence that is known to be functionally important in splicing is the so-called **branch site** which is usually located very close to the end of the intron, at most 40 nucleotides before the terminal AG dinucleotide (*Figure 1.15*).

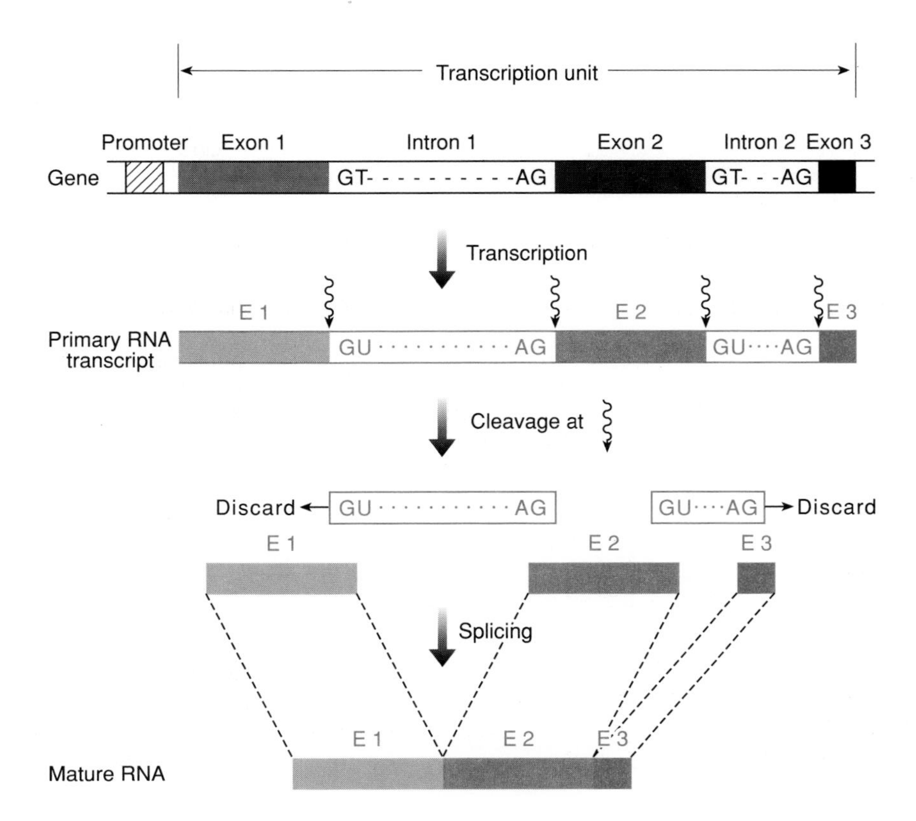

Figure 1.14: RNA splicing involves endonucleolytic cleavage and removal of intronic RNA segments and splicing of exonic RNA segments.

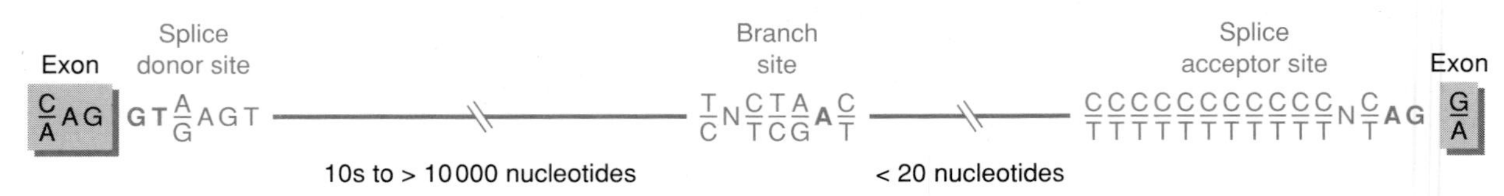

Figure 1.15: Consensus sequences at the splice donor, splice acceptor and branch sites in introns of complex eukaryotes.

Highlighted nucleotides are essentially invariant. Others represent the majority nucleotide found at this particular position. *Note* that in cases where pyrimidines (C/T or T/C) are preferred, no significance should be attached to which base comes first. For example, the consensus sequence of the branch site is written so as to highlight the similarity to the consensus branch site in yeast introns (TACTAAC), but the sequence given for the splice acceptor site does not signify a preference for C as opposed to T.

The splicing mechanism involves the following sequence:

(i) cleavage at the 5′ splice junction;

(ii) nucleolytic attack by the terminal G nucleotide of the splice donor site at the invariant A of the branch site to form a **lariat**-shaped structure;

(iii) cleavage at the 3′ splice junction, leading to release of the intronic RNA as a lariat, and splicing of the exonic RNA segments (*Figure 1.16*).

The above reactions are mediated by a number of complexes containing several snRNP particles, which are composed of **snRNA** molecules attached to specific proteins. As detailed in *Figure 1.16*, individual snRNAs involved in this process may have specific sequences which permit base pairing with a conserved intronic sequence (the 5′ splice site or the branch site) and/or with recognition sequences in other snRNA molecules. These snRNA–protein complexes (snRNPs), together with some protein splicing factors, form a large particulate complex, the **spliceosome**, which manages the splicing process.

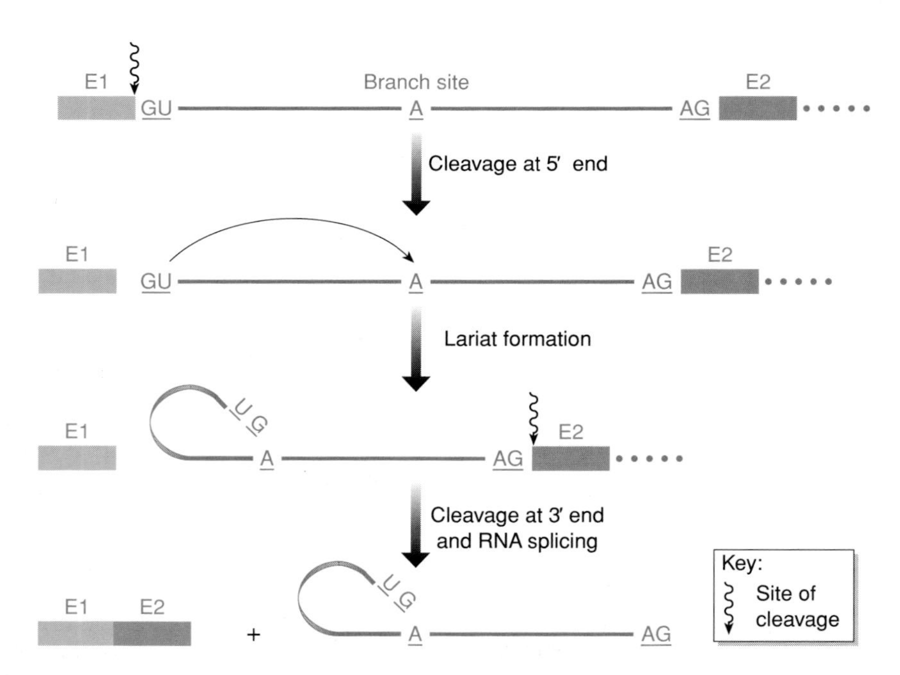

Figure 1.16: Mechanism of RNA splicing.

Small nuclear ribonucleoprotein particles (snRNPs) are directly involved in the splicing reaction. U1 snRNP recognizes and binds to the splice donor site by base pairing; the 5′ terminus of the U1 snRNA has a sequence UACUUAC which base pairs with the splice donor consensus (GUAAGUA). After the U1 snRNP has bound, U2 snRNA recognizes the branch site by a similar base-pairing reaction and, subsequently, interaction between U1 snRNP and U2 snRNP brings the two splice junctions close together. Thereafter a multi-snRNP particle, containing U4, U5 and U6 snRNAs, associates with the U1–U2 snRNP complex.

The spliceosome is envisaged to act in a processive manner: once a 5′ splice site is recognized, it scans the DNA sequence until it meets the next 3′ splice site (which would be signalled as a target by the branch site consensus sequence located just before it). However, the order in which intronic sequences are removed and their flanking exonic sequences spliced is not governed by their linear order in the RNA transcript; instead, the conformation of the RNA is thought to influence the accessibility of 5′ splice sites.

Specialized nucleotides are added to the 5′ and 3′ ends of most RNA polymerase II transcripts

In addition to RNA splicing, RNA polymerase II transcripts are subject to two additional RNA processing events:

Capping

This occurs shortly after transcription. In the case of primary transcripts which will be processed to give mRNA, a methylated nucleoside, 7-methylguanosine (m^7G) is linked to the first 5′ nucleotide of the RNA transcript by a special 5′ – 5′ phosphodiester bond. As this bond effectively bridges the 5′ carbon of the m^7G residue to the 5′ carbon of the first nucleotide, the 5′ end is said to be blocked or capped (*Figure 1.17*). Transcripts of the snRNA genes are also capped, but their caps may undergo additional modification. The cap has been envisaged to have several possible functions:

(i) to protect the transcript from 5′ → 3′ exonuclease attack (decapped mRNA molecules are rapidly degraded);

(ii) to facilitate transport from the nucleus to the cytoplasm;

(iii) to facilitate RNA splicing;

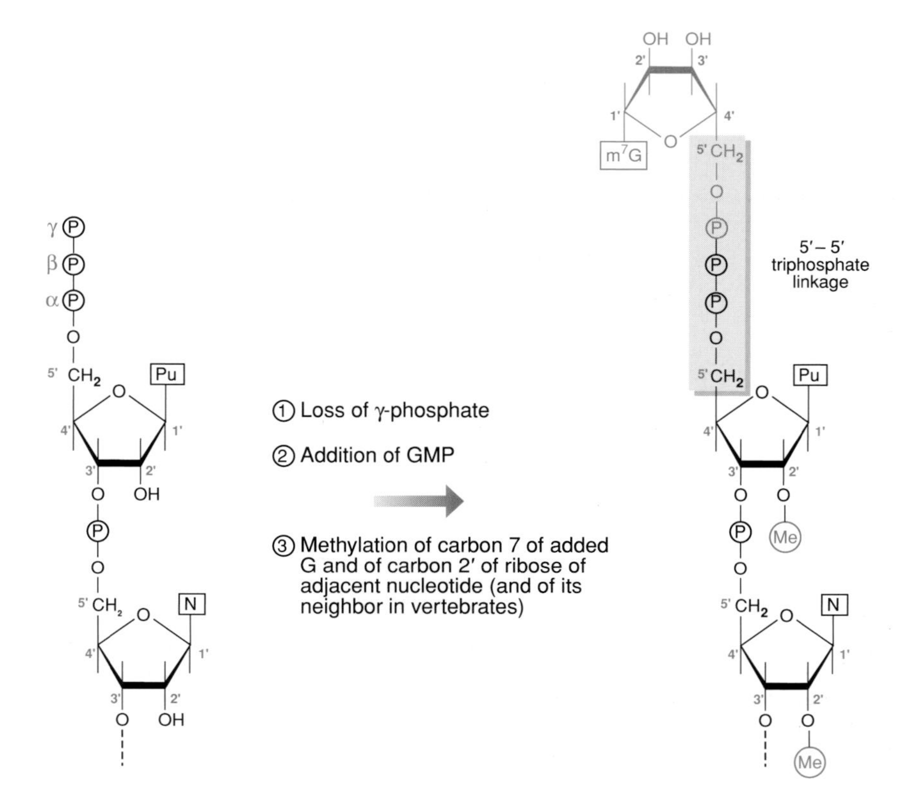

Figure 1.17: The 5′ end of eukaryotic mRNA molecules is protected by a specialized nucleotide (capping).

After the original gamma phosphate of the terminal 5′ nucleotide is removed, a new GMP residue is provided by a GTP precursor which forms a specialized **5′–5′ triphosphate** linkage with what was the terminal 5′ nucleotide. Subseqent reactions lead to methylation of carbon atom 7 of the terminal G, and, in vertebrates, of the 2′ carbon atom of the ribose of each of the two adjacent nucleotides. N, any nucleotide; Pu, purine.

(iv) to play an important role in the attachment of the 40S subunit of the cytoplasmic ribosomes to the mRNA (see below).

Polyadenylation

Transcription by both RNA polymerase I and III is known to stop after the enzyme recognizes a specific transcription termination site. However, identifying possible termination sites for transcription by RNA polymerase II is difficult because the 3′ ends of mRNA molecules are determined by a post-transcriptional cleavage reaction. The sequence AAUAAA is a major element that signals 3′ cleavage for the vast majority of polymerase II transcripts (transcripts from histone genes and snRNA genes are notable exceptions). Cleavage occurs at a specific site located 15–30 nucleotides downstream of the AAUAAA element (*Figure 1.18*).

Following the cleavage point, transcription can continue for hundreds or thousands of nucleotides until termination occurs at one of several later sites. Once cleavage has occurred downstream of the AAUAAA element, about 200 **adenylate** (i.e. AMP) residues are sequentially added in mammalian cells by the enzyme poly(A) polymerase to form a poly(A) tail. The poly(A) tail has been envisaged to have several possible functions:

(i) to facilitate transport of the mRNA molecules to the cytoplasm;

(ii) to stabilize at least some of the mRNA molecules in the cytoplasm [shortening of poly(A) tracts is associated with mRNA degradation, but some mRNA species (e.g. actin mRNA) remain stable with little or no poly(A)];

(iii) it may facilitate translation by permitting enhanced recognition of the mRNA by the ribosomal machinery.

Termination of transcription of histone genes also involves 3′ cleavage of the primary transcript. This reaction is dependent on secondary structure in the RNA transcript: it is dependent on a conserved upstream hairpin sequence and a short downstream sequence which base-pairs with a short sequence at the 5′ end of the U7 snRNA.

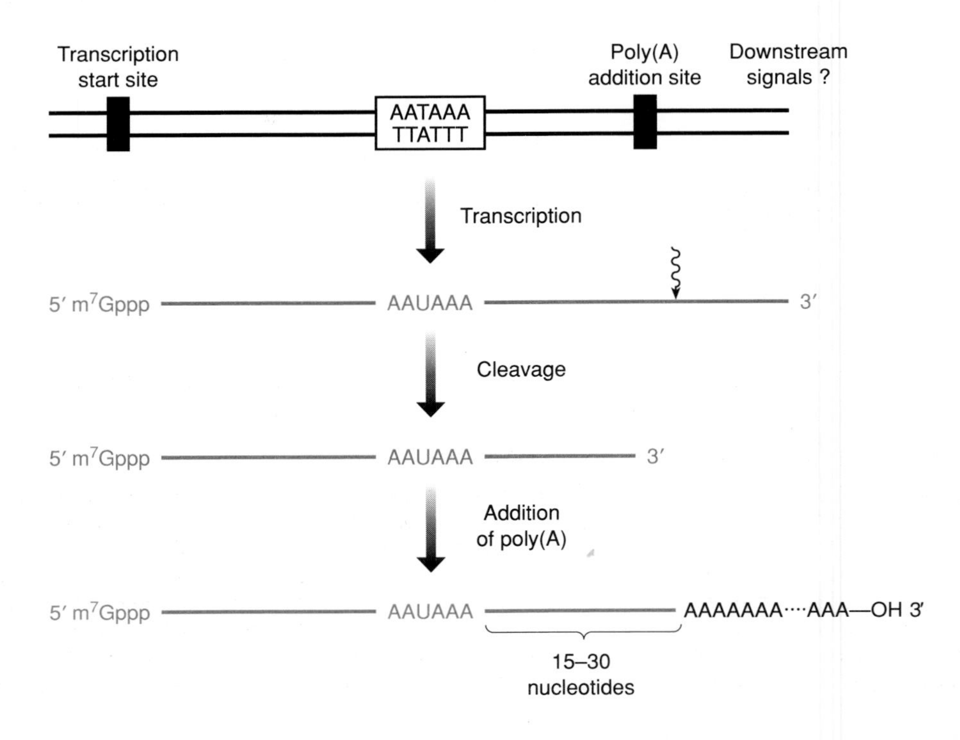

Figure 1.18: The 3′ end of most eukaryotic mRNA molecules is polyadenylated.

The end of transcription of RNA polymerase II transcripts is signalled by a 3′ cleavage in the transcribed RNA. In most mRNA species, this is achieved by an upstream AAUAAA signal in concert with, as yet, unidentified downstream signals. Cleavage occurs normally about 15–30 nucleotides downstream of the AAUAAA element and AMP residues are subsequently added by poly(A) polymerase to form a poly(A) tail. Histone mRNA undergoes a different 3′ cleavage reaction (see text).

Translation, post-translational processing and protein structure

Translation is the process whereby mRNA is decoded on ribosomes to specify the synthesis of polypeptides

Following post-transcriptional processing, mRNA transcribed from genes in nuclear DNA migrates to the cytoplasm, where it engages with ribosomes and other components to direct the synthesis of specific polypeptides. The mitochondria also have ribosomes and the capacity for protein synthesis which, however, is restricted to a few polypeptide-encoding genes located in mitochondrial DNA (see below). Only the central segment of a typical eukaryotic mRNA molecule is translated to specify the synthesis of a polypeptide. The flanking sequences, **5′ and 3′ untranslated sequences** (**5′ UTS; 3′ UTS**), are originally copied from sequence derived from the 5′ and 3′ terminal exons and, like the 5′ cap and 3′ poly(A) tail, assist in binding and stabilizing the mRNA on the ribosomes where translation of the central segment occurs (*Figure 1.19*).

Ribosomes are large RNA–protein complexes composed of two subunits. In eukaryotes, cytoplasmic ribosomes have a large 60S subunit and a smaller 40S subunit (**S values** are a measure of how fast large molecular structures will sediment in the ultracentrifuge and are governed by both molecular mass and shape). The 60S sub-

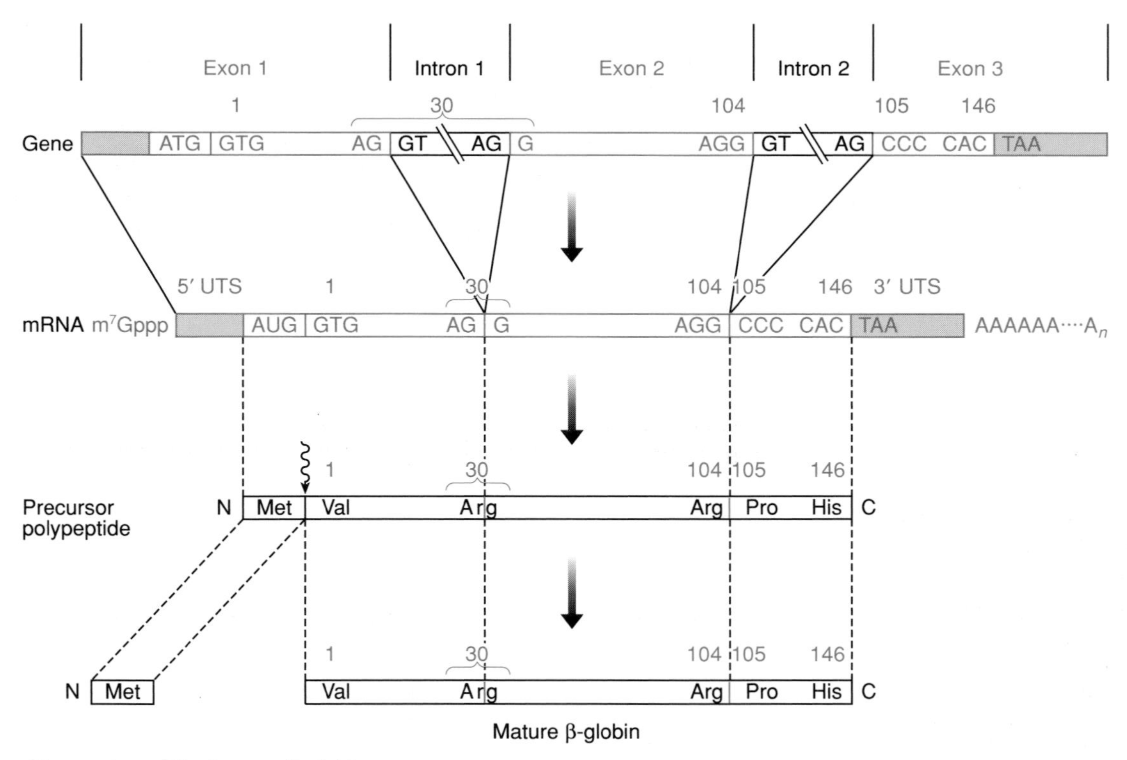

Figure 1.19: Expression of the human β-globin gene.

Exons 1 and 3 each contain noncoding sequences (shaded bars) at their extremities, which are transcribed and are present at the 5′ and 3′ ends of the β-globin mRNA, but are not translated to specify polypeptide synthesis. Such 5′ and 3′ untranslated sequences (**5′ UTS** and **3′ UTS**), however, are thought to be important in ensuring high efficiency of translation (see text). The stop codon TAA represents the first three nucleotides of the 3′ untranslated sequence. *Note* that the initial translation product has 147 amino acids, but that the N-terminal methionine is removed by post-translational processing to generate the mature β-globin polypeptide. The first two bases of the codon specifying Arg30 are encoded by exon 1 and the third base is encoded by exon 2 (i.e. intron 1 separates the second and third bases of the codon, an example of a **phase 2 intron**; see page 225). The second intron separates codons 104 and 105 and is an example of a **phase 0 intron**; see page 225, and also *Figure 1.23* for an example of a **type I intron**.

unit contains three types of rRNA molecule: 28S rRNA, 5.8S rRNA and 5S rRNA, and about 50 ribosomal proteins. The 40S subunit contains a single 18S rRNA and over 30 ribosomal proteins. Ribosomes provide a structural framework for polypeptide synthesis in which the RNA components are predominantly responsible for the catalytic function of the ribosome; the protein components are thought to enhance the function of the rRNA molecules, and a surprising number of them do not appear to be essential for ribosome function.

The assembly of a new polypeptide from its constituent amino acids is governed by a **triplet genetic code**. Successive groups of three nucleotides (**codons**) in the linear mRNA sequence are decoded sequentially in order to specify individual amino acids. The decoding process is mediated by a collection of tRNA molecules, to each of which a specific amino acid has been covalently bound (at the free 3′ hydroxyl group of the tRNA; see *Figure 1.20*) by a specific amino acyl tRNA synthetase.

Different tRNA molecules bind different amino acids. Each tRNA has a specific trinucleotide sequence, called the **anticodon**, at a crucially important site located in the center of one arm of the tRNA (*Figure 1.6*). This site provides the necessary specificity to interpret the genetic code: for an amino acid to be inserted in the growing polypeptide chain, the relevant codon of the mRNA molecule must be recognized via base pairing with a suitably complementary anticodon of the appropriate tRNA molecule (*Figure 1.20*).

One model of translation envisages that the 40S ribosomal subunit initially recognizes the 5′ cap via the participation of proteins that specifically bind to the cap. It then scans along the mRNA until it encounters the initiation codon, which is almost always AUG, specifying methionine (a few cases are known where ACG, CUG or GUG are used instead). Usually, though not always, the first AUG encountered will

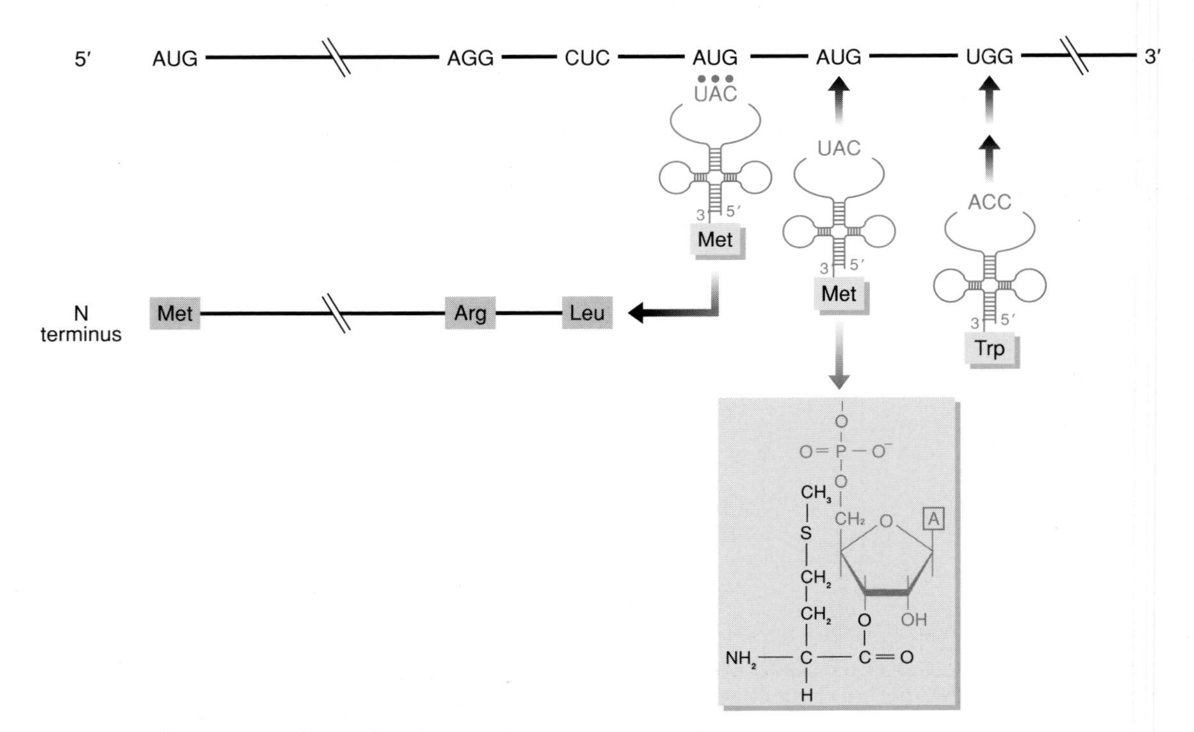

Figure 1.20: The genetic code is deciphered by codon–anticodon recognition.

The sequence of nucleotides in the mRNA sequence is interpreted from a translational start point (normally marked by the sequence AUG) and continues in the 5′ → 3′ direction until a stop codon is reached in that reading frame (see *Figure 1.19*). Each codon in the mRNA is recognized by the complementary anticodon sequence of a tRNA molecule to which a specific amino acid is covalently bonded to the adenosine at the 3′ end (see insert).

be the initiation codon. However, the AUG is recognized efficiently as an initiation codon only when it is embedded in a suitable sequence, the optimal being the sequence: GCC**A/G**CC<u>AUG</u>**G**. The most important determinants in this sequence are the G following the AUG codon, and the purine preceding it by three nucleotides.

Subsequently, successive amino acids are incorporated into the growing polypeptide chain by a condensation reaction: the amino group of the incoming amino acid reacts with the carboxyl group of the last amino acid to be incorporated, resulting in a **peptide bond** between successive residues (*Figure 1.21*). This is catalyzed by a **peptidyl transferase** activity which resides in the RNA component of the large ribosomal subunit.

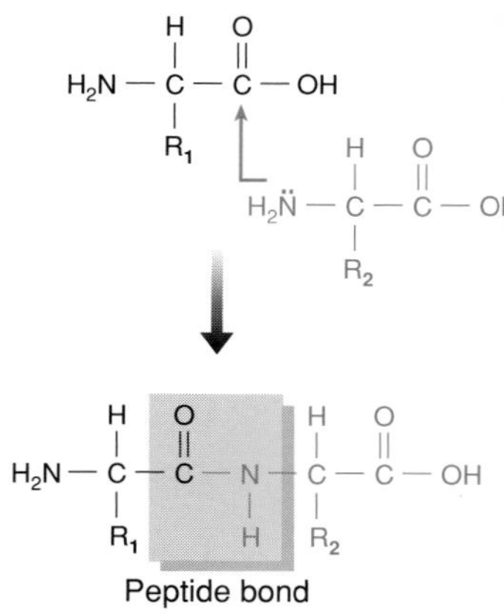

Due to there being 64 (=4^3) possible codons, but only 20 different amino acids, the genetic code is degenerate: each amino acid is specified on average by about three different codons. However, certain amino acids (such as methionine or tryptophan) can be specified by only a single codon, whereas others, including arginine, leucine and serine are specified by six codons (*Figure 1.22*).

Although there are 64 codons, the corresponding number of tRNA molecules with different anticodons is less: just over 30 types of cytoplasmic tRNA and only 22 types of mitochondrial RNA. The interpretation of all 64 codons on both cytoplasmic and mitochondrial ribosomes is possible because the normal base pairing rules are relaxed when it comes to codon–anticodon recognition. The **wobble hypothesis** states that pairing of codon and anticodon follows the normal A–U and G–C rules for the first two base positions in a codon, but that exceptional 'wobbles' occur at the third position and G–U base pairs are also admitted (see *Table 1.4*).

Translation continues until a **termination codon** is encountered (i.e. UAA, UAG or UGA in the case of nuclear-encoded mRNA; UAA, UAG, AGA or AGG in the case of mitochondrial-encoded mRNA; *Figure 1.22*). The backbone of the primary translation product will therefore have at one end a methionine with a free amino group (the **N-terminal end**) and at the other end an amino acid with a free carboxyl group (the **C-terminal end**). Note that although codons are translated in a specific **translational reading frame**, overlapping genes are occasionally found in eukaryotes, in which different translational reading frames are used (see *Figure 7.3*).

$H_2N-CH(R_1)-COOH + H_2\ddot{N}-CH(R_2)-COOH \rightarrow H_2N-CH(R_1)-CO-NH-CH(R_2)-COOH$

Peptide bond

Figure 1.21: Polypeptides are synthesized by peptide bond formation between successive amino acids.

Codon	Amino acid	Codon	Amino acid	Codon	Amino acid	Codon	Amino acid
AAA	Lys	CAA	Gln	GAA	Glu	UAA	STOP
AAG	Lys	CAG	Gln	GAG	Glu	UAG	STOP
AAC	Asn	CAC	His	GAC	Asp	UAC	Tyr
AAU	Asn	CAU	His	GAU	Asp	UAU	Tyr
ACA	Thr	CCA	Pro	GCA	Ala	UCA	Ser
ACG	Thr	CCG	Pro	GCG	Ala	UCG	Ser
ACC	Thr	CCC	Pro	GCC	Ala	UCC	Ser
ACU	Thr	CCU	Pro	GCU	Ala	UCU	Ser
AGA (boxed)	Arg / STOP	CGA	Arg	GGA	Gly	UGA (boxed)	STOP / Trp
AGG (boxed)	Arg / STOP	CGG	Arg	GGG	Gly	UGG	Trp
AGC	Ser	CGC	Arg	GGC	Gly	UGC	Cys
AGU	Ser	CGU	Arg	GGU	Gly	UGU	Cys
AUA (boxed)	Ile / Met	CUA	Leu	GUA	Val	UUA	Leu
AUG	Met	CUG	Leu	GUG	Val	UUG	Leu
AUC	Ile	CUC	Leu	GUC	Val	UUC	Phe
AUU	Ile	CUU	Leu	GUU	Val	UUU	Phe

Figure 1.22: The nuclear and mitochondrial genetic codes are similiar but not identical.

Colored alternatives represent differences in interpretation of the genetic code of nuclei (black) and of mammalian mitochondria (red). *Note* that degeneracy of the code most often involves the third base of the codon. Sometimes any base may be substituted (GGN = glycine, CCN = proline, etc., where N is any base). In other cases, any purine (Pu) or any pyrimidine (Py) will do (AAPu = lysine, AAPy = asparagine, CAPu = glutamine, CAPy = histidine, etc.).

Table 1.4: Codon–anticodon pairing admits relaxed base-pairing (wobbles) at the third base position of codons

Base at 5′ end of tRNA anticodon	Base recognized at 3′ end of mRNA codon
A	U only
C	G only
G	C or U
U	A or G

The predominant step in the control of translation is ribosome binding. In addition to the 5′ cap, the 5′ UTS (usually < 100 bp) and 3′ UTS (usually much longer than the 5′ UTS) both play critical roles in mRNA recruitment for translation. Several *cis*-acting elements that are involved in this process have been characterized and, in addition, a few *trans*-acting factors which bind to these elements have been identified. It is possible that the 5′ and 3′ UTS sequences interact to enhance translation.

The mitochondrial protein synthesis machinery is similar to that found in cytoplasmic ribosomes. For example, mitochondrial ribosomes contain two subunits, a large subunit with a 23S rRNA and several proteins, and a small subunit with a 16S rRNA and several ribosomal proteins. However, the components are assembled from the products of both mitochondrial and nuclear genes: the RNA components are synthesized in the mitochondria (the two rRNA molecules and 22 types of tRNA and a few, usually about 10, types of mRNA; see page 147 for a description of the human mitochondrial genome) whereas the proteins are encoded by nuclear genes which have been translated on cytoplasmic ribosomes and imported into the mitochondrion (RNA polymerase, 22 amino acyl tRNA synthetases and about 80 ribosomal proteins).

Post-translational modifications of proteins are frequent and can involve addition of specific chemical groups to specific amino acids and cleavage of the primary transcript

Modification reactions

Primary translation products often undergo a variety of modification reactions, involving the addition of chemical groups which are attached covalently to the polypeptide chain at the translational and post-translational levels. This can involve simple chemical modification (hydroxylation, phosphorylation, etc.) of the side chains of single amino acids (see *Table 1.5*).

A futher level of complexity is introduced in many cases by the presence of the varied carbohydrate side chains of **glycoproteins**, or the glycosaminoglycans which are found in **proteoglycans**. Few proteins in the cytosol are glycosylated and those that are carry a single sugar residue, *N*-acetylglucosamine, covalently linked to a serine or threonine residue. By contrast, those proteins which are secreted from cells or exported to lysosomes, the Golgi apparatus or the plasma membrane are glycosylated.

Oligosaccharide components of glycoproteins are largely preformed and added *en bloc* to polypeptides. Two major types of glycosylation are recognized:

(i) **N-glycosylation:** involves, in most cases, initial transfer of a common oligosaccharide sequence to the side chain NH_2 group of an Asn residue within the endoplasmic reticulum (ER; see *Table 1.5*). Subsequent trimming of residues and replacement with different monosaccharides occurs in the Golgi apparatus;

(ii) **O-glycosylation:** see *Table 1.5*.

Post-translational cleavage

The primary translation product may also undergo internal cleavage to generate a smaller mature product. Occasionally the initiating methionine is cleaved from the primary translation product, as during the synthesis of β-globin (*Figure 1.19*). More substantial polypeptide cleavage is observed in the case of the maturation of many proteins, including plasma proteins, polypeptide hormones, neuropeptides, growth factors, etc. For example, all secreted polypeptides and also polypeptides which are

Table 1.5: Major types of modification of polypeptides

Type of modification (group added)	Target amino acids	Comments
Phosphorylation (PO_4^-)	Tyrosine, serine, threonine	Achieved by specific kinases. May be reversed by phosphatases
Methylation (CH_3)	Lysine	Achieved by methylases and undone by demethylases
Hydroxylation (OH)	Proline, lysine, aspartic acid	Hydroxyproline and hydroxylysine are particularly common in collagens
Acetylation (CH_3CO)	Lysine	Achieved by an acetylase and undone by deacetylase
Carboxylation (COOH)	Glutamate	Achieved by γ-carboxylase
N-glycosylation (complex carbohydrate)	Asparagine, usually in the sequence: **Asn**-X-Ser/Thr	Takes place initially in the endoplasmic reticulum; X is any amino acid other than proline
O-glycosylation (complex carbohydrate)	Serine, threonine, hydroxylysine	Takes place in the Golgi apparatus; less common than N-glycosylation

transported across intracellular membranes (e.g. those synthesized in the cytoplasm and transported into the mitochondria) are synthesized initially as precursors, in which a **signal sequence** (sometimes called a **leader sequence**) at the N-terminal end acts as a recognition signal for transport across cellular membranes (see below). Thereafter the signal peptide is cleaved from the main polypeptide and degraded. Additionally, in some cases, a single mRNA molecule may specify more than one functional polypeptide chain as a result of proteolytic cleavage of a large precursor polypeptide (*Figure 1.23*).

Secretion of proteins and export to specific intracellular locations requires specific localization signals in the coding sequence

Proteins synthesized on mitochondrial ribosomes are required to function within the mitochondria. However, the numerous proteins that are synthesized on cytoplasmic ribosomes have diverse functions which may require them to be secreted from the cell where they were synthesized (as with hormones and other intercellular signalling molecules) or to be exported to specific intracellular locations, such as the nucleus (histones, DNA and RNA polymerases, transcription factors, RNA-processing proteins, etc.), the mitochondrion (mitochondrial ribosomal proteins, many respiratory chain components, etc.), peroxisomes, and so on. To do this, a specific **localization signal** needs to be embedded in the structure of the polypeptide so that it can be sent to the correct address. Usually, the localization signal takes the form of a short peptide sequence. Often, but not always, this constitutes a so-called **signal sequence** (or **leader sequence**) which is removed from the protein by a specialized **signal peptidase** once the sorting process has been achieved.

Signals for export to the endoplasmic reticulum and extracellular space

In the case of secreted proteins the signal peptide comprises the first 20 or so amino acids at the N-terminal end and always includes a substantial number of hydrophobic amino acids (see *Table 1.6*). The signal sequence is guided to the ER by a **signal recognition particle (SRP)** which binds both the growing polypeptide chain and the ribosome and directs them to a receptor protein on the cytosolic side surface of the rough endoplasmic reticulum (RER) membrane. Thereafter a polypeptide can pass into the lumen of the ER, destined for export from the cell, unless there are additional hydrophobic segments which stop the transfer process, as in the case of **transmembrane proteins**.

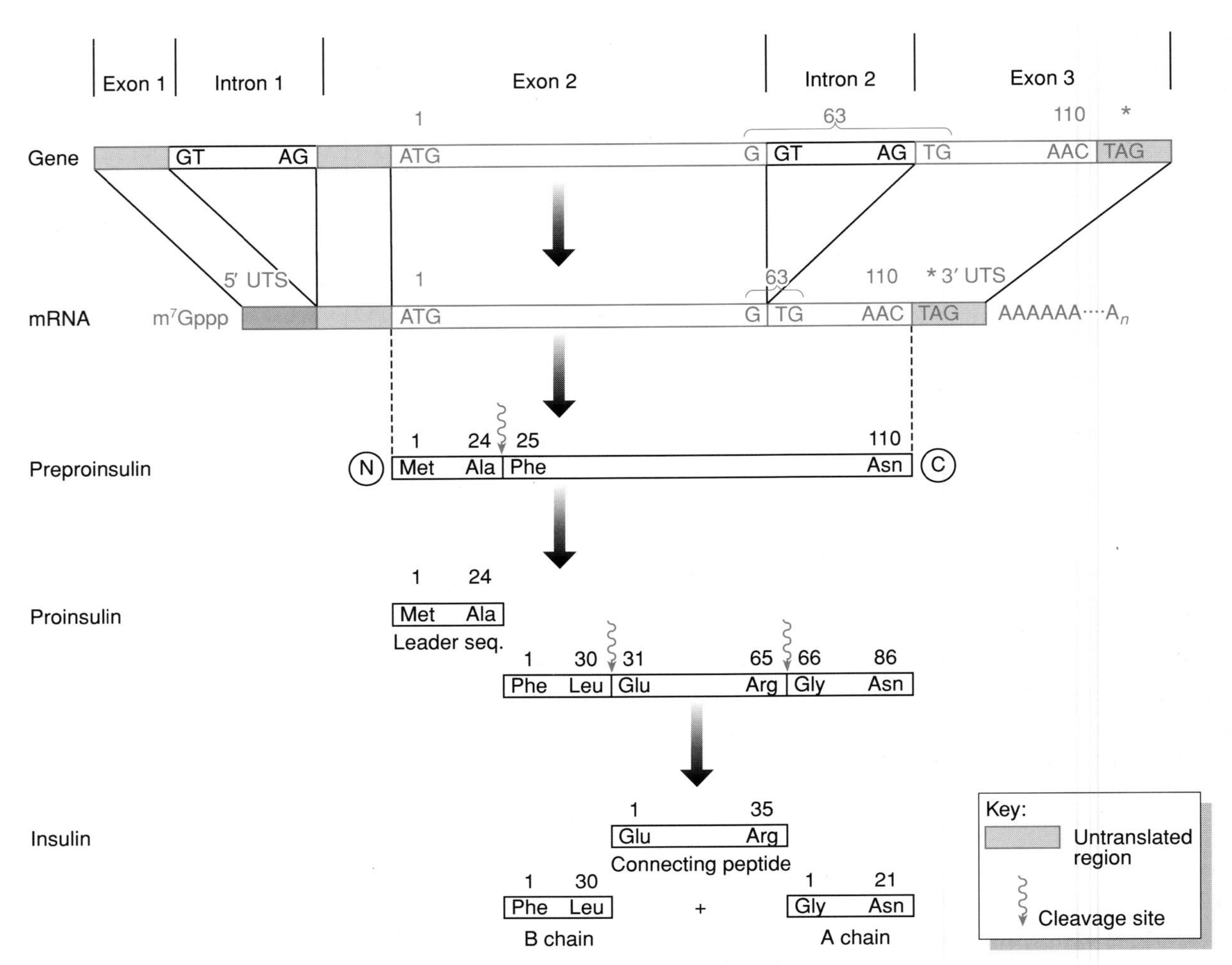

Figure 1.23: Insulin synthesis involves multiple post-translational cleavages of polypeptide precursors.

The first intron interrupts the 5′ untranslated region; the second intron interrupts base positions 1 and 2 of codon 63 and is classified as a phase I intron (see page 225, and also *Figure 1.19* for other intron phases). The primary translation product, preproinsulin, has a **leader sequence** of 24 amino acids which is required for the protein to cross the cell membrane, and is thereafter discarded. The proinsulin precursor contains a central segment, the **connecting peptide**, which is thought to be important in maintaining the conformation of the A and B chain segments so that they can form disulfide bridges (see *Figure 1.25*).

Other signals

Like the ER signal, an N-terminal signal sequence is required to traverse the mitochondrial membranes, and is subsequently cleaved. Typically, a **mitochondrial signal peptide** has, in addition to many hydrophobic amino acids, several positively charged amino acids, usually spaced at intervals of about four amino acids. This structure is thought to form an *amphipathic α-helix*, a helical structure with charged amino acids on one surface and hydrophobic amino acids on another (see below and *Figure 1.24*).

Nuclear localization signals can be located just about anywhere within the polypeptide sequence and typically consist of a stretch of four to eight positively charged amino acids, together with neighboring proline residues. Often, however, the signal is bipartite, and the positively charged amino acids are found in two blocks of two to four residues, separated by about 10 amino acids (see *Table 1.6*).

Table 1.6: Examples of signal peptides

Destination of protein	Location and form of signal peptide	Examples
Endoplasmic reticulum and secretion from cell	N-terminal 20 or so amino acids; very hydrophobic	Human insulin – 24 amino acid, highly hydrophobic signal peptide: N-Met-Ala-Leu-Trp-Met-Arg-Leu-Leu-Pro-Leu-Leu-Ala-Leu-Leu-Ala-Leu-Trp-Gly-Pro-Asp-Pro-Ala-Ala-Ala
Mitochondria	N-terminal; α-helix with positively charged residues on one face and hydrophobic ones on the other	Human mitochondrial aldehyde dehydrogenase – N-terminal 17 amino acids: N-Met-Leu-**Arg**-Ala-Ala-Ala-**Arg**-Phe-Gly-Pro-**Arg**-Leu-Gly-**Arg**-**Arg**-Leu-Leu
Nucleus	Internal; often a string of basic amino acids plus prolines; may be bipartite	SV40 T antigen – continuous: Pro-Pro-**Lys**-**Lys**-**Lys**-**Arg**-**Lys**-Val p53 – bipartite: **Lys**-**Arg**-Ala-Leu-Pro-Asn-Asn-Thr-Ser-Ser-Ser-Pro-Gln-Pro-**Lys**-**Lys**-**Lys**

Protein structure is varied and complex and is not easily predicted from the amino acid sequence of polypeptides

Proteins are composed of one or more polypeptides, each of which can be subject to post-translational modification. They can interact with specific **co-factors** (for example, divalent cations such as Ca^{2+}, Fe^{2+}, Cu^{2+}, Zn^{2+} or small molecules which are required for functional enzyme activity, e.g. NAD^+) or **ligands** (any molecule which a protein specifically binds), each of which can be powerful influences on the conformation of the protein. At least four different levels of structural organization have been distinguished for proteins (see *Table 1.7*).

Within a single polypeptide chain, there is ample scope for hydrogen bonding between different residues; irrespective of the side chains, the oxygen of a peptide bond's carbonyl (CO) group can hydrogen bond to the hydrogen of the NH group of another peptide bond. Fundamental structural units defined by hydrogen bonding between closely neighboring amino acid residues of a single polypeptide include:

(i) *the α-helix*. This involves formation of a rigid cylinder. The structure is dominated by hydrogen bonding between the carbonyl oxygen of a peptide bond with the hydrogen atom of the amino nitrogen of a peptide bond located four amino acids away (see *Figure 1.24*). Note that the DNA-binding domains of transcription factors are usually α-helical (see page 164). An **amphipathic α-helix** has charged residues on one surface and hydrophobic ones on another (*Figure 1.24*). Identical α-helices with a repeating arrangement of nonpolar side

Table 1.7: Levels of protein structure

Level	Definition	Comment
Primary	The linear sequence of amino acids in a polypeptide	Can vary enormously in length from a small peptide to thousands of amino acids long
Secondary	The path that a polypeptide backbone follows in space	May vary locally, e.g. as α-helix or β-pleated sheet, etc.
Tertiary	The overall three-dimensional structure of a polypeptide	Can vary enormously, e.g. globular, rod-like, tube, coil, sheet, etc.
Quaternary	The overall structure of a multimeric protein, i.e. of a combination of protein subunits	Often stabilized by disulfide bridges and by binding to ligands, etc.

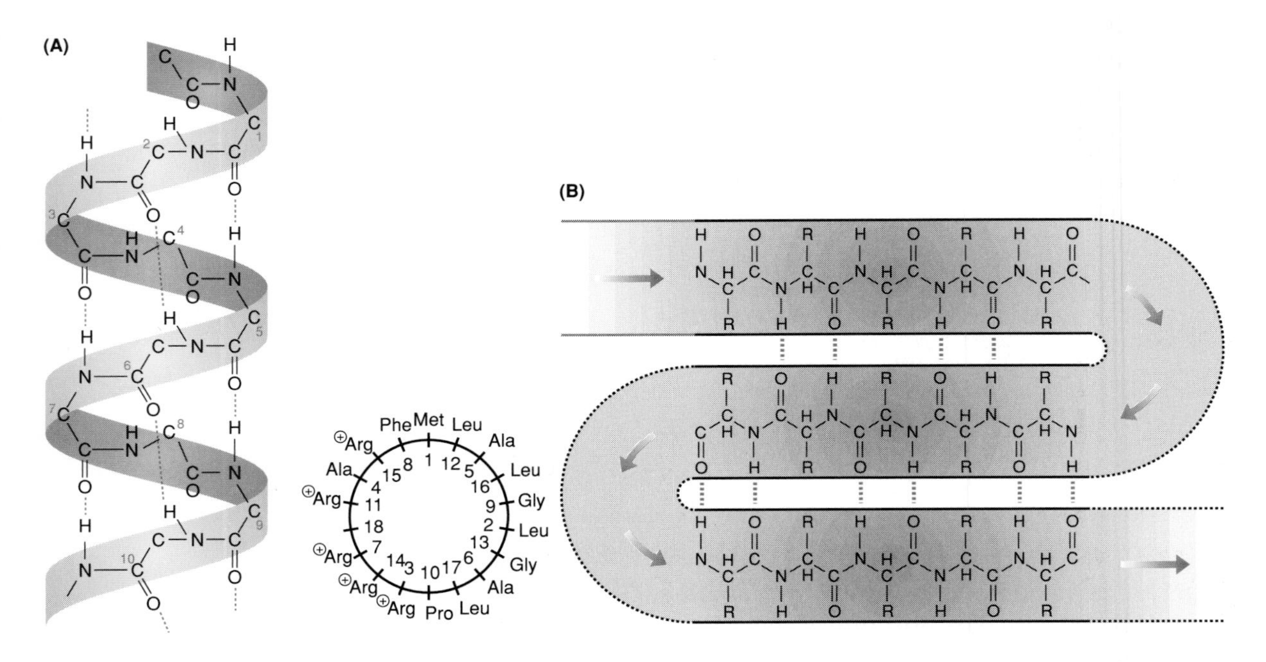

Figure 1.24: Regions of secondary structure in polypeptides are often dominated by intrachain hydrogen bonding.

(A) Structure of an α-helix. *Left:* only the backbone of the polypeptide is shown for clarity. The carbonyl (CO) oxygen of each peptide bond is hydrogen bonded to the hydrogen on the peptide bond amide group (NH) of the fourth amino acid away, so that the helix has 3.6 amino acids per turn. *Note* that for clarification purposes some bonds have been omitted. The side chains of each amino acid are located on the outside of the helix and there is almost no free space within the helix. *Right:* charged amino acids and hydrophobic amino acids are located on different surfaces in an amphipathic α-helix. The sequence shown is that for the signal peptide sequence for the mitochondrial aldehyde dehydrogenase (see *Table 1.6*).
(B) Structure of a β-pleated sheet. *Note* that hydrogen bonding occurs between the CO oxygen and NH hydrogen atoms of peptide bonds on adjacent parallel segments of the polypeptide backbone. The example shows a case of bonding between antiparallel segments of the polypeptide backbone (**antiparallel β-sheet**) and enforced abrupt change of direction between antiparallel segments is often accomplished using β-turns (see text). Arrows mark the direction from N terminus to C terminus. *Note* that parallel β-pleated sheets with the adjacent segments running in the same direction are also commonly found.

chains can coil round each other to form a particular stable structure called a **coiled coil**. Long rod-like coiled coils are found in many fibrous proteins, such as the α-keratin fibers of skin, hair and nails or fibrinogen of the blood clot.

(ii) *The β-pleated sheet.* This features hydrogen bond formation between opposed peptide bonds in parallel (often really antiparallel) segments of the same polypeptide chain (see *Figure 1.24*). β-Pleated sheets form the core of most, but not all globular proteins

(iii) *The β-turn.* Hydrogen bonding between the peptide bond CO group of amino acid residue n of a polypeptide with the peptide bond NH group of residue n+3 results in a hairpin turn. By permitting the polypeptide to reverse direction abruptly, compact globular shapes can be achieved. β-Turns are so named because they also often connect antiparallel strands in β-pleated sheets (see *Figure 1.24*).

More complex structural motifs consisting of combinations of the above structural modules constitute **protein domains**; compact regions of a protein formed by folding back of the primary structure so that elements of secondary structure can be stacked next to each other. Such domains often represent functional units involved in binding other molecules. In addition, covalent **disulfide bridges** are often formed between the sulfhydryl (–SH) groups of pairs of cysteine residues occurring within the same polypeptide chain or on different polypeptide chains (see *Figure 1.25*).

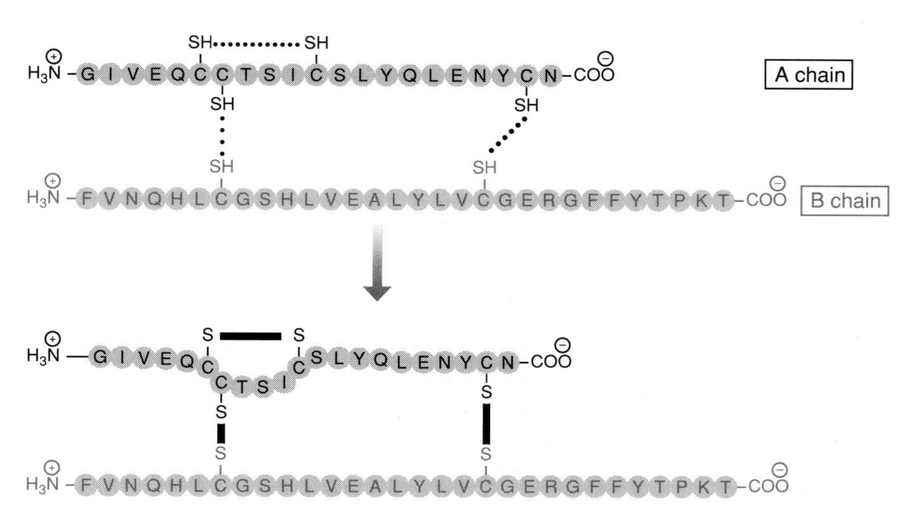

Figure 1.25: Intrachain and interchain disulfide bridges in human insulin.

The disulfide bridges (–S–S–) are formed by a condensation reaction between opposed sulfhydryl (–SH) groups of cysteine residues 6 and 11 of the A chain or between indicated residues of the different chains.

Clearly, the tertiary or quaternary structure of proteins is determined by the primary amino acid sequence. However although secondary structure motifs such as α-helices, β-pleated sheets and β-turns can be predicted by analyzing the primary sequence, the overall three-dimensional structure cannot, at present, be accurately predicted. In addition to the structural complexity of simple polypeptides, many proteins are organized as complex aggregates of multiple polypeptide subunits.

Mutation and DNA repair

Mutations in DNA provide the raw fuel for evolutionary novelty and can result in disease

Mutation is the process which produces heritable changes in DNA. It encompasses both large changes (loss, duplication or rearrangement of whole chromosomes or chromosome segments; see Chapter 2) and so-called **point mutations** (changes involving loss, duplication or alteration of small segments of DNA, often only a single nucleotide). Clearly such alterations have the possibility of causing disease. Mutations in the **germline** (the haploid gametes and those diploid cells from which they are formed) can be transmitted to subsequent generations, whereas mutations in other body cells (**somatic cells**) result in disease being confined to the individual in whom the mutation arises. However, mutations also offer the ability for acquiring new or improved functions which, if accompanied by increased reproductive success, can be fostered by natural selection. The usually low level of mutation may therefore be viewed as a balance between permitting occasional evolutionary novelty at the expense of causing disease or death in a proportion of the members of a species.

While the rate of mutation can be enhanced by exposure to environmental mutagenic agents (e.g. ionizing radiation and chemical mutagens acquired through dietary contact or occupational/environmental exposure), the basal mutation rate reflects inevitable spontaneous errors in chromosome segregation at meiosis, DNA replication and DNA repair, and spontaneous chemical attack.

The two most frequent spontaneous chemical reactions are hydrolytic reactions:

(i) *Depurination*. This involves removal of an adenine or a guanine residue – the N-glycosidic bond linking the purine residue to the carbon 1′ of the deoxyribose is hydrolyzed and the purine is replaced by an hydroxyl group at carbon 1′. Such events are extremely common: approximately 5000 purine bases are lost each day from the DNA of each human cell;

(ii) *Deamination*. Spontaneous deamination of cytosine to uracil (see *Figure 1.13*) occurs at a daily rate of about 100 bases per human cell.

DNA bases are also subject to alteration by reactive metabolites (including reactive forms of oxygen) and by ultraviolet light which can induce **pyrimidine dimers,** the formation of covalent linkages between two neighboring pyrimidines (C–C, C–T or T–T) on the same DNA strand. Most of the chemical changes would be expected to lead to a deletion of one or more base pairs or a base substitution.

DNA repair ensures prompt recognition and repair of the vast majority of altered bases

The very high frequency of alteration in DNA would have potentially lethal consequences if not checked in some way. Fortunately, the vast majority of mutations are promptly corrected by DNA repair enzymes which are constantly scanning the DNA in order to detect and replace damaged nucleotides. As a result, the fidelity with which DNA sequences are maintained in even very complex genomes is high: in 1 year a mammalian germline cell with a genome size of 3×10^9 bp normally suffers alterations to a total of only 10–20 bp.

Two major DNA repair pathways are known:

(i) *base excision repair*. This pathway commences with a DNA glycosylase which excises the relevant base. Thereafter the enzymes AP endonuclease and a phosphodiesterase act in turn to remove the sugar phosphate from which a base has been excised. The gap of a single nucleotide is then filled by the concerted action of DNA polymerase and DNA ligase (see *Figure 1.26*);

(ii) *Nucleotide excision repair*. This pathway is capable of removing almost any type of DNA damage that creates a large change in the DNA duplex. One cut is made on either side of the lesion spanning a distance of nearly 30 nucleotides in human DNA and an associated DNA helicase removes the entire portion of the damaged strand. The large gap is again sealed by a combination of DNA polymerase and DNA ligase activities (see *Figure 1.26*).

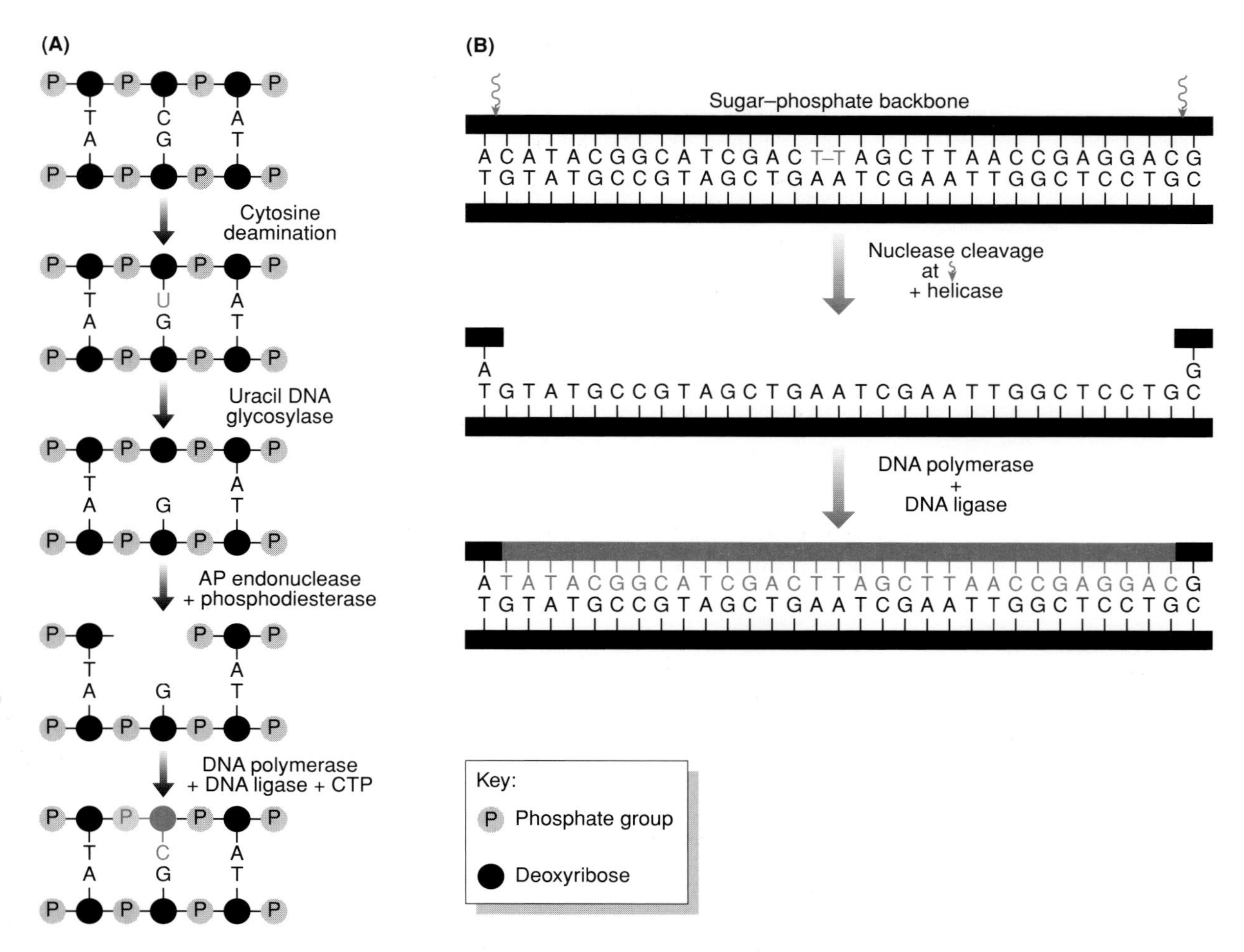

Figure 1.26: Major DNA repair pathways involve excision and repair of a single base and of a polynucleotide tract containing altered bases.

(A) Base excision repair. *Note* that repair of depurination involves the base excision pathway without the initial DNA glycosylase step. The AP endonuclease is so called because it recognizes any site in the DNA duplex that contains a deoxyribose sugar without its base, and such sites naturally arise through loss of a purine (apurinic sites) or of a pyrimidine (apyrimidinic sites).
(B) (Poly)nucleotide excision repair. The example shown is a thymidine dimer which is excised as part of a polynucleotide segment which is subsequently replaced in its entirety.

Further reading

Adams RPL, Knowler JT, Leader DP. (1992) *Biochemistry of the Nucleic Acids*, 11th Edn. Chapman & Hall, London.

Alberts B, Bray D, Lewis J, Raff M, Roberts K, Watson JD. (1994) *Molecular Biology of the Cell*, 3rd Edn. Garland Publishing, New York.

Bird A. (1986) CpG-rich islands and the function of DNA methylation. *Nature*, **321**, 209–213.

Lewin B. (1994) *Genes V*. Oxford University Press, Oxford.

Lodish H, Baltimore D, Berk A, Zipursky L, Matsudaira P, Darnell J. (1995) *Molecular Cell Biology*, 3rd Edn. Scientific American Books, New York.

Chromosome structure and function

Chapter 2

Chromosome structure and organization

The chromosome number and DNA content of eukaryotic cells varies between species and between the cells of a single organism

Bacterial cells typically contain a single small circular chromosome. By contrast, eukaryotic cells have several, usually very large, linear chromosomes in their cell nuclei whose number and DNA content can vary considerably between species (*Table 2.1*). Generally, the genome size parallels the complexity of the organism. However, there are some anomalies: the cells of an onion and a lily contain respectively about five and 30 times as much DNA as that of a typical human cell.

The chromosome number and DNA content of the individual cells of a complex eukaryotic organism can also vary. For example, the **gametes** (sperm and egg cells) of mammals are specialized sex cells which have half the number of chromosomes of somatic cells. The gametes are described as being **haploid**; somatic cells are **diploid**. The gametes arise from certain somatic cells in the testis and ovary which undergo a specialized form of reductive cell division (**meiosis**) (see page 40). In human cells, this results in the production of sperm and egg cells, each with 23 chromosomes, comprising a **sex chromosome** (chromosome which determines sex – either X or Y) and 22 different **autosomes** (nonsex chromosomes).

Table 2.1: Interspecific variation in chromosome number and genome size

Species	Haploid chromosome number	Haploid genome size (Mb)
Saccharomyces cerevisiae (yeast)	16	14
Dictyostelium discoideum (slime mold)	7	70
Caenorhabditis elegans (nematode)	11/12	100
Drosophila melanogaster (fruit fly)	4	170
Gallus domesticus (chicken)	39	1200
Mus musculus (mouse)	20	3000
Xenopus laevis (toad)	18	3000
Homo sapiens (human)	23	3000
Zea mays (maize)	10	5000
Allium cepa (onion)	8	15 000

The diploid somatic cells owe their ultimate origin to the fusion of two haploid cells. During fertilization of an egg cell by a sperm cell, the haploid genomes of the male and female **pronuclei** fuse to give a diploid genome in the resulting fertilized egg cell (**zygote**) (*Figure 2.1*). The diploid zygote and all diploid cells in the resulting individual contain pairs of **homologous chromosomes**, chromosomes which carry almost identical DNA sequences, one member (**homolog**) of each pair being donated by each parent. For example, human somatic cells normally contain 46 chromosomes: 22 homologous pairs of autosomes and two sex chromosomes. The two sex chromosomes may be fully homologous (XX) or partially homologous (XY) (see page 49).

Diploid somatic cells originate by a process of binary cell division, comprising nuclear division (**mitosis**) and cytoplasmic division (**cytokinesis**). Such mitotic cell divisions are conservative because the diploid status is maintained in the daughter cells: the chromosomes of the parental cell duplicate prior to cell division and are then distributed equally to the two daughter cells (see page 39). Ultimately, all diploid cells owe their ancestry to mitotic cell divisions starting from cleavage of the zygote.

During development, and in the subsequent mature organism, constituent cells of organisms will undergo many rounds of mitotic cell divisions prior to terminal cell differentiation or cell death. However, as somatic cells differentiate, the number of chromosomes and the DNA content can be altered. Some differentiated cells have no chromosomes, for example red blood cells and platelets lack a nucleus, and terminally differentiated skin cells lack all the organelles. Other somatic cells are multinucleated, as in the case of muscle fibers where large cells are formed by aggregation of many small cells. Some cells are naturally **polyploid** because they have

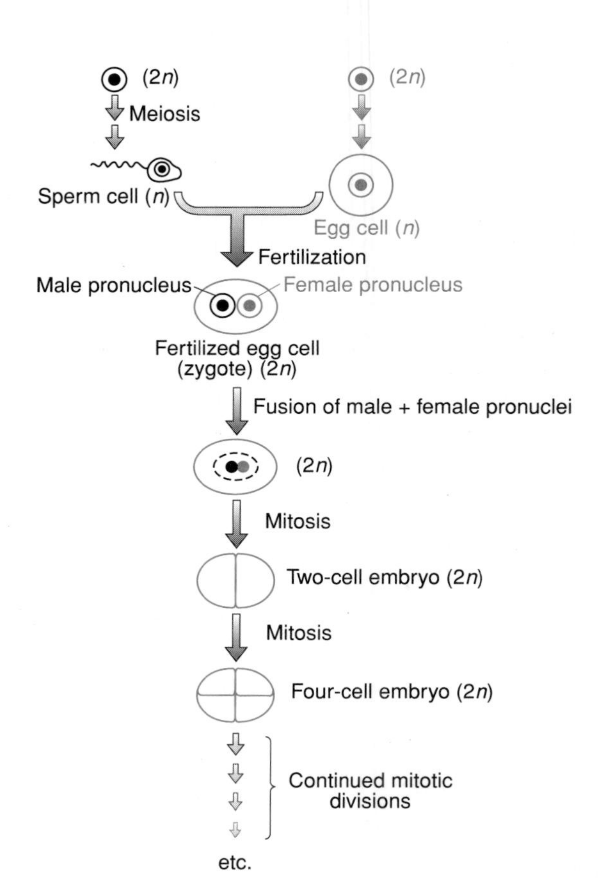

Figure 2.1: Fusion of haploid (*n*) sperm and egg cells generates a diploid (2*n*) zygote which undergoes successive binary divisions.

The haploid sperm and egg cells originate by meiosis from diploid precursors (see *Figure 2.9*). Successive mitotic cleavages of the zygote generate the diploid somatic cells during development, growth and cell turnover.

additional copies of the chromosome set due to additional rounds of DNA duplication prior to cell division (**endomitosis**). Examples include regenerating cells of the liver and other tissues which are tetraploid. The giant megakaryocytes of the bone marrow provide a conspicuous example of naturally occurring polyploid cells: they usually contain eight, 16 or 32 times the haploid DNA content and individually give rise to thousands of platelet cells.

Finally, although the chromosome number of normal undifferentiated somatic cells does not change during the mitotic cell cycle, the DNA content clearly does. The cell cycle comprises a very short stage of cell division, the **M phase** (mitosis phase) and a long intervening **interphase** which comprises three phases: **S phase** (DNA synthesis phase), **G1 phase** (gap between M phase and S phase) and **G2 phase** (gap between S phase and M phase) (see *Figure 2.2*). Cells normally enter S phase only if they are committed to mitosis; nondividing cells remain in a modified G1 stage, sometimes called G0. If the total genomic DNA content of a haploid cell is defined as 1C, the DNA content of diploid cells can be 2C or 4C, depending on which stage of the cell cycle is being considered. This is so because, although the DNA content of cells duplicates during S phase, the centromere does not, and accordingly the number of chromosomes is constant. The standard mitotic metaphase chromosomes, as normally illustrated, consist of two sister chromatids joined at the centromere. From the anaphase stage of mitosis right through until DNA duplication in S phase, a chromosome of a diploid cell contains two DNA strands and the total DNA content is 2C. Thereafter, up until the end of metaphase of the next mitosis, each chromosome effectively has four DNA strands, and the cell has a DNA content of 4C (*Figure 2.2*).

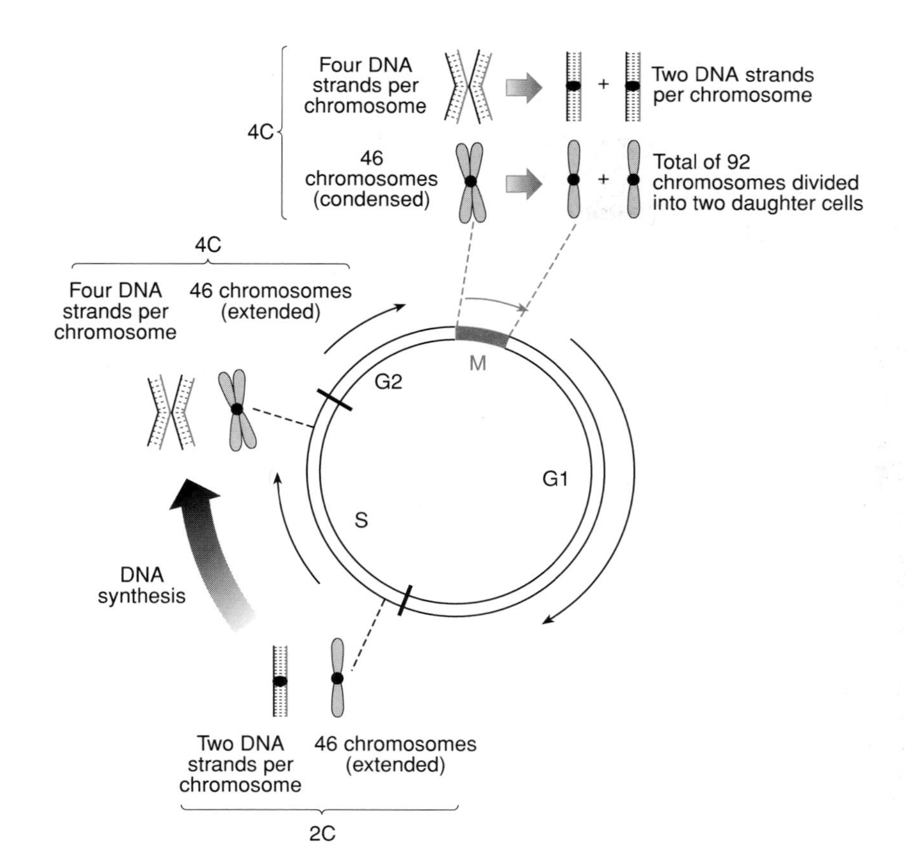

Figure 2.2: Human chromosomal DNA content during the cell cycle.

Note that interphase comprises G1 + S + G2. Chromosomes are normally defined by the presence of a single centromere. They contain one DNA double helix from the anaphase stage of mitosis right through until the stage in S phase when the DNA has duplicated. From this stage until the end of the metaphase stage of mitosis, the chromosome consists of two *chromatids* each containing a DNA duplex, making a total of four DNA strands (two double helices) per chromosome. The DNA content of a cell before S phase is equal to twice the DNA content of haploid cells (i.e. **2C**), while between S phase and mitosis it is 4C.

Packaging of DNA into chromosomes requires multiple hierarchies of DNA folding

The basic material of eukaryotic chromosomes is called **chromatin** and consists of DNA complexed with basic proteins, notably histones, and acidic proteins. In the cell the structure of each chromosome is highly ordered (Manuelidis, 1990) and compaction of the chromosomal DNA (2 nm in diameter) is achieved by complexing with DNA-binding proteins. The most fundamental unit of packaging is the **nucleosome** which consists of a central core complex of eight basic **histone** proteins (two each of histones H2A, H2B, H3 and H4) around which a stretch of 146 bp of double-stranded DNA is coiled in 1.75 turns (*Figure 2.3*). Adjacent nucleosomes are connected by a short length of spacer DNA. The elementary fiber of linked nucleosomes is approximately 10 nm in diameter and is in turn coiled into a **chromatin fiber** of 30 nm diameter which can be resolved by electron microscopy.

At the metaphase stage of cell division, the chromosomes become even more condensed, with linear lengths in human cells of close to 10^{-5} of the expected lengths of the constituent DNA molecules. Each chromatid consists of loops of chromatin fiber, containing approximately 75 kb of DNA per loop. The loops are attached to a central **scaffold** of nonhistone acidic protein, notably topoisomerase II, an enzyme which permits uncoiling of the two DNA strands of the DNA duplexes. Topoisomerase II and some other chromatin proteins are known to bind to AT-rich sequences. Among these, so-called **scaffold attachment regions** (SARs), stretches of several hundred base pairs of highly AT-rich (>65%) DNA, are candidate elements for the regions to which the chromatin loops are attached. The resulting

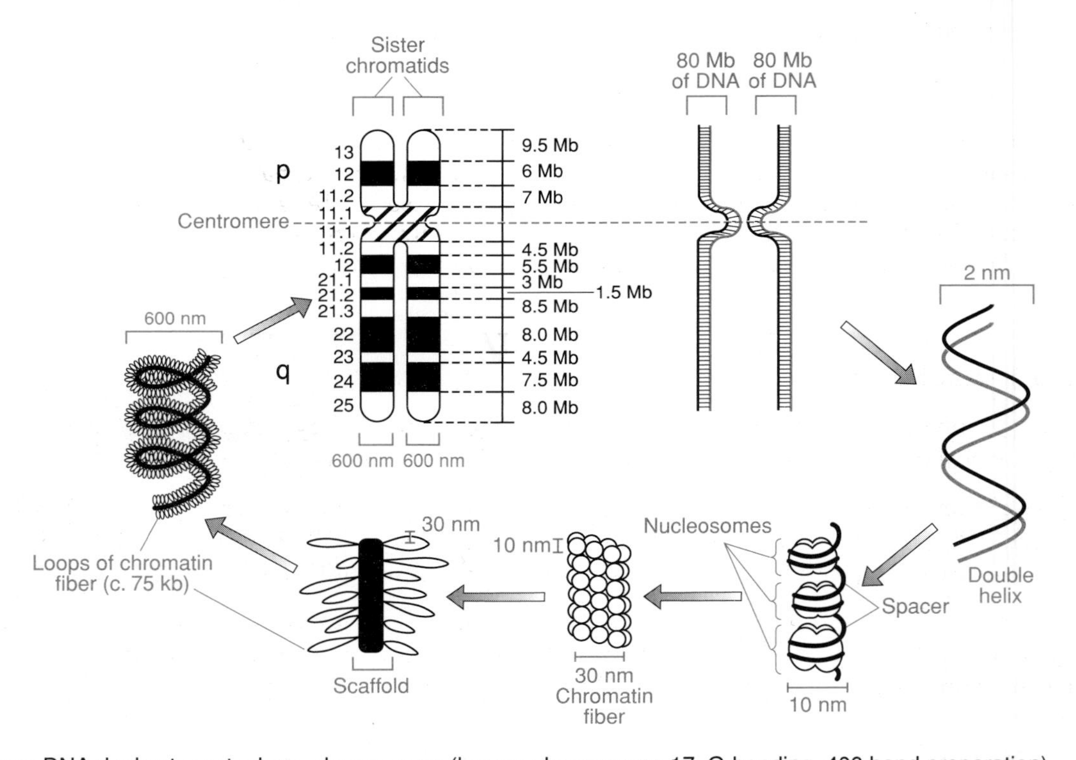

Figure 2.3: From DNA duplex to metaphase chromosome (human chromosome 17, G-banding, 400 band preparation).

Estimated packaging ratios (the degree of compaction of the linear DNA duplex) for human chromosomes are as follows: 1:6 for nucleosomes, 1:36 for the 30 nm fiber and 1: approximately 50 000–100 000 for the metaphase chromosome. Presently, it is uncertain whether the DNA at the centromere of the metaphase chromosome has been delayed in its replication unlike the rest of the chromatid, or whether full DNA replication has occurred in the S phase and the constriction at the centromere is due to some other cause.

loop–scaffold complex is compacted further by coiling to generate a chromatid (see *Figure 2.3*).

Chromosome banding permits structural definition and differentiation of chromosomes

A variety of treatments involving denaturation and/or enzymatic digestion of chromatin, followed by incorporation of a DNA-specific dye, can cause mitotic chromosomes of complex organisms to appear as a series of transverse alternating light and dark staining bands (see *Box 2.1*; Craig and Bickmore, 1993). Banding reflects variations in the longitudinal structure of chromatids; each chromatid may be viewed as a series of stacked disks where each disk differs from that of its nearest neighbor(s) in base composition, time of replication, chromatin conformation, and also in the density of genes and repetitive sequences.

Such banding can permit accurate differentiation of chromosomes – previously the only way of doing this was by examining the sizes of chromosomes and the positions of the centromeres. Additionally, chromosome banding permits more accurate definition of translocation breakpoints, and subchromosomal deletions, etc.

Banding resolution can be increased by obtaining more elongated chromosomes, for example prometaphase chromosome preparations can be used instead of the more conventional metaphase chromosomes. Typical high-resolution banding procedures for human chromosomes can resolve a total of 400, 550 or 850 bands (*Figure 2.4*). The representation of the entire chromosome set as a series of banded chromosomes is called a **karyogram** (often mistakenly described as a **karyotype**, a term which really should be restricted to describing the chromosome complement as a character set, e.g. 46,XX or 46,XY in the case of normal human chromosome preparations). See *Box 2.2* for details of chromosome nomenclature.

G bands have long been known to be late replicating and contain relatively highly condensed chromatin, while R bands generally replicate early in S phase, and have less condensed chromatin structures. Late replicating, highly condensed DNA is relatively transcriptionally inactive and genes are known to be concentrated mostly in the R bands (see below). The structural basis of differential banding, however, has

Box 2.1: Chromosome banding

G-banding – the chromosomes are subjected to controlled digestion with trypsin before staining with Giemsa, a DNA-binding chemical dye. Dark bands are known as G bands. Pale bands are G negative.

Q-banding – the chromosomes are stained with a fluorescent dye which binds preferentially to AT-rich DNA, such as Quinacrine, DAPI (4′, 6-diamidino-2-phenylindole) or Hoechst 33258, and viewed by UV fluorescence. Fluorescing bands are called Q bands and mark the same chromosomal segments as G bands.

R-banding – is essentially the reverse of the G-banding pattern. The chromosomes are heat-denatured in saline before being stained with Giemsa. The heat treatment denatures AT-rich DNA and R bands are Q negative. The same pattern can be produced by binding GC-specific chromomycin dyes such as chromomycin A_3, olivomycin or mithramycin.

T-banding – identifies a subset of the R bands which are especially concentrated at the telomeres. The T bands are the most intensely staining of the R bands and are visualized by employing either a particularly severe heat treatment of the chromosomes prior to staining with Giemsa, or a combination of dyes and fluorochromes.

C-banding – is thought to demonstrate constitutive heterochromatin. The chromosomes are typically exposed to denaturation with a saturated solution of barium hydroxide, prior to Giemsa staining.

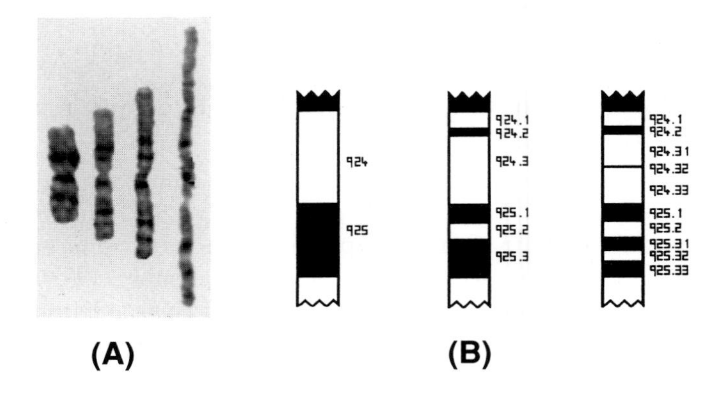

Figure 2.4: Different chromosome banding resolutions can resolve bands, sub-bands and sub-sub-bands.

(A) G-banded chromosome 1 at different banding resolutions.
(B) Numbering of bands, sub-bands, and sub-sub-bands.
Reproduced from Wolstenholme (1992) by permission of Oxford University Press.

only recently been clarified. G and R bands had long been presumed to be AT-rich and GC-rich respectively: quinacrine, which produces the Q-banding pattern, preferentially binds to AT-rich DNA, while the R-banding pattern can be elicited by using the dye chromomycin which preferentially binds GC-rich DNA. However, it is now known that human G/Q band DNA is only a few per cent richer in AT content than R band DNA.

The above paradox has been resolved by the development of new staining techniques which afford an updated scaffold–loop model (Saitoh and Laemmli, 1994). According to this model, the highly AT-rich SARs are differentially folded along the longitudinal length of chromatids: regions of tight packing of SARs are associated with G/Q bands, those with more unfolded SARs are associated with R bands. In this model, Giemsa is thought to stain the loop base selectively, while selective staining of the loop body establishes an R banding pattern (see *Figure 2.5*).

Box 2.2: Chromosome nomenclature

The classification of human chromosomes is decided by the Standing Committee of Human Cytogenetic Nomenclature who regularly update the system (most recently in 1995 – see Further reading). Their report is known as the **International System for Human Cytogenetic Nomenclature (ISCN)**. The basic terminology for banded chromosomes was decided at the meeting in Paris in 1971 and is often referred to as the Paris nomenclature. Short arm locations are labeled p (*petit*) and long arms q (*queue*). At different levels of microscopic resolution, different numbers of bands are seen. At the lowest resolutions, only a few major bands are identified and these are labeled p1, p2, p3; q1, q2, q3, etc., counting outwards from the centromere. Sub-bands are p11 (one-one, not eleven!), p12, etc, and sub-sub-bands p11.1, p11.2, etc. (see above and *Figures 7.4* and *9.25*. The centromere is designated 'cen' and the telomere 'ter'.

Note that it is conventional to refer to relative subchromosomal localizations in terms of *proximity to the centromere*. For example, *proximal* Xq means the segment of the long arm of the X which is closest to the centromere while *distal* 2p means the portion of the short arm of chromosome 2 which is most distant from the centromere, and therefore closest to the telomere.

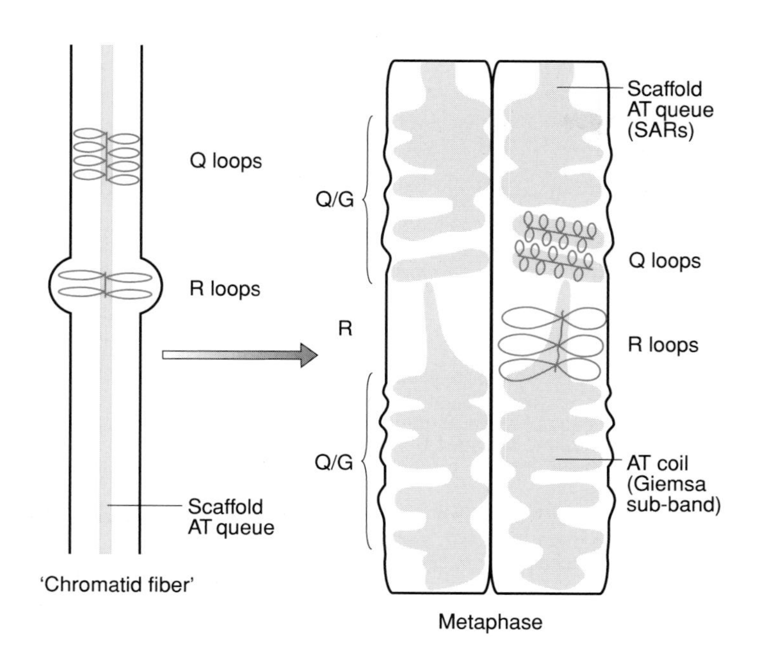

Figure 2.5: The structural basis of chromosome banding, in the model of Saito and Laemmli (1994).

Chromatin loops are attached to the chromosome scaffold at special scaffold attachment regions (SARs). DNA which forms G bands (Q bands) contains more closely spaced SARs, giving tighter loops (Q loops). There are more SARs per unit length of chromosome in G bands than in R bands. Stains with an affinity for the AT-rich SARs, such as Giemsa and quinacrine, give the G-banding pattern. Reproduced from Saitoh and Laemmli (1994) with permission from Cell Press.

Types of cell division – mitosis and meiosis

Mitosis is the normal process of cell division

As the organism develops and grows through the embryonic, fetal, infant and adult stages, mitotic cell divisions are needed to generate the required large numbers of cells. Additionally, as many cells have a limited life span, there is a continuous requirement in the mature organism to generate new cells by mitotic cell division. Accordingly, mitosis is the normal process of cell division from cleavage of the zygote to death of the organism. Large numbers of divisions are involved – approximately 10^{17} in the lifetime of a human (see page 242).

The M phase of cell division consists of the various stages of nuclear division [**mitosis**, i.e. prophase, prometaphase, metaphase, anaphase and telophase, and the process of cell division (**cytokinesis**) which overlaps the final stages of mitosis (*Figure 2.6*)]. In preparation for cell division, the previously highly extended chromosomes undergo a considerable degree of condensation so that, by the **metaphase** stage of mitosis, they are readily visible under the microscope. Even though DNA replication has occurred some time previously, it is only at the prometaphase stage of mitosis that individual chromosomes are seen to comprise two **chromatids** which are attached at the centromere. Each of the two sister chromatids of a single chromosome contains a parental DNA strand and a newly synthesized DNA strand, so that a chromosome with two chromatids in effect has the equivalent of two DNA duplexes or four DNA strands (*Figure 2.2*).

A crucial component during mitotic division is the **mitotic spindle** which is formed from fibers comprised of tubulin-based **microtubules** and microtubule-associated proteins. Two important classes of spindle fibers are the **polar fibers**, which extend from the two poles of the spindle towards the equator, and the **kinetochore fibers**, which are attached to the **kinetochore** (a large multiprotein structure which is attached to the centromere of each chromatid) and extend in the direction of the spindle poles (*Figure 2.7*).

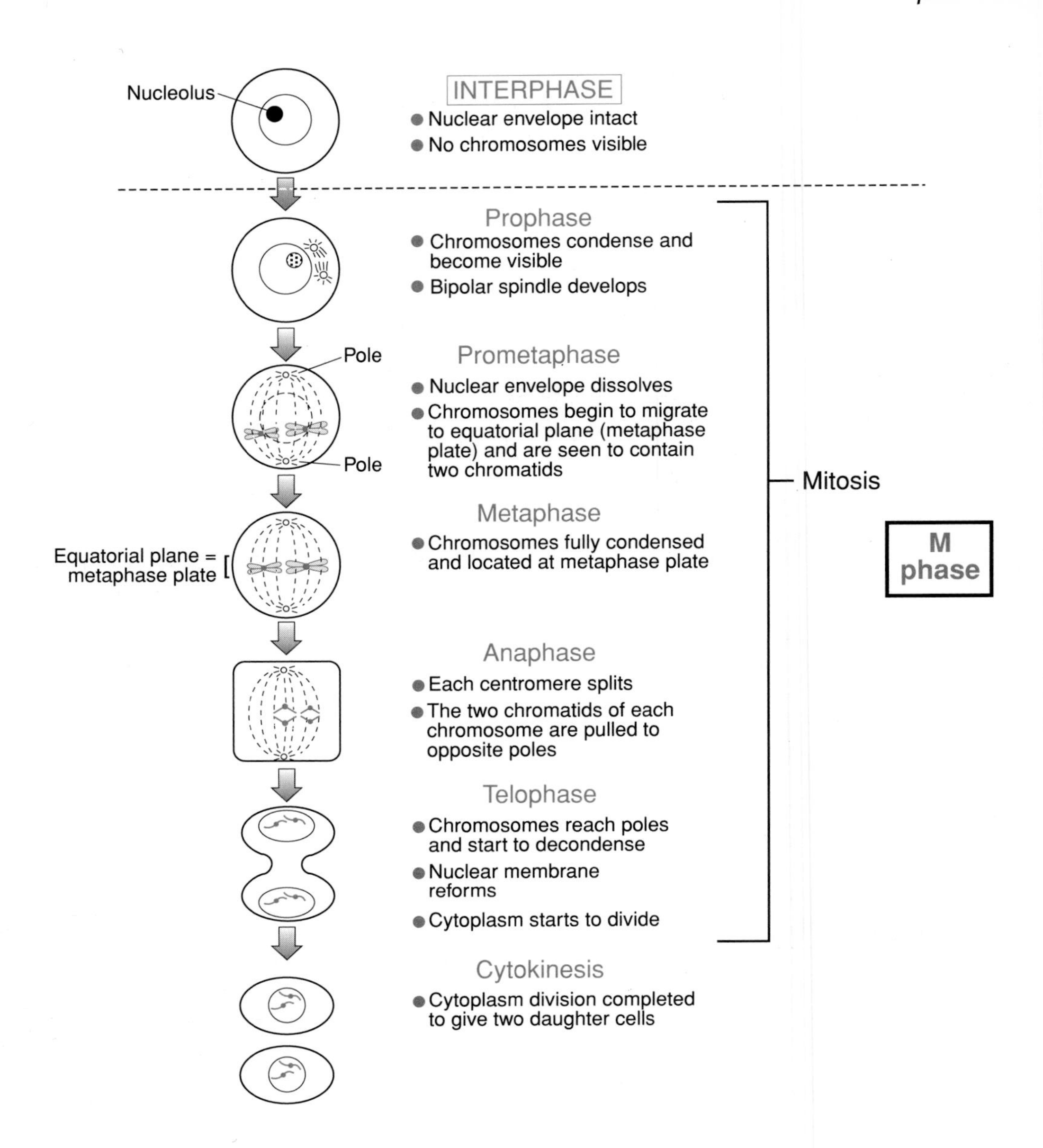

Figure 2.6: From interphase through M phase.

Although polar fibers develop at prophase while the nuclear membrane is still intact, kinetochore fibers do not develop until prometaphase. The interaction between the different spindle fibers pulls the chromosomes towards the center, and by metaphase each chromosome is independently aligned at the equatorial plane (**metaphase plate**). This means that paternal and maternal homologs are normally well separated at mitotic metaphase (*Figure 2.8*). Following centromere division at anaphase, the spindle fibers then pull the separated chromatids of each chromosome to opposing poles (*Figure 2.8*).

Meiosis is a specialized form of cell division giving rise to sperm and egg cells

Meiosis involves two successive cell divisions of specialized diploid cells in the ovary (**primary oocytes**) or testis (**primary spermatocytes**), each of which are differentiated products of diploid precursor cells (**oogonia** in females, **spermatogonia** in males) (*Figure 2.9*). In males, the product is four spermatozoa; in females, however, the cytoplasm divides unequally at each stage: the products of meiosis I (the first

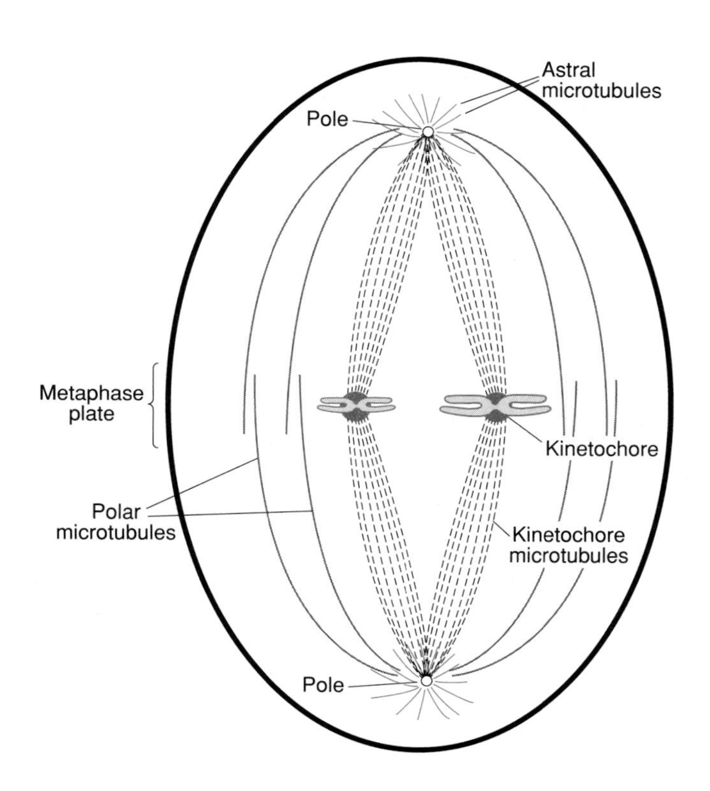

Figure 2.7: The mitotic spindle is composed of different classes of microtubule.

Spindle microtubules are of three classes: astral microtubules radiate from each pole; polar microtubules form attachments that link the two poles; and kinetochore microtubules link metaphase chromosomes at the equatorial plane (metaphase plate) to each of the poles, by attaching to the pair of kinetochores of each chromosome. For the sake of clarity, only a small fraction of the microtubules is shown.

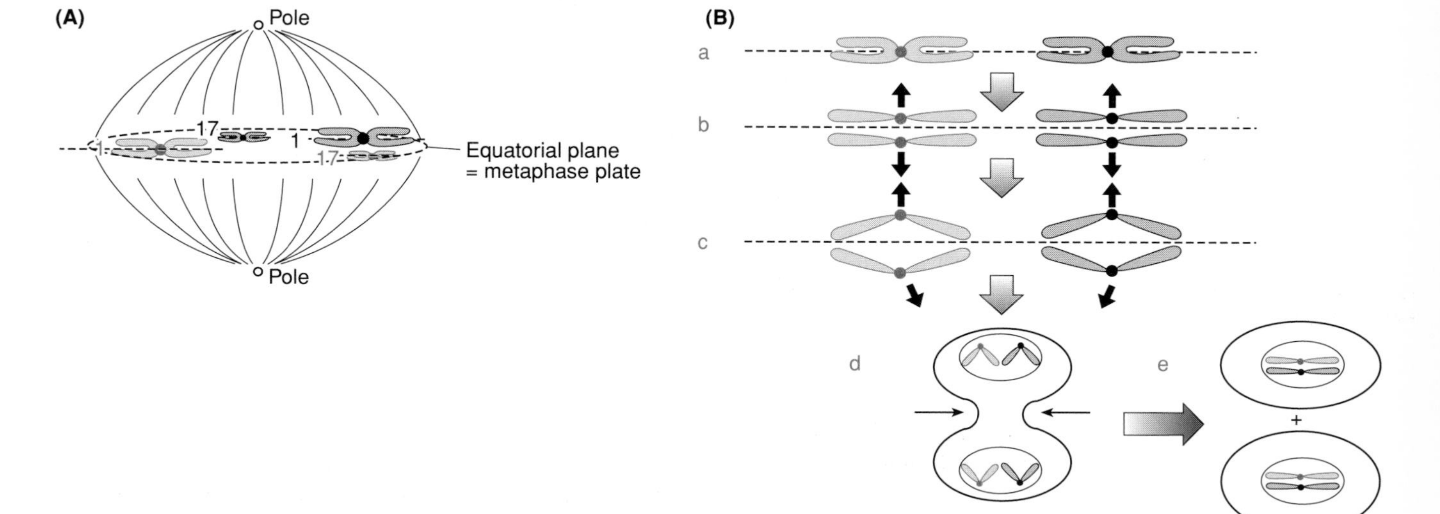

Figure 2.8: Independent alignment of homologous chromosomes at mitotic metaphase and migration of separated chromatids to the opposing poles.

(A) Physical separation of the paternal (black) and maternal (red) homologs of each chromosome pair (e.g. human chromosome 1 and chromosome 17) is meant to indicate that they are independently aligned at the metaphase plate.
(B) Independently aligned maternal and paternal homologs of each chromosome pair (a) subsequently undergo centromeric duplication and separation of chromatids at the anaphase stage (b,c). Solid arrows indicate spindle-derived forces pulling on the kinetochores of the centromeres, eventually driving the sister chromatids of each chromosome to the opposing poles. At this stage (telophase) they become enclosed in a nuclear envelope (d), prior to cytokinesis (e).

meiotic division) are a large **secondary oocyte** and a small cell (**polar body**). The secondary oocyte then gives rise to the large mature egg cell and a second polar body (*Figure 2.9*). A number of other fundamental differences between meiosis and mitosis are summarized in *Table 2.2*.

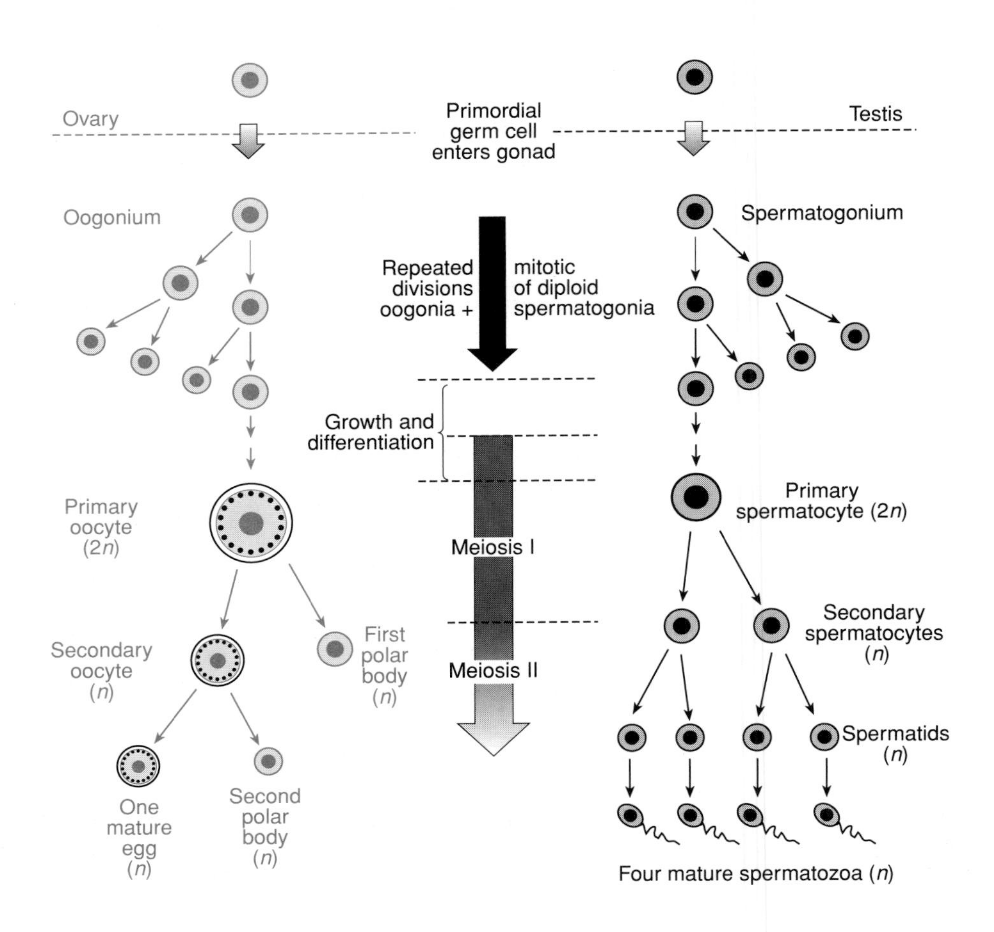

Figure 2.9: Eggs and sperm cells are generated by mitotic division of precursor gonadal cells, followed by meiotic division of primary oocytes and spermatocytes.

The first meiotic division (meiosis I) may take up to several years. For example, in human oogenesis, meiosis I is arrested at the prophase stage from the fetal stage right through to puberty. During this time, the primary oocytes complete their growth phase, acquiring an outer jelly coat, cortical granules, ribosomes, mRNA, yolk, etc. At puberty, hormones induce the completion of meiosis I in a small number of primary oocytes, and the resulting secondary oocytes undergo meiosis II. The secondary spermatocytes and oocytes are haploid because meiosis I does not involve a DNA duplication stage, unlike meiosis II.

Table 2.2: Differences between meiosis and mitosis

	Mitosis	Meiosis
Location	All tissues	Only in testis and ovary
Products	Diploid somatic cells	Haploid sperm and egg cells
DNA replication and cell division	Normally one round of replication per cell division	Only one round of replication (in meiosis I); but two cell divisions
Extent of prophase	Short (~30 min in human cells)	Long and complex in meiosis I; can take years to complete
Pairing of homologs	None	Yes (in meiosis I)
Recombination	Rare and abnormal	Normally at least once for each pair of homologs
Relationship between daughter cells	Genetically identical	Different (recombination and independent assortment of homologs)

A crucial difference between meiosis and mitosis lies in the distribution of genetic material to daughter cells. The mechanism of cell divisions at mitosis and meiosis II are equivalent: both act to ensure that each daughter cell receives identical copies of the DNA. By complete contrast, meiosis I has the function of ensuring that daughter cells receive different genetic material. This is achieved by two means, as follows.

Independent assortment of paternal and maternal homologs

Unlike mitosis or meiosis II, meiosis I is a reductive division because the cells divide without prior chromosome duplication. The mechanism of division (see below) ensures that the paternal homolog of each homologous pair of chromosomes will enter one daughter cell and the maternal homolog will enter the second daughter cell. However, for each homologous chromosome pair, the choice of which homolog enters which cell is random. Because the chromosomes show independent assortment, and because the paternal and maternal homologs normally show some genetic differences, the chromosome complement of individual sperm cells or egg cells shows enormous genetic variability. On the basis of independent assortment alone, a single individual can produce about 8.4×10^6 different gametes (*Figure 2.10*).

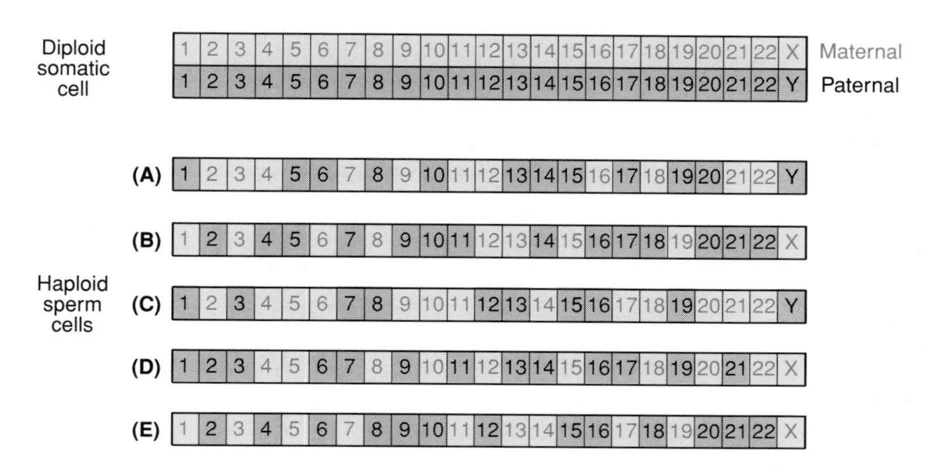

Figure 2.10: Independent assortment of maternal and paternal homologs at meiosis I generates enormous variation.

Diploid somatic cells contain 23 chromosome pairs, one set of 23 being maternally inherited and the other set being paternal in origin. However, haploid gametes, such as sperm cells, only contain one each of each pair of homologs, chosen at random. Ignoring the possibility of recombination, different sperm cells (or eggs) from a single individual would be expected to show numerous chromosome combinations (sperm cells A–E show five combinations out of a possible total of 2^{23} or 8.4 million).

Recombination

During prophase of meiosis I, the maternal and paternal homologs of each chromosome pair form a **bivalent** by pairing together (**synapsis**) (*Figure 2.11*). At the zygotene stage, each pair of homologs begins to form a **synaptonemal complex** consisting of the two homologs in close apposition, separated by a long linear protein core. Completion of this complex marks the start of the pachytene stage, the stage at which recombination (**crossing-over or crossover**) occurs. This involves breaking the double helix in each of two nonsister chromatids (i.e. a paternal chromatid and a maternal chromatid) and exchanging the fragments in a reciprocal fashion.

Currently, the mechanism allowing alignment of the homologs is not understood. However, it is thought that such close apposition is required for recombination. **Recombination nodules**, very large multiprotein assemblies located at intervals on the synaptonemal complex, are thought to mediate the recombination events. The two homologs can be seen to be physically connected at specific points. Each such connection is described as a **chiasma** (plural **chiasmata**) and represents the physical expression of a crossover point.

In addition to their role in recombination, chiasmata are thought to fulfill an essential function in chromosome segregation at meiosis I: by holding the maternal and paternal homologs of each chromosome pair together on the spindle until anaphase I (*Figures 2.11* and *2.12*), they have a role that is analogous to that of the centromeres in mitosis and meiosis II. The importance of this role is reflected by the frequency of crossover: each pair of homologs is thought to undergo at least one crossover, and often several crossovers during meiosis. Even the human X and Y chromosomes, which show only partial homology, are known to undergo one obligatory crossover during meiosis (see pages 211–212).

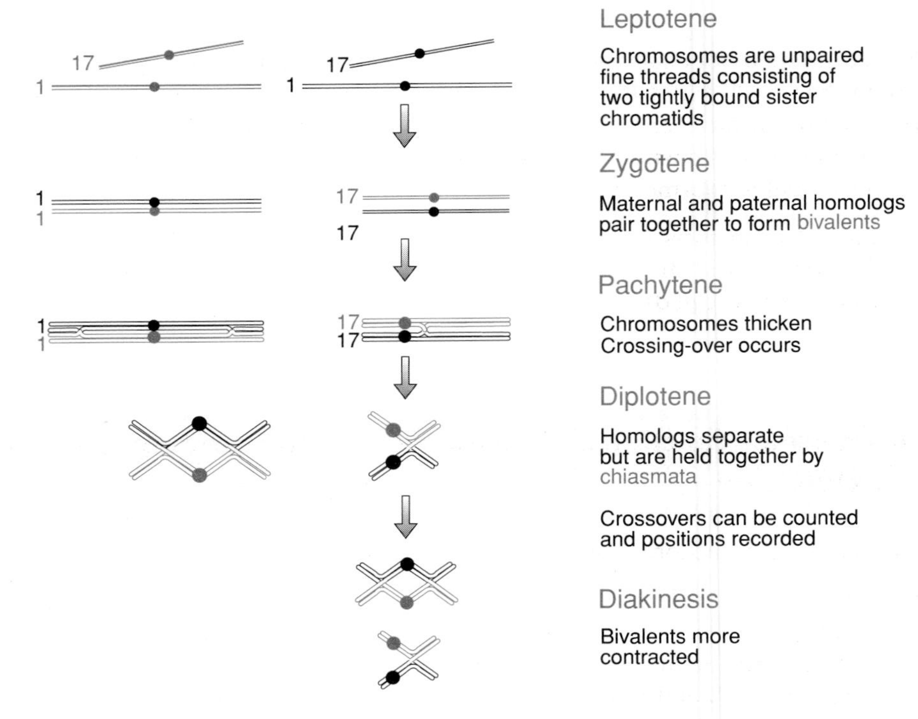

Figure 2.11: The five stages of prophase in meiosis I.

Two representative pairs of homologs are shown. Of these, the chromosome 1 pair are envisaged to undergo two crossovers; the chromosome 17 pair are shown to undergo only a single crossover. For the sake of clarity, in each case the crossover is envisaged to involve only a single paternal chromatid and a single maternal chromatid. *Note*, however, that the number of crossovers is likely to be higher in practice and may involve three or even all four chromatids in a bivalent.

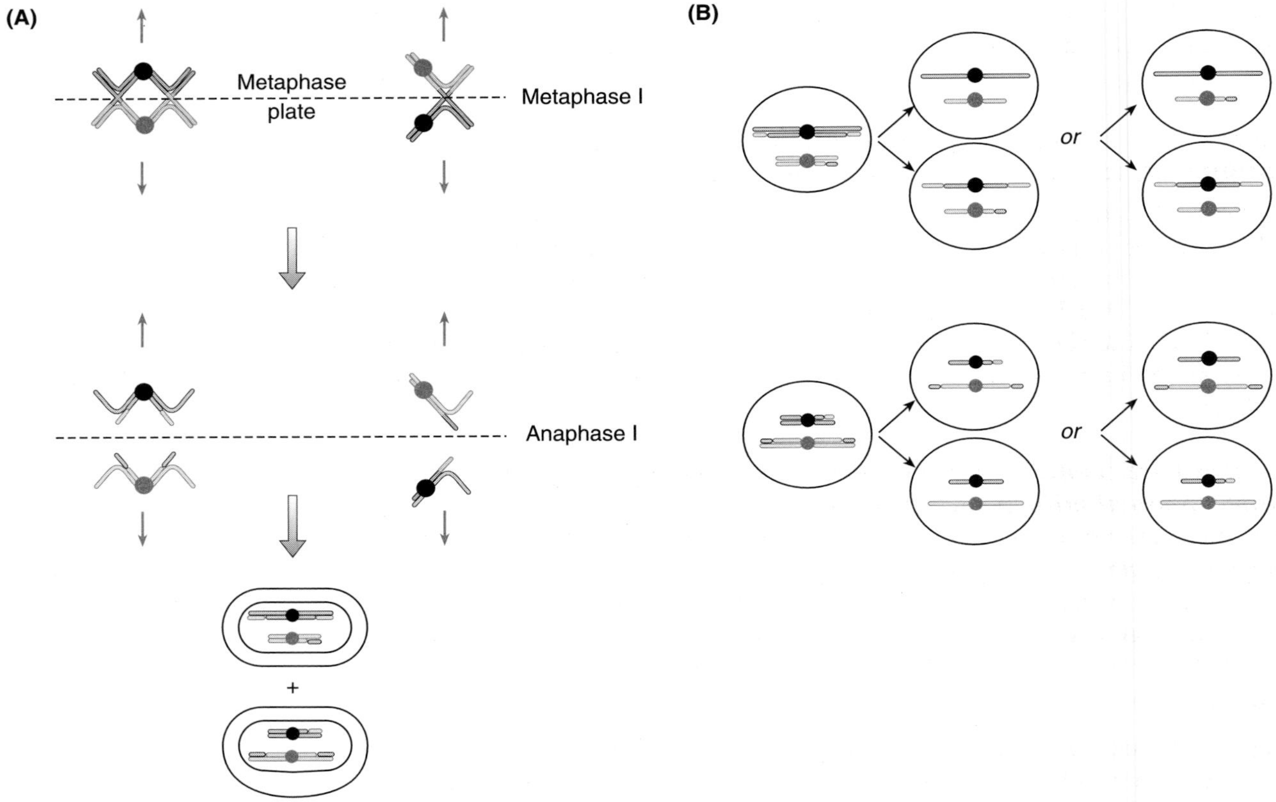

(A) Metaphase plate — Metaphase I — Anaphase I

(B) or — or

Figure 2.12: From metaphase to the products of meiosis II.

(A) From metaphase to cell division in meiosis; **(B)** from meiosis I to meiosis II. The figure follows on from *Figure 2.11* to show possible fates of two pairs of homologs that have undergone recombination at metaphase. In the example shown, the maternal chromosome 1 and paternal chromosome 17 segregate to one cell during meiosis I, while the paternal chromosome 1 and maternal chromosome 17 segregate to the other cell. Because the recombination events of *Figure 2.11* are envisaged to involve only two nonsister chromatids in each case, the final meiotic products will contain some nonrecombinant chromosomes as well as recombinant chromosomes. Although the two sister chromatids of each chromosome will segregate to different daughter cells, the different chromosomes behave independently and so different combinations are possible, as shown.

The requirement for chiasmata to ensure correct segregation during meiosis I makes certain that shuffling of the genetic material occurs by genetic recombination. The combination of independent assortment of homologs plus obligate recombination between homologs ensures that the genetic variation between the gametes of a single individual is enormous.

By contrast to meiosis I, the second meiotic division, **meiosis II**, involves a completely symmetrical distribution of genetic information to the resulting daughter cells, as in mitotic cell divisions.

Chromosome function and relationship of transcriptional activity to chromosome architecture

Mammalian chromosomes have two biological functions:

- *to perpetuate the hereditary material during development of an individual.* By replicating and ensuring that duplicated DNA is divided equally between daughter cells during mitotic cell division, they perpetuate the DNA within the cells of an individual during development, growth and cell turnover.
- *To shuffle the hereditary material between successive generations.* As detailed on pages 42–45, this is achieved by meiotic mechanisms entailing independent assortment of and recombination between parental homologs.

Chromosomal function is dependent on three kinds of element: centromeres, telomeres and replication origins

The biological functions of any eukaryotic chromosome are crucially dependent on only three classes of DNA sequence element: centromeres, telomeres and origins of replication. This simple requirement has been verified by the successful construction of artificial chromosomes in yeast: large foreign DNA fragments can be artificially ligated to short sequence elements that specify a functional centromere, two telomeres and a replication origin, provided in a specialized cloning vector (see *Figure 4.15*).

Centromeres are cis*-acting DNA elements responsible for segregation of chromosomes at mitosis and meiosis*

Normal chromosomes each have a single centromere which can be identified cytologically as the **primary constriction**, the region at which sister chromatids are associated. The centromere is essential for segregation at mitosis (and also meiosis): chromosome fragments that lack a centromere (**acentric** fragments) do not become attached to the mitotic spindle, and so fail to be included in the nuclei of either of the daughter cells.

During the late prophase stage of mitosis, a pair of kinetochores forms at each centromere, one attached to each sister chromatid. Multiple kinetochore microtubules attach to each kinetochore and provide physical links between the centromere of a chromosome and the two spindle poles (*Figure 2.7*). At anaphase, the kinetochore microtubules serve to pull the two sister chromatids toward opposite poles of the spindle. Kinetochores play a central role in this process, by controlling assembly and disassembly of the attached microtubules and, through the presence of motor molecules, by ultimately driving chromosome movement.

Since the position of the centromere on a specific chromosome is fixed, centromere structure and/or function has been presumed to be specified by DNA sequences at the centromere. In simple eukaryotes, the sequences that specify centromere function are very short. For example, in yeast cells the centromere element (CEN) is about 110 bp long and includes two highly conserved flanking elements of 9 bp and 11 bp and a central segment of about 80–90 bp which is AT-rich (see *Figure 2.13*). The centromeres of such cells are interchangeable – a CEN fragment derived from one yeast chromosome can replace the centromere of another with no apparent consequence. However, in complex eukaryotes such as mammals, centromeres are more complicated in structure with some chromosome-specific DNA sequence elements.

Telomeres seal the ends of chromosomes and confer chromosome stability

Telomeres are specialized structures, comprising DNA and protein, which cap the ends of eukaryotic chromosomes. They have been envisaged to have several functions:

- to maintain the structural integrity of a chromosome. If a telomere is lost, the resulting chromosome end is highly unstable and has a tendency to fuse with the ends of other broken chromosomes, to be involved in recombination events and to be degraded. Telomere-binding proteins recognize the overhanging 3′ end of a telomere (see below) and can protect the terminal DNA *in vitro*. Accordingly, they may provide the 'cap' that protects the telomere *in vivo*.
- To ensure complete replication of the extreme ends of chromosome termini. During DNA replication, the synthesis of the lagging strand is discontinuous and requires the presence of some DNA ahead of the sequence which is to be copied to serve as the template for an RNA primer (see *Figure 1.8*). However, at

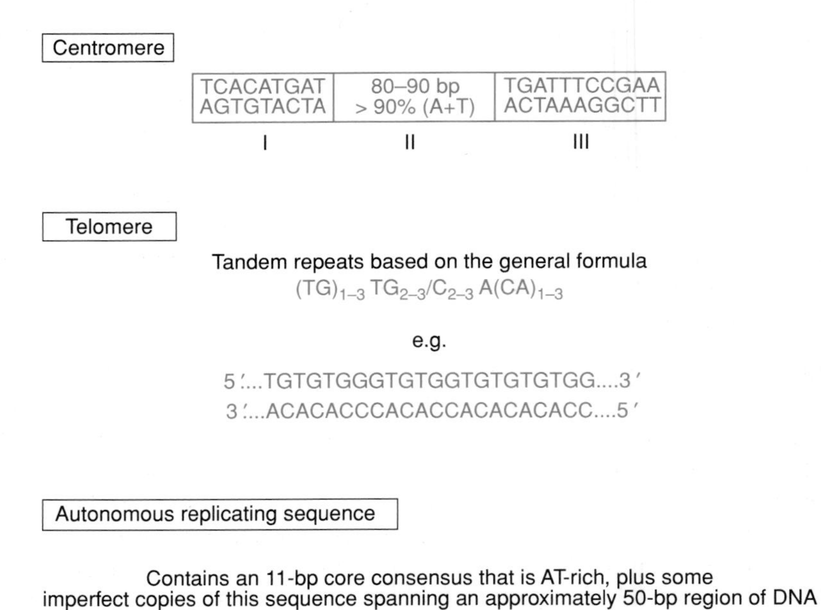

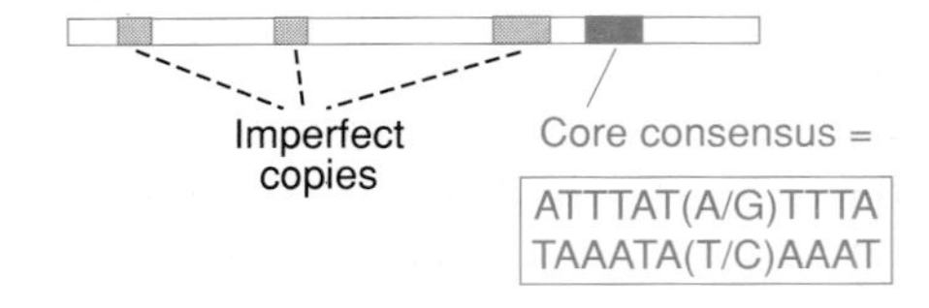

Figure 2.13: DNA sequence composition of the three elements required for chromosome function in yeast.

the extreme terminus of a linear molecule, there can never be such a template, and a different mechanism is required to solve the problem of replicating the ends of a linear DNA molecule (see below).

- To have a role in establishing the three-dimensional architecture of the nucleus and/or in chromosome pairing. This speculative role is suggested by the observation that the ends of chromosomes are located at the nuclear periphery.

The structure of eukaryotic telomeres consists of a long array of tandem repeats. One DNA strand of the telomere contains TG-rich sequences and terminates in the 3′ end; the complementary strand is CA-rich. Unlike centromeres, the sequence of telomeres has been highly conserved in evolution – there is considerable similarity in the simple sequence repeat, for example TTGGGG (*Paramecium*), TAGGG (*Trypanosoma*) TTTAGGG (*Arabidopsis*) and TTAGGG (*Homo sapiens*) (see also *Figure 2.13*).

The problem of replicating the ends of a chromosome has been solved by extending the synthesis of the leading strand using a specialized enzyme, **telomerase**. This RNA–protein complex carries within its RNA component a short sequence which will act as a template to prime extended DNA synthesis of telomeric DNA sequences on the leading strand. Further extension of the leading strand provides the necessary template for completing synthesis of the lagging strand by DNA polymerase α (*Figure 2.14*).

The continued reliance on discontinuous synthesis of the lagging strand means that although telomere synthesis permits completion of 3′ end synthesis of nontelomeric DNA, the telomere itself will have a protruding 3′ end. This provides a single-stranded DNA target for binding by telomere-specific proteins. However, the actual nature of the telomere sequence may not be important. The telomere length is known to be highly variable and is subject to genetic control.

Just internal to the essential telomeric repeats, eukaryotic chromosomes also have a more complex set of repeats called subtelomeric or telomere-associated sequences. Their sequences are not conserved in eukaryotes and their function is unknown.

Replication origins are required to initiate DNA replication and maintain chromosome copy number

The DNA in most diploid cells normally replicates only once per cell cycle (see page 37 for some exceptions). The control of the initiation of replication is governed by key *cis*-acting sequences located in the immediate proximity of the physical locations at which DNA synthesis is initiated. They have been envisaged to be sites at which *trans*-acting proteins bind.

Eukaryotic origins of replication have been most comprehensively studied in yeast cells, where the presence of a putative replication origin can be tested by a genetic assay. To do this, defective yeast cells lacking an essential gene are transformed with a bacterial plasmid containing the essential yeast gene that the host cell lacks, plus a random fragment of yeast DNA which is to be tested for its ability to promote autonomous replication. This is so because the bacterial replication origin in the plasmid does not function in yeast cells; the few plasmids which transform at high efficiency must therefore possess a sequence within the inserted yeast fragment that confers the ability to replicate extrachromosomally at high efficiency, that is an **autonomously replicating sequence** (ARS) element.

ARS elements are thought to derive from authentic origins of replication and, in some cases, this has been confirmed by mapping a specific ARS element to a specific chromosomal location and demonstrating that DNA replication is indeed

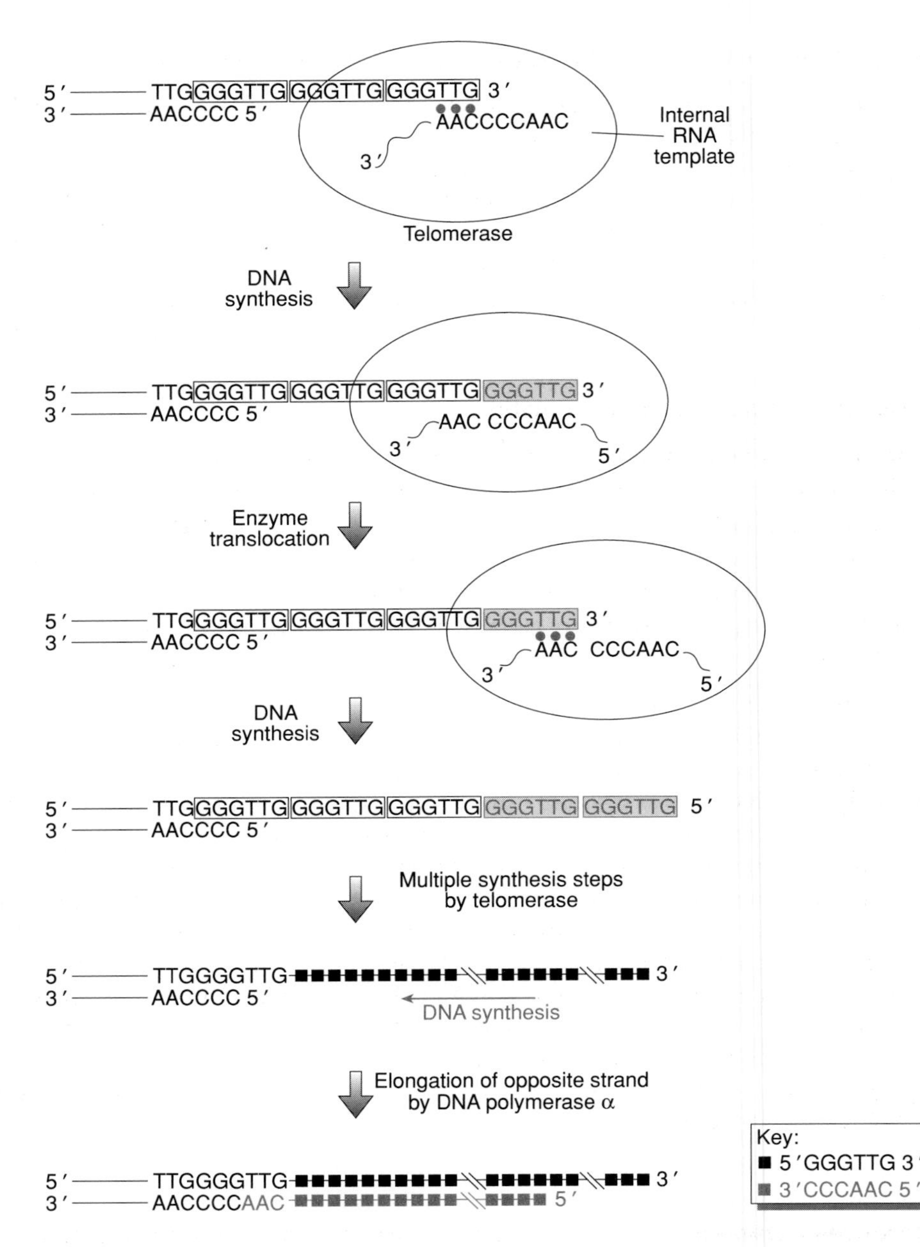

Figure 2.14: Telomerase extends the TG-rich strand of telomeres by DNA synthesis using an internal RNA template.

initiated at this location. ARS elements extend for only about 50 bp and consist of an AT-rich region which contains a conserved core consensus and some imperfect copies of this sequence (*Figure 2.13*). In addition, the ARS elements contain a binding site for a transcription factor and a multiprotein complex is known to bind to the origin.

Mammalian replication origins have, however, been much less well defined because of the absence of a genetic assay. Some initiation sites have been studied. However, such studies have not been able to identify a unique origin of replication, leading to speculation that replication can be initiated at multiple sites over regions tens of kilobases long. Recently, however, computer analysis of the sequences of the regions encompassing several eukaryotic origins of replication, including examples from human DNA and the DNA of other mammals, identified a consensus DNA sequence WAWTTDDWWWDHWGWHMAWTT where W = A or T; D = A or G or T; H = A or C or T; and M = A or C (Dobbs *et al.*, 1994).

Transcriptional activity is related to chromosome architecture

During mitosis, the chromosomes condense and are essentially transcriptionally inactive. However, during interphase, most of the chromatin fibers are much less densely packed than in the mitotic chromosomes. Such material is called **euchromatin**, and is dispersed throughout most of the nucleus and stains diffusely. The remaining chromatin is comprised of sets of fibers which are highly condensed throughout the cell cycle and form dark-staining regions (**heterochromatin**). Whereas euchromatin contains transcriptionally active regions interspersed with nontranscribed DNA sequences, heterochromatin is essentially transcriptionally inert.

There are two classes of heterochromatin:

(i) **facultative heterochromatin** can be either genetically active or inactive, as in the specialized case of mammalian X-chromosome inactivation (see page 50);

(ii) **constitutive heterochromatin** is always in the inactive state. Constitutive heterochromatic regions are composed largely of certain repetitive DNA sequences which are found in and around the centromeres of chromosomes and in certain other regions (e.g. in the human genome, see karyogram in *Figure 7.4*).

Clearly, condensation of chromatin is associated with loss of gene expression. However, although location within euchromatin is virtually essential for gene expression, it is clearly not sufficient because there is a large amount of transcriptionally inactive DNA within euchromatic regions. Given that G bands are known to contain more condensed chromatin than R bands, it was widely expected that if G and R bands showed a similar differential degree of condensation in interphase the latter would be associated with a greater degree of transcriptional activity. In support of this view, early experiments suggested that human poly(A)$^+$ mRNA preferentially hybridized to R bands, and *in situ* hybridization mapping experiments (see page 126) have suggested that 80% of genes in the human genome map to R bands, and only 20% to G bands. Very recently, CpG islands, regions of vertebrate genomes which are marked by a high degree of transcriptional activity (see page 16) have been shown to be located preferentially in R bands, notably in the subset of R bands that are revealed by T-banding (see *Figure 7.5*).

Differences between the X and Y chromosomes of mammals limit their ability to pair at meiosis and require special mechanisms to ensure correct gene expression

Unlike homologous pairs of autosomes, the two sex chromosomes of mammals, the X chromosome and the Y chromosome, are structurally rather different from one another. For example, the human X chromosome is a large submetacentric chromosome with numerous genes, whereas the corresponding Y chromosome is a small acrocentric chromosome (see human karyogram, *Figure 7.5*) in which only a very few active genes are thought to occur. Males normally carry one copy of each sex chromosome (XY); females inherit two X chromosomes, one from each parent.

The major male-determining gene is carried on the Y chromosome and the presence of a Y chromosome is normally sufficient to ensure maleness, irrespective of the number of X chromosomes. For example, *Klinefelter's syndrome* is a rare disorder in which individuals have a Y chromosome together with multiple X chromosomes (e.g. 47, XXY, 48, XXXY and 49, XXXXY), and are male. In contrast, individuals lacking a Y chromosome but with an aberrant number of X chromosomes are female [e.g. individuals with Turner's syndrome (45, X) and those with multiple X chromosomes (47, XXX or 48, XXXX)]. Some XY individuals, however, are also female. One cause of XY females is that the male-determining gene on the Y chromosome (*SRY* in humans) is absent or defective in expression. Similarly, XX males have also been identified. In some cases this occurs as a result of rare translocations in which the *SRY* gene has been translocated from a Y chromosome on to an X chromosome, so that any individual who inherits the X^{SRY} chromosome is male.

The considerable difference between the X and Y chromosomes could be expected to lead to difficulties in two major areas: meiotic pairing and gene dosage.

Meiotic pairing of the X and Y chromosomes

In female meiosis, each of the chromosomes has a partner which is fully homologous with it. However, in male meiosis, pairing is required between the only partly homologous X and Y chromosomes. Despite their differences in structure, however, pairing is made possible by a small region of homology between the X and Y chromosome at one end of the two chromosomes. In humans, this region of homology is found at the tips of the short arms of the X and Y chromosomes, and is the location of an obligatory crossover during male meiosis (pages 211–212). Because of the obligatory crossover, genes in these regions will not show the normal X-linked or Y-linked patterns of inheritance, and consequently this region is known as the **major pseudoautosomal region** (see page 211).

X inactivation and dosage compensation

The importance of correct gene expression is illustrated by the result of accidental gain or loss of whole chromosomes (see next section). For example, in human zygotes, the absence of a single chromosome is almost always incompatible with life (a notable exception is the presence of a single X in a female zygote – see below). Additionally, the presence of an additional copy of a specific chromosome most frequently results in pregnancies failing to reach term, or if they do, as is sometimes the case with chromosomes 13, 18, 21 and the X in humans, they usually result in major developmental abnormalities. Significantly, these four chromosomes are known to be comparatively lacking in transcribed genes (see *Figure 7.5*). This is consistent with the idea that there is a requirement to keep the expression levels (**gene dosage**) of at least some important genes constant, a requirement that is particularly acute in chromosomes with large numbers of actively transcribed genes.

Both the paternally inherited and maternally inherited copies of almost all functional autosomal genes are expressed (the few exceptions are **imprinted genes** in which either the paternal copy or the maternal copy, but not both, are expressed – see page 170). However, because males have one X chromosome whilst females have two, there will be an imbalance in the ratio of autosomal to X-linked genes in the two sexes: 2:1 in males but 1:1 in females. In order to compensate for the differential expression due to differences in gene number, one randomly selected X chromosome in each cell of a female mammal is inactivated (**X inactivation** or **lyonization**, see page 172). The inactive X chromosome is transcriptionally inactive and assumes a highly condensed heterochromatic form. Unlike the active X, the inactive X can be observed by light microscopy during interphase, where it assumes a distinct structure, the **Barr body**, which is located near the nuclear membrane and replicates late in S phase.

Chromosome abnormalities

Types of chromosomal abnormality

Mutations affecting genomic DNA sometimes result in loss or gain of whole chromosomes, or of subchromosomal segments. Additionally, large-scale rearrangements without any net loss or gain of genetic material are often observed (e.g. balanced translocations, inversions, etc. – see below). Traditionally, if the mutation produces an effect that is large enough to be visible under the light microscope, it is described as a **chromosomal abnormality** or **chromosome aberration**. The resolution of this technique is such that the smallest deletions or additions that can be visualized span about 4 megabases (Mb) of DNA.

Chromosomal abnormalities may be present in cells throughout the body (**constitutional abnormalities**), or may be present in a small subset of cells or tissues (**somatic** or **acquired abnormalities**). Constitutional abnormalities must be present very early in development, and result from inheritance of an abnormal sperm or egg cell, abnormal fertilization or an abnormal event in the early embryo. By contrast, an individual with a somatic abnormality is a **genetic mosaic**, containing cells with two different chromosome constitutions, with both cell types deriving from the same zygote. Such abnormalities include many which would be lethal if constitutional.

Normal chromosome constitutions in mammals are described by a **karyotype** that states the total number of chromosomes and the sex chromosome constitution (e.g. human females and males are respectively 46, XX and 46, XY). When there is a chromosomal abnormality the karyotype also describes the type of abnormality and the chromosome bands or sub-bands affected (see *Box 2.3*).

Chromosomal abnormalities, constitutional or somatic, mostly fall into one of two categories: numerical abnormalities (see below) and structural abnormalities (page 54). However, abnormalities have also been identified in which cells appear to have a normal karyotype, but in which the chromosome set can be shown to arise following unequal contributions from the two parents. Either a conceptus arises in which all chromosomes derive from a single parent (**uniparental diploidy**), or the two homologs of a specific chromosome pair can be shown to have originated from only one of the two parents (**uniparental disomy**) (page 58).

Numerical chromosomal abnormalities involve a change in the number of chromosomes, without chromosome breakage

Three classes of numerical chromosomal abnormalities can be distinguished: polyploidy, aneuploidy and mixoploidy (*Figure 2.15*).

Polyploidy

The most common polyploidy is **triploidy** which is caused by two sperm fertilizing a single egg (**dispermy**) or by fertilization involving an abnormal diploid gamete (see *Figure 2.16*). **Tetraploidy** is usually due to failure to complete the first zygotic division: the DNA has replicated to give a content equal to four times that in the haploid sperm or egg cells, but cell division subsequently has not taken place as normal.

Although constitutional polyploidy is rare, it should be noted that in all individuals some cells are naturally polyploid. For example, cells of regenerating liver and other tissues are naturally tetraploid because of reduplication of DNA, prior to mitosis. An extreme example is provided by megakaryocytes, bone marrow cells with very

Box 2.3: Nomenclature of chromosome abnormalities

Numerical abnormalities:

Triploidy	69,XXX, 69,XXY, 69,XYY
Trisomy	e.g. 47,XX, +21
Monosomy	e.g. 45,X
Mosaicism	e.g. 46,XX/47,XXX

Structural abnormalities:

Deletion	e.g. 46,XY, del(4) (p16.3→pter)
Inversion	e.g. 46,XY, inv(11) (p11p15)
Duplication	e.g. 46,XX, dup(3) (q26→qter)
Insertion	e.g. 46,XX, ins(13) (q13)
Ring	e.g. 46,XY, –22, +r(22)
Marker	e.g. 47,XX, +mar[a]
Translocation, reciprocal	e.g. 46,XX, t(2;6) (q35;p21.3)[b]
Translocation, Robertsonian	e.g. 45,XY, –14, –21, +rob(14q; 21q)[c]
	e.g. 46,XX, –14, +rob(14q;21q)[d]

Notes:
[a]Karyotype of a cell which contains an extra unidentified chromosome (a *marker*).
[b]A balanced reciprocal translocation with breakpoints in 2q35 and 6p21.3.
[c]A balanced carrier of a 14q–21q Robertsonian translocation.
[d]A patient with translocation Down syndrome carrying, in addition to one normal chromosome 14 and two normal chromosome 21s, a Robertsonian translocation 14q–21q chromosome.

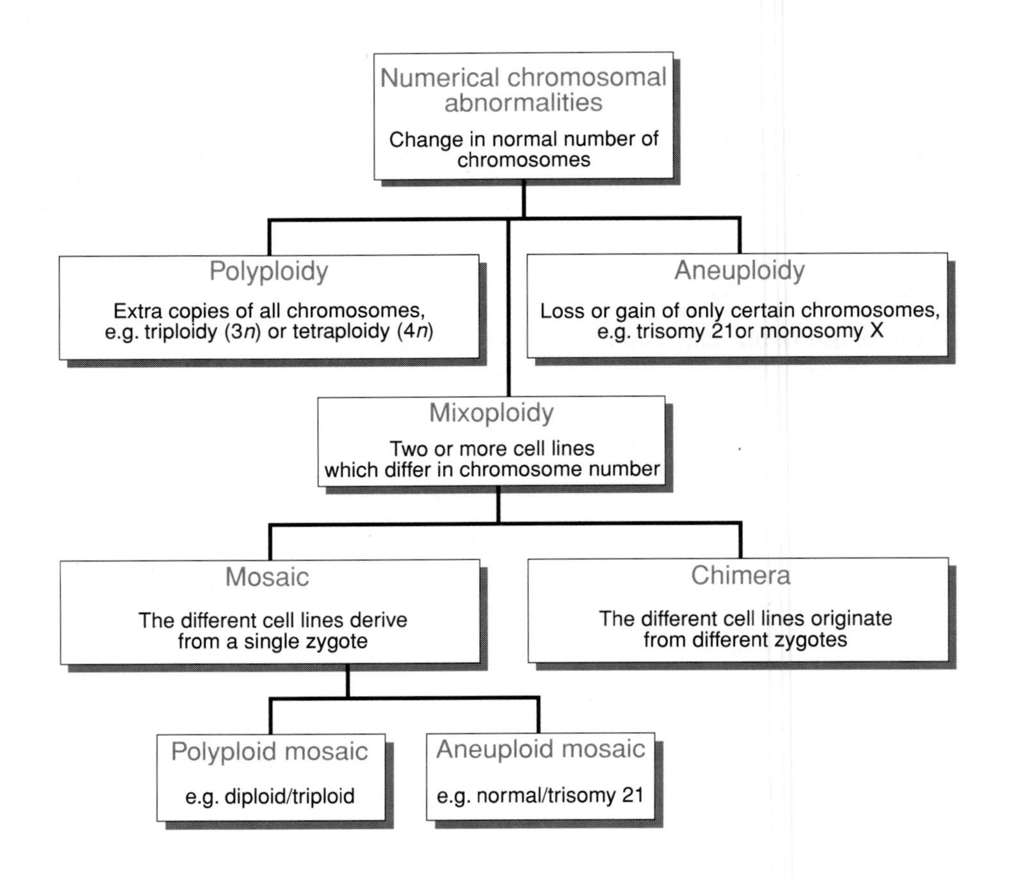

Figure 2.15: Types of numerical chromosomal abnormality.

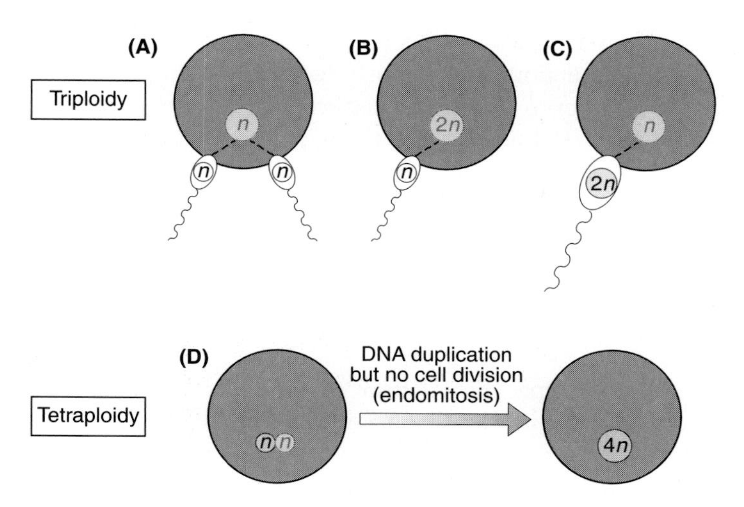

Figure 2.16: Origins of triploidy and tetraploidy.

About two-thirds of human triploidies arise by fertilization of a single egg by two sperm **(A)**. The remaining cases are due to fertilization between a normal haploid gamete and an abnormal diploid gamete, which results because of a failure of one of the maturation divisions in oogenesis or spermatogenesis. The cases which result from fertilization by a diploid sperm **(C)** are considerably more common in humans than those involving a diploid egg **(B)**. Constitutional tetraploidy results from a failure of the first cell division of a zygote following DNA duplication, and is incompatible with survival of the embryo. However, a certain proportion of the cells of all individuals are normally tetraploid because of failure of cell division following DNA duplication.

large nuclei. They have 8–16 times the normal haploid chromosome number and act as precursors to numerous platelet cells. Like many other terminally differentiated cells (e.g. red blood cells, squamous epithelial cells, etc.), platelets lack a nucleus and are therefore **nulliploid**.

Aneuploidy

Aneuploidy often results from extra copies of a specific chromosome in addition to the normal two homologs in diploid cells (e.g. **trisomy**), or loss of a homolog (**monosomy**). Cancer cells, in particular, often show extreme aneuploidy, with multiple chromosomal abnormalities.

Aneuploid cells arise as a result of either:

(i) **nondisjunction**. This describes the failure of paired chromosomes to separate (disjoin) at meiosis I, or of paired sister chromatids to disjoin either at meiosis II or at mitosis. The two conjoined chromosomes or chromatids migrate to one pole and will be included in one daughter cell, whereas the other daughter cell will lack genetic material.

(ii) **Anaphase lag**. Strictly speaking, this term describes the failure of incorporation of a chromosome into one of the daughter nuclei following cell division. It occurs as a result of delayed movement (lagging) of the chromosome during anaphase, and the chromosome is subsequently lost.

Mixoploidy

Mixoploidy can occur as a result of **mosaicism** (an individual is a genetic mosaic by possessing two or more genetically different cell lines all derived from a single zygote) or, rarely, as a result of **chimerism** (a chimera has two or more genetically different cell lines originating from different zygotes).

Aneuploidy mosaics are common. For example, mosaicism resulting in a proportion of normal cells and a proportion of aneuploid (e.g. trisomic) cells can be ascribed to nondisjunction or chromosome lag occurring in one of the mitotic divisions of the early embryo (any monosomic cells that are formed usually die out after a short interval). Polyploidy mosaics (e.g. human diploid/triploid mosaics) are occasionally found. As gain or loss of a haploid set of chromosomes by mitotic nondisjunction is most unlikely, human diploid/triploid mosaics most probably arise by fusion of the second polar body with one of the cleavage nuclei of a normal diploid zygote.

Structural chromosomal abnormalities result from chromosome breakage

Different types of structural chromosomal abnormalities can result, partly depending on the number of chromosome breakpoints (*Figure 2.17*). When a chromosome breaks at a single point, the two ends at the breakpoint are normally rejoined rapidly by repair enzymes. However, on rare occasions, the distal fragment (i.e. the fragment which is remote from the centromere and by definition is **acentric**, i.e. lacks a centromere) may be lost (**terminal deletion**). Although the large remaining fragment contains a centromere, the absence of a functional telomere at one end usually means that the chromosome is unstable and is often degraded.

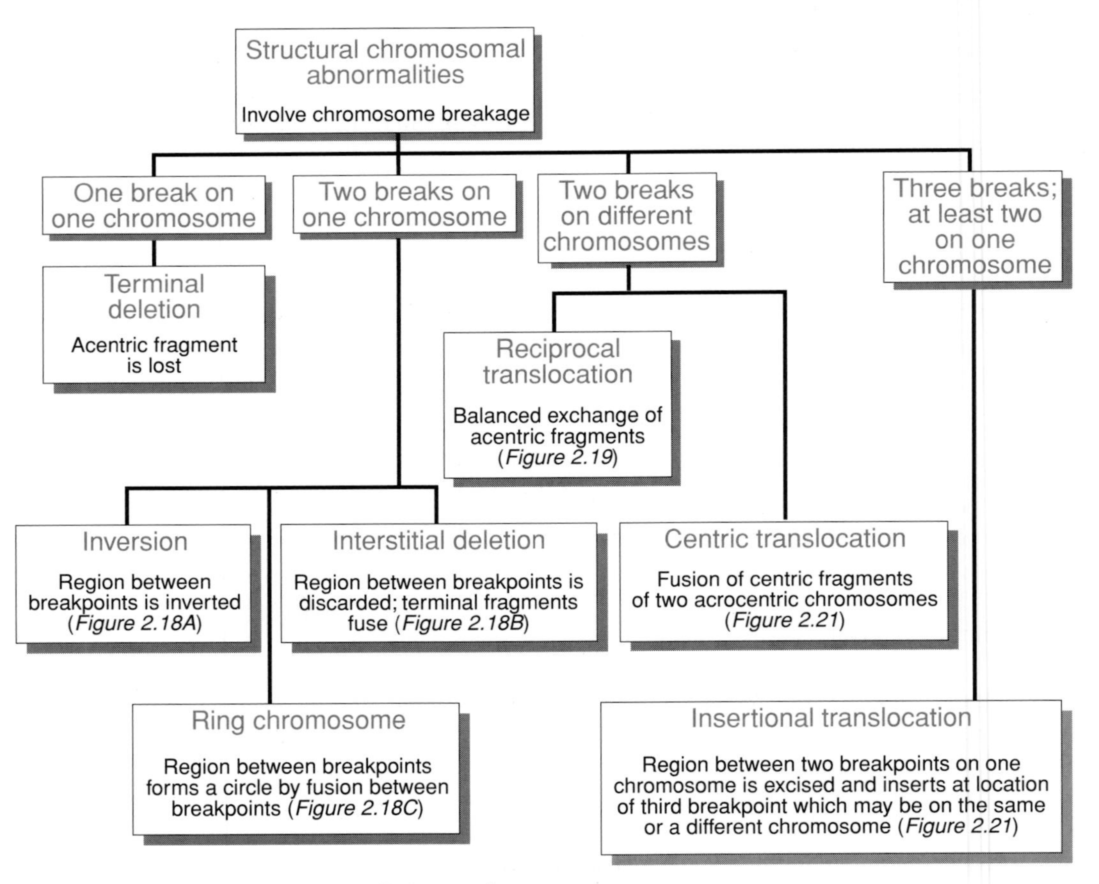

Figure 2.17: Major types of structural chromosomal abnormality.

Note that, in addition to the above, other structural abnormalities are known. For example, **isochromosomes** are abnormal metacentric chromosomes (on rare occasions, dicentric), which consist of the duplication of the long arm or of the short arm of a chromosome. They are thought to arise by aberrant division of the centromere, so that one daughter cell receives both long arms and the other gets both short arms.

If breakage occurs at two or more locations, the repair enzymes have difficulty in distinguishing the different broken ends, and a variety of structural chromosomal aberrations can result (see below). Structural chromosomal abnormalities may be **balanced** if there is no net gain or loss of chromosomal material, or **unbalanced** if there is net gain or loss of chromosomal material.

Two breaks in a single chromosome can result in an inversion, an interstitial deletion or a ring chromosome

If two breaks occur in a single chromosome, then one of three outcomes is usually observed (see *Figure 2.18*):

(i) a **chromosomal inversion** results when the chromosomal segment between the breakpoints is inverted prior to resealing the breakpoints. This is a balanced rearrangement with no net loss of chromosomal material. The inverted segment may include the centromere (**pericentric inversion**) or may be confined to one chromosome arm (**paracentric inversion**, see *Figure 2.18A*).

(ii) An **interstitial deletion** results when the two terminal chromosomal fragments become joined together with exclusion of the intervening segment. Because of the requirement for a centromere for chromosome function, the only viable chromosomes with interstitial deletions are those where the deletion is confined to one arm of a chromosome. As the deleted fragment lacks a centromere (**acentric fragment**), it will be lost at the next cell division (*Figure 2.18B*).

(iii) **A ring chromosome** results when the two ends of the segment between the breakpoints are joined to form a circular fragment. Only if it carries a centromere will the ring chromosome be able to pass through cell division (*Figure 2.18C*).

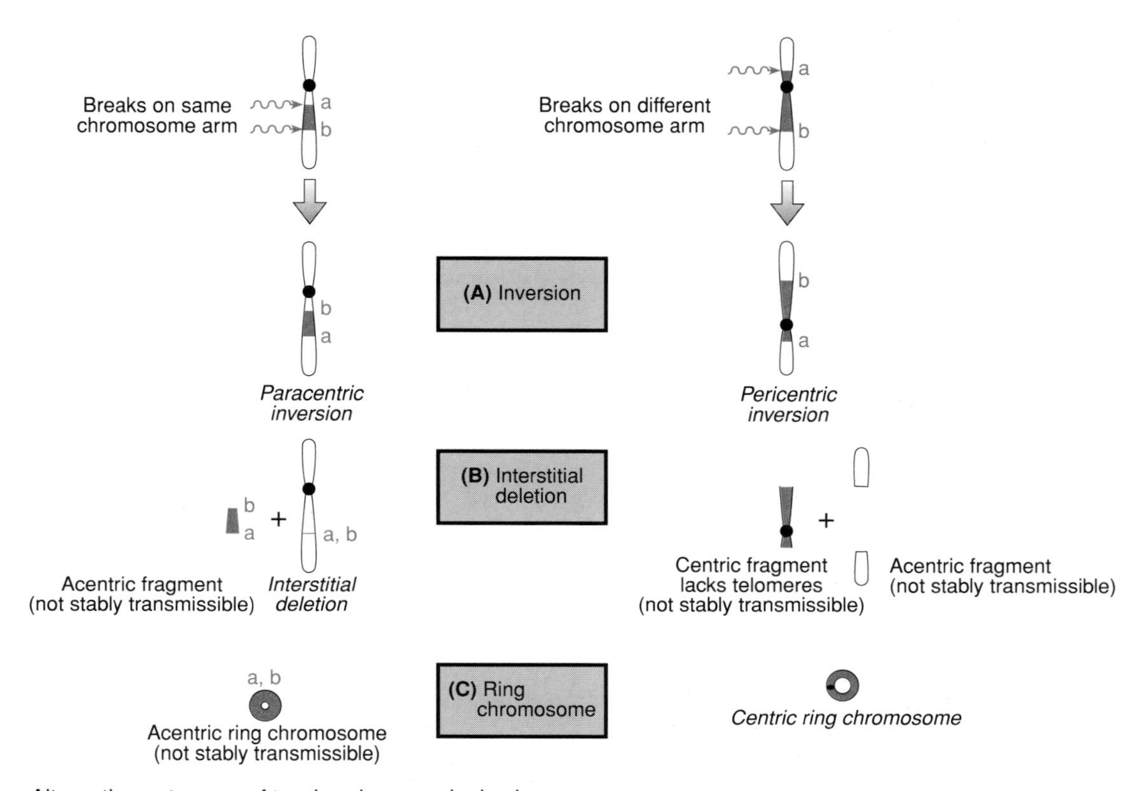

Figure 2.18: Alternative outcomes of two breaks on a single chromosome.

Breaks on more than one chromosome can result in chromosomal translocations

If breaks occur on two or more chromosomes, hybrid chromosomes may be formed by joining segments originally found on different chromosomes (**chromosomal translocation**). Three types of chromosome translocation are known:

(i) a **reciprocal translocation** is a balanced rearrangement which occurs when material distal to the breakpoints on two chromosomes (i.e. the acentric fragments) is exchanged (*Figure 2.19*). Breakpoints may occur on either the long or the short arms. Although carriers of a balanced translocation are usually asymptomatic, some cases involve interruption by one (or rarely both) translocation breakpoints of an important DNA sequence, such as that of a gene or regulatory element, leading to inactivation of gene expression or inappropriate gene expression (see *Figure 17.6*).

Even if a carrier of a balanced translocation is asymptomatic, however, offspring born to such an individual are at risk: during gametogenesis unbalanced gametes may result, which, when fertilized by a normal gamete, produce zygotes with **partial trisomy** and **partial monosomy** for defined chromosomal regions (*Figure 2.20*).

(ii) A **centric (Robertsonian) fusion** is a rearrangement which occurs when breaks occur at or near the centromere of two acrocentric chromosomes, and the large fragments of the two chromosomes fuse. Often the breaks are just above the centromere and the resulting fusion chromosome has two centromeres (**dicentric chromosome**). The acentric fragments will usually be lost at a subsequent cell division (*Figure 2.21*). Although there has been a loss of material, Robertsonian translocations are usually regarded as balanced because they cause no phenotypic change.

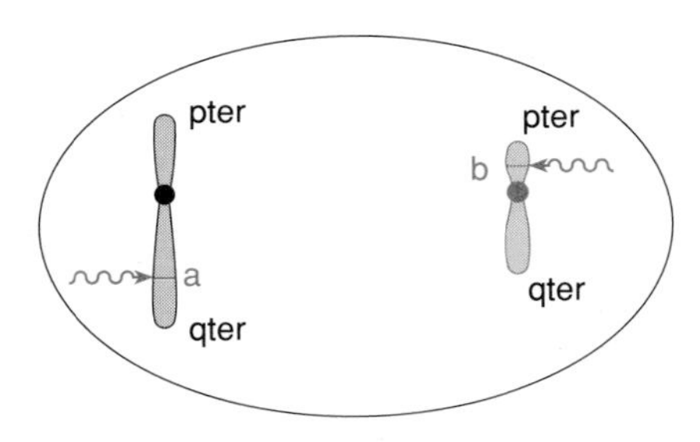

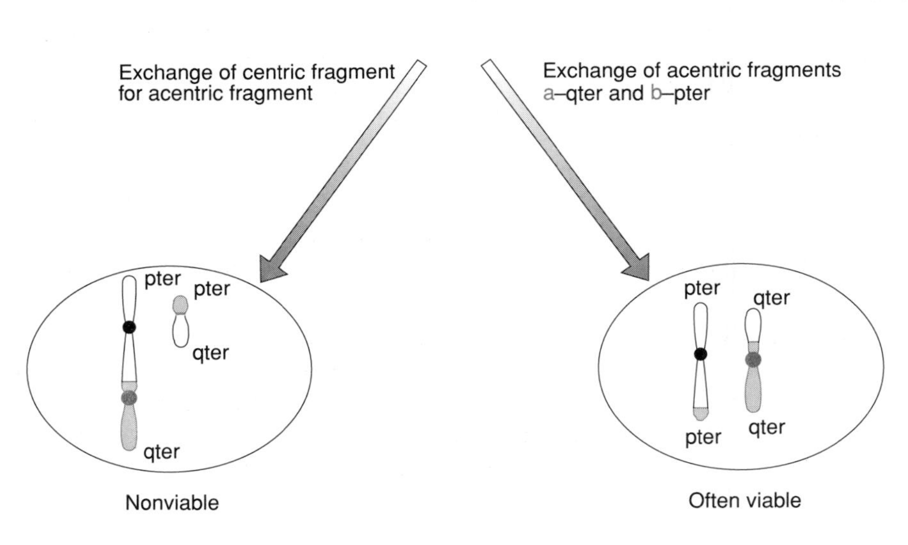

Figure 2.19: Alternative outcomes of reciprocal translocation between two chromosomes with single breaks.

Note that observed reciprocal translocations involve exchange between distal fragments (i.e. the acentric fragments), as shown on the right. The alternative possibility, shown on the left, results in the production of an acentric fragment which will not adhere to the mitotic spindle and will be lost at a subsequent cell division, and of an unstable dicentric chromosome.

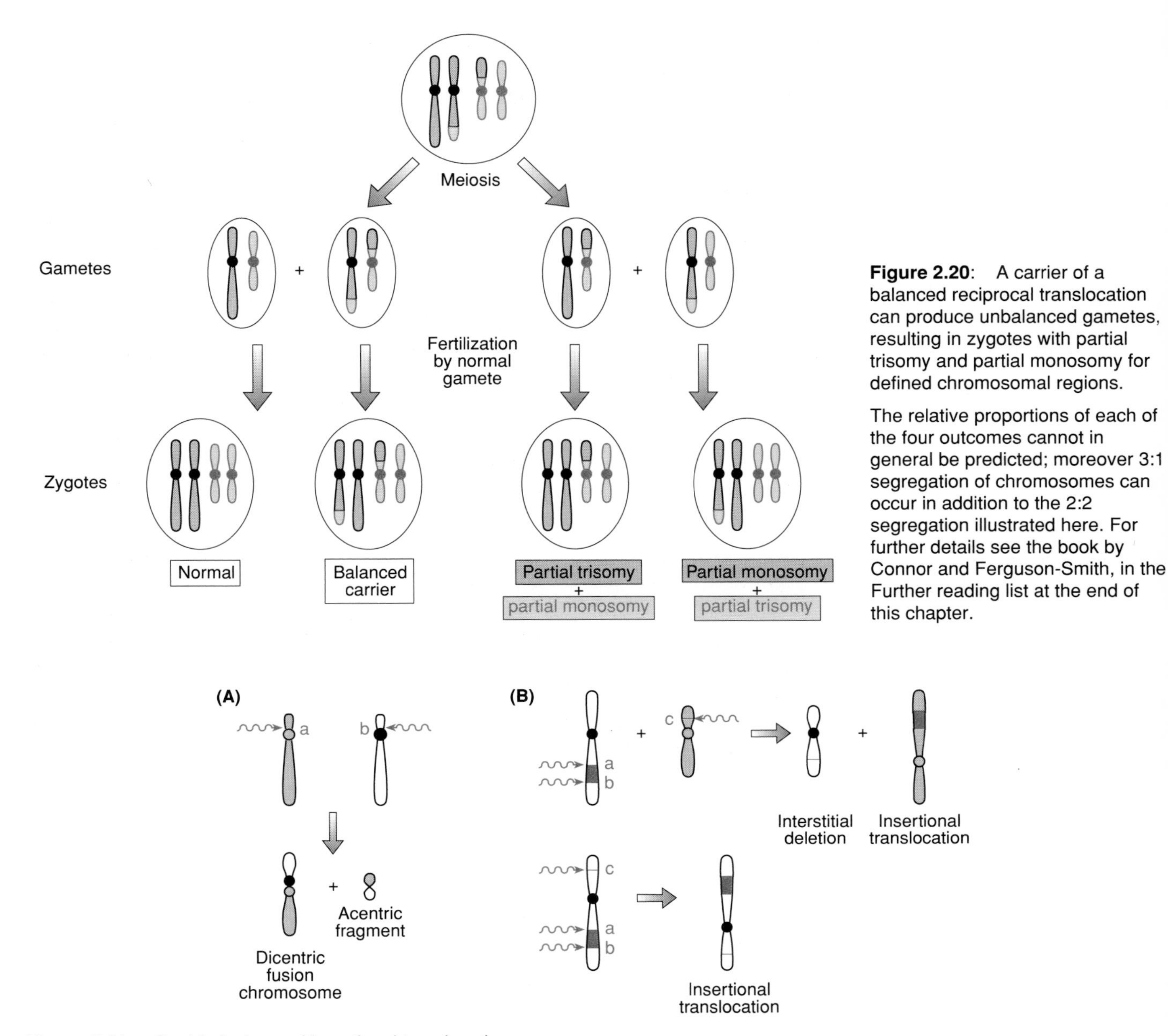

Figure 2.20: A carrier of a balanced reciprocal translocation can produce unbalanced gametes, resulting in zygotes with partial trisomy and partial monosomy for defined chromosomal regions.

The relative proportions of each of the four outcomes cannot in general be predicted; moreover 3:1 segregation of chromosomes can occur in addition to the 2:2 segregation illustrated here. For further details see the book by Connor and Ferguson-Smith, in the Further reading list at the end of this chapter.

Figure 2.21: Centric fusion and insertional translocation.

(A) Centric fusion. This involves breakpoints, usually just above the centromeres of human acrocentric chromosomes, resulting in the production of a dicentric chromosome. The reciprocal product consists of a very small acentric fragment which will be lost at a subsequent cell division. Although, technically, the loss of material results in an unbalanced translocation, the short arm of human acrocentric chromosomes contains repeated genes, and the loss of two such short arms is not particularly significant.
(B) Insertional translocation. Curly arrows denote location of breakpoints, as before.

An asymptomatic carrier of a centric fusion chromosome often produces unbalanced gametes resulting in aneuploid zygotes, which are monosomic or trisomic for one of the chromosomes involved in the centric fusion (*Figure 2.22*).

(iii) An **insertional translocation** results from three chromosome breaks and involves excision of a DNA segment located between two breakpoints on one chromosome and then insertion into an interstitial site occurring usually, but not always, on a second chromosome (*Figure 2.21*). Again, offspring of carriers may be at risk of partial monosomy or partial trisomy.

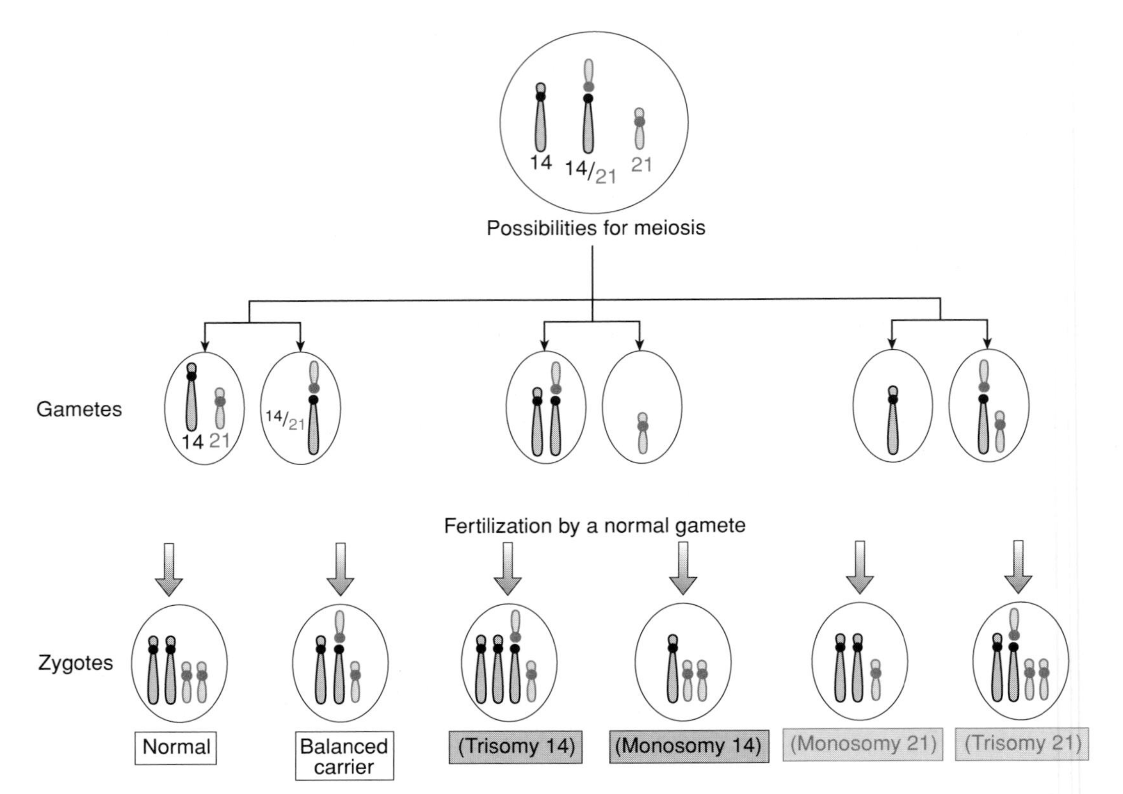

Figure 2.22: An asymptomatic carrier of a centric fusion often produces unbalanced gametes, which can result in zygotes that are monosomic or trisomic for one of the chromosomes involved in the centric fusion.

The example shown illustrates a carrier of a 14/21 Robertsonian translocation. The bracketed monosomic and trisomic outcomes indicate that conceptuses do not develop to term, unlike trisomy 21 (Down syndrome).

Uniparental diploidy and uniparental disomy

Occasionally, cells may show what cytologically appears to be the normal diploid chromosome set (e.g. 46,XX or 46,XY in humans) but in reality is an abnormal chromosome constitution because of an imbalance in the normally equal paternal and maternal chromosome contributions. An extreme case is **uniparental diploidy**, where all the chromosomes are derived from a single parent. More commonly, cases have been reported of individuals in whom two copies of a specific chromosome are inherited from a single parent, either both homologs from that parent (**uniparental disomy**) or two identical copies of a single homolog (**uniparental isodisomy**).

Uniparental diploidy results in failure of embryonic development in humans and uniparental disomy or isodisomy often contributes to disease (see following sections). This is due to inherent differences in maternally and paternally inherited chromosome homologs. Although both the paternal and maternal alleles of the vast majority of genes show equivalent potential for expression, some genes show evidence of **gametic imprinting** and either the paternal or the maternal copy is expressed, but usually not both (see page 170).

Uniparental diploidy

In humans, an **hydatidiform mole** is an abnormal conceptus derived from cells with an apparently normal 46,XX karyotype. The zygote does not develop to produce an embryo; instead the chorion villi lack fetal vasculature and become swollen. Malignant changes to the trophoblastic epithelium may then result in a choriocarcinoma. All 46 chromosomes of a hydatidiform mole are paternal in origin, and typing with polymorphic markers reveals homozygosity at all loci. Such moles are thought to result from degeneration of the female pronucleus of the fertilized egg and subsequent DNA duplication of the male pronucleus to ensure a diploid zygote. **Ovarian teratomas** by contrast have two maternal genomes, with a 46,XX karyotype. They consist of a disorganized mass of differentiated embryonic tissues, but without any of the extra-embryonic membranes of a normal conceptus.

Uniparental disomy and isodisomy

Uniparental disomy is often thought to arise by loss of an extra chromosome copy from a zygote with an inviable trisomy, thereby restoring the normal number of chromosomes. If each of the three copies has an equal chance of being lost, there will be a two in three chance of a single chromosome loss leading to the normal chromosome constitution and a one in three chance of uniparental disomy (either paternal or maternal). Uniparental isodisomy may possibly arise by selection pressure on a monosomic embryo to achieve euploidy by selective duplication of the monosomic chromosome.

Further reading

Connor JM, Ferguson-Smith MA. (1994) *Essential Medical Genetics*, 4th Edn. Blackwell Scientific Publications, Oxford.

ISCN (1995) *An International System for Human Cytogenetic Nomenclature* (ed. F. Mittelman). Karger, Basel.

Rooney DE, Czepulkowski BH. (1992) (eds) *Human Cytogeneticcs: a Practical Approach*, Vols 1 and 2. IRL Press Oxford.

Sumner AT. (1990) *Chromosome Banding*. Unwin Hyman, London.

Therman E, Sulsman M. (1992) (eds) *Human Chromosomes: Structure Behavior and Function*, 3rd Edn. Springer, New York.

Tyler-Smith C, Willard, HF. (1993) Mammalian chromosome structure. *Curr. Opin. Genet. Dev.*, **3**, 390–397.

References

Craig JM, Bickmore WA. (1993) *Bioessays*, **15**, 349–354.

Dobbs DL, Shaiu W-L, Benbow RM. (1994) *Nucl. Acid Res.*, **22**, 2479–2489.

Manuelidis L. (1990) *Science*, **250**, 1533–1540.

Saitoh Y, Laemmli UK. (1994) *Cell*, **76**, 609–622.

Wolstenholme J. (1992) In: *Human Cytogenetics: a Practical Approach*, Vol. 1, 2nd Edn (eds DE Rooney, BH Czepulkowski), pp. 1–30. IRL Press, Oxford.

Genes in pedigrees

Chapter 3

Mendelian pedigree patterns

The simplest genetic characters are those whose presence or absence depends on the **genotype** at a single **locus**. That is not to say that the character itself is programmed by only one pair of genes – expression of any human character is likely to require a large number of genes and environmental factors. However, sometimes a particular genotype at one locus is both necessary and sufficient for the character to be expressed, given the normal genetic and environmental background of the organism. Such characters are called **mendelian**.

Dominance and recessivity are properties of characters, not genes

A character is **dominant** if it is manifest in the **heterozygote** and **recessive** if not. Note that dominance and recessiveness are properties of characters, not genes. Thus sickle cell anemia is recessive because only HbS **homozygotes** manifest it. Heterozygotes for the same gene show sickling trait, which is therefore a dominant character. Most human dominant syndromes are known only in heterozygotes. Sometimes homozygotes have been described, born from matings of two heterozygous affected people, and often the homozygotes are much more severely affected. Examples are achondroplasia and type 1 Waardenburg syndrome. Nevertheless we describe achondroplasia and Waardenburg syndrome as dominant because these terms describe **phenotypes** seen in heterozygotes. Note, however, that not all geneticists would agree with this usage of the term 'dominant'. Some prefer to use the term **semi-dominant** if the heterozygote has an intermediate phenotype, reserving 'dominant' for conditions where the homozygote is indistinguishable from the heterozygote – Huntington disease for example. The question of dominance has been well reviewed recently by Wilkie (1994).

OMIM is the standard database of mendelian characters

In humans, some 5000 mendelian characters are known. A catalog of these characters is available either as a printed book updated every 2 years (McKusick, 1994) or as an on-line database (OMIM, Online Mendelian Inheritance in Man), updated weekly and accessible through the Internet (Genome Database at http://www3.ncbi.nlm.nih.gov/Omim; see *Box 13.2*). This is the essential starting point for acquiring information on human mendelian characters, both pathological and nonpathological. Each character is given a 6-digit MIM number which is widely used to identify it in genetic literature – thus achondroplasia is MIM 100800, type 1 Waardenburg syndrome is MIM 193500 and Huntington disease is MIM 143100.

There are five basic mendelian pedigree patterns

Figure 3.1 shows the symbols used for drawing pedigrees. Mendelian characters may be determined by loci on autosomes or on the X or Y sex chromosomes. Autosomal and X-linked characters can be dominant or recessive. Since there is at most one type of Y chromosome, the question of dominance or recessiveness does not arise for Y-linked characters (in the rare XYY males the two Y chromosomes are duplicates, so again there are no heterozygotes). Males are **hemizygous** for loci on the X and Y chromosomes, where they have only a single copy of each gene. Thus there are five archetypal pedigree patterns (*Figure 3.2*), but see *Figure 3.4* for complications on the basic pedigree patterns. The hallmarks of each pattern are listed in *Box 3.1*.

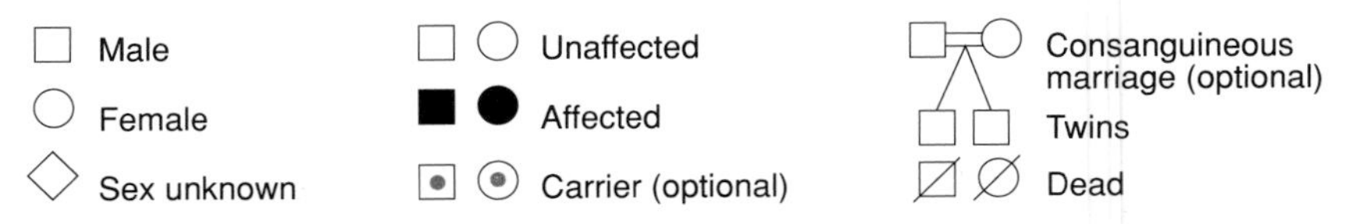

Figure 3.1: Main symbols used in pedigrees.

Generations are usually labeled in Roman numerals, and individuals within each generation in Arabic numerals; III-7 or III_7 is the seventh person from the left (unless explicitly numbered otherwise) in generation III. An arrow ↗ can be used to indicate the *propositus* (female: *proposita*) through whom the family was ascertained.

The mode of inheritance can rarely be defined unambiguously in a pedigree

Given the limited size of human families, it is rarely possible to be completely certain of the mode of inheritance of a character simply by inspecting a single pedigree. In experimental animals one would set up a test cross and check for a 1:2 or 1:4 ratio. In human pedigrees the proportion of affected children is not a very reliable indicator. This is not merely because the numbers in any one family are too small for statistical confidence. Often the way in which the family was ascertained is also important. Particularly for recessive conditions, families are normally ascertained because they already have an affected child. Families are systematically missed when both parents are carriers but, by good fortune, nobody is affected. Moreover, in many study designs, families with several affected children are more likely to be ascertained than families with only a single affected child.

Thus the ratios of affected to unaffected children in collected pedigree data are biased in ways which may be quite complex (*Figure 18.7*). The statistical technique of segregation analysis (page 491) can be used to derive an unbiased estimate from pooled family data; however, for most of the 5000 known human mendelian characters (McKusick, 1994), this has not been done. For many of the rarer conditions, the stated mode of inheritance is no more than an informed guess. Assigning modes of inheritance is important, because that is the basis of the risk estimates used in genetic counseling. However, it is important to recognize that the modes of inheritance are often working hypotheses rather than established fact. McKusick uses an asterisk to denote entries with relatively well-established modes of inheritance. Only

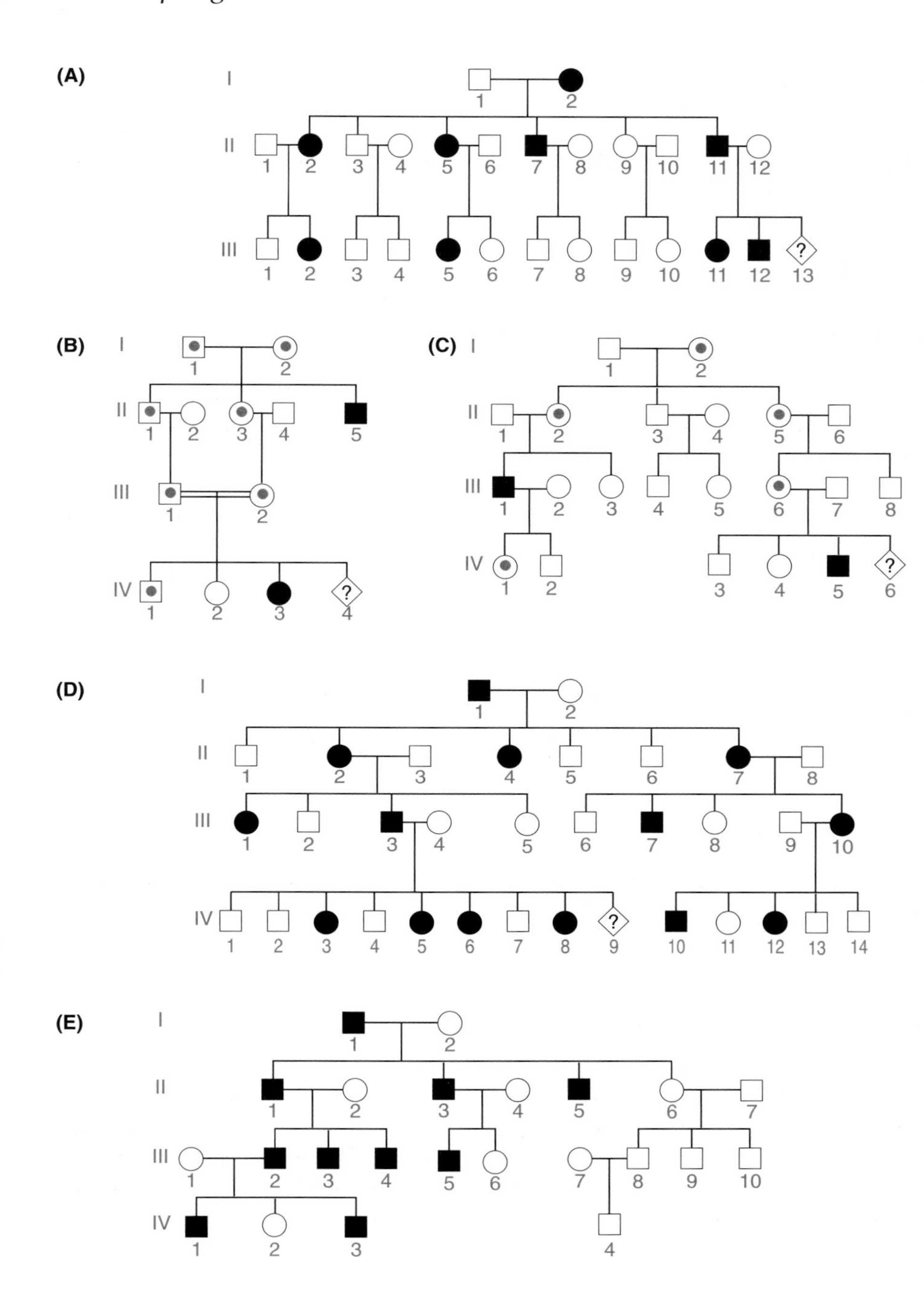

Figure 3.2: Basic mendelian pedigree patterns.

(A) Autosomal dominant; **(B)** autosomal recessive; **(C)** X-linked recessive; **(D)** X-linked dominant; **(E)** Y-linked. The risk for the individuals marked with a query are **(A)** 1 in 2, **(B)** 1 in 4, **(C)** 1 in 2 males or 1 in 4 of all offspring, **(D)** negligibly low for males, 100% for females. See page 66 and *Figure 3.4* for complications to these basic patterns.

when a cloned copy of the gene is available can the mode of inheritance be determined with complete certainty.

There is not necessarily a one-to-one correspondence between mendelian phenotypes and DNA coding sequences

The genes of classical genetics are abstract entities, hypothesized on the basis of characters which show mendelian inheritance. They do not necessarily correspond to genes as defined by molecular geneticists. If a character segregates in a mendelian pattern, then it is determined at a single chromosomal location – but the determinant may not be a gene in the molecular geneticist's sense of the word. The 'gene' for Charcot–Marie–Tooth disease type 1A (MIM 118220), when cloned,

Box 3.1: Principal features of mendelian patterns

Autosomal dominant inheritance (*Figure 3.2A*):

An affected person usually has at least one affected parent (for exceptions see *Figure 3.4*).

Affects either sex.

Transmitted by either sex.

A child of an affected × unaffected mating has a 50% chance of being affected (this assumes that the affected person is heterozygous, which is usually true for rare conditions).

Autosomal recessive inheritance (*Figure 3.2B*):

Affected people are usually born to unaffected parents.

Parents of affected people are usually asymptomatic carriers.

There is an increased incidence of parental consanguinity.

Affects either sex.

After the birth of an affected child, each subsequent child has a 25% chance of being affected.

X-linked recessive inheritance (*Figure 3.2C*):

Affects almost exclusively males.

Affected males are usually born to unaffected parents; the mother is normally an asymptomatic carrier and may have affected male relatives.

Females may be affected if the father is affected and the mother is a carrier, or occasionally as a result of nonrandom X-inactivation (page 72).

There is no male-to-male transmission in the pedigree (but matings of an affected male and carrier female can give the appearance of male-to-male transmission, see *Figure 3.4F*).

X-linked dominant inheritance (*Figure 3.2D*):

Affects either sex, but more females than males.

Females are often more mildly and more variably affected than males.

The child of an affected female, regardless of its sex, has a 50% chance of being affected.

For an affected male, all his daughters but none of his sons are affected.

Y-linked inheritance (*Figure 3.2E*):

Affects only males.

Affected males always have an affected father.

All sons of an affected man are affected.

turned out to be a tandem duplication of about 1.5 Mb of DNA on chromosome 17p11.2 (*Figure 15.6*). Insertions, microdeletions encompassing several contiguous genes, or even chromosomal translocations can produce mendelian phenotypes, and so figure in pedigree analysis as 'genes'.

In short, pedigree patterns provide the essential entry point into human genetics, but they are only a starting point for defining genes. It would be a serious error to imagine that the 5000 known mendelian characters define 5000 transcription units.

Locus heterogeneity

It commonly happens that mutations in several different genes result in the same clinical phenotype (**locus heterogeneity)**. Characters which are the end result of a long pathway are almost always genetically heterogeneous. Mental retardation is the extreme example, but even an apparently specific phenotype such as retinitis pigmentosa can turn out to be highly heterogeneous. Examples of these phenomena are discussed on page 419 and the subsequent pages.

Allelic series and clinical heterogeneity

Sometimes, several apparently distinct human phenotypes turn out to be all caused by different allelic mutations at the same locus. Mutations which partially inactivate the dystrophin gene produce Becker muscular dystrophy, whereas mutations which completely inactivate the same gene produce the more severe Duchenne muscular dystrophy (*Figure 15.2*). Inactivation of the androgen receptor gene causes androgen insensitivity (testicular feminization syndrome, MIM 313700), but expansion of a run of glutamine codons within the same gene causes a totally different disease, spinobulbar muscular atrophy or Kennedy disease (MIM 313200). Other examples are discussed in Chapter 15.

The complementation test

Often it is not clear whether two rather similar characters are determined by alleles at one locus or by two different loci. In experimental animals this would be easily resolved by a complementation test (see *Box 3.2*). Occasionally, humans 'oblige' and people with autosomal recessive profound deafness often marry each other and usually produce children with normal hearing, thus demonstrating that mutations at several different loci can cause deafness (see *Figure 15.8*).

Mitochondrial inheritance gives a nonmendelian matrilinear pedigree pattern

In addition to the mutations in genes carried on the nuclear chromosomes, mitochondrial mutations are a significant cause of human genetic disease. The mitochondrial genome (*Figure 7.2*) is small but highly mutable compared to nuclear DNA, probably because mitochondrial DNA replication is more error-prone and the number of replications is much higher. Mitochondrially encoded diseases have two unusual features, **matrilineal inheritance** and frequent **heteroplasmy** (Wallace, 1994).

Matrilineal inheritance

Inheritance is matrilineal, because fathers do not pass on mitochondria to their children (this assertion rests on limited evidence; however, paternally derived mitochondrial variants are not detected in children). Thus, a typical mitochondrially inherited condition can affect both sexes, but is passed on only by affected mothers (*Figure 3.3A*). If the mother has only abnormal mitochondria (see below), then all her children inherit only abnormal mitochondria. One of the best known mitochondrial diseases shows an unusual mode of inheritance: Leber's hereditary optic atrophy (LHON, MIM 535000; Riordan-Eva and Harding, 1995) is associated with various mitochondrial mutations and is inherited matrilinearly, but almost all affected patients are male (*Figure 3.3B*). This is unexpected – mitochondrial diseases should affect both sexes equally. One explanation could be that LHON requires both a mitochondrial and an X-linked mutation, but attempts to demonstrate an X-linked susceptibility have not been successful, so the reason for the male excess remains unknown.

Box 3.2: The complementation test to discover whether two recessive characters are determined by allelic genes

Parental cross	$a_1a_1 \times a_2a_2$	aaBB × AAbb
	↓	↓
Offspring	a_1a_2	AaBb
Phenotype	Mutant	Wild-type
	One locus	**Two loci**

Animals homozygous for the two characters are crossed and the phenotype of the offspring observed. If both animals carry mutations at the same locus the progeny will not have a wild-type allele, and so will be phenotypically abnormal. If there are two different loci, the progeny are heterozygous for each of the two recessive characters, and therefore phenotypically normal.

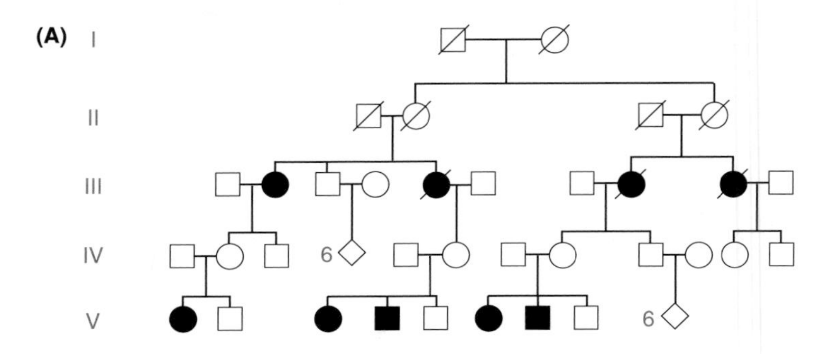

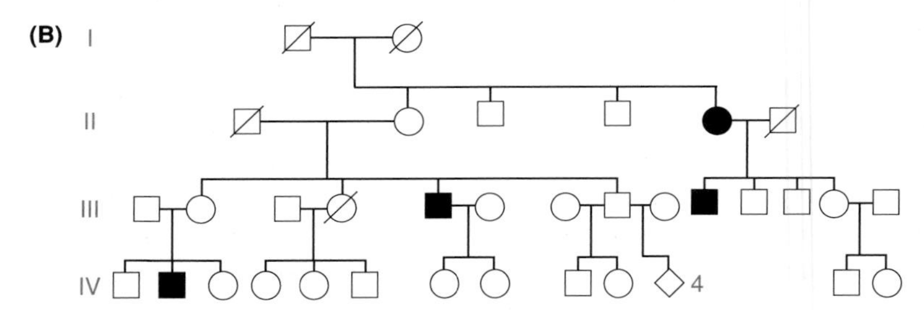

Figure 3.3: Pedigrees of mitochondrial diseases.

(A) A typical pedigree pattern, showing mitochondrially-determined hearing loss (Prezant *et al.*, 1993).
(B) The atypical pattern of Leber's hereditary optic atrophy, which affects mainly males (Sweeney *et al.*, 1992).

Heteroplasmy

Cells contain many mitochondria. In some mitochondrial diseases, every mitochondrion carries the causative mutation (homoplasmy), but in other cases a mixed population of normal and mutant mitochondria is seen within each cell (heteroplasmy). The complicated molecular pathology of mitochondrial diseases is discussed in Chapter 15.

Complications to the basic pedigree patterns

In real life, various complications often disguise a basic mendelian pattern (*Figure 3.4*). If a character is common in the population, there is a high chance that it may be brought into the pedigree independently by two or more people. A common recessive character like blood group O may be seen in successive generations because of repeated marriages of group O people with heterozygotes. This produces a pattern resembling dominant inheritance (*Figure 3.4A*). Thus the classic pedigree patterns are best seen with rare characters. However, even rare characters often show complications in the pedigree pattern.

Failure of a dominant condition to manifest is called nonpenetrance

With dominant conditions, **nonpenetrance** is a frequent complication. The *penetrance* of a character, for a given phenotype, is defined as the probability that a person who has the genotype will manifest the character. By definition, a dominant character is manifest in a heterozygous person, and so should show 100% penetrance. Nevertheless, many human characters, while generally showing dominant inheritance, occasionally skip a generation. In *Figure 3.4B*, II_2 has an affected parent and an affected child, and almost certainly carries the mutant gene, but is phenotypically normal. This would be described as a case of nonpenetrance. There is no mystery about nonpenetrance – indeed, full penetrance is the more surprising phenomenon. A fully penetrant character is wholly determined by the genotype at one locus, regardless of all the other genes and environmental factors involved in development of the organism to the stage where it can manifest the character. Not surprisingly, there are many cases where presence or absence of a character depends in the main, and in normal circumstances, on one genetic locus, but where the occasional person fails to manifest the character because of an unusual genetic background or life history.

Frequently, of course, a character depends on many factors and does not show mendelian pedigree patterns even if entirely genetic. There is a continuum of characters from fully penetrant mendelian to **multifactorial** (see page 80 and *Figure 3.10*), with increasing influence of other genetic loci and/or the environment. No logical break separates imperfectly penetrant mendelian from multifactorial characters; it is a question of which is the most useful description to apply. Nonpenetrance is a major pitfall in genetic counseling – it would be an unwise counselor who, knowing that the condition in *Figure 3.4B* was dominant and seeing that III_7 was free of signs, told her that she had no risk of having affected children. One of the jobs of genetic counsellors is to know the usual degree of penetrance of each dominant syndrome.

Late-onset diseases

A particularly important case of reduced penetrance is seen with late-onset diseases. Genetic conditions are, of course, not necessarily congenital, that is present at birth. The genotype is fixed at conception, but the phenotype may not manifest until adult life. In such cases the penetrance is age related. Huntington disease is a well-known example (*Figure 3.5*). Delayed onset might be caused by slow accumulation of a noxious substance, by slow tissue death or by inability to repair some form of environmental damage. Noncongenital cancers can be caused by a second mutation affecting a cell which already carries one mutation in a tumor suppressor gene (Chapter 17). In most cases, including Huntington disease, the reason for the late onset is unknown. Depending on the disease, the penetrance may become 100% if the person lives long enough, or there may be people who carry the gene but who

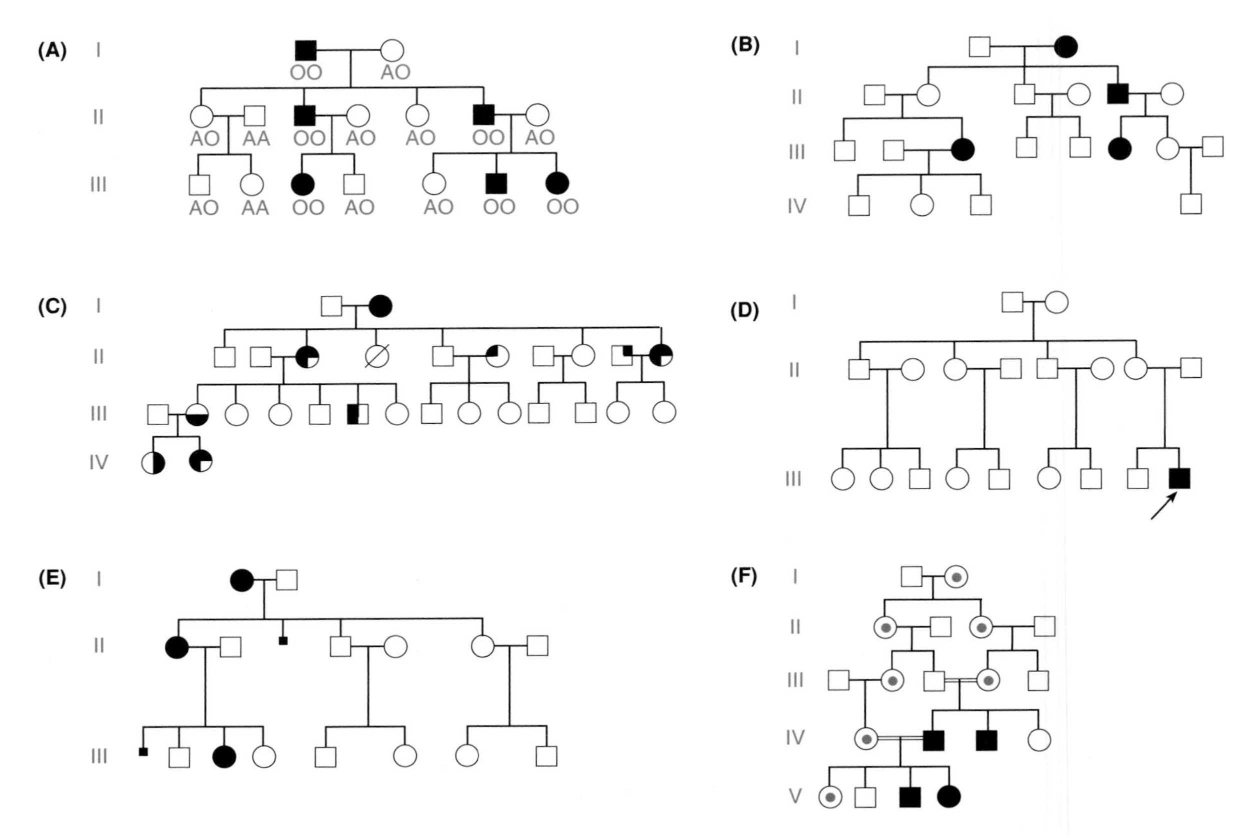

Figure 3.4: Complications to the basic mendelian patterns.

(A) A common recessive, such as blood group O, can give the appearance of a dominant pattern.
(B) Autosomal dominant inheritance with nonpenetrance in II_2.
(C) Autosomal dominant inheritance with variable expression: in this family with Waardenburg syndrome, shading of first quadrant = hearing loss; second quadrant = different colored eyes; third quadrant = white forelock; fourth quadrant = premature greying of hair.
(D) A new autosomal dominant mutation, mimicking an autosomal or X-linked recessive pattern.
(E) An X-linked dominant condition where affected males abort spontaneously (small squares).
(F) An X-linked recessive pedigree where inbreeding gives an affected female and apparent male-to-male transmission.

will never develop symptoms no matter how long they live. Age of onset curves are important tools in genetic counseling, because they enable the geneticist to estimate the chance that an at-risk but asymptomatic person will subsequently develop the disease (Harper and Newcombe, 1992).

Many conditions show variable expression

Related to nonpenetrance is the **variable expression** frequently seen in dominant conditions. *Figure 3.4C* shows an example from a family with Waardenburg syndrome. Different family members show different features of the syndrome. The cause is the same as with nonpenetrance: other genes or environmental factors have some influence on development of the symptoms. Nonpenetrance and variable expression are typically problems with dominant, rather than recessive, characters. Partly, this reflects the difficulty of detecting nonpenetrant cases in a typical recessive pedigree. However, as a general rule, recessive conditions are less variable than dominant ones, and this is probably because the phenotype of a heterozygote involves a balance between the effects of the two alleles, so that the outcome is likely to be more sensitive to outside influence than the phenotype of a homozygote. However, both nonpenetrance and variable expression are occasionally seen in recessive conditions.

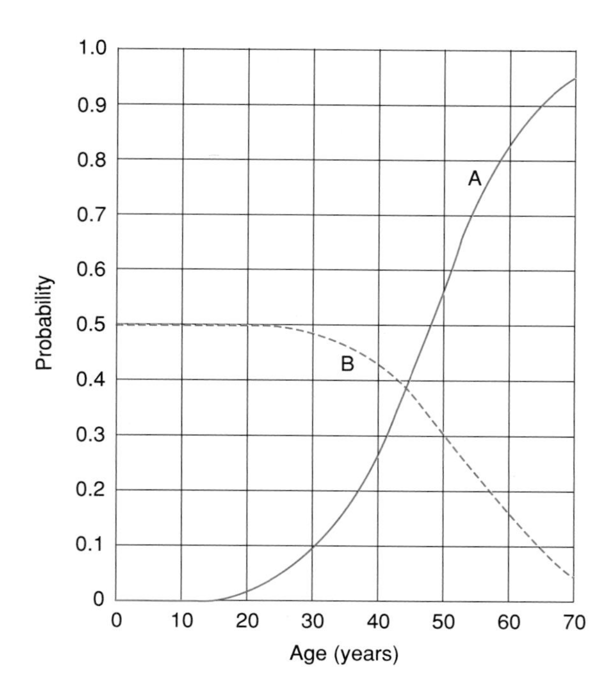

Figure 3.5: Age of onset curves for Huntingdon disease.

Curve A: probability that an individual carrying the disease gene will have developed symptoms by a given age. Curve B: risk that a healthy child of an affected parent carries the disease gene at a given age. Reproduced from Harper (1993) with permission from Butterworth-Heinemann Ltd.

These complications are much more conspicuous in humans than in plants or other animals, because laboratory animals and crop plants are far more genetically uniform than humans. What we see in human genetics is typical of a wild population. Mouse geneticists are familiar with the way expression of a mutant can change when it is bred on to a different genetic background.

Anticipation is a special type of variable expression

Anticipation describes the tendency of some variable dominant conditions to become more severe in successive generations. Until recently, most geneticists were sceptical that this ever really happened. The problem is that true anticipation is very easily mimicked by random variations in severity. A family comes to clinical attention when a severely affected child is born. Investigating the history, the geneticist notes that one of the parents is affected, but only mildly. This looks like anticipation, but may actually be just a bias of ascertainment. Had the parent been severely affected, he or she would most likely never have become a parent, and had the child been mildly affected, the family would not have come to notice. Given the lack of any plausible mechanism for anticipation, and the statistical problems of demonstrating it in the face of these biases, most geneticists were unwilling to consider anticipation seriously until molecular developments obliged them to do so.

Anticipation suddenly became respectable, even fashionable, with the discovery of unstable expanding trinucleotide repeats in fragile-X syndrome (MIM 309550), and later in myotonic dystrophy (MIM 160900) and Huntington disease (see pages 266 and 408). Severity and/or age of onset of these diseases correlate with the repeat length, and the repeat length tends to grow as the gene is transmitted down the generations. Thus these conditions show true anticipation. Now once again we see claims for anticipation being made for many diseases, and it is important to bear in mind that the old objection about bias of ascertainment remains valid. To be credible, a claim of anticipation requires careful statistical backing. Anecdotal evidence alone will not suffice.

For imprinted genes, expression depends on parental origin

Parental origin effects

We are used to the idea that the action of a given gene in particular cells of an individual may depend on the genetic background (the individual's other genes) or the environment. The idea that it might also depend on the parental origin of the gene is less familiar. Nevertheless several recent observations suggest that the maternal and paternal genomes in an individual do not always function interchangeably:

- mouse embryos manipulated so as to contain two maternal or two paternal genomes fail to develop, despite having normal diploid chromosome numbers.
- Human triploid abortuses are phenotypically different depending on whether the extra genome is maternal or paternal.
- Certain human characters are autosomal dominant but manifest only when inherited from one parent. In some families glomus tumors are inherited as an autosomal dominant character, but expressed only in people who inherit the gene from their father (Heutink *et al.*, 1992). Beckwith–Wiedemann syndrome (MIM 130650) is sometimes dominant but expressed only by people who inherit it from their mother (Viljoen and Ramesar, 1992). Example pedigrees are shown in *Figure 3.6*.
- Deletion of certain chromosomal regions produces a different phenotype when on the maternal or paternal chromosome. The best example is deletion of 15q12, which on the paternal chromosome produces Prader–Willi syndrome (mental retardation, hypotonia, gross obesity, hypogenitalism; MIM 176270), and on the maternal chromosome produces Angelman syndrome (mental retardation, growth retardation, hyperactivity, inappropriate laughter; MIM 105830). These conditions are further discussed in Chapter 15; see also Nicholls (1994).
- Allele loss in many cancers (Chapter 17) preferentially involves the paternal allele.

These and similar observations suggest that genomes bear an **imprint** of their parental origin, which in at least some cases modulates their activity. The imprint must be reversible, since a man passes on genes with his own paternal imprint even if those genes came to him with a maternal imprint from his mother.

How many genes are imprinted?

An important question concerns the universality of imprinting. Clearly most genes are not subject to imprinting, otherwise we would not see so many simple mendelian characters. Systematic surveys have been made of the mouse genome, where Robertsonian translocations allow crosses to be set up which produce offspring having both copies of one particular chromosome derived from a single parent (**uniparental disomy, UPD**, see page 58). All of the chromosomes of the mouse are acrocentric so they can all be tested in Robertsonian translocations. These reveal that UPD for some chromosomes has no phenotypic effect, for others it produces viable abnormal phenotypes (sometimes complementary for different parental origins, e.g. overgrowth in maternal UPD and undergrowth in paternal UPD) and, for some chromosomes, UPD is lethal.

Further dissection at the chromosomal and genetic level shows that imprinting is a property of a limited number of individual genes or small chromosomal regions. For example, most genes on human chromosome 11 are expressed from both chromosomes, but the *H19* gene is expressed only from the maternal chromosome

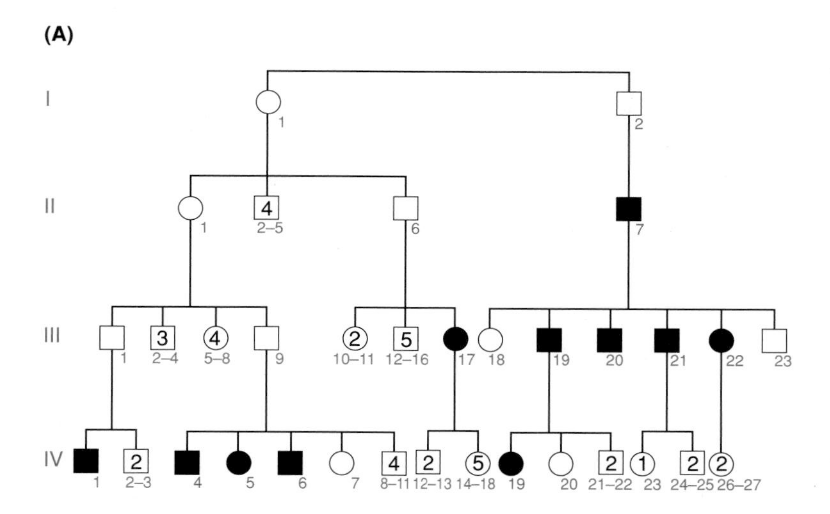

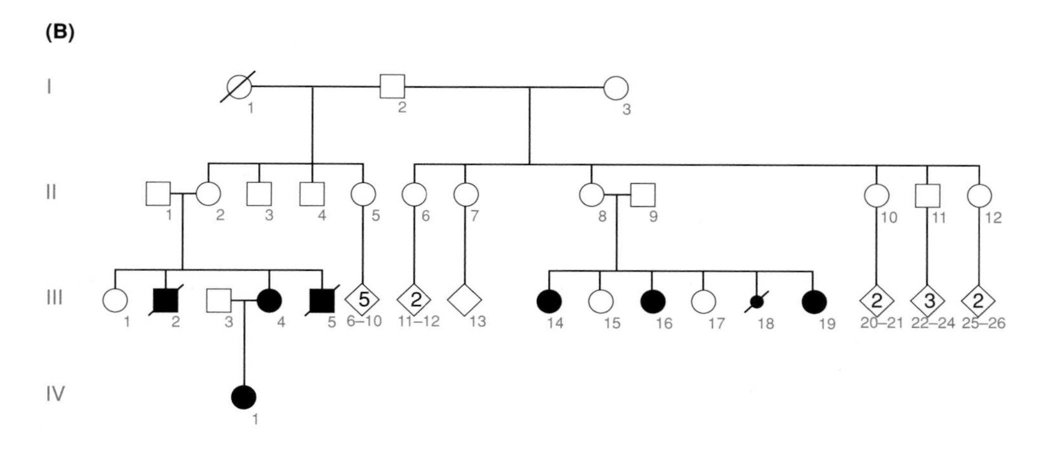

Figure 3.6: Pedigrees of imprinted characters.

(A) In some families tumors of the carotid body are inherited as an autosomal dominant character which is expressed only when inherited from the father. Redrawn from van der Mey *et al.* (1989).
(B) In some families Beckwith–Wiedemann syndrome behaves as an autosomal dominant character with high penetrance when inherited from the mother but low penetrance when inherited from the father. Redrawn from Viljoen and Ramesar (1992).

11 whilst the adjacent *IGF2* gene is expressed only from the paternal chromosome (see *Table 7.11*).

The mechanism of imprinting

Imprinting appears to operate at the transcriptional level. The mechanism involves DNA methylation (page 172), but the details are complex and poorly understood. Where an individual is heterozygous for a sequence variant which is present in the mature mRNA, mRNA from different tissues can be checked for bi-allelic expression, and the origin of each allele determined by typing the parents. It turns out that imprinting may vary between tissues, and also during development. Thus imprinting allows an extra level of control of gene expression, but it is not possible to compress its functioning into a simple uniform story.

The function of imprinting

Much debate has surrounded the biological purpose of imprinting. A popular, but unsubstantiated, theory is that imprinting is a result of evolutionary conflict between two selfish genomes in placental mammals. The paternal genome propagates itself best by creating an embryo which aggressively removes nutrients from the mother, while the maternal genome suppresses this to protect the mother. Paternal genes are preferentially expressed in the trophoblast and extra-embryonic membranes, while maternal genes are preferentially expressed in the embryo. The jury is still out on the question of whether imprinting is an important general feature of development and of genetic disease, or whether it is a rather obscure property of a small number of genes.

UPD for a number of different chromosomes has been described in humans. In some cases the individuals are phenotypically normal; when they are abnormal it can be difficult to decide whether the abnormalities result from UPD *per se* or from homozygosity for particular mutant alleles which the chromosome happened to carry.

New mutations often complicate pedigree interpretation

Many cases of severe genetic disease are caused by fresh mutations, striking without warning in a family with no previous history of the disease. People with severe genetic diseases seldom reproduce, so they do not pass on their mutant genes. On the assumption that, averaged over time, new mutations exactly replace the disease genes lost through natural selection, there is a simple relationship (described on page 78) between the rate at which natural selection is removing disadvantageous genes, the rate at which new mutation is creating them, and their frequency in the population. Mechanisms which affect the population frequency of alleles are discussed on page 245.

This mutation–selection dynamic has different effects on pedigrees, depending on the mode of inheritance. Autosomal recessive pedigrees are not significantly affected – any new mutations probably happened many generations ago, and we can safely assume that the parents of an affected child are both carriers. For dominant conditions, however, the turnover of disease genes is much faster, because they are constantly exposed to selection. People with severe dominant conditions often have unaffected parents and no previous family history of the condition. Lethal fully penetrant dominants necessarily always occur as fresh mutations, so the pedigree pattern gives no clues about their cause. Serious X-linked recessives also show a significant proportion of fresh mutations, because the gene is exposed to natural selection whenever it is in a male.

When a normal couple with no relevant family history have a child with severe abnormalities (*Figure 3.4D*), deciding the mode of inheritance and recurrence risk can be very difficult – the problem might be autosomal recessive, autosomal dominant, X-linked recessive (if the child is male) or nongenetic. Germinal mosaicism (see page 74) further complicates pedigree interpretation with possible new mutations.

Male lethality and lyonization may complicate X-linked pedigrees

X-Linked conditions lethal in males

For some X-linked dominant conditions the absence of the normal allele is lethal before birth. Thus affected males are not born, and we see a condition which affects only females, who pass it on to half their daughters but none of their sons (*Figure 3.4E*). There may be a history of miscarriages, but the family is rarely big enough to prove that the number of sons is only half the number of daughters. An example is incontinentia pigmenti (MIM 308310).

X-Inactivation (lyonization)

A further complication in X-linked pedigrees stems from **lyonization**, the inactivation in a female of one of the two X chromosomes, picked at random (page 172).

Female carriers may be heterozygous at the whole organism level, but functionally each cell expresses either the normal or the abnormal allele. Thus the distinction between dominant and recessive is much less clear-cut than with autosomal conditions. Carriers of 'recessive' X-linked conditions often manifest some signs, while heterozygotes for 'dominant' conditions are usually more mildly and variably affected than affected males. Where the phenotype depends on a circulating product, as in hemophilia A and B, there is an averaging effect between the normal and abnormal cells. Female carriers have an intermediate phenotype, usually clinically unaffected but biochemically abnormal. Where the phenotype is a localized property of individual cells, as in hypohidrotic ectodermal dysplasia (MIM 305100: missing sweat glands, abnormal teeth and hair), female carriers show patches of normal and abnormal tissue.

There are no Y-linked diseases

In one respect, mendelian pedigrees may be simpler than shown in *Figure 3.2*: there are no known Y-linked diseases. Y-Linked characters are known, but not diseases. Conceivably, such a disease may exist undiscovered, but this is unlikely for two reasons. First, the pedigree pattern would be strikingly noticeable, especially in societies which trace family through the male line, yet they have not been noted (claims for 'porcupine men' are dubious, see MIM 146600). Second, Y-linked diseases are inherently implausible because the existence of a 'disease gene' usually implies the existence of a normal gene carrying out some important function, such that if the function is not performed, disease results; yet females are perfectly normal without any Y-linked genes. Thus any Y-linked genes must code either for nonessential characters or for male-specific functions, and defects are unlikely to cause diseases apart from male infertility.

Mosaicism and chimerism

Mosaics and chimeras have two (or more) genetically different cell lines

Traditional pedigree-based genetics sees genes as factors transmitted through gametes to the offspring, but molecular genetics sees genes as nuclear DNA sequences which program cells, and allow us to study the genes of somatic cells as well as of gametes. The mechanism of mitosis is designed to ensure that (with trivial exceptions; see *Box 18.1*) every cell of a person's body contains a faithful and complete replica of the genes present in the original fertilized egg. However, mutations may occur during development of an organism. Such post-zygotic mutations produce **mosaics** with two (or more) genetically distinct cell lines (*Figure 3.7A*). Mosaicism can affect the somatic or germline tissues. The older literature on human mosaicism refers only to chromosomal mosaicism; mosaicism for single gene mutations is at least as important but, until DNA diagnostics was developed, it remained undetectable in humans.

Mosaics are presumed (though rarely proven) to derive from a single fertilized egg. **Chimeras**, on the other hand, are the result of aggregation of two zygotes into a single embryo (the reverse of twinning), or alternatively of limited colonization of one twin by cells from a nonidentical co-twin (*Figure 3.7B*). Chimerism is proved by the presence in pooled tissue samples of too many parental alleles at several loci (if just one locus were involved, one would suspect mosaicism for a single mutation). Blood-grouping centers occasionally discover chimeras among normal donors, and

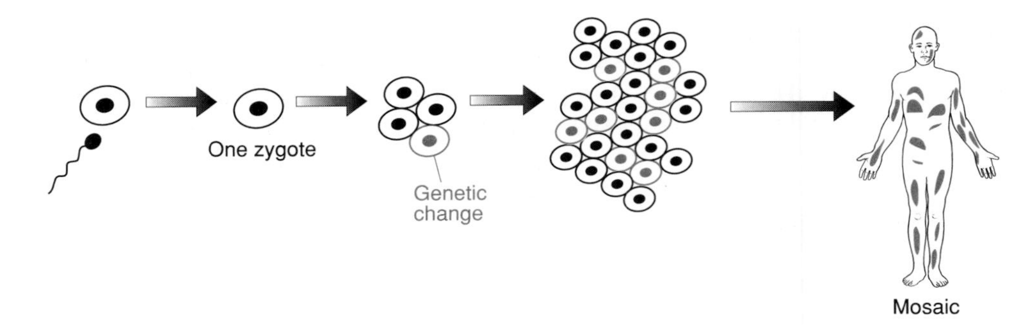

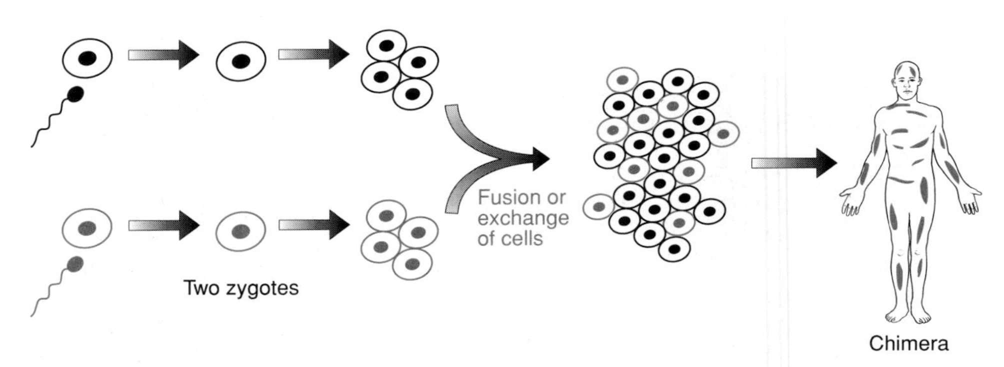

Figure 3.7: Mosaics and chimeras.

Mosaics have two or more genetically different cell lines derived from a single zygote. The genetic change indicated may be a gene mutation, a numerical or structural chromosomal change, or in the special case of lyonization, X-inactivation. A chimera is derived from two zygotes, which are usually both normal but genetically distinct.

some intersex patients turn out to be XX/XY chimeras. Chimerism is an exceptional phenomenon and rarely needs to be considered; but mosaicism is common and frequently causes difficulties in counselling and pedigree interpretation.

Germline mosaicism complicates pedigree interpretation

Pure somatic mosaicism, where the mutant clone is not present in the germline, is not relevant to inherited disease, although it is the cause of a large class of nonheritable genetic disease, particularly noninherited cancer. The rare inherited forms of cancer also require a second, somatic, mutation in addition to the inherited mutation. This whole topic is considered in detail in Chapter 17. **Germinal mosaicism** (sometimes called **gonadal mosaicism**) is most likely to be seen in those autosomal dominant or X-linked diseases where affected people are reproductively disadvantaged. As mentioned above (page 72), a substantial proportion of all cases of these diseases result from new mutations. These may occur in the affected person or in a close ancestor. The question then arises, when a person who was normal at conception produces a gamete carrying a mutation, at what point during his or her development did the mutation occur?

The timing of the mutation

A common assumption is that an entirely normal person produced a single mutant gamete. We cannot automatically assume that this is correct, unless there is something about the mutational process which means it can occur only during gametogenesis. If mutations can occur in the germline earlier in development (that is, in precursor cells of the gametes), they will produce persons who harbor a clone of mutant germline cells, and who are liable repeatedly to produce mutant gametes. The pedigree mimics recessive inheritance because normal parents, with no previous family history, produce more than one affected child. Even if the correct, dominant, mode of inheritance is realized, it is very difficult to calculate a recurrence risk to use in counseling the parents. Usually an empiric risk (see page 81) is quoted.

Molecular evidence of mosaicism

Molecular studies can be a great help in these cases. Sometimes it is possible to demonstrate directly that a normal father is producing a proportion of mutant sperm (*Figure 3.8*). Direct testing of the germline is not possible in women, but other accessible tissues such as fibroblasts or hair roots can be examined for evidence of mosaicism. A negative result on somatic tissues does not rule out germline mosaicism, but a positive result, in conjunction with an affected child, proves it.

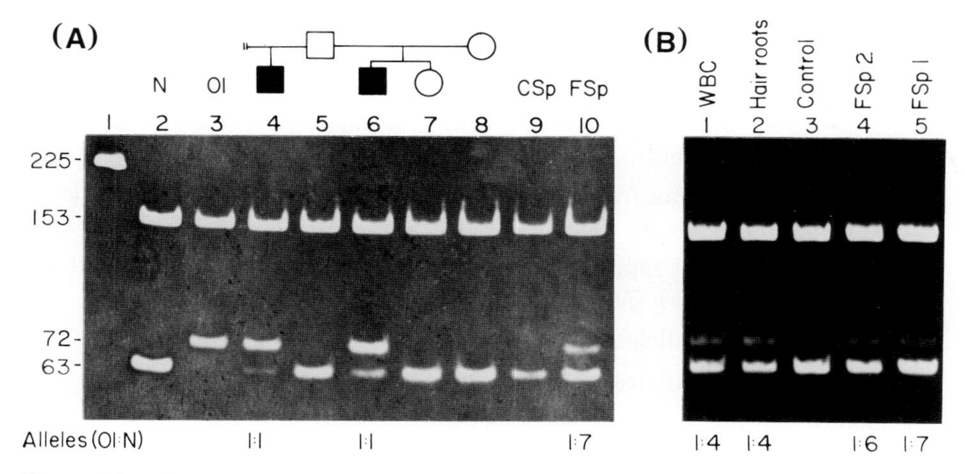

Figure 3.8: Germinal mosaicism in autosomal dominant osteogenesis imperfecta (see page 420).

The father, though phenotypically normal, carries a mutation in the *COL1A1* gene, demonstrable by PCR amplification of sperm. The normal allele gives the 63-bp band and the mutant allele the 72-bp band. Both affected sons are heterozygous with a 1:1 ratio of bands. The blood sample from the father gives only the normal band **(A** lane 5**)** but a sperm sample **(A** lane 10**)** contains both alleles with a 1:7 ratio of mutant to normal. A sperm sample from a normal control **(A** lane 9**)** gives only the normal band, as expected. **(B)** The ratio of mutant to normal alleles observed in various samples from the father. FSp, sperm of father; CSp, control sperm; WBC, white blood cells. Reproduced from Nicholls (1994) with permission from The University of Chicago Press.

Where there is no direct test for the mutation, or the relevant family members are not available, gene tracking by linked markers (page 440) can be used to identify the origin of the mutated chromosome. Such studies cannot tell us at what point on its journey through the pedigree the chromosome picked up the mutation, but sometimes the clinical data provide a clue. *Figure 3.9* shows an example. The X chromosome is mutant in III_1 but was nonmutant in the egg which produced his grandfather I_1. This produces a dilemma for II_3, II_4 and II_5. Their carrier risk may be small in percentage terms, but it is hard to quantify precisely and, with a catastrophic disease like Duchenne muscular dystrophy, even a small risk may be too high to ignore (van der Meulen *et al.*, 1995).

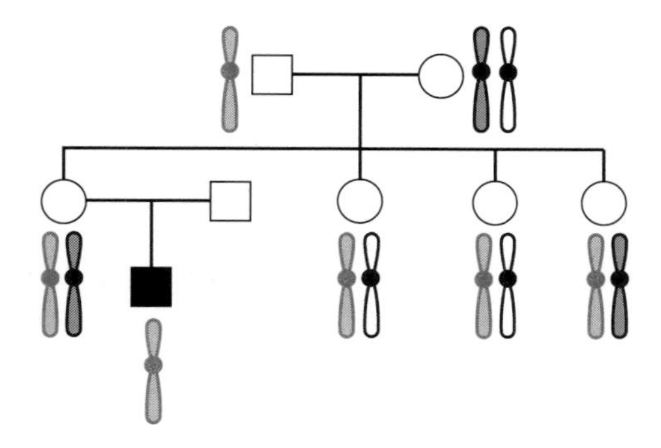

Figure 3.9: A new mutation in X-linked recessive Duchenne dystrophy.

Colors indentify the three grandparental X chromosomes (ignoring recombination). III_1 has the grandpaternal X, which has acquired a mutation at some point in the pedigree. There are four possible points at which this could have happened: (i) if III_1 carries a new mutation, the recurrence risk for all family members is very low; (ii) if II_1 is a germinal mosaic, there is a significant risk (but hard to quantify) for her future children, but not for her sisters; (iii) if II_1 was the result of a single mutant sperm, she has the standard recurrence risk for X-linked recessives, but her sisters are free of risk. If I_1 was a germinal mosaic all the sisters have a significant risk, which is hard to quantify.

Factors affecting gene frequencies

There can be a simple relationship between gene frequencies and genotype frequencies

A thought experiment: picking genes from the gene pool

Over a whole population there may be many different alleles at a particular locus, although each individual person has just two alleles, which may be identical or different. We can imagine a **gene pool**, consisting of all alleles at the A locus in the population. The **gene frequency** of allele A_1 is the proportion of all alleles in the gene pool which are A_1. Consider two alleles, A_1 and A_2 at the A locus. Let their gene frequencies be p and q respectively (p and q are each between 0 and 1). Let us perform a thought experiment:

- Pick an allele at random from the gene pool. There is a chance p that it is A_1 and a chance q that it is A_2.
- Pick a second allele at random. Again the chance of picking A_1 is p and the chance of picking A_2 is q (we assume the gene pool is very large).
- The chance that both alleles were A_1 is p^2.
- The chance that both alleles were A_2 is q^2.
- The chance that the first allele was A_1 and the second A_2 is pq.
- The chance that the first was A_2 and the second A_1 is qp.
- Overall, the chance of picking one A_1 and one A_2 allele is $2pq$.

The Hardy–Weinberg distribution

If we pick a person at random from the population, this is equivalent to picking two genes at random from the gene pool. The chance the person is A_1A_1 is p^2, the chance they are A_1A_2 is $2pq$, and the chance they are A_2A_2 is q^2. This simple relationship between gene frequencies and genotype frequencies (the **Hardy–Weinberg distribution**, see *Box 3.3*) holds whenever a person's two genes are drawn independently and at random from the gene pool. A_1 and A_2 may be the only alleles at the locus (in which case $p + q = 1$) or there may be other alleles and other genotypes ($p + q < 1$). For X-linked loci males, being hemizygous (only one allele), are A_1 or A_2 with frequencies p and q respectively, while females can be A_1A_1, A_1A_2 or A_2A_2 (see *Box 3.3*).

Limitations of the Hardy–Weinberg distribution

These simple calculations break down if the underlying assumption, that a person's two genes are picked independently from the gene pool, is violated. In particular, there is a problem if there has not been random mating. **Assortative mating** can take several forms, but the most generally important is **inbreeding**. If you marry a

Box 3.3: Hardy-Weinberg equilibrium genotype frequencies for allele frequencies p (A_1) and q (A_2)

	Autosomal locus			X-linked locus				
				Males		Females		
Genotype	A_1A_1	A_1A_2	A_2A_2	A_1	A_2	A_1A_1	A_1A_2	A_2A_2
Frequency	p^2	$2pq$	q^2	p	q	p^2	$2pq$	q^2

Note that these genotype frequencies will be seen whether or not A_1 and A_2 are the only alleles at the locus.

relative you are marrying somebody whose genes resemble your own. This increases the likelihood of your children being homozygous and decreases the likelihood that they will be heterozygous. Rare recessive conditions are strongly associated with parental consanguinity, and Hardy–Weinberg calculations which ignore this will overestimate the carrier frequency in the population at large. Population stratification (page 499) is a form of hidden inbreeding.

Use of the Hardy–Weinberg distribution in genetic counseling

Gene frequencies or genotype frequencies are essential inputs into many forms of genetic analysis, such as linkage analysis (page 320) and segregation analysis (page 491), and they have a particular importance in calculating genetic risks. *Box 3.4* gives examples.

Box 3.4: The Hardy–Weinberg distribution can be used (with caution) to calculate carrier frequencies and simple risks for counseling

An autosomal recessive condition affects one newborn in 10 000. What is the expected frequency of carriers?

Phenotypes:	Unaffected		Affected
Genotypes:	AA	Aa	aa
Frequencies:	p^2	$2pq$	$q^2 = 1/10\,000$

q^2 is 10^{-4}, and therefore $q = 10^{-2}$ or 1/100.
1 in 100 genes at the A locus are a, 99/100 are A.
The carrier frequency, $2pq$, is $2 \times 99/100 \times 1/100$, very nearly 1 in 50.

If a parent of a child affected by the above condition remarries, what is the risk of producing an affected child in the new marriage?

To produce an affected child, both parents must be carriers, and the risk is then 1 in 4. Thus the overall risk is:

	(the parent's carrier risk)	×	(the new spouse's carrier risk)	× 1/4
=	1	×	1/50	× 1/4
=	1/200			

This assumes there is no family history of the same disease in the new spouse's family.

X-Linked red–green color blindness affects one in 12 British males; what proportion of females will be carriers and what proportion will be affected?

	Males		Females		
Genotypes:	A_1	A_2	A_1A_1	A_1A_2	A_2A_2
Frequencies:	p	$q = 1/12$	p^2	$2pq$	q^2

$q = 1/12$, therefore $p = 11/12$,
$2pq = 2 \times 1/12 \times 11/12 = 22/144$
$q^2 = 1$ in 144. Thus this single-locus model predicts that 15% of females will be carriers and 0.7% will be affected.

Genotype frequencies can be used (with caution) to calculate mutation rates

Mutant genes are being created by fresh mutation but being removed by natural selection (page 245). For a given level of selection we can calculate the mutation rate which would be required to replace the genes lost. If we assume that there is an equilibrium in the population between the rates of loss and of replacement, the calculation tells us the present mutation rate. We can define the **coefficient of selection(s)** as the relative chance of reproductive failure of a genotype due to natural selection (the fittest type in the population has $s = 0$, a genetic lethal has $s = 1$).

- For an **autosomal recessive** condition, a proportion q^2 of the population are affected, and so the loss of disease genes each generation is sq^2. This is balanced by mutation at the rate of $\mu(1-q^2)$ where μ is the mutation rate per gene per generation. At equilibrium $sq^2 = \mu(1-q^2)$, or approximately (if q is small) $\mu = sq^2$.
- For a rare **autosomal dominant** condition only the heterozygotes are significantly frequent in the population. Heterozygotes occur with frequency $2pq$ (frequency of disease gene = p). Only half the genes lost through their reproductive failure are the disease gene, so the rate of gene loss is very nearly sp. Again, this is balanced by a rate of new mutation of μq^2, which is approximately μ if q is almost 1. Thus $\mu = sp$.
- For an **X-linked recessive** disease, the rate of gene loss through affected males is sq. This is balanced by a mutation rate 3μ, since all X chromosomes in the population are available for mutation, but only the one third of X chromosomes which are in males are exposed to selection. Thus $\mu = sq/3$.

Estimates derived using these formulae can be compared with the general expectation, from studies in many organisms, that mutation rates are typically 10^{-5}–10^{-7} per gene per generation.

Heterozygote advantage can be much more important than recurrent mutation for maintaining a disease gene at high frequency

The formula $\mu = sq^2$ gives an unexpectedly high mutation rate for some autosomal recessive conditions. Cystic fibrosis (CF) is an example. Until very recently, virtually nobody with cystic fibrosis lived long enough to reproduce, therefore $s = 1$. CF affects about one birth in 2000 in the UK. Thus $q^2 = 1/2000$, and the formula gives $\mu = 5 \times 10^{-4}$. This would be a strikingly high mutation rate for any gene – but there is evidence that new CF mutations are in fact very rare. This follows from the uneven ethnic distribution of CF and the existence of strong linkage disequilibrium (*Table 12.3*).

The missing factor is **heterozygote advantage**. CF carriers have, or had in the past, some reproductive advantage over normal homozygotes. There has been much debate over what this advantage might be. The CF gene encodes a membrane chloride channel, and a popular but unproven theory is that the lower number of chloride channels in carriers makes them more resistant to chloride-losing diarrhea. This could have helped them survive cholera epidemics (Gabriel *et al.*, 1994; Rodman and Zamudio, 1991). Whatever the cause of the heterozygote advantage, if s_1 and s_2 are the coefficients of selection against the AA and aa genotypes respectively, then at equilibrium, the ratio of the gene frequencies of A and a, p/q, is s_2/s_1. *Box 3.5* illustrates the calculation for CF, and shows that a heterozygote advantage too small to observe in population surveys can have a major effect on gene frequencies.

Box 3.5: Selection in favor of heterozygotes for cystic fibrosis

For CF, the disease frequency in the UK is about one in 2000 births.

Phenotypes:	Unaffected		Affected
Genotypes:	AA	Aa	aa
Frequencies:	p^2	$2pq$	$q^2 = 1/2000$

q^2 is 5×10^{-4}, therefore $q = 0.022$ and $p = 1 - q = 0.978$
$p/q = 0.978/0.022 = 43.72 = s_2/s_1$
If $s_2 = 1$ (affected homozygotes never reproduce), $s_1 = 0.023$

The present CF gene frequency will be maintained if Aa heterozygotes have on average 2.3% more surviving children than AA homozygotes.

It is worth remembering that the medically important mendelian diseases are those which are both common and serious, and they must all have one or another special trick to remain common in the face of selection. This trick may be an exceptionally high mutation rate (Duchenne muscular dystrophy), or propagation of non-pathological but unstable pre-mutations (fragile X; see page 266), or onset of symptoms after reproductive age (Huntington disease; see *Figure 3.5*), but for common serious recessive conditions it is most often heterozygote advantage.

Nonmendelian characters

Research into simple and complex traits has long defined two separate traditions within human genetics

By the time Mendel's work was rediscovered in 1900, a rival school of genetics was well established in the UK and elsewhere. Francis Galton, the remarkable and eccentric cousin of Charles Darwin, devoted much of his vast talent to systematizing the study of human variation. Starting with an article on *'Hereditary Talent and Character'* published the same year, 1865, as Mendel's paper (and expanded in 1869 to a book, *Hereditary Genius*), he spent many years investigating family resemblances.

Galton was devoted to quantifying observations and applying statistical analysis. His Anthropometric Laboratory, established in 1884, recorded from his subjects (who paid him threepence for the privilege) their weight, sitting and standing height, arm span, breathing capacity, strength of pull and of squeeze, force of blow, reaction time, keenness of sight and hearing, color discrimination and judgements of length. Except for color blindness, these are quantitative, continuously variable characters. In one of the first applications of statistics, he compared physical attributes of parents and children, and established the degree of correlation between relatives. By 1900 he had established a large body of knowledge about the inheritance of such attributes, and a tradition (*biometrics*) of their investigation.

A historical controversy

When Mendel's work was rediscovered, a controversy arose. The claims of the Mendelians, championed by Bateson, were resisted by biometricians. Biometricians allowed that mendelian genes might explain a few rare abnormalities or insignificant variants, but pointed out that most of the characters likely to be important in evolution (body size, build, strength, skill in catching prey or finding food) were continuously variable quantitative characters and not amenable to mendelian analysis. You cannot follow their inheritance by drawing pedigrees and marking in the affected people, because we all have these characters, only to different degrees. The controversy ran on, heatedly at times, until 1918. That year saw a seminal paper by R.A. Fisher demonstrating that characters governed by a large number of independent mendelian factors (**polygenic characters**) would display the quantitative variation and family correlations described by the biometricians.

The two traditions in human genetics

In principle, Fisher's description of polygenic inheritance unified genetics. This was indeed generally true for the genetics of experimental organisms or farm animals. In human genetics, however, the studies of mendelian and quantitative characters have tended to continue as separate traditions, and few investigators feel at home in both worlds. The spectacular advances of 1970–1990 were entirely in mendelian genetics. Until very recently, investigation of nonmendelian characters was limited to descriptive studies of family resemblances. Geneticists from the mendelian tradition were often reluctant to get involved in these studies, partly because of the complex statistical methodology and perhaps also because of a feeling that they were a poor investment of research effort compared to mapping and cloning genes for mendelian characters. Also, many studies concerned sensitive areas of behavioral genetics such as the heritability of IQ, where violent controversies and a distastefully confrontational style of argument often reigned.

One could perhaps characterize the two traditions as 'bottom-up' and 'top-down'. mendelian genetics is essentially reductionist: the analysis starts with the most basic simple units of inheritance, and aims to build up understanding of complex characters from the bottom by combining the effects of identified single genes. Biometricians and their intellectual descendants observe phenotypes whose causation may be immensely complex, and hope to unravel the complexity by statistical dissection, going down from the phenotype towards the genes.

Modern approaches to analysis of nonmendelian characters

Recent developments are promising at last to bring together the study of single genes and complex phenotypes. Automated gel analysis is allowing genes and markers to be analyzed on a scale unimaginable 10 years ago. This has had two consequences. Virtually every human gene will have been identified before the turn of the century, at least as an expressed sequence tag (EST, see page 303), so that molecular geneticists are looking for fresh fields to conquer. At the same time, marker studies can now be large enough to have the statistical power to detect individual loci underlying complex phenotypes. Given the overwhelming preponderance of nonmendelian conditions in the genetics of human disease, molecular dissection of complex phenotypes is widely seen as the next frontier in medical genetics. The whole subject is considered in detail in Chapter 18.

Nonmendelian characters can be polygenic, oligogenic or multifactorial

The further away a character is from the primary gene action, the less likely it is to show a simple mendelian pedigree pattern. DNA sequence variants are virtually always mendelian. That is part of their attraction as genetic markers. Protein variants (electrophoretic mobility or enzyme activity) are usually mendelian but can depend on more than one locus because of post-translational modification (page 24). Clinical syndromes usually depend on more than just a single genetic locus. They result from failure or malfunction of some physiological or developmental process, usually involving a complex of interacting factors. Thus the common birth defects (cleft palate, congenital dislocation of the hip, congenital heart disease, etc.) are rarely mendelian. Behavioral traits like IQ test performance or schizophrenia are still less likely to be mendelian. This does not however mean that they may not be genetic, either partly or entirely.

Nonmendelian characters may depend on two, three or many genetic loci, with greater or smaller contributions from environmental factors (*Figure 3.10*). We use **multifactorial** here as a catch-all term covering all these possibilities. More specifically, the genetic determination may involve a small number of loci (**oligogenic**) or many loci, each individual locus having only a small effect (**polygenic**), or there may be a single major locus with a multifactorial background. Early workers on quantitative genetics thought that so-called polygenes might be a special class of genes, but now we see them as standard mendelian genes which happen to have only a marginal effect on the particular character being considered. The same gene could appear as the mendelian determinant of one character and a polygene for another. Genetics of nonmendelian characters is discussed in detail in Chapter 18.

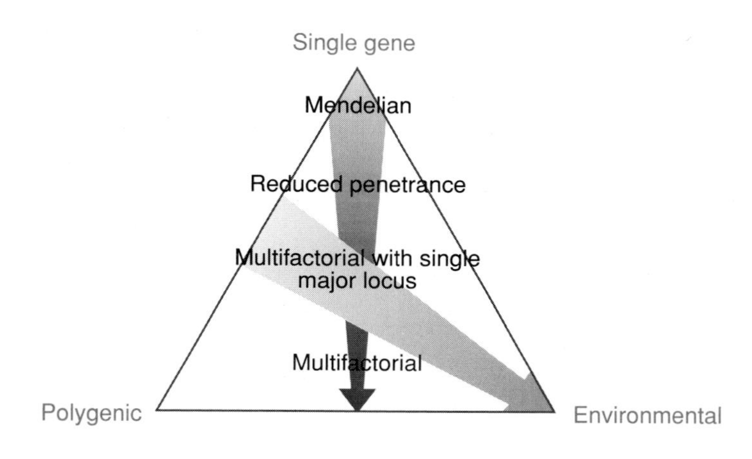

Figure 3.10: The spectrum of human characters.

Few characters are purely mendelian, purely polygenic or purely environmental. Most depend on a mixture of major and minor genetic determinants, together with environmental influences. The mix of factors determining any character could be represented by a point located somewhere within the triangle.

Counseling in nonmendelian conditions uses empiric risks

In genetic counseling for nonmendelian conditions, risks are not derived from polygenic theory; they are **empiric risks** obtained through population surveys (see, for example *Table 18.4*). This is fundamentally different from mendelian conditions, where the one in two, one in four, etc., risks come from theory. The effect of family history is also quite different. If a couple are both carriers of CF, the risk of their next child being affected is one in four. This remains true regardless of how many affected or normal children they have already produced. If they have had a baby with a neural tube defect, the recurrence risk is about one in 25 in the UK – but if they have already had two affected babies, the recurrence risk is about one in 12. It is not that having a second affected baby has caused their recurrence risk to increase, but it has enabled us to recognize them as a couple who always had been at particularly high risk. A cynic would say it involves the counselor being wise after the event – but the practice accords with our understanding based on threshold theory (page 486), as well as with epidemiological data, and it represents the best we can offer in an imperfect state of knowledge.

Further reading

Bittles AH, Neel JV. (1994) The costs of human inbreeding and their implications for variations at the DNA level. *Nature Genetics,* **8**, 117–121.

Connor JM, Ferguson-Smith MA. (1994) *Essential Medical Genetics,* 4th Edn. Blackwell Scientific Publications, Oxford.

Forrest DW. (1974) *Francis Galton: the Life and Work of a Victorian Genius*. Elek, London.

Wolf U. (1995) The genetic contribution to the phenotype. *Hum. Genet.* **95**, 127–148.

References

Gabriel SE, Brigman KN, Koller BH, Boucher RC, Stutts MJ. (1994) *Science,* **266**, 107–109.

Harper PS, Newcombe PG. (1992) *J. Med. Genet.,* **29**, 239–242.

Harper PS. (1993) *Genetic Counselling,* 4th Edn. Butterworth Heinemann, Oxford.

Heutink P, van der Mey AG, Sandkujl LA *et al.* (1992) *Hum. Mol. Genet.,* **1**, 7–10.

McKusick VA. (1994) (ed.) *Mendelian Inheritance in Man,* 11th Edn. Johns Hopkins University Press, Baltimore, MD.

Nicholls RD. (1994) *Am. J. Hum. Genet.,* **54**, 733–740.

Prezant TR, Agapian JV, Bohlman MC *et al.* (1993) *Nature Genetics,* **4**, 289–294.

Riordan-Eva P, Harding AE. (1995) *J. Med. Genet.,* **32**, 81–87.

Rodman DM, Zamudio S. (1991) *Med. Hypotheses,* **36**, 253–258.

Sweeney MG, Davis MB, Lashwood A, Brockington M, Toscano A, Harding AE. (1992) *Am. J. Hum. Genet.,* **51,** 741–748.

van der Meulen MA, van der Meulen MJP, te Meerman GJ. (1995) *J. Med. Genet.,* **32,** 102–104.

van der Mey AGL, Maaswinkel-Mooy PD, Cornelisse CJ, Schmidt PH, van de Kamp JJP. (1989) *Lancet,* **ii**, 1291–1294.

Viljoen D, Ramesar R. (1992) *J. Med. Genet.,* **29**, 221–225.

Wallace DC. (1994) *Proc. Natl Acad. Sci. USA,* **91**, 8739–8746.

Wilkie AOM. (1994) *J. Med. Genet.,* **31**, 89–98.

Cell-based DNA cloning

Chapter 4

Principles of DNA cloning

Because mammalian genomes are complex, any specific gene or DNA fragment of interest normally represents only a tiny fraction of the total DNA in a cell. For example, the β-globin gene comprises only 0.00005% of the 3000 megabases (Mb) of human genomic DNA, and even the massive 2.5 Mb dystrophin gene, the largest gene that has been identified, accounts for only about 0.08% of human genomic DNA. Enrichment for gene sequences is possible by isolating total RNA, or poly(A)$^+$ messenger RNA (mRNA) from suitable cells and converting this to **complementary DNA (cDNA)**. In some cases, this can result in a profound enrichment for specific exonic DNA sequences when the relevant genes are known to be expressed at very high levels in a specific cell type. In most cases, however, the desired gene sequences still represent only a tiny proportion of the total cDNA population. In order to study a specific DNA sequence within a complex DNA population, two major approaches have been applied (see *Figure 4.1*).

- **DNA cloning**. The desired fragment must be *selectively amplified* so that it is purified essentially to homogeneity. Thereafter, its structure and function can be comprehensively studied, for example by DNA sequencing, *in vitro* expression studies, etc.
- **Molecular hybridization**. The fragment of interest is not amplified, but instead is *specifically detected* within a complex mixture of many different sequences. Its chromosomal location can be determined in this way and some information can be gained regarding its structure. If expressed, the sequence of interest can be detected within a complex RNA or cDNA population from specific cells, enabling comprehensive analysis of its expression patterns.

The amplification step in DNA cloning involves a programmed large increase in copy number of selected DNA sequences. In practice, this involves multiple rounds of DNA replication catalyzed by a DNA polymerase acting on one or more types of template DNA molecule. This can be achieved by essentially two different approaches, involving DNA replication *in vivo* or *in vitro*.

(i) **Cell-based DNA cloning.** This is an *in vivo* DNA cloning method whose first step involves attaching foreign DNA fragments *in vitro* to DNA sequences which are capable of independent replication. The hybrid DNA fragments are then transferred into suitable host cells where they can be propagated

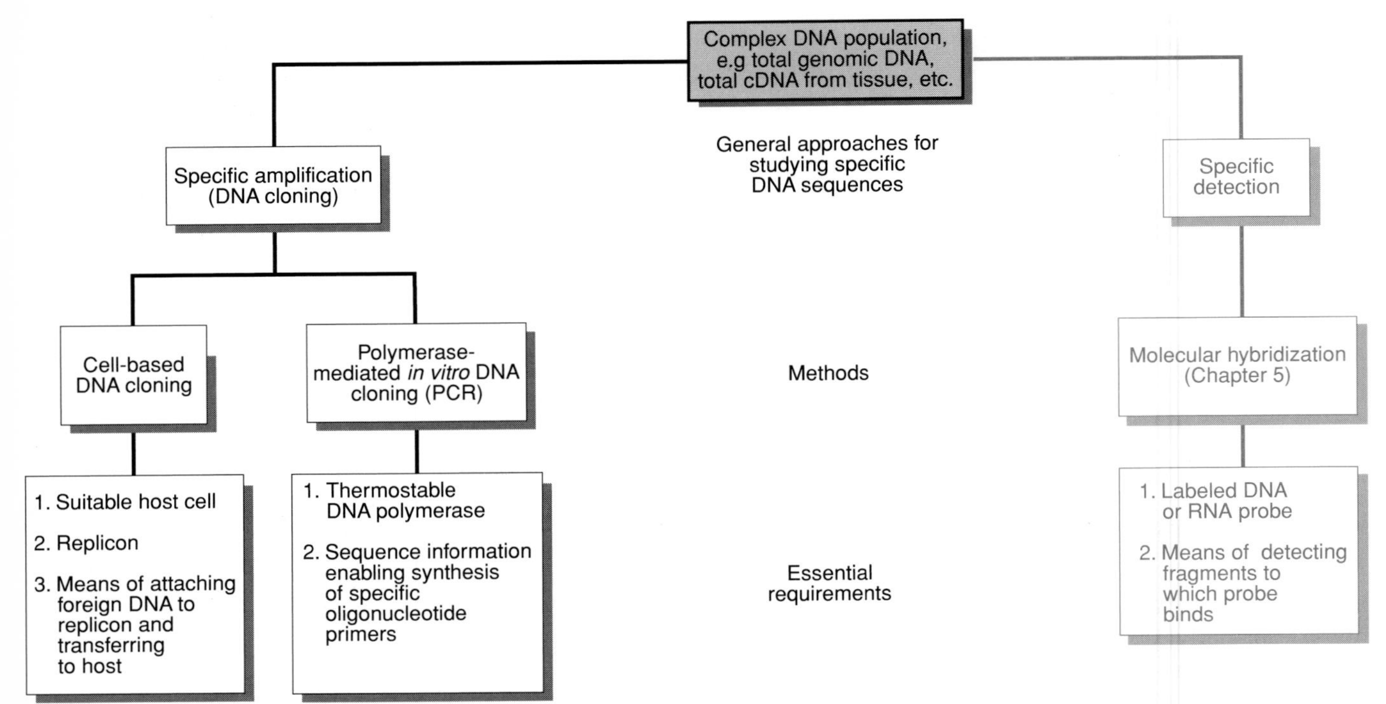

Figure 4.1: General approaches for studying specific DNA sequences in complex DNA populations.

selectively. In the past, the term DNA cloning has been used exclusively to signify this particular approach.

(ii) **Cell-free DNA cloning.** The **polymerase chain reaction (PCR)** is a newer form of DNA cloning which is enzyme mediated and is conducted entirely *in vitro*.

Principles of cell-based DNA cloning

Cell-based DNA cloning requires attachment *in vitro* of DNA fragments to purified replicons and propagation in suitable host cells

The essence of cell-based DNA cloning involves four steps (see *Figure 4.2*):

(i) *construction of recombinant DNA molecules* by *in vitro* covalent attachment (**ligation**) of the desired DNA fragments (**target DNA**) to a **replicon** (any sequence capable of independent DNA replication). This step is facilitated by cutting the target DNA and replicon molecules with specific restriction nucleases (page 87) before joining the different DNA fragments using the enzyme DNA ligase.

(ii) *Transformation* by transfer of the recombinant DNA molecules into host cells (often bacterial or yeast cells) in which the chosen replicon can undergo DNA replication independently of the host cell chromosome(s).

(iii) *Selective propagation of cell clones* involves two stages. Initially the transformed cells are plated out by spreading on an agar surface in order to encourage the growth of cell colonies, which are **cell clones** (populations of identical cells all descended from a single cell). Subsequently, individual colonies can be picked from a plate and the cells can be further expanded in liquid culture.

(iv) *Isolation of recombinant DNA clones* by harvesting expanded cell cultures and selectively isolating the recombinant DNA.

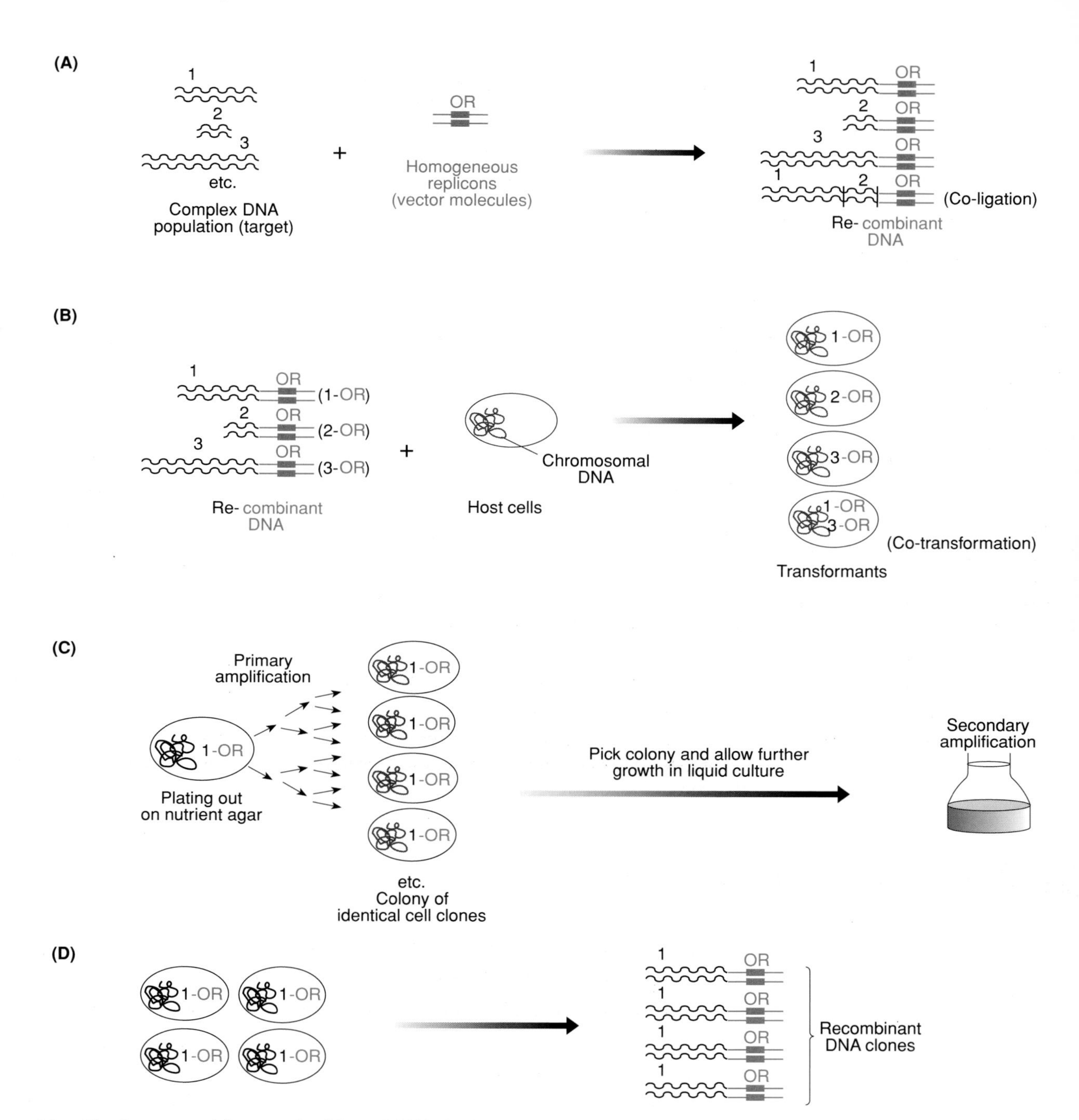

Figure 4.2: The four essential steps of cell-based DNA cloning.

(A) Formation of recombinant DNA. *Note* that, in addition to simple vector–target ligation products, co-ligation events may occur whereby two unrelated target DNA sequences may be ligated in a single product. OR, origin of replication.
(B) Transformation. *Note* that occasionally co-transformation events are observed, such as the illustrated cell which has been transformed by two different DNA molecules.
(C) Amplification to produce numerous cell clones.
(D) Isolation of recombinant DNA clones.

Replicons and host cells

Cell cloning requires that the foreign DNA fragments which are introduced into a host cell must be able to replicate; otherwise, they would soon be diluted out as the host cell undergoes many rounds of cell division. However, foreign DNA fragments will generally lack an origin of replication that will function in the host cell. They require, therefore, to be attached to an independent *replicon* so that their replication is controlled by the replicon's origin of replication.

In principle, the necessary replicon could be provided in two ways. One possibility is to introduce the fragments into the host cells in such a way that in each cell one, or a very few, fragments integrate into a host cell chromosome and are propagated under the control of a host cell replicon. This is the way that retroviruses, for example, integrate their DNA into host cell chromosomes. However, as a general method for cloning, this approach suffers from numerous disadvantages, including the difficulty in retrieving the inserted DNA. Instead, cell-based cloning very largely relies on a different approach: the foreign DNA fragments are attached *in vitro* to a purified replicon and the resulting hybrid molecules are transferred into host cells, where they replicate *independently of the host cell chromosomes*. Because the foreign DNA fragments (target DNA) can be viewed as passengers of the replicon, replicons used for cloning are described as **vector** molecules. Notice that, although the resulting replication of the introduced DNA fragments is independent of the host chromosomes, the vector may have an origin of replication that originates from either a natural extrachromosomal replicon or, in some cases, a chromosomal replicon (see page 103).

Although some DNA cloning systems involve human and other mammalian cells as hosts, the great bulk of cell-based DNA cloning has used modified bacterial or fungal host cells. The former cells are widely used because of their capacity for rapid cell division. Bacterial cell hosts have a single circular double-stranded chromosome with a single origin of replication. Replication of the host chromosome subsequently triggers cell division so that each of the two resulting daughter cells contains a single chromosome like their parent cell (i.e. the copy number is maintained at one copy per cell). However, the replication of extrachromosomal replicons is not constrained in this way: many such replicons go through several cycles of replication during the cell cycle and can reach high copy numbers. Two basic types of extrachromosomal replicon are found in such cells:

(i) **plasmids** are small circular double-stranded DNA molecules which individually contain very few genes. Their existence is intracellular, being vertically distributed to daughter cells following host cell division, but can be transferred horizontally to neighboring cells during bacterial conjugation events. Natural examples include plasmids which carry the sex factor (F) and those which carry drug-resistance genes.

(ii) **Bacteriophages** are viruses which infect bacterial cells. DNA-containing bacteriophages often have genomes containing double-stranded DNA which may be circular or linear. Unlike plasmids, they can exist extracellularly. The mature virus particle (**virion**) has its genome encased in a protein coat so as to facilitate adsorption and entry into a new host cell.

In order for naturally occurring replicons to be used as vector molecules for cell-based DNA cloning, various modifications need to be made. Similarly, the host cells that are used for cloning are specialized cells whose genotype has been selected to optimize their use in DNA cloning. Typically, cloning systems are constructed so as to ensure that joining of the foreign DNA fragment occurs at a unique location in the vector molecule. Additionally, they have in-built selection systems so that cells which contain the relevant vector molecule can be specifically selected. In many cases, there are additional screening systems to ensure detection and propagation of cells containing recombinant DNA (see page 95).

Restriction endonucleases enable the target DNA to be cut into pieces of manageable size and facilitate ligation to similarly cut vector molecules

A major boost to the development of cell-based DNA cloning was the discovery and exploitation of **type II restriction endonucleases**, enzymes which normally cleave DNA at all locations which contain a small, specific recognition sequence, usually 4–8 base pairs (bp) long (see *Box 4.1* and *Table 4.1*).

Box 4.1: Restriction endonucleases and modification–restriction systems

Bacteriophages that are liberated from a bacterium of a particular strain can infect other bacteria of the same strain but not those of a different strain. This is because the phage DNA has the same **modification** pattern as the DNA of bacterial strains it can infect; the phage is **'restricted'** to that strain of bacteria. The restriction is not an absolute one: some phages can escape restriction and can acquire the modification pattern of the new host.

The basis of modification–restriction systems is now known to involve two types of enzyme activity:

(i) a sequence-specific **DNA methylase** activity provides the basis of the modification pattern;

(ii) a sequence-specific **restriction endonuclease** activity underpins the restriction phenomenon by cleaving phage DNA whose modification (methylation) pattern is different from that of the host cell DNA.

The bacterial strain possesses a *DNA methylase activity with the same sequence specificity as the corresponding restriction nuclease activity*. As a result, cellular restriction endonucleases will not cleave the appropriately methylated host cell DNA but may cleave incoming phage DNA, if not methylated appropriately. *Note*, however, that some plasmids and bacteriophages possess genes for modification and restriction systems so that their presence in a bacterium determines its specificity.

The recognition sequences for the vast majority of type II restriction endonucleases are normally **palindromes**, that is the sequence of bases is the same on both strands when read in the $5' \rightarrow 3'$ direction, as a result of a twofold axis of symmetry. In some cases, the cleavage points occur exactly on the axis of symmetry, giving products (**restriction fragments**) which are **blunt-ended** (*Figure 4.3*). In most cases, however, the cleavage points do not fall on the symmetry axis, so that the resulting restriction fragments possess so-called **5′ overhangs** or **3′ overhangs**.

Overhanging ends generated by cleavage with a restriction nuclease are often described as **sticky ends** or **cohesive termini** because the two overhanging ends of each fragment are complementary in base sequence, and will have a tendency to associate with each other, or with any other similarly complementary overhang, by forming base pairs. Different fragments with the same sequences in their overhanging ends can be generated by: (i) cutting with the same restriction nuclease; (ii) cutting with different restriction nucleases that happen to recognize the same target sequence (**isoschizomers**); or (iii) by cutting with enzymes which have different recognition sequences but happen to produce compatible sticky ends, for example *Bam*HI and *Mbo*I (*Figure 4.3*).

The termini of restriction fragments which have the same type of overhanging ends can associate in a variety of different ways, either intramolecularly (**cyclization**), or between molecules to form linear **concatemers** or circular compound molecules.

Table 4.1: Restriction endonucleases

Enzyme	Source	Sequence cut	Average expected fragment size (kb) in human DNA[a]
*Alu*I	*Arthrobacter luteus*	AGCT	0.3
*Hae*III	*Hemophilus aegyptus*	GGCC	0.6
*Taq*I	*Thermus aquaticus*	TCGA	1.4
*Mnl*I	*Moraxella nonliquefaciens*	CCTC/GAGG	0.4
*Hind*III	*Hemophilus influenzae Rd*	AAGCTT	3.1
*Eco*RI	*Escherichia coli* R factor	GAATTC	3.1
*Bam*HI	*Bacillus amyloliquefaciens H*	GGATCC	7.0
*Pst*I	*Providencia stuartii*	CTGCAG	7.0
*Mst*I	*Microcoleus* species	CCTNAGG[c]	7.0
*Sma*I	*Serratia marcescens*	CCCGGG	78
*Bss*HII	*Bacillus stearothermophilus*	GCGCGC	390[b]
*Not*I	*Norcadia otitidis-caviarum*	GCGGCCGC	9766[b]

[a] Assuming 40% G+C, and a CpG frequency 20% of that expected.
[b] Observed average sizes are considerably lower than these estimates.
[c] N = A, C, G or T.
Note: Names are normally derived from the first letter of the genus and the first two letters of the species name, e.g. *Pst*I is the first restriction nuclease to have been isolated from *Providencia stuartii*. *Mnl*I is an example of an enzyme whose recognition sequence is not palindromic. So-called 'rare-cutters' often have recognition sequences containing one or more CpG dinucleotides and cut vertebrate DNA comparatively infrequently.

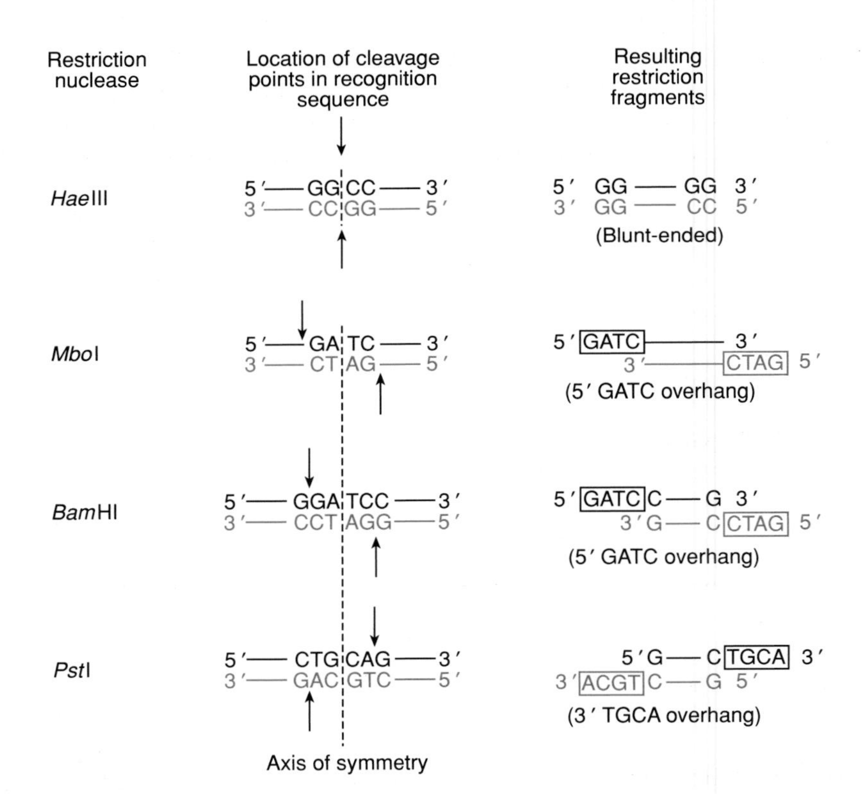

Figure 4.3: Restriction nucleases can generate blunt-ended fragments or fragments with 5′ or 3′ overhanging ends.

Intermolecular reactions occur most readily at high DNA concentrations. At very low DNA concentrations, however, individual termini on different molecules have less opportunity of making contact with each other, and intramolecular cyclization is favored.

Because the overhanging ends generated by restriction endonucleases are very short (typically four nucleotides or less), hydrogen bonding between complementary overhanging ends provides a rather weak contact between two molecules, and can only be maintained at low temperatures. However, it does facilitate subsequent covalent bonding between the two associated molecules (**DNA ligation**). This is performed using the enzyme DNA ligase. Ligation of blunt-ended fragments is also possible, although less efficient than sticky end ligation.

Generally, ligation reactions are designed to promote the formation of recombinant DNA (by ligating target DNA to vector DNA), although vector cyclization, vector–vector concatemers and target DNA–target DNA ligation are also possible (see *Figure 4.4*). To achieve this, the vector molecules are often treated so as to prevent or minimize their ability to undergo cyclization. Two common ways of achieving this are:

(i) *cutting of vectors with two different restriction endonucleases.* Often, vector molecules have multiple unique restriction sites, in which foreign DNA can be cloned, occurring in a short segment of the molecule. It is often convenient, therefore, to cut the vector with two restriction endonucleases which do not produce complementary overhanging ends (e.g. *Eco*RI and *Bam*HI), and remove the small vector fragment between the sites, resulting in a vector molecule whose two ends cannot religate. However, if target DNA is cut with the same enzyme combination, recombinant DNA can easily be formed.

(ii) *Vector dephosphorylation.* During DNA ligation *in vitro*, the enzyme DNA ligase will catalyze the formation of a phosphodiester bond only if one nucleotide contains a 5′ phosphate group and the other contains a 3′ hydroxyl group. If, therefore, the 5′ phosphate groups at both ends of the vector DNA are removed by treatment with alkaline phosphatase, the tendency for the vector DNA to recircularize will be minimized. A foreign DNA insert can, however,

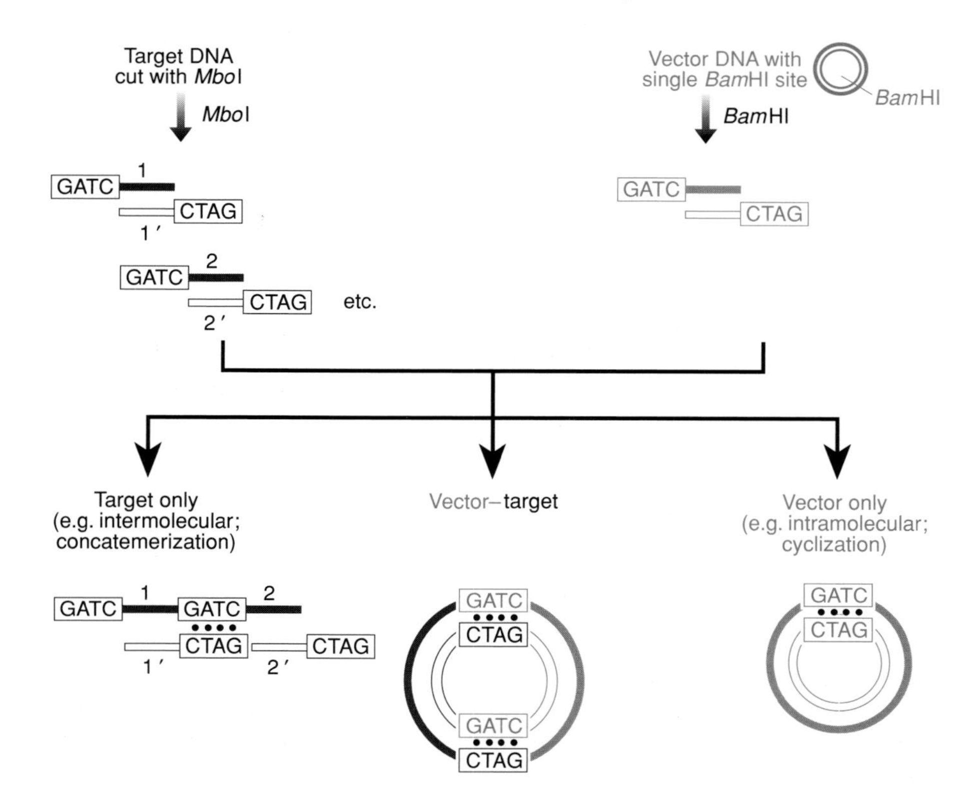

Figure 4.4: Cohesive termini can associate intramolecularly and intermolecularly.

Note that only some of the possible outcomes are shown. For example, vector molecules may also form intermolecular concatemers, multimers can undergo cyclization and co-ligation events can involve two different target sequences being included with a vector molecule in the same recombinant DNA molecule (see *Figure 4.2A*). The tendency towards cyclization of individual molecules is more pronounced when the DNA is at low concentration and the chances of collision between different molecules with complementary sticky ends is reduced.

provide 5′-terminal phosphates which can then be joined to the 3′ hydroxyl groups provided by the vector. This method, therefore, increases the frequency of cells containing recombinant DNA.

Introducing recombinant DNA into recipient cells provides a method for fractionating a complex starting DNA population

The plasma membrane of cells is selectively permeable and does not normally admit large molecules such as long DNA fragments. However, cells can be treated in certain ways (e.g. by exposure to certain high ionic strength salts, short electric shocks, etc.) so that the permeability properties of the plasma membranes are altered. As a result, a fraction of the cells become **competent**, that is capable of taking up foreign DNA from the extracellular environment. Only a small percentage of the cells will take up the foreign DNA (**DNA transformation**), but those that do often take up only a single molecule (which can, however, subsequently replicate many times within a cell). *This is the basis of the critical fractionation step in cell-based DNA cloning*; the population of transformed cells can be thought of as a sorting office in which the complex mixture of DNA fragments is sorted by depositing individual DNA molecules into individual recipient cells (*Figure 4.5*).

Because circular DNA (even nicked circular DNA) transforms much more efficiently than linear DNA, most of the cell transformants will contain cyclized products rather than linear recombinant DNA concatemers and, if an effort has been made to suppress vector cyclization (e.g. by dephosphorylation, see page 89), most of the transformants will contain recombinant DNA. Note, however, that **cotransformation** events (the occurrence of more than one type of introduced DNA molecule within a cell clone, see *Figure 4.2B*) are quite common in some cloning systems, notably yeast artifical chromosome cloning (see page 103).

The transformed cells are allowed to multiply. In the case of cloning using plasmid vectors and a bacterial cell host, a solution containing the transformed cells is simply spread over the surface of nutrient agar in a petri dish (**plating out**). This usually results in the formation of **bacterial colonies** which consist of **cell clones** (identical progeny of a single ancestral cell). Picking an individual colony into a tube for subsequent growth in liquid culture permits a secondary expansion in the number of cells which can be scaled up to provide very large yields of cell clones, all identical to an ancestral single cell (*Figure 4.5*). If the original cell contained a single type of foreign DNA fragment attached to a replicon, then so will the descendants, resulting in a huge amplification in the amount of the specific foreign fragment.

Expanded cultures representing cell clones derived from a single cell can then be processed to recover the recombinant DNA. To do this, the cells are lysed, and the DNA is extracted and purified using procedures that result in recovery of the recombinant DNA by taking advantage of differences between it and the host chromosomal DNA. In the case of bacterial cells, the bacterial chromosome is circular double-stranded DNA, like any plasmids containing introduced foreign DNA. However, the chromosomal DNA is relatively very large (~ 4.3 Mb) compared with most recombinant DNA molecules (often only a few kilobases long). During cell lysis and subsequent extraction procedures, the very large bacterial chromosomal DNA, but not the small recombinant plasmid DNA, will undergo nicking and shearing, generating linear DNA fragments with free ends. This difference can be exploited by subjecting the isolated DNA to a denaturation step, often as a result of exposure to alkaline pH in the 12.0–12.5 range. Following this treatment, the linearized host cell DNA readily denatures, but the strands of **covalently closed circular (CCC)** plasmid DNA are unable to separate. After normal conditions have

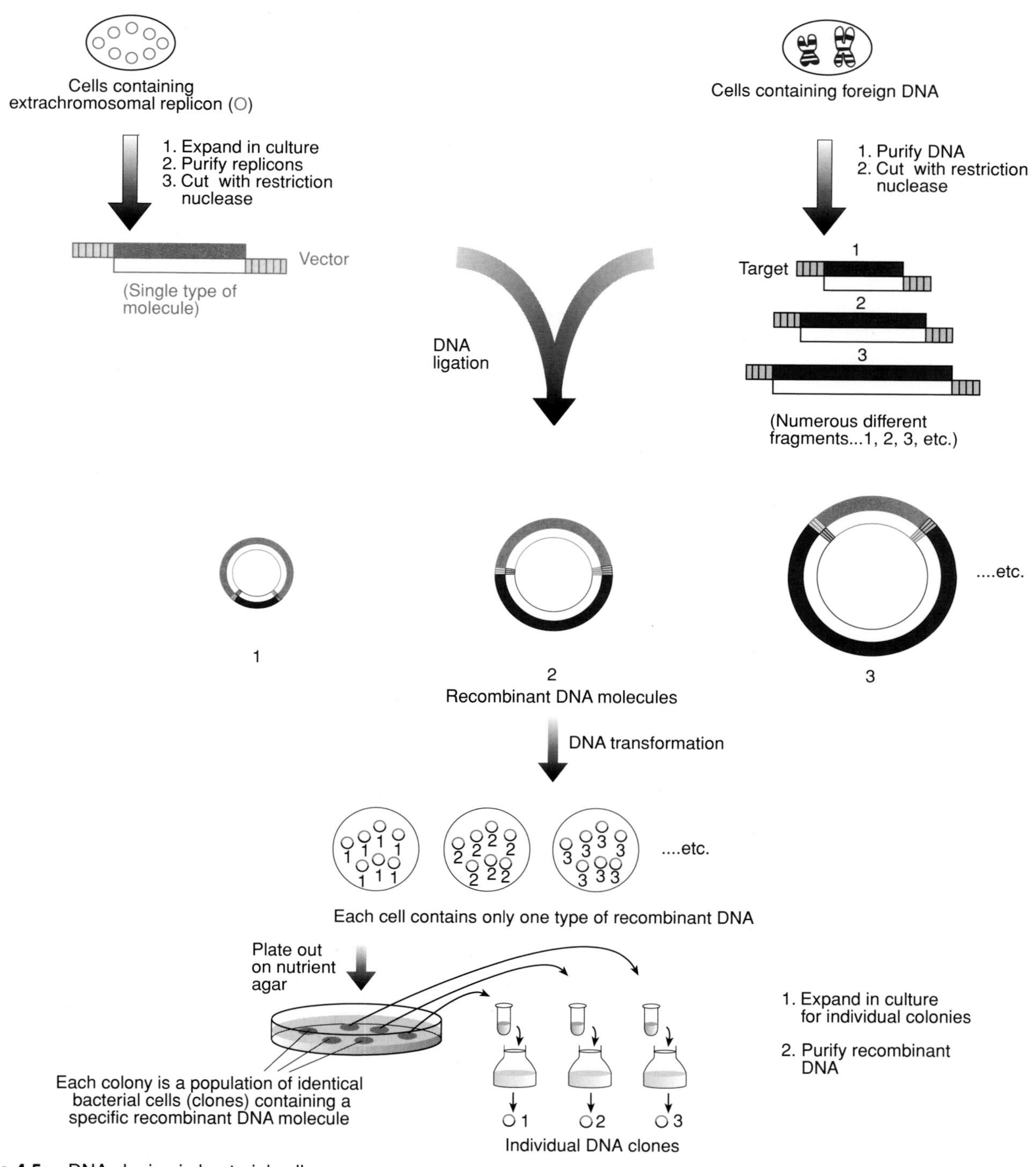

Figure 4.5: DNA cloning in bacterial cells.

The example illustrates cloning of genomic DNA but could equally be applied to cloning cDNA (see *Figure 4.7*).

been restored, the two strands of the CCC DNA rapidly re-align in perfect register to form native **superhelical** molecules or so-called **supercoiled DNA** (this higher order form of twisting occurs in any CCC DNA because the tension introduced by twisting of the double helix cannot be relaxed, unlike in DNA with one or two free ends where relaxation is possible by rotation of a free end). The denatured host cell DNA precipitates out of solution, leaving behind the CCC plasmid DNA.

If required, further purification is possible by **equilibrium density gradient centrifugation (isopycnic centrifugation)**: the partially purified DNA is centrifuged to equilibrium in a solution of cesium chloride containing ethidium bromide (EtBr). EtBr binds by *intercalating* between the base pairs, thereby causing the DNA helix to unwind. Unlike chromosomal DNA, a CCC plasmid DNA has no free ends and can only unwind to a limited extent, which limits the amount of EtBr it can bind. EtBr–DNA complexes are denser when they contain less EtBr, so CCC plasmid DNA will band at a lower position in the cesium chloride gradient than chromosomal DNA and plasmid circles that are open, enabling separation of the recombinant DNA from host cell DNA. The resulting recombinant DNA molecules will normally be identical to each other (representing a single target DNA fragment) and are referred to as **DNA clones**.

DNA libraries are a comprehensive set of DNA clones representing a complex starting DNA population

The first attempts at cloning human DNA fragments in bacterial cells concentrated on target sequences which were highly abundant in a particular starting DNA population. For example, most human cells contain much the same complex collection of DNA sequences in the nucleus, but the mRNA populations of different cell types can be quite different. Although the mRNA population in each cell is complex, some cells are particularly devoted to synthesizing a specific type of protein and so their mRNA populations have a few predominant mRNA species (e.g. much of the mRNA made in erythrocytes is α and β globin mRNA). The enzyme **reverse transcriptase** (RNA-dependent DNA polymerase) can be used to make a DNA copy that is complementary in base sequence to the mRNA, so-called cDNA. Hence, cDNA from erythrocytes is greatly enriched in globin cDNA, facilitating its isolation.

Modern DNA cloning approaches offer the possibility of making comprehensive collections of DNA clones (**DNA libraries**) from extremely complex starting DNA populations (such as total human genomic DNA). This approach enables DNA sequences which are very rare in the starting population to be represented in a library of DNA clones, whence they can be isolated individually by selecting a suitable host cell colony and amplifying it. Two basic varieties of this method have popularly been undertaken, depending on the nature of the starting DNA: genomic DNA libraries and cDNA libraries.

Newly constructed libraries are said to be *unamplified*, although this is a misleading term because the initially transformed cells have been amplified to form separated cell colonies. Often, cell colony formation is allowed to proceed on top of membranes that are overlaid on to the surface of nutrient agar in sterile culture dishes. Copies of the library can then be made by *replica plating* on to a similar sized membrane prior to overlaying on to a nutrient agar surface and colony growth. More recently, individually picked cell colonies have been spotted in gridded arrays on to suitable membranes or into the wells of microtiter dishes where they can be stored for long periods at –70°C in the presence of a cell-stabilizing medium such as glycerol. For multiple distribution, *amplified libraries* are required. The cells from representative primary filters are washed off into cell culture medium, diluted and stabilized by the presence of glycerol, or some alternative stabilizing agent. Individual aliquots can then be plated out at a later stage to regenerate the library. This additional amplification step, however, may result in distortion of the original representation of cell clones because during the amplification stage there may be differential rates of growth of different colonies.

Genomic DNA libraries

In the case of complex eukaryotes, such as mammals, all nucleated cells have much the same DNA content, and it is often convenient to prepare a genomic library from easily accessible cells, such as blood cells. The starting material is genomic DNA which has been fragmented in some way, usually by digesting with a restriction endonuclease. Typically, the genomic DNA is digested with a 4-bp cutter such as *Mbo*I which recognizes the sequence GATC. This sequence will occur about every 275 bp on average in human genomic DNA, so that there will be few DNA sequences which lack a recognition site for this enzyme. Clearly, complete digestion of the starting DNA with this enzyme would produce very small fragments. Instead, the extent of enzymic cleavage is deliberately restricted: under conditions of **partial restriction digestion** (low enzyme concentration, short time of incubation, etc.), cleavage will occur at only a small number of the potential restriction sites. This clearly has the benefit of being able to produce large fragments for cloning. Importantly, it also allows *random fragmentation* of the DNA. Thus, for a specific sequence location, the pattern of cutting will be different on different copies of the same starting DNA sequence (*Figure 4.6*). Such random fragmentation ensures that the library will contain as much representation as possible of the starting DNA. Additionally, it has the advantage that it results in clones with overlapping inserts. As a result, after characterization of the insert of one clone, attempts can be made to access clones from the same general region by identifying those with inserts showing some similiarities to that of the original clone (see *Figure 11.13*).

The **complexity** (number of independent DNA clones) of a genomic DNA library can be defined in terms of **genome equivalents (GE)**. A GE of 1 is obtained when the number of independent clones = genome size/average insert size. For example, for a human genomic DNA library with an average insert size of 40 kilobases (kb), 1 GE = 3000 Mb/40 kb = 75 000 independent clones. Because of sampling variation, however, the number of GEs must be considerably greater than 1 to have a high chance of including any particular sequence within that library. Consequently, attempts are normally made to prepare libraries with several GEs.

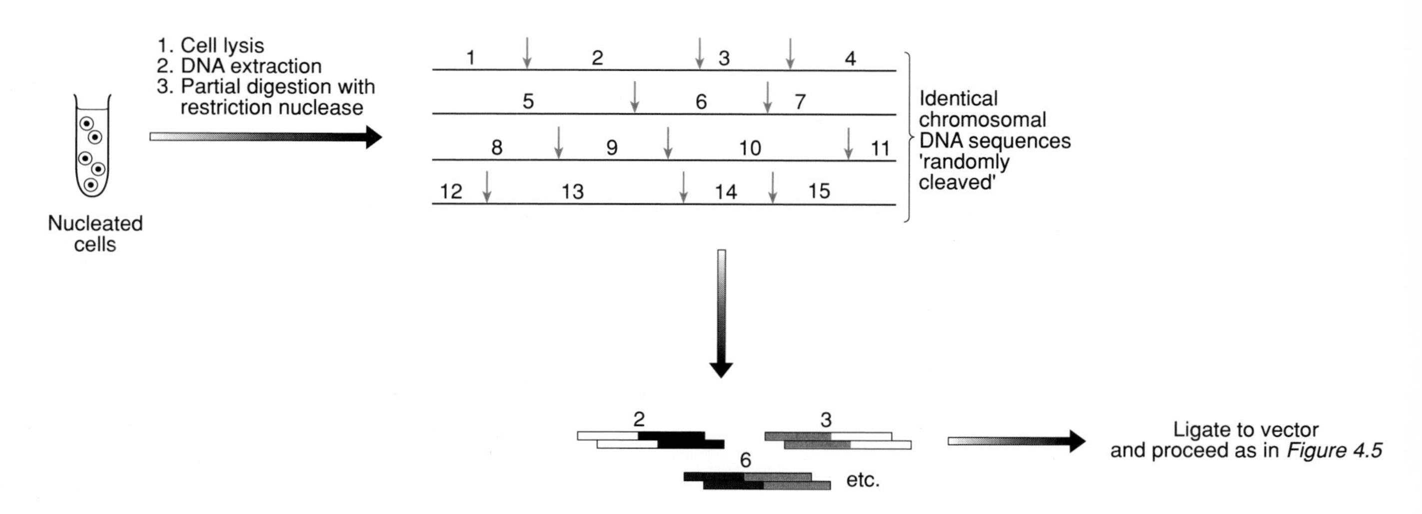

Figure 4.6: Making a genomic DNA library.

All nucleated cells of an individual will have the same genomic DNA content so that any easily accessible cells (e.g. blood cells) can be used as source material. Because DNA is extracted from numerous cells with identical DNA molecules, the isolated DNA will contain large numbers of identical DNA sequences. However, partial digestion with a restriction endonuclease will cleave the DNA at only a small subset of the available restriction sites, and the pattern will differ between individual molecules, resulting in almost random cleavage. This will generate a series of restriction fragments which, if they derive from the same locus, may share some common DNA sequence (e.g. fragment 6 partially overlaps fragments 2 and 3, as shown, and also fragments 9 and 10, and 13 and 14).

Recently, subgenomic DNA libraries have also been a valuable resource: the ability to fractionate individual human chromosomes or specific chromosome bands has permitted construction of chromosome-specific libraries (see page 284) and chromosome band-specific libraries (see page 374).

cDNA libraries

As gene expression can vary in different cells and at different stages of development, the starting material is usually total RNA from a specific tissue or specific developmental stage of embryogenesis. From this, poly(A)$^+$ mRNA can be selected by specific binding to a complementary single-stranded oligonucleotide or polynucleotide which is bound to a solid matrix. For example, chromatography columns containing poly(U) bound to Sepharose or oligo(dT) bound to cellulose have been widely used: the poly(A)$^+$ mRNA selectively binds to the poly(U) or oligo(dT) components and subsequently can be eluted using buffers of high ionic strength to disrupt the hydrogen bonding. The isolated poly(A)$^+$ mRNA can then be converted, using reverse transcriptase, to a double-stranded cDNA copy. To assist cloning, **oligonucleotide linkers** which contain suitable restriction sites are ligated to each end of the cDNA (see *Figure 4.7*). The double-stranded linkers are synthesized chemically as complementary single-stranded oligonucleotides, then allowed to anneal.

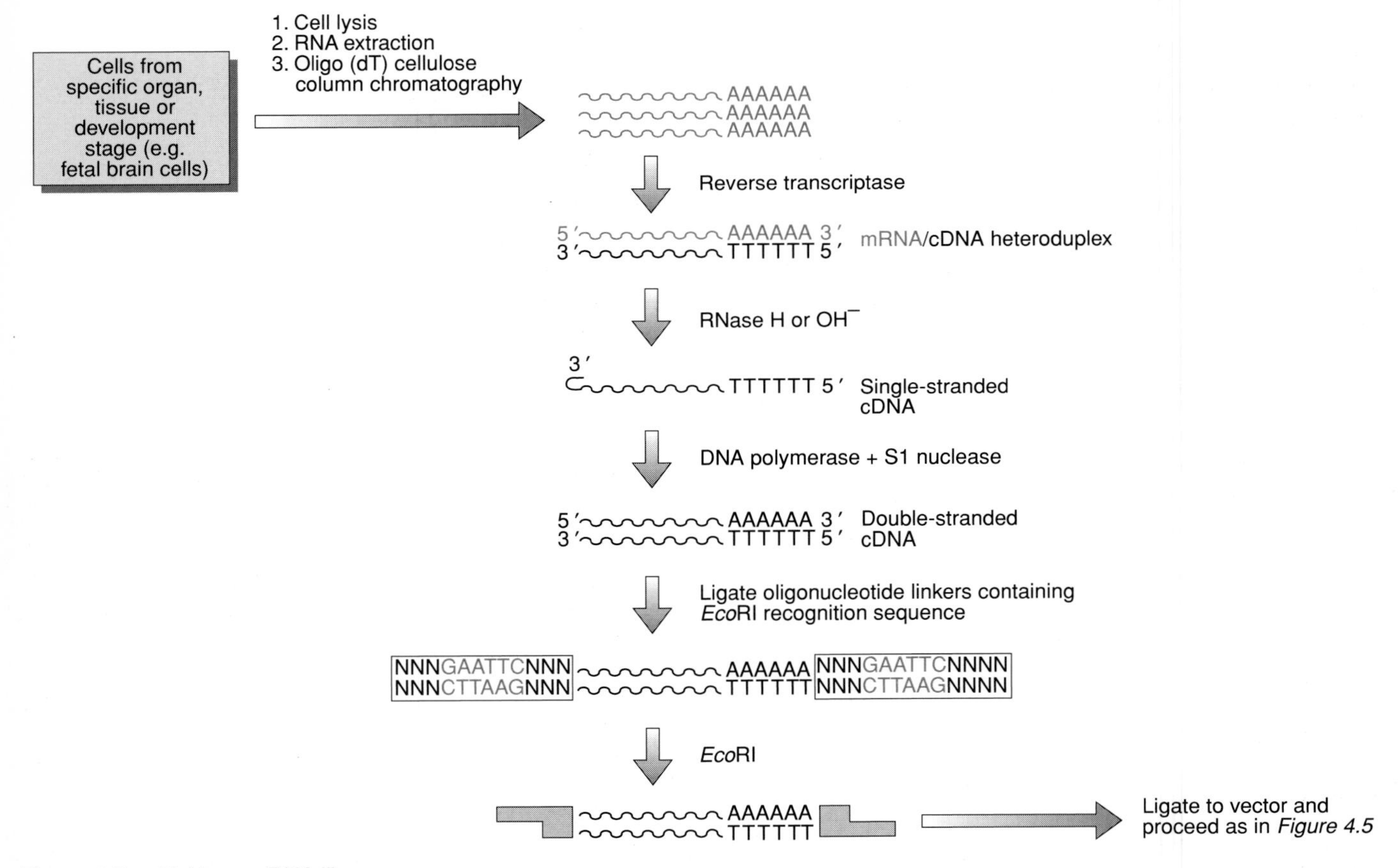

Figure 4.7: Making a cDNA library.

The reverse transcriptase step often uses an oligo(dT) primer to prime synthesis of the cDNA strand. More recently, mixtures of random oligonucleotide primers have been used instead to provide a more normal representation of sequences. RNase H will specifically digest RNA that is bound to DNA in an RNA–DNA hybrid. The 3′ end of the resulting single-stranded cDNA has a tendency to loop back to form a short hairpin. This can be used to prime second strand synthesis by DNA polymerase and the resulting short loop connecting the two strands can then be cleaved by the S1 nuclease which specifically cleaves regions of DNA that are single stranded.

Recombinant screening is often achieved by insertional inactivation of a marker gene

An essential requirement for cell-based DNA cloning systems is a method of detecting cells containing the appropriate vector molecule, and the subset which contain exogenous DNA **inserts** in the vector molecule. Identification of cells containing the vector molecule requires engineering or selection of the vector molecule to contain a suitable marker gene whose expression provides a means of identifying cells containing it. For example, many vectors contain a gene that confers resistance to an antibiotic, such as ampicillin, and are used with strains of host cells that are sensitive to the relevant antibiotic. Others may contain a fragment of the β-galactosidase gene, which, in the presence of host cells containing a complementary fragment of the same gene, produces functional β-galactosidase activity which can be assayed by conversion of a colorless substance, Xgal (5-bromo, 4-chloro, 3-indolyl β-D-galactopyranoside), to a blue product.

Identification of cells containing vector molecules with recombinant DNA is often accomplished by designing the vector molecule to have a multiple cloning site located within the marker gene. The number of nucleotides in the polylinker sequence containing the multiple cloning sites is a multiple of three, so that its insertion in the marker gene does not result in a shift in the translation reading frame. As a result of maintaining the reading frame, and the very small size of the polylinker, the activity of the marker gene is maintained despite the insertion of the polylinker. However, if the vector then contains recombinant DNA inserted in the middle of the polylinker, the resulting large insertion causes loss of expression of the marker gene (**insertional inactivation**). In the case of a marker β-galactosidase gene, insertional inactivation means that cells containing recombinant DNA are colorless in the presence of Xgal, while cells containing nonrecombinant vector are blue.

Another common screening system involves the use of **suppressor tRNA** genes, mutant tRNA genes which can recognize a normal termination codon signal and, in response, introduce an amino acid in the polypeptide chain. This is possible because the mutation in such tRNA genes (often mutated $tRNA^{Glu}$ or $tRNA^{Tyr}$ genes) results in an altered anticodon sequence that is complementary to one of the normal termination codons: UAA (**ochre**), UAG (**amber**) or UGA (**opal**) (see *Box 4.2*). In such systems, the host cell often carries a defective marker gene that is designed to have

Box 4.2: Nonsense suppressor mutations

	Glu	**Amber**
Codon	5′ GAG 3′	5′ UAG 3′
	...	...
Anticodon	3′ CUC 5′ →	3′ AUC 5′
of:	$tRNA^{Glu}$ →	$tRNA^{Glu*}$ (amber suppressor)
	Glu	**Ochre**
Codon	5′ GAA 3′	5′ UAA 3′
	...	...
Anticodon	3′ CUU 5′ →	3′ AUU 5′
of:	$tRNA^{Glu}$ →	$tRNA^{Glu*}$ (ochre suppressor)

Base changes in the anticodon of a tRNA may enable it to insert an amino acid in response to a stop codon.

a premature stop codon (i.e. an amber, ochre or opal mutation) resulting in a phenotype that can be easily scored. If the vector carries a suppressor tRNA gene that suppresses the effect of this termination codon, the marker gene activity will be recovered and the wild-type phenotype will be restored. However, if the suppressor tRNA gene is inactivated by insertional inactivation, a mutant phenotype will again be produced (see *Figure 4.15* for an example of one such system).

If there is already available a DNA probe that is closely related to the desired recombinant DNA, colonies containing the latter can be detected directly by colony hybridization (see *Figure 5.15*).

Initial characterization of recombinant DNA clones involves restriction mapping

Restriction mapping of DNA clones involves cutting the DNA with one or more of a series of different restriction nucleases and separating the resulting fragments according to size by agarose gel electrophoresis. Because of the conspicuous deficiency in the CpG dinucleotide in vertebrate genomes (see page 15), recognition sites that are GC-rich will occur comparatively less frequently than expected in vertebrate DNA, but will not be so rare in bacterial cell DNA (see *Table 4.1*).

Double digests (cleavage by two different enzymes) and **partial digests** (reduced digestion so that not every cleavage site is actually cut) help in relating the different restriction fragments to each other. The resulting information can be used to construct a **restriction map**, a linear map of the relative positions of recognition sites for a variety of restriction endonucleases (*Figure 4.8*). If the restriction maps of two independently isolated DNA fragments show extensive sharing of restriction sites,

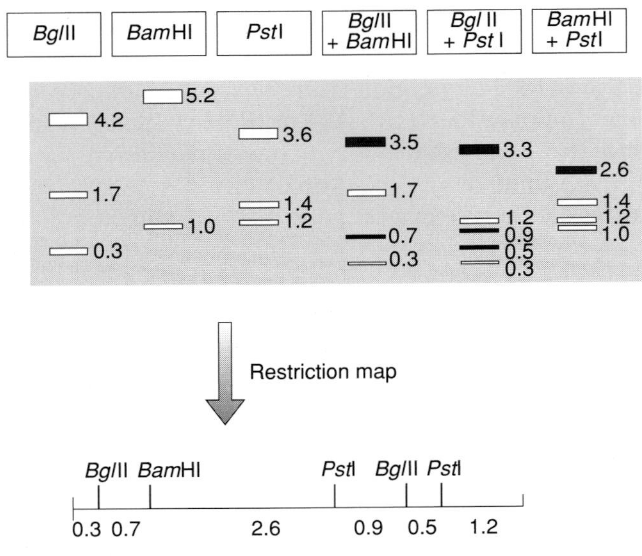

Figure 4.8: Generating a restriction map.

The size patterns from double digests provide information on the relative locations of restriction sites. The example shows size fractionation by agarose gel electrophoresis of restriction fragments following incubation of a 6.2 kb DNA fragment with the indicated enzymes. New bands in the double digests (i.e. not found in the original single digests) are indicated by black boxes. In the *Bgl*II + *Bam*HI double digest, the original 1.7-kb and 0.3-kb bands from the *Bgl*II digest alone are maintained, suggesting that these fragments do not have a *Bam*HI site, while the 4.2-kb *Bgl*II fragment is replaced by 3.5-kb and 0.7-kb fragments, suggesting that there is a *Bam*HI site within 0.7 kb from one end of the 4.2-kb *Bgl*II fragment. Similarly, in the *Bam*HI + *Pst*I double digest, the 1.4-kb and 1.2-kb fragments seen in the *Pst*I digest alone are maintained, suggesting that they lack a *Bam*HI site, while the 3.6-kb *Pst*I fragment is replaced by a 2.6-kb + 1.0-kb fragment, as a result of possession of an internal *Bam*HI site located 1.0 kb from one end. By comparing all three patterns of double digestion, the restriction map at the bottom can be deduced. *Note* that restriction mapping is often helped by the use of partial digests and also by end-labeling (page 109).

it is highly likely that the two fragments contain overlapping DNA sequences or are closely related members of a repeated DNA sequence family. Restriction mapping of a recombinant DNA clone immediately provides details of the length of the insert DNA and some information on the location of unique restriction sites which may be useful for **subcloning** purposes (where fragments of a recombinant DNA are cloned into a different vector molecule, often so as to study some aspect of its structure or function).

Vector systems for cloning different sizes of DNA fragments

Cell-based DNA cloning has been used widely as a tool for producing quantities of pure DNA for physical characterization and functional studies of individual genes, gene clusters or other DNA sequences of interest. However, the size of different DNA sequences of interest can vary enormously (e.g. human gene sizes are known to vary between 0.1 kb and 2.5 Mb). The first cell-based cloning systems to be developed could clone only rather small DNA fragments. Recently, however, there have been rapid developments in cloning systems that permit cloning of very large DNA fragments.

Plasmid vectors provide a simple way of cloning small DNA fragments in bacterial (and simple eukaryotic) cells

In order to adapt natural plasmid molecules as cloning vectors, several modifications are normally made:

- insertion of a **multiple cloning site polylinker**. This is a short (~30 bp) synthetic sequence which contains unique restriction sites for a variety of common restriction nucleases (pre-existing restriction sites for these enzymes will be deleted from the plasmid if necessary to ensure the presence of unique cloning sites).
- Insertion of an **antibiotic resistance gene**. The host cells that are used must naturally be sensitive to the antibiotic in question so that any vector molecule which transforms a host cell can confer antibiotic resistance. By plating transformed cells on a medium containing the antibiotic, only those cells that have been transformed by vector molecules survive.
- Insertion of a **selection system for screening for recombinants**. Typically this involves arranging for the multiple cloning site polylinker to be inserted into an expressible gene or gene fragment within the plasmid (see page 95).

The plasmid vector pUC19 contains a polylinker with unique cloning sites for multiple restriction nucleases and an ampicillin resistance gene to permit identification of transformed cells (*Figure 4.9*). In addition, selection for recombinants is achieved by insertional inactivation of a component of the β-galactosidase gene, a complementary portion of this gene being provided by using a specially modified *E. coli* host cell.

Lambda and cosmid vectors provide an efficient means of cloning moderately large DNA fragments in bacterial cells

The major disadvantage of plasmid vectors is that their capacity for accepting large DNA fragments is severely limited: most inserts are a few kilobases in length and inserts larger than 5–10 kb are very rare. Additionally, standard methods of

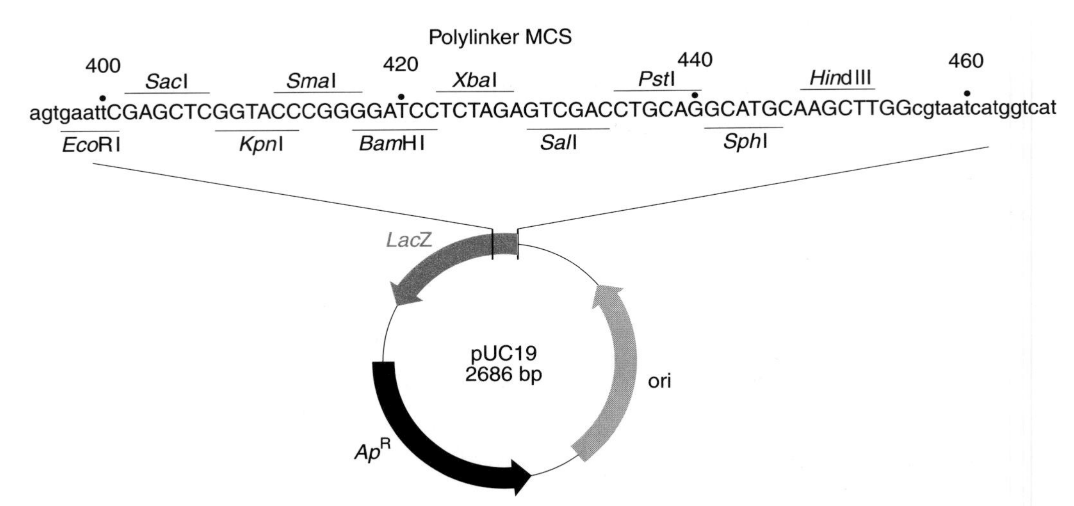

Figure 4.9: Map of plasmid vector pUC19.

The origin of replication (ori) was derived originally from a ColE1-like plasmid, pMB1. The ampicillin resistance gene (Ap^R) was derived originally from the plasmid RSF 2124 and permits selection for cells containing the vector molecule. A portion of the *lacZ* gene is included and is expressed to give an amino-terminal fragment of β-galactosidase. This is complemented by a mutant *lacZ* gene in the host cell: the products of the vector and host cell *lacZ* sequences, although individually inactive, can associate to form a functional product. The 54-bp polylinker multiple cloning site (capital letters) is inserted into the vector *lacZ* (lower case letters) component in such a way as to preserve the reading frame and functional expression. However, cloning of an insert into the multiple cloning site (MCS) will cause insertional inactivation, and absence of β-galactosidase activity.

transformation of bacterial cells with plasmid vectors are relatively inefficient. To address these difficulties, attention was focused at an early stage on the possibility of using bacteriophage **lambda** as a cloning vector. The wild-type λ virus particle (**virion**) contains a genome of close to 50 kb of linear double-stranded DNA packaged within a protein coat and has evolved a highly efficient mechanism of infecting *E. coli* cells. After the λ virion attaches to the bacterial cell, the coat protein is discarded and the λ DNA is injected into the cell. At the extreme termini of the λ DNA are overhanging 5′ ends which are 12 nucleotides long and complementary in base sequence. Because these large 5′ overhangs can base-pair, they are effectively sticky ends, similar to, but more cohesive than, the small sticky ends generated by some restriction nucleases (see page 87). Such cohesive properties are recognized in the name given to this sequence – the **cos** sequence. Once inside the bacterial cell, the cos sequences base-pair, and sealing of nicks by cellular ligases results in the formation of a double-stranded circular DNA. Thereafter the λ DNA can enter two alternative pathways (*Figure 4.10*):

(i) **the lytic cycle**. The λ DNA replicates, initially bidirectionally, and subsequently by a rolling circle model which generates linear multimers of the unit length. Coat proteins are synthesized and the λ multimers are snipped at the cos sites to generate unit lengths of λ genome which are packaged within the protein coats. Some of the λ gene products lyse the host cell, allowing the virions to escape and infect new cells.

(ii) **The lysogenic state**. The λ genome possesses a gene *att* which has a homolog in the *E. coli* chromosome. Apposition of the two *att* genes can result in recombination between the λ and *E. coli* genomes and subsequent integration of the λ DNA within the *E. coli* chromosome. In this state, the λ DNA is described as a **provirus** and the host cell as a **lysogen** because, although the λ DNA can remain stably integrated for long periods, it has the capacity for excision from the host chromosome and entry into the lytic cycle (*Figure 4.10*). Genes required for lysogenic function are located in a central segment of the λ genome (*Figure 4.11*).

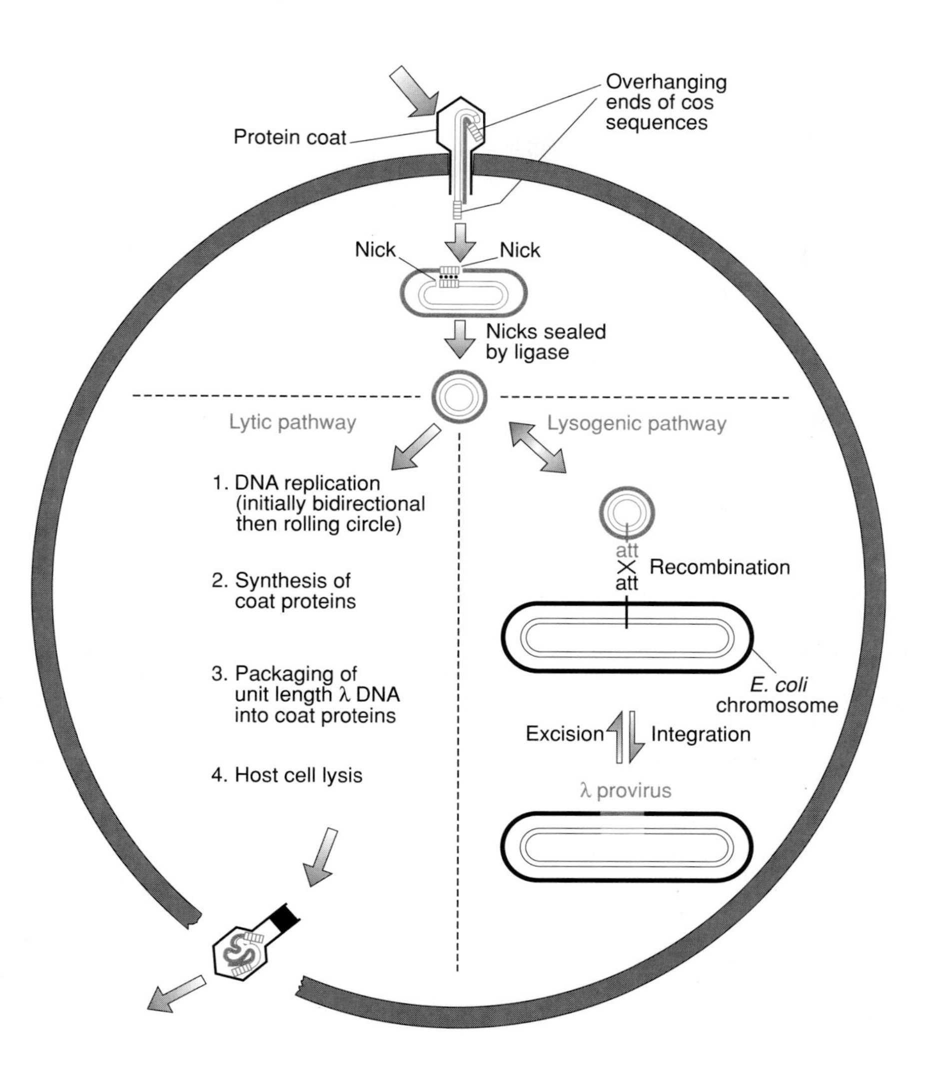

Figure 4.10: Phage λ can enter both lytic and lysogenic pathways.

The decision to enter the lytic cycle or the lysogenic state is controlled by two regulatory genes, *c*I and *cro*. These two genes are mutually antagonistic: in the lytic state the *cro* protein dominates, leading to repression of *c*I, whereas in the lysogenic state the *c*I repressor dominates and suppresses transcription of other λ genes including *cro*. In normally growing host cells, the lysogenic state is favored and the λ genome replicates along with the host chromosomal DNA. Damage to host cells favors a transition to the lytic cycle, enabling the virus to escape the damaged cell and infect new cells.

In order to design suitable cloning vectors based on λ, it was necessary to design a system whereby foreign DNA could be attached to the λ replicon *in vitro* and for the resultant recombinant DNA to be able to transform *E. coli* cells at high efficiency. The latter requirement was achieved by developing an ***in vitro*** **packaging system** which mimicked the way in which wild-type λ DNA is packaged in a protein coat, resulting in high infection efficiency (*Figure 4.12*).

Several major types of cloning vector that have been developed by modifying phage λ, or utilizing the size selection imposed by cos sequences, are described in the following sections.

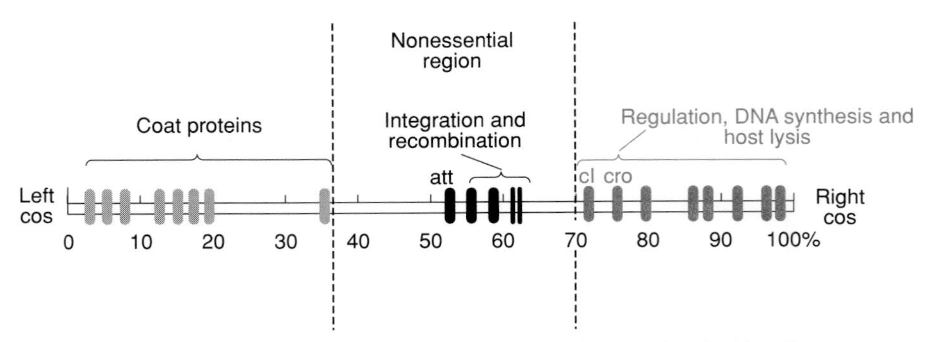

Figure 4.11: Map of the λ genome, showing positions of genes (vertical bars).

In λ replacement vectors, the nonessential region is removed by restriction endonuclease digestion, leaving a left λ arm and a right λ arm. A foreign DNA fragment, can be ligated to the two arms in place of the original 'stuffer' fragment, providing maximal insert sizes of over 20 kb.

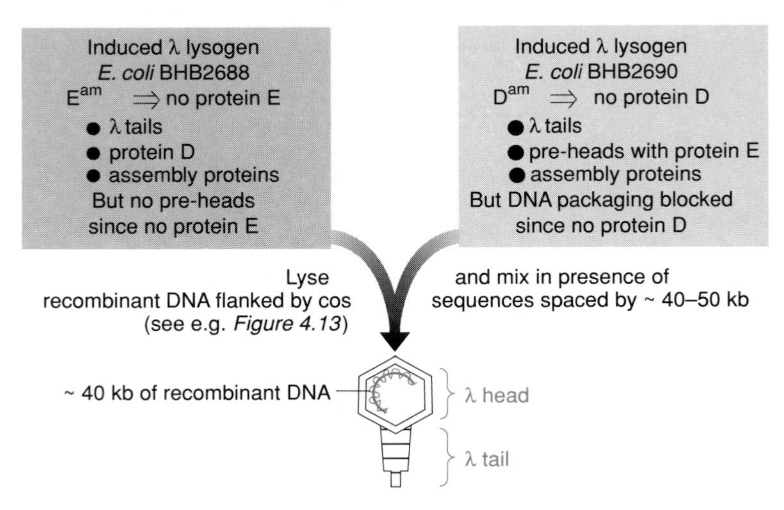

Figure 4.12: *In vitro* DNA packaging in a phage λ protein coat can be performed using a mixed lysate of two mutated λ lysogens.

Normal *in vivo* packaging of λ DNA involves first making **pre-heads**, structures composed of the major capsid protein encoded by gene *E*. A unit length of λ DNA is inserted in the pre-head, the unit length being prepared by cleavage at neighboring cos sites. A minor capsid protein D is then inserted in the pre-heads to complete head maturation, and the products of other genes serve as **assembly proteins**, ensuring joining of the completed tails to the completed heads. A defect in producing protein E, resulting from an amber mutation introduced into gene E (E^{am}), prevents pre-heads being formed by BHB2688. An amber mutation in gene D(D^{am}) prevents maturation of the pre-heads, with enclosed DNA, into complete heads. The components of the BHB2688/BHB2690 mixed lysate, however, complement each other's deficiency and provide all the products for correct packaging.

Replacement λ vectors

Only DNA molecules from 37 to 52 kb in length can be stably packaged into the λ particle. The central segment of the λ genome contains genes that are required for the lysogenic cycle but are not essential for lytic function. As a result, it can be removed and replaced by a foreign DNA fragment. Using this strategy, it is possible to clone foreign DNA up to 23 kb in length, and such vectors are normally used for making genomic DNA libraries.

Insertion λ vectors

Lambda vectors used for making cDNA libraries do not require a large insert capacity (most cDNAs are <5 kb long). Design of insertion vectors often involves modification of the λ genome to permit insertional cloning into the *c*I gene.

Cosmid vectors

Cosmid vectors contain cos sequences inserted into a small plasmid vector. Large (~30–44 kb) foreign DNA fragments can be cloned using such vectors in an *in vitro* packaging reaction because the total size of the cosmid vector is often about 8 kb (see *Figure 4.13*).

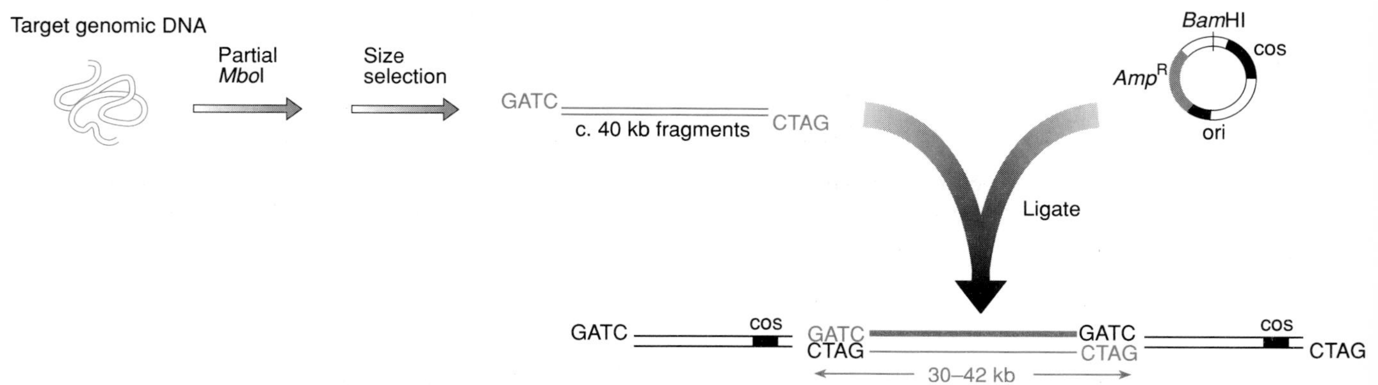

Figure 4.13: Ligation to cleaved cosmid vector molecules can produce vector–target concatemers, resulting in a large exogenous DNA fragment flanked by cos sequences.

Large DNA fragments can be cloned in bacterial cells using vectors based on bacteriophage P1 and F factor plasmids

Because the human and other mammalian genomes are so large, and because many individual genes can be very large (see *Figure 7.6*), there was a need for the development of new cloning vectors that could accept large DNA inserts. A number of such vectors have been developed recently (see *Table 4.2*) and have found immediate uses in general physical mapping of genomes and in permitting the characterization and expression of large genes or gene complexes. Examples of such vectors are discussed below.

Bacterial artificial chromosome (BAC) vectors

Many vectors which are popularly used for DNA cloning in bacterial cells contain high to medium copy number replicons. The advantage of vectors which contain such replicons is the high yield of DNA they afford: each cell in which a vector molecule is propagated will have several to multiple copies of the vector molecule, depending on the replication efficiency of its replicon. However, an important disadvantage is that such vectors often show structural instability of inserts, resulting in deletion or rearrangement of portions of the cloned DNA. Such instability is particularly common in the case of DNA inserts of eukaryotic origin where repetitive sequences occur frequently and, as a result, it is difficult to clone and maintain intact large DNA in bacterial cells.

Table 4.2: Sizes of inserted DNA commonly obtained with different cloning vectors

Cloning vector	Size of insert
Standard high copy number plasmid vectors	0–10 kb
Bacteriophage λ insertion vectors	0–10 kb
Bacteriophage λ replacement vectors	9–23 kb
Cosmid vectors	30–44 kb
Bacteriophage P1	70–100 kb
PAC (P1 artificial chromosome) vectors	130–150 kb
BAC (bacterial artificial chromosomes) vectors	up to 300 kb
YAC (yeast artificial chromosomes) vectors	0.2–2.0 Mb

In order to overcome this limitation, attention has recently been focused on vectors based on low copy number replicons, such as the *E. coli* fertility plasmid, the F-factor. This plasmid contains two genes, *parA* and *parB*, which maintain the copy number of the F-factor at 1–2 per *E. coli* cell. Vectors based on the F-factor system are able to accept large foreign DNA fragments (>300 kb). The resulting recombinants can be transferred with considerable efficiency into bacterial cells using **electroporation** (a method of exposing cells to high voltages in order to relax the selective permeability of their plasma membranes). However, because the resulting **bacterial artificial chromosomes (BACs)** contain a low copy number replicon, only very low yields of recombinant DNA can be recovered from the host cells (Shizuya *et al.*, 1992).

Bacteriophage P1 vectors and PACs

Certain bacteriophages have relatively large genomes, thereby affording the potential for developing vectors that can accommodate large foreign DNA fragments. One such is bacteriophage P1 which, like phage λ, packages its genome in a protein coat, and 110–115 kb of linear DNA is packaged in the P1 protein coat. P1 cloning vectors have therefore been designed in which components of P1 are included in a circular plasmid. The P1 plasmid vector can be cleaved to generate two vector arms to which up to 100 kb of foreign DNA can be ligated and packaged into a P1 protein coat *in vitro*. The recombinant P1 phage can be allowed to adsorb to a suitable host, following which the recombinant P1 DNA is injected into the cell, circularizes and can be amplified (Sternberg, 1992) (see *Figure 4.14*).

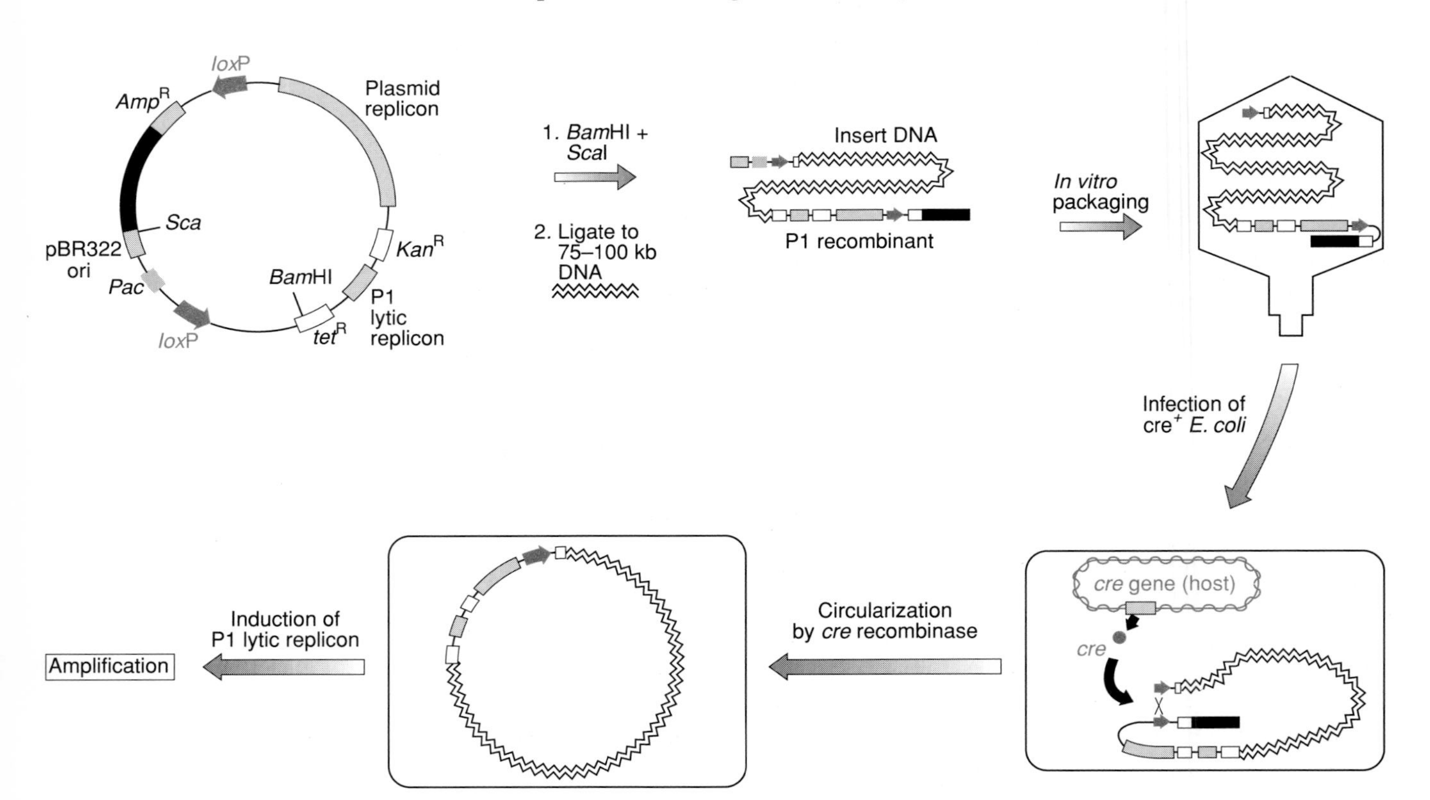

Figure 4.14: The phage P1 vector system allows DNA fragments of up to 100 kb to be cloned.

The P1 plasmid vector Ad10 incorporates various elements of the P1 genome, notably *pac*, the P1 packaging site, and two *loxP* sites, which are the sites naturally recognized by the phage recombinase, the product of the *cre* gene. The vector is digested so as to generate two arms, a short arm and a long arm to which 85–100 kb size-selected foreign DNA fragments are ligated. Packaging of the recombinant DNA occurs *in vitro* using P1 packaging extracts: pacase cleaves the recombinant DNA at the *pac* site and then works with other components to insert the DNA (maximum of 115 kb) into phage heads. Tail proteins are attached and the recombinant phage is allowed to adsorb to a *cre*+ strain of *E. coli*. The host cell *cre* product acts on the *loxP* sites so as to produce a circular plasmid which is maintained at low copy number (by the plasmid replicon) but can be amplified by inducing the P1 lytic replicon.

An improvement on the size range of inserts accepted by the basic P1 cloning system has been the use of bacteriophage T4 *in vitro* packaging systems with P1 vectors which enables the recovery of inserts up to 122 kb in size. More recently, features of the P1 and F-factor systems have been combined to produce **P1-derived artificial chromosome (PAC)** cloning systems (Iouannou *et al.*, 1994).

Yeast artificial chromosomes (YACs) enable cloning of megabase fragments

The most popularly used system for cloning very large DNA fragments involves the construction of yeast artificial chromosomes (YACs; see Schlessinger, 1990). Cloning in yeast cells offers, in principle, some advantages over cloning in bacterial cells. Certain eukaryotic sequences, notably those with repeated sequence organizations, are difficult, or impossible, to propagate in bacterial cells which do not have such types of DNA organization, but would be anticipated to be tolerated in yeast cells which are eukaryotic cells. However, the main advantage offered by YACs has been the ability to clone very large DNA fragments. Such a development proceeded from the realization that the great bulk of the DNA in a chromosome is not required for normal chromosome function. As detailed on pages 45–48, the essential functional components of yeast chromosomes are threefold:

(i) *centromeres* are required for disjunction of sister chromatids in mitosis and of homologous chromosomes at the first meiotic division.

(ii) *Telomeres* are required for complete replication of linear molecules and for protection of the ends of the chromosome from nuclease attack.

(iii) *Autonomous replicating sequence* (ARS) elements are required for autonomous replication of the chromosomal DNA and are thought to act as specific replication origins.

In each case, the DNA segment necessary for functional activity *in vivo* in yeast is limited to at most a few hundred base pairs of DNA (*Figure 2.13*). As a result, it became possible to envisage a novel cloning system based on the use of chromosomal replicons (ARS elements) as an alternative to the ubiquitous use of cloning vectors based on extrachromosomal replicons (those found in plasmids and bacteriophages).

To make a YAC, it is simply necessary to combine four short sequences that can function in yeast cells: two telomeres, one centromere and one ARS element, together with a suitably sized foreign DNA fragment to give a linear DNA molecule in which the telomere sequences are correctly positioned at the termini (*Figure 4.15*). The resulting construct cannot be transfected directly into yeast cells. Instead, yeast cells have to be treated in such a way as to remove the external cell walls. The resulting yeast **spheroplasts** can accept exogenous fragments but are osmotically unstable and need to be embedded in agar. The overall transformation efficiency is very low and the yield of cloned DNA is low (about one copy per cell). Nevertheless, the capacity to clone large exogenous DNA fragments (up to 2 Mb) has made YACs a vital tool in physical mapping (see page 340ff.).

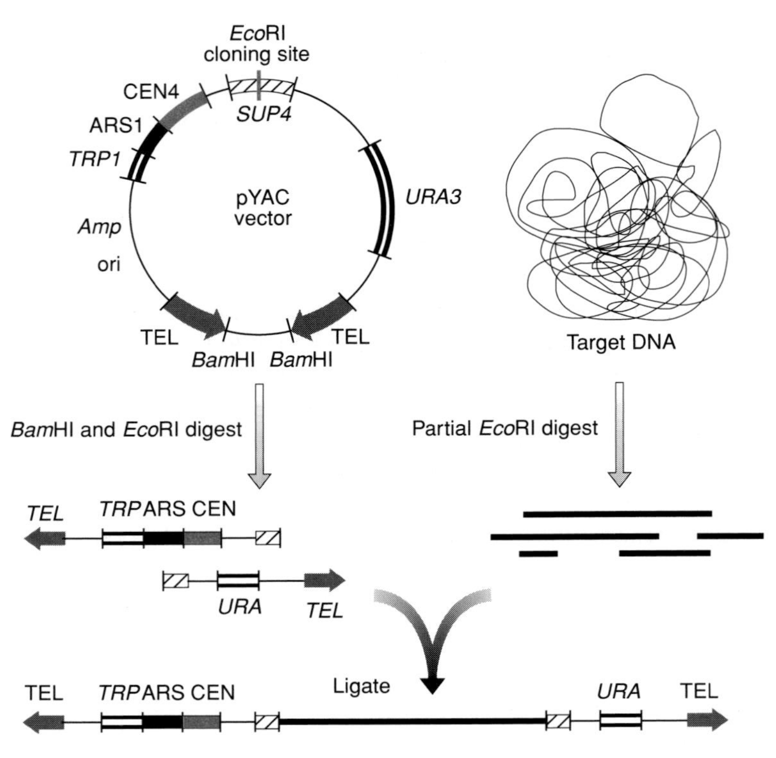

Figure 4.15: Making YACs.

Vector DNA sequences include: CEN1 – centromere sequence; TEL – telomere sequences; ARS1 – autonomous replicating sequence; *Amp* – gene conferring ampicillin-resistance; ori – origin of replication for propagation in an *E. coli* host. The vector is used with a specialized yeast host cell, AB1380, which is red colored because it carries an ochre mutation in a gene, *ade-2*, involved in adenine metabolism, resulting in accumulation of a red pigment. However, the vector carries the *SUP4* gene, a suppressor tRNA gene (see page 95) which overcomes the effect of the *ade-2* ochre mutation and restores wild-type activity, resulting in colorless colonies. The host cells are also designed to have recessive *trp1* and *ura3* alleles which can be complemented by the corresponding *TRP1* and *URA3* alleles in the vector, providing a selection system for identifying cells containing the YAC vector. Cloning of a foreign DNA fragment into the *SUP4* gene causes insertional inactivation of the suppressor gene function (see page 95), restoring the mutant (red color) phenotype.

Expression cloning

Principles of expression cloning

Many cell-based cloning systems are geared simply towards amplifying the introduced DNA to obtain sufficient quantities for a variety of subsequent structural and functional studies. In addition, however, many plasmid-based systems are designed to allow synthesis of large quantities of a strand-specific RNA transcript from the introduced DNA by an *in vitro* RNA synthesis reaction (*Figure 5.4*).

In some cases, the cloning system has been designed to promote polypeptide expression by the introduced gene (**expression cloning**). This affords the chance of producing large amounts of a specific protein, which may be of medical importance (e.g. insulin, interferon, viral antigens for producing vaccines, etc., or for raising specific antibodies; see page 551). More generally, it permits comprehensive studies of the structure and function of proteins, which are particularly useful in the case of proteins which are rare cellular components or which are difficult to isolate. A large variety of expression cloning vectors have been designed to be used in different host cell systems ranging from bacterial cells to mammalian cells, with specific vectors being engineered to be useful in specific host cell types.

Expressing eukaryotic genes in bacterial cells

Eukaryotic genes cannot normally be expressed directly in bacterial cells, because the bacterium lacks the necessary apparatus for expression of such genes (e.g. factors for RNA splicing, etc.). As a result, expression of eukaryotic polypeptides in bacterial cells is normally accomplished by first obtaining an appropriate cDNA clone (which need not be complete but must at least contain the full coding sequence), and then cloning that into a suitable bacterial expression vector. In this case, the inserted cDNA is simply a provider of genetic information required to specify a polypeptide; the control of expression is provided externally. Expression of important eukaryotic genes in bacterial cells can provide an endless bulk supply of important, medically relevant compounds. As the production of very large amounts of a protein (which are often foreign to the host cell) can be detrimental to host cell growth, expression systems are often designed with **inducible promoters**. In such cases, cells with the required recombinant DNA can be selected and grown up in large quantities without expression of the foreign gene. Expression of the inserted gene can be switched on by exposure to an inducing agent and cells can be harvested shortly thereafter. One major problem in some cases, however, concerns the importance of post-translational processing in the production of many proteins. Because of differences between such systems in prokaryotes and eukaryotes, eukaryotic proteins produced by expression of cDNA clones in bacterial cells may be unstable, or show limited or no biological activity.

Expression of eukaryotic proteins in bacterial cells can also be used as a screening system to identify a novel uncharacterized gene. Although there may be no information concerning the coding sequence of a gene, partial purification of its protein product may allow a specific antibody to be raised, enabling **antibody screening of a cDNA library**. To do this, a suitable eukaryotic cDNA library is constructed using a vector such as λgt11, which permits expression of the foreign inserts in bacterial cells. Filters containing individual bacterial colonies (**colony filters**, see *Figure 5.15*) can be screened by exposure to the antibody (see *Figure 19.1*). Positively reacting bacteria can then be propagated to isolate the cDNA clone, and the isolated cDNA clone can, in turn, be used to screen a genomic library to identify the cognate gene.

Expressing eukaryotic genes in eukaryotic cells

Expression cloning systems where the host cells are from eukaryotes allow the expression of a specific eukaryotic gene or DNA segment to be studied in a *heterologous* eukaryotic host cell. This may provide an environment that is similar, but not identical, to the original host cell, enabling specific detection and, in some cases, purification of the expressed protein. A variety of different expression systems have therefore been designed to permit study of transient or stable expression patterns enabling dissection of functional activity (see, for example, *Figure 19.8*). Such systems are also useful for identifying the functional status of members of multigene families: occasionally the sequence of pseudogenes is so similar to that of functional members of the same family that it is helpful to have a functional assay for discriminating between different genomic clones.

Eukaryotic cloning systems have also become much more popular as methods for bulk preparation of a specific eukaryotic protein. Many eukaryotic proteins require specific post-translational processing events (e.g. specific glycosylation patterns) that are critically important for their stability or functional activity. As heterologous eukaryotic hosts often have identical or similar post-translational processing systems, eukaryotic expression systems, such as the insect baculovirus system, are preferred in many cases for bulk preparation of specific eukaryotic proteins.

Further reading

Berger SL, Kimmel AR. (1987) *Guide to Molecular Cloning Techniques, Methods in Enzymology,* Vol. 152. Academic Press, San Diego.

Glover D, Hames BD. (1995) (eds) *DNA Cloning: a Practical Approach,* Vol. 1, *Core Techniques.* IRL Press, Oxford.

Glover D, Hames BD. (1995) (eds) *DNA Cloning: a Practical Approach,* Vol. 2, *Expression Systems.* IRL Press, Oxford.

Old RW, Primrose SB. (1994) *Principles of Gene Manipulation: an Introduction to Genetic Engineering,* 5th Edn. Blackwell Scientific Publications, Oxford.

Sambrook J, Fritsch EF, Maniatis T. (1989) *Molecular Cloning: a Laboratory Manual,* 2nd Edn. Cold Spring Harbor Laboratory Press, Cold Spring Harbor, NY.

References

Burke DT, Carle GF, Olson MV. (1987) *Science,* **236**, 806–812.

Iouannou PA, Amemiya CT, Garnes J, Kroisel PM, Shizuya H, Chen C, Batzer MA, de Jong P. (1994) *Nature Genetics,* **6**, 84–89.

Shizuya H, Birren B, Kim U-J, Mancino V, Slepak T, Tachiiri Y, Simon M. (1992) *Proc. Natl Acad. Sci. USA,* **89**, 8794–8797.

Schlessinger D. (1990) *Trends Genet.,* **6**, 248–258.

Sternberg N. (1992) *Trends Genet.,* **8**, 11–16.

DNA hybridization assays

Chapter 5

Nucleic acid probes

A **DNA** (or **RNA**) **probe** is any piece of DNA (or RNA) which has been labeled in some way and which is used in a hybridization assay (see page 114) to identify other DNA or RNA sequences which are closely related to it in base sequence. Nucleic acid probes may be made as single-stranded or double-stranded molecules (see *Figure 5.1*), but the working probe must be in the form of single strands.

Nucleic acids used as probes are normally obtained by DNA cloning or chemical synthesis

Conventional DNA probes

Conventional DNA probes are isolated by cell-based DNA cloning (see Chapter 4) or by PCR-based DNA cloning (see Chapter 6). In the former case, the starting DNA may range in size from 0.1 kb to hundreds of kilobases in length and is usually (but not always) originally double-stranded. PCR-derived DNA probes have often been less than 10 kb long and are usually, but not always, originally double-stranded. Conventional DNA probes are usually labeled during an *in vitro* DNA synthesis reaction (see pages 108–110).

RNA probes

RNA probes can conveniently be generated from DNA which has been cloned in a specialized plasmid vector (Melton *et al.*, 1984). Such vectors normally contain a phage promoter sequence immediately adjacent to the multiple cloning site, and specific RNA transcripts can be obtained from the cloned insert using the relevant phage RNA polymerase (see page 111).

Oligonucleotide probes

Oligonucleotide probes are short (typically 15–50 nucleotides) single-stranded pieces of DNA made by chemical synthesis: mononucleotides are added, one at a time, to a starting mononucleotide, conventionally the 3′ end nucleotide, which is bound to a solid support. Generally, oligonucleotide probes are designed with a specific sequence chosen in response to prior information about the target DNA. Sometimes, however, oligonucleotide probes are used which are **degenerate** in

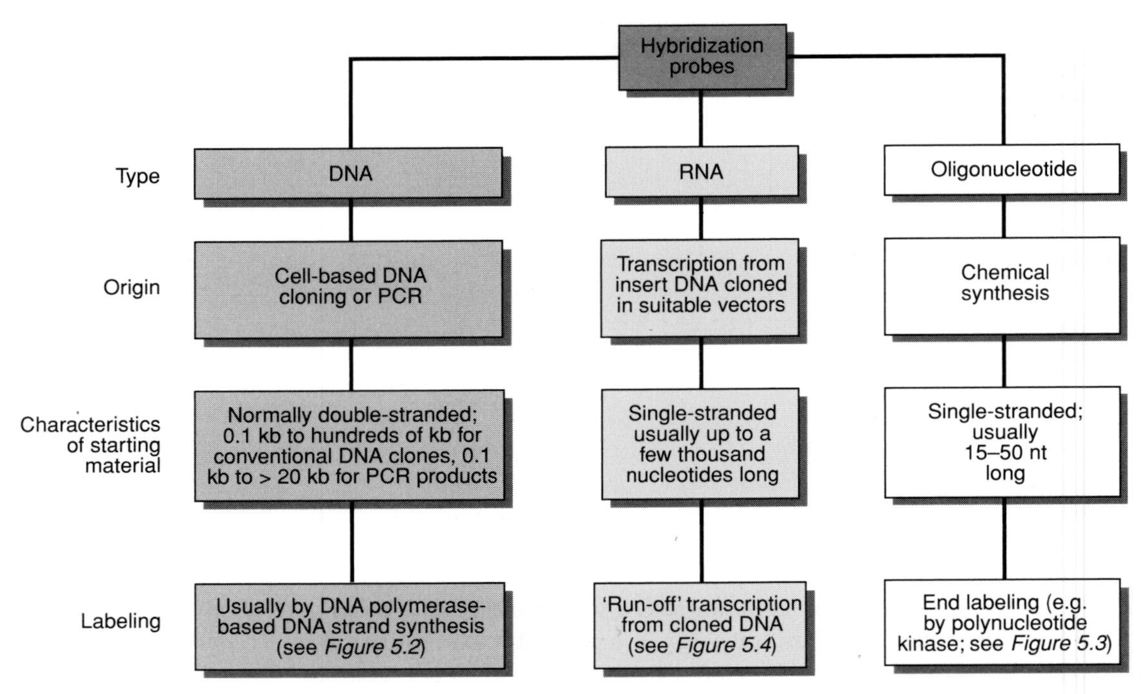

Figure 5.1: Origin and characteristics of nucleic acid hybridization probes.

sequence. Typically this involves parallel syntheses of a set of oligonucleotides which are identical at certain nucleotide positions but different at others. Oligonucleotide probes are often labeled by incorporating a ^{32}P atom at the 5′ end (see opposite).

DNA and RNA can conveniently be labeled *in vitro* by incorporation of nucleotides (or nucleotide components) containing a labeled atom or chemical group

Although, in principle, DNA and RNA can be labeled *in vivo*, by supplying labeled deoxynucleotides to tissue culture cells, this procedure is of limited general use; it has been restricted largely to preparing labeled viral DNA from virus-infected cells, and studying RNA processing events. A much more versatile method involves extracting or cloning DNA and RNA, and labeling the purified DNA or RNA *in vitro* by incorporating labeled nucleotides or nucleotide components using a suitable enzyme. Two major types of procedure have been widely used to label DNA *in vitro*:

(i) **DNA polymerase-based strand labeling.** In this type of procedure, the DNA is labeled in such a way that numerous labeled nucleotides are incorporated during a DNA strand synthesis reaction. Usually the DNA synthesis reaction involves ensuring that at least one of the four deoxyribo-nucleoside triphosphates (dNTPs) carries a labeled group.

(ii) **End-labeling.** This type of procedure involves addition of a labeled group to one or a few terminal nucleotides. It is less widely used, but is useful for a number of procedures, including labeling of single-stranded oligonucleotides (see below) and restriction mapping (page 96). Inevitably, because only one or a very few labeled groups are incorporated, the **specific activity** (the amount of radioactivity incorporated divided by the total mass) of the labeled DNA is much less than that for probes in which there has been incorporation of multiple labeled nucleotides along the length of the DNA.

Preparation of RNA probes is now most conveniently done by cloning the appropriate DNA in a plasmid vector that is designed to permit synthesis of strand-specific transcripts from the inserted DNA. The transcripts can be labeled by including one or more labeled NTPs in the *in vitro* transcription reaction (see below).

Labeling DNA by nick translation

The **nick-translation** procedure involves introducing single-strand breaks (nicks) in the DNA, leaving exposed 3′ hydroxyl termini and 5′ phosphate termini. The nicking can be achieved by adding a suitable endonuclease such as pancreatic deoxyribonuclease I (DNase I). The exposed nick can then serve as a start point for introducing new nucleotides at the 3′ hydroxyl side of the nick using the DNA polymerase activity of *E. coli* DNA polymerase I at the same time as existing nucleotides are removed from the other side of the nick by the 5′ → 3′ exonuclease activity of the same enzyme. As a result, the nick will be moved progressively along the DNA ('translated') in the 5′ → 3′ direction (see *Figure 5.2A*). If the reaction is carried out at a relatively low temperature (about 15°C), the reaction proceeds no further than one complete renewal of the existing nucleotide sequence. Although there is no net DNA synthesis at these temperatures, the synthesis reaction allows the incorporation of labeled nucleotides in place of the previously existing unlabeled ones.

Random primed DNA labeling

The **random primed DNA labeling** method (sometimes known as **oligolabeling**) (Feinberg and Vogelstein, 1983) is based on hybridization of a mixture of all possible hexanucleotides: the starting DNA is denatured and then cooled slowly so that the individual hexanucleotides can bind to suitably complementary sequences within the DNA strands. Synthesis of new complementary DNA strands is primed by the bound hexanucleotides. The synthesis is catalyzed by the Klenow subunit of DNA polymerase I (which contains the polymerase activity in the absence of associated exonuclease activities) in the presence of the four dNTPs, at least one of which has a labeled group (see *Figure 5.2B*). This method produces labeled DNAs of high specific activity. Because all sequence combinations are represented in the hexanucleotide mixture, binding of primer to template DNA occurs in a random manner, and labeling is uniform across the length of the DNA.

End-labeling of DNA

Labeling of single-stranded oligonucleotides is usually achieved by an end-labeling reaction involving polynucleotide kinase. This enzyme can catalyze the exchange of a labeled γ-phosphate group from ATP with the 5′-terminal phosphates on single- or double-stranded DNA (see *Figure 5.3A*). Typically, the label is provided in the form of a ^{32}P at the γ-phosphate position of ATP. The same procedure can also be used for labeling double-stranded DNA, as can *fill-in end-labeling (Figure 5.3B)* using the Klenow subunit of *E. coli* DNA polymerase (which has DNA polymerase activity, but lacks the 5′ → 3′ exonuclease activity of the whole enzyme). In either case, fragments carrying label at one end only can then be generated by cleavage at an internal restriction site, generating two differently sized fragments which can be separated by gel electrophoresis and purified.

Labeling of RNA

The preparation of labeled RNA probes (**riboprobes**) is most easily achieved by *in vitro* transcription of insert DNA cloned in a suitable plasmid vector. The vector is

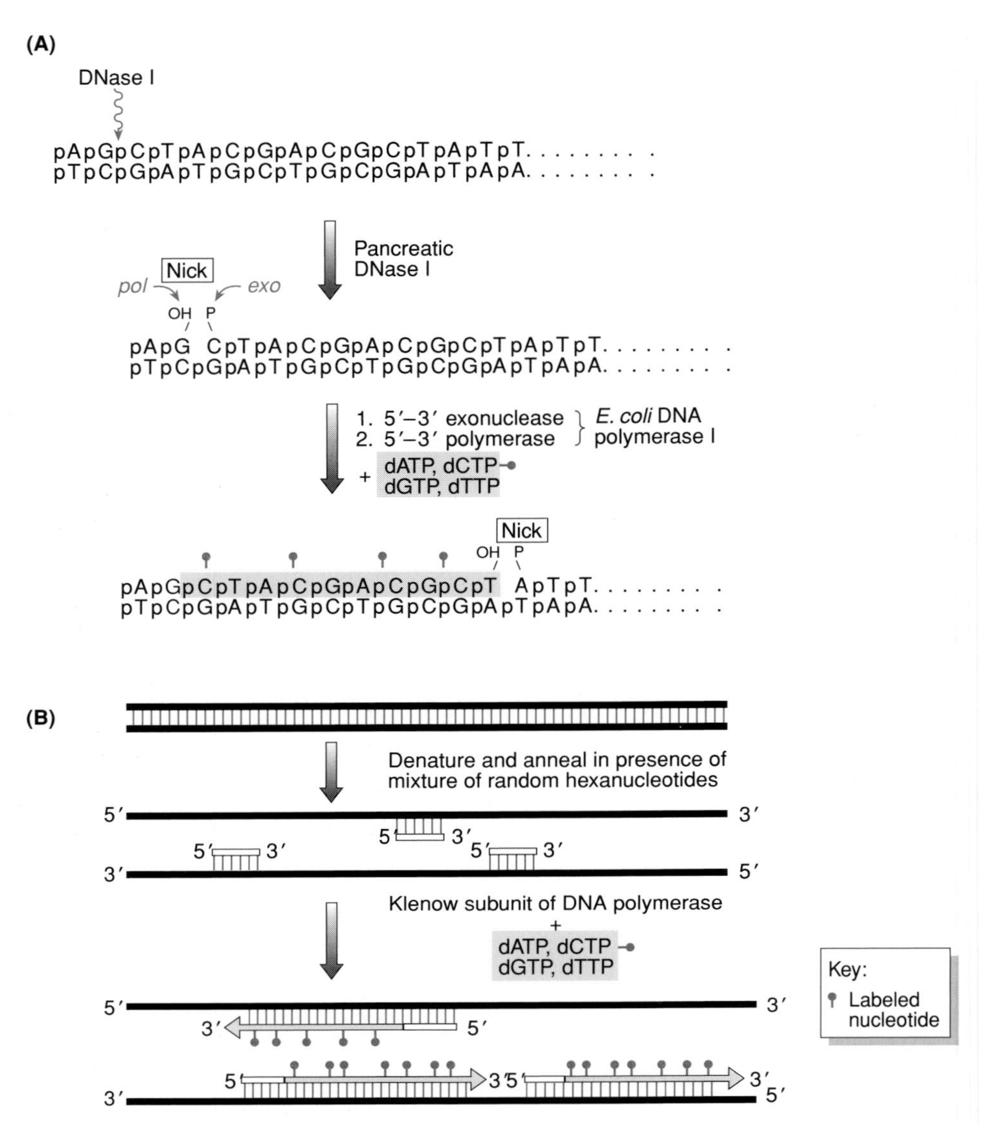

Figure 5.2: DNA labeling by *in vitro* DNA strand synthesis.

(A) Nick translation. Pancreatic DNase I introduces single-stranded nicks by cleaving internal phosphodiester bonds (p), generating a 5′ phosphate group and a 3′ hydroxyl terminus. Addition of the multisubunit enzyme *E. coli* DNA polymerase I contributes two enzyme activities: (i) a 5′ → 3′ exonuclease attacks the exposed 5′ termini of a nick and sequentially removes nucleotides in the 5′ → 3′ direction; (ii) a DNA polymerase adds new nucleotides to the exposed 3′ hydroxyl group, continuing in the 5′ → 3′ direction, thereby replacing nucleotides removed by the exonuclease and causing lateral displacement (translation) of the nick.
(B) Random primed labeling. The Klenow subunit of *E. coli* DNA polymerase I can synthesize new radiolabeled DNA strands using as a template separated strands of DNA, and random hexanucleotide primers.

designed so that adjacent to the multiple cloning site is a phage promoter sequence, which can be recognized by the corresponding phage RNA polymerase. For example, the plasmid vector pSP64 contains the bacteriophage SP6 promoter sequence immediately adjacent to a multiple cloning site. The SP6 RNA polymerase can then be used to initiate transcription from a specific start point in the SP6 promoter sequence, transcribing through any DNA sequence that has been inserted into the multiple cloning site. By using a mix of NTPs, at least one of which is labeled, high specific activity radiolabeled transcripts can be generated (*Figure 5.4*). Bacteriophage T3 and T7 promoter/RNA polymerase systems are also used commonly for generating riboprobes. Labeled sense and antisense riboprobes can be generated

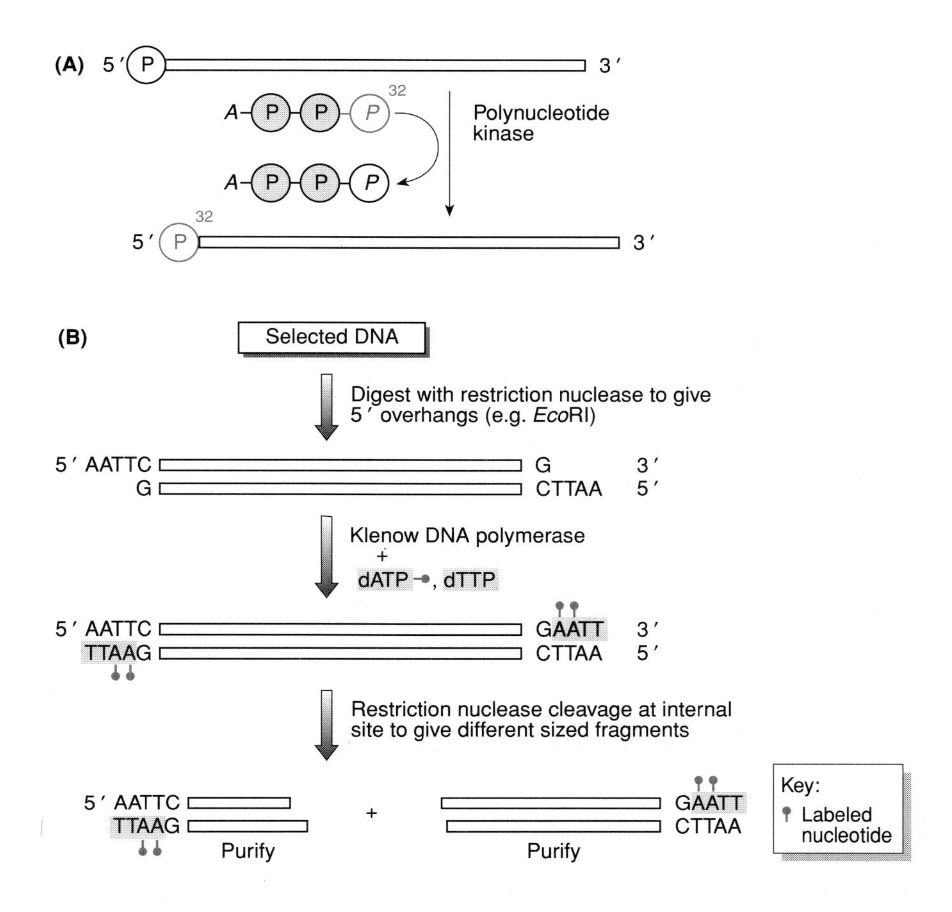

Figure 5.3: End-labeling of DNA.

(A) Kinase end-labeling of oligonucleotides. The 5′-terminal phosphate of the oligonucleotide is replaced in an exchange reaction by the ^{32}P-labeled γ-phosphate of [γ-^{32}P]ATP. The same procedure can be used to label the two 5′ termini of double-stranded DNA. **(B)** Fill-in end-labeling by Klenow. The DNA of interest is cleaved with a suitable restriction nuclease to generate 5′ overhangs. The overhangs act as a primer for Klenow DNA polymerase to incorporate labeled nucleotides complementary to the overhang. Fragments labeled at one end only can be generated by internal cleavage with a suitable restriction site to generate two differently sized fragments which can easily be size-fractionated.

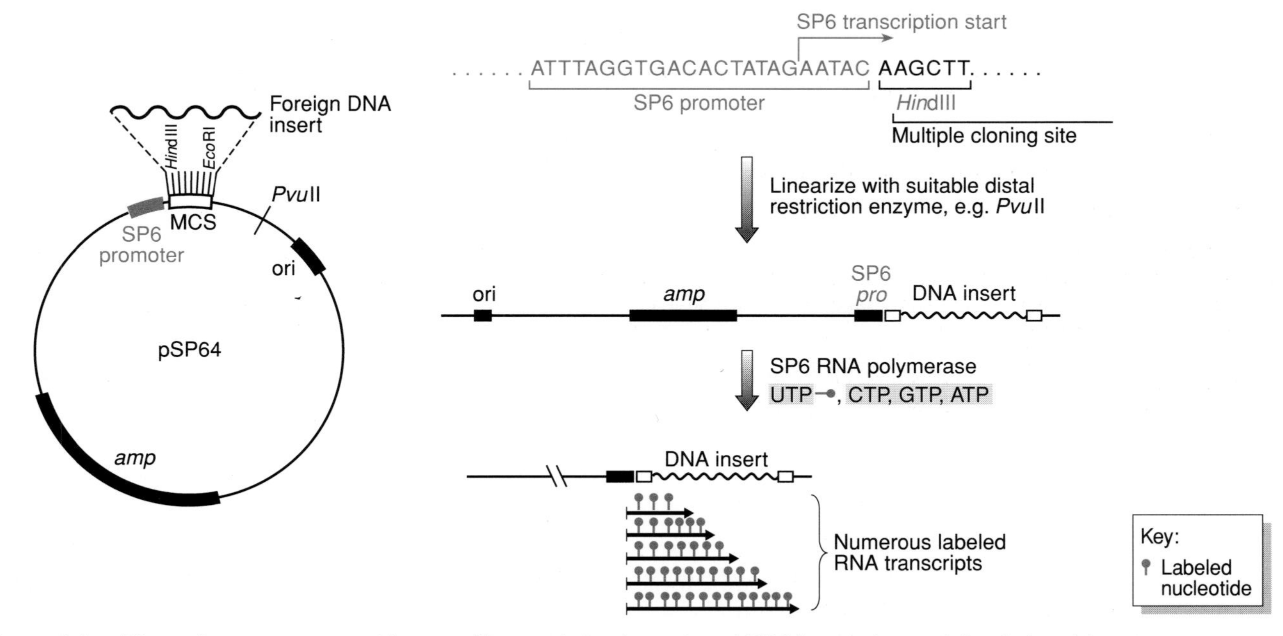

Figure 5.4: Riboprobes are generated by run-off transcription from cloned DNA inserts in specialized plasmid vectors.

The plasmid vector pSP64 contains a promoter sequence for phage SP6 RNA polymerase linked to the multiple cloning site (MCS) in addition to an origin of replication (ori) and ampicillin resistance gene (*amp*). After cloning a suitable DNA fragment in one of the 11 unique restriction sites of the MCS, the purified recombinant DNA is linearized by cutting with a restriction enzyme at a unique restriction site just distal to the insert DNA. Thereafter labeled insert-specific RNA transcripts can be generated using SP6 RNA polymerase and a cocktail of NTPs, at least one of which is labeled.

from any gene cloned in such vectors (the gene can be cloned in either of the two orientations) and are widely used in tissue *in situ* hybridization (page 126).

Nucleic acids can be labeled by isotopic and nonisotopic methods

Isotopic labeling and detection

Radiolabeled probes contain nucleotides with a radioisotope (often ^{32}P, ^{35}S or ^{3}H), which can be detected specifically in solution or, much more commonly, within a solid specimen (**autoradiography** – see *Box 5.1*).

The intensity of an autoradiographic signal is dependent on the intensity of the radiation emitted by the radioisotope, and the time of exposure, which may often be long (1 or more days, or even weeks in some applications). ^{32}P has been used widely in Southern blot hybridization, dot–blot hybridization, colony and plaque hybridization (see below) because it emits high energy β-particles which afford a high degree of sensitivity of detection. It has the disadvantage, however, that it is relatively unstable (see *Box 5.2*). Additionally, its high energy β-particle emission can be a disadvantage under circumstances when fine physical resolution is required to interpret the resulting image unambiguously. For this reason, radionuclides which provide less energetic β-particle radiation have been preferred in certain procedures, for example ^{35}S-labeled nucleotides for DNA sequencing and tissue *in situ* hybridization, and ^{3}H-labeled nucleotides for chromosome *in situ* hybridization (see pages 304 and 126 respectively). ^{35}S and ^{3}H have moderate and very long half-lives respectively, but the latter isotope is disadvantaged by its low energy β-particle emission, necessitating very long exposure times.

^{32}P-labeled nucleotides used in DNA strand synthesis labeling reactions have the radioisotope at the α-phosphate position, because the β- and γ-phosphates from dNTP precursors are not incorporated into the growing DNA chain. Kinase-mediated end-labeling, however, uses [γ-^{32}P]ATP (see *Figure 5.3A*). In the case of ^{35}S-labeled nucleotides which are incorporated during the synthesis of DNA or RNA strands, the NTP or dNTP carries a ^{35}S isotope in place of the O^- of the α-phosphate group. ^{3}H-labeled nucleotides carry the radioisotope at several positions.

Specific detection of molecules carrying a radioisotope is most often performed by autoradiography.

Nonisotopic labeling and detection

Nonisotopic labeling systems involve the use of nonradioactive probes. Although only developed comparatively recently, such systems are finding increasing applications in a variety of different areas (Kricka, 1992). Most such systems feature the chemical coupling of a modified *reporter molecule* to a nucleotide precursor. When incorporated into DNA, the reporter groups can be specifically bound by a protein or other ligand which has a very high affinity for the reporter group. Attached to this *affinity group* is a *marker group* which can be detected in a suitable assay (*Figure 5.5*).

A widely used nonisotopic labeling system utilizes the extremely high affinity of two ligands: **biotin** (a naturally occurring vitamin) and the bacterial protein steptavidin, which demonstrate an affinity constant of 10^{-14}, one of the strongest known in biology. Biotinylated probes can be made easily by including a suitable biotinylated nucleotide in the labeling reaction. A popular alternative is to use the plant steroid **digoxigenin** (obtained from *Digitalis* plants) as a label. A specific antibody for this compound permits detection of molecules containing it. Both digoxigenin- and biotin-labeled nucleotides are designed to have an intermediate spacer so that, once

Box 5.1: Principles of autoradiography

Autoradiography is a procedure for localizing and recording a radiolabeled compound within a solid sample, which involves the production of an image in a photographic emulsion. In molecular genetic applications, the solid sample often consists of size-fractionated DNA or protein samples that are embedded within a dried gel, fixed to the surface of a dried nylon membrane or nitrocellulose filter, or located within fixed chromatin or tissue samples mounted on a glass slide. The photographic emulsions consist of silver halide crystals in suspension in a clear gelatinous phase. Following passage through the emulsion of a β-particle or a γ-ray emitted by a radionuclide, the Ag^+ ions are converted to Ag atoms. The resulting latent image can then be converted to a visible image once the image is **developed**, an amplification process in which entire silver halide crystals are reduced to give metallic silver. The **fixing** process results in removal of any unexposed silver halide crystals, giving an autoradiographic image which provides a two-dimensional representation of the distribution of the radiolabel in the original sample.

Direct autoradiography involves placing the sample in intimate contact with an X-ray film, a plastic sheet with a coating of photographic emulsion; the radioactive emissions from the sample produce dark areas on the developed film. This method is best suited to detection of weak to medium strength β-emitting radionuclides (e.g. ^{3}H, ^{35}S, etc.). However, it is not suited to high energy β-particles (e.g. from ^{32}P): such emissions pass through the film, resulting in the wasting of the majority of the energy. **Indirect autoradiography** is a modification in which the emitted energy is converted to light by a suitable chemical (**scintillator** or **fluor**) using either of the following methods:

- **fluorography** – here the sample is impregnated with a liquid scintillator. The energy transferred from the radioactive emissions causes the scintillator molecules to emit photons, thereby exposing the photographic emulsion. This method is largely used to improve the detection of weak emitters such as ^{3}H or ^{35}S. If, for example, the sample occurs within a gel, the weak energy of the ^{3}H or ^{35}S radiolabel would result in a significant amount of the energy being absorbed by the sample, unless the gel is impregnated with a fluor such as PPO (2,5-diphenyl-oxazole).
- **Intensifying screens** – these are sheets of a solid inorganic scintillator which are placed behind the film in the case of samples emitting high energy radiation, such as ^{32}P. Those emissions which pass through the photographic emulsion are absorbed by the screen and converted to light. By effectively superimposing a photographic emission upon the direct autoradiographic emission, the image is intensified.

Box 5.2: Characteristics of radioisotopes commonly used for labeling DNA and RNA probes

Radioisotope	Half-life	Type/energy of emission
^{3}H	12.4 years	β^-/0.018 MeV
^{32}P	14.3 days	β^-/1.71 MeV
^{35}S	87.4 days	β^-/0.167 MeV

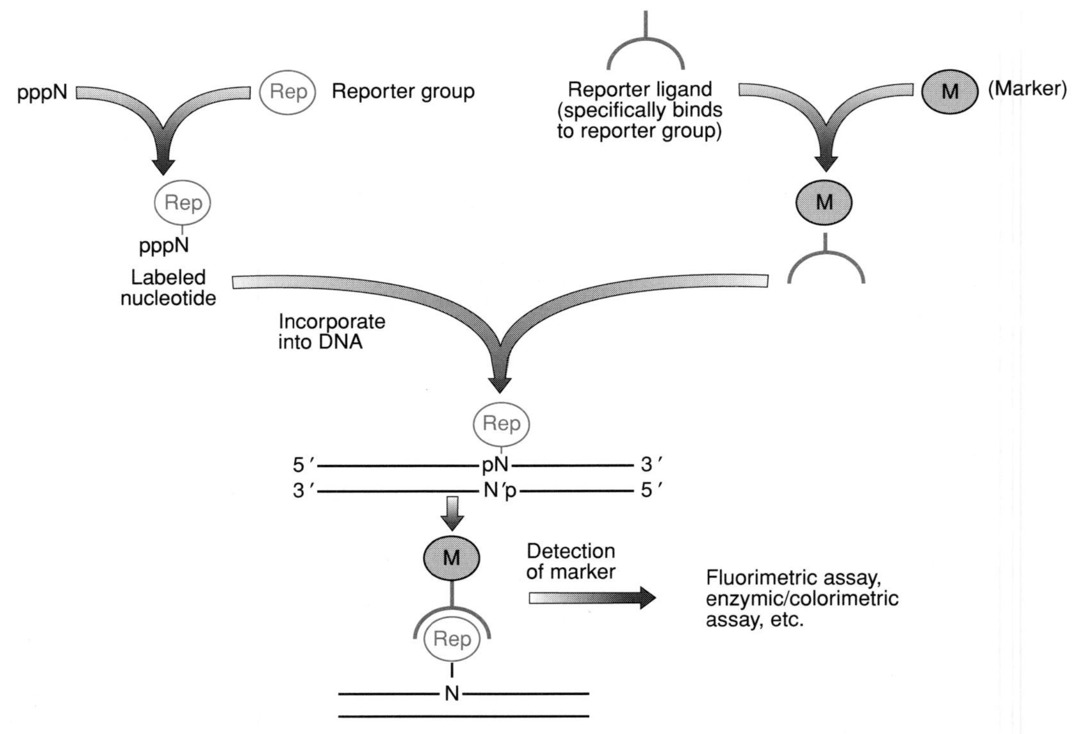

Figure 5.5: General principles of nonisotopic labeling.

The protein recognizing the reporter group is often a specific antibody, as in the digoxigenin system, or any other ligand that has a very high affinity for a specific group, such as streptavidin in the case of biotin labeling (see *Figure 5.6)*. The marker can be detected in various ways. If it carries a specific fluorescent dye, it can be detected in a fluorimetric assay. Alternatively, it can be an enzyme such as alkaline phosphatase which can be coupled to an enzyme assay, yielding a product that can be measured colorimetrically.

introduced into a labeled nucleic acid probe, the attached reporter groups can protrude sufficiently far from the nucleic acid backbone so as to facilitate their detection by streptavidin or the digoxigenin-specific antibody respectively (see *Figure 5.6*).

Principles of molecular hybridization

Molecular hybridization is a method for identifying nucleic acid molecules which are closely related in base sequence to a labeled nucleic acid probe

Definition of molecular hybridization

Molecular hybridization involves mixing single strands of two sources of nucleic acids, a labeled nucleic acid probe and an unlabeled target DNA. The two DNA strands of a double-stranded DNA probe must therefore be separated before being used (**DNA denaturation** or **melting**). If the target DNA is also double-stranded, it too must be denatured, generally by heating or by alkaline treatment, to separate the two DNA strands. After mixing single strands of probe DNA with single strands of target DNA, DNA strands with complementary base sequences can be allowed to reassociate (**DNA renaturation** or **annealing**). Complementary probe DNA strands can reanneal to form **homoduplexes**, as can complementary target DNA strands. However, it is the annealing of a probe DNA strand and a complementary target DNA strand to form a labeled **heteroduplex** that defines the usefulness of a

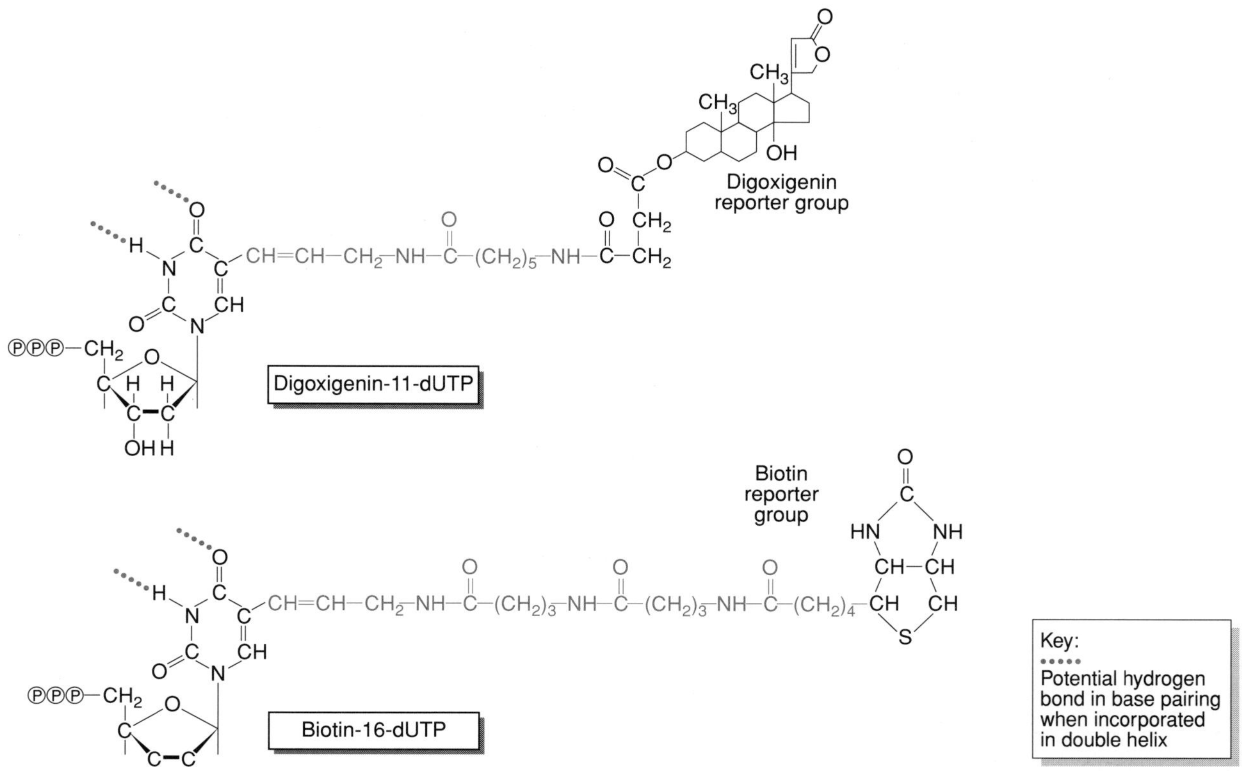

Figure 5.6: Nonisotopic labeling uses labeled nucleotides where the reporter group is attached to a suitable nucleotide by a long spacer group.

Spacer groups are shown in red.

molecular hybridization assay. Usually the probe consists of one specific small sequence, while the target is a heterogeneous mixture of many different sequences. The rationale of the method, therefore, is to use the probe to identify any DNA fragments in the complex target DNA which may be related in sequence to the probe DNA (*Figure 5.7*).

Melting temperature and hybridization stringency

Denaturation of probe DNA is generally achieved by heating a solution of the labeled DNA to a temperature which disrupts the hydrogen bonds that hold the two complementary DNA strands together. The energy required to separate two perfectly complementary DNA strands is dependent on a number of factors, notably:

- **strand length** – long homoduplexes contain a large number of hydrogen bonds and require more energy to separate them; because the labeling procedure typically results in short DNA probes, this effect is negligible above an original (i.e. prior to labeling) length of 500 bp;
- **base composition** – because GC base pairs have one more hydrogen bond than AT base pairs (see *Figure 1.2*), strands with a high % GC composition are more difficult to separate than those with a low % GC composition;
- **chemical environment** – the presence of monovalent cations (e.g. Na^+ ions) stabilizes the duplex, whereas chemical denaturants such as formamide and urea destabilize the duplex by chemically disrupting the hydrogen bonds.

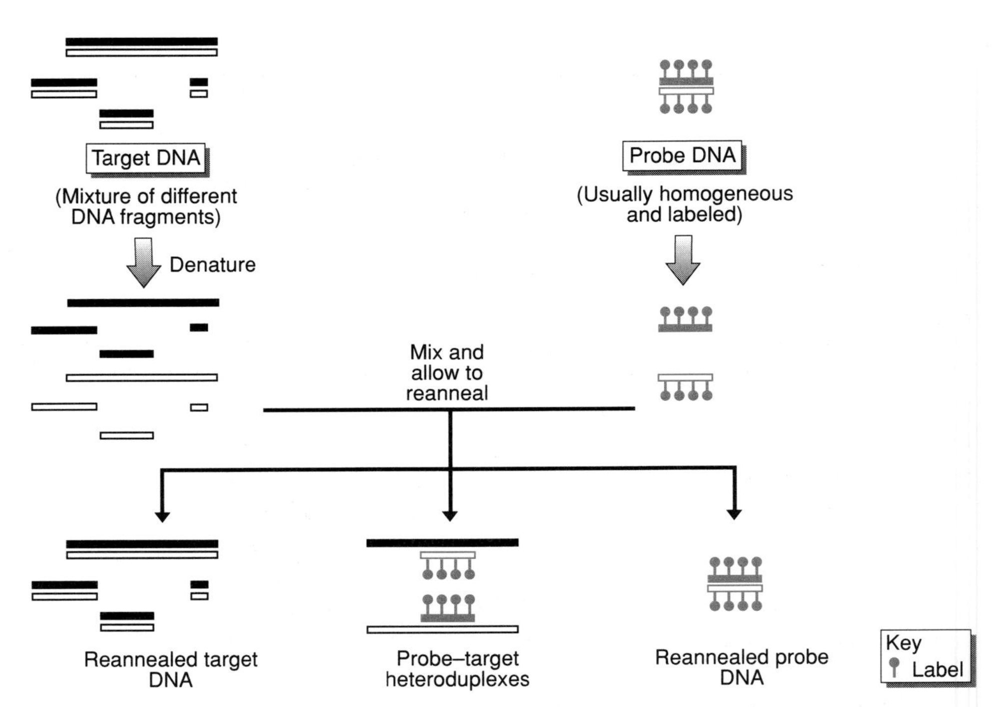

Figure 5.7: A DNA hybridization assay requires the formation of heteroduplexes between labeled single-stranded probe DNA and complementary target DNA.

The probe is envisaged to be strongly related in sequence to a central segment of one of the many types of DNA molecule in the target DNA. Mixing of denatured probe DNA and denatured target DNA will result in reannealed homoduplexes of probe DNA and of target DNA sequences, and also in heteroduplexes formed between probe DNA and any target DNA molecules that are significantly related in DNA sequence. If a method is available for removing the probe DNA that is not bound to target DNA, the heteroduplexes can easily be identified by methods that can detect the label.

A useful measure of the stability of a nucleic acid duplex is the **melting temperature (T_m)**. This is the temperature corresponding to the mid-point in the observed transition from double-stranded form to single-stranded form. Conveniently, this transition can be followed by measuring the optical density of the DNA. The bases of the nucleic acids absorb 260 nm ultraviolet (UV) light strongly. However, the adsorption by double-stranded DNA is considerably less than that of the free nucleotides. This difference, the so-called **hypochromic effect,** is due to interactions between the electron systems of adjacent bases, arising from the way in which adjacent bases are stacked in parallel in a double helix. If duplex DNA is gradually heated, therefore, there will be an increase in the light absorbed at 260 nm (the **optical density$_{260}$** or **OD_{260}**) towards the value characteristic of the free bases. The temperature at which there is a mid-point in the optical density shift is then taken as the T_m (see *Figure 5.8*).

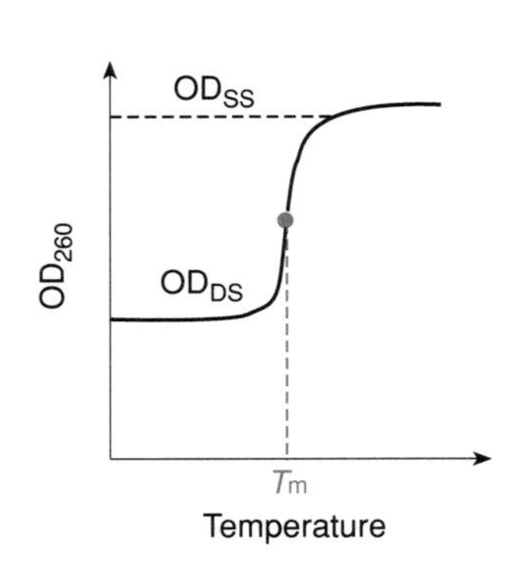

Figure 5.8: Denaturation of DNA results in an increase in optical density.

OD_{SS} and OD_{DS} indicate the optical density of single-stranded and double-stranded DNA respectively. The difference between them represents the hypochromic effect (see text).

For mammalian genomes, with a base composition of about 40% GC, the DNA denatures with a T_m of about 87°C under approximately physiological conditions. The T_m of perfect hybrids formed by DNA, RNA or oligonucleotide probes can be determined according to the formulae in *Table 5.1*. Often, hybridization conditions are chosen so as to promote heteroduplex formation and the hybridization temperature is often as much as 25°C below the T_m. However, after the hybridization and removal of excess probe, hybridization washes may be conducted under more stringent conditions so as to disrupt all duplexes other than those between very closely related sequences. Probe–target heteroduplexes are most stable thermodynamically when the region of duplex formation contains perfect base matching. Mismatches

Table 5.1: Equations for calculating T_m

Hybrids	T_m (°C)
DNA–DNA	$81.5 + 16.6(\log_{10}[Na^+]^a) + 0.41(\%GC^b) - 500/L^c$
DNA–RNA or RNA–RNA	$79.8 + 18.5(\log_{10}[Na^+]^a) + 0.58(\%GC^b) + 11.8\,(\%GC^b)^2 - 820/L^c$
oligo–DNA or oligo–RNA[d]	2 × no. of AT base pairs + 4 × no. of GC base pairs

[a]Or for other monovalent cation, but only accurate in the 0.01–0.4 M range.
[b]Only accurate for %GC from 30 to 75%.
[c]L = length of duplex in base pairs.
[d]Oligo = oligonucleotide; only accurate for lengths up to 20 nucleotides (low GC content) or up to 40 nucleotides (if GC content is high).
Note that for each 1% formamide, the T_m is reduced by about 0.6°C, while the presence of 6 M urea reduces the T_m by about 30°C.

between the two strands of a heteroduplex reduce the T_m: for normal DNA probes, each 1% of mismatching reduces the T_m by approximately 1°C. Although probe–target heteroduplexes are usually not as stable as reannealed probe homoduplexes, a considerable degree of mismatching can be tolerated if the overall region of base complementarity is long (>100 bp; see *Figure 5.9*). Increasing the concentration of NaCl and reducing the temperature reduces the **hybridization stringency**, and enhances the stability of mismatched heteroduplexes. This means that comparatively diverged members of a multigene family or other repetitive DNA family (see Chapter 8) can be identified by hybridization using a specific family member as a probe. Additionally, a gene sequence from one species can be used as a probe to identify homologs in other comparatively diverged species, provided the sequence is reasonably conserved during evolution (see pages 297 and 509).

Conditions can also be chosen to maximize hybridization stringency (e.g. lowering the concentration of NaCl and increasing the temperature), so as to encourage dissociation (denaturation) of mismatched heteroduplexes. If the region of base complementarity is small, as with oligonucleotide probes (typically 15–20 nucleotides), hybridization conditions can be chosen such that a single mismatch renders a heteroduplex unstable (see pages 119–120).

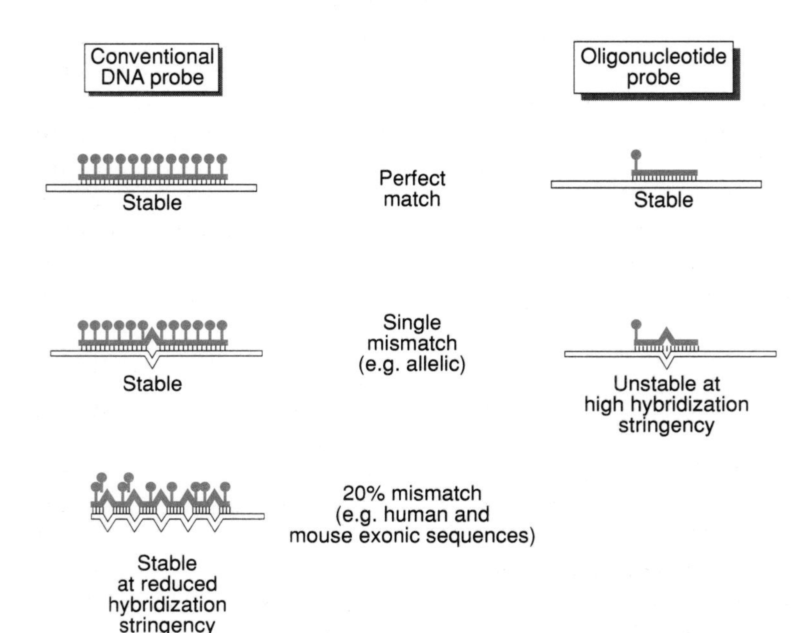

Figure 5.9: DNA hybridization can identify target sequences considerably diverged from a conventional DNA probe, or identical to an oligonucleotide probe.

The kinetics of DNA reassociation are defined by the product of DNA concentration and time ($C_o t$)

Reassociation kinetics describes the speed at which a single strand sequence is able to find a complementary sequence and base-pair with it. It is determined by two major parameters: the starting concentration (C_o) of the specific DNA sequence in moles of nucleotides per liter and the reaction time (t) in seconds. Since the rate of reassociation is proportional to C_o and to t, the $C_o t$ value (often loosely referred to as the **cot** value) is a useful measure. Because the $C_o t$ value will also vary depending on the temperature of reassociation and the concentration of monovalent cations, it is usual to use fixed reference values: a reassociation temperature of 65°C and an [Na^+] concentration of 0.3 M NaCl. Because the rate of reassociation is dependent on DNA concentration, and because the probe DNA normally is homogeneous while the target DNA is usually extremely heterogeneous, an excess of target DNA over probe DNA is used in order to encourage probe–target heteroduplex formation.

In the case of a complex genome such as the human genome, individual DNA sequences which occur as a single copy in the genome may constitute a tiny fraction of the DNA (e.g. the β-globin gene represents only 0.00005% of human genomic DNA). The speed of reassociation of such **single copy** DNA sequences present in human genomic DNA is rather slow. By contrast, certain other sequences are highly repeated in genomic DNA (see Chapter 8), and this greatly elevated DNA concentration results in a comparatively rapid reassociation time. Because the amount of target DNA bound by a probe depends on the copy number of the recognized sequence, hybridization signal intensity is proportional to the copy number of the recognized sequence: single copy genes give weak hybridization signals, highly repetitive DNA sequences give very strong signals. If a particular probe is heterogeneous and contains a low copy sequence of interest, such as a specific gene, mixed with a highly abundant DNA repeat, the weak hybridization signal obtained with the former will be completely masked by the strong repetitive DNA hybridization signal. This effect can, however, be overcome by competition hybridization (see *Box 5.3*).

Box 5.3: Competition hybridization and Cot-1 DNA

Competition (or **suppression**) **hybridization** involves blocking a potentially strong repetitive DNA signal which can be obtained when using a complex DNA probe. The labeled probe DNA is denatured and allowed to reassociate in the presence of unlabeled total genomic DNA in solution, or preferably a fraction that is enriched for highly repetitive DNA sequences. In either case, the highly repetitive DNA within the unlabeled DNA is present in large excess over the repetitive elements in the labeled probe. As a result, such sequences will readily associate with complementary strands of the repetitive sequences within the labeled probe, thereby effectively blocking their hybridization to target sequences.

Instead of using total genomic DNA as a blocking agent in hybridization, it is more effective to use a fraction of total genomic DNA that is enriched for highly repetitive DNA sequences, such as the *Alu*, LINE-1 and *THE* repeats of human DNA. For human DNA, and other mammalian DNA where the genome size is much the same as that of the human genome, the latter usually involves preparing a fraction of DNA known as **Cot-1 DNA** (i.e. DNA with a cot value of 1.0). Total purified human genomic DNA is sonicated to an average length of about 400 bp, denatured by heating, then allowed to renature in 0.3 M NaCl at 65°C at a starting concentration of x moles of nucleotides per liter for a time of t sec, where $xt = 1.0$. For example, since 1 mole of nucleotide = 330 g on average, then a starting DNA concentration of 1 mg ml^{-1} (3 μmol l^{-1}) and a renaturation time of about 5.5 min (330 sec) will produce a cot value of about 1.0 in mol nucleotides sec^{-1} l^{-1}.

Methods and applications of molecular hybridization

Early experiments in molecular hybridization utilized **solution hybridization**, involving mixing of aqueous solutions of probe and target nucleic acids. However, the very low concentration of single copy sequences in complex genomes meant that reassociation times were inevitably slow. One widely used way of increasing the reassociation speed is to artificially increase the overall DNA concentration in aqueous solution by abstracting water molecules (e.g. by adding high concentrations of polyethylene glycol).

Another approach has been to facilitate detection of reassociated molecules by immobilizing the target DNA on a solid support, such as a membrane made of nitrocellulose or nylon, to both of which single-stranded DNA binds readily. Attachment of labeled probe to the immobilized target DNA can then be followed by removing the solution containing unbound probe DNA, extensive washing and drying in preparation for detection.

Dot–blot and slot–blot hybridization are convenient methods of identifying nucleic acid sequences in an unfractionated nucleic acid sample

In this procedure, an aqueous solution of target DNA, for example total human genomic DNA, is simply spotted on to a nitrocellulose or nylon membrane and allowed to dry (**dot–blot**), or applied as above but through an individual slot in a suitable template (**slot–blot**). The target DNA sequences are denatured, either by previously exposing to heat, or by exposure of the filter containing them to alkali. The denatured target DNA sequences now immobilized on the membrane are exposed to a solution containing single-stranded labeled probe sequences. After allowing sufficient time for probe–target heteroduplex formation, the probe solution is decanted, and the membrane is washed to remove excess probe that may have become nonspecifically bound to the filter. It is then dried and exposed to an autoradiographic film.

Allele-specific oligonucleotide (ASO) hybridization can discriminate between alleles differing at a single nucleotide position

A general method for distinguishing between alleles that differ by even a single nucleotide substitution involves constructing **allele-specific oligonucleotide (ASO)** probes from sequences spanning the variant nucleotide site. ASO probes are typically 15–20 nucleotides long and are normally employed under hybridization conditions at which the DNA duplex between probe and target is stable only if there is perfect base complementarity between them: a single mismatch between probe and target sequence is sufficient to render the short heteroduplex unstable (*Figure 5.9*). Typically, this involves designing the oligonucleotides so that the single nucleotide difference between alleles occurs in a central segment of the oligonucleotide sequence, thereby maximizing the thermodynamic instability of a mismatched duplex. Such discrimination can be employed for a variety of research and diagnostic purposes. Although ASOs can be used in conventional Southern blot hybridization (see below), it is more convenient to use them in dot–blot assays (see *Figure 5.10*).

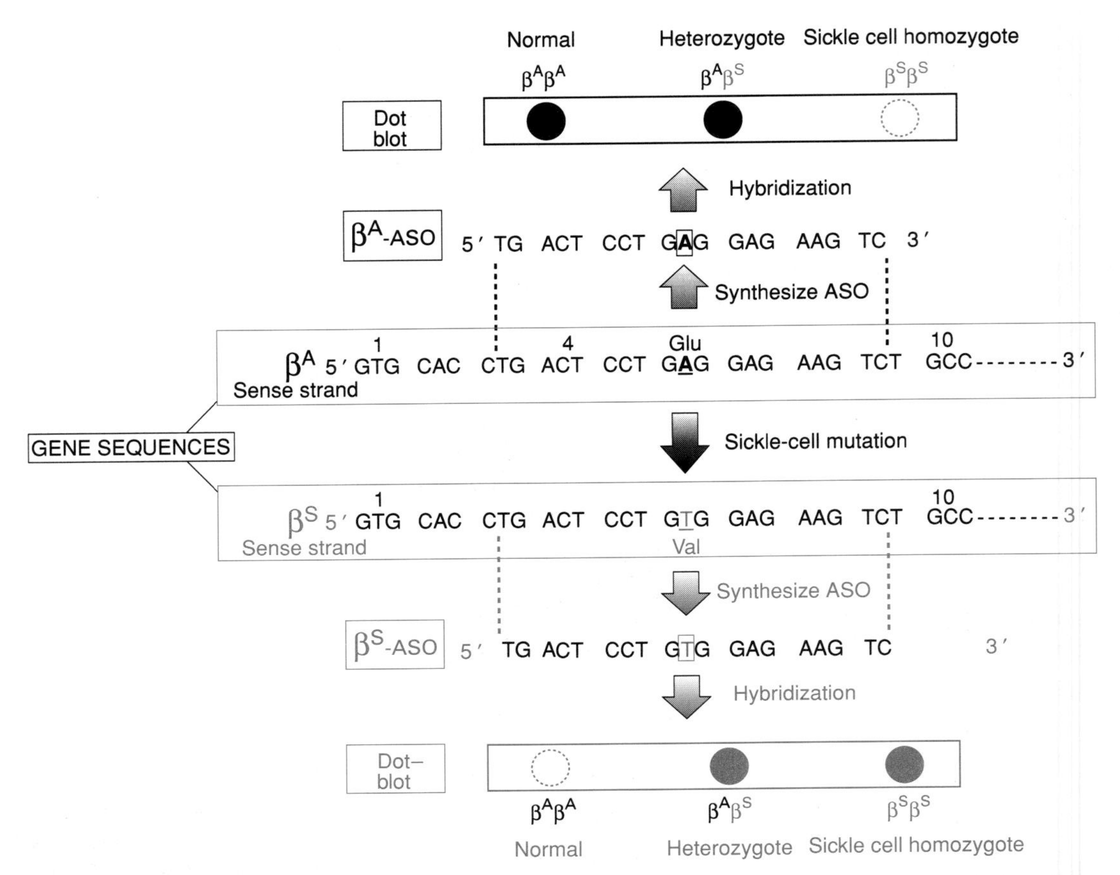

Figure 5.10: Allele-specific oligonucleotide (ASO) dot–blot hybridization can identify individuals with the sickle cell mutation.

The sickle cell mutation is a single nucleotide substitution (A → T) at codon 6 in the β-globin gene, resulting in a GAG (Glu) → GTG (Val) substitution. The example shows how one can design ASOs: one specific for the normal (β^A) allele and identical to a sequence of 19 nucleotides encompassing codons 3–9 of this allele, and one specific for the mutant (β^S) allele, being identical to the equivalent sequence of the mutant allele. The labeled ASOs can be individually hybridized to denatured genomic DNA samples on dot–blots (see page 119). The β^A- and β^S-specific ASOs can hybridize to the complementary antisense strand of the normal and mutant alleles respectively, forming perfect 19-bp duplexes. However, duplexes between the β^A-specific ASO and the β^S allele, or between the β^S-specific ASO and the β^A allele have a single mismatch and are unstable at high hybridization stringency (see *Figure 5.9*).

Southern and Northern blot hybridizations detect target DNA and RNA fragments that have been size-fractionated by gel electrophoresis

Southern blot hybridization

In this procedure, the target DNA is digested with one or more restriction endonucleases, size-fractionated by agarose gel electrophoresis, denatured and transferred to a nitrocellulose or nylon membrane for hybridization (*Figure 5.11*). During the electrophoresis, DNA fragments, which are negatively charged because of the phosphate groups, are repelled from the negative electrode towards the positive electrode, and sieve through the porous gel. Smaller DNA fragments move faster. For fragments between 0.1 and 20 kb long, the migration speed depends on fragment length, but scarcely at all on the base composition. Thus, fragments in this size range are fractionated by size in a conventional agarose gel electrophoresis system. To achieve efficient size-fractionation of large fragments (40 kb to several megabases), a more specialized system is required, such as a pulsed-field gel electrophoresis apparatus (see page 286).

Following electrophoresis, the test DNA fragments are denatured in strong alkali. As agarose gels are fragile, and the DNA in them can diffuse within the gel, it is

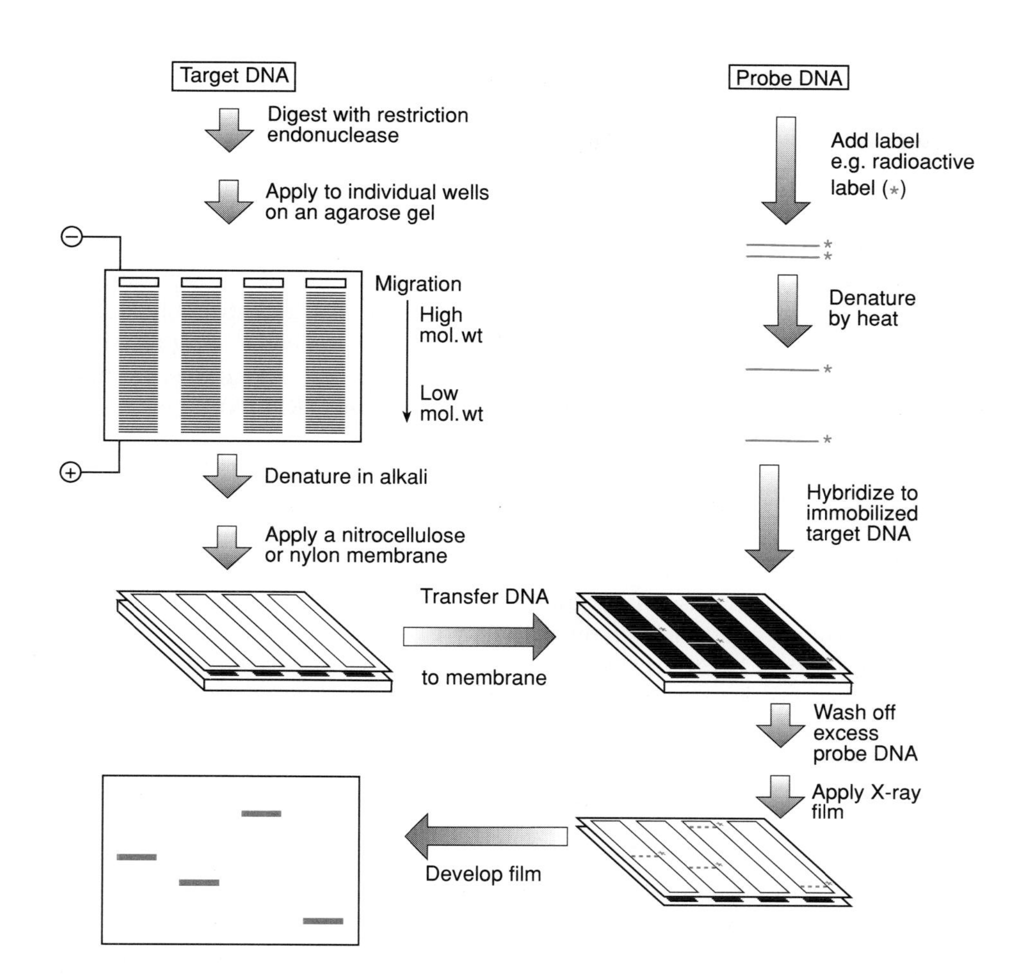

Figure 5.11: Southern blot hybridization detects target DNA fragments that have been size-fractionated by gel electrophoresis.

usual to transfer the denatured DNA fragments by blotting on to a durable nitrocellulose or nylon membrane, to which single-stranded DNA binds readily. The individual DNA fragments become immobilized on the membrane at positions which are a faithful record of the size separation achieved by agarose gel electrophoresis. Subsequently, the immobilized single-stranded target DNA sequences are allowed to associate with labeled single-stranded probe DNA. The probe will bind only to related DNA sequences in the target DNA, and their position on the membrane can be related back to the original gel in order to estimate their size.

Northern blot hybridization

Northern blot hybridization is a variant of Southern blotting in which the target nucleic acid is RNA instead of DNA. A principal use of this method is to obtain information on the expression patterns of specific genes. Once a gene has been cloned, it can be used as a probe and hybridized against a Northern blot containing, in different lanes, samples of RNA isolated from a variety of different tissues (see *Figure 19.7*). The data obtained can provide information on the range of cell types in which the gene is expressed, and the relative abundance of transcripts. Additionally, by revealing transcripts of different sizes, it may provide evidence for the use of alternative promoters, splice sites or polyadenylation sites.

Southern blot hybridization permits assay of RFLPs and moderately small scale mutations, and detection of families of related DNA sequences

Southern blot hybridization has been used extensively in molecular genetic studies as a means of genomic restriction mapping: a labeled DNA probe from one genome can be used to infer the structure of related sequences in the same or different genomes. Because the genomic DNA samples are fractionated by separation of restriction fragments according to size, mutations which alter a restriction site, and significantly large insertions or deletions occurring between neighboring restriction sites, can be typed. Such mutations will result in altered restriction fragment lengths, that is **restriction fragment length polymorphisms (RFLPs)**.

Direct detection of pathogenic point mutations by restriction mapping

Very occasionally, a pathogenic mutation directly abolishes or creates a restriction site, enabling direct screening for the pathogenic mutation. For example, the sickle cell mutation is a single nucleotide substitution (A $\rightarrow$ T) at codon 6 in the β-globin gene, which causes a missense mutation (Glu $\rightarrow$ Val), and at the same time abolishes an *Mst*II restriction site which spans codons 5–7. The nearest flanking restriction sites for *Mst*II, located 1.2 kb upstream in the 5′-flanking region and 0.2 kb downstream at the 3′ end of the first intron, are well conserved. Consequently, a β-globin DNA probe can differentiate the normal β^A-globin and the mutant β^S-globin alleles in *Mst*II-digested human DNA: the former exhibits 1.2-kb + 0.2-kb *Mst*II fragments, whereas the sickle cell allele exhibits a 1.4-kb *Mst*II fragment (*Figure 5.12*).

Detection of conventional RFLPs

The great majority of mutations are not associated with disease; instead, they often occur within noncoding DNA sequences. As a large number of recognition sequences are known for type II restriction endonucleases, many point mutation polymorphisms will be characterized by alleles which possess or lack a recognition site for a specific restriction endonuclease and therefore display **restriction site**

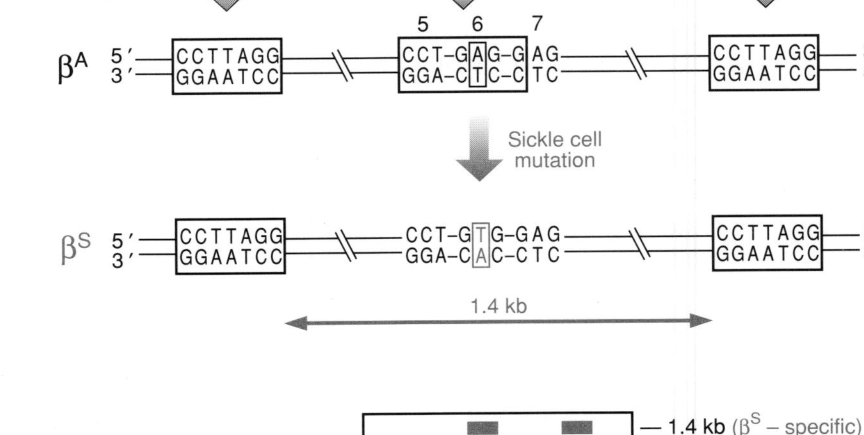

Figure 5.12: The sickle cell mutation destroys an *Mst*II site and generates a disease-specific RFLP.

The *Mst*II restriction nuclease recognizes the sequence CCTNAGG where N = A, C, G or T. A restriction site for *Mst*II is found in the normal β^A-globin allele but is destroyed by the sickle cell mutation. The nearest flanking *Mst*II sites are located, respectively, 1.2 kb upstream in the 5′-flanking region of the β-globin gene and 0.2 kb downstream at the 3′ end of the first intron. Conservation of these flanking sites results in the β^A-associated (1.2 kb + 0.2 kb) *Mst*II RFLP and the sickle cell-associated 1.4-kb *Mst*II RFLP.

polymorphism (RSP). Accordingly, individual RSPs normally have two detectable alleles (one lacking and one possessing the specific restriction site). RSPs can be assayed by digesting genomic DNA samples with the relevant restriction endonuclease and identifying specific restriction fragments whose lengths are characteristic of the two alleles, so-called RFLPs (*Figure 5.13*).

VNTR-based RFLPs and DNA fingerprinting

DNA probes can also be used to monitor **VNTR polymorphisms** where alleles differ by a variable number of tandem repeats. To do this, genomic DNA samples are digested with a restriction endonuclease which recognizes well-conserved restriction sites flanking a specific VNTR locus. The resulting restriction fragments are separated according to size on agarose gels, transferred to a suitable membrane and hybridized with a probe representing a unique sequence from the corresponding locus. The resulting pattern of locus-specific RFLPs does not reflect RSP: instead, the differences in sizes of the restriction fragments represent integral numbers of the tandemly repeated unit.

Although the term VNTR could, in theory, encompass a wide range of repeat lengths, in practice the term is usually reserved for moderately large arrays of a repeat unit which is typically in the 5–64 bp region (so-called *hypervariable minisatellite DNA*, see page 196), distinguishing it from simple tandem repeat polymorphism (where the repeat unit length is from 1 to 4 bp; i.e. microsatellite DNA, see page 197) and tandem repeat polymorphism associated with very large arrays of satellite DNA (see page 194).

If the VNTR locus is a member of a repeated DNA family, the use of a VNTR repeat probe, rather than a unique flanking probe, will produce a complex polymorphic pattern. For example, hypervariable minisatellite DNA clones have been used as probes against Southern blots of appropriately digested genomic DNA. Cross-hybridization of such probes with the other members of this highly repeated DNA family results in a pattern of hybridization bands representing the summed contributions of two alleles at each of many hypervariable loci scattered throughout the genome. Consequently, the overall polymorphism of the multilocus hybridization patterns is uniquely high. Because it permits distinction between any two individuals who are not identical twins, probing with hypervariable minisatellites has been termed 'DNA fingerprinting' (see page 446).

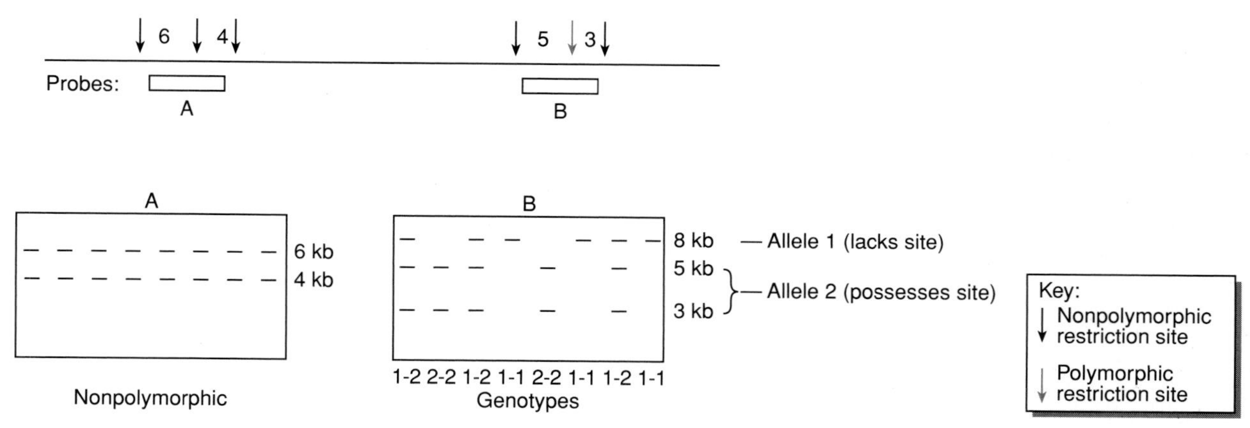

Figure 5.13: Assay of conventional (RSP-based) RFLPs.

Detection of gene deletions by restriction mapping

Certain diseases are associated with a high frequency of deletion of all or part of a gene. If a partial restriction map has been established for the gene under investigation, deletions can be screened by Southern blot hybridization using an appropriate intragenic DNA probe. If the deletion is a small one, for example a few hundred base pairs, it is often apparent as a consistent reduction in size of normal restriction fragments in the gene. An individual who is homozygous for this mutation, or is a heterozygote with one normal allele and another with a small deletion, can easily be identified by detecting the aberrant size restriction fragments.

Large deletions will lead to absence of specific restriction fragments. Homozygous deletion of large DNA segments can easily be detected as complete absence of appropriate restriction fragments associated with the gene. If, however, an individual is heterozygous for a relatively large gene deletion, the deletion may still be detected by demonstrating comparatively reduced intensity of specific gene fragments. For example, patients with 21-hydroxylase deficiency often have deletions of about 30 kb of the 21-hydroxylase/*C4* gene cluster. Such pathological deletions eliminate the functional 21-hydroxylase gene, *CYP21*, and an adjacent *C4B* gene, leaving the related *CYP21P* pseudogene and *C4A* genes. Patients with homozygous deletions will show absence of diagnostic restriction fragments associated with *CYP21* and *C4B*, while carriers of the deletion will show a 2:1 ratio of *CYP21P*:*CYP21* and of *C4A*:*C4B* (Collier *et al.*, 1989; *Figure 5.14*).

Identifying families of related DNA sequences

An important application of Southern blot hybridization in mammalian genetics is the ability to identify a given DNA probe as a member of a repetitive DNA family. Many important mammalian genes belong to multigene families, and many other DNA sequences show varying degrees of repetition (Chapter 8). Once a newly isolated probe is demonstrated to be related to other uncharacterized sequences, attempts can then be made to isolate the other members of the family by screening genomic DNA libraries (page 508). Additionally, screening can also be conducted on genomic DNA samples from different species to identify interspecific related sequences. An important route to identifying coding DNA involves identifying sequences that are highly conserved in evolution (see page 296).

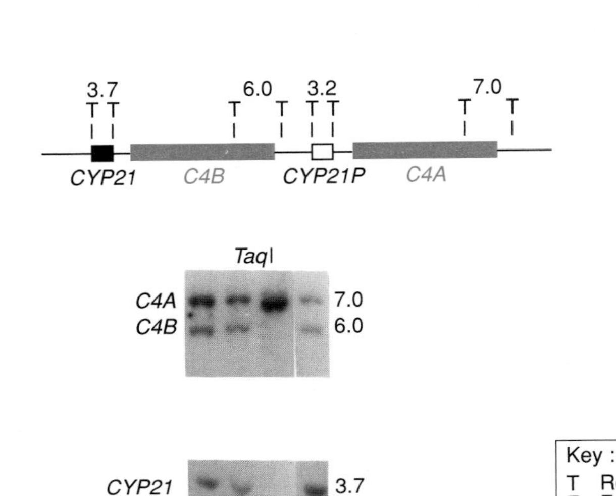

Figure 5.14: Detection of heterozygous and homozygous gene deletions associated with 21-hydroxylase deficiency.

Numbers indicate the size in kilobases of indicated *Taq*I restriction fragments. Southern blots of *Taq*I-digested genomic DNA from family members were hybridized with a complement *C4* gene probe (recognizing the duplicated *C4A* and *C4B* genes equally) (middle panel) or with a 21-hydroxylase (*CYP21*) gene probe (lower panel).

Colony blot and plaque lift hybridization are methods for screening separated bacterial colonies or plaques following bacteriophage infection of bacteria

As described on page 95 colonies of bacteria or other suitable host cells which contain recombinant DNA can generally be selected or identified by the ability of the insert to inactivate a marker vector gene (e.g. β-galactosidase, or an antibiotic-resistance gene). A specific approach can be used when the desired recombinant DNA contains a DNA sequence that is closely related to an available nucleic acid probe. In the case of bacterial cells used to propagate plasmid recombinants, the cell colonies are allowed to grow on an agar surface and then transferred by surface contact to a nitrocellulose or nylon membrane, a process known as **colony blotting** (see *Figure 5.15*). Alternatively the cell mixture is spread out on a nitrocellulose or nylon membrane placed on top of a nutrient agar surface, and colonies are allowed to form directly on top of the membrane. In either approach, the membrane is then exposed to alkali to denature the DNA prior to hybridizing with a labeled nucleic acid probe. After hybridization, the probe solution is removed, and the filter is washed extensively, dried and submitted to autoradiography using X-ray film. The position of strong radioactive signals is related back to a master plate containing the original pattern of colonies, in order to identify colonies containing DNA related to the

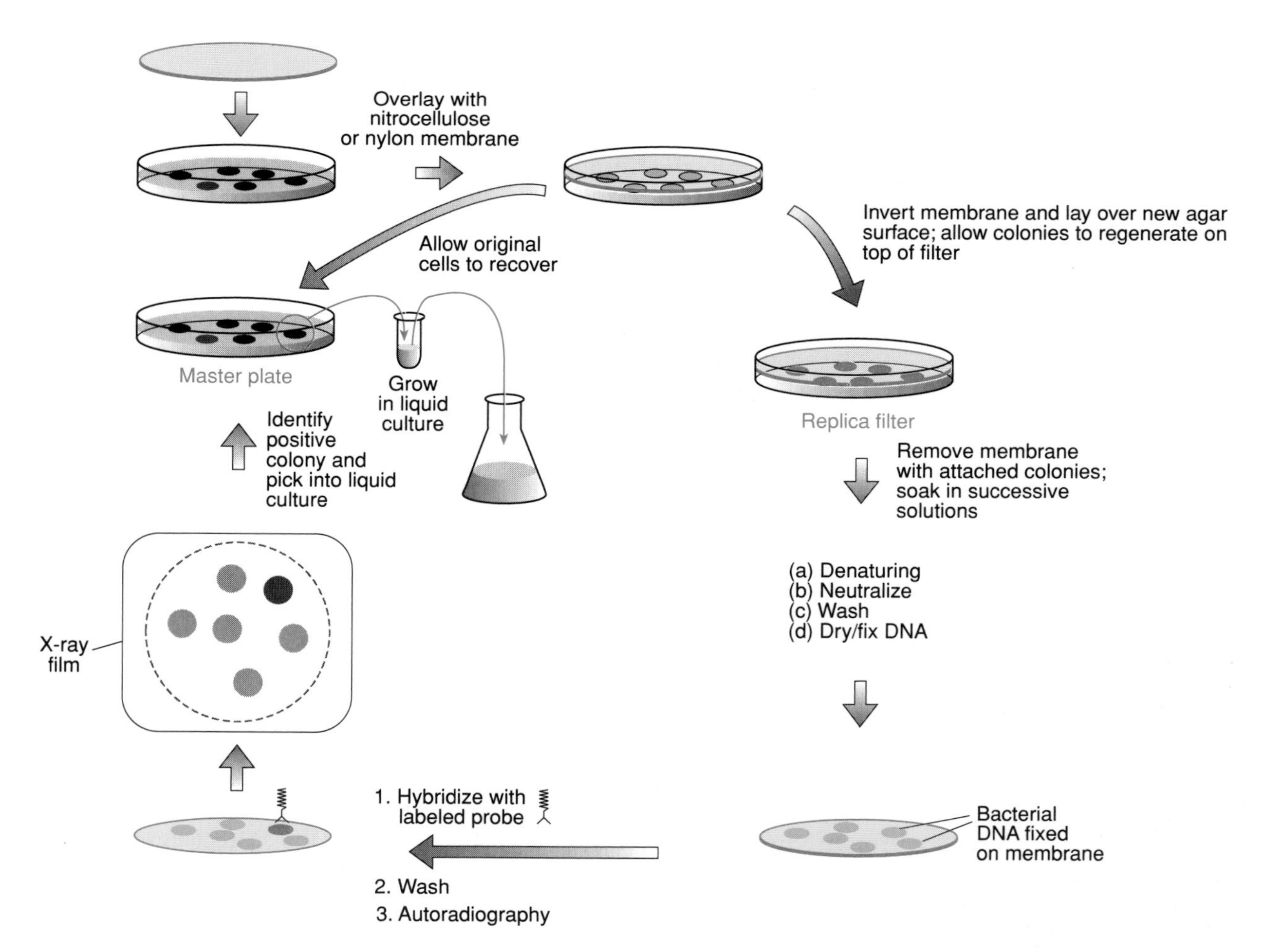

Figure 5.15: Colony hybridization involves replicating colonies on to a durable membrane prior to hybridization with a labeled nucleic acid probe.

This method is popularly used to identify colonies containing recombinant DNA, should a suitable labeled probe be available.

probe. These can then be individually picked and amplified in culture prior to DNA extraction and purification of the recombinant DNA.

A similar process is possible when using phage vectors. The plaques which are formed following lysis of bacterial cells by phage will contain residual phage particles. A nitrocellulose or nylon membrane is placed on top of the agar plate in the same way as above, and when removed from the plate will constitute a faithful copy of the phage material in the plaques, a so called **plaque-lift**. Subsequent processing of the filter is identical to the scheme in *Figure 5.15*).

In situ hybridization usually involves hybridizing a nucleic acid probe to the denatured DNA of a chromosome preparation or the RNA of a tissue section fixed on a glass slide

Chromosome in situ *hybridization*

A simple procedure for mapping genes and other DNA sequences is to hybridize a suitable labeled DNA probe against chromosomal DNA that has been denatured *in situ*. To do this, an air-dried microscope slide preparation of metaphase or prometaphase chromosomes is made, usually from peripheral blood lymphocytes or lymphoblastoid cell lines. Treatment with RNase and proteinase K results in partially purified chromosomal DNA, which is denatured by exposure to formamide. The denatured DNA is then available for *in situ* hybridization with an added solution containing a labeled nucleic acid probe, overlaid with a coverslip. Depending on the particular technique that is used, chromosome banding of the chromosomes can be arranged either before or after the hybridization step. As a result, the signal obtained after removal of excess probe can be correlated with the chromosome band pattern in order to identify a map location for the DNA sequences recognized by the probe (see also pages 280–286).

Tissue in situ *hybridization*

In this procedure, a labeled probe is hybridized against RNA in tissue sections (Wilkinson, 1992). Tissue sections are made from either paraffin-embedded or frozen tissue using a cryostat, and then mounted on to glass slides. A hybridization mix including a labeled nucleic acid probe is applied to the section on the slide and covered with a glass coverslip. Typically, the hybridization mix has formamide at a concentration of 50% in order to reduce the hybridization temperature and minimize evaporation problems. Although double-stranded cDNAs have been used as probes, single-stranded **complementary RNA (cRNA)** probes are preferred: the sensitivity of initially single-stranded probes is generally higher than that of double-stranded probes, presumably because a proportion of the denatured double-stranded probe renatures to form probe homoduplexes. cRNA riboprobes that are complementary to the mRNA of a gene can be obtained by cloning a gene in the 'reverse orientation' in a suitable vector such as pSP64 (see *Figure 5.4*). In such cases, the phage polymerase will synthesize labeled transcripts from the opposite DNA strand to that which is normally transcribed *in vivo*.

Following hybridization and washing to remove excess labeled probe, the signal is visualized using autoradiographic procedures. The localization of the silver grains is often visualized using only **dark-field microscopy** (direct light is not allowed to reach the objective; only scattered light). However **bright-field microscopy** provides the best detection of the intracellular localization of an mRNA (see *Figure 5.16*).

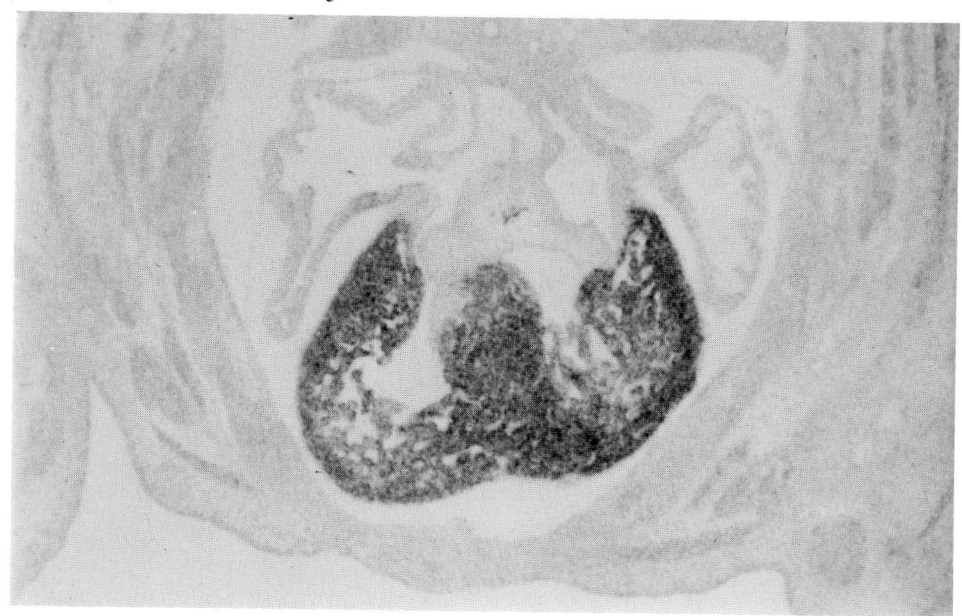

Figure 5.16: Tissue *in situ* hybridization.

The example shows the pattern of hybridization produced using a ^{35}S-labeled β-myosin heavy chain antisense riboprobe against a transverse section of tissue from a 13-day embryonic mouse. The dark areas represent strong labeling, notably in the ventricles of the heart. Kindly supplied by Dr David Wilson, University of Newcastle upon Tyne.

Further reading

Berger SL, Kimmel AR. (1987) *Guide to Molecular Cloning Techniques, Methods in Enzymology,* Vol. 152. Academic Press, San Diego.

Hames BD, Higgins SJ. (1985) *Genes Probes: A Practical Approach.* IRL Press, Oxford.

Sambrook J, Fritsch EF, Maniatis T. (1989) *Molecular Cloning: a Laboratory Manual,* 2nd Edn. Cold Spring Harbor Laboratory Press, Cold Spring Harbor, NY.

Old RW, Primrose SB. (1994) *Principles of Gene Manipulation: an Introduction to Genetic Engineering,* 5th Edn. Blackwell Scientific Publications, Oxford.

References

Collier S, Sinnott PJ, Dyer PA, Price DA, Harris R, Strachan T. (1989) *EMBO J.*, **8**, 1393–1402.

Feinberg AP, Vogelstein B. (1983) *Anal. Biochem.*, **132**, 6–13.

Kricka LJ. (1992) *Nonisotopic DNA Probing Techniques.* Academic Press, San Diego, CA.

Melton DA, Krieg PA, Rebagliati MR, Maniatis T, Zinn K, Green MR. (1984) *Nucleic Acids Res.*, **12**, 7035–7056.

Wilkinson, D. (1992) In Situ *Hybridization: a Practical Approach.* IRL Press, Oxford.

PCR-based DNA cloning and DNA analyses

Chapter 6

Principles of PCR

PCR is a cell-free method of DNA cloning

The standard PCR reaction

PCR is a rapid and versatile *in vitro* method for amplifying defined target DNA sequences present within a source of DNA. Usually, the method is designed to permit *selective* amplification of a specific target DNA sequence or sequences within a heterogeneous collection of DNA sequences (e.g. total genomic DNA or a complex cDNA population). To permit such selective amplification, some prior DNA sequence information from the target sequences is required, enabling the construction of two oligonucleotide primer sequences (often 15–30 nucleotides long). Such so-called **amplimers**, when added to denatured genomic DNA, will bind specifically to complementary DNA sequences immediately flanking the desired target region. They are designed so that, in the presence of a suitably heat-stable DNA polymerase and DNA precursors (the four deoxynucleoside triphosphates, dATP, dCTP, dGTP and dTTP), they can initiate the synthesis of new DNA strands which are complementary to the individual DNA strands of the target DNA segment, and which will overlap each other (*Figure 6.1*).

PCR is a chain reaction because newly synthesized DNA strands will act as templates for further DNA synthesis in subsequent cycles. After about 30 cycles of DNA synthesis, the products of the PCR will include, in addition to the starting DNA, about 10^5 copies of the specific target sequence, an amount which is easily visualized as a discrete band of a specific size when submitted to agarose gel electrophoresis. A heat-stable DNA polymerase is used because the reaction involves sequential cycles composed of three steps:

(i) Denaturation – typically at about 93–95°C for human genomic DNA;

(ii) Reannealing – at temperatures usually from about 50°C to 70°C, depending on the T_m (see page 115) of the expected duplex (the annealing temperature is typically about 5°C below the calculated T_m);

(iii) DNA synthesis – typically at about 70–75°C.

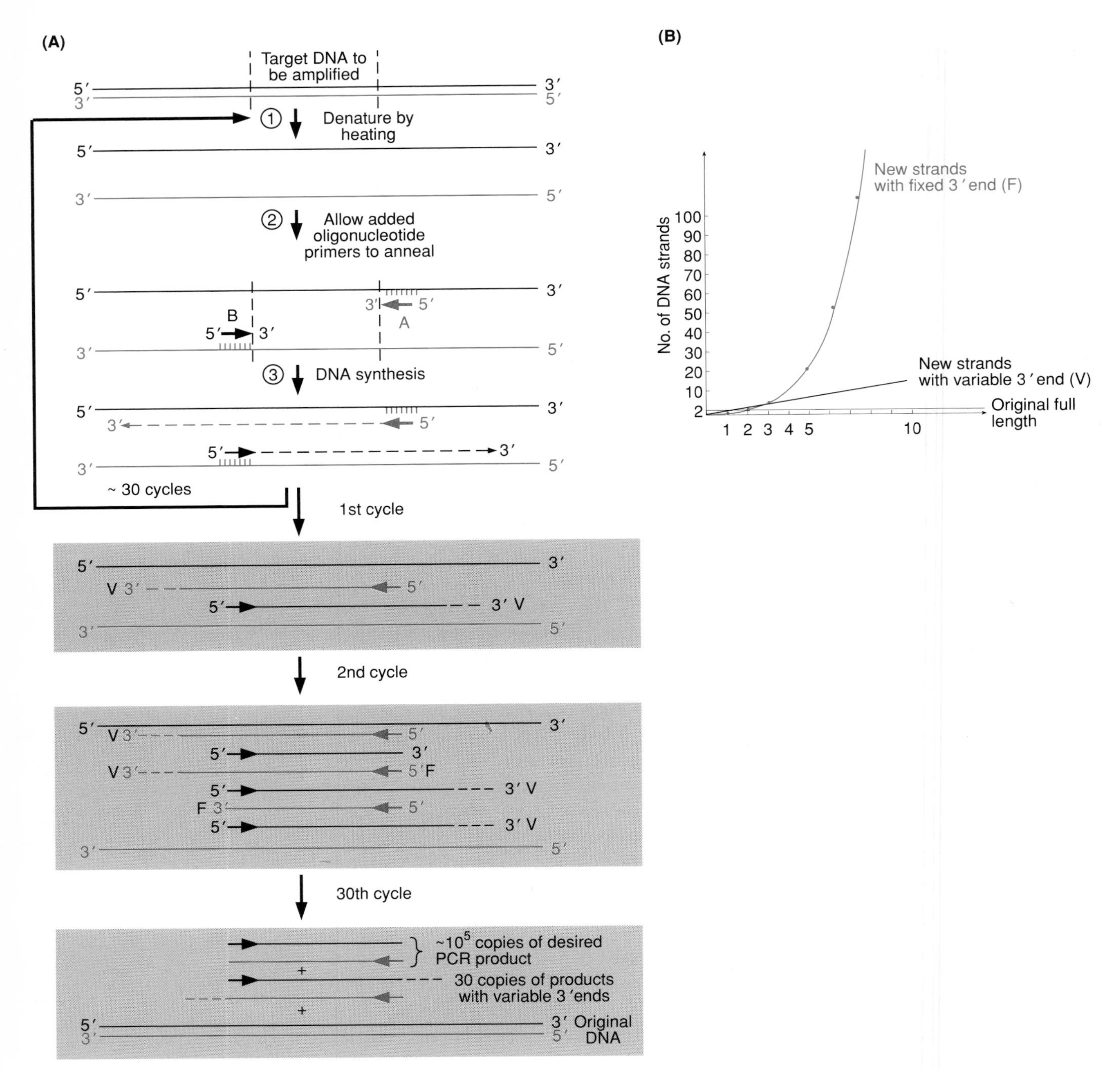

Figure 6.1: PCR is an *in vitro* method for amplifying DNA sequences using defined oligonucleotide primers.

Oligonucleotide primers A and B are complementary to DNA sequences located on opposite DNA strands and flanking the region to be amplified. Annealed primers are incorporated into the newly synthesized DNA strands. The first cycle will result in two new DNA strands whose 5′ end is fixed by the position of the oligonucleotide primer but whose 3′ end is variable ('ragged' 3′ ends). The two new strands can serve in turn as templates for synthesis of complementary strands of the desired length (the 5′ ends are defined by the primer and the 3′ ends are fixed because synthesis cannot proceed past the terminus of the opposing primer). After a few cycles, the desired fixed length product begins to predominate.

Suitably heat-stable DNA polymerases have been obtained from microorganisms whose natural habitat is hot springs. For example, the widely used *Taq* polymerase is obtained from *Thermus aquaticus* and is thermostable up to 94°C, with an optimum working temperature of 80°C.

Specificity of amplification and primer design

The specificity of amplification necessarily depends on the extent to which the designed primers can recognize different target DNA sequences other than the sequence which was intended to be amplified. For complex DNA sources, such as total genomic DNA from a mammalian cell, it is often sufficient to design two primers that are about 20 nucleotides long. This is because the chance of an accidental perfect match elsewhere in the genome for either one of the primers is extremely low, and for both sequences to occur by chance in close proximity in the specified direction is normally exceedingly low. Although conditions are usually chosen to ensure that only strongly matched primer–target duplexes are stable, spurious amplification products can nevertheless be observed. This can happen if one or both chosen primer sequences contain part of a repetitive DNA sequence, and primers are usually designed to avoid matching to known repetitive DNA sequences, including large runs of a single nucleotide (*Figure 6.2*).

Accidental matching at the 3′ end of the primer is critically important: spurious products may derive from substantially mismatched primer–target duplexes unless the 3′ end of the primer shows perfect matching. One way of minimizing the amplification of spurious products involves the use of **nested primers**. The products of an initial amplification reaction are diluted and used as the target DNA source for a second reaction in which a different set of primers is used, corresponding to sequences located close (but internal) to those used in the first reaction.

The major advantages of PCR as a cloning method are its rapidity, sensitivity and robustness

Because of its simplicity, PCR is a popular technique with a wide range of applications which depend on essentially three major advantages of the method.

Length	Usually about 20 nt for target sequences in complex genomic DNA; can be much less if target DNA is less complex
Base composition	Substantial tandem repeats of one or more nucleotides to be avoided Overall %GC plus length to be chosen so that the T_m of each oligonucleotide (*Table 5.1*) should be equal or nearly identical
3 ′end	Base complementarity of the two bases at the extreme 3′ end of the two primers to be avoided. Otherwise primer dimers can result, reducing amplification efficiency

Figure 6.2: PCR primer design.

Speed and ease of use of PCR

DNA cloning by PCR can be performed in a few hours, using relatively unsophisticated equipment. Typically, a PCR reaction consists of 30 cycles containing a denaturation, synthesis and reannealing step, with an individual cycle typically taking 3–5 min in an automated thermal cycler. This compares favorably with the time required for cell-based DNA cloning, which may often be weeks or even months. Clearly, some time is also required for designing and synthesizing oligonucleotide primers, but this has been simplified by the availability of computer software for primer design and rapid commercial synthesis of custom oligonucleotides. Once the conditions for a reaction have been tested, the reaction can then be repeated simply.

Sensitivity of the PCR reaction

PCR is capable of amplifying sequences from minute amounts of target DNA, even the DNA from a single cell. Such exquisite sensitivity has afforded new methods of studying molecular pathogenesis and has found numerous applications in forensic science, in diagnosis (page 429), in genetic linkage analysis using single-sperm typing (page 328) and in molecular paleontology studies, where samples may contain minute numbers of cells. However, the extreme sensitivity of the method means that great care has to be taken to avoid contamination of the sample under investigation by external DNA, such as from minute amounts of cells from the operator.

Robustness of PCR

PCR can permit amplification of specific sequences from material in which the DNA is badly degraded or embedded in a medium from which conventional DNA isolation is problematic. As a result, it is again very suitable for molecular anthropology and paleontology studies, for example the analysis of DNA recovered from mummified individuals and the quest for identifying DNA from fossil samples containing minute amounts of cells from long extinct creatures. It has also been used successfully to amplify DNA from formalin-fixed tissue samples, which has important applications in molecular pathology and, in some cases, genetic linkage studies.

The major disadvantages of PCR are the requirement for prior target sequence information, short product sizes and infidelity of DNA replication

Despite its huge popularity, PCR has certain limitations as a method for selectively cloning specific DNA sequences.

Need for target DNA sequence information

In order to construct specific oligonucleotide primers that permit *selective amplification* of a particular DNA sequence, some prior sequence information is necessary. This normally means that the DNA region of interest has been partly characterized previously, often following cell-based DNA cloning. However, a variety of techniques have been developed that reduce or even exclude the need for prior DNA sequence information concerning the target DNA, when certain aims are to be met. For example, previously uncharacterized DNA sequences can sometimes be cloned using PCR with degenerate oligonucleotides if they are members of a gene or repetitive DNA family at least one of whose members has previously been characterized (page 509). In some cases, PCR can be used effectively without any prior sequence information concerning the target DNA to permit *indiscriminate amplification* of DNA sequences from a source of DNA that is present in extemely limited quantities (page 139). Therefore, although PCR can be applied to ensure whole genome amplification, it does not have the advantage of cell-based DNA cloning in offering a way of separating the individual DNA clones comprising a genomic DNA library.

Short PCR product size

A clear disadvantage of PCR as a DNA cloning method has been the size range of the DNA sequences that can be cloned. Unlike cell-based DNA cloning where the size of cloned DNA sequences can approach 2 Mb (page 103), reported DNA sequences cloned by PCR have typically been in the 0–5 kb size range, often at the lower end of this scale. Although small segments of DNA can usually be amplified easily by PCR, it becomes increasingly more difficult to obtain efficient amplification as the desired product length increases. Recently, however, conditions have been identified for effective amplification of longer targets, including a 42-kb product from the bacteriophage λ genome (see Cheng *et al.*, 1994). Often, the conditions for **long range PCR** involve a combination of modifications to standard conditions with a two-polymerase system to provide optimal levels of DNA polymerase and $3' \rightarrow 5'$ exonuclease activity which serves as a proofreading mechanism (see *Box 6.1*).

Box 6.1: Proofreading by DNA polymerase-associated $3' \rightarrow 5'$ exonuclease activity

The fidelity of DNA replication *in vivo* is extremely high: during replication of mammalian genomes, for example, only one base in about 3×10^9 is copied incorrectly. Misincorporation occurs at a low frequency, dependent on the relative free energies of correctly and incorrectly paired bases. Very minor changes in helix geometry can stabilize G–T base pairs (with two hydrogen bonds; note the frequent occurrence of G–U base pairing in RNA, see page 7). *In vivo* copying normally shows an error rate much lower than these thermodynamic limitations would imply. This is achieved by proofreading mechanisms, one of which is a common property of DNA polymerases. Unlike RNA polymerases, DNA polymerases *absolutely* require the 3′ hydroxyl end of a base-paired primer strand as a substrate for chain extension. Additionally, DNA molecules with a mismatched nucleotide at the 3′ end of the primer strand are not effective templates for DNA synthesis. Many DNA polymerases, including that of *E. coli*, contain an integral $3' \rightarrow 5'$ exonuclease activity. When an incorrect base is inserted during DNA synthesis, DNA synthesis does not proceed. Instead, the $3' \rightarrow 5'$ exonuclease activity removes one nucleotide at a time from the 3′ hydroxyl terminus until a correctly base-paired terminus is obtained, enabling DNA synthesis to proceed again. As a result, DNA polymerases are usually self-correcting, removing errors made by the DNA polymerase activity during DNA synthesis.

Infidelity of DNA replication

Cell-based DNA cloning involves DNA replication *in vivo*, which is associated with a very high fidelity of copying because of proofreading mechanisms (see *Box 6.1*). However, when DNA is replicated *in vitro* the copying error rate is considerably greater. Of the heat-stable DNA polymerases required for PCR, the most widely used is *Taq* polymerase derived from *T. aquaticus*. This DNA polymerase, however, has no associated $3' \rightarrow 5'$ exonuclease to confer a proofreading function, and the error rate due to base misincorporation during DNA replication is rather high: for a 1-kb sequence that has undergone 20 effective cycles of duplication, approximately 40% of the new DNA strands synthesized by PCR using this enzyme will contain an incorrect nucleotide resulting from a copying error. This means that, even if the PCR reaction involves amplification of a single DNA sequence, the final product will be a mixture of extremely similar, but not identical DNA sequences. However, DNA sequencing of the total PCR product may nevertheless give the correct sequence. This is because, although individual DNA strands in the PCR product often contain incorrect bases, the incorporation of incorrect bases is essentially random. As a result, *for each base position*, the contribution of one incorrect base on one or more strands is overwhelmed by the contributions from the huge majority of strands which will have the correct sequence. What it does mean, however, is that further analysis of the product may be difficult. For example, functional studies frequently require that the DNA be cloned into a suitable expression vector. However, subsequent cell-based cloning of PCR products runs into the problem that transformation selects for a single molecule, and the cell clones chosen to be amplified will contain identical molecules, each the same as a single starting molecule which may well have the incorrect DNA sequence because of a copying error during PCR amplification. As a result, several individual clones may need to be sequenced in order to determine the correct (consensus) sequence, before selecting one with the authentic sequence for subsequent experiments.

Recently, the problem of infidelity of DNA replication during the PCR reaction has been considerably reduced by using alternative heat-stable DNA polymerases which have associated 3′ → 5′ exonuclease activity. For example, the *Pyrococcus furiosus* DNA polymerase is becoming more widely used because of the proof-reading conferred by its associated 3′ → 5′ exonuclease activity. The resulting PCR product has a much lower level of mutations introduced by copying errors: for a 1-kb segment of DNA that has undergone 20 effective cycles of duplication, about 3.5% of the DNA strands in the product carry an altered base.

Applications of PCR

Although successful PCR was first developed only 10 years ago, the simplicity and the versatility of the technique have ensured that it is among the most ubiquitous of molecular genetic methodologies, with a wide range of general applications (*Figure 6.3*).

PCR is frequently used to assay for polymorphisms and pathogenic mutations

Assay of restriction site and tandem repeat polymorphisms

RSPs are polymorphisms resulting in alleles possessing or lacking a specific restriction site. Such polymorphisms can be typed using Southern blot hybridization by hybridizing a DNA probe representing the locus against genomic DNA samples that have been digested with the appropriate restriction enzyme and size-fractionated by agarose gel electrophoresis. The resulting RFLPs have two alleles corresponding to the presence or absence of the restriction site (see page 122). As a convenient alternative to RFLPs, PCR can type RSPs by simply designing primers using sequences which flank the polymorphic restriction site, amplifying from genomic DNA, then cutting the PCR product with the appropriate restriction enzyme and separating the fragments by agarose gel electrophoresis (*Figure 6.4*).

Short tandem repeat polymorphisms (**STRPs**), also called microsatellite markers, consist of a short sequence, typically from one to four nucleotides long, that is tandemly repeated several times, and often characterized by many alleles. For example, $(CA)_n/(TG)_n$ repeats are often polymorphic when n exceeds 12, and have been widely used as polymorphic markers in the human genome (see below). Increasingly, however, trinucleotide and tetranucleotide marker polymorphisms are being typed. In each case the STRPs can be typed conveniently by PCR. Primers are designed from sequences known to flank a specific STRP locus, permitting PCR amplification of alleles whose sizes differ by integral repeat units (*Figure 6.5*). The PCR products can then be size-fractionated by polyacrylamide gel electrophoresis. The PCR normally includes a radioactive or fluorescent nucleotide precursor which becomes incorporated into the small PCR products and facilitates their detection. To ensure adequate size fractionation of alleles, the PCR products are denatured prior to electrophoresis. An example of the use of a CA repeat marker is shown in *Figure 6.6*.

Typing genetic markers	RFLPs (see *Figure 6.4*); STRPs (see *Figures 6.5* and *6.6*)
DNA templates for mutation screening	Genomic mutation screening and RT-PCR (see *Figure 6.7*)
Detecting point mutations	Mutations changing restriction site–same principle as *Figure 6.4*. Other mutations by allele-specific amplification (ARMS; *Figure 6.8*)
cDNA cloning	From amino acid sequence by DOP-PCR (*Figure 6.9*); cloning of the ends of cDNA by RACE (*Figure 19.4*)
Genomic DNA cloning	Cloning of new members of a DNA family by DOP-PCR (see page 510). Whole genome amplification or subgenomic amplification (e.g. microdissected chromosome bands) by DOP-PCR or linker-primed PCR (*Figure 6.10*)
Genome walking	Inverse PCR (see *Figure 6.11*); bubble linker (vectorette) PCR (see *Figure 6.12* and page 141), IRE-PCR (see *Figure 6.13* and page 142)
DNA templates for DNA sequencing	Single-stranded DNA by asymmetric PCR (page 143). Double-stranded DNA for direct sequencing or for conventional cloning then sequencing (pages 142–144)
In vitro mutagenesis	Using 5′ add-on mutagenesis to create a recombinant PCR product (see *Figure 6.14A*). Using mispaired primers to change a single predetermined nucleotide (see *Figure 6.14B*)

Figure 6.3: PCR has numerous general applications.

The figure illustrates general applications. Specific applications are described in separate chapters. Genome walking means accessing uncharacterized DNA starting from a neighboring characterized sequence.

Mutation screening

Because of its rapidity and simplicity, PCR is ideally suited to providing numerous DNA templates for mutation screening. Partial DNA sequences, at the genomic or the cDNA level, from a gene associated with disease, or some other interesting phenotype, immediately enable gene-specific PCR reactions to be designed. Amplification of the appropriate gene segment then enables rapid testing for the presence of associated mutations in large numbers of individuals. By contrast, cell-based DNA cloning of the gene from numerous different individuals is far too slow and labor-intensive to be considered as a serious alternative.

Typically, the identification of exon/intron boundaries and sequencing of the ends of introns of a gene of interest offers the possibility of *genomic mutation screening*: individual exon-specific amplification reactions are developed by designing primers which recognize intronic sequences located close to the exon/intron boundary (see *Figure 6.7A*). The resulting PCR products are then analyzed by rapid mutation-screening methods, for several of which the optimal size for mutation screening is about 200 bp (see page 395). Conveniently, the average size of a human exon is about 180 bp but, in the case of very large exons, it is usual to design a series of primers to generate overlapping exonic products.

Figure 6.4: Restriction site polymorphisms can easily be typed by PCR as an alternative to laborious RFLP assays.

Alleles 1 and 2 are distinguished by a polymorphism which alters the nucleotide sequence of a specific restriction site for restriction nuclease R: allele 1 possesses the site, but allele 2 has an altered nucleotide(s) X, X′ and so lacks it. PCR primers can be designed simply from sequences flanking the restriction site to produce a short product. Digestion of the PCR product with enzyme R and size-fractionation can result in simple typing for the two alleles.

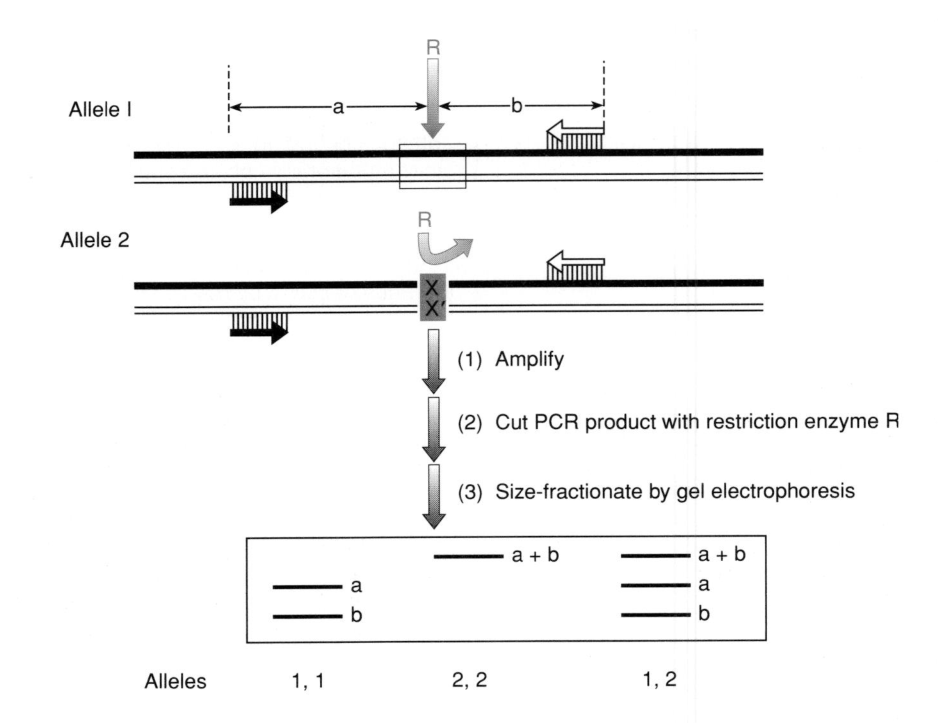

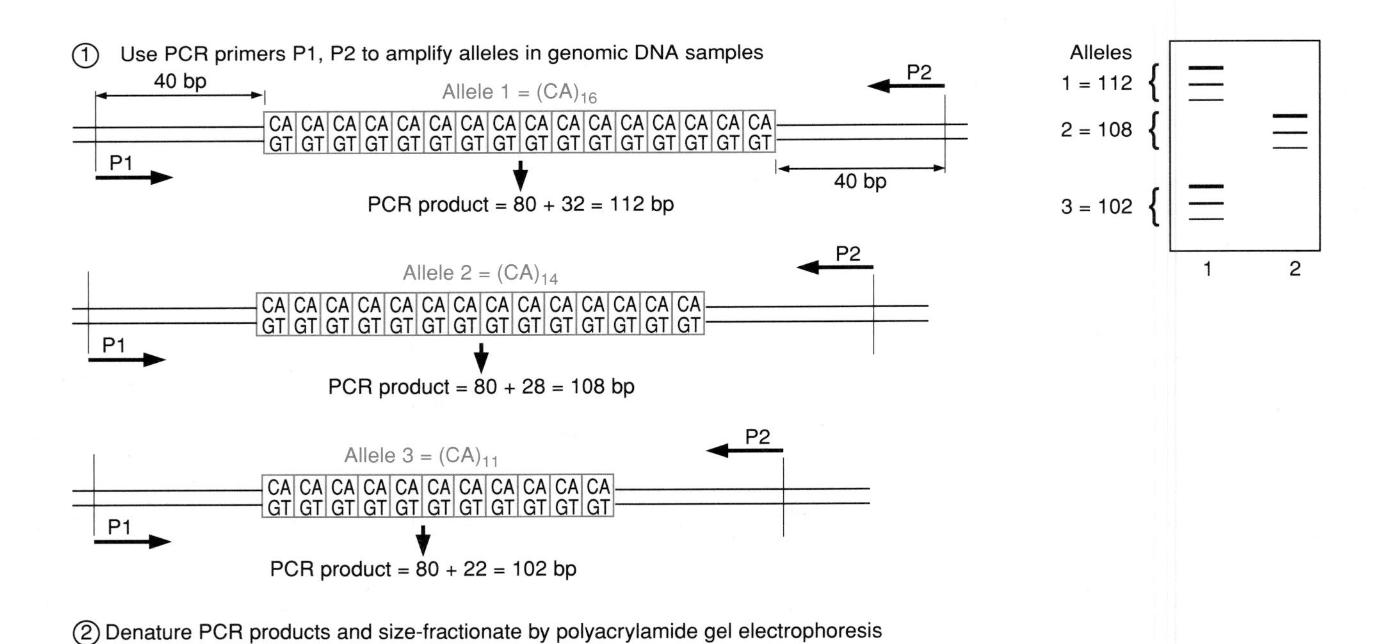

② Denature PCR products and size-fractionate by polyacrylamide gel electrophoresis

③ Autoradiography

Figure 6.5: PCR can be used to type short tandem repeat polymorphisms (STRPs).

The example illustrates typing of a (CA)/(TG) dinucleotide repeat polymorphism which has three alleles as a result of variation in the number of the (CA) repeats. On the autoradiograph each allele is represented by a major upper band and two minor 'shadow bands' (see *Figure 6.6*). Individuals 1 and 2 have genotypes (in brackets) as follows: 1 (1,3); 2 (2,2).

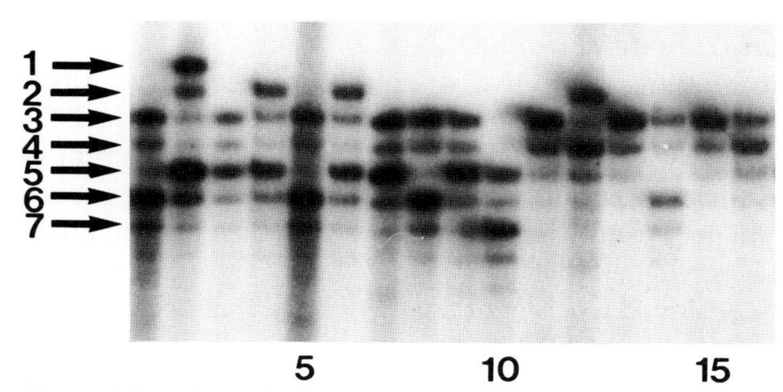

Figure 6.6: Example of typing for a CA repeat.

The example illustrated shows typing of members of a large family with the (CA)/(TG) marker *D17S800*. Arrows to the left mark the top (main) band seen in different alleles 1–7. *Note* that individual alleles show a strong upper band followed by two lower 'shadow bands', one of intermediate intensity immediately underneath the strong upper band, and one that is very faint and is located immediately below the first shadow band. For the indicated individuals, the genotypes (in brackets) are as follows: 1 (3,6); 2 (1,5); 3 (3,5); 4 (2,5); 5 (3,6); 6 (2,5); 7 (3,5); 8 (3,6); 9 (3,5); 10 (5,7); 11 (3,3); 12 (2,4); 13 (3,3); 14 (3,6); 15 (3,3); 16 (3,4). *Note* that in the latter case, the middle band is particularly intense because it contains both the main band for allele 4 plus the major shadow band for allele 3. Slipped strand mispairing (see page 254) is thought to be the major mechanism responsible for producing shadow bands at tandem dinucleotide repeats (Hauge and Litt, 1993).

PCR can also quickly provide amplified cDNA sequences for mutation screening. Such *cDNA mutation screening* may be the only way in which mutations can be screened if the exon/intron organization of a gene has not been established. To do this, mRNA is isolated from a convenient source of tissue, such as blood cells, converted into cDNA using reverse transcriptase (see page 94) and the cDNA is used as a template for a PCR reaction. This version of the standard genomic PCR reaction is consequently often referred to as **RT-PCR** (**reverse transcriptase-PCR**; see *Figure 6.7B*). Clearly, the method is ideally suited to genes that are expressed at high levels in easily accessible cells, such as blood cells. However, as a result of low level *ectopic transcription* of genes in all tissues, it has also been applied to transcript analysis of genes which are not significantly expressed in blood cells, such as the dystrophin (*DMD*) gene (Chelly *et al.*, 1989).

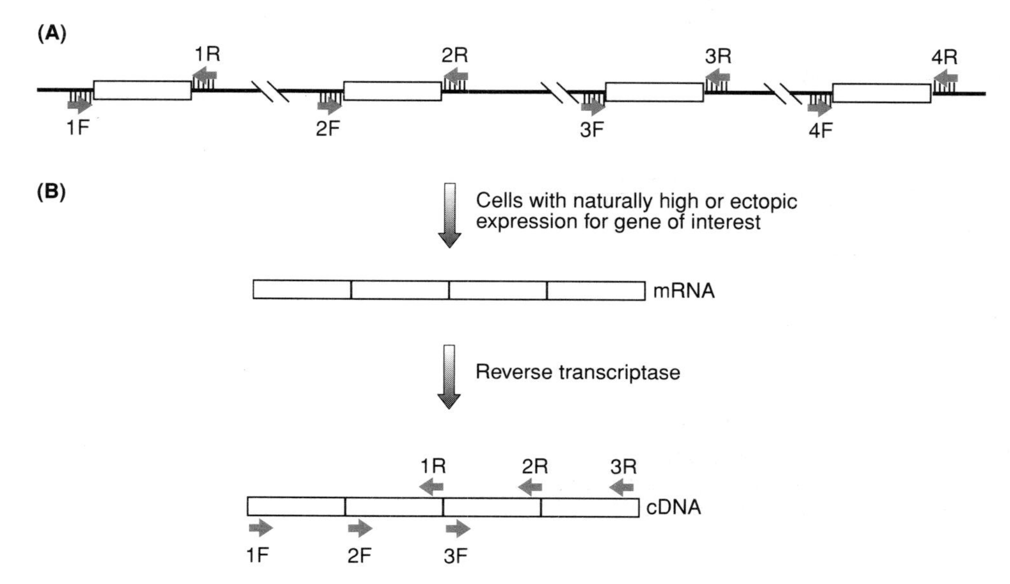

Figure 6.7: PCR products for gene mutation screening are obtained from genomic DNA using intron-specific primers flanking exons or by RT-PCR.

(A) Genomic DNA. Exons 1–4 can be amplified separately from genomic DNA using pairs of intron-specific primers 1F + 1R, 2F + 2R, etc.

(B) RT-PCR. This relies on at least some mRNA being present in easily accessible cells such as blood cells, permitting conversion to cDNA. The cDNA can then be used as a template for pairs of exon-specific primers (1F+1R, 2F+2R, etc.) to generate overlapping DNA fragments.

Allele-specific PCR and mutation detection

Oligonucleotide primers can be designed so as to discriminate between target DNA sequences that differ by a single nucleotide in the region of interest. This is the PCR equivalent of the allele-specific hybridization which is possible with ASO probes (see page 109). In the case of allele-specific hybridization, alternative ASO probes are designed to have differences in a central segment of the sequence (to maximize thermodynamic instability of mismatched duplexes). However, in the case of allele-specific PCR, ASO primers are designed to differ at the nucleotide that occurs at the extreme 3′ terminus. This is so because the DNA synthesis step in a PCR reaction is crucially dependent on correct base pairing at the 3′ end (see *Figure 6.8*). This method can be used to type specific alleles at a polymorphic locus, but has found particular use as a method for detecting a specific pathogenic mutation, the so-called **amplification refractory mutation system** (**ARMS**; Newton *et al.*, 1989).

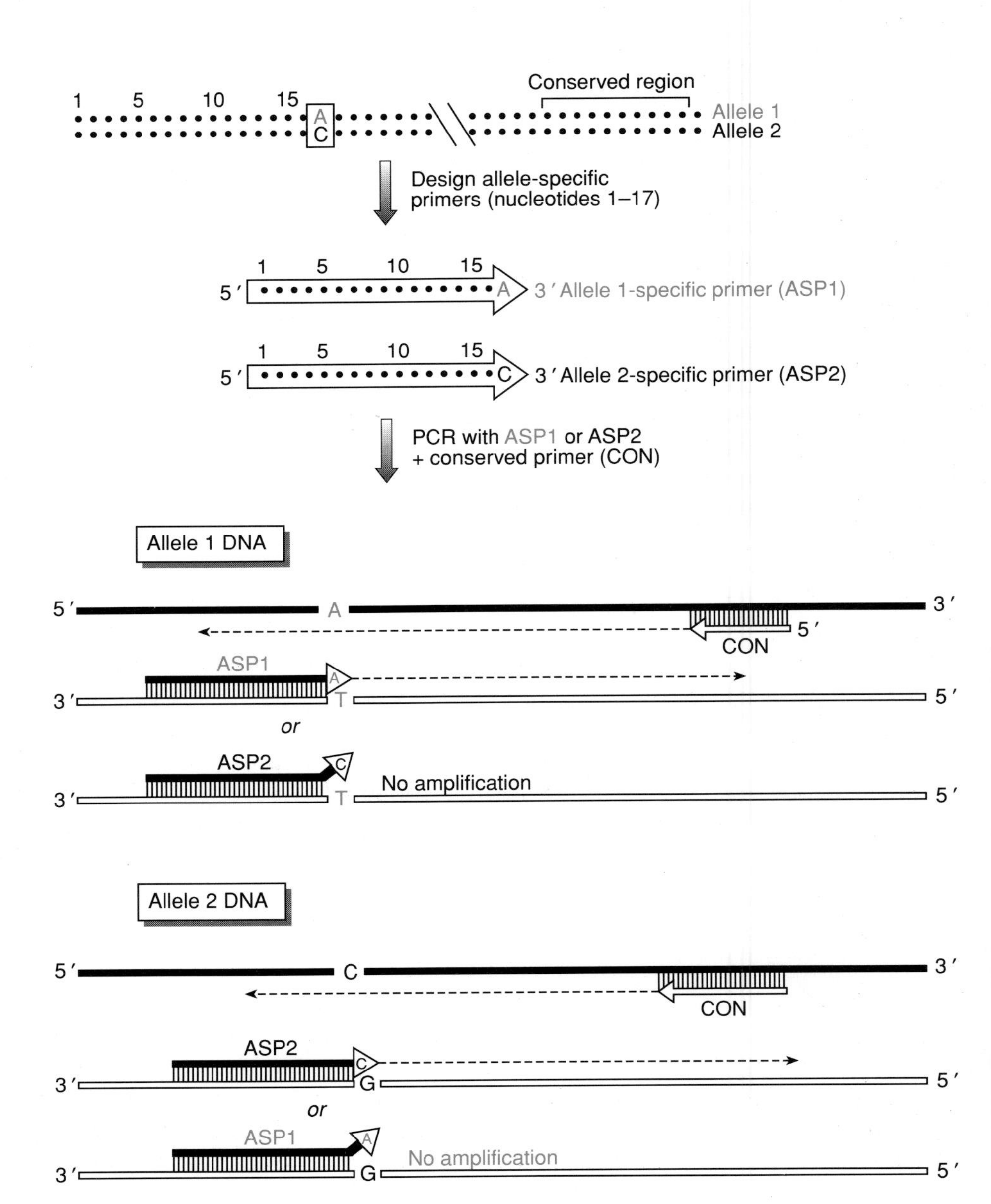

Figure 6.8: Correct base pairing at the 3′ end of PCR primers is the basis of allele-specific PCR.

The allele-specific oligonucleotide primers ASP1 and ASP2 are designed to be identical to the sequence of the two alleles over a region preceding the position of the variant nucleotide, up to and terminating in the variant nucleotide itself. ASP1 will bind perfectly to the complementary strand of the allele 1 sequence, permitting amplification with the conserved primer. However, the 3′-terminal C of the ASP2 primer mismatches with the T of the allele 1 sequence, making amplification impossible. Similarly ASP2 can bind perfectly to allele 2 and initiate amplification, unlike ASP1.

PCR with degenerate oligonucleotide primers can permit rapid cDNA cloning and cloning of uncharacterized members of DNA families

DOP-PCR (degenerate oligonucleotide-primed PCR) is a form of PCR which is deliberately designed to permit possible amplification of several products. Typically, the two primers are partially degenerate oligonucleotides, comprised of panels of oligonucleotide sequences that have the same base at certain nucleotide positions, but are different at others. As a result, there may be many primer binding sites in the source DNA. This provides a means of searching for a new or uncharacterized DNA sequence that belongs to a family of related sequences either within or between species. Additionally, it provides a way of cloning a gene when only a limited portion of amino acid sequence is known for the product (*Figure 6.9*).

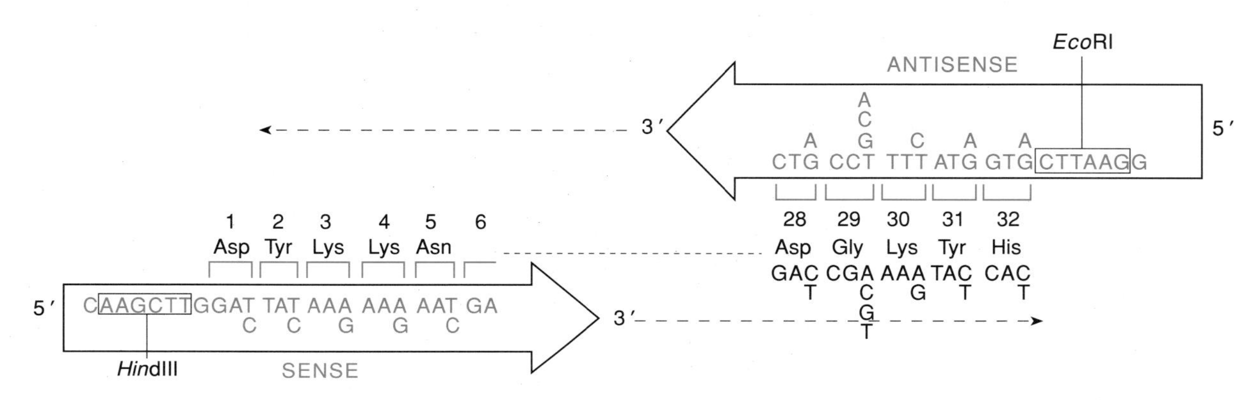

Figure 6.9: DOP-PCR can permit cDNA cloning using degenerate oligonucleotides.

The figure illustrates cloning of a cDNA for porcine urate oxidase using degenerate oligonucleotides corresponding to a known amino acid sequence. The sense primer was constructed to correspond to the first five codons plus the first two bases of codon 6, and the antisense primer corresponded to codons 28–32 (Lee *et al.*, 1988). The amino acid sequences chosen for constructing primers were selected on the basis of their high content of amino acids which were specified by only two codons (Asp, Tyr, Lys, Asn, His, see *Figure 1.22*). The primers have 5′ extensions containing recognition sequences for restriction nucleases, in order to facilitate subsequent cell-based cloning (see page 143).

PCR can be used to permit indiscriminate amplification of DNA sequences from a tiny amount of source DNA

As mentioned above, a major limitation of the standard PCR reaction is that it requires prior sequence information concerning the target DNA to be amplified. However, certain PCR applications involve amplification of target sequences without any prior sequence information concerning the target DNA. Instead of the standard reaction which involves *selective amplification* of specific target DNA sequences (*Figure 6.1*), comparatively *indiscriminate* amplification of target DNA is also possible. This is particularly advantageous when a source DNA is originally present in tiny amounts (e.g. extracts from ancient DNA samples, microdissected chromosome bands, single cell typing, etc.), and PCR amplification of essentially all sequences increases the amount of DNA for study.

Linker-primed PCR (ligation adaptor PCR)

One way of enabling amplification of essentially all DNA sequences in a complex DNA mixture involves first ligating a known sequence to all fragments. To do this, the target DNA population is digested with a suitable restriction nuclease, and double-stranded oligonucleotide *linkers* (*adaptors*) with a suitable *overhanging end* are ligated to the ends of target DNA fragments. Amplification is then conducted using oligonucleotide primers which are specific for the linker sequences. In this way, all fragments of the DNA source which are flanked by linker oligonucleotides can be amplified (*Figure 6.10*).

Whole genome amplification by DOP-PCR

An extension of the DOP-PCR method (see above) involving the use of completely random oligonucleotides, permits an alternative to linker-primed PCR as a means of obtaining indiscriminate amplification: completely degenerate oligonucleotide primers permit **whole genome amplification**, which has been useful for preimplantation diagnosis (page 429).

PCR can be used to clone uncharacterized regions of DNA flanking a previously characterized region

A major limitation of the standard PCR reaction is that it allows amplification of a DNA sequence flanked between two *convergent* primers, each of which prime DNA chain extension in the direction of the other primer. Often, however, it is desirable to be able to access uncharacterized DNA sequences flanking a region for which sequence information is available. To do this, a variety of methods have been developed.

Inverse PCR

This is a general method for amplifying DNA flanking a previously characterized region (Ochman *et al.*, 1988). It utilizes primers derived from the extremities of a characterized region which are *divergent*, that is the 5′ → 3′ direction of each primer

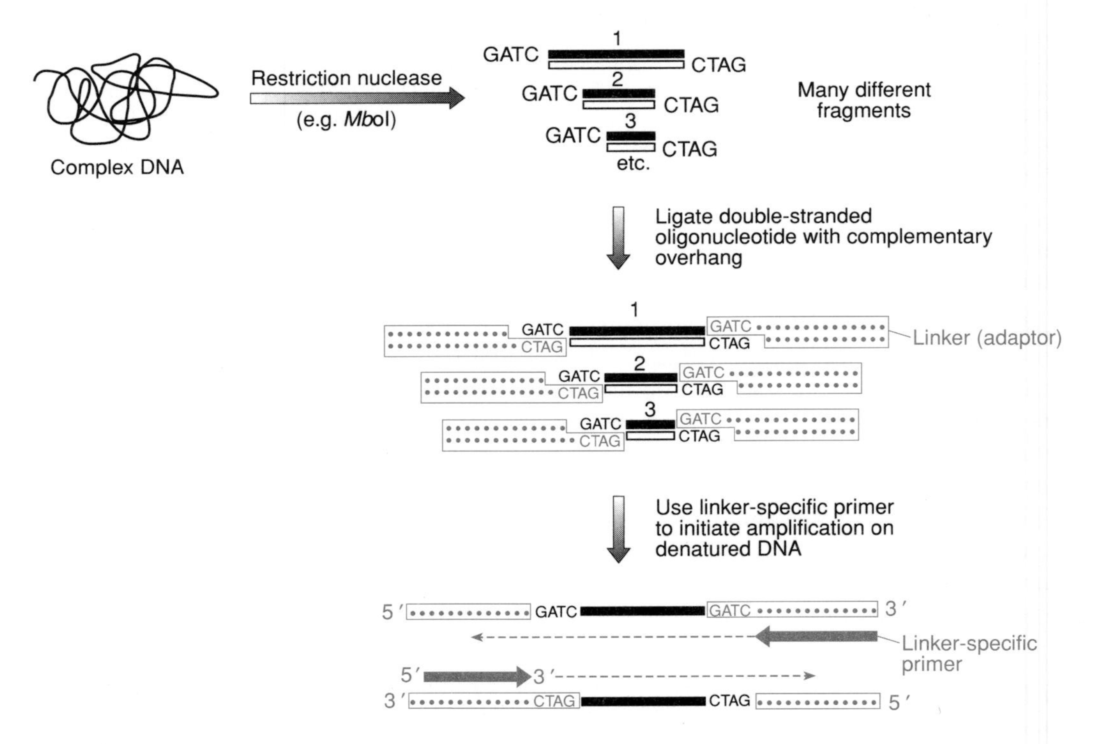

Figure 6.10: Linker-primed PCR permits indiscriminate amplification of DNA sequences in a complex target DNA.

The linker (adaptor) molecule is a double-stranded oligonucleotide formed by ligating two single-stranded oligonucleotides which are complementary in sequence except that one possesses a 5′ overhang compatible with a restriction nuclease overhang (in this case, the 5′ GATC overhang produced by *Mbo*I). After ligation of the linker to the target restriction fragments, a linker-specific primer can result in amplification of all fragments by binding to two flanking linker molecules.

points away from the other primer, instead of towards it, as is customary. Normally, PCR amplification would be impossible with such primers. However, inverse PCR involves first cutting the DNA so that the desired region is present on a small restriction fragment and then encouraging the DNA to circularize (by conducting the ligations at very low DNA concentrations, the two ends of a single molecule are more likely to come into contact than those on different molecules). Amplification of the uncharacterized flanking regions is then possible using standard conditions (see *Figure 6.11*).

Bubble linker PCR

Bubble linker PCR (**vectorette PCR**) is an alternative method to inverse PCR, enabling amplification of uncharacterized sequences flanking a region for which DNA sequence information is available (Riley *et al.*, 1990). The method is partially a variant of linker-primed PCR (see page 139) in that a double-stranded oligonucleotide linker is ligated to the target DNA. However, in bubble linker PCR, amplification occurs using a primer derived from the target sequence as well as one derived from the linker sequence. The double-stranded linker does not contain perfectly complementary oligonucleotides (as is normal, except for overhangs designed to help ligation to target DNA). Instead, the two oligonucleotides of the linker are designed to be complementary at the ends, but not in the middle; the lack of pairing in the middle segment of the linker has then been termed a 'bubble'. Because the two sequences in the region of the bubble are noncomplementary, it is possible to design a linker *strand-specific* primer which, with a convergent target DNA-specific primer, permits amplification of the uncharacterized sequence flanking one side of the characterized DNA (see *Figure 6.12*). Vectorette PCR has been particularly useful for amplifying exon/intron boundaries to determine gene structure (page 514) and for isolating the ends of YACs (page 291).

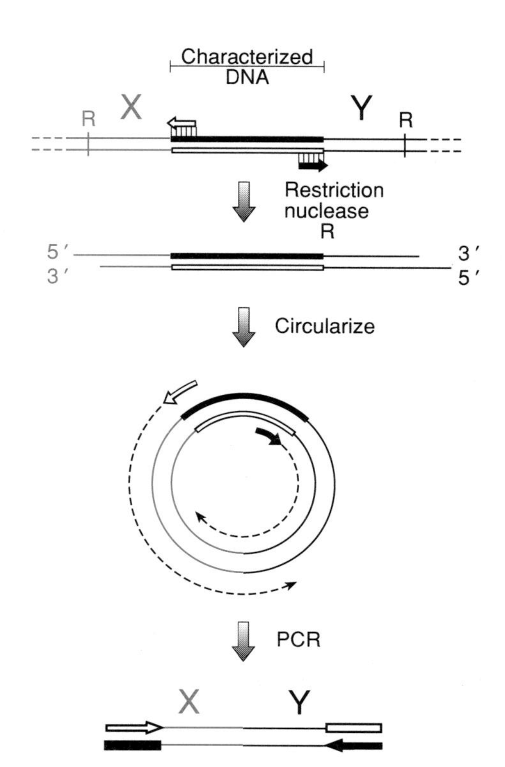

Figure 6.11: Inverse PCR permits cloning of flanking DNA sequences from a circularized template.

The two primer sequences are designed to bind to the characterized DNA sequence but with their 3′ ends facing away from each other, instead of towards each other. After the characterized sequence and its two flanking regions have been recovered on a short restriction fragment, the two flanking regions can be brought next to each other by forcing the fragment to circularize (by ligation at very low DNA concentrations). The PCR primers can now permit amplification of the flanking regions.

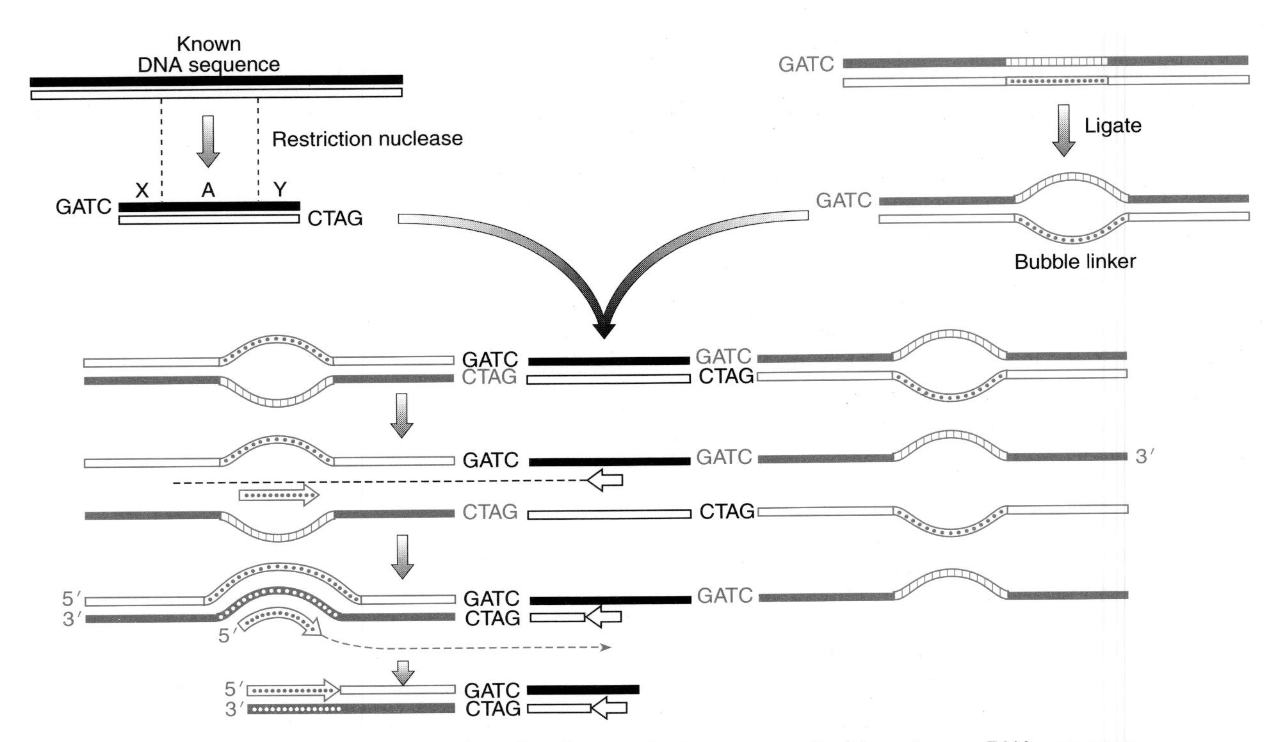

Figure 6.12: Bubble linker PCR permits amplification of uncharacterized sequences flanking a known DNA sequence.

Note that the bubble linkers ('vectorettes') contain complementary sequences at their ends but unrelated sequences within the bubble. A restriction fragment containing a known sequence A flanked by uncharacterized regions X and Y is ligated to a bubble linker containing a suitable overhang at one end. PCR amplification of the uncharacterized sequence X is then possible using a primer specific for the known DNA sequence (A) and a primer specific for one of the strands of the bubble in the bubble linker (B). The bubble linker primer B cannot prime DNA synthesis initially as there is no sequence to which it can bind: it is identical, not complementary in sequence to one strand of the bubble, and unrelated in sequence to the other. However, primer A initiates synthesis of a complementary DNA strand which will contain a sequence *complementary to one of the unique sequences within the bubble linker*. As a result, primer B can bind to this newly synthesized strand and initiate new strand synthesis to start a PCR reaction. The flanking sequence Y can similarly be isolated in another reaction using a suitable A-specific primer and a bubble-specific primer derived from the strand opposite to that used for making primer B.

IRE-PCR (interspersed repeat element PCR)

This method is intended to amplify uncharacterized sequence located between two members of a highly repetitive interspersed DNA family, examples of which have been particularly well characterized in mammalian DNA (see pages 197–202). If such a family has a high enough copy number so that examples of family members occur every few kilobases, then amplification is possible using a *single primer* whose sequence is determined by a consensus sequence of the repetitive DNA family. Such single primer PCR is clearly only possible in the case of pairs of closely spaced repeats in the opposite orientation, enabling binding of convergent primers (*Figure 6.13*). In a complex genome, IRE-PCR would be expected to give a very large number of products, and its use has been limited largely to subcloning of large cloned inserts, such as those from YACs. In the human genome, this method has been widely used in the case of the highly repetitive *Alu* repeat family (i.e. ***Alu*-PCR**; Nelson *et al.*, 1989, and see page 294).

PCR-amplified products are often used for DNA sequencing

Asymmetric PCR

The most widely used form of DNA sequencing requires the DNA to be in a single-stranded form before use (see page 304). Although this can be accomplished by DNA denaturation of PCR products prior to DNA sequencing, the quality of the

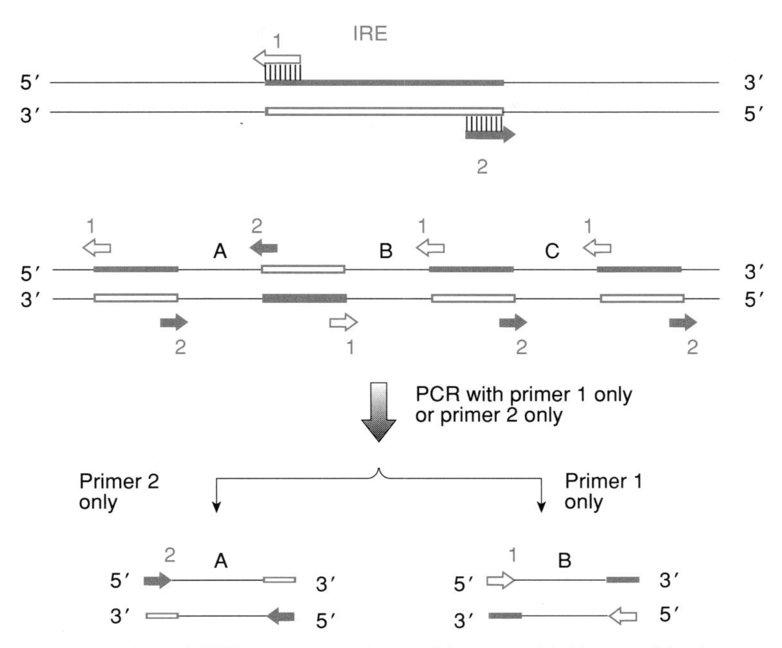

Figure 6.13: IRE-PCR results in amplification of DNA sequences located between highly repetitive interspersed repeat elements.

Primers for a specific class of interspersed repeat element (IRE) are designed to hybridize to sequences at either terminus of the repeat element, with the 5′ → 3′ direction pointing away from the repeat element. A single such primer can be used in a PCR reaction to amplify sequences located between two related repeat elements which are in opposite orientation, and arranged such that the two bound primers are convergent (e.g. PCR with primer 2 alone can amplify sequence A but not B; primer 1 alone can amplify sequence B but not A). A combination of primers 1 and 2 should, in theory, permit amplification of sequences between repeats in the same orientation. In practice, PCR amplification may be difficult or impossible if the distance between the repeats is too great (see page 133).

DNA sequence is greater if an originally single-stranded DNA template is prepared. **Asymmetric PCR** (or *single-stranded PCR*) has been designed as a way of amplifying DNA to generate ultimately a single-stranded DNA product. To do this, the method uses unequal (asymmetric) primer concentrations: typically the concentration of one primer is set to be 100 times that of the other primer. During the initial stages (up to the 15th to 25th cycles), most of the product generated is double-stranded and accumulates exponentially. Thereafter, as the low concentration primer becomes depleted, further cycles generate an excess of one of the two strands, and single-stranded product accumulates linearly.

Cell-based DNA cloning of PCR products

The quality of DNA sequences obtained from denatured double-stranded PCR products or from single-stranded PCR products obtained by asymmetric PCR is often less than optimal. As an alternative, PCR products are often cloned in conventional cell-based cloning systems. However, cloning of the PCR product into a unique restriction site in a vector molecule is not straightforward. This is because a proportion of the PCR products are not *blunt ended*. Some thermostable DNA polymerases, such as Taq polymerase possess a terminal deoxynucleotidyl transferase activity which adds an adenylate residue to the 3′ end of PCR products, notably to those containing a 3′-terminal cytosine.

Various vectors have been designed to take advantage of the 3′-terminal A. For example, one has been designed with a 3′ extension of a complementary T residue which can base-pair with the 3′ A overhang of the PCR product (Mead *et al.*, 1991). Another approach is to use a 5′ extension to the primers (see below) in order to introduce a suitable restriction site at the ends of the PCR product during amplification (see *Figure 6.14A* for the basic principle and *Figure 6.9* for an example).

PCR can be used to produce specific pre-determined mutations in DNA sequences

Because the major critical feature for successful specific amplification using oligonucleotide primers is correct base pairing at the 3′ end of the primers, considerable flexibility is available in primer design. As a result, PCR methods have been developed to permit oligonucleotide-directed ***in vitro*** **site-specific mutagenesis**, as an alternative to several other such methods (see pages 526–528). Two major approaches have been taken towards PCR mutagenesis:

(i) **mismatched primer mutagenesis** – the primers are deliberately synthesized to be similar but not identical to the target sequence.

(ii) **5′ add-on mutagenesis** – the primers are designed to have a novel protruding sequence at the 5′ end. The extra 5′ sequence does not participate in the first annealing step of the PCR reaction (only the 3′ part of the primer forms a duplex with the target DNA), but it subsequently becomes incorporated into the amplified product, thereby generating a recombinant product (*Figure 6.14*). The extra sequence is often designed to contain a convenient restriction site which may help in subsequent cell-based DNA cloning. Alternatively, it can contain some functional component such as a promoter sequence for adding to the 5′ end of a DNA fragment from which an expres-

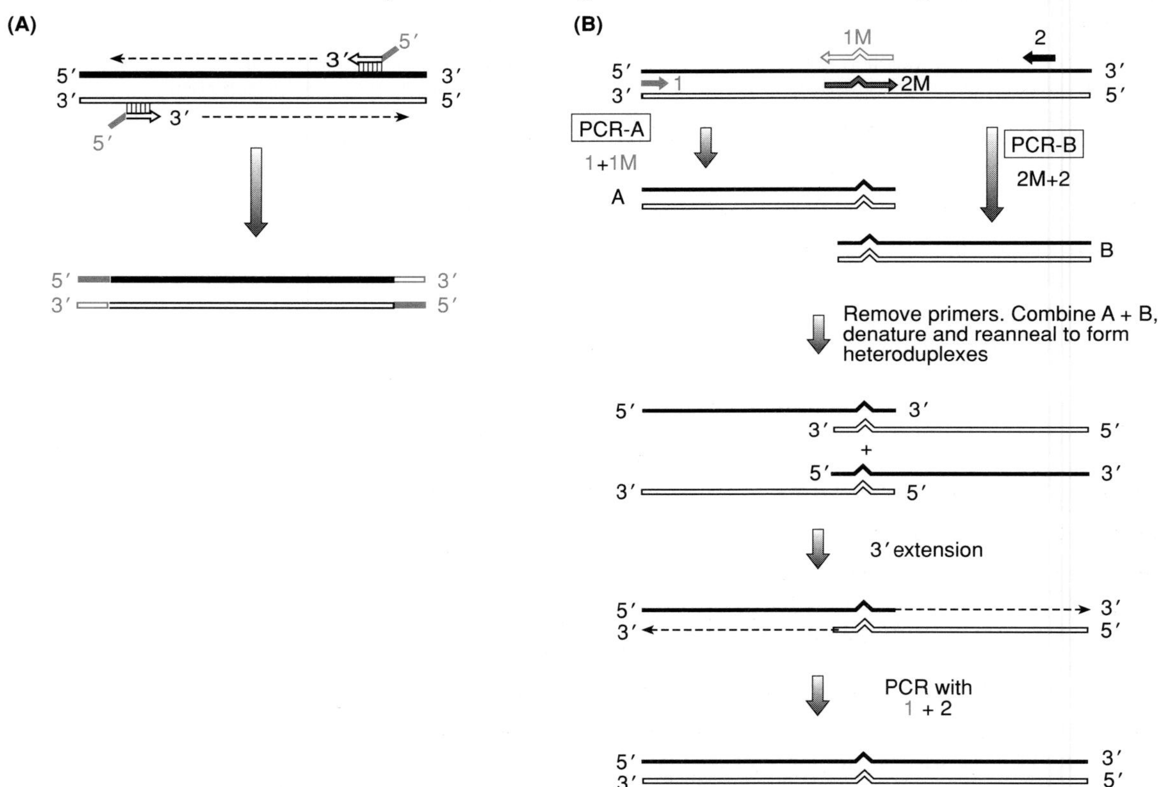

Figure 6.14: PCR mutagenesis.

(A) 5′ add-on mutagenesis. Primers can be modified at the 5′ end to introduce, for example, a labeled group (Figure 11.20), a sequence containing a suitable restriction site (*Figure 19.8*) or phage promoter (*Figure 14.16*).

(B) Site-specific mutagenesis. The mutagenesis shown can result in an amplified product with a specific pre-determined mutation located in a central segment.

PCR reactions A and B are envisaged as amplifying overlapping segments of DNA containing an introduced mutation (by deliberate base mismatching using a mutant primer – 1M or 2M). After the the two products are combined, denatured and allowed to reanneal, the DNA polymerase can extend the 3′ end of heteroduplexes with recessed 3′ ends. Thereafter, a full length product with the introduced mutation in a central segment can be amplified by using the outer primers 1 and 2 only.

sion product is required (e.g. for transcription by a phage RNA polymerase without the need for the normal cloning steps shown in *Figure 5.5;* see *Figure 14.16*). Another application is to ensure that a labeled group, such as a biotinylated group, becomes incorporated at the end of a PCR product (see *Figure 11.20* for an example).

Although PCR mutagenesis involves using altered primer sequences to introduce the mutation, the method is not limited to altering sequences at the extreme end of a chosen region. Instead, the mutation can be introduced at any point within a chosen sequence using mismatched primers. This is most easily accomplished by designing two mutagenic reactions in which the two separate PCR products have partially overlapping sequences containing the mutation, and then combining the denatured products to generate a larger product with the mutation in a more central location (Higuchi, 1990; see *Figure 6.14B*).

Further reading

Ehrlich HA. (1989) *PCR Technology. Principles and Applications for DNA Amplification.* Stockton Press, New York.

Ehrlich HA, Gelfand D, Sninsky JJ. (1991) Recent advances in the polymerase chain reaction. *Science*, **252**, 1643–1651.

Innis MA, Gelfand DH, Sninsky JJ, White TJ. (1990) *PCR Protocols. A Guide to Methods and Applications.* Academic Press, San Diego.

McPherson MJ, Taylor GR, Quirke P. (1991) *PCR: a Practical Approach.* IRL Press, Oxford.

Newton CR, Graham A. (1994) *PCR.* BIOS Scientific Publishers, Oxford

References

Chelly J, Concordet JP, Kaplan JC, Kahn A. (1989) *Proc. Natl Acad. Sci. USA*, **86**, 2617–2621.

Cheng S, Chang S-Y, Gravitt P, Respess R. (1994) *Nature*, **369**, 684–685.

Hauge Y, Litt M. (1993) *Hum. Mol. Genet.*, **2**, 411–415.

Higuchi R. (1990) In: *PCR Protocols. A Guide to Methods and Applications* (eds MA Innis, DH Gelfand, JJ Sninsky, TJ White), pp. 177–183. Academic Press, San Diego.

Lee CC, Wu X, Gibbs RA, Cook RG, Muzny DM, Caskey CT. (1988) *Science*, **239**, 1288–1291.

Mead DA, Pey NK, Herrnstadt C, Marcil RA, Smith LM. (1991) *Biotechnology*, **9**, 657–663.

Nelson DL, Ledbetter SA, Corbo L, Victoria MF, Ramirez-Solis R, Webster TD, Ledbetter DH, Caskey CT. (1989) *Proc. Natl Acad. Sci. USA*, **86**, 6686–6690.

Newton CR, Graham A, Heptinstall LE, Powell SJ, Summers C, Kalsheker N, Smith JC, Markham AF. (1989) *Nucleic Acids Res.*, **17**, 2503–2516.

Ochman H, Gerber AS, Hartl DL. (1988) *Genetics*, **120**, 621–623.

Riley J, Butler R, Ogilvie D, Finniear R, Jenner D, Powell S, Anand R, Smith JC, Markham AF. (1990) *Nucleic Acids Res.*, **18**, 2887–2890.

Organization and expression of the human genome

Chapter 7

Organization of the human genome

The **human genome** is the term used to describe the total genetic information (DNA content) in human cells. It really comprises two genomes: a complex nuclear genome and a simple mitochondrial genome (*Figure 7.1*). The nuclear genome provides the great bulk of essential genetic information, most of which is ultimately decoded to specify polypeptide synthesis on cytoplasmic ribosomes. Mitochondria possess their own ribosomes and the few polypeptide-encoding genes in the mitochondrial genome produce mRNAs which are translated on the mitochondrial ribosomes. However, the mitochondrial genome specifies only a very small portion of the specific mitochondrial functions; the bulk of the mitochondrial polypeptides are encoded by nuclear genes and are synthesized on cytoplasmic ribosomes, before being imported into the mitochondria (*Figure 1.10*). In many ways, the human genome is typical of mammalian genomes, both in terms of genome organization and expression.

The mitochondrial genome consists of a single type of DNA duplex which is densely packed with genetic information

General structure and inheritance of the mitochondrial genome

The human mitochondrial genome is defined by a single type of circular double-stranded DNA whose complete nucleotide sequence has been established (Anderson *et al.*, 1981). It is 16 569 bp in length and is 44% (G + C). The two DNA strands have significantly different base compositions: the **heavy (H) strand** is rich in guanines, the **light (L) strand** is rich in cytosines. Although the mitochondrial DNA is principally double-stranded, a small section is defined by a triple DNA strand structure due to the additional synthesis of a segment of mitochondrial DNA, 7S DNA [see *Figure 7.2,* and Clayton (1992) for a general review of transcription and replication of animal mitochondrial DNAs]. Human cells usually contain thousands of copies of the double-stranded mitochondrial DNA molecule. Accordingly, although a single mitochondrial DNA duplex has only about 1/8000 as much DNA as an average sized chromosome, the total mitochondrial DNA complement can account for up to about 0.5% of the DNA in a nucleated somatic cell.

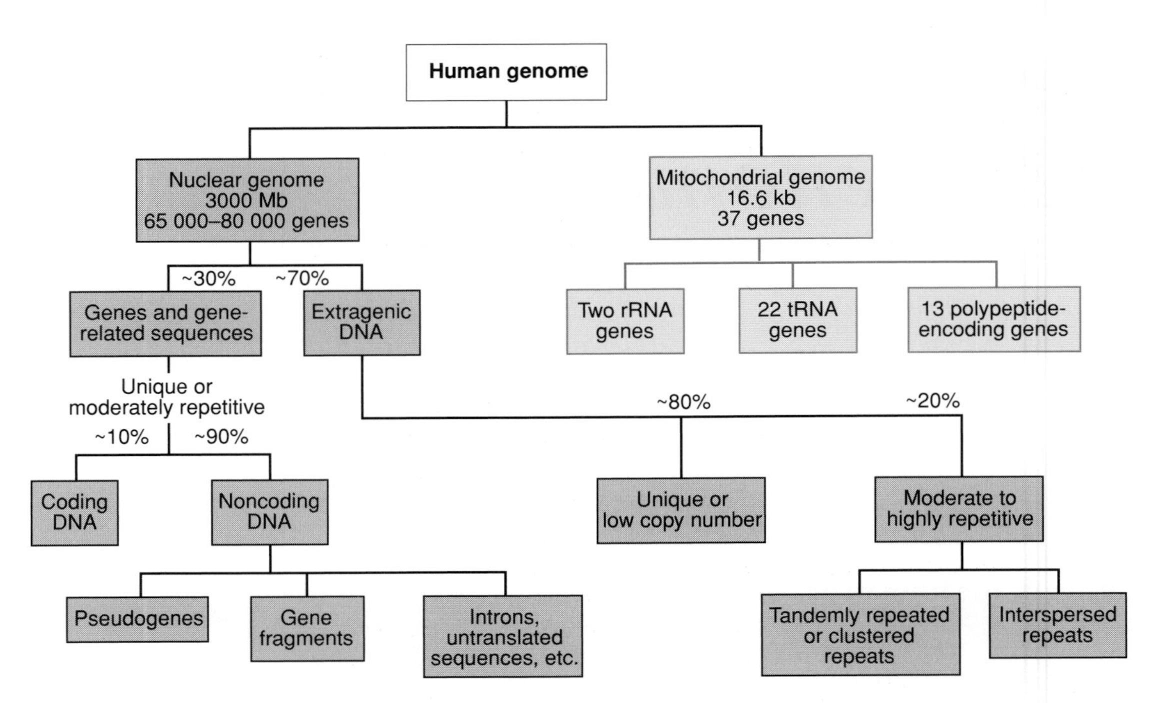

Figure 7.1: Organization of the human genome.

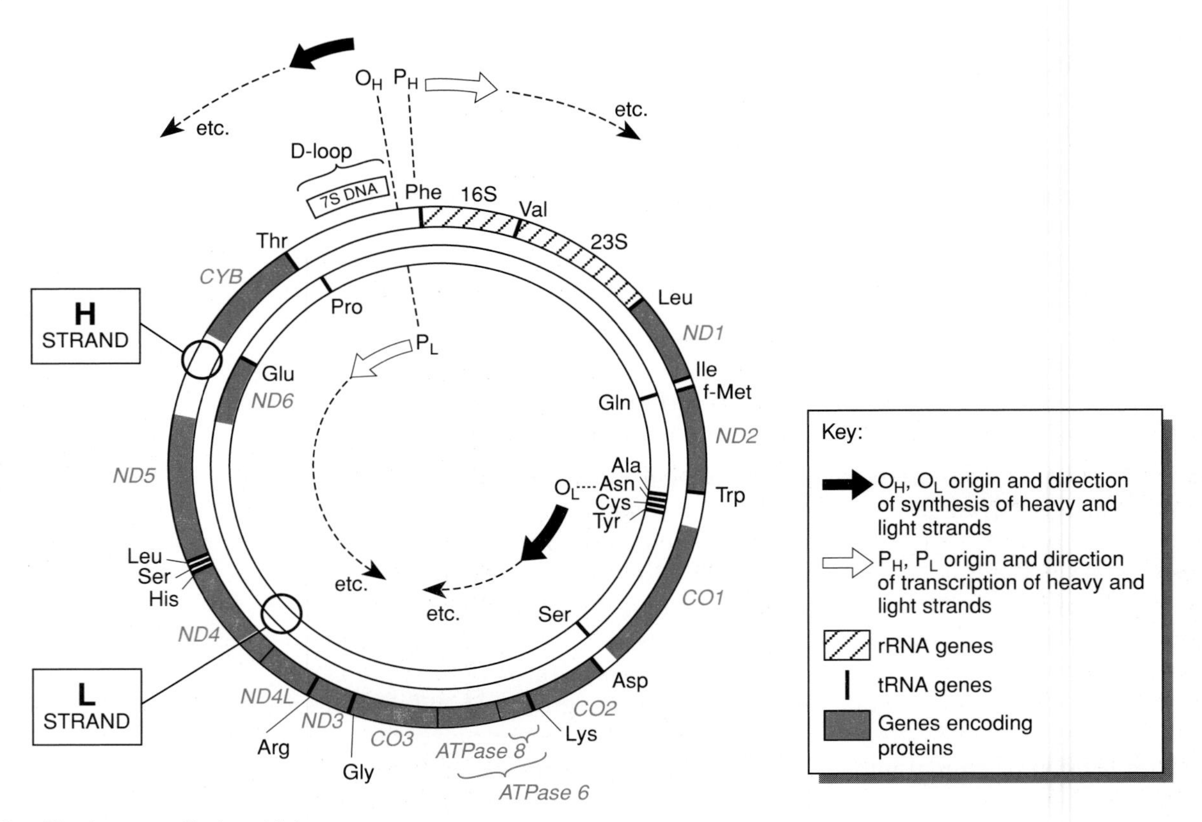

Figure 7.2: The human mitochondrial genome.

Note that transcription of the heavy (H) strand actually originates from two closely spaced promoters located in the D loop region, which for the sake of clarity are grouped as P_H. Transcription from these promoters runs clockwise round the circle; that from the light strand promoter P_L runs anticlockwise. In both cases, the resulting large transcripts are then cleaved to generate RNAs for individual genes. The high coding sequence density results from absence of introns in all genes and close apposition of genes, including one case of overlapping genes: the *ATPase 8* gene partly overlaps the *ATPase 6* gene (see *Figure 7.3*). Other polypeptide-encoding genes specify NADH dehydrogenase subunits 1–6 (*ND1–ND6*); cytochrome *c* oxidase subunits 1–3 (*CO1–CO3*) and cytochrome *b* (*CYB*).

During zygote formation, a sperm cell contributes its nuclear genome, but not its mitochondrial genome, to the egg cell. Consequently, the mitochondrial genome of the zygote is determined exclusively by that originally found in the unfertilized egg. The mitochondrial genome is therefore maternally inherited: males and females both inherit their mitochondria from their mother, whereas males cannot transmit their mitochondria to subsequent generations. Thus mitochondrially encoded genes or DNA variants give the pedigree pattern shown in *Figure 3.3A*. During mitotic cell division, the mitochondrial DNA molecules of the dividing cell segregate in a purely random way to the two daughter cells.

Mitochondrial genes

The human mitochondrial genome contains 37 genes: 28 are encoded by the heavy strand, and nine by the light strand (*Figure 7.2*). A total of 13 out of the 37 genes encode polypeptides which are synthesized on mitochondrial ribosomes. They provide a few subunits for four of the five *respiratory complexes*, the multichain enzymes of oxidative phosphorylation which are engaged in the production of ATP. However, all other human mitochondrial proteins are encoded by the nuclear genome and are translated on cytoplasmic ribosomes before being imported into the mitochondrion (see *Figure 1.10*). The remaining 24 mitochondrial genes encode 22 types of tRNA and two rRNA molecules, which constitute part of the mitochondrial protein synthesis apparatus; other components are encoded exclusively by nuclear genes (see *Box 7.1*)

The mitochondrial genetic code

The genetic code employed to decipher mitochondrially encoded mRNA on mitochondrial ribosomes differs slightly from that used to decipher nuclear encoded mRNA on cytoplasmic ribosomes. Eight of the 22 human mitochondrial tRNA molecules have anticodons which are each able to recognize families of four codons differing only at the third base, and 14 recognize pairs of codons which are identical at

Box 7.1: The limited autonomy of the mitochondrial genome

	Encoded by mitochondrial genome	Encoded by nuclear genome
Components of oxidative phosphorylation system	13 subunits	>80 subunits
I NADH dehydrogenase	7 subunits	>41 subunits
II Succinate CoQ reductase	0 subunits	4 subunits
III Cytochrome *b*–*c*1 complex	1 subunit	10 subunits
IV Cytochrome *c* oxidase complex	3 subunits	10 subunits
V ATP synthase complex	2 subunits	14 subunits
Components of protein synthesis apparatus	2 rRNAs 22 tRNAs (13 mRNAs)	All mitochondrial ribosomal proteins (~70 in total)
Other mitochondrial proteins	None	All, e.g. mitochondrial DNA polymerase, RNA polymerase plus numerous other enzymes, structural and transport proteins, etc.

the first two base positions and share either a purine or a pyrimidine at the third base. Between them, therefore, the 22 mitochondrial tRNA molecules can recognize a total of 60 codons [$(8 \times 4) + (14 \times 2)$]. The remaining four codons, UAG, UAA, AGA and AGG cannot be recognized by mitochondrial tRNA and act as stop codons (see *Figure 1.22*). In addition, the mitochondrial and nuclear genomes differ in many other aspects of their organization and expression (*Table 7.1*).

Table 7.1: The human nuclear and mitochondrial genomes

	Nuclear genome	Mitochondrial genome
Size	3000 Mb	16.6 kb
No. of different DNA molecules	23 (in XX) or 24 (in XY) cells, all linear	One circular DNA molecule
Total no. of DNA molecules per cell	23 in haploid cells; 46 in diploid cells	Several $\times 10^3$
Associated protein	Several classes of histone and nonhistone protein	Largely free of protein
No. of genes	~65 000–80 000	37
Gene density	~1/40 kb	1/0.45 kb
Repetitive DNA	Large fraction, see *Figure 7.1.*	Very little
Transcription	The great bulk of genes are transcribed individually	Continuous transcription of multiple genes
Introns	Found in most genes	Absent
% of coding DNA	~3%	~93%
Codon usage	See *Figure 1.22*	See *Figure 1.22*
Recombination	At least once for each pair of homologs at meiosis	None
Inheritance	Mendelian for sequences on X and autosomes; paternal for sequences on Y	Exclusively maternal

Coding and noncoding DNA

Unlike its nuclear counterpart, the human mitochondrial genome is extremely compact: approximately 93% of the DNA sequence represents coding sequence. The genes all lack introns and they are tightly packed (on average one per 0.45 kb). The coding sequences of some genes (notably those encoding the sixth and eighth subunits of the mitochondrial ATPase) show some overlap (*Figures 7.2* and *7.3*) and, in most other cases, the coding sequences of neighboring genes are contiguous or separated by one or two noncoding bases. Some genes even lack termination codons; to overcome this deficiency, UAA codons have to be introduced at the post-transcriptional level (Anderson *et al.*, 1981; see legend to *Figure 7.3*).

The only significant region lacking any known coding DNA is the **displacement (D) loop** region. This is the region in which a triple-stranded DNA structure is generated

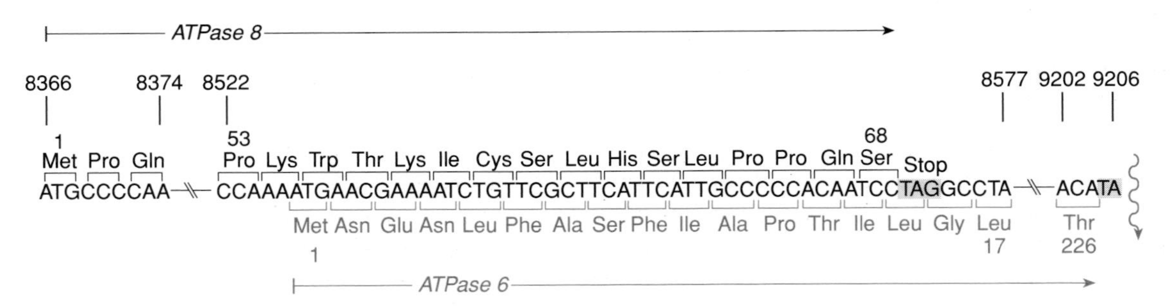

Figure 7.3: The genes for mitochondrial ATPase subunits 6 and 8 are partially overlapping and translated in different reading frames.

Note that the overlapping genes share a common sense strand, the H strand. Coding sequence co-ordinates are as follows: *ATPase* subunit 8, 8366–8569; *ATPase* subunit 6, 8527–9204.The C terminus of the *ATPase 6* subunit gene is defined by the post-transcriptional introduction of a UAA codon: following transcription the RNA is cleaved after position 9206 and polyadenylated, resulting in a UAA codon where the first two nucleotides are derived ultimately from the TA at positions 9205–9206 and the third nucleotide is the first A of the poly(A) tail. Other human genes are known to be overlapping but are often transcribed from opposite strands.

by synthesizing an additional short piece of the H-strand DNA, the 7S DNA. The replication of both the H and L strands is unidirectional and starts at specific origins. In the former case, the origin is in the D loop and only after about two-thirds of the daughter H strand has been synthesized (by using the L strand as a template and displacing the old H strand) does the origin for L strand replication become exposed. Thereafter, replication of the L strand proceeds in the opposite direction, using the H strand as a template (*Figure 7.2*). The D loop also contains the predominant promoter for transcription of both the H and L strands. Unlike transcription of nuclear genes, in which individual genes are almost always transcribed separately using individual promoters, transcription of the mitochondrial DNA starts from the promoters in the D loop region and continues, in opposing directions for the two different strands, round the circle to generate large multigenic transcripts (see *Figure 7.2*). The mature RNAs are subsequently generated by cleavage of the multigenic transcripts.

The nuclear genome is distributed between 24 different types of DNA duplex which show considerable regional variation in base composition and gene density

Size and banding patterns of human chromosomes

The nucleus of a human cell contains more than 99% of the cellular DNA. The nuclear genome is distributed between 22 types of autosome and two types of sex chromosome which can easily be differentiated by chromosome banding techniques (*Figure 7.4*), and have been classified into groups largely according to size and, to some extent, centromere position (*Table 7.2*). In addition to the primary constriction (centromere) present on each chromosome, the long arms of chromosomes 1, 9 and 16 possess so-called **secondary constrictions** (light staining, apparently uncoiled chromosomal regions) which, like the centromeres, are composed of constitutive heterochromatin (see page 49). By comparison with the size of a mitochondrial DNA molecule, an average size human chromosome has an enormous amount of DNA, approximately 130 Mb on average, but varying between approximately 50 and 250 Mb (*Table 7.3*). In a 550-band metaphase chromosome preparation (*Figure 7.4*), an average band corresponds to about 6 Mb of DNA.

Differences between dark and light chromosome bands and regional differences in gene density

The alternating pale and light bands may reflect the compartmentalization of the human genome into **isochores** (Bernardi, 1993). These are defined large chromosomal regions (about >300 kb) which may reflect regional variations in base composition, or alternatively variable spacing of *scaffold attachment regions* (see *Figure 2.5*). The dark G bands show a number of features which distinguish them from the pale bands obtained with standard Giemsa staining (*Table 7.4*).

Gene density can vary substantially between chromosomal regions and also between whole chromosomes. For example, heterochromatic regions are known to be very largely composed of repetitive noncoding DNA, and the centromeres and large regions of the Y chromosome, in particular, are notably devoid of genes. Recently, insight into gene distribution along the lengths of the different chromosomes has been obtained by hybridizing purified CpG island fractions of the genome (which are associated with perhaps about 56% of human genes; Antequera and Bird, 1993) to metaphase chromosomes (Craig and Bickmore, 1994). On this basis, it is clear that gene density is high in subtelomeric regions and that some chromosomes (e.g. 19 and 22) are gene rich while others (e.g. 4 and 18) are gene poor (see *Figure 7.5*).

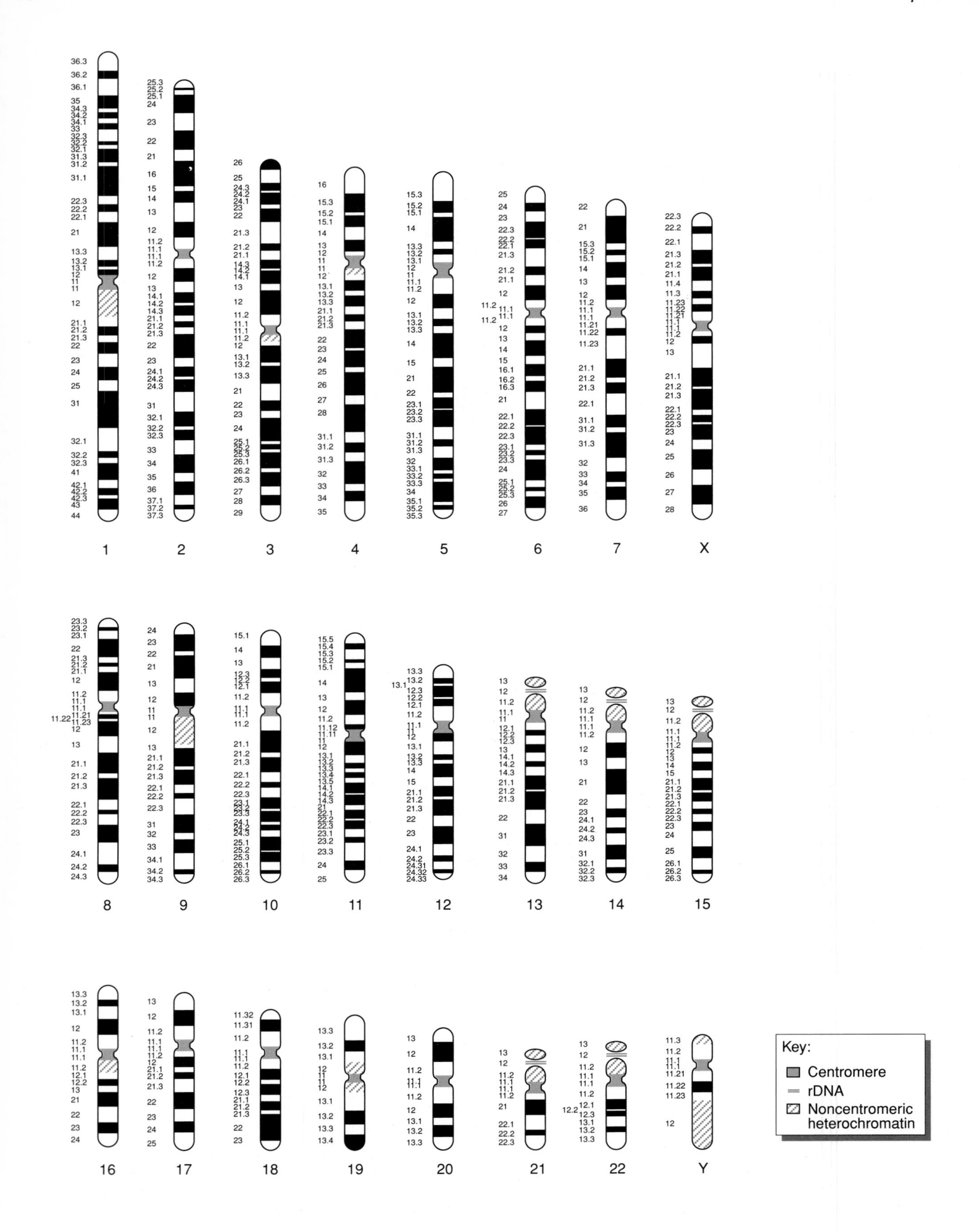
1
2
3
4
5
6
7
X
8
9
10
11
12
13
14
15
16
17
18
19
20
21
22
Y
Key:
Centromere
rDNA
Noncentromeric heterochromatin

Table 7.2: Human chromosome groups

Group	Chromosomes	Description
A	1–3	Largest; 1 and 3 are metacentric but 2 is submetacentric
B	4,5	Large; submetacentric with two arms very different in size
C	6–12, X	Medium size; submetacentric
D	13–15	Medium size; acrocentric with satellites
E	16–18	Small; 16 is metacentric but 17 and 18 are submetacentric
F	19, 20	Small; metacentric
G	21, 22, Y	Small; acrocentric, with satellites on 21 and 22 but not on the Y

Note that numbering of autosomes is in decreasing order in size, except that chromosome 21 is now known to be smaller than chromosome 22

Table 7.3: DNA content of human chromosomes[a]

Chromosomes	Percentage of total length	Amount of DNA (Mb)	Chromosome	Percentage of total length	Amount of DNA (Mb)
1	8.3	250	13	3.6	110
2	7.9	240	14	3.5	105
3	6.4	190	15	3.3	100
4	6.1	180	16	2.8	85
5	5.8	175	17	2.7	80
6	5.5	165	18	2.5	75
7	5.1	155	19	2.3	70
8	4.5	135	20	2.1	65
9	4.4	130	21	1.8	55
10	4.4	130	22	1.9	60
11	4.4	130	X	4.7	140
12	4.1	120	Y	2.0	60

[a]The DNA content is given for chromosomes prior to entering the S (DNA replication) phase of cell division (see *Figure 1.3*).
Data abstracted from Stephens *et al.* (1990).

Table 7.4: Properties of chromosome bands seen with standard Giemsa staining

Dark bands (G bands)	Pale bands (correspond to R bands – see *Box 2.1*)
Stain strongly with dyes that bind preferentially to AT-rich regions, such as Giemsa and Quinacrine	Stain weakly with Giemsa and Quinacrine
May be comparatively AT-rich	May be comparatively GC-rich
DNase insensitive	DNase sensitive
Condense early during the cell cycle but replicate late	Condense late during cell cycle but replicate early
Gene poor. Genes may be large because exons are often separated by very large introns	Gene rich. Genes are comparatively small because of close clustering of exons
LINE rich, but may be poor in *Alu* repeats (see pages 199 and 202)	LINE poor, but may be enriched in *Alu* repeats (see pages 199 and 202)

Figure 7.4: Banding pattern of human chromosomes (G-banding, 550-band karyogram).

Note that chromosome 21 is in fact smaller than chromosome 22 in size. The observed metaphase chromosome lengths range between 2 μm (chromosome 21) and 10 μm (chromosome 1), whereas the fully uncoiled DNA strands would be expected to measure between 1.7 and 8.5 cm respectively. *Note* the presence of extensive heterochromatic regions on the Y chromosome, at the secondary constrictions of chromosomes 1q, 9q and 16q, and on the short arms of the acrocentric chromosomes 13, 14, 15, 21 and 22.

Differences between pale and dark G bands, which are, respectively, gene rich and gene poor, are illustrated by the contrast between the human leukocyte antigen (HLA) complex and the dystrophin gene (*DMD*). The former is located in the pale G band, 6p21.3: at the time of writing, 170 genes have been identified in a span of 4 Mb, and the total is expected to be at least 200. By contrast, a full 2.4 Mb of DNA in the dark G band region, Xp21, appears to be devoted to just the dystrophin gene (which can, however, be expressed alternatively to produce a variety of different transcripts – see *Figure 7.11*).

The nuclear genome contains the vast majority of genes in the human genome, a figure that is currently estimated to be about 65 000–80 000

The number of genes in the human genome has been the subject of much speculation. The small mitochondrial genome is known to have precisely 37 genes, but the number of genes in the nuclear genome remains unknown. Theoretical calculations based on the *mutational load* that a genome can tolerate and observed average mutation rates of human genes ($\sim 10^{-5}$ per gene per generation) suggest an upper limit of 100 000. A variety of different approaches have been used to obtain more precise estimates of the total gene number. Three approaches suggest a best estimate of about 65 000–80 000 genes:

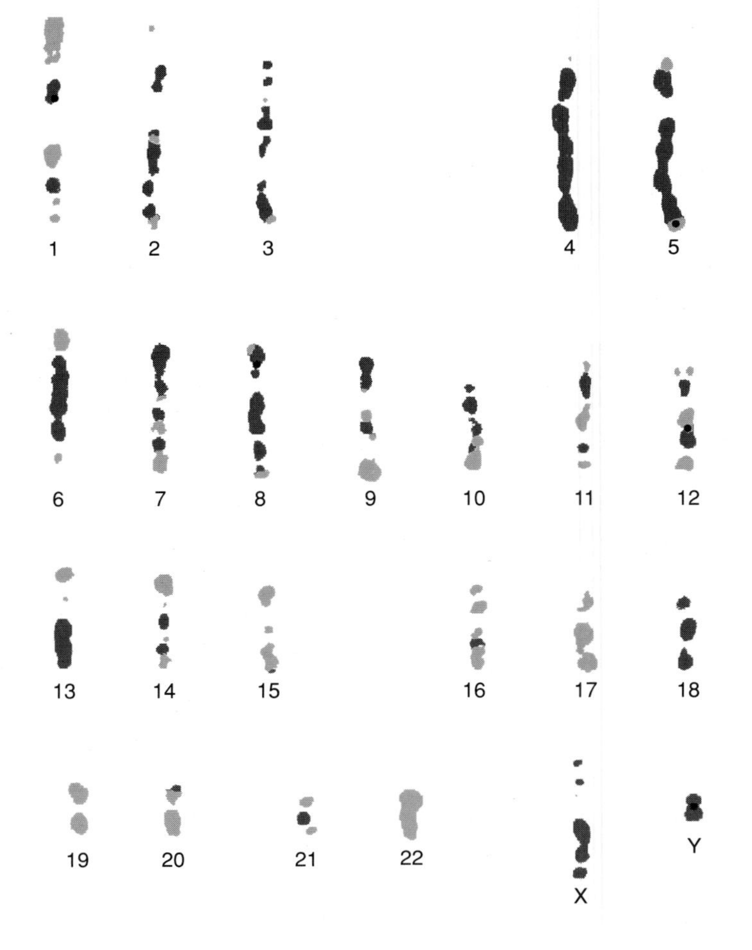

Figure 7.5: Clustering of CpG islands in the human genome.

The diagram represents FISH of a CpG island fraction from human DNA to human metaphase chromosomes (Craig and Bickmore, 1994). The Texas Red signal is derived from the CpG island probe, while the fluorescein isothiocyanate (FITC) green signal (shown here as gray) represents late replicating regions (which are mostly transcriptionally inactive), recognized by incorporation of bromodeoxyuridine (BrdU). Black regions represent overlap of the signals, indicating hybridization of the CpG island fraction to late replicating DNA. There is no counterstain, so that early replicating regions of the genome which do not have high densities of CpG islands are invisible, as are centromeres (where the anti-BrdU cannot get access). In addition to the rDNA clusters on the short arms of chromosomes 13–15, 21 and 22, high CpG island density is found on chromosomes 1, 9, 15–17, 19, 20 and 22. Adapted from Craig and Bickmore (1994) with permission from Nature America Inc.

- **genomic sequencing.** Extrapolation from sequencing of large chromosomal regions may suggest that there are about 70 000 genes (Fields *et al.*, 1994). This is based on the observation that gene-rich regions (e.g. the HLA complex) have an average gene density of close to one per 20 kb, but gene-poor regions have a much lower density, say one-tenth of this density, and that the genome is split 50:50 into gene-rich and gene-poor regions.
- **CpG island number.** Restriction enzyme analysis using the methylation-sensitive enzyme *Hpa*II suggests that the total number of *CpG islands* (see page 15) in the human genome is 45 000 (Antequera and Bird, 1993). Using an estimate that approximately 56% of genes are associated with CpG islands, these authors have suggested a total of about 80 000 human genes.
- **EST analysis.** Large-scale random sequencing of cDNA clones provides so-called *expressed sequence tags* (ESTs, see page 346). Comparison of known human EST sequences with a large set of different human genomic coding DNA sequences listed in sequence databases has suggested a figure of about 65 000 human genes (Fields *et al.*, 1994).

Such values suggest that the genes in the nuclear genome represent about 99.98% of the total number of cellular genes. If the average size of a human gene is taken to be about 10–15 kb, this would mean that if the genes did not show overlaps, about 25–35% of the total nuclear DNA would be occupied by genes. As the vast majority of nuclear genes encode polypeptides and the coding sequence required for an average size human polypeptide is taken to be about 500 codons, that is 1.5 kb, only about 3% of the nuclear genome (80–100 Mb of the 3000 Mb) would be expected to have a coding function.

Organization of human genes

Functionally similar genes are occasionally clustered in the human genome, but are more often dispersed over different chromosomes

Functional diversity of human genes

The great majority of human genes ultimately specify polypeptides which collectively carry out numerous diverse functions, including structural proteins (membrane components, cytoskeleton proteins, etc.), transport proteins, hormones, receptors, enzymes, regulatory proteins, signaling molecules, etc. Only a small minority specify a mature RNA product, but there is increasing recognition of the functional diversity of genes encoding RNA products. In addition to the many genes involved in assisting protein synthesis (rRNA genes, tRNA genes, etc.) there are numerous other RNA genes which encode a diverse group of *small nuclear RNAs* (snRNA) and small cytoplasmic RNAs which have been identified recently as participating in RNA splicing and a variety of other roles. A wider role for RNA-encoding genes has been suggested by the recent characterization of two genes.

- **The *XIST* gene** is thought to be the major gene involved in initiating the process of X chromosome inactivation, being expressed exclusively from inactivated X chromosomes. No long open reading frames can be identified, and gene function is thought to be carried out through an RNA product by a mechanism that remains obscure (see page 173).
- **The *H19* gene** appears to encode an RNA that acts as tumor suppressor (Wrana, 1994).

Organization of functionally similar human genes

Although many human genes are *single copy genes* (present at a single locus and distinct from other genes), many others are members of *gene families* which have arisen by gene duplication events from initially a single gene (see Chapter 8). Members of a gene family often have identical or closely related functions and may be clustered at specific chromosomal locations or dispersed in the genome. In addition, genes can be related functionally to other genes by a variety of routes. For example, they may encode different subunits of a single multisubunit complex; they may encode products that interact directly with each other, for example signaling molecules plus their receptors, etc.; they may encode products that are functionally related through participating in a common metabolic pathway, etc. Often, however, such products are encoded by genes that are dispersed in the genome (see *Table 7.5*).

Table 7.5: Distribution of genes encoding functionally related products

Genes which encode	Organization	Examples
The same product	Often clustered	Genes encoding rDNA, histones, etc.
Tissue-specific protein isoforms or isozymes	Sometimes clustered; sometimes non-syntenic	Clustering of pancreatic and salivary amylase genes (1p21); nonsynteny of α-actin genes expressed in skeletal (1p) and cardiac (15q) muscle
Isozymes specific for different subcellular compartments	Usually nonsyntenic	Cytoplasmic and mitochondrial isozymes of aldehyde dehydrogenase (*ALDH1*-9q, *ALDH2*-12q), aconitase (*ACO1*-9p; *ACO2*-22q), thymidine kinase (*TK1*-17q; *TK2*-16), etc.
Enzymes in the same metabolic pathway	Usually nonsyntenic	Genes for steroidogenic enzymes e.g. steroid 11-hydroxylase-8q, steroid 17-hydroxylase-10, steroid 21-hydroxylase-6p
Subunits of the same protein or enzyme	Usually nonsyntenic	Hemoglobin: α-16p; β-11p; collagen: α(1)I-7q; α(2)I-17q; ferritin: H-11q; L-22q; class I HLA: H-6p; L-15q; immunoglobulins: H-14q; L-2p or 22q
Ligand plus associated receptor	Usually nonsyntenic	Genes encoding interferons (*IFNA, IFNB* both 9p, *IFNG*-12q) and their receptors (*IFNAR*, *IFNBR*, both 21q, *IFNGR*-18); insulin gene *INS*-11p, but insulin receptor *INSR*-19p.

Human genes show enormous variation in size and internal organization

Size diversity

Genes in simple organisms such as bacteria are comparatively similar in size, and usually very short. By contrast, complex organisms such as mammals show wide variation in gene size, a feature found especially in human genes which can vary in length from hundreds of nucleotides to several megabases (*Figure 7.6*). The enormous size of some human genes means that transcription can be time consuming. For example, the human dystrophin gene requires about 16 h to be transcribed, and transcripts undergo splicing before transcription is completed (Tennyson *et al.*, 1995).

As one would expect, there is a direct correlation between the size of a gene and the size of its product, but there are some striking anomalies. For example, the apolipoprotein B gene is about 50 times smaller than the 2.4 Mb dystrophin gene but encodes a product with 4563 amino acids, significantly larger than the dystrophin polypeptide.

Diversity in internal organization

There is an inverse correlation between gene size and the proportion of the gene length which is expressed at the RNA level (*Figure 7.6*). A very small minority of human genes lack introns and are generally very small genes (see *Table 7.6* for examples). For those that do possess introns, the exon content as a percentage of gene length tends to be very small in large genes. This does not arise because exons in large genes are smaller than those in small genes: the average exon size in human genes is about 170 bp and, although very large exons are known (see *Box 7.2*), exon size is comparatively independent of gene length (*Table 7.7*). Instead, the explanation is due to the huge variation in intron lengths: large genes tend to have very large introns (*Table 7.7*). However, the relationship between gene and intron length is not without anomalies: the human type 7 collagen gene (*COL7A1*) is an intermediate size gene (31 kb) but has a total of 118 exons (the highest number for any characterized gene and significantly more than the giant dystrophin gene), and an average intron size of only 188 bp (Christiano *et al.*, 1994).

Many human genes, like other eukaryotic genes, contain intragenic repeated sequences which may involve different forms of repeated structure, often in coding DNA, and different modes of repetition. Repeated coding DNA can result in internal

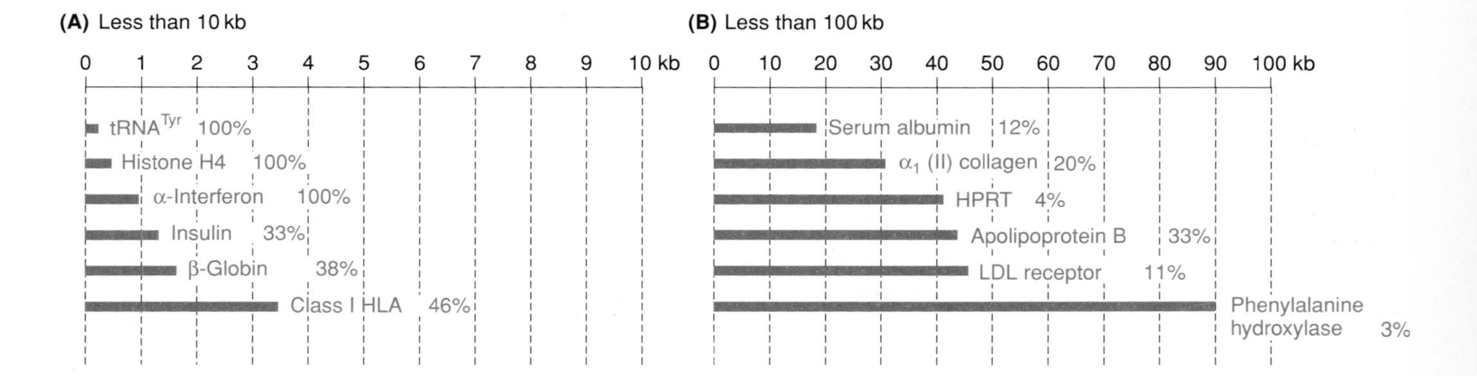

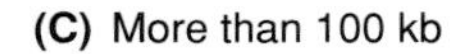

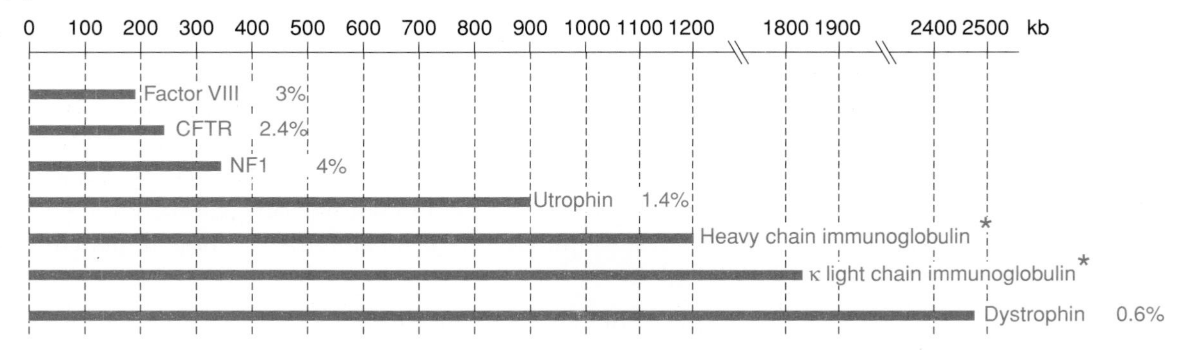

Figure 7.6: Human genes vary enormously in size and exon content.

Exon content is shown as a percentage of the lengths of indicated genes. *Note* the generally inverse relationship between gene length and percentage of exon content. Asterisks emphasize that the lengths given for the indicated Ig heavy chain and light chain loci correspond to the germline organizations. Immunoglobulin and T-cell receptor genes have unique organizations, requiring cell-specific somatic rearrangements in order to be expressed in B or T lymphocytes respectively (see page 177). CFTR, cystic fibrosis transmembrane regulator; HPRT, hypoxanthine phosphoribosyl transferase; NF1, neurofibromatous type 1.

Table 7.6: Examples of human genes with uninterrupted coding sequences

All 37 mitochondrial genes
Histone genes
Many genes encoding small RNA, e.g. most tRNA genes
Various hormone receptor genes, e.g. dopamine D1 and D5 receptors, 5-HT$_{1B}$ serotonin receptor, angiotensin II type 1 receptor, formyl peptide receptor, bradykinin B2 receptor, α2 adrenergic receptor
Autosomal processed copies of intron-containing X-linked genes Typically have testis-specific expression patterns, e.g. *PGK2* (phosphoglycerate kinase), *GK* (glycerol kinase), *MYCL2* (*myc* family member), *PDHA2* (pyruvate dehydrogenase E1a), *GLUD2* (glutamate dehydrogenase)
Others, e.g. IFN-α, thrombomodulin, *SRY* and many SOX (SRY HMG box-related) genes, *XIST*

repetition within the amino acid sequences of encoded polypetides. In some cases, there may be a very high degree of sequence homology between the repeats. In many cases, a repeat sequence may correspond to a *protein domain,* but the degree of sequence homology between such repeats can vary considerably (*Table 7.8*).

Rare examples of overlapping genes and genes within genes are known in the human genome

Partially overlapping genes

The genes of simple organisms are generally more clustered than those in complex organisms. The average gene density in the human genome is about one per 40–45 kb of DNA which, assuming an average gene size of, say, 10–15 kb means that there is about 30 kb of DNA on average between human genes. By contrast, average gene densities in simple organisms are very much higher: roughly one per 1, 2 and 5 kb, respectively, for *E. coli, Saccharomyces cerevisiae* and *Caenorhabditis elegans*. Simple genomes such as those of certain phages and bacteria often show examples of partially overlapping genes which use different reading frames, sometimes from a common sense strand. One such example is known in the simple human mitochondrial genome (see *Figure 7.3*). However, reported occurrences of overlapping genes in the complex nuclear genomes of mammals are rare and, where they do occur, the overlapping genes are often transcribed from the two different DNA strands. As noted on page 151, the degree of gene clustering in the nuclear genome is largely dependent on the chromosomal region, and in regions of high density occasional examples of overlapping genes have been noted. For example, the 4 Mb human HLA complex at 6p21.3, which has an average gene density of about one gene per 20 kb, is known to contain a few examples of overlapping genes.

Genes within genes

The genes of complex genomes such as the human nuclear genome often contain large introns. Recently, a few examples have been identified of large human introns containing whole small genes.

- *The NF1 gene.* Intron 27 of the *NF1* (neurofibromatosis type I) gene spans about 40 kb and contains three small genes, each with two exons which are transcribed from the opposite strand to that used for the *NF1* gene (Viskochil *et al.*, 1991; see *Figure 7.7*).
- *The factor VIII gene.* Intron 22 of the blood clotting factor VIII gene (*F8*) contains a CpG island from which two internal genes, *F8A* and *F8B* are transcribed in opposite directions (Levinson *et al., 1992*). *F8A* is transcribed from the opposite strand to that used by the factor VIII gene. *F8B* is transcribed in the same direction as the factor VIII gene to give a short mRNA containing a new exon spliced on to exons 23–26 of the factor VIII gene (see *Figure 10.17*).

Box 7.2: Human gene organization

Gene number

Total genome	Mitochondrial genome: 37 (page 149). Nuclear genome: 65 000–80 000 (page 154).
Chromosome	Average of about 3000; but dependent on chromosome length and also on chromosome type (see *Figure 7.5*).
Chromosome band	Average of about 130 per band in a 550-band chromosome preparation.
Gene density	Averages of about one per 0.45 kb in the mitochondrial genome one per 40–45 kb in the nuclear genome.
Gene size	Average 10–15 kb, but enormous variation (see *Figure 7.6*).
Intergenic distance	On average, 25–30 kb in nuclear genome.
Exon number	Generally correlated with gene length, and shows wide variation from small genes with a single exon (see *Table 7.7*) to large genes with numerous exons, e.g. the dystrophin gene (*DMD*) has 79 exons.
Exon size	Average size about 170 bp with comparatively little length variation and independent of gene length. However, some very long exons are known, e.g. exon 26 of the apoB gene (*APOB*), 7.6 kb exon 15 of the adenomatous polyposis coli gene (*APC*), 6.5 kb exon 11 of the *BRCA1* breast cancer gene, 3.4 kb.
Intron size	Enormous variation. Strong direct correlation with gene size: small genes tend to have small introns and large genes tend to have large introns. Examples of typical intron sizes are as follows: β-globin gene (*HBB*; 1.6 kb) 0.5 kb myoglobin gene (*MB*; 10.4 kb) 4.7 kb dystrophin gene (*DMD*; 2.5 Mb) 30.0 kb
mRNA size	Average of about 2.2 kb, but considerable variation
5′ UTS	Average of about 0.1 kb
coding DNA	Average of about 1.5 kb (500 codons)
3′ UTS	Average of about 0.6 kb

Table 7.7: Average sizes of exons and introns in human genes

Gene product	Size of gene	Number of exons	Average size of exon (bp)	Average size of intron (kb)
$tRNA^{tyr}$	0.1	2	50	0.02
Insulin	1.4	3	155	0.48
β-Globin	1.6	3	150	0.49
Class I HLA	3.5	8	187	0.26
Serum albumin	18	14	137	1.1
Type VII collagen	31	118	77	0.19
Complement C3	41	29	122	0.9
Phenylalanine hydroxylase	90	26	96	3.5
Factor VIII	186	26	375	7.1
CFTR (cystic fibrosis)	250	27	227	9.1
Dystrophin	2400	79	180	30.0

Table 7.8: Examples of intragenic repetitive coding DNA (see also page 266)

Gene product	Size of encoded repeat in amino acids	No. of copies	Nucleotide sequence homology between copies
Ubiquitin (*UbB* and *UbC* genes)	76	3 (*UbB*) 9 (*UbC*)	High homology
Involucrin	10	59	High homology for central 39 repeats
Apolipoprotein (a)	114 = kringle 4-like repeat[a]	37	High homology; 24 of the repeats are identical in sequence
Plasminogen	~75–80	5	Low homology but conserved protein domains (kringles[a])
Collagen	18	57	Low homology but conserved amino acid motifs based on $(\text{Gly-X-Y})_6$
Serum albumin	195	3	Low homology
Proline-rich protein genes	16–21	5	Low homology
Tropomyosin α-chain	42	7	Low homology
Immunoglobulin ε-chain, C region	108	4	Low homology
Dystrophin	109	24	Low homology

[a]A kringle is a cysteine-rich sequence that contains three internal disulfide bridges and forms a pretzel-shaped structure.

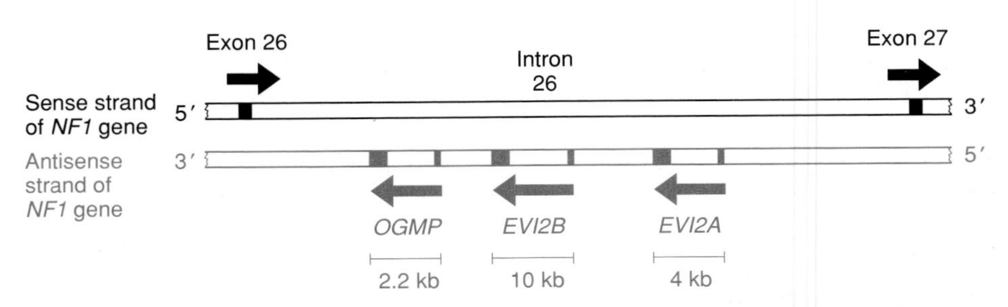

Figure 7.7: Genes within genes: intron 26 of the gene for neurofibromatosis type I (NF1) contains three internal genes each with two exons.

Note that the three internal genes are transcribed from the opposing strand to that used for transcription of the *NF1* gene. Genes are: *OGMP*, oligodendrocyte myelin glycoprotein; *EVI2A* and *EVI2B*, human homologs of murine/genes thought to be involved in leukemogenesis, and located at ecotropic viral integration sites.

Human gene expression

The range of expression patterns of human genes and of the control systems used to regulate their expression are fundamentally similar to those found in other mammals, and generally resemble those in eukaryotes in general. As in other eukaryotes, the control of human gene expression is largely exerted at the transcriptional level. Increasingly, however, the importance of post-transcriptional regulation of gene expression is being appreciated. Recent studies of gene expression in humans (and other complex eukaryotes) have emphasized the need for a wider definition of the term *gene*. For example, many cases are known where a single human transcription unit can give rise to a variety of gene products. In some cases, two or more functional products are generated by cleavage of a larger precursor expression product, a mechanism that is commonly employed in prokaryotic cells. However, in humans and other eukaryotes, there is also a variety of additional post-transcriptional mechanisms which can generate many distinct products with overlapping sequences from a single gene.

As in other complex eukaryotes, transcriptionally active human chromatin is characterized by an open structure, which is comparatively free of methylation and is accessible to transcription factors and to cleavage by nucleases such as DNase I (page 15). Several types of control mechanism can regulate human gene expression at specific developmental stages and in fully differentiated cells of the mature organism. Of these, transcriptional control mechanisms are particularly prevalent (*Table 7.9*). Because developmental stage-specific expression often involves sequential transcriptional activation and inactivation of members of a gene family, the mechanisms involved in regulating such expression will be examined in Chapter 8, as will other mechanisms that regulate tissue-specific gene expression in multigene families. The control of tissue-specific gene expression as it affects isolated human genes, be they members of multigene families or not, will be discussed in the present chapter.

Human genes have complex sets of *cis*-acting transcriptional control sequences

Cell type-specific transcription factors are known to play an important role in activating cell-specific functions. The transcription factors are *trans*-acting elements which recognize and bind specific *cis*-acting sequence elements (*Table 7.10*). Often, the recognition sequences for both prokaryotic and eukaryotic transcription factors are only about 8–10 nucleotides long. In simple prokaryotic genomes, sequence

Table 7.9: Overview of the regulation of gene expression in human cells

Selective expression mechanism	Examples
A. ***Epigenetic mechanisms and long-range control of gene expression by chromatin structure***	Cell position-dependent short-range signaling (page 175) Imprinting of certain autosomal genes (page 170 and *Figure 7.15*) Imprinting of the *XIST* gene (page 173) DNA rearrangements in B and T lymphocytes which produce cell-specific immunoglobulins and T-cell receptors (page 177ff. and *Figures 7.18* and *7.19*) Competition for enchancers or silencers (e.g. in globin expression; see page 190) and the imprinting of *H19* and *IGF2* genes (page 171) Classic position effects (e.g. possibly in the case of facioscapulohumeral dystrophy (page 174) Suppression of gene expression by chromatin domains (e.g. of the *PAX6* gene in aniridia and the *SOX9* gene in campomelic dysplasia; (page 174) Inactivation by the *XIST* gene product of many genes on the one X chromosome on which it is expressed in female cells (*Figure 9.11*)
B. ***Transcriptional***	
Binding of tissue-specific transcription factors to *cis*-acting elements of a single gene	See *Table 7.10*
Direct binding of hormones, growth factors or intermediates (e.g. cAMP) to response elements in inducible transcription elements	CREB and cAMP response elements; retinoic acid and response elements (see *Table 1.3*)
Use of alternative promoters in a single gene	See *Figure 7.11* for dystrophin gene. Also applies to many other genes
C. ***Post-transcriptional***	
Tissue-specific alternative RNA processing of a single gene	Alternative splicing (see *Figure 7.12*) Alternative polyadenylation (see *Figure 7.12*)
Tissue-specific RNA editing	Editing of apoB mRNA (*Figure 7.13*)
Translational control mechanisms	Binding of IBF to iron-response elements in ferritin and transferrin mRNA (*Figure 7.14*)

Table 7.10: Examples of tissue-restricted and tissue-specific transcription factors

Transcription factor	Binding sequence	Expression patterns
AP2	CCC(A/C)N(C/G)$_3$	Especially neural crest lineage, keratinocytes?
GATA-1,-2, etc.	(A/T)GATA(A/G)	Erythroid cells
NF-E2	TGACTCAG	Erythroid cells
HNF-1	GTTAATNATTAAC (= PE element)	Differentiated liver, kidney, stomach, intestine, spleen
HNF-5	T(G/A)TTTG(C/T)	Liver
Ker1	GCCTGCAGGC	Keratinocytes
MBF-1	(C/T)TAAAAATAA(C/T)$_3$	Myocytes
MEF-2	(C/T)TA(A/T)AAATA(A/G)	Myocytes
MyoD	CAACTGAC	Myoblasts + myotubes
OTF-2	ATGCAAAT	Lymphoid cells
TCF-1	(C/A)A(C/A)AG	T cells

discrimination has not been a problem. However, because the human genome and the genomes of other complex eukaryotes are large, the chance of inappropriate expression is increased: the probability of unintended chance occurrence of elements with the same sequence as dedicated *cis*-acting control elements is so much higher. Partly because of this, and also because of the general need for more sophisticated control systems imposed by having very large numbers of interacting genes, control elements in complex eukaryotic genomes are quite elaborate. Often, human genes show evidence of three or more classes of *cis*-acting regulatory elements. Each class is typically made up of multiple short sequence elements distributed over a few hundred base pairs. However, the different classes which regulate a single gene may be located at considerable distances. Recognized classes of *cis*-acting elements for individual genes include:

- **Promoters** – these are usually located in the immediate upstream region of the gene (often within 200 bp from the transcription start site) and serve to initiate transcription. They consist of combinations of short sequence elements (often CCAAT box, TATA box, GC box, see *Table 1.3*) that can be recognized by ubiquitous transcription factors. In addition, promoters of genes that show tissue-specific expression patterns often include *cis*-acting sequence elements that can be recognized by tissue-specific transcription factors (*Table 7.10*).
- **Response elements (REs)** – these are found only in selected genes whose expression is controlled by an external factor, such as a hormone, growth factor or by an internal signaling molecule such as cAMP. They are often located a short distance upstream of the promoter elements (often within 1 kb of the transcription start site). Depending on the particular gene, a wide variety of elements are possible, including hormome REs such as glucocorticoid and retinoic acid REs, growth factor REs, cAMP REs, etc. High level expression of the relevant gene can be induced by binding of the appropriate signaling factor to a suitable RE.
- **Enhancers** – these are positive regulatory elements whose functions, unlike those of promoters, are independent of both their orientation and, to some extent, their distance from the genes they regulate. They often contain, within a span of only 200–300 bp, elements recognized by ubiquitous transcription factors as well as elements recognized by tissue-specific transcription factors. They serve to increase the level of transcription which is initiated by the promoter elements and are found in a wide variety of human genes.
- **Silencers** – these negative regulatory elements have been less well studied. Where studied in human genes, silencer elements have been reported close to

the promoter, both upstream and also within introns (see page 227). However, the evidence for such sequences often relies on *in vitro* DNA binding studies and their significance *in vivo* remains largely unclear [see Clark and Docherty (1993) for a general review of the negative regulation of transcription in eukaryotes].

In addition to the above groups of elements, positive *cis*-acting regulatory elements may occasionally co-ordinate the expression of several genes in a gene cluster. Such elements may be located far upstream of the genes whose expression they control. For example, the expression of the human α-globin and β-globin genes is controlled predominantly by positive regulatory sequences located about 50–60 kb upstream (see pages 190–192). As in other genes expressed in erythroid cells, the sequences involved in regulating globin gene expression often contain multiple recognition sequences for the erythroid-specific transcription factors GATA-1 and NF-E2 (*Figure 7.8*).

Transcription factors often contain conserved structural motifs that permit DNA binding

Transcription factors recognize and bind a short nucleotide sequence, usually as a result of extensive complementarity between the surface of the protein and surface features of the double helix in the region of binding. Although the individual interactions between the amino acids and nucleotides are weak (usually hydrogen bonds, ionic bonds and hydrophobic interactions), the region of DNA–protein binding is typically characterized by about 20 such contacts, which collectively ensure that the binding is strong and specific.

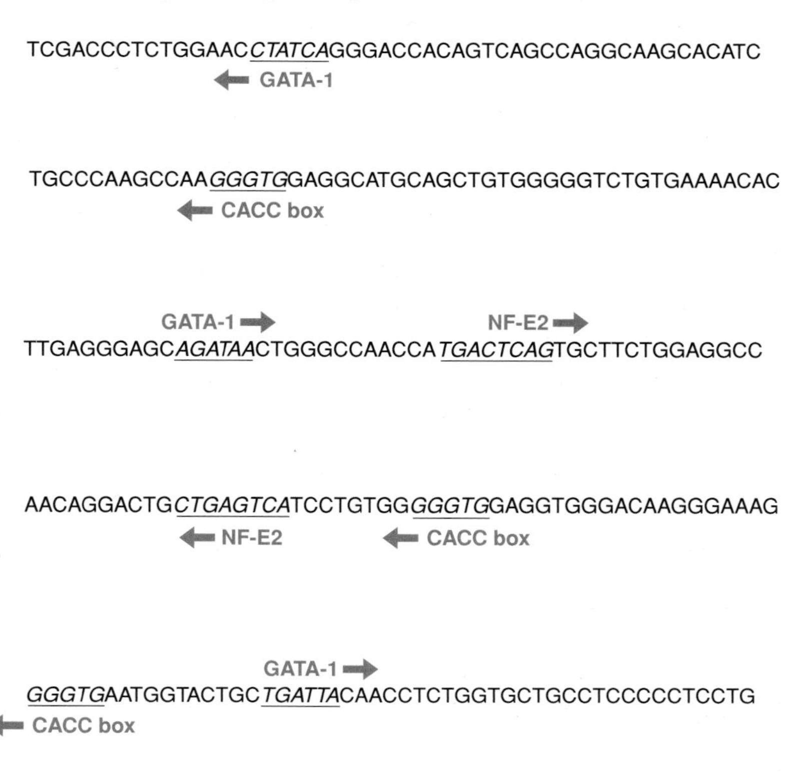

Figure 7.8: The HS-40 α-globin regulatory site contains many recognition elements for erythroid-specific transcription factors.

Note that the HS-40 site is a *locus control region* for the α-globin gene cluster (see page 191).

In human and other eukaryotic transcription factors, a number of conserved structural motifs have been identified which are known to assist in this process and are common to many different transcription factors with quite different specificities. Each of the motifs uses α-helices (or occasionally β-sheets; see *Figure 1.24*) to bind to the major groove of DNA. Often, two distinct functions can be identified in a transcription factor: a **DNA-binding domain** and an **activation domain**. Clearly, although the motifs in general provide the basis for DNA binding, the precise collection of sequence elements in the DNA-binding domain will provide the basis for the required sequence-specific recognition. Most transcription factors bind to DNA as homodimers, with the DNA-binding region of the protein usually distinct from the region responsible for forming dimers.

The leucine zipper motif

The **leucine zipper** is a helical stretch of amino acids rich in leucine residues (typically occurring once every seven amino acid residues, i.e. once every two turns of the helix – see *Figure 7.9*), which readily forms a dimer. Each monomer unit consists of an *amphipathic α-helix* (hydrophobic side groups of the constituent amino acids face one way; polar groups face the other way, see *Figure 1.24*). The two α-helices of the individual monomers join together over a short distance to form a *coiled-coil* (see page 28) with the predominant interactions occurring between opposed hydrophobic amino acids of the individual monomers. Beyond this region the two α-helices separate, so that the overall dimer is a Y-shaped structure. The dimer is thought to grip the double helix much like a clothes peg grips a clothes line (*Figure 7.10*). In addition to forming homodimers, leucine zipper proteins can occasionally form heterodimers depending on the compatibility of the hydrophobic surfaces of the two different monomers. Such heterodimer formation provides an important combinatorial control mechanism in gene regulation.

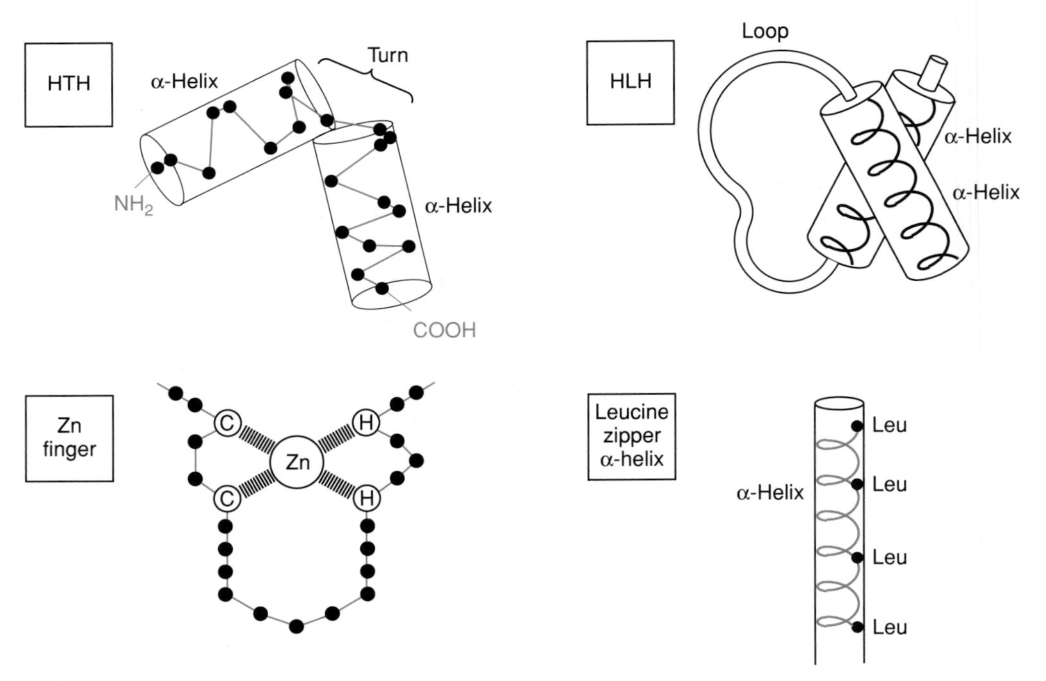

Figure 7.9: Structural motifs commonly found in transcription factors and DNA-binding proteins.

Abbreviations are: HTH, helix–turn–helix; HLH, helix–loop–helix.
Note that the leucine zipper monomer is *amphipathic* [i.e. has hydrophobic residues (leucines) consistently on one face of the helix]. Two such helices can align with their hydrophobic faces in opposition to form a coiled-coil structure.

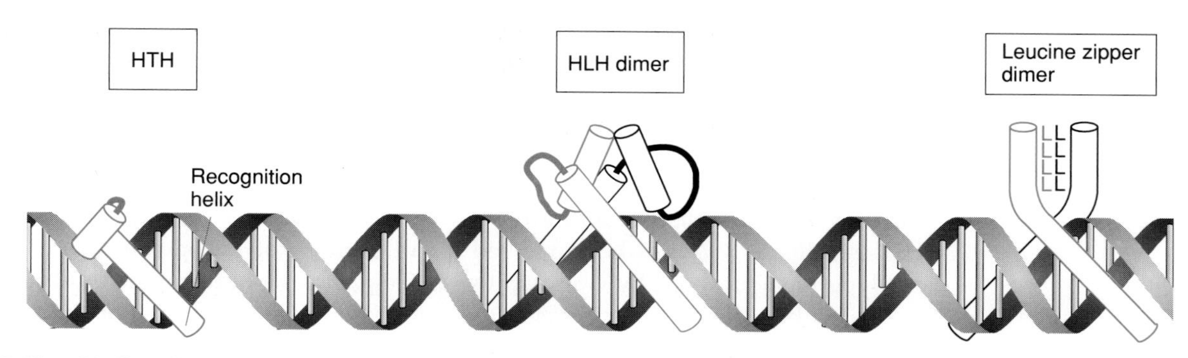

Figure 7.10: Binding of conserved structural motifs in transcription factors to the double helix.

Note that the individual monomers of the helix–loop–helix (HLH) dimer and the leucine zipper dimer are colored differently to permit distinction, but may be identical (homodimers). HLH heterodimers and leucine zipper heterodimers may provide a higher level of regulation (see text).

The helix–loop–helix motif

The **helix–loop–helix (HLH) motif** is related to the leucine zipper and should be distinguished from the **helix–turn–helix (HTH)** motif described in the next section. It consists of two α-helices, one short and one long, connected by a flexible loop. Unlike the short turn in the HTH motif, the loop in the HLH motif is flexible enough to permit folding back so that the two helices can pack against each other, that is the two helices lie in planes that are parallel to each other, in contrast to the two helices in the HTH motif (*Figure 7.9*). The HLH motif mediates both DNA binding and protein dimer formation (*Figure 7.10*) and it permits occasional heterodimer formation. In the latter case, however, heterodimers form between a full-length HLH protein and a truncated HLH protein which lacks the full length of the α-helix necessary to bind to the DNA. The resulting hetero-dimer is unable to bind DNA tightly. As a result, HLH dimers are thought to act as a control mechanism, by enabling *inactivation* of specific gene regulatory proteins.

The helix–turn–helix motif

The **HTH motif** is a common motif found in homeoboxes, and a number of other transcription factors. It consists of two short α-helices separated by a short amino acid sequence which induces a turn, so that the two α-helices are orientated differently (i.e. the two helices do not lie in the same plane, unlike those in the HLH motif; *Figure 7.9*). The structure is very similar to the DNA-binding motif of several bacteriophage regulatory proteins such as the λ cro protein whose binding to DNA has been intensively studied by X-ray crystallography. In the case of both the λ cro protein and eukaryotic HTH motifs, it is thought that while the HTH motif in general mediates DNA binding, the more C-terminal helix acts as a specific **recognition helix** because it fits into the major groove of the DNA (*Figure 7.10*), controlling the precise DNA sequence which is recognized.

The zinc finger motif

The **zinc finger motif** involves binding of a zinc ion by four conserved amino acids so as to form a loop (finger), a structure which is often tandemly repeated. Although several different forms exist, common forms involve binding of a Zn^{2+} ion by two conserved cysteine residues and two conserved histidine residues, or by four conserved cysteine residues. The resulting structure may then consist of an α-helix and a β-sheet held together by co-ordination with the Zn^{2+} ion, or of two α-helices. In either case, the primary contact with the DNA is made by an α-helix binding to the major groove. The so-called Cys_2/His_2 finger typically comprises about 23 amino acids with neighboring fingers separated by a stretch of about seven or eight amino acids (*Figure 7.9*).

A single gene or transcription unit can give rise to a variety of different gene products

Genes versus transcription units

Classically, a *gene* has been viewed as an entity that encodes a single RNA or polypeptide product. By contrast, the concept of several different products being encoded by a single *transcription unit* (a segment of DNA that is continuously transcribed into RNA) has long been familiar in simple genomes, such as bacterial genomes, where so-called *polycistronic mRNAs* are common. These arise by continuous transcription through several adjacent genes (the term *cistron* is essentially an old-fashioned word for gene). Such multigenic transcripts are then cleaved to generate individual gene transcripts. In such cases, the term *transcription unit* is clearly not functionally equivalent to the term *gene* because each transcription unit corresponds to several genes. In complex genomes, such as the human genome, however, the vast majority of genes are transcribed individually and, in these cases, the term *gene* and *transcription unit* are essentially equivalent. Nevertheless, rare examples are known in complex genomes of multigenic transcription units (see *Box 7.3*). In addition, the genes of complex eukaryotes are known to undergo a variety of alternative processing events, resulting in a variety of different products. Note that the immunoglobulin and T-cell receptor genes provide additional complexity: the individual genes are very large in germline DNA (see *Figures 7.6* and *7.17*), but the corresponding transcription units are much smaller and of variable size because of cell-specific DNA rearrangements in B and T lymphocytes (see *Figures 7.18* and *7.19*). The classical view of a gene is no longer valid (see *Box 7.3*).

Box 7.3: Individual human transcription units and genes can encode several products

Mechanisms	Frequency and examples
Multigenic transcription units	Rare. Examples include 18S, 28S and 5.8S rRNA genes (see *Figure 8.3*) and mitochondrial genes (see page 151)
Use of alternative promoters	Quite common. See below and *Figure 7.11*
Alternative splicing	Very frequent. See page 167 and *Figure 7.12*
Alternative polyadenylation	Quite common. See *Figure 7.12*
RNA editing	Extremely rare. See page 167 and *Figure 7.13*
Post-translational cleavage	Rare. May generate functionally related polypeptides as in the case of several hormones, e.g. human insulin. See *Figure 1.23*

Use of alternative promoters

Several human genes are known to have two or more alternative promoters, which can result in cell type-specific expression patterns.

For example, transcription of the human dystrophin gene is known to result in a variety of different products as a result of the alternative use of different promoter elements which are activated in different cell types. In addition to a muscle-specific promoter which defines the conventional start point of the gene, a brain-specific promoter activates transcription at a location which is more than 100 kb upstream of the muscle-specific promoter, whereas an alternative promoter located over 100 kb downstream of the muscle-specific promoter is used in Purkinje cells of the cerebellum (*Figure 7.11*). Each of these promoters uses a different first exon, resulting in a different N-terminal amino acid sequence. In addition, at least four other alternative internal promoters can be used, resulting in smaller proteins (D'Souza *et al.*, 1995; see *Figure 7.11*).

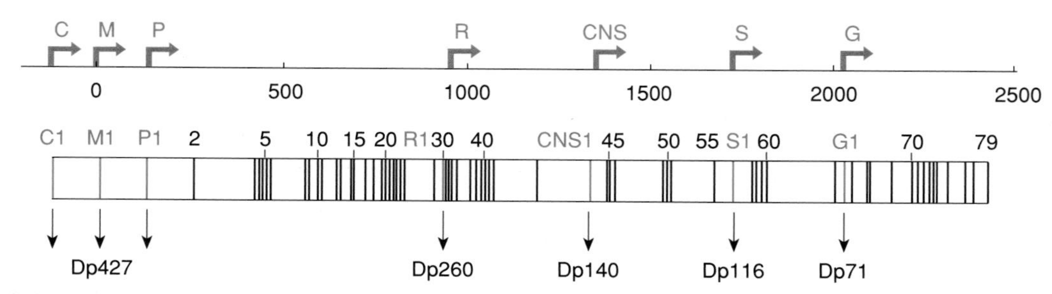

Figure 7.11: At least seven distinct promoters can be used to generate cell type-specific expression of the dystrophin gene.

The positions of the seven alternative promoters are illustrated at the top: C, cortical; M, muscle; P, Purkinje; R, retinal; CNS, central nervous system; S, Schwann cell; G, general or glial cell. The approximate positions of the exons are illustrated below. *Note* that each promoter uses its own first exon (in red: C1, M1, P1, R1, CNS1, S1 and G1) together with downstream exons (in black). All 78 downstream exons are used in the case of the full-length C-, M- and P-dystrophins to generate a product which in each case is about 427 kDa (Dp427). The other promoters are located immediately upstream of indicated exons as follows: R, exon 30; CNS, exon 45; S, exon 56; G, exon 63. *Note also* that initiation of translation for the Dp140 isoform does not occur until exon 51, although the promoter is thought to be in intron 44. In addition to the diversity of isoforms generated by usage of differential promoters, alternative splicing is known to occur, notably at the 3′ end.

Alternative splicing and polyadenylation

Many individual human genes undergo **alternative splicing** to yield different mRNA sequences encoding protein isoforms which may be tissue specific. In some cases, tissue-specific products of divergent function may also derive from a single gene. For example, a combination of alternative splicing and **alternative polyadenylation** of the calcitonin gene (*CALC*) results in the synthesis in the thyroid of calcitonin, a circulating Ca^{2+} homeostatic hormone, and in the hypothalamus of calcitonin gene-related peptide (CGRP), which may have both neuromodulatory and trophic activities (*Figure 7.12*). Alternative splicing can also regulate protein localization by generating soluble forms of membrane proteins, as in the case of many members of the Ig superfamily, including class I and II HLA genes, IgM and the *CD8* gene. Approaches to defining alternative splicing mechanisms have recently been reviewed by Hodges and Bernstein (1994).

RNA editing

RNA editing is a form of post-transcriptional processing which can involve enzyme-mediated insertion or deletion of nucleotides or subsititution of single nucleotides at the *RNA level*. Insertion or deletion RNA editing appears to be a peculiar property of gene expression in mitochondria of kinetoplastid protozoa and slime molds. **Substitution RNA editing** is frequently employed in some systems, such as the mitochondria and chloroplasts of vascular plants where indivdual mRNAs may undergo multiple C $\rightarrow$ U or U $\rightarrow$ C editing events, and has also been observed in a few mammalian genes such as the Wilm's tumor susceptibility gene (*WT1*) and genes encoding apoB lipoprotein, glutamate receptor and a very few other products (see Scott, 1995). In addition to C $\rightarrow$ U and U $\rightarrow$ C substitutions, A $\rightarrow$ I (*inosine*, a base which is produced by hydrolytic deamination of adenosine and which is thought to be read as G by the translation apparatus) substitutions have also been observed.

Structure of inosine

This is a deaminated form of adenosine in which the amino (NH_2) group at carbon 6 of adenosine is replaced by a carbonyl (C=0) group (see *Figure 1.2*)

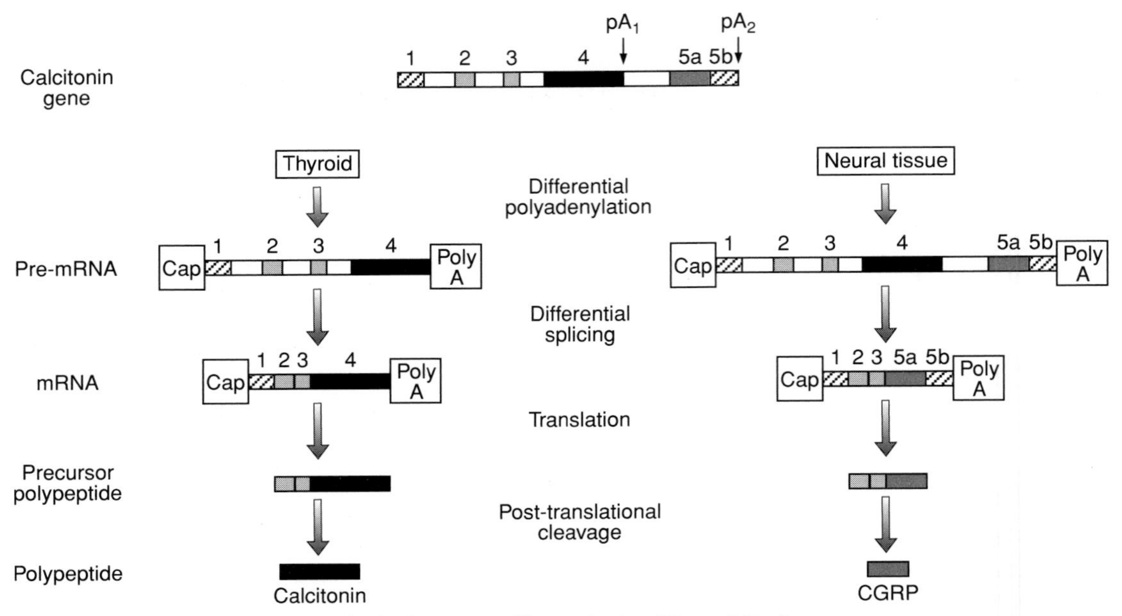

Figure 7.12: Differential RNA processing results in tissue-specific products of the calcitonin gene.

pA_1 and pA_2 represent alternative polyadenylation signals which are employed in thyroid and neural tissue respectively. Exon 1 and the 3′ part of exon 5 (5b) encode 5′ and 3′ untranslated sequences respectively. Calcitonin is encoded by exon 4 sequences in the thyroid, while the calcitonin gene-related peptide (CGRP), which is synthesized in neural tissue is encoded by the 5′ part of exon 5 (5a) as a result of alternative splicing.

RNA editing of human *APOB* lipoprotein gene has been well studied. In the liver the *APOB* gene encodes a 14.1 kb mRNA transcript and a 4536 amino acid product, apoB100. However, in the intestine the same gene encodes a 7 kb mRNA which contains a premature stop codon not present in the gene and encodes a product, apoB48, which is identical in sequence to the first 2152 amino acids of apoB100 (see Hodges and Scott, 1992). The premature stop codon is generated enzymatically in the intestinal mRNA by a single C $\rightarrow$ U change at nucleotide 6666 (*Figure 7.13*).

The mechanism of *APOB* mRNA editing most likely involves a sequence-specific cytidine deamination. The *APOB* mRNA editing enzyme is a multisubunit protein whose catalytic subunit, p27, is a member of the cytidine deaminase family. This family includes bacterial enzymes whose substrates are the mononucleotide cytidine, and recent data suggest that the p27 cytidine deaminase evolved from a simple enzyme of this type which acquired an RNA-binding motif at its active site, enabling it to recognize large polyribonucleotide substrates (Navaratnam *et al.*, 1995).

The expression of some human genes is controlled predominantly at the translational level

Although the control of human gene expression is exerted predominantly at the transcriptional level, as is the case for eukaryotic genes in general, the expression of some human genes is known to be controlled primarily at the translational level. For example, increased iron levels stimulate the synthesis of the iron-binding protein, ferritin, without any corresponding increase in ferritin mRNA. Conversely, decreased iron levels stimulate the production of transferrin receptor (*TFR*) without any effect on the production of transferrin receptor mRNA. The 5′-untranslated sequences of both ferritin heavy chain mRNA and light chain mRNA contain a single **iron-response element (IRE)**, a specific sequence which forms a hairpin structure, and the 3′ untranslated sequence of the transferrin receptor mRNA contains several

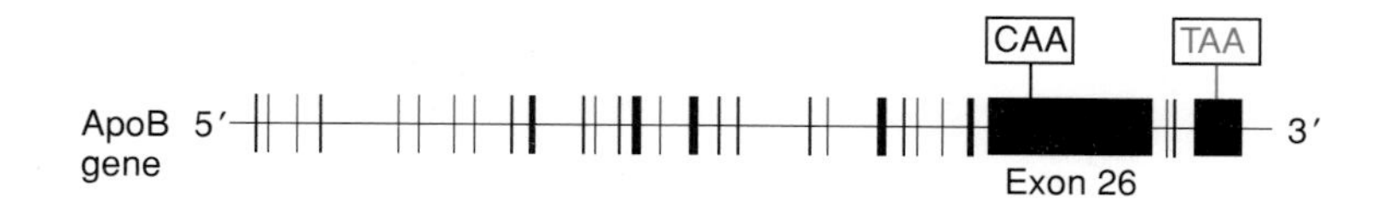

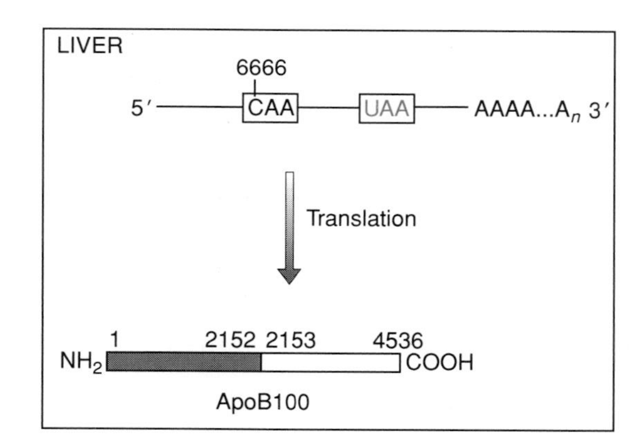

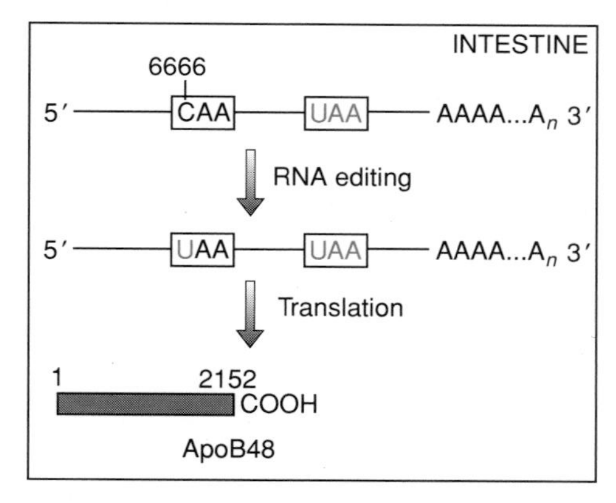

Figure 7.13: Expression of the human apolipoprotein B gene in the intestine involves tissue-specific RNA editing.

Note that codon 2153 specified by the CAA triplet at nucleotide positions 6666–6668 specifies glutamine in the ApoB100 product synthesized in liver, but because of RNA editing signals a translation stop in the intestine.

IREs (see Klausner *et al.*, 1993). Regulation is exerted by binding of IREs by a specific IRE-binding protein which is activated at low iron levels (see *Figure 7.14*).

Some human genes show selective expression of only one of the two parental alleles

X-linked genes in females and all autosomal genes are *biallelic* because both father and mother normally contribute one allele each. In males possessing one X chromosome and one Y chromosome, the great majority of sex-linked genes are *mono-allelic*: most of the many genes on the X do not have a functional homolog on the Y chromosome; and some of the few genes on the Y chromosome are known to be Y-specific, for example *SRY*, the major male sex-determining locus. A few genes on the Y chromosome do have functional homologs on the X chromosome and so are biallelic. In some cases of X–Y homologous loci, both homologs are normally functional (see page 211).

We are accustomed to assuming that both the paternal and maternal alleles of biallelic genes are expressed, unless one or both copies have sustained mutations which affect expression. Clearly the expression can be tissue specific so that in some cells both parental alleles are strongly expressed; in others, both gene copies are not apparently expressed. Thus, although there may be cell-specific differences in expression, there is no discrimination between the capacity of the two parental alleles to be expressed, other than that due to genetic (mutational) differences between them. However, in humans and other mammals, several biallelic genes are known where, at least in some cells, only one parental allele, either the paternal or the maternal allele but not both, is normally expressed. In such cases there is **functional hemizygosity**: only one half of the maximum gene product is obtained. This occurs naturally *even although both parental alleles are perfectly capable of normal gene expression and may even be identical.* There are three classes of mechanism known to result in monoallelic expression of biallelic genes in human and mammalian genomes (see *Box 7.4*).

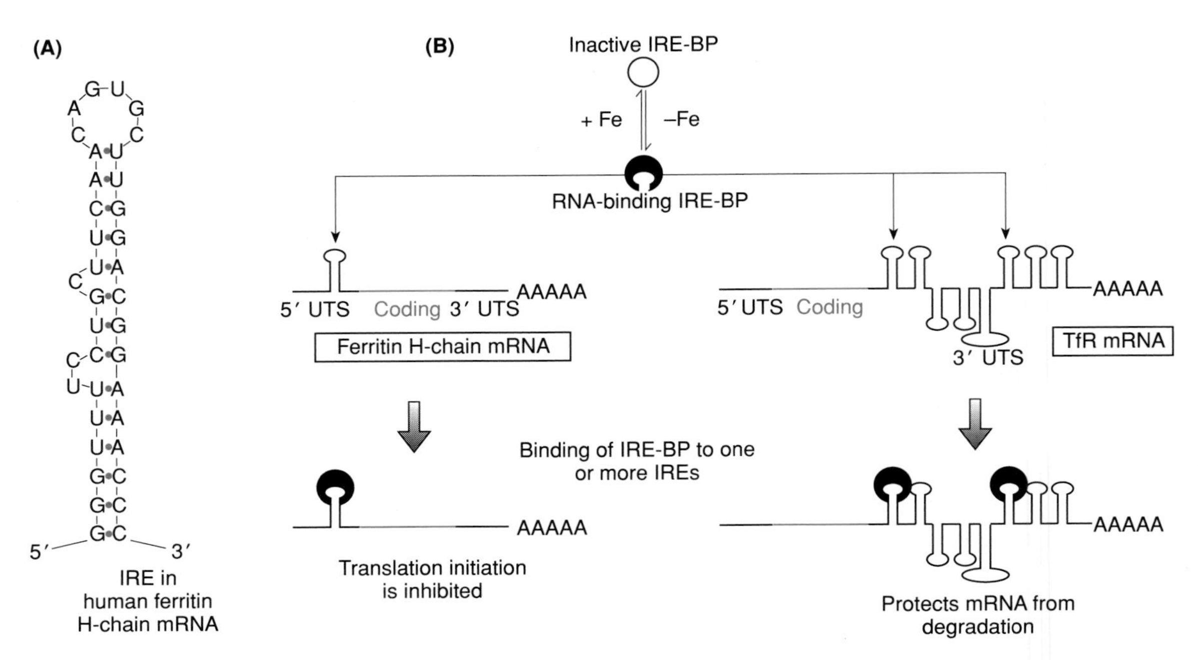

Figure 7.14: The IRE-binding protein regulates the production of ferritin heavy chain and transferrin receptor by binding to iron-response elements (IREs) in the 5′- or 3′-untranslated regions.

(A) Structure of the IRE in the 5′-UTR of the ferritin heavy chain.
(B) Binding of the IRE-binding protein to ferritin and transferrin mRNAs has contrasting effects on protein synthesis.

Box 7.4: Mechanisms resulting in monoallelic expression from biallelic genes in human (mammalian) cells

Mechanism	Relevant genes and cellular location of monoallelic expression
Genomic imprinting	A small number of autosomal genes. Cellular locations depend on where an individual gene is expressed but note (page 172) that some imprinted genes show monoallelic expression in some cell types but biallelic expression in others.
X chromosome inactivation (page 172)	Confined to certain X-linked genes in females only. Monoallelic expression in all cells in which genes are expressed. Includes: (i) numerous genes which are expressed only on the active X chromosome (see *Figure 9.11*); (ii) the *XIST* gene which uniquely is expressed only on the inactive X chromosome
Allelic exclusion (page 178)	Ig gene expression in B lymphocytes; TCR gene expression in T lymphocytes

Genomic imprinting describes differences in the expression of paternal and maternal alleles of certain autosomal genes in mammals

For the great majority of mammalian genes, the expression of an allele does not depend on whether that allele has been inherited from the mother or from the father. However, a few mammalian autosomal genes are unusual in that the expression of an allele depends on its parental origin: for some genes, the maternal allele but not the paternal allele is expressed in certain cells; for others it is always the paternal allele which is expressed. This phenomenon is known as **genomic imprinting**. Disease can occur if the normally expressed allele receives a deleterious mutation, or as a result of *uniparental disomy* (where an individual receives two paternal copies of a specific chromosome, or two maternal homologs instead of one maternal and one paternal homolog). If two copies of the chromosome with the nonexpressed allele are inherited, gene function will be abolished by what is essentially an *epigenetic* mechanism (both gene copies are genetically capable of function). See pages 70–72 for an overview of genomic imprinting, its relationship to certain inherited disorders and ideas on the rationale for imprinting.

The above observations suggest that some mechanism must be able to distinguish between maternally and paternally inherited alleles: as chromosomes pass through the male and female germlines they must acquire some *imprint* to signal a difference between paternal and maternal alleles in the developing organism. During development, this imprint would be expected to be stably inherited at least for many rounds of DNA duplication (but see below). Clearly, there must also be a mechanism for erasing the imprints, as required when, for example, a man passes on an allele which he had inherited from his mother (see *Figure 7.15*). Of the small number of human genes that are known to be imprinted, examples are known of loci where only the paternal allele is expressed and loci where only the maternal allele is expressed. They include two closely neighboring genes, *H19* and *IGF2*, where complementary patterns of uniparental expression are seen (*Table 7.11*). The murine *H19* and *Igf2* homologs show the same patterns of imprinting and are located with two other imprinted genes in a stretch of 350 kb of DNA. Recently, evidence has been obtained for long-range transcriptional control mechanisms in this chromatin domain: sequences around the mouse *H19* gene participate in the regulation of expression of two other imprinted genes located over 90 kb upstream (Leighton *et al.*, 1995; see also page 174)

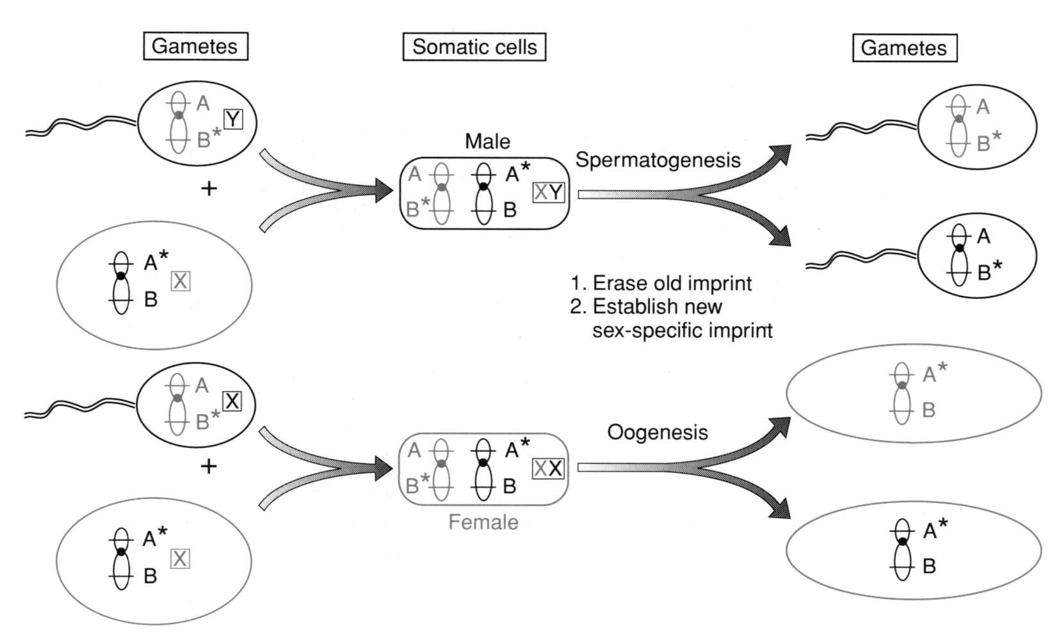

Figure 7.15: Genomic (gametic) imprinting.

The diagram illustrates the fate of a chromosome carrying two genes, A and B, which are subject to imprinting: A is imprinted in the female germline, B is imprinted in the male germline, as indicated by asterisks. As a result, in diploid somatic cells A is imprinted when present on a maternally inherited chromosome and B is imprinted when present on a paternally inherited chromosome. An individual chromosome may pass through the male and female germlines in successive generations: a man may transmit a chromosome inherited from his mother and a woman can transmit a chromosome inherited from her father, as indicated by the gametes in the left panel. As a result, there must be a mechanism whereby the old imprint is erased from the germline prior to establishing a new sex-specific imprint.

Table 7.11: Examples of imprinted human genes

Gene and location	Expressed allele	Differences in expression patterns
IGF2 (insulin-like growth factor type 2) at 11p15	Paternal	Adult liver is unusual in showing biallelic expression
H19 (tumor suppressor RNA) at 11p15	Maternal	
SNRPN (small nuclear ribonucleoprotein polypeptide N) at 15q12	Maternal	
WT1 (Wilms' tumor)	(Rarely – maternal)	Biallelic expression in kidney but frequently imprinted in cells of placenta and brain, showing expression of maternal allele

Currently, the imprinting mechanism remains obscure (Barlow, 1994). The imprint would be expected to be acquired when the parental genomes are separate (most likely in the germline, or in the early zygote before fusion of male and female pronuclei), but should be present thereafter on one of the parental chromosomes. Allele-specific DNA methylation differences have been found in the endogenously imprinted genes, suggesting that DNA methylation may be involved in the mechanism. In addition, imprinted genes often have short tandem direct repeats embedded in GC-rich sequences subject to monoparental methylation (Neumann *et al.*, 1995). Because of the association of hypermethylated regions of DNA with transcriptional inactivity, the imprinting signal has been widely expected to be essentially a negative one. Surprisingly, however, in at least one murine locus, it is the expressed allele which has been found to be methylated in the region thought to contain the primary gametic imprint (Barlow, 1994). Additionally, tissue-specific differences are known in some cases, such as the Wilms' tumor gene and the insulin-like growth factor type 2 gene (*IGF2; Table 7.10*). This suggests that a mechanism must exist whereby certain somatic cells can either relax or override the initial imprint, resulting in both alleles being expressed. According to one theory, the evolution of the mammalian imprinting occurred as a result of the competing interests of paternal and maternal genes in pregnancy (see pages 71–72)

X chromosome inactivation in mammals is designed to overcome differences in the ratio of X-linked genes and autosomal genes in male and female cells

X chromosome inactivation is a process that occurs in all mammals, resulting in selective inactivation of alleles on one of the two X chromosomes in females (Migeon, 1994). It provides a mechanism of **dosage compensation**. This overcomes sex differences in the expected ratio of autosomal gene dosage to X chromosome gene dosage (2:1 in males: 1:1 in females): males with a single X chromosome are *constitutionally* **hemizygous** for X chromosome genes, but females become *functionally hemizygous* by inactivating one of the parental X chromosome alleles (see page 50).

X chromosome inactivation occurs in female mammals at an early stage in development, being initiated at the late blastula stage in mice, and most likely also in humans. In each cell that will give rise to the female fetus, one of the two parental X chromosomes is *randomly* inactivated. Thereafter, however, the inactive chromosome usually remains inactive in all progeny cells, that is the X chromosome inactivation pattern is clonally inherited (*Figure 7.16*). This means that female mammals are mosaics, comprising mixtures of cell lines in which the paternal X is inactivated

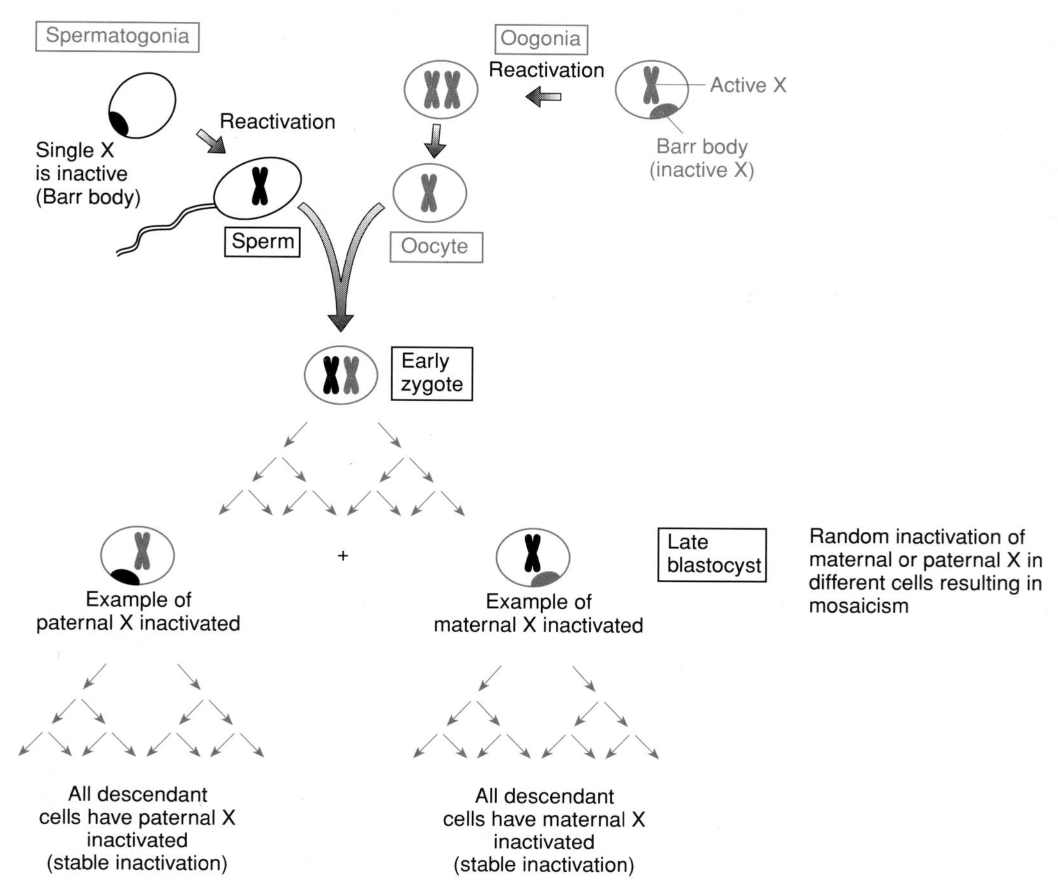

Figure 7.16: The process of X chromosome inactivation in mammals.

Note that during spermatogenesis and oogenesis the X chromosome is transiently inactivated, forming the characteristic condensed X chromosome *(Barr body)*, prior to reactivation. In the early female zygote, both X chromosomes are active but, in the late blastocyst, a choice is made randomly in each cell to inactivate either the paternal or the maternal X chromosome. Thereafter, the pattern of X chromosome inactivation remains essentially stable so that the resulting organism will have two populations of cells, one with the paternal X inactivated, the other with the maternal X inactivated. Adapted from Migeon (1994) with permission from Elsevier Trends Journals.

and cell lines where the maternally inherited X is inactivated. In addition to X chromosome inactivation in female somatic cells, the X chromosome is known to be inactivated transiently during gametogenesis in both males and females.

The process of X chromosome inactivation is complex, and distinct molecular mechanisms are involved in initiation of inactivation and maintenance of the inactivation. The basic mechanism involves inhibition of transcription, with initiation of transcriptional inactivation probably affecting the entire chromosome, while subsequent maintenance of the inhibition is thought to occur locus by locus. The gene that is directly responsible, *XIST*, is expressed exclusively from the inactivated X chromosome. Thus, like the genes that are inactivated on the X chromosome, it too is subject to monoallelic expression but in this unique case the allele on the inactive X chromosome is the one that is expressed in female cells. It appears to encode a functional RNA with several conserved tandem repeats (Brown *et al.*, 1992) and is thought to be required for initiation of X chromosome inactivation, but not for its maintenance (Brown and Willard, 1994).

Chromatin structure may exert long-range control over gene expression

Promoters and related upstream elements typically control expression of a single gene with a transcription start point located within 1 kb of the control element. Some *cis*-acting elements, however, exert long-range control over a much larger chromosomal region. Several different mechanisms have been described, and probably many others will be discovered as large-scale mapping of the genome proceeds.

Competition for enhancers or silencers

Sometimes long-range control of gene expression appears to depend on competition between clustered genes for an enhancer. The control of globin gene expression by the locus control regions is described in Chapter 8 (see page 190). Another example comes from the *H19* and *IGF2* genes located at 11p15. These adjacent genes are imprinted in opposite senses, so that only one of them is expressed on any one chromosome, depending on the parental origin (see page 171). Studies of deletions (summarized by Eden and Cedar, 1995) suggest that the mechanism is, at least in part, competition for a single enhancer.

Position effects

Studies of chromosomal rearrangements in *Drosophila* have shown that proximity to centromeres, telomeres or heterochromatic blocks may suppress gene expression, presumably by altering the structure of a large chromatin domain. Fascio-scapulohumeral muscular dystrophy (FSHD; MIM 158900) is a possible example of a similar **position effect** in man. The gene for this autosomal dominant progressive neuromuscular disease maps close to the telomere of chromosome 4q. When Southern blots of *Eco*RI-digested DNA are hybridized to a subtelomeric probe p13H-11, a very large hybridizing fragment of over 30 kb is seen. DNA from FSHD patients consistently shows smaller bands of 14–28 kb. These patients have a reduced copy number of a 3.2 kb repetitive sequence that is recognized by the probe, and *de novo* deletion of stretches of the repeats have been observed by FSHD patients with no previous family history.

Hopes that the 3.2 kb sequence would contain the *FSHD* gene, however, have been disappointed. Even though it contains part of a homeodomain (Hewitt *et al.*, 1994), there is no evidence that any part of it is transcribed or expressed. Most probably the *FSHD* gene is located proximally to the tandem 3.2 kb repeats, and the deletions move it closer to the 4q telomere, where it is silenced by a position effect.

Chromatin domains

Evidence of long-range effects controlling gene expression over large chromosomal domains has emerged from research aimed at cloning disease genes. Once a disease gene has been mapped, positional cloning (pages 372–387) can be greatly helped by discovering an affected patient who has a chromosomal breakpoint at the location of the disease gene. Hopefully, the patient is affected because the break has disrupted the disease gene. Cloning the breakpoint will then identify the gene, usually with much less effort than alternative strategies.

Such an approach has succeeded many times, but sometimes cloning apparently correctly located breakpoints fails to reveal the disease gene. Examples are aniridia (AN1; MIM 106210), which is caused by loss-of-function mutations of the *PAX6* gene on 11p13, and campomelic dysplasia (CMPD1; MIM 211970), which is caused by mutations in the *SOX9* gene on 17q24. In each case typically affected patients are known who have clearly causative chromosomal breaks, but the breakpoints lie 50–100 kb outside the gene and do not physically disrupt it (Fantes *et al.*, 1995;

Foster *et al.*, 1994). It seems likely that expression of the gene is suppressed by a long-range effect analogous to the classic position effects described above, reflecting the novel chromosomal environment created by the translocation. Probably these effects are related to the isochore structure of chromosomes which underlies the banding patterns (page 151).

Prader–Willi and Angelman syndromes (pages 407 and 409) bring together position effects, imprinting and DNA methylation. A *cis*-acting sequence analogous to the globin locus control region has been identified which governs parent-specific methylation and gene expression of a megabase-size chromosomal region at 15q11.

X inactivation

X inactivation must be mediated by some sort of long-range chromatin structural change. This is so because a diffusible agent would not be able to affect just the X chromosome on which the *XIST* gene (page 173) is expressed.

Selective gene expression in cells of mammalian embryos most likely develops in response to short range cell–cell signaling events

Because nucleated human cells contain basically identical genetic instructions, the question arises as to how selective gene expression at different developmental stages or in differentiated cells ever develops in the first place. Something must set up an asymmetry or axis in the fertilized egg cell. In *Drosophila*, the egg is inherently asymmetrical because of transfer of gene products from asymmetrically sited nurse cells. The embryo develops initially as a multinucleate *syncytium* (effectively one big cell) and regionalization depends on the response of individual nuclei to long-range gradients of regulatory molecules.

In mammals, the egg cell is relatively small and early embryonic development creates an apparently symmetrical aggregate of individual cells. Nevertheless, development becomes asymmetric. The initial positional clue may be provided by the point of entry of the sperm. It is likely that short range intercellular signaling molecules provide a means of identifying cell position, and triggering differential gene expression. For example, if an intercellular signaling molecule has a range of, say, one cell diameter, then the cells at the outside of the blastula will receive different signals from those surrounded by neighbors on all sides, and the different positional cues may be translated into differential gene expression. As particular cell systems develop during, for example, organogenesis (mostly accomplished between the 4th and 9th embryonic weeks), particular cell type growth or differentiation factors may then induce the expression of developmental stage- and/or tissue-specific transcription factors.

The unique organization and expression of Ig and TCR genes

The organization and expression of immunoglobulin (Ig) and T-cell recepter (TCR) genes is in many ways quite different from that of other genes. This is so because of the need for each individual to produce a huge variety of different Igs and TCRs. An individual B or T lymphocyte is *monospecific* and produces a single type of Ig or TCR; it is the population of different B and T cells in any one individual that enables the synthesis of so many different types of these molecules. B and T lymphocytes need to be extremely diverse because they represent the cells that provide antibody responses or cell-mediated responses to foreign antigen: by providing a large repertoire of Igs and TCRs, the possibilities for being able to recognize and bind very many different types of foreign antigen are greatly increased.

Ig and TCR genes exhibit a unique organization: multiple gene segments can encode each of several different regions of the polypeptide

Polypeptide structure

An Ig molecule consists of four polypeptide chains, two identical heavy chains and two identical light chains (see *Figure 8.2*). The light chains fall into two classes: kappa (κ) and lambda (λ) light chains, which are functionally equivalent. At the N-terminal segments of each type of chain are the so-called **variable (V) regions,** which need to bind foreign antigen; the remaining C-terminal segments are **constant (C) regions**. In the case of the heavy chains, there are different alternatives for the constant region which specify the tissues in which the Ig will be expressed and dictate the **immunoglobulin class** (*Table 7.12*). Similarly, TCRs, which provide cell-mediated immune responses to foreign antigens, consist of two types of chain. Each such chain has Ig-like variable regions which bind foreign antigen, and constant regions which anchor the molecule to the cell surface (see *Figure 8.2*). The most frequently occurring TCRs have a β and a γ chain; a minor population consists of an α chain and a δ chain.

Table 7.12: Ig classes and subclasses

Class (and subclass)	Type of heavy chain	Location
IgA (IgA1,IgA2)	α (α_1,α_2)	Predominant Ig in seromuccous secretions, e.g. saliva, milk, etc.
IgD	δ	Low in serum but present in large quantities on surface of many circulating B cells
IgE	ε	Especially on suface membrane of basophils and mast cells
IgG (IgG1,IgG2,IgG3,IgG4)	γ (γ_1,γ_2,γ_3,γ_4)	Major serum Ig
IgM	μ	Predominant 'early' antibody

Gene structure

The genes which encode the different types of chain in Igs and TCRs are located on different chromosomes and are organized as clusters of numerous gene segments (*Table 7.13*). Each such cluster is unusual in that the coding sequences for *specific segments* of each chain are often present in numerous different copies that are sequentially repeated. For example, although the constant region of human κ light chain Ig is encoded by a single C_κ sequence, the variable regions are encoded by a combination of a V_κ segment (which encodes most of the variable region) and a short J_κ segment (joining segment; encodes a small part at the C-terminal end of the variable region) which are selected from a total of about 76 alternative V_κ segments and five alternative J_κ segments. Although the λ light chain is similarly encoded by V_λ, J_λ and

Table 7.13: Functional human Ig and TCR loci

Locus	Location	Number of gene segments			
		V	D	J	C
IGH	14q32.3	86	30	9	11
IGK	2p12	76	0	5	1
IGL	22q11	52	0	7	7
TCRA	14q11–12	60	0	75	1
TCRB	7q32–33	70–100	2	13	2
TCRG	7p15	8	0	5	2
TCRD	14q11–12	6	3	3	1

C_λ segments, the heavy chain Ig locus shows some differences. The variable region is encoded by a combination of a V_H gene segment, a J_H gene segment and also a D_H gene segment (encoding a diversity segment), each selected from many repeated gene segments. Additionally, there are a variety of different C_H sequences which specify the *class* of the Ig (see above). In total this cluster comprises about 140 gene segments, of which about one-third are known to be incapable of expression, and spans about 1200 kb (see *Figure 7.17*).

As each Ig gene cluster or TCR gene cluster in an individual B or T lymphocyte only ever gives rise to at most one Ig or TCR polypeptide, an entire cluster can functionally be regarded as a single, albeit unusual, type of gene. However, the individual gene segments cannot be regarded as the functional equivalent of *classical exons.* This is so because individual gene segments in these clusters are sometimes composed of coding DNA and noncoding DNA. For example, each of the human C_H sequences is itself composed of three or four classical exons separated by introns: after transcription into RNA, the intronic sequences are discarded, and only the exonic sequences are retained in the mRNA.

Programmed DNA rearrangements at the Ig and TCR loci occur during the maturation of B and T lymphocytes, respectively

The unique arrangement of gene segments in the Ig and TCR gene clusters reflects the very unusual way in which somatic recombinations are required in B and T lymphocytes before functional Ig and TCR genes can be assembled and then expressed (see below). Such somatic recombinations result in bringing together different

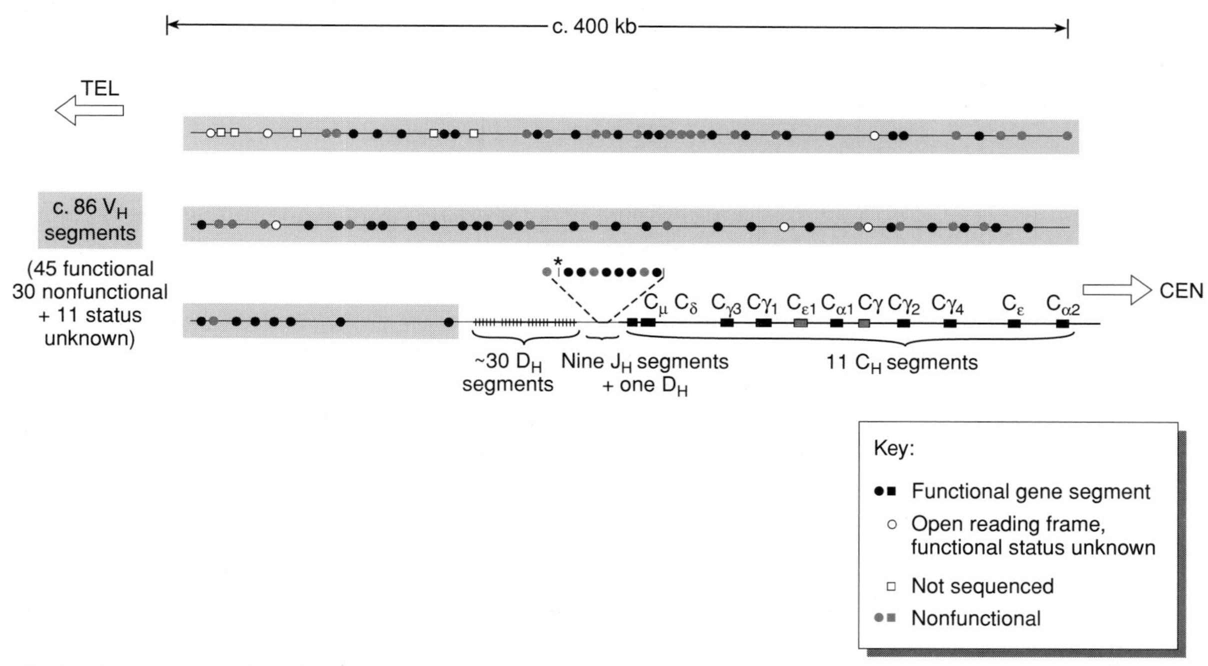

Figure 7.17: The Ig heavy chain locus on 14q32 contains about 86 variable (V) region sequences, 30 diversity (D) segments, nine joining (J) segments and 11 constant region (C) sequences.

The entire locus spans about 1200 kb of 14q32.3 and, for clarification, is shown as three segments of 400 kb from the telomeric end (top) to the centromeric end (bottom). Although the D_H segments are mostly located in a few clusters separating the V_H and J_H segments, at least one such segment is located in the J_H segment region. Segments indicated by open circles have the required open reading frames but have not been observed in productive rearrangements and so their functional status is unknown. Segments which are known to be nonfunctional are indicated in red, and account for approximately one third of all the segments. *Note* that although this is the only functional human heavy chain locus, small clusters of V_H and D segments are also located on 15q11.2 and 16p11.2. Adapted from data in Cook *et al.* (1994) with permission from Nature America Inc.

combinations of the different gene segments in different individual lymphocytes. Consequently, they can be regarded as both *tissue-specific (confined to B and T lymphocytes) and cell-specific* ***alternative DNA splicing*** *events* (as opposed to alternative RNA splicing which brings about different combinations of exons *at the RNA level*; see page 16). As a result, the original germline gene organization is altered: gene segments that were distant in the germline are spliced together at the DNA level. Because the choice of which of the many repeated gene segments are recombined to give a functional V–J or V–D–J unit is *cell specific*, individual B and T cells produce different Igs and TCRs. This means that, in a sense, every individual is a mosaic with respect to the organization of the Ig and TCR genes in B and T lymphocytes; even identical twins will diverge genetically.

The rearrangements which lead to the production of functional light chains and heavy chains of Igs are slightly different.

- **Making a light chain**. In order to generate a functional κ light chain Ig, for example, a somatic recombination event brings together a specific combination of one of the $V_κ$ gene segments and one of the $J_κ$ gene segments (**V–J joining**). Thereafter, splicing to the single $C_κ$ sequence occurs *at the RNA level* (*Figure 7.18A*).
- **Making a heavy chain**. Two successive somatic recombinations are required, resulting first in D_H–J_H joining, and then V_H–D_H–J_H joining. Subsequently the resulting V_H–D_H–J_H coding sequence is spliced *at the RNA level* to the nearest C_H sequence, initially $C_μ$ (*Figure 7.18B*).

Because there are three types of functional Ig gene loci in human cells (heavy chain, κ light chain and λ light chain), and because these occur on both maternal and paternal homologs, there are six chromosomal segments in which DNA rearrangments can result in production of an Ig chain. However, an *individual* B cell is *monospecific*: it produces only one type of Ig molecule with a single type of heavy chain and a single type of light chain. This is so for two reasons:

- *allelic exclusion.* A light chain or a heavy chain can be synthesized from a maternal chromosome or a paternal chromosome in any one B cell, but not from both parental homologs. As a result, there is *monoallelic* expression at the heavy chain gene locus in B cells. This phenomenon also applies to TCR gene clusters.
- *Light chain exclusion.* A light chain synthesized in a single B cell may be a κ chain, or a λ chain, but never both. As a result of this requirement plus that of allelic exclusion, there is monoallelic expression at one of the two functional light chain gene clusters *and no expression at the other.*

The decision to choose which of the two heavy chain alleles and which of the four possible segments can make a light chain appears to be random. Most likely, in each B-cell precursor, productive DNA rearrangements are attempted at all six Ig alleles but the chances of productive arrangements in more than one light chain cluster or more than one heavy chain allele may not be high. Additionally, however, there appears to be some kind of *negative feedback regulation*: a functional rearrangement at one of the heavy chain alleles suppresses rearrangements occurring in the other allele, and a functional rearrangement at any one of the four regions capable of encoding a light chain suppresses rearrangements occurring in the other three.

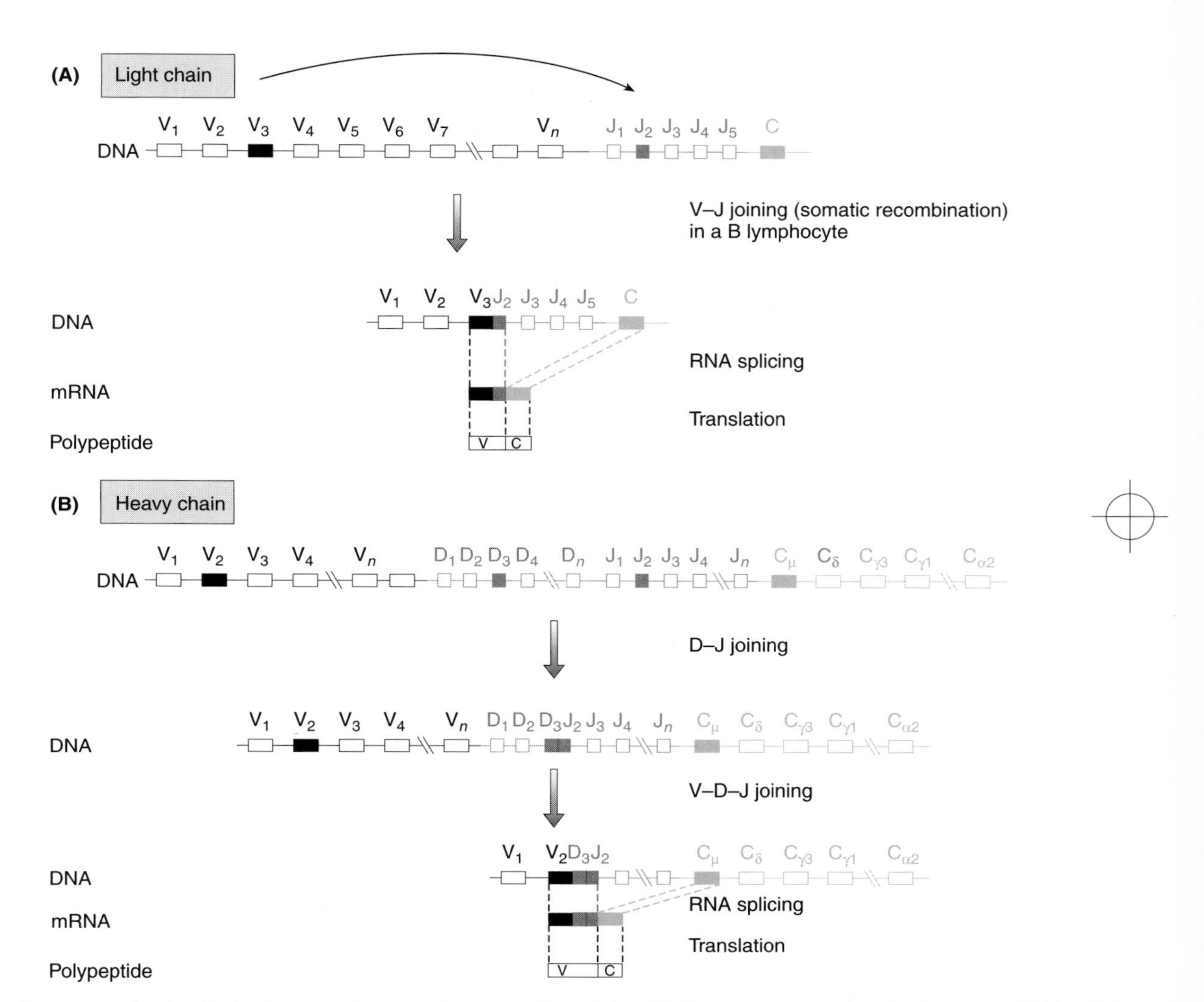

Figure 7.18: Igs are synthesized following somatic recombination of V and J, or V, D and J segments and subsequent RNA splicing to C sequences.

(A) Light chain synthesis. Somatic recombination (*DNA splicing*) results in joining of a specific variable (V) segment to a specific joining (J) segment; the example shows a V_3–J_2 joining which is only one of many possibilities. The VJ unit is then spliced to the constant region (C) sequence by *RNA splicing*.
(B) Heavy chain synthesis. Two sequential somatic recombinations produce first D–J joining, then a VDJ unit. Subsequent RNA splicing results in splicing of the VDJ sequence to the C_μ sequence. As the B cell matures, however, subsequent somatic recombinations result in joining of the previously selected VDJ unit to different C genes (*heavy chain switch*, see text and *Figure 7.20*).

V–J and V–D–J joining is often achieved by intrachromatid deletions, and also by megabase inversions in the former case

The genetic mechanism leading to V–J and V–D–J joining often involves large-scale deletions which are thought to occur by an intrachromatid recombination event, similar to those used in V–D–J–C joining (see next section). In addition, V–J joining often occurs as a result of megabase inversions. The human κ light chain gene locus spans about 1840 kb on 2p12 and includes about 76 V_κ segments, mostly comprising pairs of duplicated V gene segments, organised as two clusters: a *proximal cluster* located adjacent to the J_κ segments and to the C_κ segment, and a *distal cluster*. This occurs as a result of an inverted repeat structure: V gene segments in the proximal V_κ cluster usually have a corresponding duplicate in a distal V_κ cluster which is separated from the proximal cluster by about 800 kb and in the opposite orientation (*Figure 7.19*). Depending on which V segment cluster is involved, V–J joining occurs by two possible routes:

- *V segments in the distal cluster* are joined to J segments by inversions.
- *V segments in the proximal cluster* are joined to J segments by deletions *Figure 7.19*).

Note that the joining process is imprecise, and so can also introduce a measure of variability in the sequence at the junctions of joined segments.

Class switching of heavy chains involves differential joining of a single VDJ unit to alternative DNA segments encoding constant regions

Although a B cell produces only one type of Ig molecule, the **heavy chain class** (see *Table 7.12*) can change during the cell lineage (**class switching** or **isotype switching**). Such switching involves differential joining of the same VDJ unit that was brought together by two successive somatic recombinations (see *Figure 7.18B*) to different segments encoding alternative constant regions. The initial joining of a VDJ sequence to constant region segments is accomplished *at the RNA level*. However, subsequently, class switching involves joining the same VDJ unit *at the DNA level* to alternative constant regions by yet more somatic recombination events (**V–D–J–C joining**). Class switching involves the following progression:

- *initial synthesis of IgM only by immature B cells.* This occurs because the VDJ unit is spliced at the RNA level to a C_μ sequence (*Figure 7.18B*).
- *Later synthesis of both IgM and IgD by immature B cells.* The partial switch to making IgD occurs because the VDJ unit can be spliced at the RNA level to a C_δ sequence, as a result of alternative RNA splicing (*Figure 7.20*).
- *Synthesis of IgG, IgE or IgA by mature B cells.* Class switching events involve splicing the same VDJ unit to a C_γ, C_ε or C_α sequence, respectively, at the DNA level as a result of a somatic recombination event (VDJ–C joining). The mechanism involves deletion of the intervening sequence by intrachromatid recombination (*Figure 7.20*).

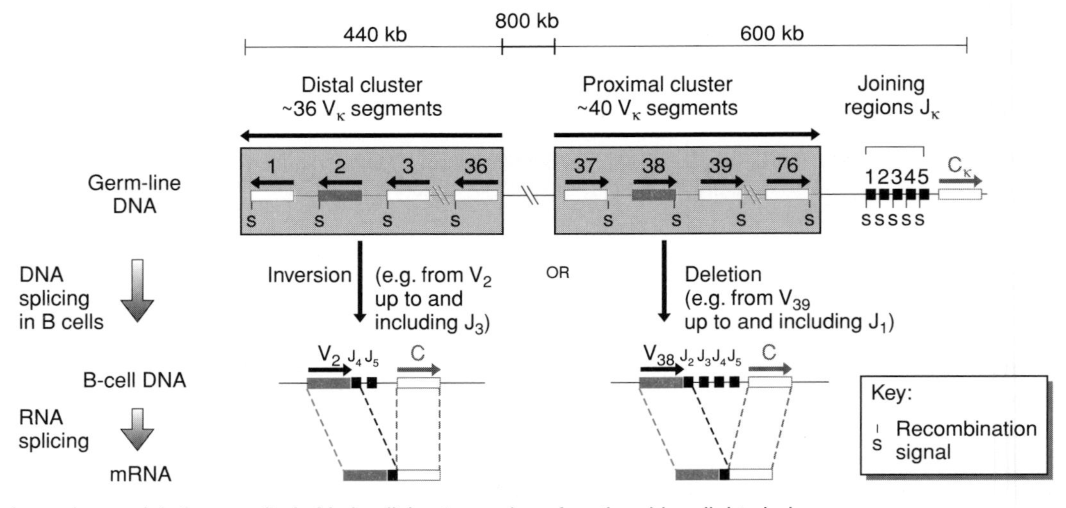

Figure 7.19: Inversion or deletion results in V–J splicing to produce functional Ig κ light chain genes.

The human κ light chain gene cluster contains about 76 V_κ segments arranged in two large clusters, in opposite orientations. V segments in the distal cluster have the opposite orientation to the J_κ segments and the single C_κ sequence. As a result, the DNA rearrangements used to splice distal V_κ segments to a J_κ segment are megabase inversions (Weichhold *et al.*, 1990). Those in the proximal cluster can undergo V–J joining by a somatic recombination resulting in a deletion of the intervening chromosomal segment, most likely through an intrachromatid recombination event such as that used in class switching (see *Figure 7.20*).

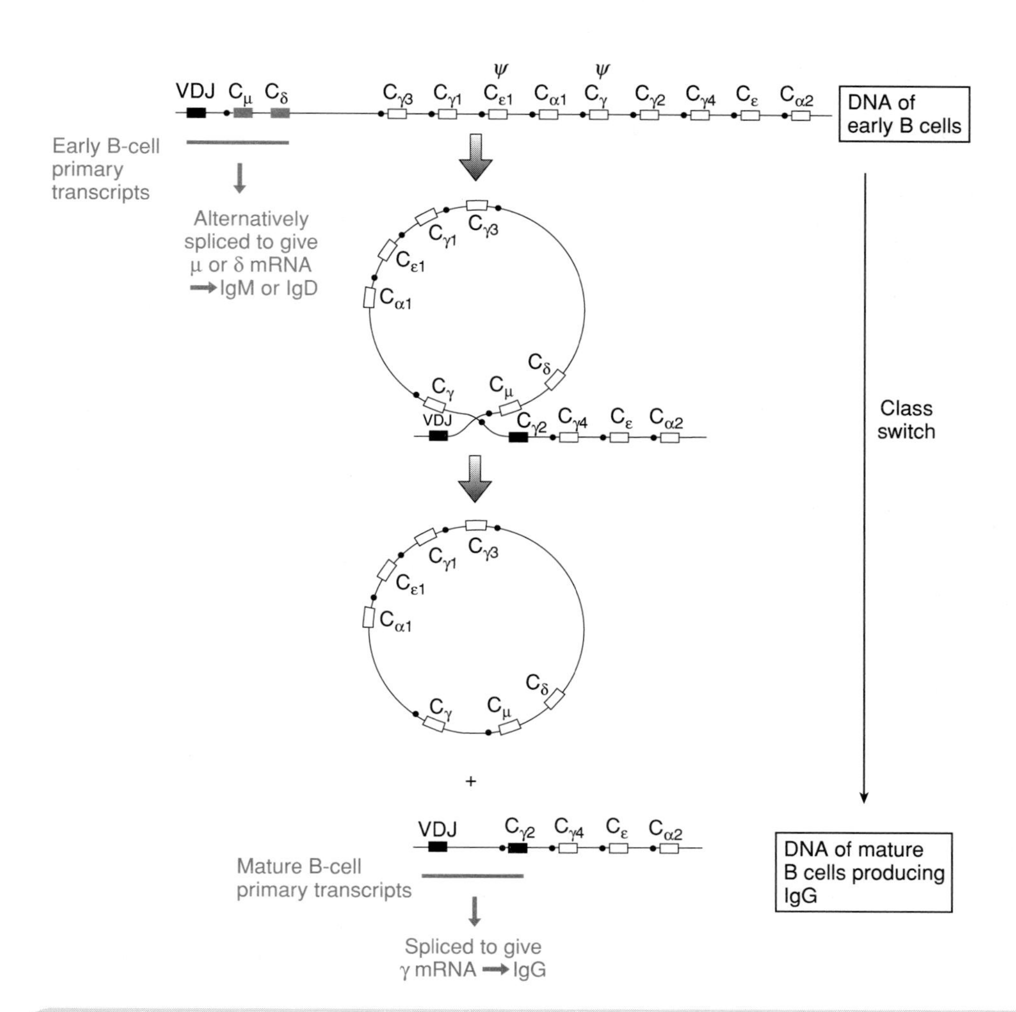

Figure 7.20: Ig heavy chain class switching is mediated by intrachromatid recombination.

Note that joining of the same VDJ unit to a C_μ or a C_δ sequence occurs at the level of RNA splicing to generate heavy chains for IgM and IgD respectively. In contrast, class switching to generate IgA, IgE or IgG involves joining of the same VDJ unit at the DNA level to, respectively, a C_α, C_ε or, as illustrated in the figure, a C_γ sequence.

Further reading

Gardiner K. (1995) Human genome organization. *Curr. Opin. Genet. Dev.*, **5**, 315.

Kao F-T. (1985) Human genome structure. *Int. Rev. Cytol.* **96**, 51.

Roitt I, Brostoff J, Male D. (1985) *Immunology*. Gower Medical Publishing, London.

Singer M, Berg P. (1991) *Genes and Genomes*. University Science Books, Mill Way, CA.

Travers A. (1993) *DNA–Protein Interactions*. Chapman & Hall, London.

References

Anderson S, Bankier AT, Barrell BG *et al.* (1981) *Nature*, **290**, 457–465.

Antequera F, Bird A. (1993) *Proc. Natl Acad. Sci. USA*, **90**, 11995–11999.

Barlow DP. (1994) *Trends Genet.*, **10**, 194–199.

Bernardi G. (1993) *Gene*, **135**, 57–66.

Brown CJ, Willard, H. (1994) *Nature*, **368**, 154–156.

Brown CJ, Hendrich BD, Rupert JL, Lafreniere RG, Xing Y, Lawrence J, Willard H. (1992) *Cell*, **71**, 527–542.

Christiano A, Hoffman GG, Chung-Honet LC, Lee S, Cheng W, Uitto J, Greenspan DS. (1994) *Genomics*, **21**, 169–179.

Clark AR, Doherty K. (1993) *Biochem. J.*, **296**, 521–541.

Clayton DA. (1992) *Int. Rev. Cytol.*, **141**, 217–232.

Cook GP, Tomlinson IM, Walter G, Riethman H, Carter NP, Buluwela L, Winter G, Rabbitts TH. (1994) *Nature Genetics*, **7**, 162–168.

Craig JM, Bickmore WA. (1994) *Nature Genetics*, **7**, 376–381.

D'Souza VN, Man NT, Morris GE, Karges W, Pillers D-AM, Ray PN. (1995) *Hum. Mol. Genet.*, **4**, 837–842.

Eden S, Cedar H. (1995) *Nature*, **375**, 16–17.

Fantes J, Redcker B, Breen M *et al.* (1995) *Hum. Mol. Genet.*, **4**, 415–422.

Fields C, Adams MD, White O, Venter JC. (1994) *Nature Genetics*, **7**, 345–346.

Foster, JW, Dominquez-Steglich MA, Guioli C *et al.* (1994) *Nature*, **372**, 525–530.

Hewitt JE, Lyle R, Clark LN. (1994) *Hum. Mol. Genet.*, **3**, 1287–1295.

Hodges D, Bernstein SI. (1994) *Adv. Genet.*, **31**, 207–281.

Hodges P, Scott J. (1992) *Trends Biochem. Sci.*, **17**, 77–81.

Klausner RD, Rouault TA, Harford JB. (1993) *Cell*, **72**, 19–28.

Leighton PA, Ingram RS, Eggenschwiler J, Efstratiadis A, Tilghman SM. (1995) *Nature*, **375**, 34–39.

Levinson B, Kenwrick S, Gamel P, Fisher K, Gitschier J. (1992) *Genomics*, **14**, 585–589.

Migeon BR. (1994) *Trends Genet.*, **10**, 230–235.

Navaratnam N, Bhattacharya S, Fujuno T, Patel D, Jarmuz AL, Scott J. (1995) *Cell*, **81**, 187–195.

Neumann B, Kubicka P, Barlow DP. (1995) *Nature Genetics*, **9**, 12–13.

Scott J. (1995) *Cell*, **81,** 833–836.

Serfling E, Jasin M, Schaffner W. (1985) *Trends Genet.*, **1**, 224–230.

Stephens JC, Cavanaugh ML, Gradie MI, Mador ML, Kidd K. (1990) *Science*, **250**, 237–244.

Tennyson CN, Klamut HJ, Warton RG. (1995) *Nature Genetics*, **9**, 184–190.

Viskochil D, Cawthorn R, O'Connell P *et al.* (1991) *Mol. Cell Biol.*, **11**, 906–912.

Weichhold GM, Klobeck H-G, Ohnheiser R, Combriato G, Zachau HG. (1990) *Nature*, **347**, 90–92.

Wrana JL. (1994) *Bioessays*, **16**, 89–90.

Xu H, Wei H, Tassone F, Graw S, Gardiner K, Weissman SM. (1995) *Genomics*, **27**, 1–8.

Human multigene families and repetitive DNA

Chapter 8

Principle of repetitive DNA and multigene families

DNA sequences in the nuclear diploid genome usually exist as two allelic copies (on paternal and maternal homologous chromosomes). In addition to this degree of repetition, approximately 40% of the human nuclear genome in both haploid and diploid cells is composed of sets of closely related *nonallelic* DNA sequences (**DNA sequence families** or **repetitive DNA**). Within the considerable variety of different repetitive DNA sequences are DNA sequence families whose individual members include functional genes (**multigene families**), and also many examples of nongenic repetitive DNA sequence families.

The reassociation kinetics of human DNA suggest three broad classes of DNA sequence

Reassociation kinetics (page 118) first suggested that complex genomes, such as the human genome, comprise different sequence classes on the basis of the *copy number*. Typically, this involves randomly shearing human DNA (e.g. by sonication) to give fragments whose average size is about 500 bp and denaturing the sheared DNA by heating to separate the complementary strands of each fragment. Thereafter the DNA is cooled, typically to a temperature of about 20–30°C below the *melting temperature*, T_m (which marks the mid-point of the transition between the double-stranded and single-stranded states of DNA heated in solution, see page 115). The cooled DNA renatures but the rate of reassociation depends not only on time (t) but also on the initial concentration (C_0) of that sequence (i.e. the C_0t value; see page 118). This type of analysis has suggested that the human genome consists of roughly three broad sequence components:

- **Single copy** (or **very low copy number**) **DNA** (60%). Reassociates very slowly because a single strand from a single copy sequence will require some considerable time to find a complementary partner strand, given that the vast majority of DNA fragments are unrelated to it.
- **Moderately repetitive** (30%). Intermediate speed of reassociation.
- **Highly repetitive** (10%). Reassociates very rapidly because there are numerous copies of the same sequence and the chances of quickly finding complementary partners within the mass of different fragments are high.

Members of DNA sequence families can be identified by a variety of different approaches

Irrespective of the nature of the DNA sequences, the operational definition of a DNA sequence family is the comparatively high level of DNA sequence similarity (**sequence homology**) between whole family members, or components of the family members. Members of a DNA sequence family can be identified and actively sought by a variety of methods:

- **DNA sequencing** can identify a particular sequence as belonging to a DNA sequence family when there is prior sequence information for other family members. The sequence analysis also allows direct calculations on the degree of sequence relatedness of family members (see page 515).
- **DNA hybridization assays** are a generally quick and convenient way of identifying a particular DNA fragment as belonging to a family of related sequences. A cloned gene or sequence of interest can be identified as belonging to a DNA sequence family by using it as a hybridization probe: against Southern blots of genomic DNA samples it will reveal characteristically complex band patterns. Identification of new members of small gene families may involve hybridization screening of monochromosomal hybrids (page 277) followed by screening chromosome-specific genomic DNA libraries (page 284). Note that for some multigene families the hybridization probe used may be a small part of a gene which is particularly highly conserved, such as a homeobox, etc.
- **PCR assays** are extremely valuable for identifying novel members of a gene family by designing primers corresponding to amino acid sequences that are known to be very highly conserved between the products of known members of a gene family.

When two members of a repetitive DNA sequence family exhibit a high degree of sequence homology, a recent common evolutionary origin is indicated. As detailed in the following sections, DNA sequence families show considerable variation in the number of different repeat unit members in the family, the size of the repeating unit, chromosomal location, mode of repetition and capacity for expression.

Multigene families

Human gene families vary in the overall sequence relatedness of different family members and the extent to which particularly conserved subgenic sequences define the family

A large percentage of actively expressed human genes are members of families of DNA sequences which show a high degree of sequence similarity. However, the extent of sequence sharing and the organization of family members can vary widely. Some family members may be nonfunctional (see page 187) and rapidly accumulate sequence differences, leading to marked sequence divergence.

Classical gene families

Classical gene families are distinguished by members which exhibit a high degree of sequence homology over most of the gene length or, at least, the coding DNA component, a feature which automatically identifies such sequences as being closely related evolutionarily as well as functionally. In some cases, such as the individual rRNA gene families and individual histone gene families, there is an extremely high degree of sequence similarity between family members. Many other large gene families show a high degree of sequence similarity between family members.

Gene families encoding products with large, highly conserved domains

In some gene families there is particularly pronounced homology within specific strongly conserved regions of the genes. Often such families encode products that have an important developmental function for which the presence of a very highly conserved domain is crucially important. Examples include genes possessing large highly conserved sequence motifs, such as the paired box of *PAX* genes (~390 bp) or the homeobox (~180 bp). The sequence similarity between the full lengths of family members may be low. However, if a sufficiently long conserved sequence can be isolated from one family, it can be used as a probe in a DNA hybridization assay in order to identify other members of the family.

Gene families encoding products with very short conserved amino acid motifs

The members of some gene families may not be very obviously related at the DNA sequence level, but nevertheless encode gene products that are characterized by a common general function and the presence of very short conserved sequence motifs. For example, the DEAD box gene family contains several different genes whose products appear to function as RNA helicases and are characterized by the presence of eight short amino acid sequence motifs, including the DEAD box (i.e. the sequence Asp–Glu–Ala–Asp as represented by the one-letter amino acid code). The WD repeat gene family is an example of a family where there is considerable diversity in product function, although all the products appear to have a regulatory function. Each of the products is, however, characterized by tandem repeats in which a central core is a fixed length and contains small conserved amino acid motifs, including the WD sequence (tryptophan–aspartate; see *Figure 8.1*).

Gene superfamilies

In some types of gene family, the genes encode products that are known to be functionally related in a general sense, and show only very weak sequence homology over a large segment, without very significant conserved amino acid motifs.

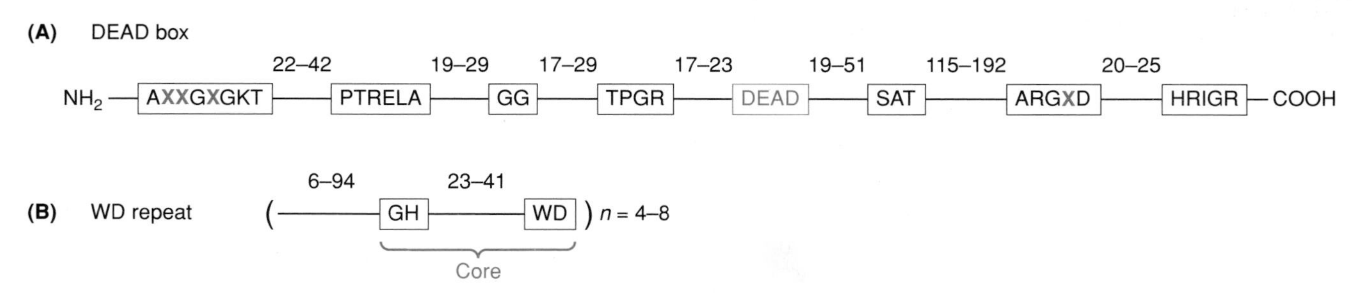

Figure 8.1: Some gene families are defined by functionally related gene products bearing very short conserved amino acid motifs.

(A) Motifs in the DEAD box family. This gene family encodes products implicated in cellular processes involving alteration of RNA secondary structure, such as translation initiation and splicing. Eight very highly conserved amino acid motifs are evident, including the DEAD box (Asp–Glu–Ala–Asp). Numbers refer to frequently found size ranges for intervening amino acid sequences (see Schmid and Linder, 1992). X = any amino acid. See inside front cover for the one-letter amino acid code.
(B) WD repeat family. This gene family encodes products that are involved in a variety of regulatory functions, such as regulation of cell division, transcription, transmembrane signaling, mRNA modification, etc. The gene products are characterized by between four and eight tandem repeats containing a core sequence of fixed length (from 27–45 amino acids, terminating in the dipeptide WD, i.e. Trp-Asp) preceded by a unit whose length can vary between repeats (see Neer *et al.*, 1994).

Instead, there may be some evidence for general common structural features. Such genes, which appear to be evolutionarily related but more distantly than those in a classical or conserved domain/motif gene family, have been considered to be members of a **gene superfamily.** For example, in addition to the immunoglobulin gene family, other related genes such as the *HLA* genes, TCR genes, *T4* and *T8* genes are known to encode products with an immune system function and a domain structure that resembles that of immunoglobulins. Although, therefore, the level of sequence homology between such genes may be very low, the similarities in function and general domain structure have suggested the existence of a so-called Ig *superfamily,* in which there appears to be a distant common evolutionary relationship (*Figure 8.2*; see also *Figure 9.15* for the globin superfamily).

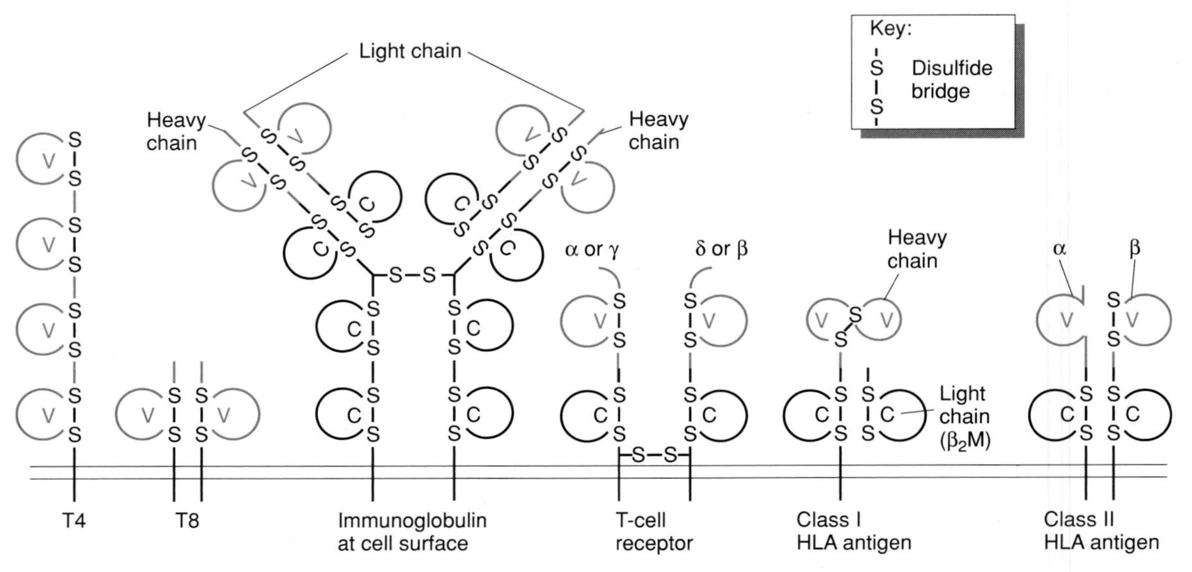

Figure 8.2: Members of the Ig superfamily are surface proteins with similar types of domain structure.

Most members of the Ig superfamily are dimers consisting of extracellular variable domains (V) located at the N-terminal ends and constant (C) domains, located at the C-terminal (membrane-proximal) ends. The light chain of class I HLA antigens has a single constant domain and does not span the membrane. It associates with the transmembrane heavy chain which has two variable and one constant domain, giving an overall structure similar to that of the class II HLA antigens.

Human gene families can occur as closely clustered genes at specific subchromosomal localizations, or as widely dispersed genes

Although gene families often contain members that are nonfunctional (see below), the expressed genes in a specific family occasionally have virtually identical functions. For example, individual genes in a particular histone gene family encode essentially identical products, and the human *HBA1* and *HBA2* genes encode identical α-globins. Usually, however, duplicated genes have related but divergent expression patterns, and in some cases divergent functions. For example, individual genes of some gene families are expressed at different developmental stages (e.g. those of the β-globin and the α-globin clusters, see below), in different tissues (e.g. the alkaline phosphatase gene family; see page 219) or in different subcellular compartments (see examples in page 219). Two major classes of sequence organization are seen in multigene families: clustered gene families and interspersed gene families.

Clustered gene families

Such families consist of one or more gene clusters located at specific subchromosomal locations. The genes in individual clusters often arise by *tandem gene duplication* events, often giving rise to nonfunctional *pseudogenes* (see *Box 8.1*). In some cases, the genes in a cluster can show very close clustering. For example, the major ribosomal genes are arranged in large tandemly repeated units (see page 188). Close clustering of genes in the α-globin cluster, and also in the β-globin cluster, facilitates co-ordination of their expression during development: a single *locus control region* located upstream of the genes in each cluster co-ordinates switching on or off of individual genes at different developmental stages (see page 190 and *Figure 8.6*).

Box 8.1: Pseudogenes and gene fragments

The processes that give rise to gene families often result in the formation of nonfunctional copies of a gene or a fragment of a gene, either a **pseudogene** (a nonfunctional copy of most or all of a gene, or at least its coding DNA), or **truncated genes** and **gene fragments** (nonfunctional copies of a segment of the gene). Various different classes of such sequences exist.

- **Nonprocessed (conventional) pseudogenes.** These are nonfunctional copies of the genomic DNA sequence of a gene, and so contains sequences corresponding to exons and introns, and often to flanking sequences. Nonprocessed pseudogenes can often be recognized by the presence of inappropriate termination codons in sequences corresponding to exons of functional gene homologs. Such pseudogenes are common in clustered gene families (page 189) and are thought to arise by tandem gene duplication events (see *Figure 9.2*).
- **Expressed nonprocessed pseudogene**. Nonprocessed pseudogenes originate from gene duplication events. Immediately after a gene duplication event, both gene copies will be functional, but thereafter one copy may pick up deleterious mutations and eventually lose its original function and even the capacity to be expressed (see *Figure 9.2*). Early in this process, therefore, there may be a transition stage where the gene is no longer functional but continues to be expressed at the RNA level, and possibly in some cases even at the polypeptide level. For example, the θ-globin gene (*HBQ1*, see *Figure 8.6*) is known to be expressed, but there is no evidence that a θ-globin polypeptide becomes incorporated in a functional hemoglobin (Clegg, 1987). The human chorionic somatomammotropin-like gene *CS-L* may be another example.
- **Processed pseudogenes**. Processed pseudogenes are nonfunctional copies of the exonic sequences of an active gene and are often found in interspersed gene families. They are thought to arise by integration into chromosomes of a natural complementary DNA sequence generated by a reverse transcriptase from an RNA transcript (see *Figure 8.7*). Processed pseudogenes derived from genes transcribed by RNA polymerase II (including all genes encoding polypeptides, see page 14) are normally incapable of expression because they lack promoter sequences. Processed pseudogenes derived from genes transcribed by RNA polymerase III, such as the *Alu* repeat, can reach very high copy numbers (see *Box 8.2* and page 200).
- **Expressed processed pseudogene.** A processed pseudogene may originate by integrating into a chromosomal DNA site which just happens, by chance, to be adjacent to a promoter which can drive expression of the processed gene copy. In some cases, selection pressure may ensure that the expressed processed gene copy continues to be expressed. A variety of such processed genes are known to have testis-specific expression patterns (see the example of the pyruvate dehydrogenase gene, *PDHA2*, page 216).

  Another class of expressed processed pseudogene results from processed copies of genes transcribed by RNA polymerase III, which are unusual because the promoter is internal to the gene (see *Figure 1.12*). As the processed copy in this case contains promoter elements, it may be expressed (see the example of expressed *Alu* sequences, page 200).
- **Truncated genes and gene fragments.** Sometimes genomic sequences closely resemble a small component of a functional gene [e.g. a 5′ fragment or a 3′ fragment (*truncated genes*), or a very small segment of the gene, even a single exon from a multiexon gene (*gene fragment*)]. These sequences are often found in clustered gene families (see *Figure 8.4*) and are likely to have originated by unequal crossover or unequal sister chromatid exchanges (page 255).

In other clustered gene families, however, the physical relationship between genes in a cluster may be less close and a cluster of related genes may also contain within it genes that are unrelated in sequence and function. For example, the HLA complex on 6p21.3 is dominated by families of genes which encode class I and II HLA antigens and various serum complement factors, but individual family members may be separated by functionally unrelated genes such as members of the steroid 21-hydroxylase gene family, etc.

Often sequence homology is greater within a cluster than between clusters. For example, genes in the β-globin cluster are more related to each other than they are to genes in the α-globin cluster. However, in the *HOX* gene family sequence identity can be greater between genes in different clusters than between genes in the same cluster (see *Figure 9.5*).

Interspersed gene families

In these families, there is no obvious physical relationship between family members, which are usually dispersed over several chromosomes. The family members may show considerable sequence divergence unless their dispersion has been a relatively recent event, or there has been considerable selection pressure to maintain sequence conservation. They often contain *processed pseudogenes* (see *Box 8.1*).

RNA-encoding gene families often have numerous family members

Human genes whose final expression product is an RNA molecule often belong to high copy number multigene families. For example, genes which specify the more than 30 different types of cytoplasmic tRNA are members of different but closely related multigene families; individual tRNA gene families have copy numbers between 10 and 100 and collectively they comprise about 1300 genes. Genes encoding the different classes of snRNA molecule also belong to large gene families. One common characteristic of each of these gene families is that they include a considerable number of pseudogenes and truncated genes/gene fragments (see *Box 8.1*).

The genes which encode rRNA are also members of multigene families. The 28S, 5.8S and 18S rRNA genes are clustered together in large arrays containing about 60 copies of each gene. Such clusters span about 2 Mb of DNA (so-called **ribosomal DNA** or **rDNA**) and are located on the short arms of each of the five acrocentric chromosomes, at the **nucleolar organizer regions** (see *Figure 7.4*). The other major RNA constituent of ribosomes, 5S rRNA, is transcribed from genes which constitute part of a very large gene family which is clustered on the long arm of chromosome 1. Including spacer units (see below), the rRNA gene families account for about 0.4% of the total DNA in the human genome.

The 28S, 5.8S and 18S rRNA genes are unusual in that unlike the vast majority of nuclear genes, which are individually transcribed, they are initially expressed as multigenic transcripts, in much the same way as the mitochondrial genes (see page 151). A common 13-kb transcription unit is expressed to give a precursor 45S rRNA, which then undergoes a variety of cleavage reactions to generate the mature 28S, 5.8S and 18S rRNA species (see *Figure 8.3*). Like the products of the mitochondrial genome, the individual rRNA products of the rDNA clusters are clearly inter-related components with a single overall function. In this sense, therefore, the unusual use of polygenic RNA transcripts is, in principle, no different from the way in which a single primary translation product is occasionally cleaved to generate two or more polypeptides with a common function. For example, the insulin molecule contains two polypeptide chains which are known to be derived from a single translation product (see *Figure 1.23*). Again, this way of generating related nonoverlapping

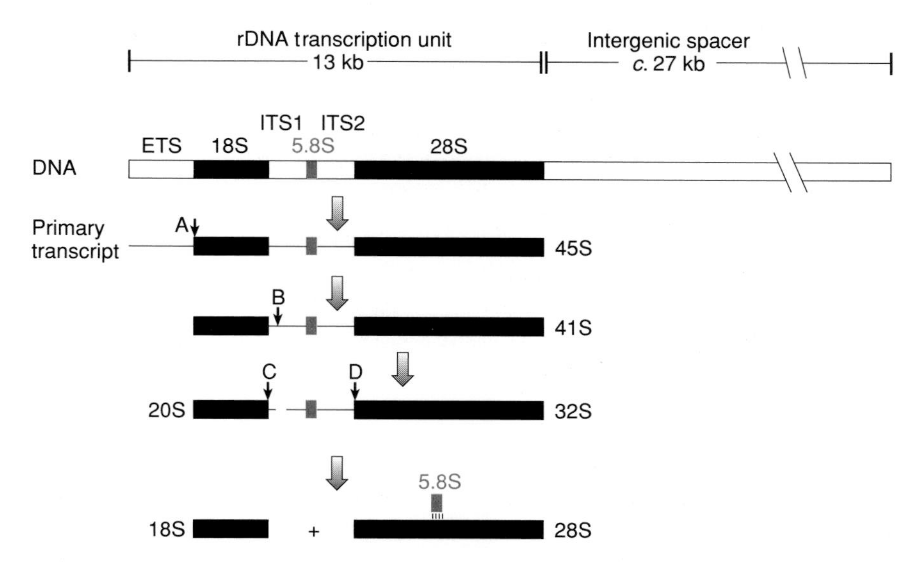

Figure 8.3: The major human rRNA species are synthesized by cleavage from a common 13-kb transcription unit which is part of a 40-kb tandemly repeated unit.

Small arrows indicated by letters A–D signify positions of endonuclease cleavage of RNA precursors. Cleavage of the 41S precursor at B generates two products: 20S + 32S. Following cleavage of the 32S precursor at D, and excision of the small 5.8S rRNA, hydrogen bonding takes places between the 5.8S rRNA and a complementary central segment of the 28S rRNA. The approximately 6 kb of RNA sequence originating from the external and internal transcribed spacer units (ETS, ITS1 and ITS2) are degraded in the nucleus. S is the sedimentation coefficient, a measure of size.

products from a single precursor molecule is relatively rare: the individual components of the great majority of human multisubunit proteins are encoded by distinct genes, often on different chromosomes (see *Table 7.5*).

Human gene clusters often contain conventional pseudogenes and truncated genes or gene fragments

Genes which encode polypeptides are often members of multigene families whose copy number can vary widely. In some cases, such as the genes which encode an individual type of histone molecule, the sequences of the different members of a high copy number gene family are nearly identical. Other multigene families may exhibit considerable sequence divergence between the family members, resulting in divergent expression patterns and sometimes functions. The most extreme example of functional divergence is illustrated by the frequent occurrence of nonfunctional family members. For example, nonprocessed (conventional) pseudogenes (see *Box 8.1*) are often found in clustered multigene families (see *Table 8.1*). Other gene members may be functionless truncated genes or gene fragments, following loss of one or more exons, probably as a result of unequal crossover or of unequal sister chromatid exchange (see page 255). For example, the class I *HLA* gene family on 6p21.3 has several such family members. Although the number of class I *HLA* genes can vary on different chromosome 6s, comprehensive analysis of one of these identified 17 family members clustered over about 2 Mb, of which six are known to be expressed, four are conventional full-length pseudogenes and seven represent nonfunctional truncated genes or gene fragments (Geraghty *et al.*, 1992; see *Figure 8.4*).

Expression of individual genes in gene clusters may be co-ordinated by a common locus control region

Some human gene clusters show evidence of co-ordinated expression of the individual genes in the cluster. For example, individual genes in the α-globin, β-globin

Table 8.1: Examples of clustered multigene families

Family	Copy no.	Organization	Chromosome location
Complement C4 gene clusters	2	On tandem compound repeats ~30 kb in length; both expressed	6p21.3
Growth hormone gene cluster	5	Clustered within 67 kb; one conventional pseudogene	17q22–24
α-Globin gene cluster	7	Clustered over ~50 kb; three functional genes, one expressed gene of unknown function, three conventional pseudogenes (*Figure 8.6*)	16p13.3
Class I HLA heavy chain genes	~20	Clustered over 2 Mb; six are expressed, four are conventional pseudogenes, seven are truncated genes or gene fragments (*Figure 8.4*)	6p21.3
HOX genes	38	Organized in four clusters (*Figure 9.5*)	2p, 7, 12, 17
Histone genes	100	Clustering at a few locations, especially compound cluster on chromosome 1p21	1p21, 6, 12q

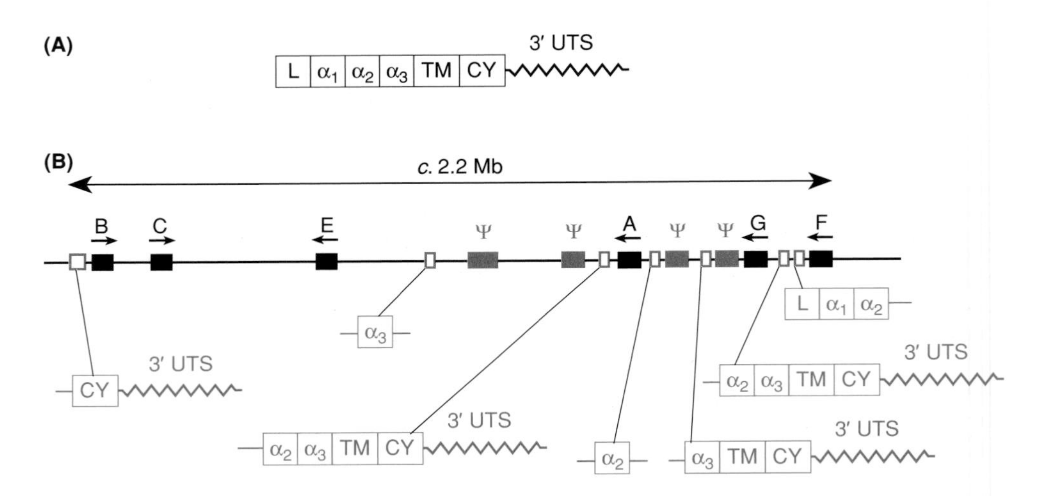

Figure 8.4: Clustered gene families often contain nonprocessed pseudogenes and truncated genes or gene fragments: example of the class I HLA gene family.

(A) Structure of a class I HLA heavy chain mRNA. The full-length mRNA contains a polypeptide-encoding sequence – blocks represent different domains as follows: L – leader sequence; α_1, α_2, α_3 – extracellular domains; TM – transmembrane sequence; CY – cytoplasmic tail) and a 3′-untranslated sequence (3′-UTS). The three extracellular domains α_1–α_3 are each encoded essentially by a single exon. The very small 5′-UTS is not shown.
(B) The class I HLA heavy chain gene cluster. The cluster is located at 6p21.3 and comprises about 20 genes. They include six expressed genes (black), four full-length nonprocessed pseudogenes (filled red blocks, ψ) and a variety of nonfunctional truncated genes or gene fragments (open red blocks). Some of the latter are truncated at the 5′ end (e.g. the one next to *HLA-B*), some are truncated at the 3′ end (e.g. the one next to *HLA-F*) and some contain single exons (e.g. the one next to *HLA-E*).

and the four *HOX* gene clusters are activated sequentially in a temporal sequence that corresponds exactly with their linear order on the chromosome. In the case of the globin genes, there is a clear developmental stage-specific expression: different genes can be active at the embryonic, fetal or adult stages to generate slightly different forms of hemoglobin (**hemoglobin switching;** *Figure 8.5*).

Recently, it has become apparent that the expression of the genes in each of the two globin gene clusters is co-ordinated by a dominant control region, the **locus control region (LCR)** which is located some distance upstream of the gene cluster (see Grosveld *et al.*, 1993). Such cluster-specific LCRs are thought to organize the cluster into an **active chromatin domain** and to act as enhancers of globin gene transcription. The open conformation of transcriptionally active chromatin domains makes them more accessible to cleavage by the enzyme DNase I. Consistent with this relationship, the β-globin LCR has been considered to comprise short sequences at three major erythroid-specific DNase I-hypersensitive sites (HS2, HS3 and HS4) clustered over a 15 kb region located about 50–60 kb upstream of the β-globin gene,

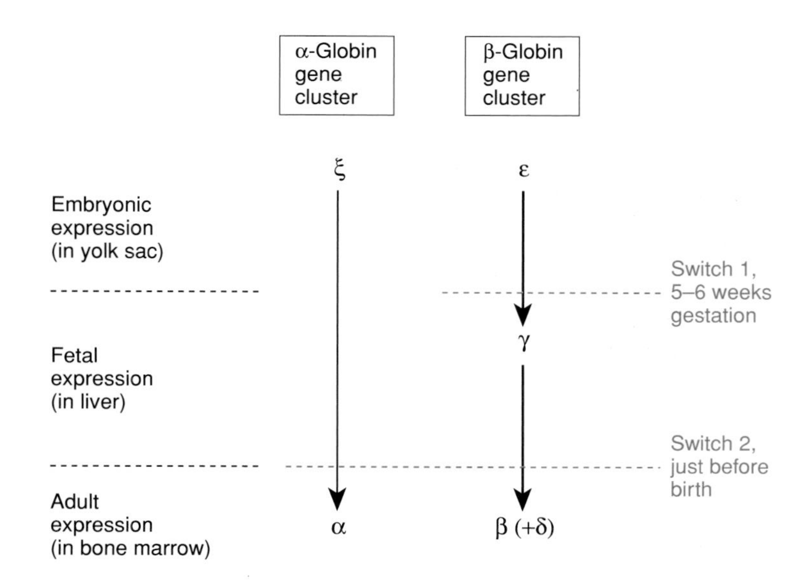

Figure 8.5: Human hemoglobin switching occurs at two distinct developmental stages.

while the α-globin LCR has been identified to occur at an erythroid-specific DNase-hypersensitive site, HS-40, located 60 kb upstream of the α-globin gene (*Figure 8.6*). Each site marks the location of what is effectively an *enhancer* sequence (see pages 14 and 162) of about 200–300 bp of DNA which contains short *cis*-acting sequence elements, including multiple sequence elements recognized by erythroid-specific transcription factors (see *Figure 7.8*). Without the respective LCRs, globin gene expression is negligible and, in the case of the β-globin LCR, it appears that the HS2, HS3 and HS4 elements interact with each other to form a larger complex that interacts with the individual globin genes.

Other DNase I-hypersensitive sites are located at the promoters of the globin genes, but show developmental stage specificity. For example, in fetal liver, the promoters of the two γ genes, the β and δ genes, are marked by DNase I-hypersensitive site but, in adult bone marrow, the two γ genes are no longer transcriptionally active and their promoters no longer reveal DNase I-hypersensitive sites. Developmental stage-specific switching in globin gene expression is then thought to be accomplished by competition between the globin genes for interaction with their respective LCR and stage-specific activation of gene-specific silencer elements. For example, transcription of the ε-globin gene (*HBE1*) is preferentially stimulated by the neighboring LCR at the embryonic stage. In the fetus, however, ε-globin expression is suppressed following activation of a silencer and γ-globin expression becomes dominant (*Figure 8.6*).

Interspersed human gene families often contain many processed pseudogenes

In addition to clustered gene families, there are numerous examples of interspersed gene families which show little or no evidence of clustering (*Table 8.2*). Some of the interspersed gene families contain only a few members, all or most of which are functional, and may individually encode highly related products which can, however, have different patterns of tissue- or cell compartment-restricted expression. Such related genes almost certainly arose as a result of ancient genome or gene duplication events (pages 205–206). However, the great majority of the interspersed gene families contain one or a small number of active genes and a large number of processed pseudogenes (see *Box 8.1*). In addition to consistently lacking any sequences corresponding to the introns or promoters of functional gene members,

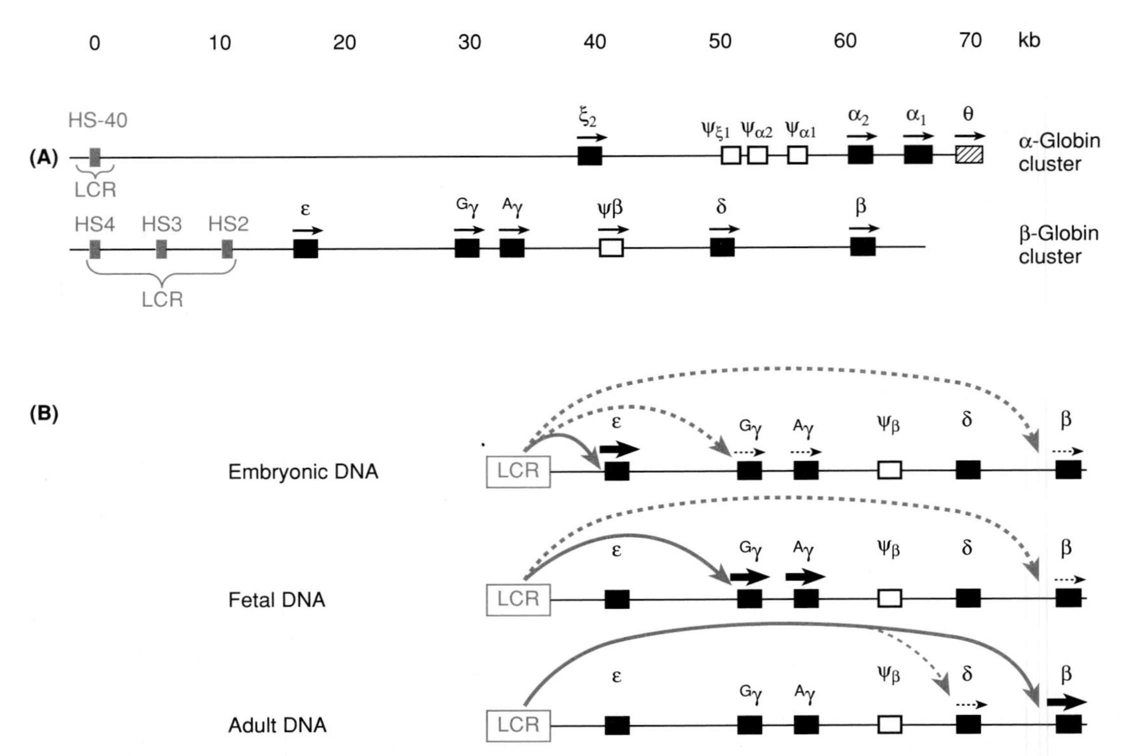

Figure 8.6: Gene expression in the α- and β-globin gene clusters is controlled by common locus control regions.

(A) Organization of the human α- and β-globin gene clusters. The locus control regions (LCRs) consist of one or more erythroid-specific DNase I-hypersensitive sites (HS-40, etc.) located upstream of the cluster. Arrows mark the direction of transcription of expressed genes. The functional status of the θ-globin gene is uncertain: it is expressed, but may be an expressed pseudogene (see *Box 8.1*). **(B)** Regulation of gene expression by the β-globin LCR. The strong red arrows indicate a powerful enhancer effect by the LCR on the indicated genes, resulting in a high expression level; dotted red arrows indicate correspondingly weak effects.

Table 8.2: Examples of interspersed multigene families

Family	Copy no.	Features
Aldolase	5	Three functional genes and two pseudogenes scattered on five different chromosomes
PAX	9	At least eight are expressed
Ferritin heavy chain	>15	One functional gene known on chromosome 11; most are processed pseudogenes
Glyceraldehyde 3-phosphate dehydrogenase	>18	One functional gene on 12p; many processed pseudogenes
Actin	>20	Four functional genes and many processed pseudogenes

processed pseudogenes contain at one end an oligo(dA)/(dT) sequence, which is consistent with an origin by an RNA-mediated DNA transposition mechanism (Vanin, 1985; see *Figure 8.7* for one possible mechanism).

As can be seen in *Table 8.2*, a wide variety of polypeptide-encoding genes are represented in interspersed gene families, including genes which encode enzymes, regulatory proteins, storage proteins, structural proteins, etc. In some cases, no obvious evolutionary advantage is apparent for having so many additional transcriptionally inactive gene copies. However, several examples are known of interspersed gene families where, in addition to a functional gene locus that contains introns, there are one or more processed gene copies which lack introns but appear to be expressed, often in a testis-specific fashion. Often the functional intron-containing gene locus in such families is located on the X chromosome (see page 216).

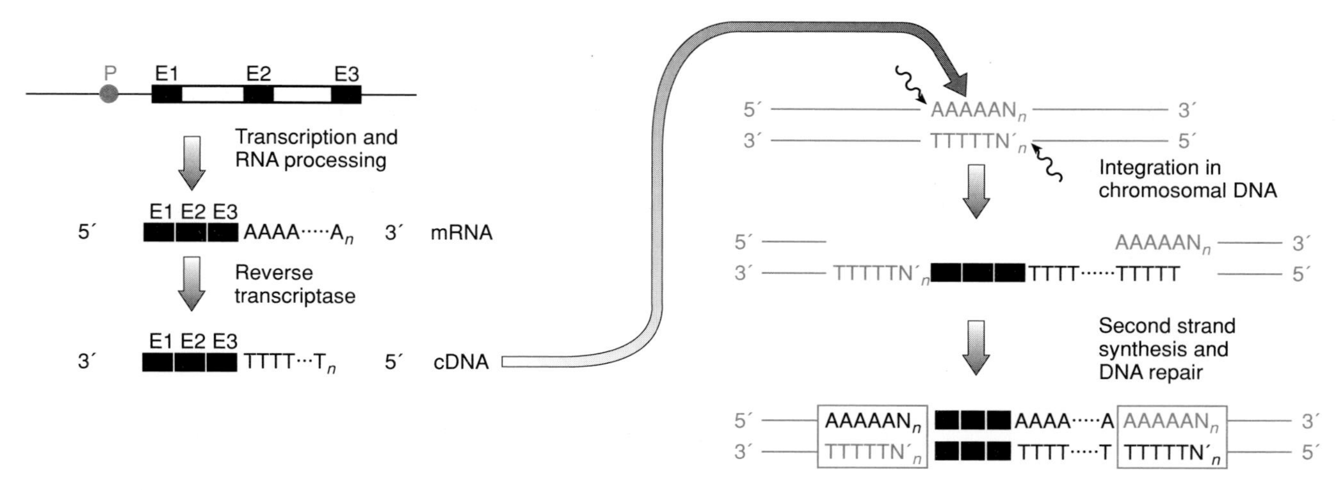

Figure 8.7: Processed pseudogenes originate by reverse transcription from RNA transcripts.

The reverse transcriptase function could be provided by LINE-1 (*Kpn*) repeats (see *Figure 8.12*). The model for integration shown in the figure is only one of several possibilities (see Vanin, 1985). This model envisages integration at staggered breaks (indicated by curly arrows) in A-rich sequences. If the A-rich sequence is included in a 5′ overhang, it could form a hybrid with the distal end of the poly(T) of the cDNA, facilitating second strand synthesis. Because of the staggered breaks during integration, the inserted sequence will be flanked by short direct repeats (boxed sequences). E1–E3 represent exons. P = promoter (shown here as for genes transcribed by RNA polymerase II, but note that genes transcribed by RNA polymerase III often contain internal promoters, see *Figure 1.12*). Some processed gene copies may possibly be functional, such as in the case of some autosomal processed copies of X-linked genes which conserve an important biological function during spermatogenesis (see page 216).

Although the size of some interspersed polypeptide-encoding gene families testifies to the success of retrotransposition as a mechanism for generating processed gene copies, the really successful (in terms of high copy number) retrotranspositions have been performed from RNA polymerase III transcripts. This is the case because, unlike polymerase I and polymerase II transcripts, RNA polymerase III transcripts often contain an internal promoter which facilitates the expression of newly transposed copies (see *Figure 1.12*).

Extragenic repeated DNA sequences and transposable elements

The human nuclear genome, like that of other complex eukaryotes, contains a large amount of highly repeated DNA sequence families which are largely transcriptionally inactive. Like multigene families, they occur in two major types of organization:

- **tandemly repeated DNA**. Such families are defined by *blocks* (or *arrays*) of tandemly repeated DNA sequences. Individual arrays can occur at a few or many different chromosomal locations. Depending on the average size of the arrays of repeat units, highly repetitive noncoding DNA belonging to this class can be grouped into three subclasses: *satellite DNA, minisatellite DNA* and *microsatellite DNA*.
- **Interspersed repetitive DNA.** The individual repeat units are not clustered, but are dispersed at numerous locations in the genome. Most of the DNA families belonging to this class contain members that are capable of undergoing *retrotransposition*, that is transposition through an RNA intermediate.

The chromosomal locations of different types of tandemly repeated DNA can show a very restricted or highly dispersed pattern, while different classes of interspersed

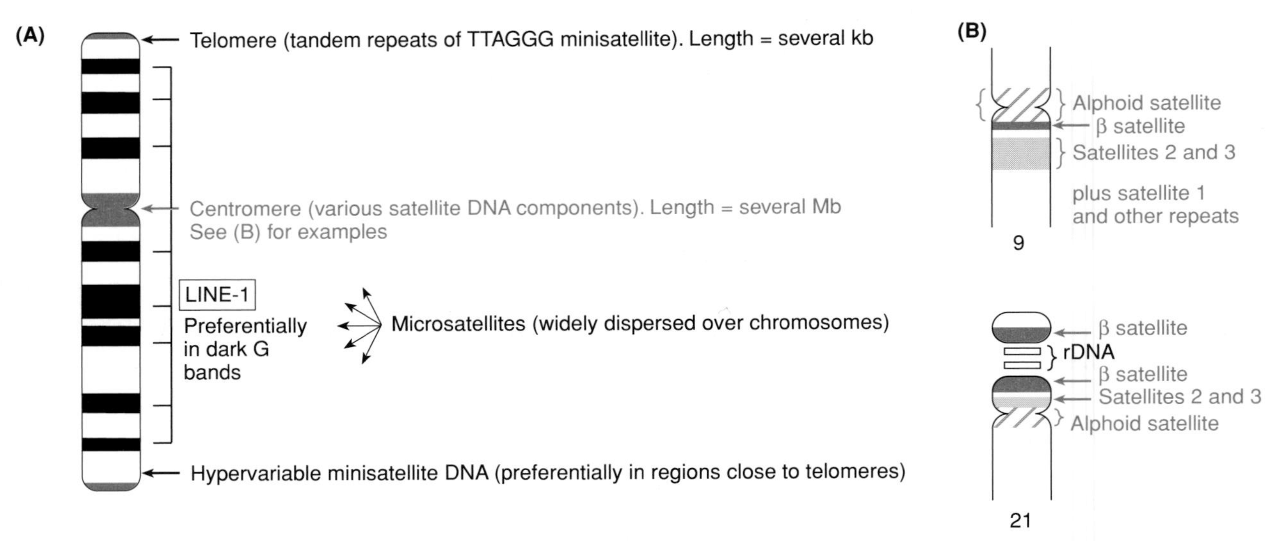

Figure 8.8: Chromosomal location of major repetitive DNA classes.

(A) General overview. *Note* the restricted locations of certain types of tandemly repeated DNA, such as satellite DNAs which are found in heterochromatin (notably at the centromeres) and minisatellite DNAs which are often found at telomeres or close to them.
(B) Satellite DNA organization at centromeres. The locations of different classes of satellite DNA is shown for chromosome 9 and for chromosome 21 (one of the five examples of an autosomal acrocentric chromosome). The illustration in this case is redrawn from Tyler-Smith and Willard (1993) with permission from Current Biology Ltd.

repeat DNA can show preferential location within different types of chromosome bands (see below and *Figure 8.8*).

Satellite DNA is composed of very long arrays of tandem repeats which can be separated from bulk DNA by buoyant density gradient centrifugation

Human **satellite DNA** is comprised of very large arrays of tandemly repeated DNA with the repeat unit being a simple or moderately complex sequence (*Table 8.3*; see Singer, 1982a). Repeated DNA of this type is not transcribed and accounts for the bulk of the heterochromatic regions of the genome. The base composition, and therefore density, of such DNA regions is dictated by the base composition of their constituent short repeat units and may diverge substantially from the overall base composition of bulk cellular DNA.

Isolation by buoyant density gradient centrifugation

When DNA is isolated from human cells by conventional methods, it is subject to mechanical shearing. Fragments are generated from the bulk DNA (with a base composition of ~42% GC) and fragments from the satellite DNA regions which may have a similar or different base composition. If the base composition is significantly different, satellite DNA sequences can be separated from the bulk DNA by buoyant density gradient centrifugation. Following centrifugation, they appear as minor (or *satellite*) bands of different buoyant density from a major band which represents bulk DNA. Typically, human DNA is complexed with Ag^+ ions and then fractionated in buoyant density gradients containing cesium sulfate, whereupon three satellite bands are identified at different densities: 1, 1.687 g cm^{-3}; 2, 1.693 g cm^{-3}; 3, 1.697 g cm^{-3}.

Each of these satellite classes includes a number of different tandemly repeated DNA sequence families (satellite subfamilies), some of which are shared between different classes. DNA sequence analysis has revealed that some of the repetitive

Table 8.3: Major classes of tandemly repeated human DNA

Class	Size of repeat unit (bp)	Major chromosomal location(s)
Satellite DNA (blocks often from 100 kb to several Mb in length)		
Satellites 2 and 3	5	Most, possibly all, chromosomes
Satellite 1 (AT-rich)	25–48	Centromeric heterochromatin of most chromosomes and other heterochromatic regions
α (alphoid DNA)	171	Centromeric heterochromatin of all chromosomes
β (*Sau*3A family)	68	Notably the centromeric heterochromatin of 1, 9, 13, 14, 15, 21, 22 and Y
Minisatellite DNA (blocks often within the 0.1–20 kb range)		
telomeric family	6	All telomeres
hypervariable family	9–24	All chromosomes, often near telomeres
Microsatellite DNA (blocks often less than 150 bp)	1–4	All chromosomes

DNA families in the satellites are based on very simple repeat units. For example, both satellite II and satellite III contain sequence arrays which are based on tandem repetition of the sequence ATTCC. Additionally, restriction mapping (page 96) has revealed satellite subfamilies which show additional higher order repeat units superimposed on the small basic repeat units. Such subfamilies are thought to arise as a result of subsequent amplification of a unit which is larger than the initial basic repeat unit and contains some diverged units (*Figure 8.9*).

Alphoid DNA and centromeric heterochromatin

Other types of satellite DNA sequence cannot easily be resolved by density gradient centrifugation. They were first identified by digestion of genomic DNA with a restriction endonuclease which typically has a single recognition site in the basic repeat unit. In addition to the basic repeat unit size (monomer), such enzymes will produce a characteristic pattern of multimers of the unit length because of occasional random loss of the restriction site in some of the repeats (Singer, 1982a). Alpha satellite (or **alphoid DNA**) constitutes the bulk of the centromeric heterochromatin

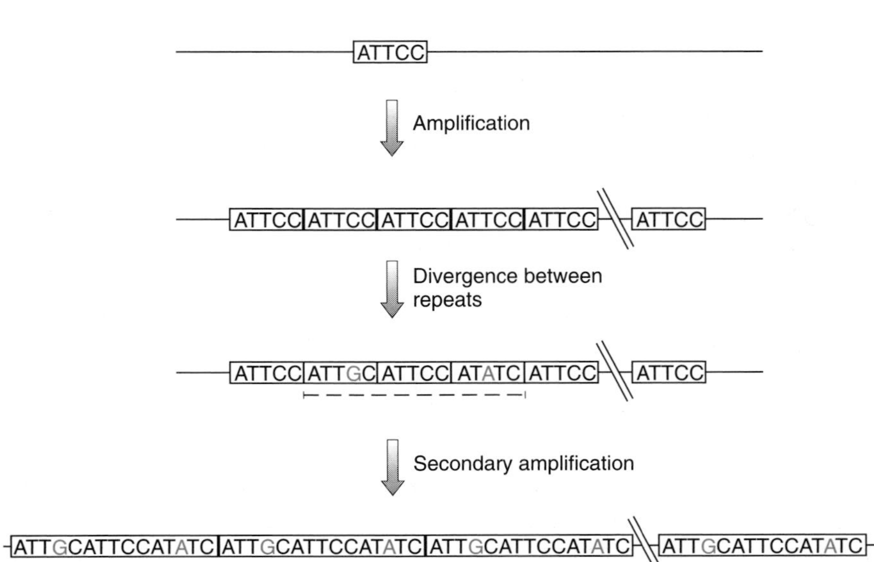

Figure 8.9: Formation of higher order repeat units in simple sequence satellite DNA.

Secondary amplification is envisaged in this case to involve a 15-bp repeat unit comprising three diverged 5-bp repeats.

and accounts for about 3–5% of the DNA of each chromosome. It is characterized by tandem repeats of a basic mean length of 171 bp, although higher order units are also seen. The sequence divergence between individual members of the alphoid DNA family can be so high that it is possible to isolate chromosome-specific subfamilies for each of the human chromosomes (Choo *et al.*, *1991*).

It is not clear at present to what extent satellite DNA can be considered 'junk' DNA. The centromeric DNA of human chromosomes largely consists of various families of satellite DNA (see *Figure 8.8B*). Of these, only the α satellite is known to be present on all chromosomes, although sequence divergence between the repeats has led to chromosome-specific subsets. Although the 171-bp repeat unit of α satellite DNA often contains a binding site for a specific centromere protein, CENP-B, there is presently no compelling evidence to suggest that centromere function depends on this association, or even on the presence of α satellite DNA.

Minisatellite DNA is composed of moderately sized arrays of tandem repeats and is often located at or close to telomeres

Minisatellite DNA comprises a collection of moderately sized arrays of tandemly repeated DNA sequences which are dispersed over considerable portions of the nuclear genome (*Table 8.3*). Like satellite DNA sequences, they are not normally transcribed (but see below).

Hypervariable minisatellite DNA

Hypervariable minisatellite DNA sequences are highly polymorphic and are organized in over 1000 arrays (from 0.1 to 20 kb long) of short tandem repeats (Jeffreys, 1987). The repeat units in different hypervariable arrays vary considerably in size, but share a common core sequence, GGGCAGGAXG (where X = any nucleotide), which is similar in size and in G content to the *chi* sequence, a signal for generalized recombination in *E. coli*. While many of the arrays are found near the telomeres, several hypervariable minisatellite DNA sequences occur at other chromosomal locations. Although the great majority of hypervariable minisatellite DNA sequences are not transcribed, some rare cases are known to be expressed. For example, the *MUC1* locus on 1q is known to encode a highly polymorphic glycoprotein found in several epithelial tissues and body fluids as a result of extensive variation in the number of minisatellite-encoded repeats (Swallow *et al.*, 1987).

The significance of hypervariable minisatellite DNA is not clear, although it has been reported to be a 'hotspot' for homologous recombination in human cells (Wahls *et al.*, 1990). Nevertheless it has found many applications. Various individual loci have been characterized and used as genetic markers, although the preferential localization in subtelomeric regions has limited their use for genome-wide linkage studies. A major application has been in *DNA fingerprinting*, in which a single DNA probe which contains the common core sequence can hybridize simultaneously to multiple minisatellite DNA loci on all chromosomes, resulting in a complex individual-specific hybridization pattern (see page 446).

Telomeric DNA

Another major family of minisatellite DNA sequences is found at the termini of chromosomes, the telomeres. The principal constituent of telomeric DNA is 10–15 kb of tandem hexanucleotide repeat units, especially TTAGGG, which are added by a specialized enzyme, telomerase (see *Figure 2.14*). By acting as buffers to protect the ends of the chromosomes from degradation and loss and by providing a mechanism for replicating the ends of the linear DNA of chromosomes, these simple repeats are directly responsible for telomere function (see *Figure 2.14* and page 46).

Microsatellite DNA is defined by the presence of short arrays of tandem simple repeat units and is dispersed throughout the human genome

Microsatellite DNA families include small arrays of tandem repeats which are simple in sequence (often 1–4 bp) and are interspersed throughout the genome. Of the mononucleotide repeats, runs of A and of T are very common (see *Figure 8.13*) and together account for about 10 Mb, or 0.3% of the nuclear genome. By contrast, runs of G and of C are very much rarer. In the case of dinucleotide repeats, arrays of CA repeats (TG repeats on the complementary strand) are very common, accounting for 0.5% of the genome, and are often highly polymorphic (see Chapter 12). CT/AG repeats are also common, occurring on average once every 50 kb and accounting for 0.2% of the genome, but CG/GC repeats are very rare. This is so because C residues which are flanked at their 3′ end by a G residue (i.e. CpG) are prone to methylation and subsequent deamination, resulting in TpG (or CpA on the opposite strand, see page 16). Trinucleotide and tetranucleotide tandem repeats are comparatively rare, but are often highly polymorphic and increasingly have been investigated to develop highly polymorphic markers.

The significance of microsatellite DNA is not known. Alternating purine–pyrimidine repeats, such as tandem repeats of the dinucleotide pair CA/TG, are capable of adopting an altered DNA conformation, Z-DNA, *in vitro*, but there is little evidence that they do so in the cell. Although microsatellite DNA has generally been identified in intergenic DNA or within the introns of genes, a few examples have been recorded within the coding sequences of genes. Tandem repeats of three nucleotides in coding DNA may be sites that are prone to pathogenic expansions (see page 266).

Highly repeated interspersed DNA families contain a small percentage of actively transposing DNA elements

Two major classes of mammalian interspersed repetitive DNA families have been discerned on the basis of repeat unit length (Singer, 1982b):

- **SINEs** are short interspersed nuclear elements. The most conspicuous human SINE is the *Alu* repeat family (so called because of early attempts at characterizing the sequence using the restriction nuclease *Alu*I). The *Alu* repeat is primate specific and has attained a very high copy number in the human genome (*Table 8.4* and page 199). Other mammals have similar types of sequences, such as the mouse B1 family.
- **LINEs** are long interspersed nuclear elements. Human LINEs are exemplified by the *LINE-1* or *L1* element (also called the *Kpn* repeat because of early attempts at characterizing this family using the restriction nuclease *Kpn*I; see pages 200–202). The LINE-1 element is also found in other mammals such as the mouse.

In addition to the human *Alu* and LINE-1 repeat families, there are many smaller families, including the THE-1 (transposable human element family), many MER (medium reiteration frequency) families and families of human endogenous retroviruses (HERV) or retrovirus-like elements (RTLV; see *Table 8.4*).

Members of many of the interspersed repeat families can be considered as **transposable elements**, unstable DNA elements which can migrate to different regions of the genome (*Figure 8.10*). Although rare examples are known of human transposable elements which undergo transposition through a DNA intermediate, the great majority of human transposable elements undergo *retrotransposition*, that is their RNA transcripts can be converted within the cell to a complementary DNA form which can reinsert back into chromosomal DNA at a variety of different locations

Table 8.4: Major classes of interspersed human repetitive DNA

Class[a]	Size of repeat unit	Total no. of copies of repeat units
Alu family	Full length is about 280 bp, but often less	~700 000–1 000 000
LINE-1 (*Kpn*) family	Full length is 6.1 kb, but many truncated copies and average size = 1.4 kb	~60 000–100 000
MER families	Variable in the different families but often a few hundred base pairs long	Collectively ~100 000–200 000; ~200–10 000 in individual families
THE-1 family	About 2.3 kb	~10 000 plus about 10 000 solitary LTRs
HERV/RTLV families	Full length elements are often 6–10 kb	Few × 1000

[a]See text.

(see Weiner *et al.*, 1986, and *Figure 8.7*). Three classes of mammalian sequence are known to be able to transpose through an RNA intermediate (see *Box 8.2*). In each case, the transposition event involves duplication of a very short sequence at the target site, causing the transposed sequence to be flanked by short repeats (see *Figures 8.11* and *8.7*).

The *Alu* repeat occurs about once every 4 kb in the human genome and includes examples that are transcribed

The ***Alu*** repeat is the most abundant sequence in the human genome, with a copy number of about 750 000 (see Deininger, 1989). *Alu* repeats have a relatively high GC content and, although dispersed mainly throughout the euchromatic regions of the genome, have been reported to be preferentially located in R chromosome bands (Korenberg and Rykowski, 1988). The latter correspond to the pale bands seen when using standard Giemsa staining (see *Box 2.1*) and represent the most transcriptionally active regions of the genome (see page 49). The full-length *Alu* repeat is about 280

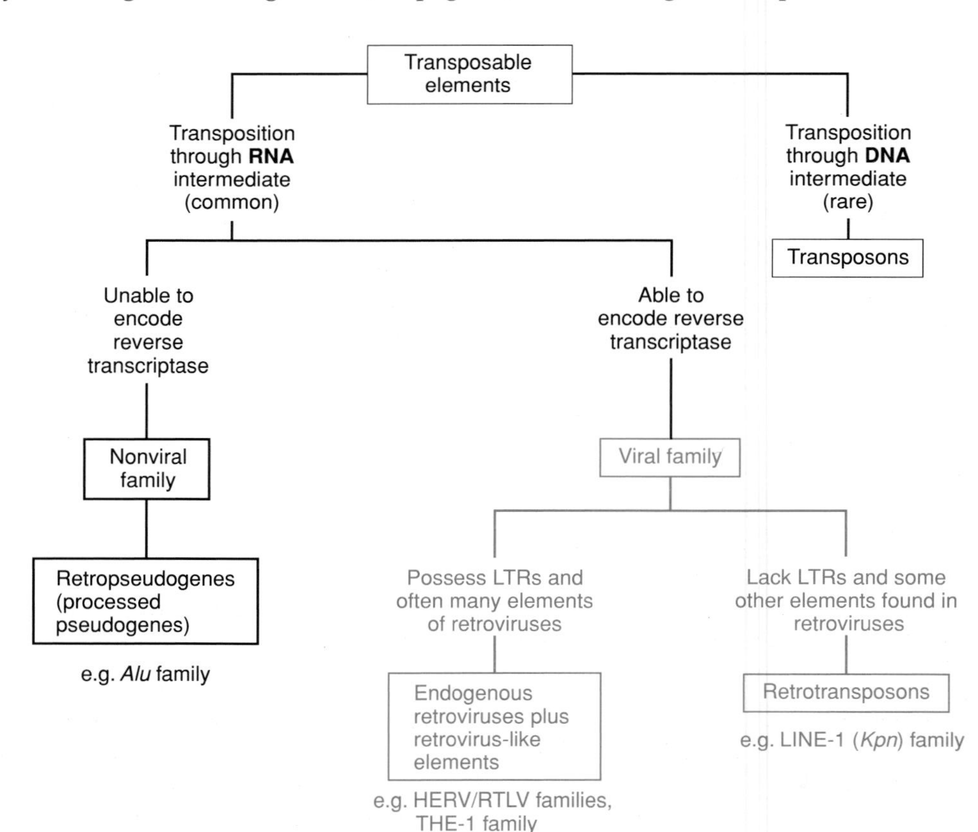

Figure 8.10: Human transposable elements.

Only a small proportion of members of any of the above families may be capable of transposing; many have lost such capacity by acquiring inactivating mutations and many are short truncated copies. See *Figures 8.11* and *8.12* for the typical structures of some human transposable elements.

Box 8.2: Classes of mammalian sequence which undergo transposition through an RNA intermediate

Endogenous retroviruses are sequences which resemble retroviruses but which cannot infect new cells and are therefore restricted to one genome. They include sequences which have the flanking *long terminal repeats (LTRs)* of retroviruses and other elements found in fully functional retroviruses including sequences which can encode a functional reverse transcriptase. Truncated and degraded retrovirus-like elements (RTLVs) are comparatively common, and can include solitary LTRs.

Retrotransposons (sometimes abbreviated as *retroposons*) lack LTRs and often other elements of retroviruses (e.g. the *env* gene of retroviruses which encodes coat proteins). They encode a reverse transcriptase and so are capable of independent transposition. They contain an $(A)_n/(T)_n$ sequence at one end. Examples include the human and mouse LINE-1 elements.

Processed pseudogenes (retropseudogenes) lack a reverse transcriptase and so are incapable of independent transposition, relying instead on the reverse transcriptase activity of other elements. They contain an $(A)_n/(T)_n$ sequence at one end. This class includes two groups:

- low copy number processed pseudogenes derived from genes transcribed by RNA polymerase II (i.e. polypeptide-encoding genes). *Note* that some rare examples of processed copies of genes may possibly be functional (see page 216);
- high copy number mammalian SINEs, such as the human *Alu* and the mouse B1 repeat families, are derived from genes transcribed by RNA polymerase III using an internal promoter (see *Figure 1.12* for examples of this general class of gene).

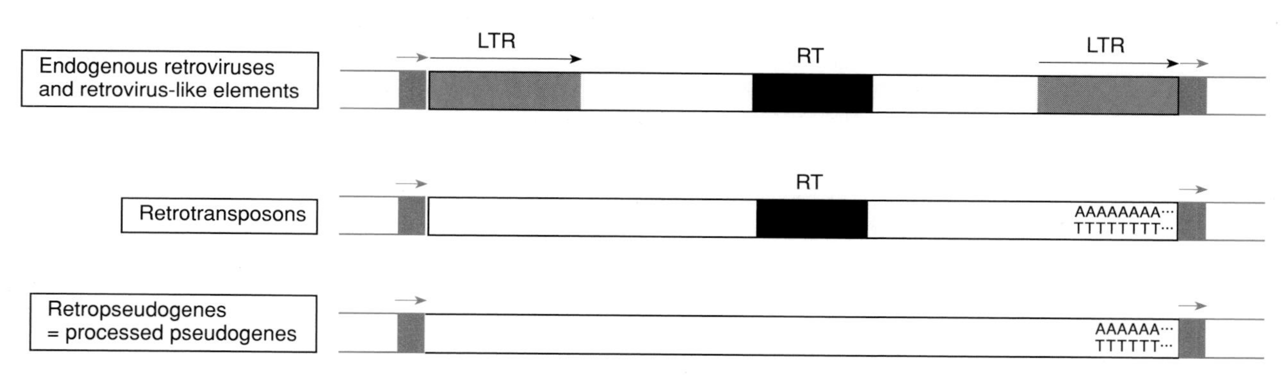

Figure 8.11: Classes of mammalian transposable element which undergo transposition through an RNA intermediate.

Red blocks flanking the sequences represent short repeats of a sequence originally present at the target site that was duplicated during the integration process (*see Figure 8.7*). LTR, long terminal repeat; RT, reverse transcriptase.

bp long and is usually flanked by short (often 6–18 bp) *direct repeats* (i.e. the repeats are in the same orientation). The typical *Alu* sequence is a tandemly repeated dimer, with the repeats sharing an approximately 120-bp sequence followed by a short sequence which is rich in A residues on one strand and T residues on the complementary strand. However, there is asymmetry between the tandem repeats: one repeat unit contains an internal 32-bp sequence lacking in the other (*Figure 8.12*). Monomers, containing only one of the two tandem repeats, and various truncated versions of dimers and monomers are also common.

The two repeated units of the *Alu* sequence show a striking resemblance to the sequence for 7SL RNA, a component of the signal recognition particle, which facilitates transport of proteins across the membrane of the endoplasmic reticulum (page 25). Because of this and the observation of the A-rich regions and the flanking direct repeats, it has been assumed widely that the *Alu* sequence has been propagated by

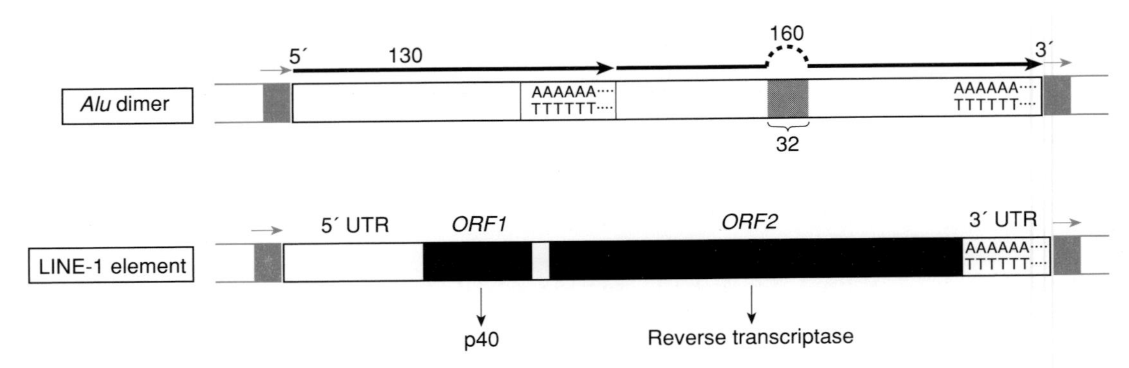

Figure 8.12: Structures of full-length *Alu* and LINE-1 repeats.

The consensus standard *Alu* dimer is shown with two similar repeats terminating in an $(A)_n$ /$(T)_n$ like sequence. They have different sizes because of the insertion of a 32-bp element within the larger repeat. *Alu* monomers also exist in the human genome, as do various truncated copies of both monomers and dimers. The consensus full-length LINE-1 element is 6.1 kb long but most LINE-1 elements are truncated and the average size is very much smaller. ORF1, ORF2, open reading frames 1 and 2.

retrotransposition from 7SL RNA, and therefore represents a processed 7SL RNA pseudogene (*Figure 9.24*). Certainly, transposition by *Alu* sequences is known to occur [presumably as a result of *trans*-acting cellular reverse transcriptases such as those encoded by LINE-1 (*Kpn*) elements (see below), and may occasionally cause clinical problems (page 271)]. Possibly, the very high copy number achieved by this processed pseudogene is related to the presence of a promoter sequence in the 7SL RNA sequence (the 7SL RNA gene, like tRNA genes is transcribed by RNA polymerase III from an internal promoter, see *Figure 1.12*). In contrast, processed pseudogenes from RNA polymerase II transcripts lack promoter sequences and their only chance of expression is if the integration event places them next to a functional promoter sequence.

Currently, the function of the *Alu* sequence, if any, is unknown. Although the average expected frequency is one copy per 4 kb, clusters of *Alu* repeats are known to occur in certain regions. Because of their ubiquity, *Alu* sequences have been considered to promote unequal recombination, a mechanism which is known to be pathogenic in some cases (see page 270) but which could be evolutionarily advantageous by promoting gene duplication (page 218). Although conspicuously absent from coding sequences, *Alu* sequences are often found in noncoding intragenic locations, notably in introns and occasionally in untranslated sequences (see *Figure 8.13*). Consequently, they are often represented in the primary transcript RNA from genes encoding polypeptides, and occasionally in mRNA. The *Alu* sequence can also be transcribed from its internal promoter by RNA polymerase III *in vitro*, and *in vivo* transcription of some *Alu* sequences can result in the accumulation of a small cytoplasmic RNA which can be specifically bound by a protein component of the signal recognition particle (Chang *et al.*, 1994).

The LINE-1 (*Kpn* repeat) family occurs about once every 50 kb in the human genome and includes examples that appear to encode a reverse transcriptase

The human **LINE-1 (L1, *Kpn* repeat) family** consists of about 50 000–10 0000 interspersed repeats (Fanning and Singer, 1987). Of these, multiple members are known to be actively transposing (Holmes *et al.*, 1994). The full-length consensus element is 6.1 kb long and has two *open reading frames* (*ORFs*), which are, however, not present in most individual sequences. ORF1 is located close to one end (conventionally known as the 5′ end) of the consensus element and encodes a protein of unknown

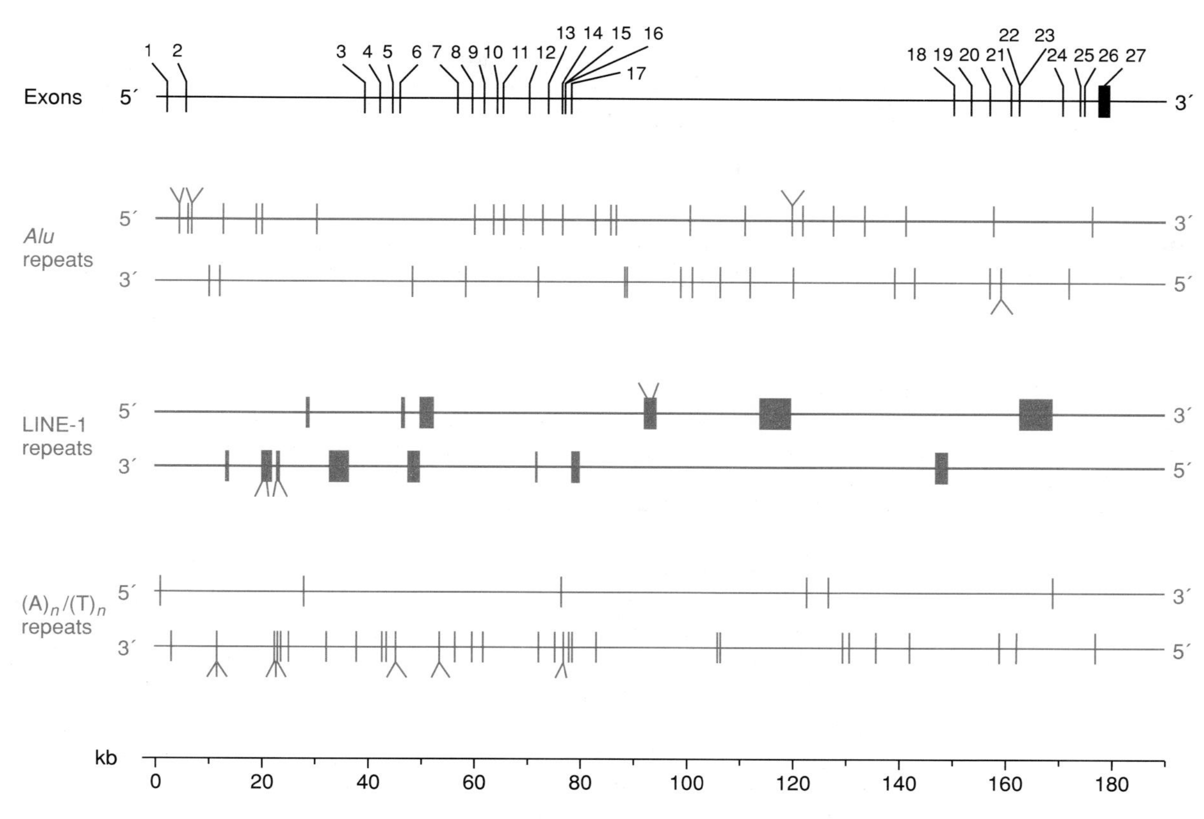

Figure 8.13: Location of *Alu*, LINE-1 and $(A)_n/(T)_n$ repeats within the human retinoblastoma gene.

The entire sequence of the 180-kb human retinoblastoma gene has been determined, enabling identification within the gene of many examples of abundant DNA repeats. The top line 5′–3′ of each pair shows the repeat elements orientated in the sense direction; the bottom line (3′–5′) shows them in the antisense orientation. There are 46 *Alu* repeats corresponding to the expected frequency of one per 4 kb. There is a particularly high frequency of LINE-1 elements in this gene (one per 11 kb), but only two of the elements approach the full 6.1 kb length. The $(A)_n/(T)_n$ sequences (n = 12 or greater) which are indicated are only those not found within the sequences of the interspersed repeats. No examples were found for $(C)_n/(G)_n$ for$_n$ = 12 or greater. The frequencies of dinucleotide repeats with n = 6 or greater were as follows: $(CG)_n/(GC)_n$ 0; (AT)n 1; $(CT)_n/(AG)_n$ 2; $(CA)_n/(TG)_n$ 4 (three of which are polymorphic). Redrawn from Toguchida *et al.* (1993) with permission from Academic Press Inc.

function, p40 (so called because its molecular weight is ~40 kDa). ORF2 has regions which are similar to various reverse transcriptases and other retroviral proteins. The full-length consensus element contains an internal promoter within a region of untranslated DNA preceding ORF1 (conventionally called the 5′-UTR) while at the other end there is an $(A)_n/(T)_n$ sequence, often described as the 3′ poly(A) tail. As in the case of other elements that transpose, the LINE-1 elements are flanked by short duplicated repeats (*Figure 8.12*).

The full-length LINE-1 element is comparatively rare (only about 3500 copies) and most repeats are truncated at the 5′ end, resulting in a population that is heterogeneous in length but sharing a common 3′ end with the poly(A) tail. LINE-1 elements are primarily located in euchromatic regions but show an inverse relationship with *Alu* repeats by appearing to be located preferentially in the dark G bands (Giemsa positive) of metaphase chromosomes. Like the *Alu* repeats, they are conspicuously absent from coding sequences but may be found in intragenic noncoding sequences (*Figure 8.13*). As a result, they may be represented in the primary RNA transcript of large genes, but they are virtually absent from mRNA.

Further reading

John B, Miklos G. (1988) *The Eukaryote Genome in Development and Evolution.* Allen & Unwin, London.

Singer M, Berg P. (1991) *Genes and Genomes.* University Science Books, Mill Way, CA.

References

Chang D-Y, Nelson B, Bilyeu T, Hsu K, Darlington GJ, Maraia RJ. (1994) *Mol. Cell. Biol.*, **14**, 3949–3959.

Choo KH, Vissel B, Nagy A, Earle E, Kalitsis P. (1991) *Nucleic Acids Res.*, **19**, 1179–1182.

Clegg JB. (1987) *Nature*, **329**, 465–467.

Deininger P. (1989) In: *Mobile DNA* (eds DE Berg, MM Howe), pp. 619–636. American Society for Microbiology, Washington, DC.

Fanning TG, Singer MF. (1987) *Biochim. Biophys. Acta*, **910**, 203–212.

Geraghty DE, Koller BH, Hansen JA, Orr HT. (1992) *J. Immunol.*, **149**, 1934–1936.

Grosveld F, Dillon N, Higgs D. (1993) *Bailliere's Clin. Haematol.*, **6**, 31–55.

Holmes SE, Dombroski BA, Krebs CM, Boehm CD, Kazazian HH Jr. (1994) *Nature Genetics*, **7**, 143–148.

Jeffreys AJ. (1987) *Biochem. Soc. Trans.*, **15**, 309–317.

Korenberg JR, Rykowski MC. (1988) *Cell*, **53**, 391–400.

Liu W-M, Maraia RJ, Rubin CM, Schmid CW. (1994) *Nucleic Acids Res.*, **22**, 1087–1095.

Neer EJ, Schmidt CJ, Nambudripad R, Smith TF. (1994) *Nature*, **371**, 297–300.

Schmid SR, Linder P. (1992) *Mol. Microbiol.*, **6**, 283–292.

Singer MF. (1982a) *Int. Rev. Cytol.*, **76**, 67–112.

Singer MF. (1982b) *Cell*, **28**, 433–434.

Swallow DM, Gendler S, Griffiths B, Corney G, Taylor-Papadimitrou J, Bramwell ME. (1987) *Nature*, **328**, 82–84.

Toguchida J, McGee TL, Paterson JC, Eagle JR, Tucker S, Yandell DW, Dryda TP. (1993) *Genomics*, **17**, 535–543.

Tyler-Smith C, Willard HF. (1993) *Curr. Opin. Genet. Dev.*, **3**, 390–397.

Vanin EF. (1985) *Biochim. Biophys. Acta*, **782**, 231–241.

Vogt P. (1990) *Hum. Genet.*, **84**, 301–336.

Wahls WP, Wallace LJ, Moore PJ. (1990) *Cell*, **60**, 95–103.

Weiner AM, Deininger PL, Efstratiadis A. (1986) *Annu. Rev. Biochem.*, **55**, 631–661.

Wong Z, Royle N, Jeffreys AJ. (1990) *Genomics*, **7**, 222–234.

Yunis JJ, Yasmineh WG. (1979) *Science*, **174**, 1200–1209.

Footprints of evolution

Chapter 9

By definition, the evolutionary origins of the human genome, and all genomes, are as old as life itself. The present chapter is not intended as an introduction to basic evolutionary theory or an overview of molecular evolutionary genetics *per se*. As a result, many fascinating areas are not covered, such as the idea that RNA used to be the primary information molecule before being superseded by DNA or the idea that the genetic code was initially a doublet code before evolving into the familiar triplet code, etc. The interested reader is advised to consult one of the more general molecular evolutionary genetics textbooks (see Further reading). Instead, the present chapter is meant to focus on how comparative analyses of present day genomes have shed light on the evolutionary origin of human DNA. Much of the data are derived from comparisons of mammalian genomes, although comparison with more distant genomes is occasionally used to explain certain *footprints of evolution*, as in the origin of mitochondrial DNA and introns. A final section considers our uniqueness when compared with mammalian models, notably the mouse (an important model for understanding early human development and also human disease – see Chapter 19) and primates, our closest living relatives.

Evolution of the human mitochondrial genome

The human mitochondrial genome probably originated following endocytosis of an aerobic eubacterium and subsequent large-scale gene transfer to the nuclear genome

The organization of the human mitochondrial genome is radically different from that of the nuclear genome (pages 147–151). Instead, it and the mitochondrial genomes of other animal cells have several features in common with most prokaryotic genomes: small size, absence of introns, a very high percentage of coding DNA, a conspicuous lack of repeated DNA sequences and comparatively small prokaryotic-like rRNA genes. Phylogenetic analyses of rRNA sequences suggest that mitochondria are particularly closely related to the α subdivision of purple bacteria. Consequently mitochondria are believed to have originated as a result of endocytosis by anaerobic eukaryotic precursor cells of an aerobic eubacterium (most likely a purple bacterium) with an oxidative phosphorylation system. According to the so-called **endosymbiont hypothesis,** this event most likely happened when oxygen started to accumulate in significant quantities in the Earth's atmosphere, probably about 1.5 billion years ago. At that time, the precursors to eukaryotes were simple anaerobic cells. The change to a progressively aerobic atmosphere provided a strong selection system for previously anaerobic cells to acquire the capacity for oxidative phosphorylation. The endocytosis of a suitably respiratory bacterial cell provided, therefore, a strong advantage for anaerobic eukaryotic precursor cells: by subverting the bacterial oxidative phosphorylation system they were able to promote their own rapid growth and evolution in the oxygen-containing atmosphere (see Gray, 1993, and *Figure 9.1*).

The size of the human mitochondrial genome (16.6 kb) and those found in other animal cells (typically less than 20 kb) is, however, much smaller than those of many

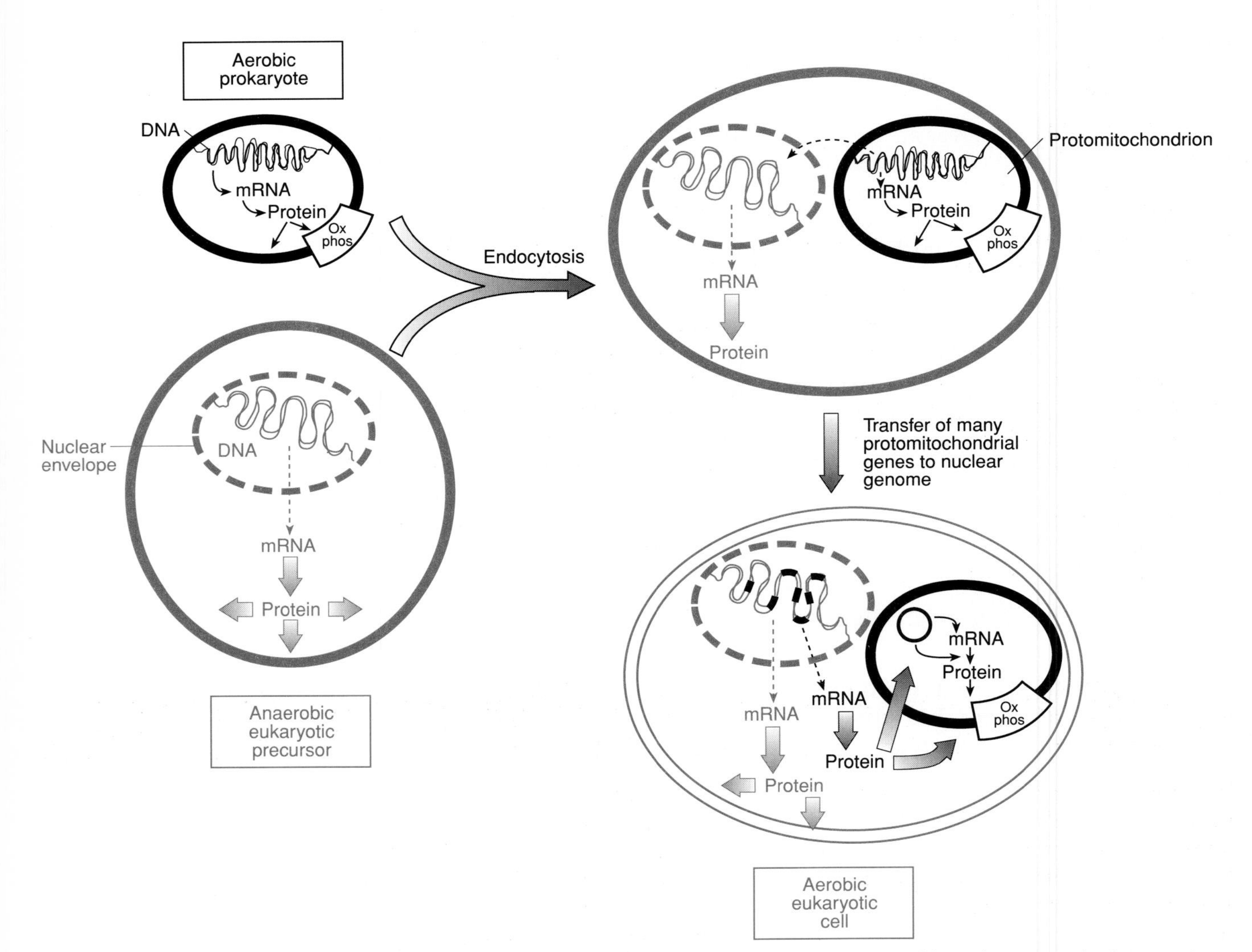

Figure 9.1: The human mitochondrial genome probably originated following endocytosis of an aerobic prokaryote and subsequent large-scale gene transfer to the nuclear genome

Although initially the engulfed aerobic prokaryote was capable of independent protein synthesis, over long evolutionary time scales most of its genes were transferred into the nuclear genome. This left a small remnant genome which is now only able to synthesize rRNA, tRNA and a few of the proteins of the oxidative phosphorylation system (Oxphos). The great majority of mitochondrial proteins, including mitochondrial enzymes, the proteins of mitochondrial ribosomes and most of the oxidative phosphorylation system proteins, are specified by nuclear genes, synthesized on cytoplasmic ribosomes and imported into the mitochondria.

bacterial genomes, such as the 4.6 Mb genome of *E. coli*. It is probable, therefore, that most of the genes donated by the eubacterial endosymbiont were gained by the nuclear genome; the mitochondrial genome presumably retained only a very small number of the eubacterial genes, including the core rRNA and tRNA genes, and a few of the many genes which encode components of the oxidative phosphorylation system. This would explain why present-day mitochondria are so heavily dependent on the products of nuclear genes (which are translated on cytoplasmic ribosomes before being imported into the mitochondria) (see page 149 and *Figure 1.10*).

The mitochondrial genetic code most likely evolved as a result of reduced selection pressure in response to a greatly diminished coding capacity

The human mitochondrial genetic code is slightly different from the 'universal' genetic code that is used in the expression of polypeptides encoded by prokaryotic genomes, eukaryotic nuclear genomes and plant mitochondrial genomes (*Figure 1.22*). In addition, although it is identical to the genetic codes of other mammalian mitochondrial genomes, it shows some differences to the nonuniversal genetic code in the mitochondria of other eukaryotes, such as *Drosophila* and yeast cells. Most probably, the evolution of the altered mitochondrial genetic codes arose in response to the progressive loss of coding potential by gene transfer from the original endosymbiont's genome to the nuclear genome; the resulting tiny mitochondrial genome in animal cells specifies the synthesis of only a few types of polypeptide (13 in the case of human mitochondria). The original bacterial endosymbiont would have been expected to display the universal genetic code. However, once most of its coding capacity was transferred to the nuclear genome, there would have been less selection pressure to conserve the original code – slight altering of the otherwise universal genetic code could be achieved without provoking disastrous consequences because only a tiny number of polypeptides were involved. It is also likely that the codons which have been altered (see *Figure 1.22*) have not been used extensively in locations where amino acid substitutions would have been deleterious. In contrast, there is a huge selection pressure to conserve the universal genetic code in large genomes, such as the nuclear genome – even slight alterations could result in lack of function or aberrant function for large numbers of vitally important gene products, resulting in cell death.

Evolution of the human nuclear genome: genome duplication and large-scale chromosomal alterations

In many ways the human nuclear genome is typical of those found in the cells of complex multicellular eukaryotes, such as mammals. The overall genome size is large, a parameter which, with some notable exceptions, parallels the complexity of the organism (*Table 2.1*). It contains a large fraction of noncoding DNA sequence whose significance is mostly obscure. In addition, a high proportion of both the coding and noncoding DNA is repetitive. In contrast, present day bacterial genomes are very much smaller – the size of the *E. coli* genome is barely 1/1000 of that of a typical mammalian genome – and they have little noncoding DNA or repeated DNA sequences. The evolution of complex multicellular organisms can therefore be expected to have involved a progressive increase in genome size. Different mechanisms are likely to have contributed to the large increase in genome size, including rare duplications of the whole genome (**tetraploidization**) and a variety of frequent subgenomic duplication events resulting from different genetic mechanisms including *unequal crossover, unequal sister chromatid exchange* and *slippage replication* (see pages 255–257). In each case, the increase in genome size must have been accomplished without initially compromising the functions of the original DNA set. Instead, by providing additional DNA, subsequent mutations could result in comparatively rapid sequence divergence: at each duplicated gene locus, one gene is surplus to requirements and so can diverge rapidly because of the absence of *selection pressure* to conserve function (see page 244). In a few cases, such diverged genes may acquire novel functions which could be selectively advantageous (see page 219). In most cases, however, the additional gene sequences would be expected to acquire deleterious mutations and degenerate into nonfunctional pseudogenes (*Figure 9.2*).

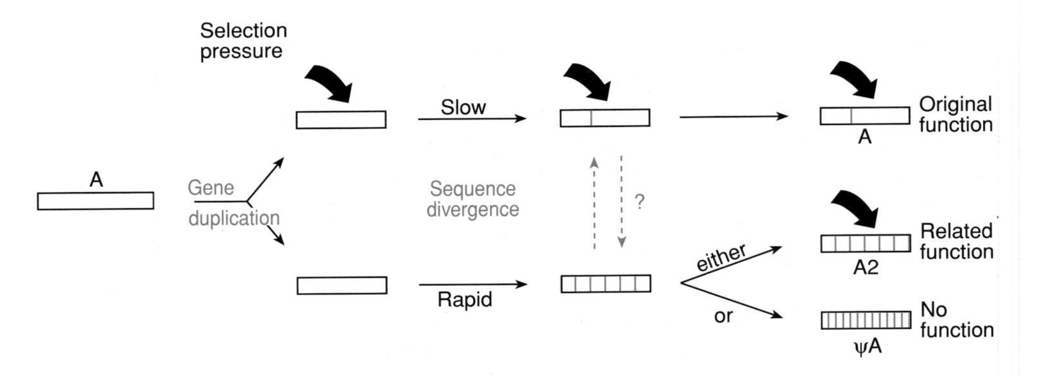

Figure 9.2: Gene duplication can lead to the acquisition of novel function or the formation of a pseudogene.

Duplication of gene A results in two equivalent gene copies. Selection pressure need be applied to only one gene copy (top) to maintain the presence of the original functional gene product. The other copy (bottom), will continue to be expressed but, in the absence of selection pressure to conserve its sequence, will accumulate mutations (vertical bars) relatively rapidly. It may acquire deleterious mutations and become a nonfunctional pseudogene which may continue to be expressed at the RNA level for some time, but which will eventually be transcriptionally silent (ψA). In some cases, however, the mutational differences may lead to a different expression pattern or other property that is selectively advantageous (A2). In the case of tandem gene duplication, subsequent sequence exchanges between the two copies (by mechanisms such as unequal crossover, see pages 255–257) will act as a brake on the rate of sequence divergence between the two gene copies.

Human genome evolution may have involved ancient genome duplication events, but the evidence has been obscured by subsequent chromosome and DNA rearrangements

Genome duplication is an effective way of increasing genome size and is responsible for the extensive polyploidy of many flowering plants. It can occur naturally when there is a failure of cell division after DNA duplication, so that a cell has double the usual number of chromosomes. Human somatic cells are normally diploid. However, if there is a failure of the first zygotic cell division, constitutional tetraploidy can result. Tetraploidy can be harmful and is often selected against. However, some animals, such as particular species of fish, are naturally tetraploid and it is quite feasible, therefore, that in the distant evolutionary past ancestral cells did undergo genome duplication events. In this way, an initially diploid cell could have undergone a transient tetraploid state; subsequent large-scale chromosome inversions and translocations, etc., could result in chromosome divergence and restore diploidy, but now with twice the number of chromosomes (*Figure 9.3*).

If ancient tetraploidization events were rare in the evolution of the human genome, much intragenomic DNA shuffling would have occurred since the last such event. This means that the original evidence for tetraploidization events would be largely

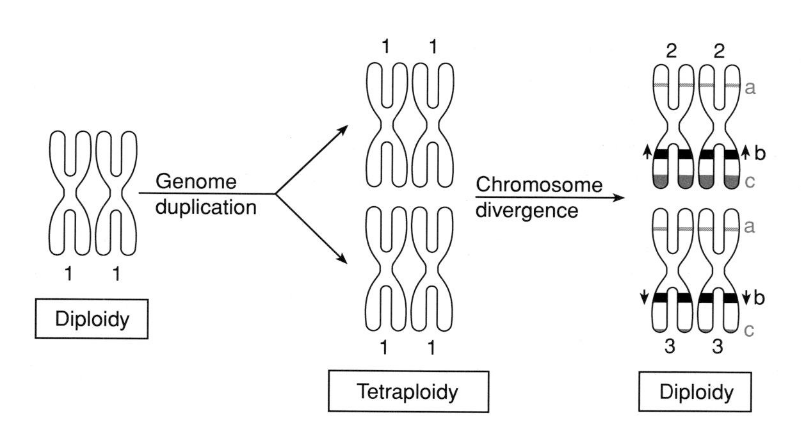

Figure 9.3: Genome duplication can lead to a transient tetraploid state before chromosome divergence restores diploidy.

Following duplication of a diploid genome, each pair of homologous chromosomes (e.g. chromosome 1) is now present as a pair of identical pairs. The resulting tetraploid state, however, can be restored to diploidy by chromosome divergence, e.g. by an interstitial deletion (upper panel, a), a terminal deletion (lower panel, c) or by an inversion (b).

obscured by subsequent chromosomal inversions, translocations, etc. Additionally, traces of gene duplication following genome duplication are likely to be reduced by silencing of one member of each duplicated gene pair which then degenerates into a pseudogene. After hundreds of millions of years without any function, the non-processed pseudogenes generated following the last proposed genome duplication would have diverged so much in sequence as to be not recognizably related to the functional gene, even assuming they have not been lost during occasional rearrangements leading to gene deletion.

At low banding resolution, certain pairs of nonhomologous human chromosomes (e.g. chromosomes 1 and 2) show apparent similarities in banding patterns which were initially suggested as evidence of ancient tetraploidy. However, the evidence for such conclusions has not been supported by subsequent high resolution chromosome banding studies. Another approach has been to identify common chromosomal locations of nonallelic genes that are obviously related. Such genes may be related in sequence (members of a gene family) or in function (participation in common biochemical pathway, receptor–ligand relationships, subunits of same protein, etc.). Using this approach, various groups of related nonhomologous chromosomes have been suggested in the human genome (Lundin, 1993), representing so-called *paralogous* chromosome segments (see *Box 9.1*)

Box 9.1: Paralogy, orthology and homology

Paralogy means close similarity of nonallelic chromosomal segments or DNA sequences *within a species*, indicative of a close evolutionary relationship which may or may not have pre-dated speciation. For example, the two different human α-globin genes (see *Figure 8.6*) are paralogous. Different degrees of paralogy can be identified in a DNA sequence family with more than two members. For example, although all 38 human *HOX* genes are clearly related and so to some extent can be considered as paralogs, there is an especially close structural and functional relationship between specific genes in the four clusters. As a result, in the *HOX* gene family, the term *paralogous* is customarily confined to sets of particularly related genes (see below and *Figure 9.5*).

Orthology means close similarity of chromosomal segments or DNA sequences *between species*. For example, the major human sex-determining gene is *SRY* and its ortholog in mouse is the *Sry* gene. In gene families with closely related members in a species, it may be difficult to recognize true orthologs between species.

Homology was initially used to describe the relationship of allelic chromosomal segments. Since then it has been widely used in a loose sense to signify any type of relatedness indicative of a common evolutionary origin, whether within or between species.

Possibly the most significant evidence for paralogous chromosome segments comes from comparisons of chromosomes 12 and 17 (HSA12 and HSA17 – from Homo sapiens) (*Figure 9.4*). These two chromosomes contain two of the four human *HOX* clusters of homeobox genes. Homeobox genes are believed to play a crucial role in establishing the anterior–posterior axis during development and, because of their fundamental importance to embryonic development, have been very strongly conserved throughout evolution. Other mammals also have four *HOX* clusters and the linear order of the genes in a cluster is thought to dictate the temporal order in which they are expressed during development and, also their anterior limits of expression along the anterior–posterior axis (see Ruddle *et al.*, 1994). The close similarity of the four clusters extends to clear identification of *paralogous HOX* genes, that is genes which occur on different clusters but which are more closely related to each other than they are to their neighbors. A single such *HOX* cluster exists in *Amphioxus*, with very close similarities to the mammalian *HOX* gene clusters (Garcia-Fernandez and Holland, 1994) (see *Figure 9.5*). As *Amphioxus* is thought to be the closest invertebrate relative of the vertebrates, the vertebrate ancestor may have had a single such cluster. During vertebrate evolution either successive duplications of the *HOX* gene cluster occurred, or possibly one or two rounds of tetraploidy contributed to the formation of the existing four clusters (Kappen *et al.*, 1989). Similar chromosomal locations suggest that the non-I integrin α chain genes (which mediate a variety of cell–cell and cell–matrix interactions) appear to have evolved in parallel with the *HOX* gene clusters in a co-ordinated fashion, presumably because of close proximity to the *HOX* genes in the ancestral cluster (Wang *et al.*, 1995).

There have been numerous major chromosome rearrangements during the evolution of mammalian genomes

In addition to whole genome duplication, a variety of different subgenomic DNA duplication events are possible, resulting from exchanges between nonhomologous

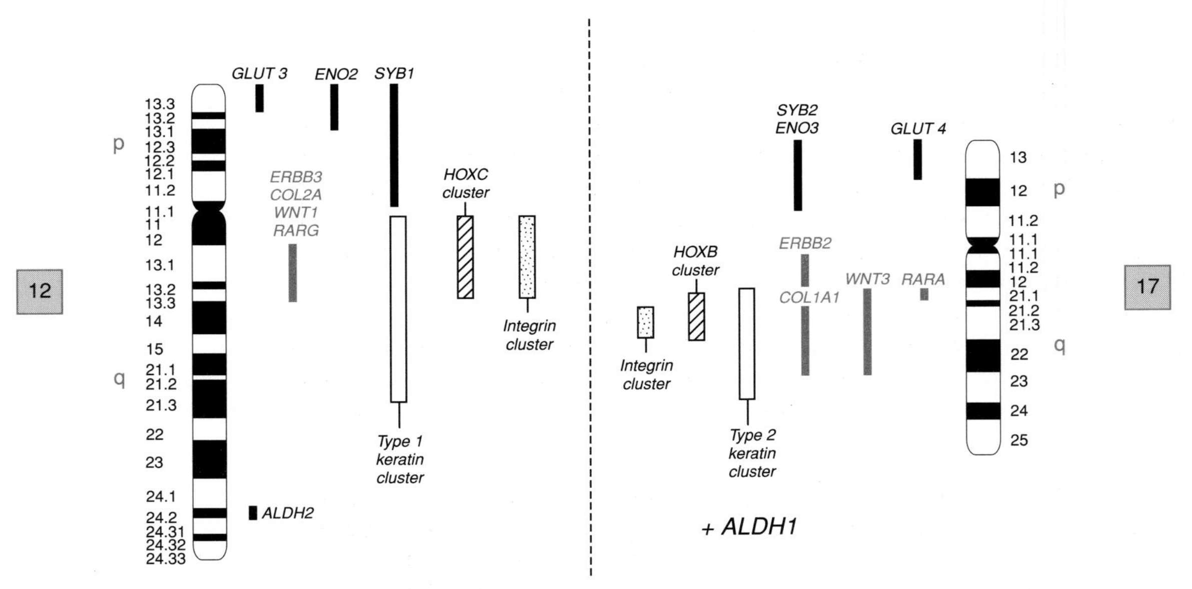

Figure 9.4: HSA12 and HSA17 appear to have paralogous chromosomal segments.

The approximate positions of some of the related genes mapping to human chromosomes 12 and 17 are indicated. The lengths of the bars marks the maximum uncertainty about the position of any member of a cluster; probably most are in fact closely clustered. *Note* that the aldehyde dehydrogenase gene *ALDH1* gene has been mapped only broadly to chromosome 17 but is clearly related to the *ALDH2* gene on 12q24.2.

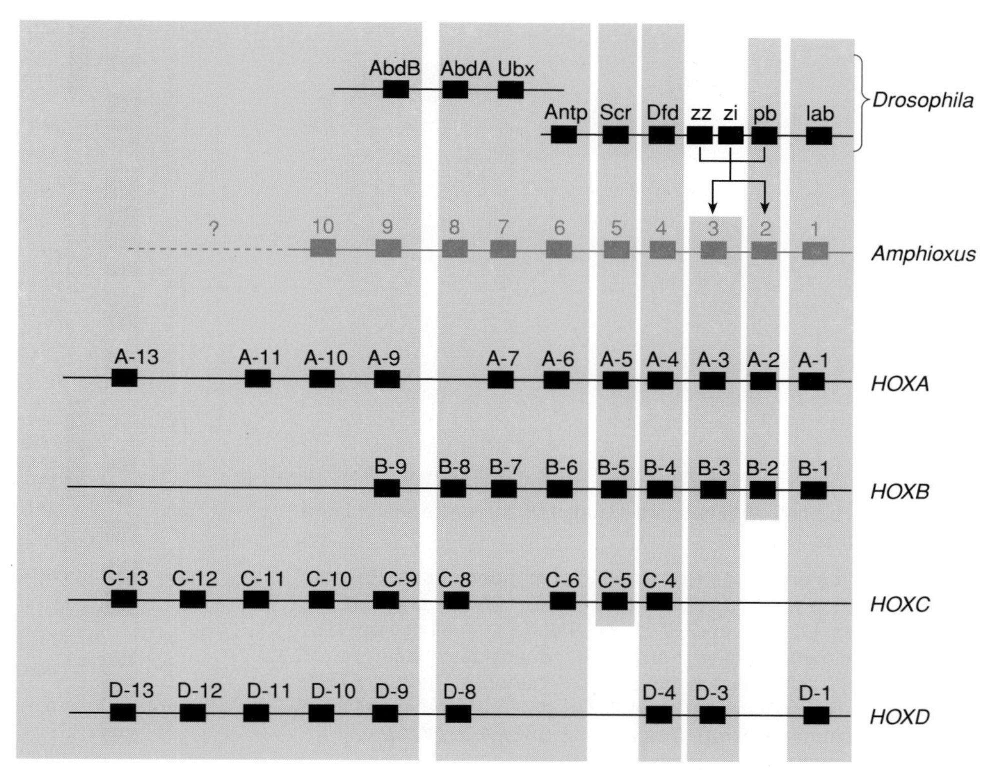

Figure 9.5: The organization of *HOX* gene clusters in mammals and *Amphioxus* suggests the possibility of one or two rounds of ancestral genome duplication.

Indicated paralogous groups consist of genes with very similar expression patterns and presumably similar functions. At the time of writing, 10 *HOX* genes had been isolated in *Amphioxus*, the invertebrate considered to be most closely related to vertebrates. The equivalent genes in *Drosophila* were presumably organized on a single cluster in an ancestral genome prior to a translocation which resulted in the *Ultrabithorax (Ubx)* and *Antennapedia (Antp)* clusters. The vertebrate ancestor presumably had 13 *HOX* genes but loss of individual genes following cluster duplications has led to each of the mammalian *HOX* clusters lacking one or more of the original genes. Boxes group genes with clearly related homeoboxes.

chromosomes (chromosomal translocations), unequal exchanges between homologous chromosomes or the sister chromatids of a single chromosome (page 255), and DNA copy transposition events (page 241). Clearly, some of these mechanisms can also result in loss of genetic material. In addition, other mechanisms (chromosome inversions, simple DNA transpositions and balanced translocations) can result in no net gain or loss of material. Mammalian genome evolution appears to have involved frequent subgenomic duplications and also rearrangements without net loss of DNA. In the human genome, for example, strong evidence exists for large paralogous regions on the two arms of chromosome 1 and to a lesser extent on the different arms of several other chromosomes (Lundin, 1993) (see *Figure 9.6*). The assumed large-scale duplications could have resulted from ancestral whole chromosome duplications by *Robertsonian fusion* (page 56) or subchromosomal duplications followed by *pericentric inversions* (page 55).

Comparisons of the present-day genome organization of humans and other mammals also suggest that large-scale rearrangements may have been frequent, and that karyotype and phenotype evolution can be uncoupled. For example, the Indian muntjac deer (*Muntiacus muntjak*) has only three types of (very large) chromosome, whereas its very close relative, the Chinese muntjac deer (*Muntiacus reevesi*), has 23 different chromosomes. The human and mouse karyograms are quite different from each other and even the highly conserved X chromosome linkage group shows numerous differences in organization between the species (see *Figure 9.10*). The

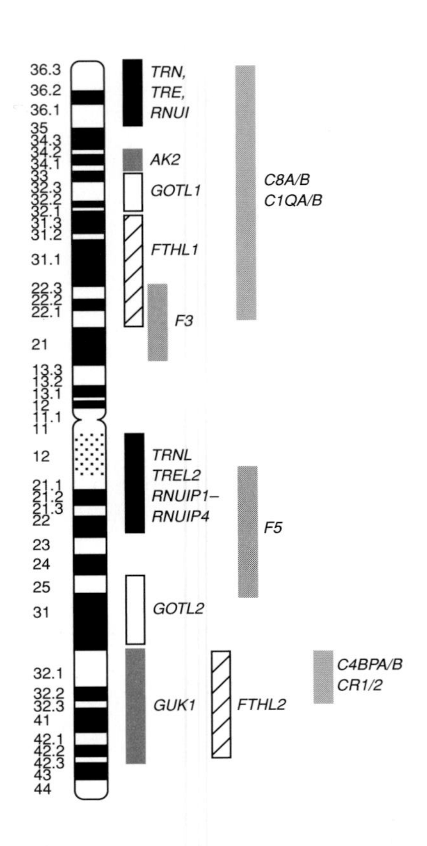

Figure 9.6: Gene mapping on human chromosome 1 suggests the presence of large duplicated regions on the two arms.

Vertical bars mark the approximate chromosomal locations of mapped genes. Paired sets are as follows: *TRN*, tRNA^{Asn} and *TRNL*, tRNA^{Asn}-like; *TRE*, tRNA^{Glu} and *TREL1*, *tRNA*^{Glu}-like; *RNU1*, small nuclear U1 RNA and *RNU1P1–RNU1P4*, small nuclear U1 RNA pseudogenes 1–4; *AK2*, adenylate kinase 2 and *GUK1*, guanylate kinase 1; *GOTL1* and *GOTL2*, glutamic–oxaloacetic acid transaminase 2-like sequences 1 and 2; *FTHL1* and *FTLH2*, ferritin heavy chain-like sequences 1 and 2; *F3* and *F5*, coagulation factors III and V; *C8A/B, C1QA/B*, complement C8 α/β, C1q α/β and *C4BPA/B, CR1/2*, complement C4-binding protein, complement component receptors 1/2.

great apes are extremely closely related to humans but show clear cytogenetic differences as a result of several inversions, a translocation that has occurred exclusively in the human lineage and another that has occurred in the gorilla lineage (see *Figure 9.25*). Old World monkeys are also closely related to humans but, with the exception of the gibbons, numerous chromosome rearrangements have occurred since divergence from the human lineage.

Evolution of the human sex chromosomes

Despite their considerable structural differences, substantial blocks of sequence homology between the human X and Y chromosomes suggest a common origin

In mammals, pairs of homologous autosomal chromosomes are structurally virtually identical (*homomorphic*); chromosome pairing at meiosis is presumed to be facilitated by the high degree of sequence identity between homologs, albeit by a mechanism that is not understood. By contrast, the X and Y chromosomes of humans and other mammalian species are *heteromorphic*. The human X chromosome is a submetacentric chromosome which contains about 165 Mb of DNA, whereas the Y is acrocentric and is much smaller (containing about 60 Mb of DNA). The human X chromosome contains numerous important genes: on the basis of its size alone, it might be expected to contain about 3000 genes, but the comparative lack of CpG islands on the X chromosome (see *Figure 7.5*) indicates that the true figure may be substantially smaller than this. In marked contrast, the Y chromosome carries only a very few functional genes, including several which are also found on the X chromosome, and some which are Y-specific, notably the primary male determinant gene,

Table 9.1: Known genes on the human Y chromosome

Gene symbol	Name/function	Chromosome location	Homolog on X
CSF2RAY	Granulocyte–macrophage colony-stimulating factor receptor	PAR1, Yp	*CSF2RA*
IL3RAY	Interleukin-3 receptor	PAR1, Yp	*IL3RA*
ASMTY	Encodes acetyl seronin *N*-methyl transferase	PAR1, Yp	*ASMT*
XE7Y	Encodes XE7 antigen; function unknown but homology to CD99	PAR1, Yp	*XE7*
MIC2Y	Encodes cell surface antigen CD99	PAR1, Yp	*MIC2*
XGRY	Encodes Xg blood group regulator	PAR1, Yp	*XGR*
XGY	Encodes Xg blood group antigen	PAR1, Yp	*XG*
ZFY	Zinc finger protein of unknown function	Distal Yp11.1	*ZFX*
RSP4Y	Encodes ribosomal protein S4	Distal Yp11.1	*RSP4X*
AMGY	Encodes amelogenin	Proximal Yp11.1	*AMGX*
SMCY	Encodes H–Y antigen	Proximal Yq	*SMCX*
SRY	Encodes the testis-determining factor	Next to PAR1, Yp	No
YRRM gene family	Regulators of spermatogenesis?	Yq11.23	No

PAR1, major pseudoautosomal region.

SRY (*Table 9.1*). The great bulk of the Y chromosome, however, is genetically inert and is composed of constitutive heterochromatin consisting of different types of highly and moderately repetitive noncoding DNA.

Despite being morphologically distinct, however, the X and Y chromosomes are able to pair during meiosis in male cells, and to exchange sequence information. Sequence exchanges occur within certain small regions of homology between the X and the Y chromosomes, known as **pseudoautosomal regions** because DNA sequences in these regions do not show strict sex-linked inheritance. In the human sex chromosomes, there are two pseudoautosomal regions (see Rappold, 1993):

- the **major pseudoautosomal region (PAR1)** extends over 2.6 Mb at the extreme tips of the short arms of the X and Y. It is the site of an *obligate crossover* during male meiosis which is thought to be required for correct meiotic segregation. Genes within this region are characteristically expressed, irrespective of whether or not the chromosome is subject to X inactivation (see page 172). This very small region is remarkable for its high recombination frequency (the sex-averaged recombination frequency is 28% which, for a region of only 2.6 Mb, is approximately 10 times the normal recombination frequency). The high figure is, of course, mostly due to the obligatory crossover in male meiosis resulting in a crossover frequency approaching 50% (see *Figure 9.7*). The boundary between the major pseudoautosomal region and the sex-specific region has been shown very recently to map within the *XG* blood group gene, with the *SRY* male determinant gene occurring only about 5 kb from the boundary on the Y chromosome (*Figure 9.8*).
- The **minor pseudoautosomal region (PAR2)** extends over 320 kb at the extreme tips of the long arms of the X and Y. Unlike the major pseudoautosomal region, crossover between the X and Y in this region is not so frequent, and is neither necessary nor sufficient for successful male meiosis.

In addition to the two pseudoautosomal regions, the human X and Y chromosomes show substantial regions of homology elsewhere, including a variety of Xp–Yq and Xq–Yp homologies, as well as Xp–Yp and Xq–Yq homologies. The existence of such homologies suggests that the two chromosomes have evolved from an ancestral homomorphic pair of chromosomes. Clearly, the two chromosomes have subsequently undergone substantial divergence, and sequences that are physically close on one chromosome may have very widely spaced counterparts on the other (*Figure 9.9*). However, primate comparisons have shown that at least some of the

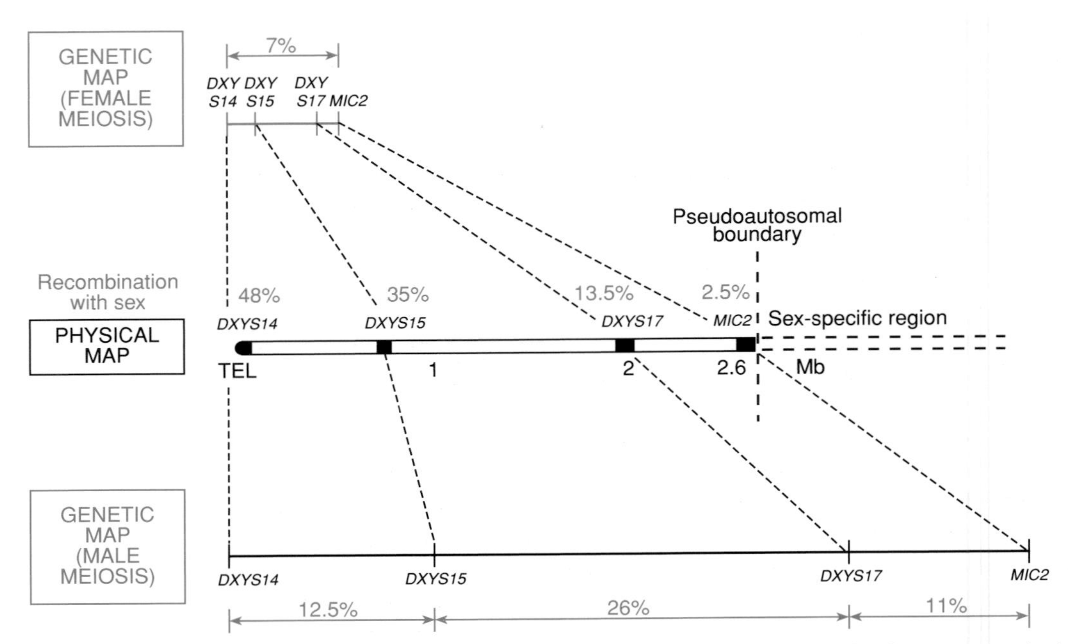

Figure 9.7: The major human pseudoautosomal region is characterized by a high overall recombination frequency and a large sex difference in recombination frequency.

The frequency of recombination with sex (i.e. of exchange between the X and Y chromosomes in male meiosis) shows a gradient along the length of the pseudoautosomal region – sequences such as *DXYS14* which are located at the telomere have an almost 50% chance of being exchanged between the X and the Y. Because of the obligate crossover in this very small region in male meiosis, the male genetic map is much larger than the corresponding female genetic map. For example, the recombination frequency between the markers *DXYS14* and *MIC2* is about 49.5% in male meiosis, but only 7% in female meiosis.

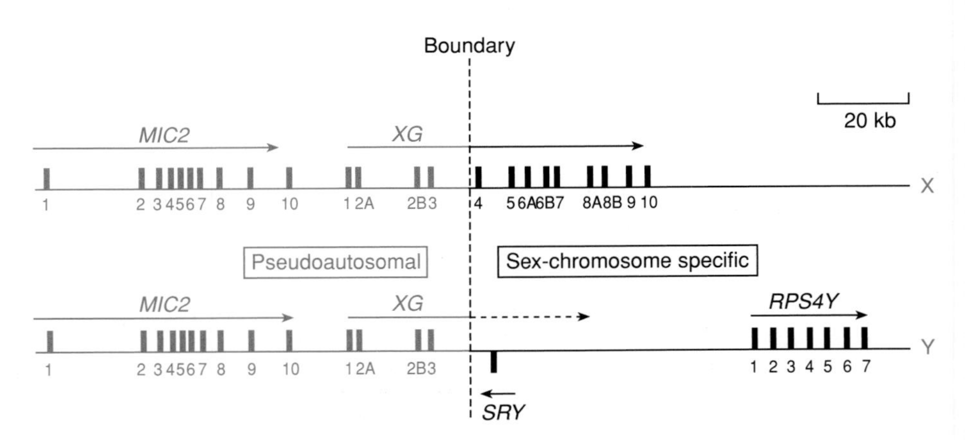

Figure 9.8: The boundary of the major human pseudoautosomal region occurs within the *XG* blood group gene.

Arrows indicate the extent and transcriptional orientation of known genes. Exons are represented by numbered vertical bars and, for the sake of clarity, are shown as being of uniform size. *Note* that the boundary occurs within the *XG* blood group gene on the X chromosome. On the Y chromosome, the *XG* gene is truncated: the promoter and exons 1, 2A, 2B and 3 are in the pseudoautosomal region, but thereafter there are Y chromosome-specific sequences, including the *SRY* gene which is transcribed from the opposite strand. Transcription from the *XG* promoter occurs across the boundary on the Y chromosome (dashed arrow) resulting in both *XG*-antisense *SRY* and *XG–RPS4Y* transcripts. *Note* that the *RSP4Y* gene has a homolog on the X chromosome at Xq13 (see *Figure 9.9*) and that the intron sizes are unknown and represented as being equal for the sake of simplicity. Another copy of the *XG* gene is known to be located on the Y chromosome: *XGPY* at Yq11.21 appears to have almost all of the exons found in the X chromosome gene and appears to be an example of an expressed pseudogene (see page 187). Reproduced from Weller *et al.* (1995) by permission of Oxford University Press.

existing X–Y homology results from very recent *duplicative transposition events* (i.e. duplication of a particular chromosomal segment and subsequent transposition of one of the two copies to another chromosome).

The evolution of the mammalian X chromosome has led to substantial species differences, both in chromosomal DNA organization and the pattern of X inactivation

Human–mouse divergence in gene order

Conservation of synteny between mouse and human is most pronounced in the case of the X chromosome: almost the entire X linkage group appears to be conserved between the two species (known exceptions include human pseudoautosomal genes with autosomal orthologs in mouse; see legend to *Figure 9.10*). This remarkable conservation of synteny for X-linked genes also applies to other mammals and appears to be evolutionarily related to the development of a special form of dosage compensation: *X chromosome inactivation* (see pages 50 and 172). Once established by evolutionary design, or accident (Ohno, 1973), X chromosome inactivation would be expected to ensure conservation of synteny on the X chromosome: X–autosome translocations would be selected against because the normal 2:1 ratio of gene dosage for autosomal and X-linked genes would be destroyed. As expected, there is extensive conservation of synteny of the genes on the mouse and human X chromosomes. Nevertheless, there are major differences in gene order: fine mapping of X-linked DNA sequences in the two species indicates regions of homology which can only have been generated by a variety of different chromosomal inversions in the lineages leading to present day mice and humans (*Figure 9.10*).

Human–mouse divergence in X inactivation patterns

The rationale for X chromosome inactivation is to act as a dosage compensation mechanism for those X chromosome genes (the vast majority) which do *not* have homologs on the Y chromosome. However, a small minority of human X-linked genes do have functional homologs on the Y chromosome. Such genes, which are common to both the X and the Y, do not show a sex difference in dosage; as a result,

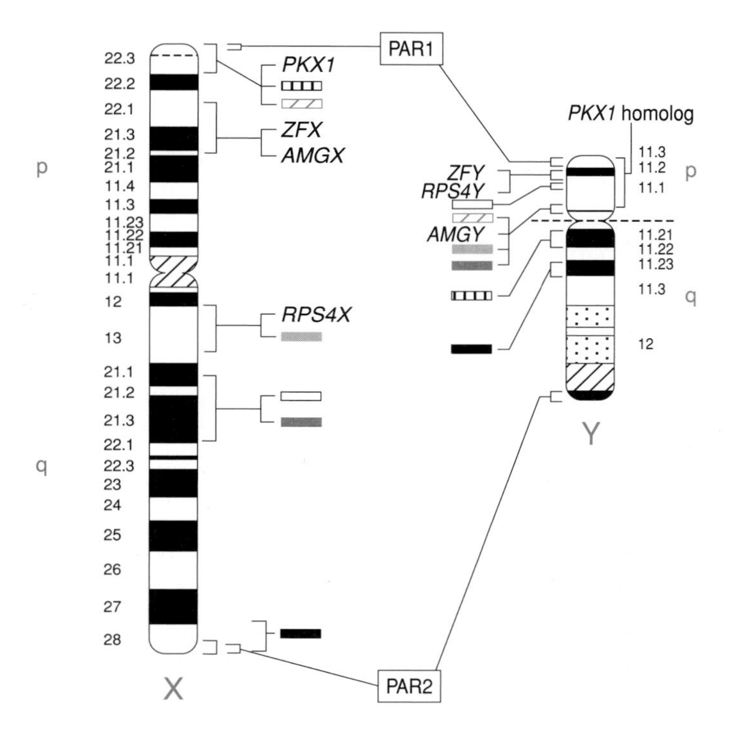

Figure 9.9: The human X and Y chromosomes show several regions of homology in addition to the common pseudoautosomal regions.

Note the contrasting spatial organization between different pairs of homology blocks. For example, a sequence block in the Yq11.21 region has a counterpart in the most distal Xp band, Xp22.3, whereas sequences in the nearby band Yq11.23 have homologs in the most distal Xq band, Xq28. The gene encoding ribosomal protein S4, *RPS4*, and the *ZFY* gene are clustered on distal Yp11.1 but have widely spaced homologs on the X, and the amelogenin gene, *AMGY*, is part of a cluster of homology regions at proximal Yp11.1 with widely scattered homologs on the X. PAR1, major pseudoautosomal region; PAR2, minor pseudoautosomal region.

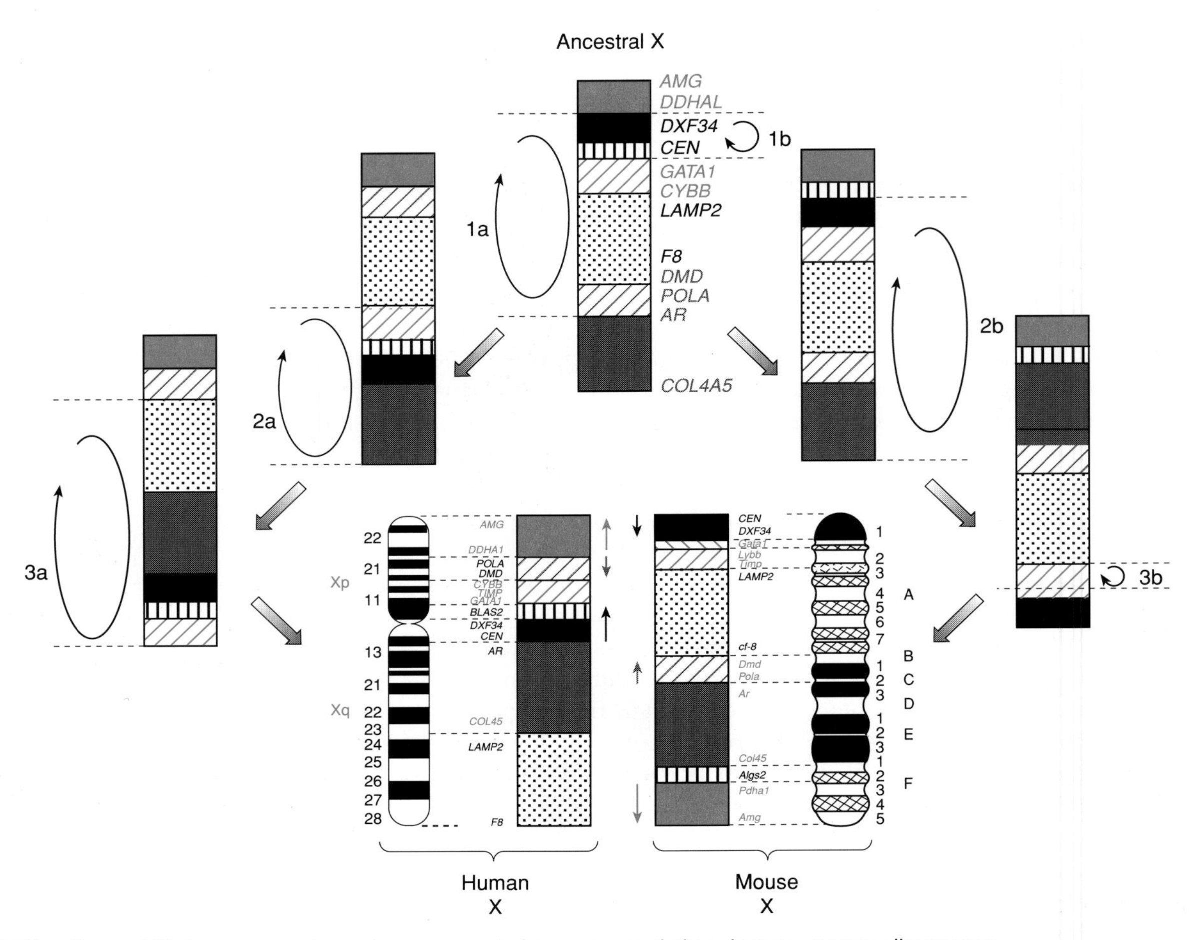

Figure 9.10: Several X chromosome inversions appear to have occurred since human–mouse divergence.

The present-day organization of human and mouse X chromosomes is shown at the bottom. A minimum of eight different homology blocks are defined by the presence in each of multiple orthologous genes and DNA sequences, of which only the most proximal and the most distal markers are shown. The remainder of the figure shows one possible explanation of how the existing chromosome organizations may have been derived by a series of inversions from a common ancestral X chromosome. Redrawn from Blair *et al.* (1994) with permission from Academic Press Inc.

they would be expected to escape X inactivation. Those genes in the major pseudoautosomal region which have been tested for X inactivation status, have all been shown to be expressed on both active and inactive X chromosomes. In addition, several other human X-linked genes are known to escape X inactivation, including not only genes which map close to the pseudoautosomal region but also genes which map to proximal Xp and proximal Xq regions, while genes known to map to intermediate locations are often subject to X inactivation (*Figure 9.11*).

As expected, many of the human X-linked genes which map outside the major pseudoautosomal region and escape X inactivation have functional homologs on the Y chromosome. Some, however, do not. For example, the *UBE1* and *SB1.8* genes escape inactivation but do not appear to have any homologs on the Y chromosome. Other genes such as the Kallman syndrome gene *KAL1*, and the steroid sulfatase gene *STS* do have homologs on the Y chromosome, but these are nonfunctional pseudogenes. It is likely, therefore, that for some genes sex difference in gene dosage is not a problem and is tolerated (see Disteche, 1995). In the mouse, however, there are considerable differences in the pattern of X inactivation. For example, the human nonpseudoautosomal genes *ZFX, RPS4X* and *UBE1* all escape inactivation,

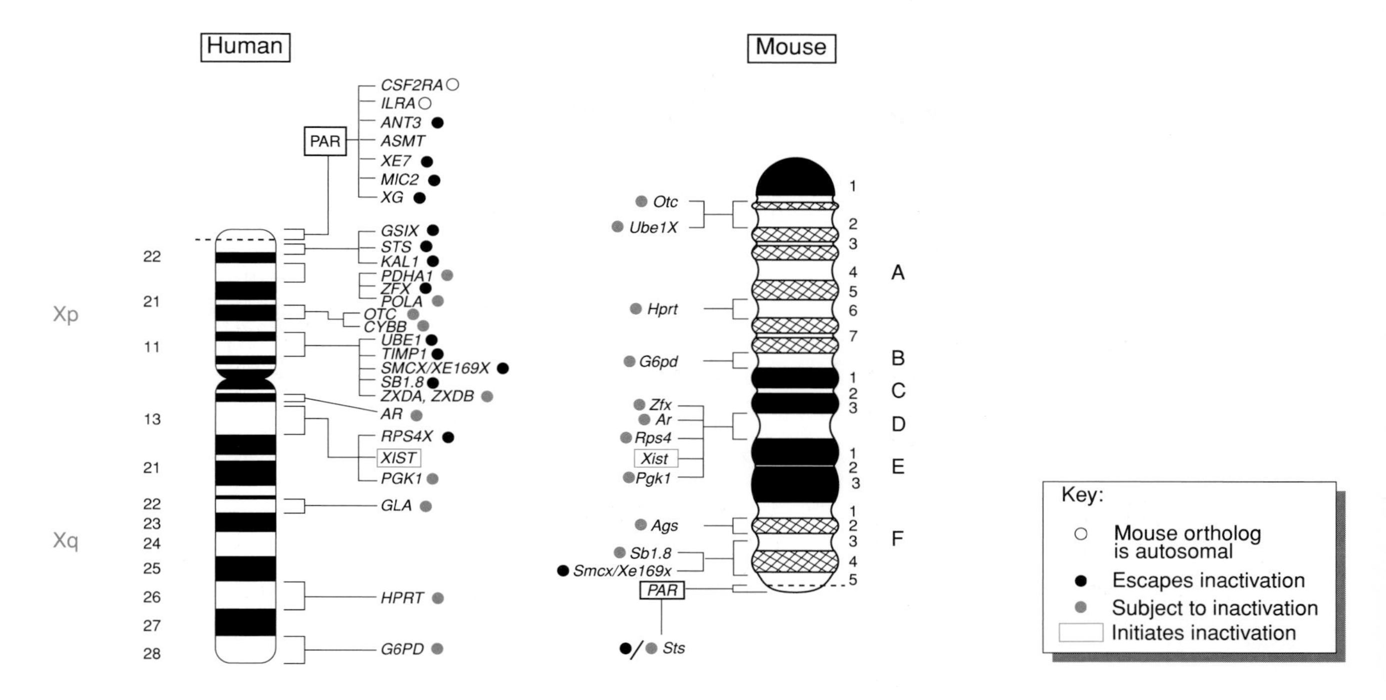

Figure 9.11: Genes that escape inactivation on the human X chromosome are widely distributed, unlike those on the mouse X.

Note that the mouse *Sts* (steroid sulfatase) locus may escape inactivation or be subject to inactivation in different mouse strains. The murine homologs of the *CSF2RA* and *ILRA* genes (located in the major human pseudoautosomal region) are autosomal. The *XIST* and *Xist* loci initiate X inactivation, and are expressed on the inactivated X, but *not* on an active X (see page 173).

but the murine homologs *Zfx, Rps4* and *Ube1X* (which unlike the human *UBE1* gene has a homolog on the Y) are all subject to X inactivation (*Figure 9.11*).

The differences in human–mouse X inactivation patterns may explain the difference in phenotype of an XO human (i.e. 45,X) and an XO mouse. Only about 1% of human XO conceptuses survive to birth, usually presenting with the characteristics of **Turner syndrome**, including short stature, gonadal dysgenesis and anatomical defects. In sharp contrast, XO mice are viable, with no prenatal lethality, and are anatomically normal and fertile. Both the embryonic lethality and Turner syndrome in human XO individuals are thought to be due to monosomy for a gene or genes common to the X and the Y: normal XX and XY individuals will be biallelic for such genes but an XO individual must be monoallelic. Because attempts to localize a Turner syndrome gene on the X have been difficult, efforts have been made to localize the homolog on the Y instead. This can be done by mapping XY females who carry different deletions of the Y: the deletions remove the major male determining gene, *SRY*, and, depending on the extent of the remaining segment that is deleted, individuals may or may not have features of Turner syndrome. As a result, the Turner phenotype has been mapped to a 90-kb segment between *SRY* and *ZFY*. Within this region maps a gene for the very highly conserved 40S ribosomal S4 protein, *RSP4Y* (see *Figure 9.8*), and as this gene and its X-linked homolog, *RPS4X*, are both known to be functionally active, *RPS4* deficiency has been suggested to have a role in Turner syndrome (Watanabe *et al.*, 1993).

Comparison of genes in distantly related mammals suggests that much of the short arm of the human X chromosome has recently been acquired by X–autosomal translocation

Mammals are classified into two subclasses *prototheria* (the monotremes or egg-laying mammals) and *theria* which in turn are subdivided into two infraclasses: **metatheria** (marsupials) and **eutheria**, a group which includes placental mammals (*Figure 9.12*). Although many eutherian X-linked genes are found to be X-linked in marsupials, genes mapping to a large part of the short arm of the human X have orthologs on autosomes of both marsupials and monotremes. Because the prototherian divergence pre-dated the metatherian–eutherian divergence, the simplest explanation is that at least one large autosomal region was translocated to the X chromosome early in the eutherian lineage.

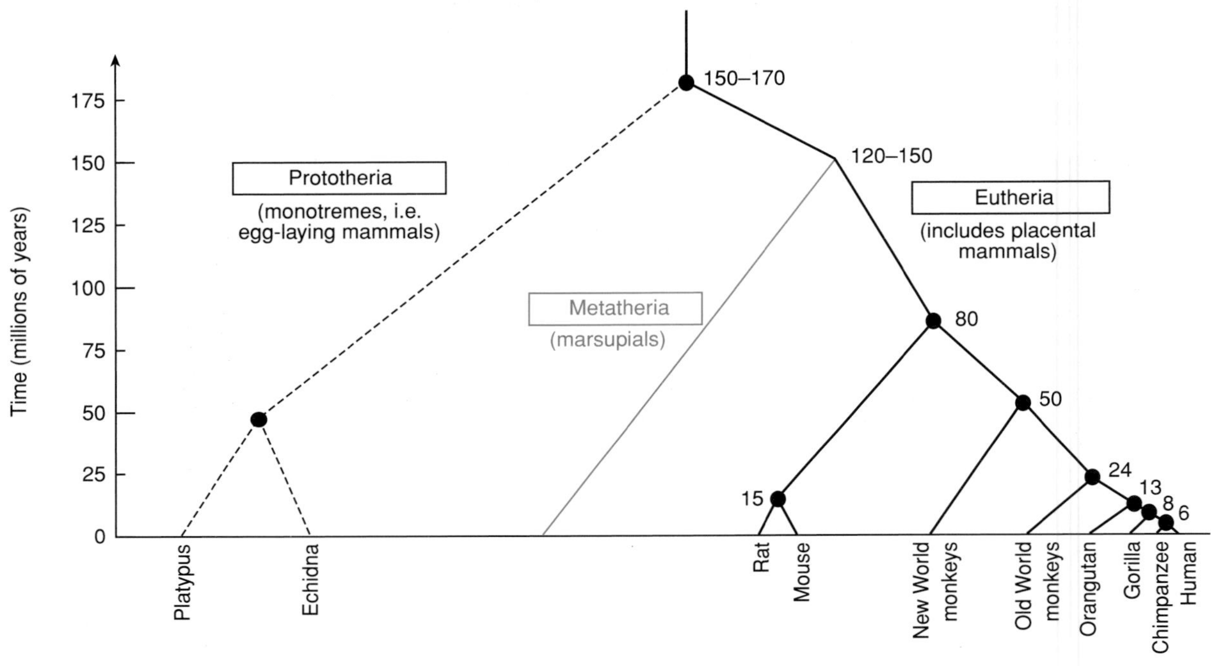

Figure 9.12: Mammalian phylogeny.

Numbers to the right refer to the approximate dates of divergence of the indicated lineages in millions of years. For example, the lineage giving rise to modern day humans is thought to have diverged from the rodent (mouse/rat) lineage at about 80 million years ago and from chimpanzees about 6 million years ago. Figures for divergence of primate lineages represent estimates based on the DNA hybridization studies of Sibley and Ahlquist (1984, 1987).

Translocation of autosomal genes on to the X chromosome will result in the formerly autosomal genes being subject to X inactivation. Not only is one X chromosome shut down in all female cells, but inactivation of the single X chromosome in male cells is required during spermatogenesis. However, certain genes on human Xp would be expected to be crucially important for cell function. For example, the *PDHA1* gene encodes the E1α subunit of the pyruvate dehydrogenase complex, an enzyme essential in aerobic energy metabolism. In marsupials the *PDHA1* gene is located on an autosome and so is expressed during spermatogenesis. In contrast, the *PDHA1* gene in humans (and other eutherian mammals) is X-linked, and is not expressed during spermatogenesis. However, a closely related gene, *PDHA2*, encodes a testis-specific human isoform which presumably has evolved in response to the silencing of X-linked genes during spermatogenesis. The *PDHA2* gene is intronless and is thought to be an example of a functional processed gene, generated by reverse transcription from the mRNA of the *PDHA1* gene (see *Figure 8.7* for the general mechanism).

The human Y chromosome most likely lost most of its former genetic capacity as a result of lack of recombination enforced by selection pressure

Distinct sex chromosomes have been independently developed in many animals with disparate evolutionary lineages, including not only mammals, but birds (where the females are XY, the *heterogametic sex,* and the males are XX, the *homogametic sex*), and certain species of fish, reptiles and insects. In each case, it is thought that the X and Y started off as virtually identical autosomes, except that one of them happened to evolve a major sex-determining locus (the *SRY* locus in humans) (*Figure 9.13*). Subsequent evolution resulted in the two chromosomes becoming increasingly dissimilar until, in many species, the Y was reduced to a tiny chromosome with only a very few functional genes. There would appear to be evolutionary pressure to adopt the strategy of having two structurally and functionally different sex chromosomes, and it seems that this pressure is gradually driving the Y chromosome to extinction. Eventually, one would expect a switch to a sex determination

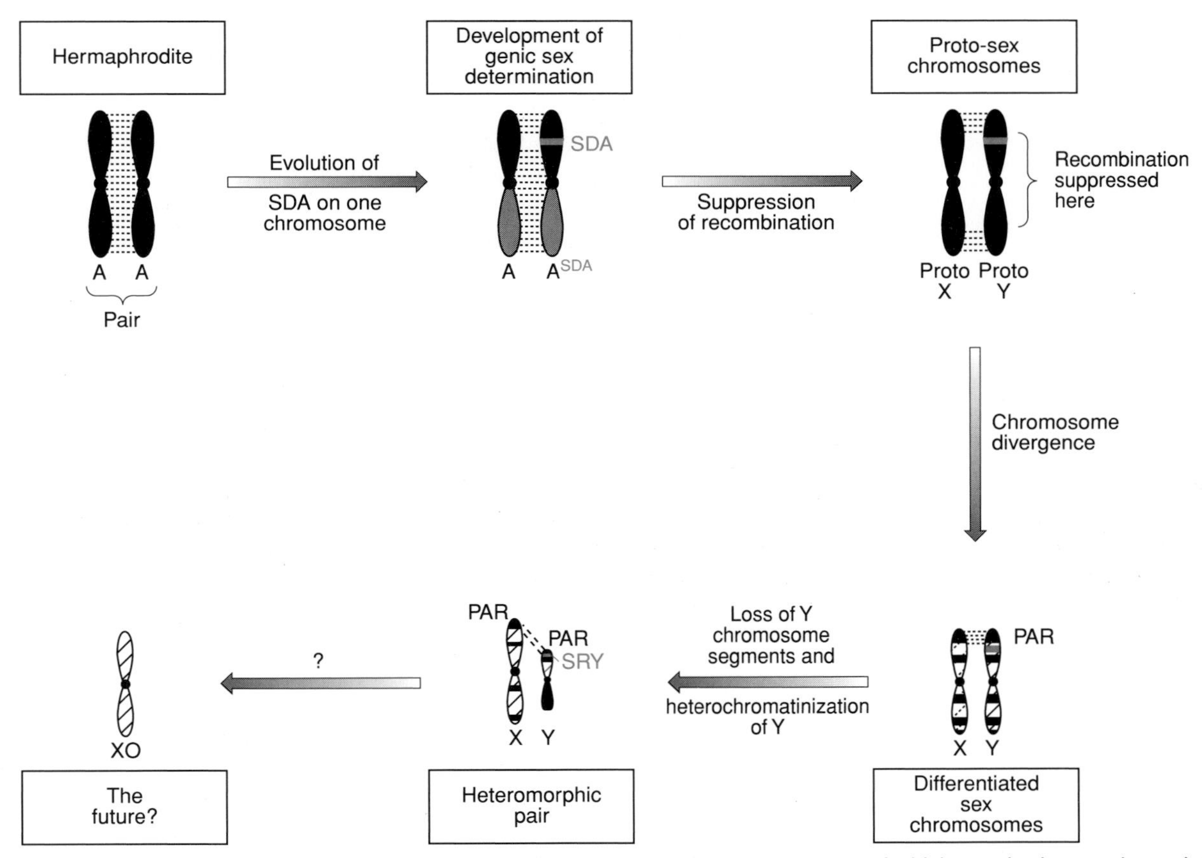

Figure 9.13: Mammalian sex chromosomes most likely evolved from a pair of autosomes, one of which acquired a sex-determining allele (SDA), leading to recombination suppression and chromosome differentiation.

One of an homologous pair of autosomes in an ancestral genome is envisaged to have evolved an SDA. Thereafter, the need to avoid exchange of the SDA and possibly the evolution of *sexually antagonistic genes* (see page 218) led to suppression of recombination between the two chromosomes, except in small regions, known as PARs. Lack of exchange between the homologs led to chromosome divergence. Because most of the Y is not involved in any recombination events, it degenerated by a series of chromosomal deletions. Present day human X and Y chromosomes retain small regions of homology outside the PARs, partly as a result of very recent X–Y transpositions (see page 211). Possibly the inexorable pressure to reduce the Y means that eventually it will be completely eliminated and a mechanism of sex determination will evolve which is based simply on X : autosome gene dosage.

system where maleness is conferred simply by X : autosome gene dosage and XO individuals are male, as in the case of *Drosophila*.

Why should the X and Y chromosomes diverge, and why should the Y degenerate? Clearly, recombination needs to be suppressed in the region of the major sex-determining locus (e.g. *SRY* in humans is located in a nonrecombining region just 5 kb proximal to the major pseudoautosomal region; see *Figure 9.8*). Additionally, environmental circumstances may have offered a selective advantage for breaking down recombination between the sex chromosomes. For example, one trigger could have been the development of *sexually antagonistic genes*, with alleles which may be of benefit to the heterogametic sex (XY), but harmful to the homogametic sex (XX). If such genes accumulate, then there will be a selective pressure to ensure that they are not transmitted to the homogametic sex, a restriction which can be met if they are present on a nonrecombining Y chromosome. Certainly, recombination between the present-day human X and Y chromosomes is very limited, being very largely confined to the tiny pseudoautosomal region at the tips of the short arms (see page 211). This means that the human Y chromosome is virtually an asexual component, a nonrecombining chromosome, within an otherwise sexual genome (the X chromosome can recombine along its length with a fully paired homolog in female meiosis). Population genetics predicts that a nonrecombining chromosome should degenerate. If the mutation rate is high, harmful mutations can gradually accumulate in genes on that chromosome over long evolutionary time scales: mutant alleles may drift to fixation as Y chromosomes with fewer mutants lost by chance, or they may 'hitchhike' along with a favorable allele in a region protected from recombination. This theory has recently received experimental verification using *Drosophila* mutants (Rice, 1994). Once mutations accumulate in the nonrecombining Y, the loss of function of genes means that there is no selective pressure to retain that DNA segment and the chromosome gradually contracts by a series of deletions.

Evolution of human DNA sequence families and DNA organization

Gene duplication is a mechanism for generating functional divergence that has frequently been used in the evolution of mammalian genomes

In addition to large-scale gene duplication events involving the whole genome or large chromosomal segments, selective duplication of specific genes can occur by other means, including copy transposition events and also tandem gene duplication.

Mechanisms resulting in gene duplication

Duplicative or **copy transposition** events involve a duplication of a DNA sequence prior to transposition. DNA transposition in mammalian genomes most often occurs through an RNA intermediate and frequently results in a moderate to large copy number interspersed repeat family. Processed copies of genes transcribed by RNA polymerase II normally lack functional regulatory sequences present in the original gene and, with few exceptions, degenerate into pseudogenes (*processed pseudogenes*; see page 187). The exceptions all appear to be sequences which are copied from X-linked genes and which show testis-specific gene expression (e.g. the pyruvate dehydrogenase genes, *PDHA2*; see page 216). Genes which are transcribed by RNA polymerase III often contain a promoter sequence within the transcribed sequence. As a result, the cDNA copies of such genes can reach very high copy numbers: the internal promoter sequence could confer transcriptional activity on the copies. This is the way in which the *Alu* repeat family appears to have evolved (see page 200 and *Figure 9.24*).

Tandem gene duplication often occurs as a result of unequal crossover events or unequal sister chromatid exchanges (see page 255). Numerous clustered human gene families show evidence of having acquired multiple members by this mechanism. In many cases, the duplicated genes degenerate into *nonprocessed pseudogenes* (see page 187). However, the transition between functioning duplicated gene and nonfunctional pseudogene may be a gradual one. This has given rise to the concept of the *expressed pseudogene,* a gene which is expressed at the mRNA level, or even at the polypeptide level, but which is nevertheless nonfunctional (see page 187). The absence of function means that selection pressure to conserve function will be relaxed and eventually mutations will accumulate, often leading to silencing of gene expression. Alternatively, the mutations may eventually result in the acquisition of different expression patterns and sometimes different functions (*Figure 9.2*).

Acquisition of different expression patterns

Some diverged duplicated genes are known to be expressed predominantly in different environments. Sequence divergence in the different genes in the α-globin gene cluster and in the β-globin gene cluster may result in encoded products with slightly different biological properties. For example, the ε-, ζ- and γ-globin chains could possibly be especially suited to binding oxygen in the comparatively oxygen-poor environment of early development, whereas the α- and β-globin chains may be the preferred polypeptides in the environment of adult tissues.

Genes encoding different tissue-specific **isoforms** (alternative forms of the same protein) or **isozymes** (alternative forms of an enzyme) also appear to have evolved by gene duplication. For example, the enzyme alkaline phosphatase is encoded by at least four different genes which show tissue-specific differences in expression. Of these, three are clustered near the telomere of 2q: *ALPI* and *ALPP* encode alternative forms of the enzyme (87% protein sequence similarity) found in intestine and placenta, respectively, and *ALPPL* encodes a placental-like isozyme. A fourth member, *ALPL,* is located near the telomere of 1p and encodes an isozyme expressed in liver, bone, kidney and some other tissues, and is more distantly related to the intestinal and placental forms (57% and 52% sequence similarity, respectively). Note, however, that duplicated genes encoding subcellular-specific forms of the same protein are often located on different chromosomes, with gene duplication possibly arising from an ancestral genome duplication event. For example, in human liver there are two major isoforms of aldehyde dehydrogenase, a cytosolic and a mitochondrial form, which show 68% sequence identity over their 500 amino acid long sequences. The cytosolic and mitochondrial forms are, respectively, encoded by the *ALDH1* gene on chromosome 9q and the *ALDH2* gene on chromosome 12q. The two genes each have 13 exons and nine out of the 12 introns occur in homologous positions in the two coding sequences, strongly suggesting a common evolutionary origin by some kind of ancient gene duplication event (see Strachan, 1992, pp. 32–33).

Concerted evolution occurs as a result of intragenomic (intraspecific) sequence exchanges within a DNA sequence family

In the case of certain gene and DNA sequence families, there may be a closer sequence relationship between individual family members in one species (paralogs) than that between orthologs in different species (**concerted evolution**). Thus, if we consider a specific gene family in two species, A and B, concerted evolution means that a family member in species A will be more closely related to other members of that family in species A than it will be to an ortholog or any other members of the same family in species B (*Figure 9.14*). Concerted evolution occurs because of various genetic mechanisms which cause sequence exchange between nonallelic DNA sequences within a genome. These mechanisms, which include *unequal crossover, unequal sister chromatid exchange* and *gene conversion*-like mechanisms (see pages 255–259), are particularly prevalent in the case of tandemly repeated DNA sequences. For example, unequal crossover and unequal sister chromatid exchange can result in a specific repeat sequence spreading through an array of tandem repeats, and eventually replacing the other repeats, thereby resulting in **sequence homogenization** (see *Figure 10.8*). Because of meiotic recombination, the resulting effect can be transmitted to other genomes in a sexual population. As a result, concerted evolution may be observed between members of a DNA family *within a species*; sequence exchange between homologous sequences in the DNA from different species is essentially nonexistent.

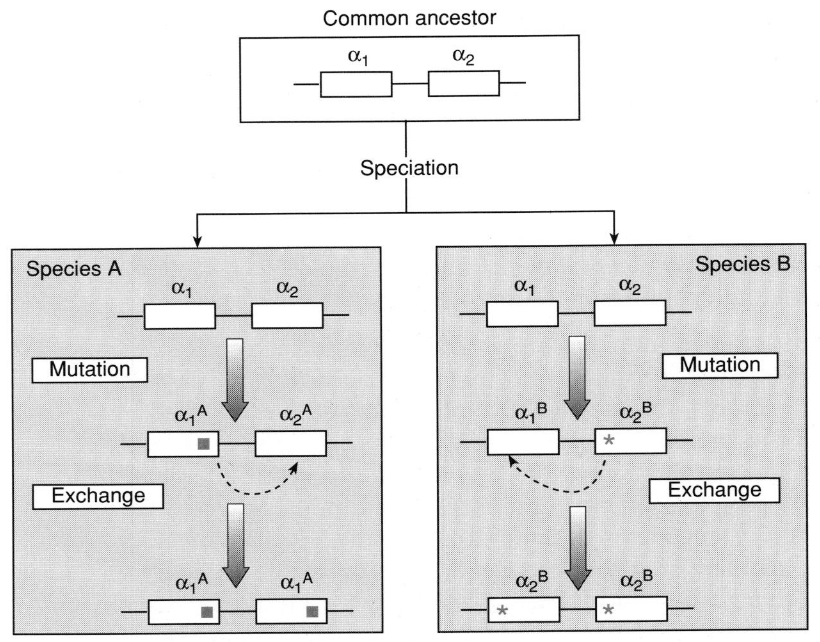

Figure 9.14: Concerted evolution occurs in DNA sequence families when a relatively high level of sequence exchange occurs between family members.

The family illustrated has two members α_1 and α_2 which arose by tandem duplication, and was inherited from a common ancestor by two species, A and B. Mutation leads to sequence divergence and the $\alpha_1^A - \alpha_2^A$ divergence would be expected to be similar to that for $\alpha_1^B - \alpha_2^B$. Indeed, if one copy were under selection pressure to maintain function, and the other diverged rapidly, one might expect the conserved copy in species A to be more similar to that in B than it would be to the other copy in A. However, if there are frequent sequence exchanges *within a species*, the copies in A will be more related to each other than to those in B and vice versa.

The rDNA genes provide a useful example of sequence exchange between repeats within a cluster but, in addition and more unusually, sequence exchanges can occur between clusters. In the human genome the rDNA genes are organized as large clusters of tandem repeats containing 50–60 copies of an approximately 40-kb repeat unit (see *Figure 8.3*). The high degree of sequence homology between such large repeats facilitates frequent sequence exchanges between nonallelic repeats. Additionally, the clusters are located on the short arms of the acrocentric chromosomes 13, 14, 15, 21 and 22 which frequently exchange sequences by nonhomologous chromosome translocations. As a result, individual human rDNA genes are more similar to each other than they are to the rDNA genes of other primates.

Some gene families do not show strong evidence of concerted evolution: sequence homologies between orthologs may be greater than between different family members in one species

In general, members of a gene family or superfamily which are located in the same gene cluster show a higher degree of sequence homology than do members present on different clusters, and the degree of sequence homology is usually greatest between closely neighboring genes within a cluster. This is so for the following reasons:

- gene duplication events leading to formation of any one cluster are often examples of recent tandem duplications, whereas the duplications that have given rise to the different clusters are often comparatively ancient and may have resulted from ancestral genome duplication events
- The evolution of gene clusters by a series of tandem duplications will tend to mean that closely neighboring genes are more likely to have originated by a recent tandem duplication than more distantly spaced genes in the same cluster.
- Following gene duplication there may be two competing forces which affect the sequence identity between the duplicated genes:

  sequence divergence (the sequences of the duplicated genes may be identical initially but during evolution will gradually become different as a result of independent accumulation of mutations in the two genes); and

  sequence homogenization (periodic sequence exchanges between the two genes will tend to result in sharing of sequences between them and therefore maintain sequence identity). Such homogenization results from genetic mechanisms (such as unequal crossover, unequal sister chromatid exchange and gene conversion (see pages 255–259) which are much more prevalent in tandemly duplicated genes than in duplicated genes which are distantly located or nonsyntenic. As a result, the sequences of distantly spaced genes or nonsyntenic genes will have a tendency to diverge more rapidly than those of tandemly duplicated genes.

The globin gene superfamily provides some useful examples. Sequence homology between the genes and gene products from different clusters (e.g. α-globin and β-globin) is much less than between genes and gene products from a single cluster (*Figure 9.15* and *Table 9.2*). This is largely so because the different clusters are presumed to have originated early in evolution while gene duplications within clusters occurred comparatively recently. In the latter case, some duplication events are presumed to have occurred very recently, leading to duplicated genes which are almost identical. For example, the two human α-globin genes *HBA1* and *HBA2* encode identical products, and the products of the two γ-globin genes *HBG1* and *HBG2* differ by a single amino acid. In other cases, the duplicated genes within a cluster are clearly more diverged in sequence, presumably because the relevant duplication events occurred some time ago.

In contrast to tandemly repeated genes, intra-cluster sequence exchanges between globin genes are likely to be infrequent (except for the very recently duplicated genes, such as *HBA1* and *HBA2*). This is so because the different globin genes are small (1.6 kb) and the chromosomal DNA separating them is not well conserved. Additionally, the stringent developmental regulation of gene expression within a cluster (see page 190) presumably imposes a functional constraint, minimizing sequence exchanges within different types of gene in a cluster. As a result, the sequence homology between orthologs in distant species such as mouse and humans may be greater than that between genes on the same human cluster. For

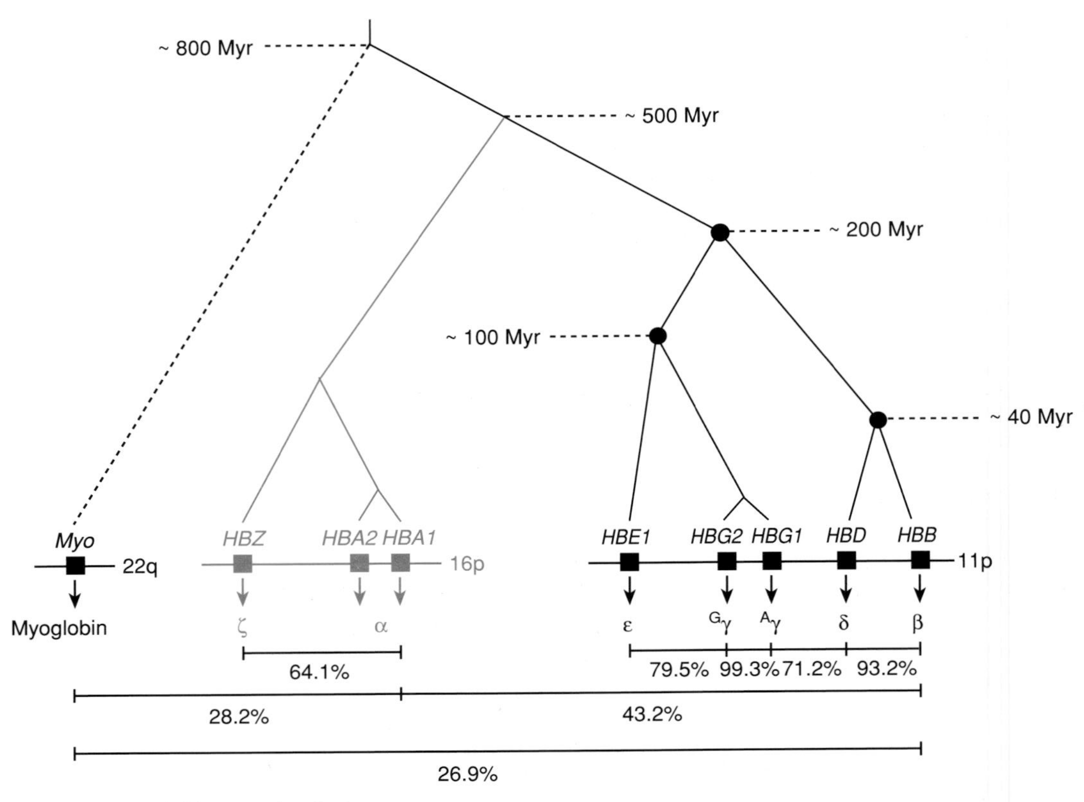

Figure 9.15: Evolution of the globin superfamily.

Globins encoded by genes within a cluster show a greater degree of sequence homology than those encoded by genes on different clusters (as do the genes themselves). The close relationship between genes within a cluster is thought to be due to comparatively recent gene duplication events, whereas the duplication events giving rise to the different clusters are much more ancient. *Note* that neighboring genes in a cluster may be very closely related as a result of very recent tandem gene duplication: the *HBA1* and *HBA2* genes encode identical α-globins, and the *HBG1* and *HBG2* encode γ-globins that differ by a single amino acid.

Table 9.2: Sequence variation in globin genes

	Sequence homology (%)			
	Coding DNA	5′ + 3′ UTS	Introns	Amino acid sequence
Human β-globin/chimp β-globin	100	100[a]	98.4	100
Human β-globin/rabbit β-globin	89.3	< 79[b]	< 67[b]	90.4
Human β-globin/mouse β-globin[c]	82.1	< 66[b]	< 61[b]	80.1
Human β-globin/human ε-globin	79.1	62	50	75.3

[a]5′ UTS only.
[b]Maximum homologies based on counting insertions or deletions of three or more nucleotides as single sites.
[c]Either β^{maj} or β^{min}.

example, the sequence homology between human and rabbit β-globin genes or even between human and mouse β-globin genes is greater than that between the human β-globin and ε-globin genes (*Table 9.2*). The β–ε-globin gene split most likely occurred some time before human–mouse divergence and the lack of frequent sequence exchanges within the human β-globin cluster has resulted in the considerable genetic distance between human β-globin and human ε-globin being maintained.

Evolution of gene structure

Eukaryotic genes are often larger and more complex than those from simple organisms. The process of gene elongation during evolution appears to have frequently involved repetition of existing amino acid sequences, often as a result of *exon duplication*. Additionally, the structure of many eukaryotic polypeptides suggests that frequent exchange of structural or functional protein domains has occurred at the gene level by *exon shuffling*, resulting in complex mosaic genes capable of specifying a variety of different protein modules.

Complex genes can evolve by intragenic duplication, often as a result of exon duplication

In addition to the other forms of DNA duplication discussed earlier, human genes, like other eukaryotic genes, often show evidence of intragenic DNA duplication which can be substantial. For example, many genes are known to encode polypeptides whose sequences are completely or mostly composed of large repeats, with sequence homology between the repeats being very high in some cases (see *Table 7.8*). In many cases, a repeat corresponds to a protein domain. Building of larger polypeptides by repetition of a previously designed protein module offers a variety of evolutionary advantages:

- *dosage repetition.* The ubiquitin-encoding genes, *UbB* and *UbC*, encode polypeptides containing, respectively, three and nine tandem repeats of the sequence for ubiquitin, a small protein with several functions, notably in proteolysis. Most of the proteins that are degraded in the cytosol are hydrolyzed in large protein complexes known as *proteasomes* but, before the proteins can be delivered to the proteasomes, they need to be tagged by covalent binding to a series of ubiquitin molecules, forming a multiubiquitin chain. Because many proteins are short lived, large amounts of ubiquitin molecules need to be synthesized. These genes may therefore have evolved to express many copies of the ubiquitin sequence by gene elongation through intragenic repetition, as opposed to the tandem gene duplication mechanism used in the case of genes encoding rRNA or histones (see pages 188–190). The large polypeptide precursor is then cleaved to generate multiple copies of the desired ubiquitin monomer.
- *Structural extension.* Repeating domains may be particularly advantageous in the case of proteins that have a major structural role. An illustrative example is provided by the 41 exons of the *COL1A1* gene which encode the part of α1(I) collagen that forms a triple helix; each exon encodes essentially an integral number of copies (one to three) of an 18 amino acid motif which itself is composed of six tandem repeats of the structure Gly–X–Y where X and Y are variable amino acids.
- *Domain divergence.* In most cases, intragenic duplication events have been followed by substantial nucleotide sequence divergence between the different repeat units. Such divergence presumably provides the opportunity of acquiring different, though related, functions. Sometimes the degree of sequence divergence between the repeats is such that the repeated structure may not be obvious at the sequence level. For example, in the case of the variable and constant domains of immunoglobulins, conservation of the secondary structure is much more apparent than that of the amino acid sequence (*Figure 8.2*). In some cases where the repeated structure is not obvious, statistical analysis can nevertheless reveal evidence for structural similarity.

Exon shuffling permits diverse combinations of structure and functional modules, and may be mediated by transposable elements

The organization of many genes reveals that they encode modules which have been considered characteristic of another type of gene. For example, the large extracellular matrix protein fibronectin contains multiple repeated domains which are encoded by individual exons or pairs of exons, indicative of classical **exon duplication** (*Figure 9.16*). A type of domain which is present in multiple copies at the N and C termini was considered to be characteristic of fibronectin, but was subsequently found in tissue plasminogen activator. Like fibronectin, tissue plasminogen activator also contains other domains – they include a structural module that is repeated extensively in, and is characteristic of, the epidermal growth factor precursor, and so-called kringle modules, which have been found in other polypeptides such as prourokinase and plasminogen, etc. (*Figure 9.16*). Such observations have suggested the possibility of mechanisms permitting **exon shuffling** between genes (see Patthy, 1994).

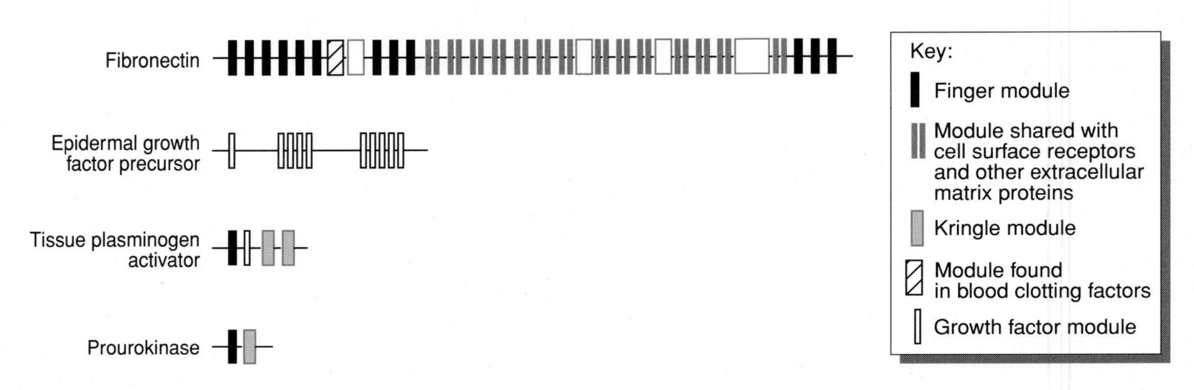

Figure 9.16: Exon duplication and exon shuffling.

The fibronectin gene contains 12 copies of an exon encoding a finger module, which is also found in the products of other genes such as the tissue plasminogen activator (TPA) gene. In addition, it contains 15 copies of a pair of exons which together specify a module shared with cell surface receptors and other extracellular matrix proteins. Similarly, the epidermal growth factor precursor gene has 10 copies of an exon encoding a growth factor module which is also found in the TPA and prourokinase (*PLAU*) genes. The latter two genes also contain exons encoding *kringle* modules. Exon duplication within a gene and exon shuffling between genes selectively use exons where the total number of nucleotides is exactly divisible by three (exon groups 0,0; 1,1 and 2,2; see *Box 9.2*). Exon duplication could be mediated by unequal crossover and unequal sister chromatid exchanges, possibly assisted by interspersed repetitive sequences within introns. Exon shuffling may be mediated by transposable elements (see *Figure 9.17*).

Intragenic exon duplication can be explained by a variety of mechanisms including unequal crossover, or unequal sister chromatid exchanges (see *Figure 10.7*). In order to avoid frameshifts, one would expect selective amplification of exons with a total number of nucleotides exactly divisible by three (i.e. exons which are flanked by introns of the same phase, such as 0,0, 1,1 and 2,2 exons; see *Box 9.2*). This is what is observed for exons that are duplicated within a gene and also for exons encoding modules shared by different genes (*Figure 9.16*).

How do genes which are not necessarily closely related come to share sequences encoding very similar protein modules? In principle, exon shuffling could take place by nonhomologous recombination between introns of nonallelic genes. However, although this type of exon shuffling mechanism could have operated in evolution, it is hardly ideal: the recombination could lead to loss of genetic material and inevitably the result would be a *fusion gene* which, in many cases, would be expected to result in loss of function. A more attractive possibility is exon shuffling

Box 9.2: Intron groups and intron phases

Introns are heterogeneous entities with different functional capacities and notable structural differences. Depending on the extent to which they rely on extrinsic factors to engage in RNA splicing and on the nature of the splicing reaction, they can be classified into different **intron groups**. The structural differences include enormous length differences (unlike exons which appear to be much more homogeneous in length; see *Figure 7.6*) and some other characteristics, such as differences in positions within coding sequences. The latter characteristic, which is particularly relevant to gene expression, is recognized in the subdivision of introns into different **intron phases** or **intron types**.

Intron groups

- **Spliceosomal introns** are the conventional introns of eukaryotic cells. They are transcribed into RNA in the primary transcript and are excised at the RNA level during RNA processing by spliceosomes (see page 17 and *Figure 1.14*). Only a few short sequences [those at the splice junctions and at the branch site (see *Figure 1.15*) and some regulatory sequences which are occasionally found] appear to be important for gene function. As a result, spliceosomal introns can tolerate large insertions and can be very long (e.g. intron 44 of the dystrophin gene is 140 kb long). Spliceosomal introns are likely to have arisen comparatively recently in evolution, and may have evolved from group II introns (see Cavalier-Smith, 1991). By tolerating the insertion of mobile elements, they facilitated exon shuffling.
- **Group I and II introns** have significant secondary structure and can catalyze their own excision without the requirement for a spliceosome (i.e. they are *self-splicing* or *autocatalytic introns*). They are found in both *eubacteria* and eukaryotes but are very restricted in their distribution, being found primarily in rRNA and tRNA genes, and in a few protein-coding genes found in some types of mitochondria, chloroplasts and bacteriophages. Both groups may also act as mobile elements, and mobile group II introns encode a reverse transcriptase-like activity which is strikingly similar to that of LINE-1 elements (see Belfort, 1993). The two groups differ in the identity of conserved splicing signals and in the nature of the splicing reaction.
- **Archaeal introns** have been found only in tRNA and rRNA genes in *archaebacteria*. They have no conserved internal structure and, unlike group I and group II introns, are not self-splicing. Although they require proteins for the splicing mechanism, they do not, unlike splicesosomal introns, require *trans*-acting RNA molecules for the splicing reaction.

Intron phases

The *phase* of an intron refers to the position at which it interrupts a coding DNA sequence which specifies polypeptides (*note* that this term is therefore irrelevant for those introns which happen to interrupt an untranslated sequence). There are three types (see *Figure 9.18*):

- **phase 0 introns** interrupt the coding sequence between adjacent codons. They are much more numerous than phase 1 or phase 2 introns (see below) and may represent the ancestral state.
- **Phase 1 introns** interrupt a codon between the first and second base positions.
- **Phase 2 introns** interrupt a codon between the second and third base positions.

Note that internal exons can be classified into various groups, depending on the phase of the two flanking introns. Exons where the total number of nucleotides is exactly divisible by three will fall into three groups (0,0; 1,1; and 2,2). *Exon duplication* within a gene and *exon shuffling* between genes involves such exons because when inserted they do not alter the translational reading frame, unlike the other six exon groups (0,1; 0,2; 1,0; 1,2; 2,0; and 2,1).

Notebox: archaebacteria and eubacteria are the two kingdoms of prokaryotes. Eubacteria appear to have branched off the universal evolutionary tree first, prior to subsequent divergence of archaebacteria and eukaryotes. Archaebacterial genomes appear to resemble eubacteria in form, but their genes are more closely related to eukaryotic genes than are those of eubacteria.

by transposons (see page 198). This could be achieved if two transposable elements that are recognized by the same *transposase* (the site-specific recombination enzyme required for transposition) have integrated in closely neighboring introns of the same gene (*Figure 9.17*). Transposases recognize the terminal repeats flanking a transposable element. However, if two such elements are integrated in close proximity then, as an alternative to the normal excision of individual transposable elements, occasional aberrant excision of the intervening chromosomal DNA segment between the two repeats could occur. The excised fragment, containing exonic sequences as well as flanking intron sequences could then transpose to a different chromosomal location where it could integrate into an intron of another gene (*Figure 9.17*).

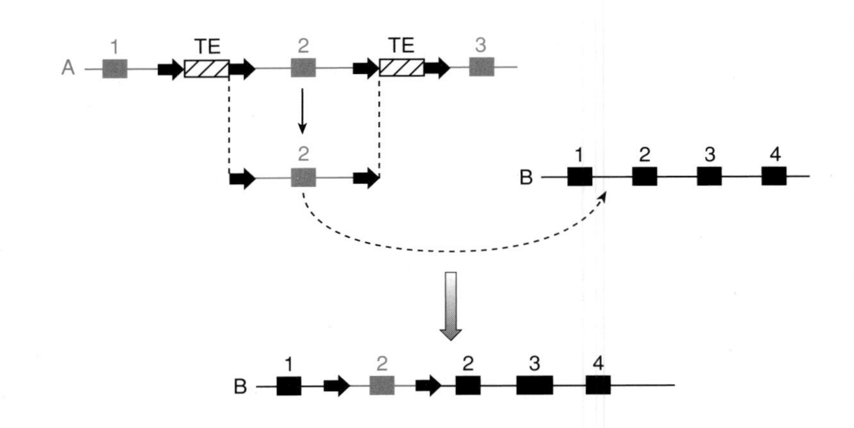

Figure 9.17: Exon shuffling between genes can be mediated by transposable elements.

Gene A contains two copies of the same transposable element (TE) flanking exon 2. An error by the relevant transposase leads to excision not of one of the TEs but instead the intervening sequence between them. The excised exon can then integrate at another chromosomal location, such as in gene B.

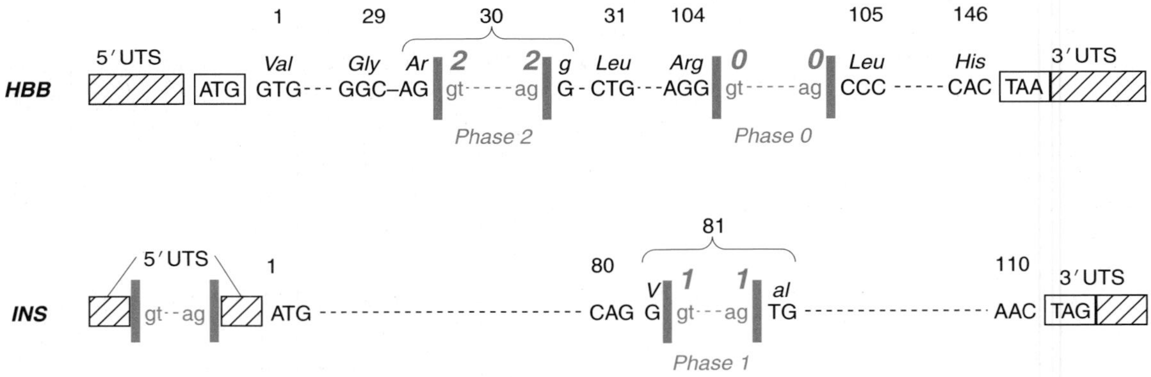

1 29 30 31 104 105 146
HBB 5′UTS ATG Val GTG--- Gly GGC–AG Ar 2 gt-----ag 2 g G - Leu CTG---Arg AGG 0 gt-----ag 0 Leu CCC ----- His CAC TAA 3′UTS
Phase 2 Phase 0

81
INS 5′UTS gt--ag 1 ATG ------------------------ 80 CAG V G 1 gt---ag 1 al TG ------------------ 110 AAC TAG 3′UTS
Phase 1

Figure 9.18: Introns within polypeptide-encoding DNA can be grouped into three phases, according to the point of insertion.

Phase 0 introns do not interrupt codons unlike phase 1 and phase 2 introns. A phase 1 intron in the human insulin gene *INS* interrupts a codon specifying valine (position 81 in the preproinsulin precursor). A phase 2 intron in the human β-globin gene *HBB* interrupts a codon specifying arginine at position 30 in the mature polypeptide (the initial product has 147 amino acids but the initiator methionine is cleaved during processing; see *Figure 1.19*). Internal exon 2 of the *HBB* gene spans codon positions 30–104 and is an example of a 2,0 exon (see *Box 9.2*). Even if it encoded a specific structure/function module, it would not be eligible for exon duplication or exon shuffling because the number of nucleotides it contains is not exactly divisible by three. *Note* that some introns are found in the untranslated sequences, such as the first intron in the *INS* gene. This means that some exons (such as exon 1 of the *INS* gene) can be composed entirely of untranslated sequence.

Most exons do not encode integral structural or functional protein modules and are flanked by introns which appear to have been inserted comparatively recently in evolution

Following the discovery of split genes in 1977, the evolution of introns and their relationship to exons have been the subject of intense debate (see Mattick, 1994). Essentially there are two alternative positions:

- *the 'introns-early' view (the exon theory of genes):* exons are the descendants of ancient minigenes and introns are the descendants of the spacers between them which were present in primordial cells. In this view, exons are considered as units which encode structural or functional domains, permitting evolution of larger genes by exon shuffling, a strategy that was favored particularly in eukaryotes. By contrast, introns are imagined to have been effectively lost from both kingdoms of prokaryotes. By constructing an exon database from available protein sequences and searching for homologous exons, Dorit *et al.* (1990) estimated that only 1000–7000 different exons were needed to construct all proteins.
- *The 'introns-late' view* does not deny that exon shuffling between genes occurs but holds that split genes have arisen as a result of comparatively recent insertion of introns into genes. Increasingly, the balance of the evidence is strongly tilting towards the view that introns have recent origins (see *Box 9.3*).

Any proposed function for introns in human genes cannot be a general one because of the small minority which lack introns (see *Table 7.6*). It is becoming apparent that some introns contain functionally important sequences, for example, important regulatory sequences. Additionally, although introns are not highly conserved in evolution, considerable sequence conservation may be found in the case of some short

Box 9.3: Evidence supporting a late rather than an early origin for introns

The lack of a correspondence between exons and structural/functional domains in the gene products

An important component of the exon theory of genes was the idea that exons in polypeptide-encoding genes represented functional or structural units. Exons consisting only of untranslated sequences (e.g. the first exon of the insulin gene; see *Figure 9.18*) could not be accommodated easily in this view. Even in the case of exons containing coding DNA, however, the evidence has been meager and, in four major examples cited as evidence for the exon theory of genes, objective methods for detecting correspondence between exons and units of protein structure have failed to identify any such correspondence (Stoltzfus *et al.*, 1994).

The general lack of conservation of intron positions

The exon theory of genes was supported by the apparent conservation of the positions of introns in genes known to have duplicated early in evolution, such as the globin genes (see *Figure 9.19*). In particular, nuclear genes encoding chloroplast and cytosolic glyceraldehyde-3-phosphate dehydrogenase are thought to have duplicated very early in evolution – they are imagined to have derived from the ancestral prokaryotic genomes that gave rise to chloroplasts and mitochondria, respectively, prior to transfer to nuclear locations. Yet, five intron positions are precisely conserved in the two genes (Kersanach *et al.*, 1994). However, there are numerous gene families (e.g. actin, myosin and tubulin families) where the locations of introns are not conserved, suggesting instead that introns have been inserted recently into the different genes at different locations. Additionally, the requirement for introns to have been present since primordial times would mean not only that many original introns have been lost subsequently from genes, but that some genes must originally have had such a large number of introns that the corresponding exon sizes must have been tiny. The discovery of autocatalytic introns which may function as mobile elements, and from which spliceosomal introns are thought to have evolved (see *Box 9.2*), also supported the idea that introns could have a relatively recent origin.

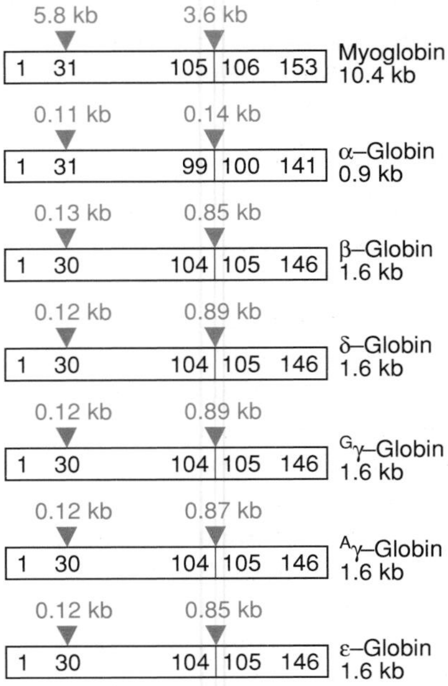

5.8 kb 3.6 kb
1 31 105 106 153 Myoglobin 10.4 kb
0.11 kb 0.14 kb
1 31 99 100 141 α–Globin 0.9 kb
0.13 kb 0.85 kb
1 30 104 105 146 β–Globin 1.6 kb
0.12 kb 0.89 kb
1 30 104 105 146 δ–Globin 1.6 kb
0.12 kb 0.89 kb
1 30 104 105 146 $^{G}\gamma$–Globin 1.6 kb
0.12 kb 0.87 kb
1 30 104 105 146 $^{A}\gamma$–Globin 1.6 kb
0.12 kb 0.85 kb
1 30 104 105 146 ε–Globin 1.6 kb

Figure 9.19: Members of the globin superfamily contain two introns which show reasonable conservation of positions, but not of size.

Boxes represent the mature polypeptides. Numbers contained within the boxes are the amino acid positions. Gene sizes are indicated to the right and intron sizes and locations are shown in red.

introns (see *Table 9.2* for an example). However, in general, the introns found in mammalian genes are large compared with those in other species and the intron sequences are not well conserved.

What makes us human? Comparative mammalian genome organization and evolution

The virtual universality of the genetic code, the high degree of conservation of key biochemical reactions, the huge evolutionary conservation of key developmental processes – these are features which emphasize the close relationship of humans to species that are morphologically quite distinct and evolutionarily distantly related. So what is it that makes us different? While many of the fundamental feaures of human cells, genome organization and gene expression are common to all eukaryotes, mammalian-specific features can be identified, such as genomic imprinting and X inactivation (pages 170–173). In addition, certain other components of the genome or aspects of its expression show still higher levels of specificity.

What makes us different from mice?

Increasingly we rely on extrapolation from mouse studies to infer the situation in humans. For example, our knowledge of gene expression patterns in early human development is minimal because of the lack of early stage embryos for study. As early stage mouse and human embryos are anatomically extremely similar, readily available mouse embryos are increasingly providing data on gene expression patterns which would be expected to be strongly conserved in human embryos. Additionally, because of the power of transgenic technology and the ability to perform gene targeting in mouse embryonic stem cells (see page 536), the mouse is the most commonly used animal model of human disease. The extrapolation from mouse studies to humans has been justified by the general assumption that genomic DNA organization and gene expression patterns of mice and humans have been highly conserved, despite the 80 million years since the two lineages diverged from a common ancestor. Increasingly, however, there is greater appreciation of differences between the two species.

General aspects of genome organization

The genome sizes are comparable (3000 Mb of DNA), and both genomes are divided into *isochores* (large chromosomal regions in which the base composition of the DNA is comparatively homogeneous but which is variable between isochores). The human isochore classes include two light (AT rich) classes L1 and L2, and three heavy (GC rich) classes H1, H2 and H3, but the mouse genome is comparatively lacking in the H3 isochore (Sabeur *et al.*, 1993). Cytogenetic analyses appear to reveal very different chromosome organizations: the mouse has 20 pairs of acrocentric chromosomes whereas there are 23 pairs of human chromosomes, most of which are metacentric or submetacentric. Nevertheless, comparison of high resolution mouse and human chromosome maps has indicated that *orthologous* chromosomal segments (e.g. those containing the major histocompatibility complex of mouse and humans) are located in regions where there is considerable similarity of cytogenetic banding patterns, albeit over relatively small chromosomal regions (Sawyer and Hozier, 1986).

Gene number

There appears to have been an erosion of CpG islands from the mouse genome: there are aproximately 45 000 CpG islands in the human genome, but only 37 000 in the equivalent sized mouse genome (Antequera and Bird, 1993). This does not simply reflect a proportional reduction in gene number in the mouse genome because analysis of the sequence databases suggests that about 56% of human genes but only about 47% of mouse genes have CpG islands. On this basis, therefore, the total number of genes in humans and mice would appear to be much the same (about 80 000 when calculated from CpG island data, but see page 155 for alternative estimates of total gene number). However, gene families often show different numbers of genes (*Figure 9.20*). Such differences are expected to reflect complex processes of gene duplication and loss, with clear evidence of interlocus sequence exchanges (*Figure 9.21*).

In some cases, human genes do not appear to have any rodent orthologs. For example, genes for eosinophil-derived neurotoxin and eosinophilic cationic protein – two host defense proteins, both of which are members of the ribonuclease superfamily – appear to have evolved very recently; orthologs cannot be detected in rodent genomes and they appear to be primate-specific (Rosenberg *et al.*, 1995). Similarly, there appear to be four human apolipoprotein (a) genes but none can be detected in rodent genomes (see Byrne and Lawn, 1994). The recent origin of the apolipoprotein (a) genes most likely occurred following a duplication of the related plasminogen locus.

Gene distribution

Gene order has not been generally well conserved between human and mouse chromosomes. As in most mammals, human–mouse comparisons show a generally strong conservation of genes on the X chromosome. However, a few genes on the human X chromosome are known to have autosomal orthologs in mouse (see legend to *Figure 9.11*). In addition, the general order of genes on the human and mouse X chromosomes is rather different, although conserved over subchromosomal regions (see *Figure 9.10*). For any one human or mouse autosome, orthologous

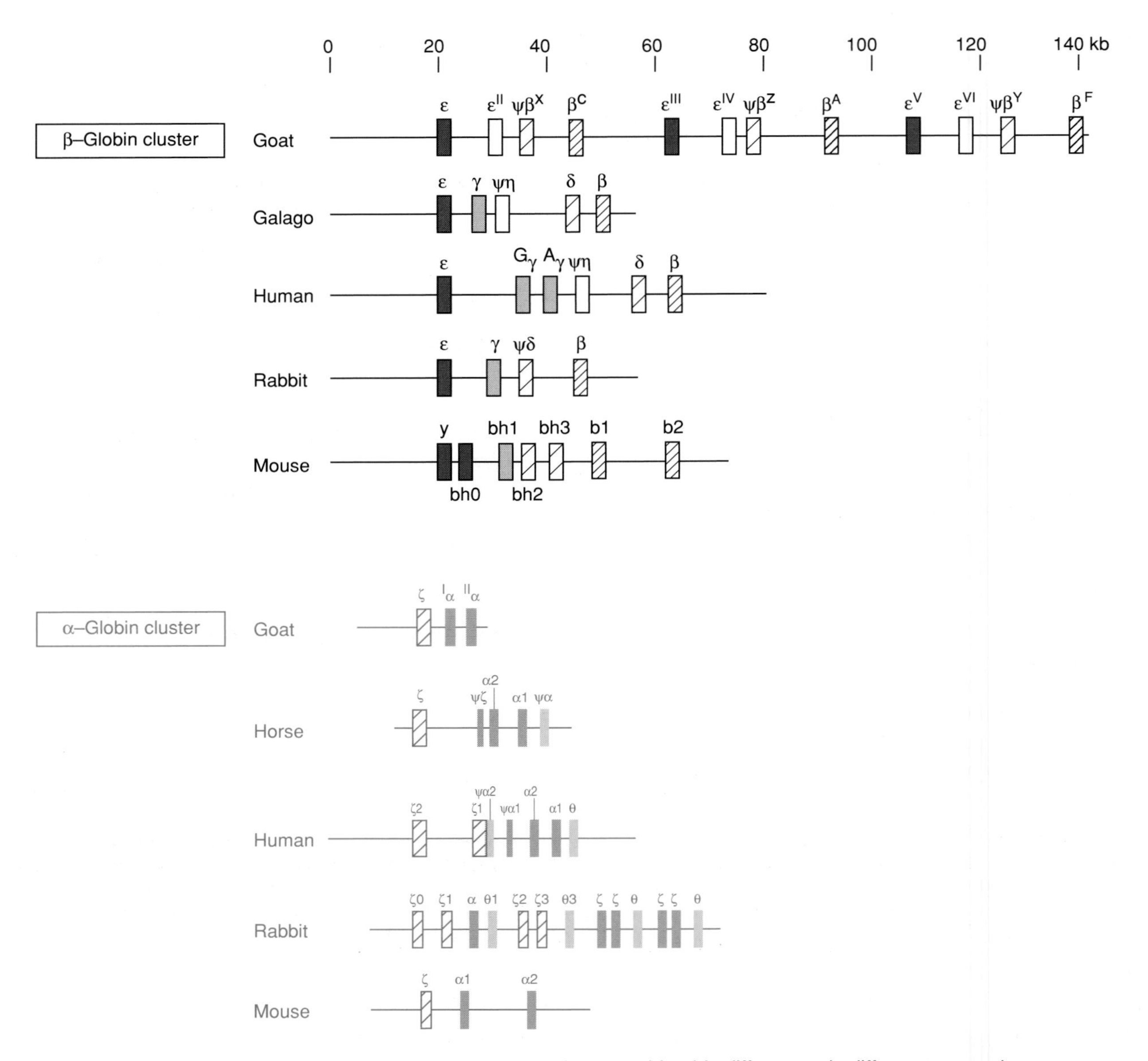

Figure 9.20: The organizations of orthologous gene families can show considerable differences in different mammals.

Shading of genes indicates proposed orthologous relationships, so that, for example, the horse ψα gene is orthologous to the θ gene of humans and rabbits. *Note* that the large number of genes in the goat β-globin cluster and the rabbit α-globin cluster reflects recent tandem triplication and quadruplication events respectively. The functional status of the θ genes is uncertain: the human θ gene may be an example of an expressed pseudogene (see *Box 8.1*) while in the rabbit the θ1 and θ2 genes are pseudogenes, and the other two θ genes are likely to be pseudogenes. Redrawn from Hardison and Miller (1993) with permission from The University of Chicago Press. 1993,© The University of Chicago Press.

regions are found on a variety of different chromosomes in the other species (*Figure 9.22*). However, again there is conservation of gene order over small to moderate sized subchromosomal regions. Such partial **conservation of synteny** (i.e. a group of linked genes in one species is paralleled by a linkage group between the orthologous genes in the other species) has proven to be very useful in identifying some human disease genes (see page 389).

Gene organization and gene expression

The sizes of coding DNA in mouse and human genes are nearly identical, with an average size of about 500 codons (Duret *et al.*, 1995). The respective polypeptide

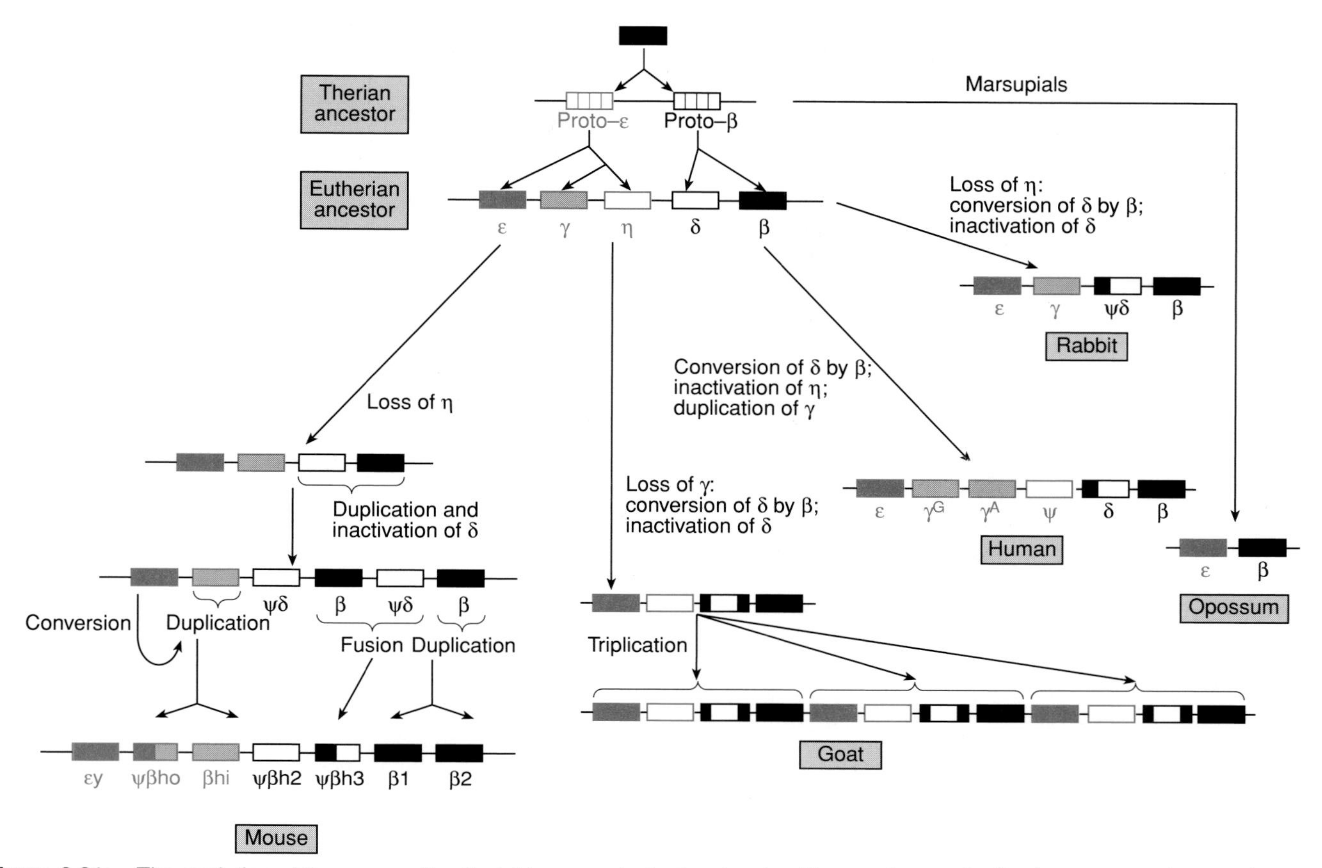

Figure 9.21: The evolution of the mammalian β-globin gene cluster has involved frequent gene duplications, conversions and gene loss or inactivation.

Note that, in addition to gene duplication and gene loss events, there are frequent examples where the sequence of one gene shows evidence of having been copied from the sequence of another gene. This is loosely described as conversion in this figure, but may have involved mechanisms other than gene conversion (see page 256) in some cases. For example, the conversion of δ by β in the lineage leading to rabbits could have involved an unequal crossover event of the type that most likely resulted in the production of the ψh3 gene in the mouse lineage. Redrawn from Tagle *et al.* (1992) with permission from Academic Press Inc.

sequences show a high degree of sequence similarity, often within the 80–95% range. However, different classes of polypeptides may be extremely conserved (such as many gene products that are important in development or some other crucially important event such as ribosomal function) while others, notably ligands and receptors that are important in host defense, can be much more divergent (*Figure 9.23*). The sequence similarity of coding DNA is generally a few per cent less than that for the polypeptide products (largely because of silent nucleotide substitutions, notably at the third base position of codons).

Species differences in gene expression include differences in RNA processing and the alternative usage of promoters. For example, the human aldolase A gene has an additional promoter which does not function in the rat ortholog (Mukai *et al.*, 1991) and similar parallels are expected for some human and mouse orthologs. Other human–mouse differences include considerable differences in the pattern of X chromosome inactivation (see *Figure 9.11*) and also in the conservation of imprinting. For example, the mouse insulin-like growth factor receptor gene *Igfr* is subject to imprinting, but its human ortholog, *IGFR*, is not (Kalscheuser *et al.*, 1993).

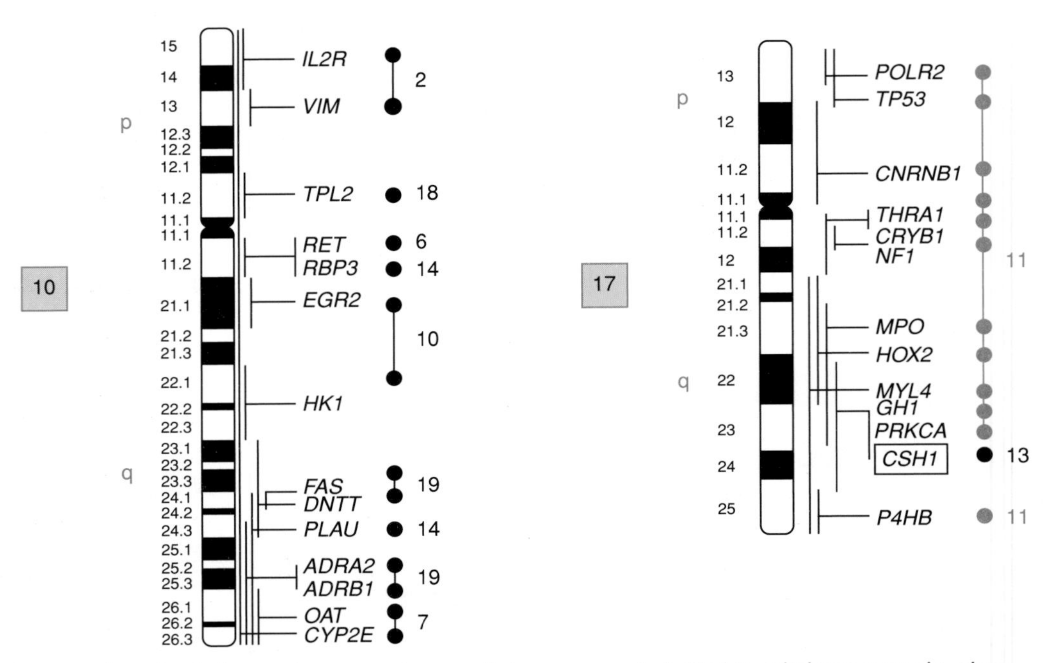

Figure 9.22: Conservation of orthologous human and mouse linkage groups is limited to subchromosomal regions.

Note that genes on human chromosome 10 (HSA10) have orthologs on at least seven different mouse chromosomes, whereas human chromosome 17 (HSA17) shows very considerable conservation of synteny with mouse chromosome 11. For a more complete representation see *Figure 14.13.* Redrawn from O'Brien *et al.* (1993) with permission from Nature America Inc.

Figure 9.23: Amino acid sequence divergence between human and rodent orthologs.

The average human–rodent sequence divergence is shown for 14 protein families, representing a total of 603 proteins compared between human and mouse/rat. *Note* that certain types of protein, such as the ribosomal proteins, are very highly conserved between mice and humans. Others, notably ligands or receptors that function in host defense, show considerable sequence divergence. Redrawn from Murphy (1993) with permission from Cell Press.

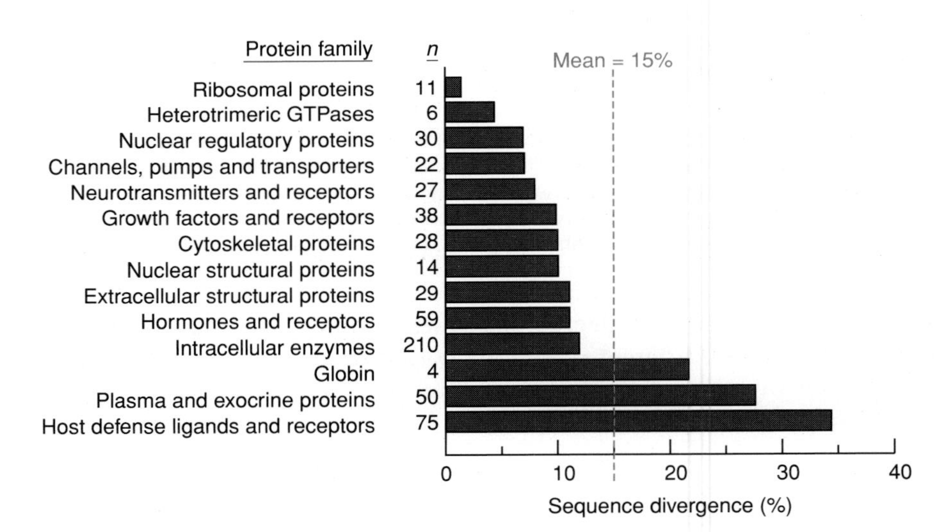

Noncoding DNA

Introns and noncoding DNA flanking genes are generally so highly diverged that alignment of orthologous sequences from the two species can be extremely difficult unless the comparison is confined to sequences which are located close to exons. Thus, very short introns (less than 200 bp) can be aligned and compared (see *Table 9.2*) but larger introns (accounting for the great majority of introns) are progressively more difficult to compare because of the very high sequence divergence. However, a striking example of the conservation of noncoding DNA occurs in the TCR cluster, where sequencing of about 100 kb in mouse and humans reveals a sequence identity of approximately 70%, even though only about 6% of the DNA is coding DNA (Koop and Hood, 1994). This is likely to be related to the very unusual mechanisms for expressing TCR and immunoglobulin genes (page 175 ff.).

Table 9.3: Comparison of genome organization and gene expression in humans and mice

Parameter	Similarities/differences in humans and mice
Genome size and gene number	About 3000 Mb in both genomes; estimate of 65 000–80 000 human genes and the number in mice is expected to be close to this figure
Chromosome number and type	Humans: 22 pairs of autosomes, X and Y; most are metacentric or submetacentric (*Figure 7.4*). Mice: 19 pairs of autosomes, X and Y, all acrocentric
CpG islands	Human genome: about 45 000. Mouse genome: about 37 000 (Antequera and Bird, 1993)
Conservation of synteny	Apart from the X chromosome (see below), is confined to subchromosomal regions which may be small or substantial (see *Figures 9.22* and *14.13*)
X chromosome genes	Almost complete conservation of synteny, but gene order can be quite different because of inversions (*Figure 9.10*)
Gene family organization	Very similar for individual families. Recent gene duplications have led to differences in gene number, e.g. there are two mouse β-globin genes and one α-globin gene, but only one human β-globin gene and two α-globin genes
Gene organization	Generally similar. Similar sizes of coding DNA (average ~ 500 codons) and intron positions often conserved
Sequence homology	Coding DNA: usually about 70–90% homology, with polypeptide products usually within the 75–95% range. Noncoding DNA: usually extremely dissimilar
Gene expression	Very similar in general. For orthologous genes, can have differences in the choice of alternative promoters and patterns of splicing. Some differences in imprinting patterns for autosomal genes (page 231) and many differences in X inactivation status for X-linked genes (*Figure 9.11*)
Tandemly repeated noncoding DNA	Telomeric repeat sequences highly conserved. Other repeats not normally conserved in sequence, but similar overall patterns (location of satellite sequences at centromeres, widespread occurrence of microsatellite sequences, etc.)
Interspersed noncoding DNA	Some sequence conservation of interspersed repeats, notably the LINE-1 element. The human *Alu* and mouse *B1* repeats are both thought to have originated from 7SL RNA copies (see *Figure 9.24*) but are considerably diverged in sequence

Telomeric minisatellite DNA is conserved between mice and humans, and indeed the TTAGGG repeats are conserved throughout vertebrates, presumably because of selection pressure to ensure continued recognition by the telomerase enzyme (*Figure 2.14*). However, highly repetitive DNA sequences in general are among the most rapidly diverging sequences because of the virtual absence of conservative selection pressure. For example satellite DNA sequences in the human and mouse genomes are quite different, and there is poor conservation of hypervariable minisatellites and microsatellites at orthologous locations in humans and mice. Nevertheless, there are several examples of apparent conservation of intragenic microsatellites located within orthologs in human and mouse/rat (Stallings, 1995).

Highly repeated interspersed elements are generally not well-conserved. The *Alu* repeat appears to have evolved as a processed pseudogene from transcripts of the 7SL RNA gene (Ullu and Tschudi, 1984; see *Figure 9.24*) and appears to be specific to primates. It does have a counterpart in the mouse genome, the *B1* repeat, which also appears to have been generated from a 7SL RNA-like gene. However, the huge amplification in copy number apppears largely to have been generated some time ago in both species and the sequence divergence between the consensus sequences for the two types of repeat family is sufficiently high that probes can be made from them which permit distinction between the two genomes. Unlike the *Alu* repeat, the LINE-1 repeats are conserved throughout mammals. This appears to be largely due to conservative selection pressure to maintain the sequence of the large *ORF2* sequence which specifies the reverse transcriptase (see Fanning and Singer, 1987).

What makes us different from the great apes?

Humans are unusual among primates in that we show much more limited genetic variability than our close relatives, the chimpanzees, and other apes. For example,

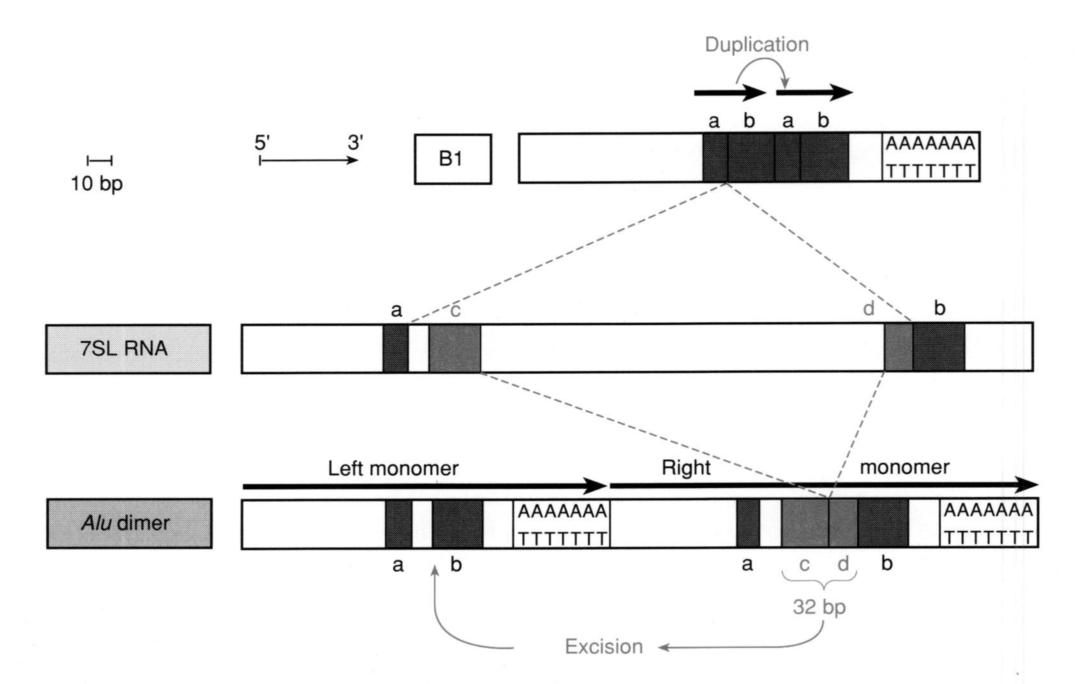

Figure 9.24: The human *Alu* repeat and the mouse *B1* repeat evolved from processed copies of the 7SL RNA gene.

Extensive homology of the *Alu* repeat sequences to the ends of the 7SL RNA sequence suggests that a polyadenylated copy of the 7SL RNA gene integrated elsewhere in the genome by a retrotransposition event (see *Figure 8.7*). In some cases, the integrated copies were able to produce RNA transcripts of their own, using the internal promoter of the 7SL RNA gene. At a very early stage, an internal segment (between c and d) was lost. Subsequently, a 32-bp central segment containing regions flanking the original deletion (c + d) was deleted to give a related repeat unit. Fusion of the two types of unit resulted in the classical *Alu* dimeric repeat, with the left (5′) monomer lacking a 32-bp sequence and the right (3′) monomer containing the 32-bp sequence. *Note* that in the human genome there also multiple copies of a *free left Alu monomer* (*FLAM*) and a *free right Alu monomer* (*FRAM*). In the mouse, a similar process of copying from the 7SL RNA gene appears to have occurred with subsequent deletion of a large internal unit (between a and b), followed by tandem duplication of flanking regions (a + b).

the sequence of an intron in the *ZFY* gene on the Y chromosome revealed no differences when the Y chromosomes of 38 different men were sampled, although there were many differences when referenced against the equivalent sequences from chimpanzees, gorillas and orangutans (Dorit *et al.*, 1995). Other studies on genes located in diverse genomic regions all confirm the low nucleotide diversity and therefore limited genetic variability of humans (see Li and Sadler, 1991). One possibility, therefore, is that at a recent time in our evolutionary past, the human population went through a 'bottleneck' (i.e. a severe reduction in effective population size), so that a large part of the previously existing genetic variability was lost.

Chimpanzees and gorillas are judged to be our closest relatives (Marks, 1992). Although still controversial, nucleotide sequence data suggest that the chimpanzee is our closest living relative (Goodman *et al.*, 1990), with recent evidence suggesting that human–chimpanzee divergence and human–gorilla divergence occurred, respectively, about 4.7 and 7 million years ago (Takahata *et al.*, 1995). The divergence into separate species may initially have been driven by small cytogenetic differences and/or mutations in key genes regulating gamete formation or regulation of early embryonic development. However, once speciation had been accomplished, the effective reproductive isolation meant that species-specific patterns of intragenomic sequence exchange (see page 219) could result in extending differences between species.

Genome organization and coding DNA

Cytogenetic comparisons of the great apes emphasize the very strong conservation of banding patterns (Yunis and Prakash, 1982). The only major structural differences are a number of pericentric and paracentric inversions, the recent fusion of two chromosomes to form human chromosome 2, and a reciprocal translocation between the gorilla chromosomes which correspond to human chromosomes 5 and 17. In addition, the extent of heterochromatinization is variable, with most gorilla chromosome arms and about half of the chimpanzee chromosome arms containing terminal heterochromatic G bands which are absent from human and orangutan chromosomes (see *Figure 9.25*). Although the present information on comparative gene mapping in primates is sketchy, the available details show evidence of strong

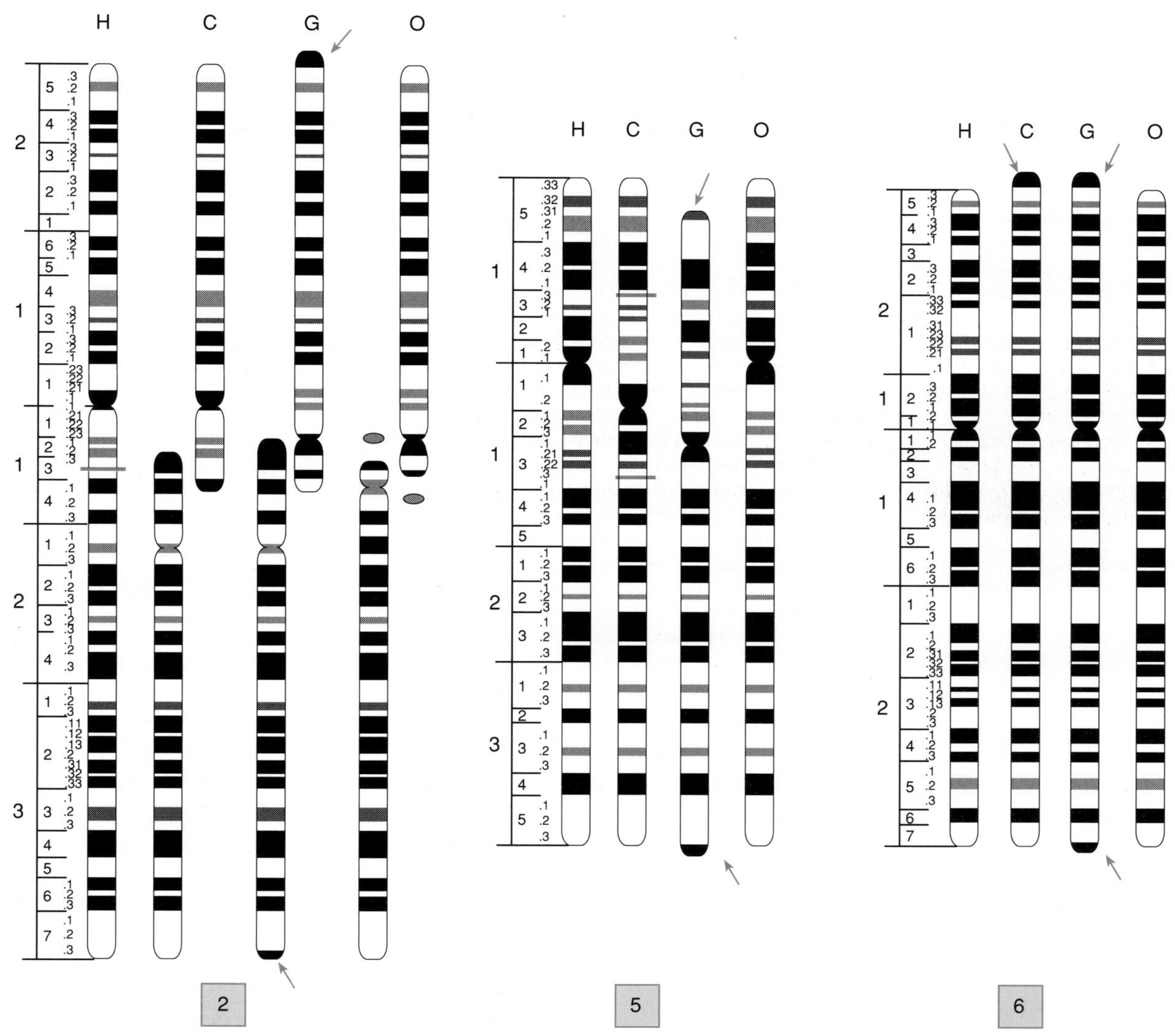

Figure 9.25: Human chromosome banding patterns are very similar to those of the great apes.

The ideograms represent selected primate chromosomes from 1000-band late prophase preparations. Human (H) chromosomes 2, 5 and 6 are illustrated together with the corresponding orthologs in chimpanzee (C), gorilla (G) and orangutan (O). Human chromosome 2 appears to have evolved by fusion of two primate chromosomes (with the point of fusion possibly at 2q13). *Note* the extremely similar structures for human chromosome 6 and its orthologs – the only readily visible differences are due to additional telomeric heterochromatin on the short arm of the chimpanzee ortholog, and on both arms of the gorilla ortholog (see arrows). In contrast, the orthologs of human chromosome 5 show appreciable differences. They include a pericentric inversion in the chimpanzee chromosome with breakpoints corresponding to human 5p13 and 5q13, and considerable differences in the gorilla ortholog (which has undergone a reciprocal translocation with a chromosome corresponding to human chromosome 17). Reprinted with permission from Yunis and Prakash (1982). © 1982, American Association for the Advancement of Science.

conservation of synteny (i.e. linked genes in humans are almost always linked in the great apes). However, large-scale organization at certain loci can differ. For example, a large part of the human Ig κ locus on 2p is duplicated (see _Figure 7.19_) but this is not the case in the corresponding chimpanzee and gorilla chromosomes (Ermert _et al._, 1995).

When orthologous human and chimpanzee sequences are compared, the coding DNA typically shows 98–100% sequence identity (see _Table 9.4_ for an example). Indeed, in some cases, specific alleles of certain human genes are more closely related to orthologs in chimpanzees than they are to other human alleles. For example, at the human HLA-DRβ locus, the alleles _HLA-DRB1*0302_ and _HLA-DRB1*0701_ are clearly closer in sequence to certain alleles of the orthologous chimpanzee (_Pan troglodytes_) gene _Patr-DRB_ than they are to each other (_Figure 9.26_). Such observations are consistent with a comparatively ancient origin for such divergent alleles, predating man–chimpanzee divergence. Although there are as yet limited data, extrapolating from known human–mouse differences suggests that there are very few human genes which lack counterparts in the chimp and gorilla genomes and

Table 9.4: Comparison of genome organization and gene expression in humans and the great apes

Parameter	Similarities/differences in humans and great apes
Genome size and gene number	About 3000 Mb in both genomes; estimate of 65 000–80 000 human genes and number in great apes expected to be extremely close to this figure
Chromosome banding patterns	Very similar. The only major differences are the fusion of two primate chromosomes to produce human chromosome 2, a reciprocal translocation in the gorilla lineage, and a few examples of paracentric and pericentric inversions (see _Figure 9.25_)
Conservation of synteny	Extensive
Gene family organization	Extremely similar for individual families. Very recent gene duplications may occasionally result in differences in gene number
Sequence homology	Coding DNA: average sequence homology about 98–100%. Noncoding DNA: usually extremely similar (~ 98%), but some sequences are restricted to humans and others which are found in some apes appear to be missing from the human genome
Gene expression	Extremely similar in general
Tandemly repeated noncoding DNA	Telomeric repeat sequences highly conserved. Other repeats not normally conserved in sequence, but similar overall patterns (location of satellite sequences at centromeres, widespread occurrence of microsatellite sequences, etc.)
Interspersed noncoding DNA	The _Alu_ repeat is found in the great apes and other primates, but human-specific subsets are known (see text)

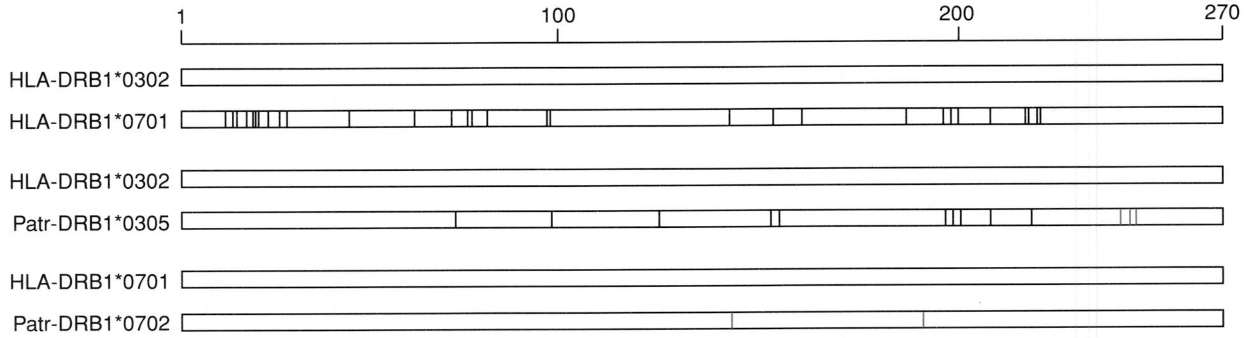

1 100 200 270

HLA-DRB1*0302
HLA-DRB1*0701

HLA-DRB1*0302
Patr-DRB1*0305

HLA-DRB1*0701
Patr-DRB1*0702

Figure 9.26: Some human alleles show greater sequence divergence than when individually compared with orthologous chimpanzee genes.

From a total of 270 amino acid positions, the _HLA-DRB1*0302_ and _HLA-DRB1*0701_ alleles show a total of 31 differences (13%). Comparison of either allele with alleles at the orthologous chimpanzee locus (_Patr-DRB1_) identifies more closely related human–chimpanzee pairs, such as _HLA-DRB1*0701_ and _Patr-DRB1*0702_ (only two amino acid differences out of 270). This suggests that some present-day _HLA_ alleles pre-date the human–chimpanzee split. See Gibbons (1995) for further details. Redrawn from Klein _et al._ (1993) with permission from Scientific American Inc.

vice versa. In just about all cases, one would expect that these human-specific genes would have arisen by very recent gene duplication events so that both gene copies may be identical, and as such be rather unlikely to provide much input into reinforcing the observed anatomical and developmental differences between humans and the great apes.

Noncoding DNA

Noncoding DNA from humans and apes can show extremely high levels of sequence homology. For example, pairwise comparisons of orthologous noncoding sequences spanning more than 12.5 kb of the β-globin gene cluster showed sequence divergence of only 1.7, 1.8 and 3.3% in the case of human–chimp, human–gorilla and human–orangutan comparisons, respectively (Goodman *et al.*, 1990). However, highly repeated DNA families appear to be undergoing a more rapid evolution. Although a common alphoid sequence is conserved in all human and great ape chromosomes (Baldini *et al.*, 1993), the vast majority of human chromosome-specific alphoid sequences do not hybridize to the centromeres of the corresponding chimpanzee and gorilla chromosomes (Archidiacono *et al.*, 1995). In addition, a subterminal satellite DNA located adjacent to the telomeres of chimpanzee and gorilla chromosomes has no counterpart in human and orangutan chromosomes (Royle *et al.*, 1994). This satellite most likely is the major component of the additional heterochromatic terminal G bands of chimpanzee and gorilla chromosomes (see *Figure 9.25*). Minisatellite and microsatellite sequences can also differ between humans and primates. Telomere sequences are conserved but hypervariable minisatellite sequences show transient evolution in the primate genomes – highly polymorphic human minisatellites often have monomorphic or minimal variability in the corresponding chromosomes of the great apes (Gray and Jeffreys, 1991). Microsatellites also show differences at orthologous positions in humans and other primates, (Rubinsztein *et al.*, 1995).

Highly repetitive interspersed DNA can also show differences. Although the *Alu* repeat family is found in other primates, several different subfamilies have been recognized (see Jurka *et al.*, 1991, for a classification) and appear to have spread at different periods of primate evolution. The average age of the oldest subfamily, the *Alu* J repeats, was estimated at about 55 million years. This family, like other old subfamilies, is characterized by considerable divergence beween the members but comparatively close resemblance of the consensus to the 7SL RNA sequence. A small number of the *Alu* sequences belong to families which are extremely recent in evolutionary origin and contain members that are actively transposing. They include the Sb1 (previously alternatively known as the PV or HS subfamily) and Sb2 families which appear, on the basis of copy number, to be very largely human-specific (see Zietkiewicz *et al.*, 1994).

Further reading

Jackson M, Strachan T, Dover GA. (1996) *Human Genome Evolution*. BIOS Scientific Publishers, Oxford.

John B, Miklos G. (1988) *The Eukaryote Genome in Development and Evolution*. Allen & Unwin, London.

Jones S, Martin R, Pilbeam D. (1992) *The Cambridge Encyclopaedia of Human Evolution.* Cambridge University Press, Cambridge.

Li W-H, Grauer D. (1991) *Fundamentals of Molecular Evolution.* Sinauer Associates, Sunderland, MA.

MacIntyre RJ. (1985) *Molecular Evolutionary Genetics.* Plenum Press, New York.

Nei M. (1987) *Molecular Evolutionary Genetics.* Columbia University Press. New York.

References

Antequera F, Bird A. (1993) *Proc. Natl Acad. Sci. USA,* **90**, 11995–11999.

Archidiacono N, Antonacci R, Marzella R, Finelli P, Lonoce A, Rocchi M. (1995) *Genomics,* **25**, 477–484.

Baldini A, Ried T, Shridhar V, Ogura K, D'Aiuto L, Rocchi M, Ward DC. (1993) *Hum. Genet.,* **90**, 577–583.

Belfort M. (1993) *Science,* **262**, 1009–1010.

Blair HJ, Reed V, Laval SH, Boyd Y. (1994) *Genomics,* **19,** 215–220.

Byrne CD, Lawn RM. (1994) *Clin. Genet.,* **46**, 34–41.

Cavalier-Smith T. (1991) *Trends Genet.,* **7**, 145–148.

Disteche C.M. (1995) *Trends Genet.,* **11**, 17–22.

Dorit RL, Schoenbach L, Gilbert W. (1990) *Science,* **250**, 1377–1382.

Duret L, Mouchiroud D, Gautier C. (1995) *J. Mol. Evol.,* **40**, 308–317.

Ermert K, Mitlohner H, Schempp W, Zachau HG. (1995) *Genomics,* **25**, 623–629.

Fanning TG, Singer MF. (1987) *Biochim. Biophys. Acta,* **910**, 203–212.

Garcia-Fernandez J, Holland PW. (1994) *Nature,* **360**, 563–566.

Gibbons A. (1995) *Science,* **267**, 35–36.

Goodman M, Tagle DA, Fitch DHA, Bailey W, Czelusniak J, Koop B, Benson P, Slightom JL. (1990) *J. Mol. Evol.,* **30**, 260–266.

Gray IC, Jeffreys AJ. (1991) *Proc. R. Soc. Lond. B.,* **243**, 241–253.

Gray MW. (1993) *Curr. Opin. Genet. Dev.,* **3**, 884–890.

Hardison R, Miller W. (1993) *Mol. Biol. Evol.,* **10**, 73–102.

Hewitt SM, Fraizer GC, Saunders GF. (1995) *J. Biol. Chem.,* **270**, 17908–17912.

Jurka J, Miloslajevic A. (1991) *J. Mol. Evol.,* **32**, 105–121.

Kalscheuser VM, Mariman EC, Schepens MT, Rehder H, Ropers H-H. (1993) *Nature Genetics,* **5**, 74–78.

Kappen C, Schughart K, Ruddle F. (1989) *Proc. Natl. Acad. Sci. USA,* **86**, 5459–5463.

Kersanach R, Brinkmann H, Liqud MF, Zhang DX, Martin W, Cerff R. (1994) *Nature,* **367**, 387–389.

Koop BF, Hood L. (1994) *Nature Genetics,* **7**, 48–53.

Li WH, Sadler LA. (1991) *Genetics,* **129**, 513–523.

Lundin LG. (1993) *Genomics,* **16**, 1–19.

Marks J. (1992) *Curr. Opin. Genet. Dev.,* **2**, 883–889.

Mattick JS. (1994) *Curr. Opin. Genet. Dev.*, **4**, 823–831.

Mukai T, Arai Y, Yatsuki H, Joh K, Hori K. (1991) *Eur. J. Biochem.*, **195**, 781–787.

Murphy PM. (1993) *Cell*, **72**, 823–826.

O'Brien SJ, Womack JE, Lyons LA, Moore KJ, Jenkins NA, Copeland NG. (1993) *Nature Genetics*, **3**, 103–112.

Ohno S. (1973) *Nature*, **244**, 259–262.

Patthy L. (1994) *Curr. Opin. Struct. Biol.*, **4**, 383–392.

Rappold GA. (1993) *Hum. Genet.*, **92**, 315–324.

Rice WR. (1994) *Science* **263**, 230–232.

Rosenberg HF, Dyer KD, Tiffany HL, Gonzalez M. (1995) *Nature Genetics*, **10**, 219–223.

Royle NJ, Baird DM, Jeffreys AJ. (1994) *Nature Genetics*, **6**, 52–56.

Rubinsztein DC, Amos W, Leggo J *et al*. (1995) *Nature Genetics*, **10**, 337–343.

Ruddle FW, Bartels JL, Bentley KL, Kappen C, Murtha MT and Pendleton JW. (1994) *Annu. Rev. Genet.*, **28**, 423–442.

Sabeur G, Macaya G, Kadi F, Bernardi G. (1993) *J. Mol. Evol.*, **37**, 93–108.

Sawyer JR, Hozier JC. (1986) *Science*, **232**, 1632–1635.

Sibley CG, Ahlquist JE. (1984) *J. Mol. Evol.*, **20**, 2–15.

Sibley CG, Ahlquist JE. (1987) *J. Mol. Evol.*, **26**, 99–121.

Stallings RL. (1995) *Genomics*, **25**, 107–113.

Stoltzfus A, Spenceer DF, Zuker M, Logsdon JM Jr, Doolittle WF. (1994) *Science*, **265**, 202–207.

Strachan T. (1992) *The Human Genome*, pp. 32–33. BIOS Scientific Publishers, Oxford.

Tagle DA, Stanhope MJ, Siemieniak DR, Benson P, Goodman M, Slightom JL. (1992) *Genomics*, **13**, 741–760.

Takahata N, Satta Y, Klein J. (1995) *Theor. Pop. Biol.*, **48**, 198–221.

Ullu E, Tschudi C. (1984) *Nature*, **312**, 171–172.

Wang W, Desai T, Ward DC, Kaufman SJ. (1995) *Genomics*, **26**, 563–570.

Watanabe M, Zinn AR, Page DC, Nishimoto T. (1993) *Nature Genetics*, **4**, 268–271.

Weller PA, Critcher R, Goodfellow PN, German J, Ellis NA. (1995) *Hum. Mol. Genet.*, **4**, 859–868.

Yunis JJ, Prakash O. (1982) *Science*, **215**, 1525–1530.

Zietkiewicz E, Richer C, Makalowski W, Jurka J, Labuda D. (1994) *Nucleic Acids Res.*, **22**, 5608–5612.

Mutation and instability of human DNA

Chapter 10

Mutation and polymorphism

As in other genomes, the DNA of the human genome is not a static entity. Instead, it is subject to a variety of different types of heritable change (**mutation**). Large-scale chromosome abnormalities involve loss or gain of chromosomes or breakage and rejoining of chromatids (see page 51 ff). Smaller scale mutations can be grouped into different mutation classes and can also be categorized on the basis of whether they involve a single DNA sequence (**simple mutations** – see page 242) or whether they involve exchanges between two allelic or nonallelic sequences (page 252). Three classes of small-scale mutation can be distinguished (see also *Table 10.1*):

- **base substitutions** – involve replacement of usually a single base; in rare cases several clustered bases may be replaced simultaneously as a result of a form of *gene conversion* (see page 256).
- **Deletions** – one or more nucleotides are eliminated from a sequence.
- **Insertions** – one or more nucleotides are inserted into a sequence. In rare cases this involves transposition from another locus. **Copy** or **duplicative transposition** involves a sequence from one locus being replicated and the copy inserted into another locus. **Noncopy transposition** involves simple transposition of a DNA sequence from one locus to another. In human and mammalian genomes, noncopy transposition is very rare: the great majority of DNA transposition occurs via an RNA intermediate so that the insertion is of a sequence copied from another locus.

New mutations arise in single individuals, in somatic cells or in the germline. If a germline mutation does not seriously impair an individual's ability to have offspring who can transmit the mutation, it can spread to other members of a (sexual) population. Allelic sequence variation is traditionally described as a **DNA polymorphism** if more than one variant (allele) at a locus occurs in a human population with a frequency greater than 0.01 (a frequency high enough such that an origin as a result of chance recurrence is highly unlikely). The **mean heterozygosity** for human genomic DNA has been calculated to be about 0.0037 (i.e. approximately 1:250 to 1:300 bases are different between allelic sequences; Cooper *et al.*, 1985). Certain genes, however, notably some HLA genes, are exceptionally polymorphic and alleles can show very substantial sequence divergence (see *Figure 9.26*). Variation in

Table 10.1: Incidence of mutation classes in the human genome

Mutation class	Type of mutation	Incidence
Base substitutions	All types	Comparatively common type of mutation in coding DNA but also common in noncoding DNA
	Transitions and transversions	Unexpectedly, transitions are commoner than transversions, especially in mitochondrial DNA
	Synonymous and nonsynonymous substitutions	Synonymous substitutions are considerably more common than nonsynonymous substitutions in coding DNA; conservative substitutions are more common than nonconservative
	Gene conversion-like events (multiple base substitution)	Rare except at certain tandemly repeated loci or clustered repeats
Insertions	Of one or a few nucleotides	Very common in noncoding DNA but rare in coding DNA where they produce frameshifts
	Triplet repeat expansions	Rare but can contribute to several disorders, especially neurological disorders (see page 266)
	Other large insertions	Rare; can occasionally get large-scale tandem duplications and also insertions of transposable elements (page 271)
Deletions	Of one or a few nucleotides	Very common in noncoding DNA but rare in coding DNA where they produce frameshifts
	Larger deletions	Rare, but often occur at regions containing tandem repeats (page 268) or between interspersed repeats (see page 254 and *Figure 10.9*)
Chromosomal abnormalities	Numerical and structural	Rare as constitutional mutations, but can often be pathogenic (see page 51ff.) Much more common as somatic mutations and often in tumor cells

allelic sequences occurs rarely as a result of new mutations: the mutation rate is comparatively low so that the vast majority of the differences between allelic sequences within an individual are inherited, rather than resulting from *de novo* mutations.

Simple mutations

Mutations due to errors in DNA replication and repair are frequent

Mutations can be induced in our DNA by exposure to a variety of mutagens occurring in our external environment or to mutagens generated in the intracellular environment (see page 30). However, by far the greatest source of mutations is from *endogenous mutation*, notably spontaneous errors in DNA replication and repair. During an average human lifetime in the order of 10^{17} cell divisions can be estimated to take place: about 2×10^{14} divisions are required to generate the approximately 10^{14} cells in the adult, and additional mitoses are required to permit cell renewal in the case of certain cell types, notably epithelial cells (see Cairns, 1975). As each cell division requires the incorporation of 6×10^9 new nucleotides, error-free DNA replication in an average lifetime would require a DNA replication-repair process whose accuracy was great enough such that the correct nucleotide was inserted on the growing DNA strands on each of about 6×10^{26} occasions.

Such a level of DNA replication fidelity is impossible to sustain; indeed, the observed fidelity of replication of DNA polymerases is very much less than this and uncorrected replication errors occur with a frequency of about 10^{-9}–10^{-11} per incorporated nucleotide (see Cooper *et al.*, 1995). As the coding DNA of an average gene is about 1.5 kb, coding DNA mutations will occur spontaneously with an average frequency of about $1.5 \times 10^{-6} - 1.5 \times 10^{-8}$ per gene per cell division. Thus, during the approximately 10^{16} mitoses undergone in an average human lifetime, each gene will be a locus for about 10^8–10^{10} mutations (but for any one gene, only a tiny minority of cells will carry a mutation). In many cases, a deleterious gene mutation in a somatic cell will be inconsequential: the mutation may cause lethality for that single cell, but will not have consequences for other cells. However, in some cases, the mutation may lead to an inappropriate continuation of cell division, causing cancer (see Chapter 17).

The frequency of individual base substitutions is nonrandom

Base substitutions are among the most common mutations and can be grouped into two classes:

(i) **transitions** are substitutions of a pyrimidine (C or T) by a pyrimidine, or of a purine (A or G) by a purine;

(ii) **transversions** are substitutions of a pyrimidine by a purine or of a purine by a pyrimidine.

When one base is substituted by another, there are always two possible choices for transversion, but only one choice for a transition. For example, the base adenine can undergo two possible transversions (to cytosine or to thymine) but only one transition (to guanine; see *Figure 10.1*). One might, therefore, expect transversions to be twice as frequent as transitions. Because the substitution of alleles in a population takes thousands or even millions of years to complete, nucleotide substitutions cannot be observed directly. Instead, they are always inferred from pairwise comparisons of DNA molecules that share a common origin, such as orthologs in different species. When this is done, the transition rate in mammalian genomes is found to be unexpectedly higher than transversion rates. For example, comparison of 337 pairs of human and rodent orthologs reveals that for substitutions which do not lead to an altered amino acid the transition rate exceeds the transversion rate by a ratio of 1.4:1; for those which result in a change of amino acid, the transition rate again exceeds the transversion rate by a ratio of more than 2:1 (Collins and Jukes, 1994).

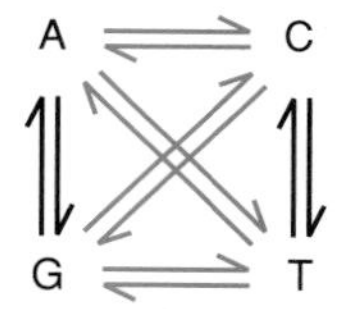

Figure 10.1: Transversions are theoretically expected to be twice as frequent as transitions. Red arrows, transversions; black arrows, transitions.

Transitions may be favored over transversions in coding DNA because they usually result in a more conserved polypeptide sequence (see below). In both coding and noncoding DNA the excess of transitions over transversions is at least partly due to the comparatively high frequency of C → T transitions, resulting from instability of cytosine residues occurring in the CpG dinucleotide. In such dinucleotides the cytosine is often methylated at the 5′ C atom and 5-methylcytosines are susceptible to spontaneous deamination to give thymine (see page 16). Presumably as a result of this, the CpG dinucleotide is a hotspot for mutation in vertebrate genomes: its mutation rate is about 8.5 times higher than that of the average dinucleotide (see Cooper *et al.*, 1995). Other factors favoring transitions over transversions are likely to include differential repair of mispaired bases by the sequence-dependent proofreading activities of the relevant DNA polymerases.

The frequency and spectrum of mutations in coding DNA differs from that in noncoding DNA

Many mutations are generated essentially randomly in the DNA of individuals. As a result, coding DNA and noncoding DNA are about equally susceptible to mutation. Clearly, however, the major consequences of mutation are largely restricted to the approximately 3% of the DNA in the human genome which is coding DNA. Mutations which occur in this component of the genome are of two types:

(i) **silent (synonymous) mutations** do not change the sequence of the gene product;

(ii) **nonsynonymous mutations** result in an altered sequence in a polypeptide or functional RNA: one or more components of the sequence are altered or eliminated, or an additional sequence is inserted into the product.

Silent mutations are thought to be effectively **neutral mutations** (conferring no advantage or disadvantage to the organism in whose genome they arise). In contrast, nonsynonymous mutations can be grouped into three classes, depending on their effect: those having a deleterious effect; those with no effect; and those with a beneficial effect (e.g. improved gene function or gene–gene interaction). Most new nonsynonymous mutations are likely to have a deleterious effect on gene expression and so can result in disease or lethality. However, the frequency of such mutation in the population is very much reduced because of **natural selection** (see *Box 10.1*). As a result, the overall mutation rate in coding DNA is much less than that in noncoding DNA. Consequently, the coding DNA component of a specific gene and the derived amino acid sequence show a relatively high degree of evolutionary conservation, as do important regulatory sequences such as the multiple elements of promoters and enhancers, and intronic sequences immediately flanking exons.

Selection pressure (the constraints imposed by natural selection) reduces both the overall frequency of surviving mutations in coding DNA and the spectrum of mutations seen. For example, deletions/insertions of one or several nucleotides are frequent in noncoding DNA but are conspicuously absent from coding DNA. This is so because often such mutations will cause a shift in the translational reading frame (**frameshift mutation**) and are potentially pathogenic. By altering the translational reading frame, a frameshift mutation results in a completely different sequence of amino acids downstream of the mutation and the polypeptide is often truncated shortly afterwards by the introduction of a premature termination codon. Even if insertions/deletions do not cause a frameshift mutation, they can often affect gene function, for example, as a result of removing a key coding sequence. Instead, coding DNA is marked by a comparatively high frequency of nonrandom base substitution occurring at locations which lead to minimal effects on gene expression (see following section).

The location of base substitutions in coding DNA is nonrandom

Nucleotide substitutions occurring in noncoding DNA usually have no net effect on gene expression, unless occurring in a promoter element or some other DNA sequence that regulates gene expression, or in important intronic sequence positions, such as at splice junctions or the splice branch site (see *Figure 1.15*). Those substitutions occurring in a segment of coding DNA which is used to specify polypeptides show a very nonrandom pattern of substitutions because of the need to conserve polypeptide sequence and biological function. In principle, base substitutions can be grouped into three classes, depending on their effect on coding potential (see *Box 10.2*).

The different classes of base substitution listed in the box show differential tendencies to be located at the first, second or third base positions of codons. Because of the design of the genetic code, different degrees of *degeneracy* characterize different sites. Base positions in codons can be grouped into three classes:

(i) **nondegenerate sites** are base positions where all three possible substitutions are nonsynonymous. They include the first base position of all but eight codons, the second base position of all codons and the third base position of

Box 10.1: Mechanisms which affect the population frequency of alleles

Individuals within a population differ from each other. Much of the basis of such differences is due to inherited genetic variation. The frequency of any mutant allele in a population is dependent on a number of factors, including natural selection, random genetic drift and sequence exchanges between nonallelic sequences.

Natural selection

Natural selection is the process whereby some of the inherited genetic variation will result in differences between individuals regarding their ability to survive and reproduce successfully. The differential reproduction is due to differences between individuals in their capacity to engage in reproduction (affected by parameters such as mortality, health and mating success) and to produce healthy offspring (differences in fertility, fecundity and viability of the offspring). The **fitness** of an organism is therefore a measure of the individual's ability to survive and to reproduce successfully. In the simplest models, the fitness of an individual is considered to be determined solely by its genetic make-up, and all loci are imagined to contribute independently to the fitness of an individual, so that each locus can be treated separately. As a result, the term *fitness* can also be applied to a genotype.

The great majority of new nonsynonymous mutations in coding DNA reduce the fitness of their carriers. They are therefore selected against and removed from the population (**negative** or **purifying selection**). Occasionally, a new mutation may be as fit as the best allele in the population; such a mutation is selectively **neutral**. Very rarely, a new mutation confers a selective advantage and increases the fitness of its carrier. Such a mutation will be subjected to **positive** or **advantageous selection**, which would be expected to foster its spread through a population. If we consider a locus with two alleles that have different fitnesses, the heterozygote may have a fitness intermediate between the two types of homozygote. The mode of selection in this case is **codominant** and the selection will be directional, resulting in an increase of the advantageous allele. In some cases, however, a new mutation may not be advantageous in homozygotes, but only in heterozygotes (**heterozygote advantage**). This situation, in which the heterozygote has a higher fitness than both the mutant homozygote *and the normal homozygote*, is a form of balancing selection known as **overdominant selection** (see page 62).

Random genetic drift

Changes in allele frequency can also occur by chance (**random genetic drift**). Even if all the individuals in a population had exactly the same fitness so that natural selection could not operate, allele frequencies would change because of *random sampling of gametes*. Only a tiny fraction of the available gametes in any generation is ever passed on to the next generation. If the number of gametes contributing to the next population is not large, certain alleles may not be transmitted to the next generation at the expected frequency, simply because of sampling variation. Because of the randomness in sampling, allele frequencies will fluctuate between generations. In the absence of new mutation and other factors affecting allele frequency, such as selection, alleles subject to random genetic drift will eventually reach **fixation** (the point at which the allele frequency is 0 or 100%). Genetic drift causes rapid changes in small populations but has little effect in large ones.

Interlocus sequence exchange

Individual genes in some gene families encode essentially the same product, but there may be sequence exchanges occurring between the different gene copies. For example, human 5.8S rRNA, 18S rRNA and 28S rRNA are encoded by numerous genes which are organized in tandemly repeated arrays (see *Figure 8.3*) and are particularly prone to sequence exchanges between the different repeats. Simply as a result of the sequence exchanges between different repeats, one type of repeat can increase in population frequency (see *Figure 10.8* for an illustration of the general principle). In such cases, where multiple loci produce essentially identical products, all the genes can effectively be considered as the equivalent of alleles, although not in the mendelian sense (which normally allows a maximum of two alleles in a diploid cell). The frequency of a specific repeat ('allele') can therefore be determined in part by the frequency with which it engages in sequence exchanges.

Box 10.2: Classes of single base substitution in polypeptide-encoding DNA

- **Synonymous ('silent') substitution** results in no change in an amino acid. This class of mutations are the most frequently observed in coding DNA, because they are neutral mutations and not subject to selection pressure. Such substitutions often occur at the third base position of a codon: *third base wobble* means that the altered codon often specifies the same amino acid (see pages 23–24) However, base substitution at the first base position can occasionally give rise to a synonymous substitution, e.g. leucine codons **C**UA ↔ **U**UA and **C**UG ↔ **U**UG, and arginine codons **A**GA ↔ **C**GA and **A**GG ↔ **C**GG. *Note* that some synonymous substitutions can be pathogenic by activating a cryptic splice site (page 261).
- A **nonsense mutation** is a nonsynonymous substitution resulting in replacement of a codon specifying an amino acid by a termination codon. Because such mutations are almost always associated with a dramatic reduction in gene function, selection pressure ensures that they are normally rare. The average human polypeptide is specified by about 500 codons, a size which would be expected to harbor about 25 termination codons, if no such functional constraints applied.
- A **missense mutation** is a nonsynonymous substitution resulting in an altered codon specifying a different amino acid. Missense mutations can be classified into two subgroups:
  (i) a *conservative substitution* results in replacement of an amino acid by another that is chemically similar to it. Often, the effect of such substitutions on protein function is minimal because the side chain of the new amino acid may be functionally similar to that of the amino acid it replaces. To minimize the effect of nucleotide substitution, the genetic code appears to have evolved so that codons specifying related amino acids are themselves related. For example, the Asp (GAC, GAT) and Glu (GAA, GAG) codon pairs ensure that third base wobble in a GAX codon (where X is any nucleotide) has a minimal effect. However, some first codon position changes can also be conservative, e.g. **C**UX (Leu) ↔ **G**UX (Val).
  (ii) A *nonconservative substitution* is a mutation that results in replacement of one amino acid by another which has a dissimilar side chain. Sometimes a charge difference is introduced; other changes may involve replacement of polar side chains by nonpolar ones and vice versa. Base substitutions at the first and second codon positions can often result in nonconservative substitutions, e.g. CGX (Arg) → **G**GX (Gly), C**C**X (Pro), C**U**X (Leu) or C**A**X (Gln/His), etc.

two codons, AUG and UGG (see *Figure 10.2*). Taking into account the observed codon frequencies in human genes, they comprise about 65% of the base positions in human codons. The base substitution rate at nondegenerate sites is very low, consistent with a strong conservative selection pressure to avoid amino acid changes (see *Figure 10.3*).

(ii) **Fourfold degenerate sites** are base positions in which all three possible substitutions are synonymous and are found at the third base position of several codons (see *Figure 10.2*). They comprise about 16% of the base positions in human codons. The substitution rate at fourfold sites is very similar to that within introns and pseudogenes, consistent with the assumption that synonymous substitutions are selectively neutral (*Figure 10.3*).

(iii) **Twofold degenerate sites** are base positions in which one of the three possible substitutions is synonymous. They are often found at the third base positions of codons, but also at the first base position in eight codons (see *Figure 10.2*). They comprise about 19% of the base positions in human codons. As expected, the substitution rate for twofold degenerate sites is intermediate (see *Figure 10.3*): only one out of the three possible substitutions, a transition, maintains the same amino acid. The other two possible substitutions are transversions which, because of the way in which the genetic code has evolved, are often conservative substitutions. For example, at the third base position of the

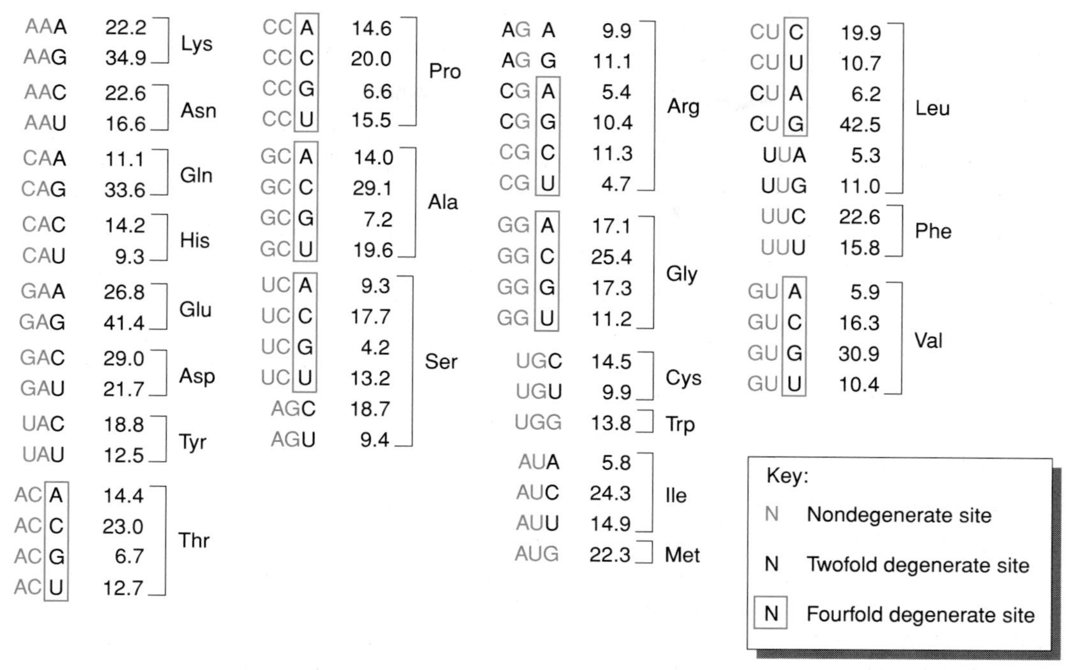

Figure 10.2: Codon frequencies in human genes and locations of nondegenerate, two- and fourfold degenerate sites.

Observed codon frequencies were derived from an analysis of 1490 human genes by Wada *et al.* (1990). *Note* that although eight of the 61 first base positions are twofold degenerate, about 96% of all possible substitutions at the first base position are nonsynonymous. Of base substitutions at the second base position, 100% are nonsynonymous and at the third base position, about 33%.

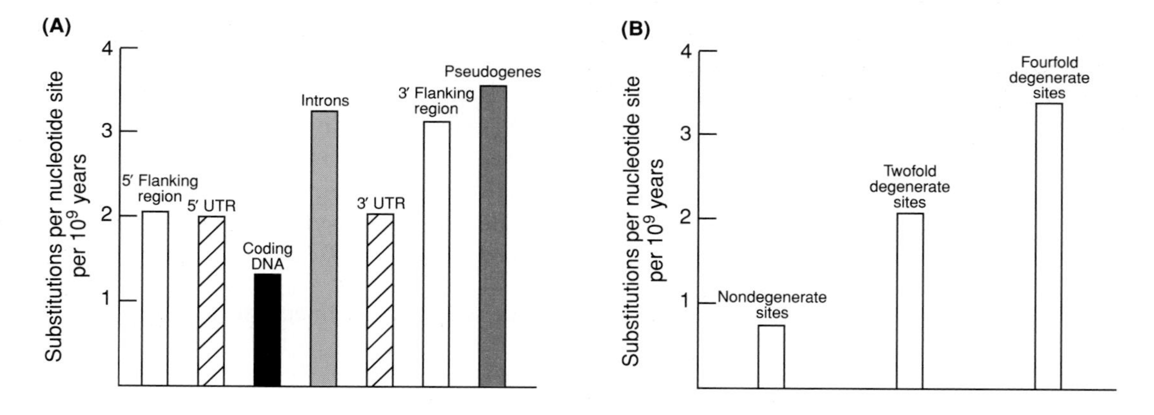

Figure 10.3: The rate of nucleotide substitution varies in different gene components and gene-associated sequences.

On the basis of the above substitution rates and the observation that an average mammalian coding DNA sequence comprises 400 codons, the coding DNA of an average human gene would be expected to undergo about one or two substitutions every million years. UTR, untranslated region. Redrawn from Figure 2 of Chapter 4 in Li and Grauer (1991) with permission from Sinauer Associates.

glutamate codon GAA, a transition A → G is silent, while the two transversions (A → C; A → T) result in replacement by a closely similar amino acid, aspartate.

The design of the genetic code and the degree to which one amino acid is functionally similar to another affect the relative mutabilities of individual amino acids. Certain amino acids may play key roles which cannot be substituted easily by others. For example, cysteine is often involved in disulfide bonding which can play a crucially important role in establishing the conformation of a polypeptide (see *Figure 1.25*). As no other amino acid has a side chain with a sulfhydryl group, there is strong selection pressure to conserve cysteine residues at many locations, and cysteine is among the least mutable of the amino acids (see Collins and Jukes, 1994). In contrast, certain other amino acids such as serine and threonine have very similar side chains, and substitutions at both the first base position of codons (**A**CX ↔**U**CX) and second base positions (A**C**Py ↔ A**G**Py) can result in serine ↔ threonine substitutions. Presumably as a result, serine and threonine are among the most mutable of the amino acids (see Collins and Jukes, 1994).

Protein-coding genes show enormous variation in the rate of nonsynonymous substitutions

The rate and type of substitution varies between different genes. At one extreme are proteins whose sequences are extremely highly conserved, such as ubiquitin, histones H3 and H4, calmodulin, ribosomal proteins, etc. For example, the ubiquitin proteins of humans, mouse and *Drosophila* show 100% sequence identity, and comparison with the yeast ubiquitin reveals 96.1% sequence identity. These genes are not especially protected from mutation, because the rate of synonymous codon substitution is typical of that for many protein-encoding genes. Instead, what distinguishes them is the extremely low rate of nonsynonymous codon substitution compared with other genes (see *Table 10.2* for some examples). Presumably, ubiquitin and the other highly conserved proteins play such crucial roles that they are under huge selection pressure to conserve the sequence. At the other extreme, the fibrinopeptides are proteins which are evolving extremely rapidly and do not appear to be subject to any selective constraint. These proteins (only 20 amino acids long) are thought to be functionless – they are fragments which are generated as part of the protein *fibrinogen* and discarded when the protein is activated to form *fibrin* during blood clotting. Another extremely rapidly evolving sequence is the major sex-determining locus, *SRY*. This gene encodes a protein which contains a central 'high mobility group' domain (*HMG box*) of about 78 amino acids. The HMG box is central to *SRY* function and is well conserved, but the flanking N- and C-terminal segments are evolving extremely rapidly, which may indicate that the majority of the *SRY* coding sequence is not functionally significant (Whitfield *et al.*, 1993). In between the two extremes in the rate of nonsynonymous substitution are the vast majority of polypeptide-encoding genes (see *Table 10.2*).

Table 10.2: Rates of synonymous and nonsynonymous substitutions in mammalian protein-coding genes

Gene	No. of codons compared	Nonsynonymous rate ($\times 10^9$)	Synonymous rate ($\times 10^9$)
Histone H3	135	0.00	6.38
Histone H4	101	0.00	6.12
Actin α	376	0.01	3.68
Aldolase A	363	0.07	3.59
HPRT	217	0.13	2.13
Insulin	51	0.13	4.02
α-Globin	141	0.55	5.14
β-Globin	144	0.80	3.05
Albumin	590	0.91	6.63
Ig V_H	100	1.07	5.66
Growth hormone	189	1.23	4.95
Ig κ	106	1.87	5.66
Interferon-β1	159	2.21	5.88
Interferon-γ	136	2.79	8.59

Data from human–rodent comparisons abstracted from Table 1 in Chapter 4 of Li and Grauer (1991).

The molecular clock can vary from gene to gene, and is different in different lineages

Synonymous substitutions have been considered to be effectively neutral from the point of view of selective constraints. As a result, the concept of a constant **molecular clock** (the idea of a constant rate of molecular evolution for a given gene or gene product) developed. However, substitution rates are now known to vary widely between different genes. For example, the genes listed in *Table 10.2* show considerable differences not only in their rates of nonsynonymous codon substitutions, but also in the rate of synonymous codon substitutions. Such differences may be governed by a number of factors:

- **timing of DNA replication**. The DNA of different genomic components is replicated at different times. Actively transcribing genes are replicated early; transcriptionally inactive DNA such as the inactivated X chromosome is replicated late. Early replicating and late replicating DNA may be subject to different intracellular concentrations of free nucleotides and of the various enzymes involved in replication and DNA repair, causing possible differences in mutation rates.
- **Differences in GC content.** This parameter is not independent of the previous one because the early replicating DNA is relatively GC-rich. The present evidence suggests that any relationship between the GC content of a mammalian gene and its mutation rate is not a simple one (see Sharp and Matassi, 1994).
- **Different genomes**. The mitochondrial DNA in mammals and many other animals is thought to be evolving at a much higher rate than nuclear DNA (see *Box 10.3*).

For a given gene, the molecular clock appears to vary depending on the species lineage. In order to estimate the relative rates of nucleotide substitutions in two lineages leading to present-day species A and B, a **relative rate test** is used, involving the use of a third reference species C which is known to have branched off earlier in evolution, before the A–B split. Pairwise comparisons of orthologs in A and C, and in B and C are then used to calculate the K value, the number of synonymous substitutions per 100 sites. The K_{AC} and K_{BC} values then provide a measure of the relative rates of mutation in the lineages leading to species A and to species B. For example, when a variety of orthologs in mouse (species A) and rat (species B) are referenced against orthologs in humans (species C), the overall K_{AC} and K_{BC} values are nearly identical (see Li and Grauer, 1991, page 82). This suggests that the base substitution rates in the lineages leading to present-day mouse and rat have been nearly equal. However, similar analyses suggest that the substitution rate appears to be lower in lineages leading to the primates and lower still in the lineage leading to modern day humans (see *Table 10.3*).

The data in *Table 10.3* may suggest that molecular evolution has effectively slowed down for organisms which have long generation times. With hindsight, perhaps this is not so surprising – most mutations arise when DNA is being replicated in gametogenesis (especially in males; see below). Rodents and monkeys have comparatively shorter generation times than humans, and so will go through more generations per unit time. In additon, it has been suggested that longer-lived animals have a greater ability to repair their DNA than do short-lived species, thereby resulting in lower mutation rates (Britten, 1986).

Box 10.3: Unresolved questions concerning mutation in mammalian mitochondrial DNA

With regard to mutation in the mitochondrial DNA in humans (and other mammalian cells), two features in particular have been difficult to resolve.

How are new mitochondrial mutations *fixed* (i.e. achieve a frequency of 100% in a population)?
There are thousands of copies of the mitochondrial DNA (mtDNA) molecule in each human somatic cell but the number can vary considerably according to cell type (some cells, such as brain and muscle cells, have particularly high oxidative phosphorylation requirements and so more mitochondria). The mtDNA is inherited from the maternal oocyte, which is an exceptional cell with many more mtDNA molecules than somatic cells (most likely ~100 000 per mammalian oocyte; see Hauswirth and Laipis, 1985). In normal individuals, ~99.9% of the mtDNA molecules are identical (**homoplasmy**). However, if a new mutation arises and spreads in the mtDNA population, there will be two significantly frequent mtDNA genotypes (**heteroplasmy**). A new mutation will arise on a single molecule and so, to be fixed in a population, the mutant molecule has to proliferate so that it replaces virtually all other mtDNA molecules in a whole population of individuals, thereby restoring the homoplasmy state. The fixation process is not facilitated by an apparent lack of recombination between mtDNA molecules in mammalian and most animal cells (it should be noted, however, that although there is no evidence for recombinational processes, possibly some kind of recombination may yet occur).

In contrast to the anticipated difficulty in fixing mtDNA mutations, fixation of alleles in the nuclear genome would be expected to be comparatively easy to achieve: for each type of nuclear DNA molecule, there are at most only two copies in a diploid somatic cell and gametogenesis generates a single type of molecule by a process (meiosis) which is specifically designed to ensure shuffling of alleles between molecules (each chromosome in a sperm or egg cell is a recombinant of maternal and paternal homologs; see *Figure 2.10*). Intuitively, therefore, one might expect that the time taken for fixation of a human mitochondrial mutation would be very much longer than that for fixation of a mutation in the nuclear genome. Paradoxically, the fixation rates for many mtDNA mutations in mammalian cells (including polymorphisms expected to be selectively neutral) is about 10 times that seen for mutations in the nuclear DNA (see below).

One possible explanation is that there is a *bottleneck* during oogenesis: the *effective number* of mtDNA molecules could be drastically reduced during oogenesis and, beyond this developmental bottleneck, over-replication of mtDNA restores the normal proportion of mtDNA molecules in somatic cells. For neutral polymorphisms, homoplasmy would be rapidly restored by random segregation following the bottleneck, but the maintenance of heteroplasmy for disease-associated mutations (see page 411) may be positively selected. As the actual number of mtDNA molecules in the oocyte is exceptionally high (Cheng *et al.*, 1995), the bottleneck could take the form of a *selective amplification* mechanism during oogenesis – a very few mtDNA molecules could be selected to serve as templates for multiple rounds of DNA replication (Hauswirth and Laipis, 1982). In addition, while there is an approximately 100-fold amplification in mtDNA molecules in oocytes, a parallel 1000-fold amplification has been claimed for the number of mitochondria (see Hauswirth and Laipis, 1985); however, this assumes that individual mitochondria exist as such, which is still uncertain (see below). This could mean that cytoplasmic partitioning in early embryogenesis and subsequent embryonic partitioning (e.g. into the three primary embryonic tissues) could provide possibilities for segregation of mitochondrial genotypes.

The bottleneck idea was initially developed from observations of polymorphisms in the mtDNA of Holstein cows. Some point mutation differences between mothers and their offspring showed that complete switching of the mtDNA type can occur in a very few generations, even a single generation, suggesting the possibility that the bottleneck could be even a single mtDNA molecule (Koehler *et al.*, 1991). Other mutations do not show such rapid switching, and observations of the segregation of heteroplasmic human polymorphisms have suggested that the bottleneck is larger than suspected (see Howell *et al.*, 1992). Rapidly and slowly segregating mtDNA genotypes could be explained if the bottleneck consisted of a single mitochondrion which was, respectively, homoplasmic or heteroplasmic. This assumes that individual mitochondria actually exist in human cells. Recent evidence suggests mitochondria are in fact part of a single dynamic cellular unit (Hayashi *et al.*, 1994). The evidence, however, was

(*continued*)

Box 10.3: (*continued*) Unresolved questions concerning mutation in mammalian mitochondrial DNA

obtained by microscopy of individual cells following *cybrid fusion*, an artificial situation whereby enucleated cells containing one kind of mitochondria are fused to cells harboring another. Finally, some polymorphisms may be able to elude the bottleneck (see Poulton, 1995). This field of research is a dynamic one and interested readers are advised to consult recent reviews.

Why is the mutation rate in animal mitochondrial DNA so high?
Plant mtDNA molecules are comparatively large (150 kb–2.5 Mb), have introns, engage in recombination and are evolving comparatively slowly. In contrast, the small mtDNA molecules of mammalian and many animal cells do not appear to recombine and appear to be evolving remarkably rapidly: mutations have been reported to be fixed at a rate which is about 10 times greater than that occurring in equivalent sequences in nuclear genomes (Brown *et al.*, 1979). This results in approximately 2–4% sequence divergence per million years. The high rate of fixation of mtDNA mutations is presumed to reflect a very high mutation rate. Presently, there remains some uncertainty as to why the mtDNA mutation rate should be so high. Various possibilities have been suggested:

- the high flux of oxygen radicals in the respiratory chain could cause substantial oxidative damage to mtDNA which, unlike nuclear DNA, is not protected by histones.
- The mtDNA has to undergo many more rounds of replication than chromosomal DNA (as a result of the additional rounds of DNA replication following the bottleneck in oogenesis – see above). This could be significant because mutations mostly arise during DNA replication (see below).
- Unlike nuclear DNA, the replication of mtDNA occurs throughout the cell cycle, is comparatively slow and is highly asymmetric. The origins of replication for the heavy and light strands are located in different regions (see *Figure 7.2*). During mtDNA replication, the parental H strand is displaced by the daughter H strand and remains in a single-stranded state for a considerable period of time, until the daughter L strand is synthesized. Single-stranded DNA is particularly prone to spontaneous mutations, e.g. the rate of spontaneous deamination of C to U is about 200 times that for double-stranded DNA. This could explain the pronounced strand asymmetry seen in human mtDNA mutation, e.g. G $\rightarrow$ A transitions on the L strand are about nine times more frequent than on the H strand, presumably because of the high frequency of spontaneous deamination of C to U on the H strand (Tanaka and Ozawa, 1994).
- Although animal mitochondria possess a highly accurate DNA polymerase γ (see page 9) with extensive $3' \rightarrow 5'$ exonucleolytic proofreading activity, they do lack the nucleotide excision repair mechanism operating in the nucleus (see *Figure 1.26*).

Table 10.3: Rates of synonymous substitution per site per year in primates and rodents

Species pair	Number of sites	Percentage divergence	Substitution rate ($\times 10^9$)[a]
Human/chimpanzee	921	1.9	1.3 (0.9–1.9)
Human/Old World monkeys	998	11.0	2.2 (1.8–2.8)
Mouse/rat	3886	23.7	7.9 (3.9–11.8)

[a]Values represent the likely mean, and in parentheses the lower and upper ranges, according to the likely mean and lower and upper ranges for the estimated times of species divergence. The latter are as follows: human/chimpanzee, 7 (5–10) million years; human/Old World monkey, 25 (20–30) million years; mouse/rat, 15 (10–30) million years. Reproduced from Li and Grauer (1991) with permission from Sinauer Press.

Higher mutation rates in males are likely to be related to the greater number of germ cell divisions

The number of human germ cell divisions is quite different in females and males. In females the number of successive cell divisions estimated from zygote to mature egg is 24. However, in males about 30 cell divisions are required from zygote to stem spermatogonia at puberty, five cell divisions are required for spermatogenesis and thereafter the spermatogenesis cycle occurs approximately every 16 days or 23 cycles per year. If an average age of 13 is taken for onset of puberty and an average of 20 for reproductive age, the total number of cell divisions is about 30 + 5 + [20–13] × 23, or about 195 divisions (*Figure 10.4*). Given that errors in DNA replication/repair provide the great majority of mutations, one might then expect that the male mutation rate would be substantially greater than that of the female. If the ratio of male to female mutation rate, α_m, were very large, then the rate of synonymous substitution in X-linked genes would be expected to be only about two-thirds of that in autosomal genes. X-linked genes often do show lower synonymous substitution rates than autosomal genes (see the example of the *HPRT* gene in *Table 10.2*). However, there appears to be considerable gene to gene variation in mutation rates and the generally low rates for X-linked genes could alternatively reflect the requirement for one X chromosome to undergo X inactivation – the inactivated X is replicated late in the cell cycle, a factor which may influence mutation rates (see above). Recent experimental verification of the higher mutation rate in males has been provided by a comparison of intronic sequences of the homologous *ZFX* and *ZFY* loci, suggesting that α_m is approximately 6 (Shimmin *et al.*, 1993).

Genetic mechanisms which result in sequence exchanges between DNA repeats

In addition to very frequent simple mutations, there are several mutation classes which involve sequence exchange between allelic or nonallelic sequences, often involving repeated sequences. For example, tandemly repetitive DNA is prone to deletion/insertion polymorphism whereby different alleles vary in the number of integral copies of the tandem repeat. Such **variable number tandem repeat (VNTR) polymorphisms** can occur in the case of repeated units that are very short (microsatellites; see page 197), intermediate (minisatellites; see page 196) or large. Note, however, that in genetic linkage analyses, the term *VNTR marker* or *VNTR polymorphism* is normally used in the very restricted sense of VNTR polymorphism involving hypervariable *minisatellites* only. Different genetic mechanisms can account for VNTR polymorphism depending on the size of the repeating unit (see the following two sections). In addition, interspersed repeats can also predispose to deletions/duplications by a variety of different genetic mechanisms.

Slipped strand mispairing can cause VNTR polymorphism at short tandem repeats (microsatellites)

Germline mutation rates at a variety of microsatellite loci exhibit considerable variation, ranging from an undetectable level up to about 8×10^{-3} (Mahtani and Willard, 1993; Weber and Wong, 1993). Novel length alleles at (CA)/(TG) microsatellites and at tetranucleotide marker loci are known to be formed without exchange of flanking markers. This means that they are not generated by unequal crossover (see below). Instead, as new mutant alleles have been observed to differ by a single repeat unit from the originating parental allele (Mahtani and Willard, 1993), the most likely mechanism to explain length variation is a form of exchange of

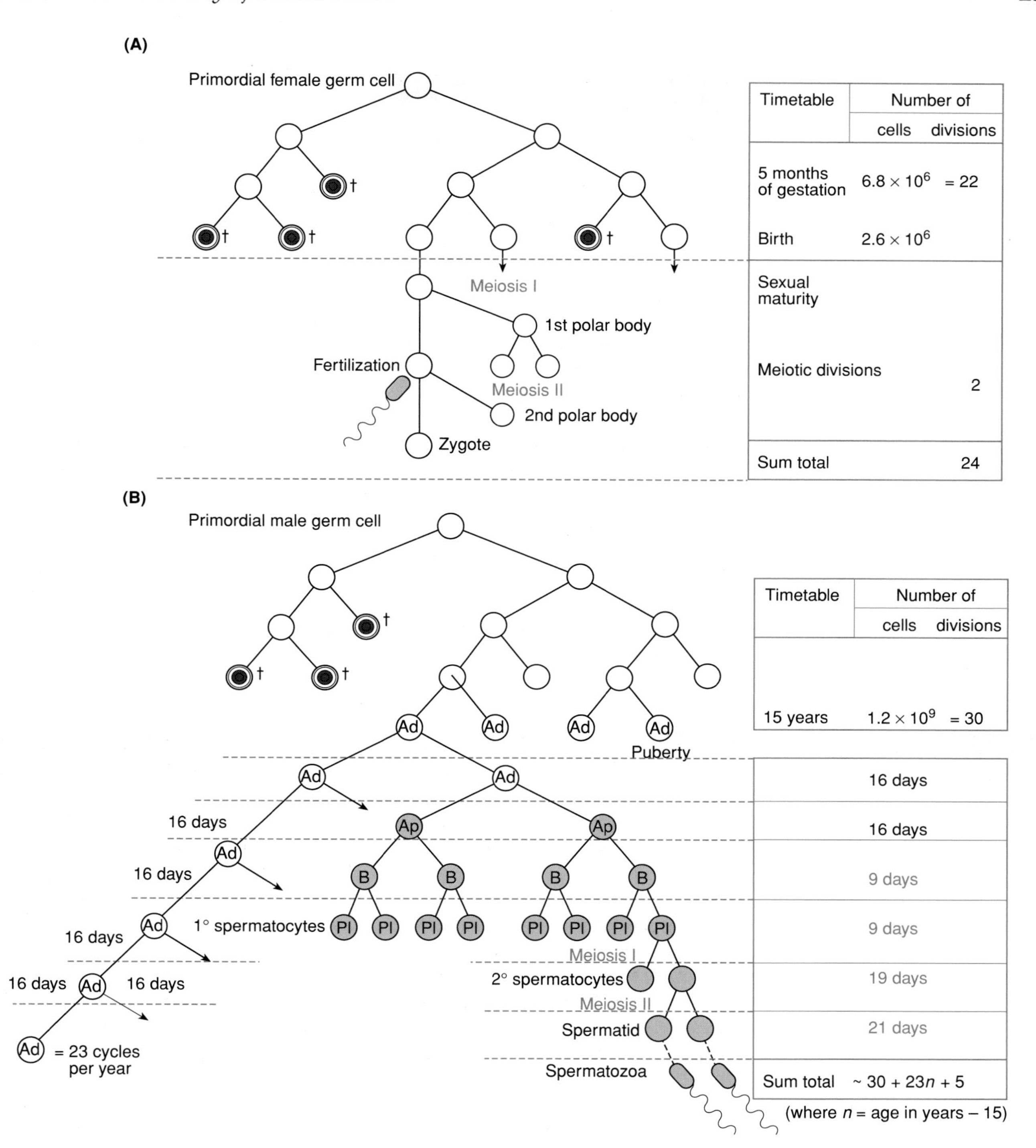

Figure 10.4: The number of cell divisions that are required to produce a human sperm cell is much greater than the number required to produce an egg cell.

(**A**) Human oogenesis occurs only during fetal life and ceases by the time of birth. The total population of germ cells in the human embryo rises to an estimated maximum of 6.8×10^6 during the fifth month. The expected number of binary divisions required to generate this number would be about 22 ($2^{22} = \sim 4 \times 10^6$). At sexual maturity, two subsequent meiotic divisions are required to produce an egg cell. †Cell atrophy.
(**B**) Human spermatogenesis continues through adult life. From the early embryonic stage up to the age of puberty, the seminiferous tubules continue to become populated by so-called *Ad spermatogonia* (**d**ark-staining A-type spermatogonia) and spermatogenesis is fully established by puberty. The number of Ad spermatogonia is estimated to be about 6×10^8 per testis, i.e. a total of about 1.2×10^9, a value which can be reached by about 30 successive cell divisions. The Ad spermatogonia then undergo a series of cell divisions, lasting 16 days. Of the two products of cell division, one prepares for the next division into two Ad cells. Because each Ad division cycle is 16 days long, following puberty a total of 23 cycles occur each year (365/16) – see bottom left. The other product of an Ad cell division divides to give two Ap (pale-staining A type spermatogonia) cells which are precursors of sperm cells (red cells on bottom right). The Ap cells give rise to B spermatogonia and then spermatocytes, which finally undergo two meiotic divisions to generate sperm cells. Modified from Vogel and Motulsky (1986) with permission from Springer Verlag.

sequence information which commences by **slipped strand mispairing.** This occurs when the normal pairing between the two complementary strands of a double helix is altered by staggering of the repeats on the two strands, leading to incorrect pairing of repeats. Although slipped strand mispairing can be envisaged to occur in nonreplicating DNA, replicating DNA may offer more opportunity for slippage and hence the mechanism is often also called **replication slippage** or **polymerase slippage** (see *Figure 10.5*). In addition to mispairing between tandem repeats, slippage replication has been envisaged to generate large deletions and duplications by mispairing between *noncontiguous repeats* and has been suggested to be a major

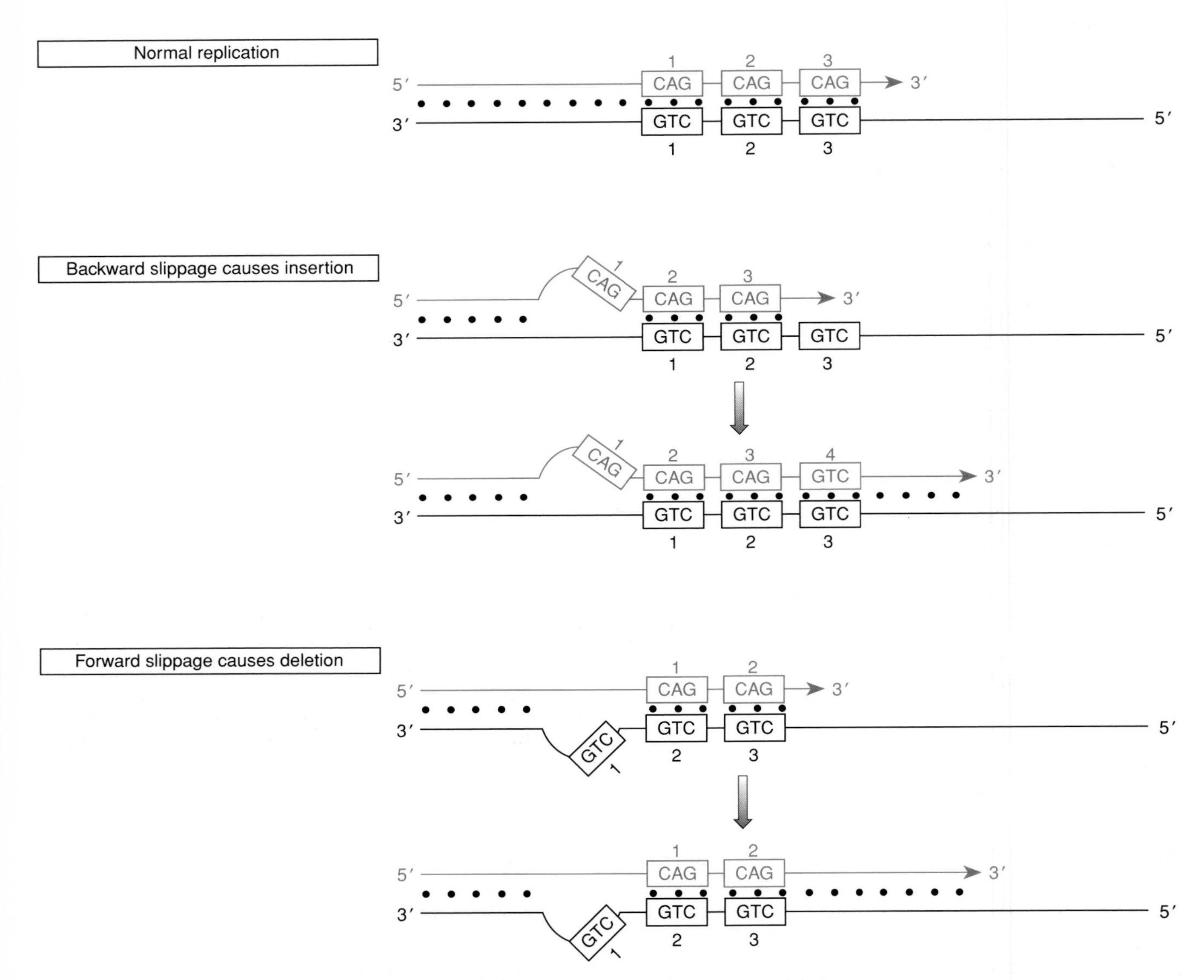

Figure 10.5: Slipped strand mispairing during DNA replication can cause insertions or deletions.

Short tandem repeats are thought to be particularly prone to slipped strand mispairing, i.e. mispairing of the complementary DNA strands of a single DNA double helix. The examples show how slipped strand mispairing can occur during replication, with the lower strand representing a parental DNA strand and the upper strand representing the newly synthesized complementary strand. In such cases, slippage involves a region of nonpairing (shown as a bubble) containing one or more repeats of the newly synthesized strand (backward slippage) or of the parental strand (forward slippage), causing, respectively, an insertion or a deletion on the newly synthesized strand. *Note* that it is conceivable that slipped strand mispairing can also cause insertions/deletions in nonreplicating DNA. In such cases, two regions of nonpairing are required, one containing repeats from one DNA strand and the other containing repeats from the complementary strand (see Levinson and Gutman, 1987).

mechanism for DNA sequence and genome evolution (Levinson and Gutman, 1987; see also Dover, 1995). The pathogenic potential of short tandem repeats is considerable (see pages 264–267).

Large units of tandemly repeated DNA are prone to insertion/deletion as a result of unequal crossover or unequal sister chromatid exchanges

Homologous recombination describes recombination (*crossover*) occurring at meiosis or, rarely, mitosis between identical or very similar DNA sequences, and usually involves breakage of nonsister chromatids of a pair of homologs and rejoining of the fragments to generate new recombinant strands. **Sister chromatid exchange** is an analogous type of sequence exchange involving breakage of individual sister chromatids and rejoining fragments that initially were on different chromatids of the same chromosome. Both homologous recombination and sister chromatid exchange normally involve *equal* exchanges – cleavage and rejoining of the chromatids occurs at the same position on each chromatid. As a result, the exchanges occur *between allelic sequences* and at corresponding positions within alleles. In the case of intragenic equal crossover between two alleles, a new allele can result which is a **fusion gene** (or **hybrid gene**), comprising a terminal fragment from one allele and the remaining sequence of the second allele (*Figure 10.6*). However, equal sister chromatid exchanges cannot normally produce genetic variation because sister chromatids have identical DNA sequences.

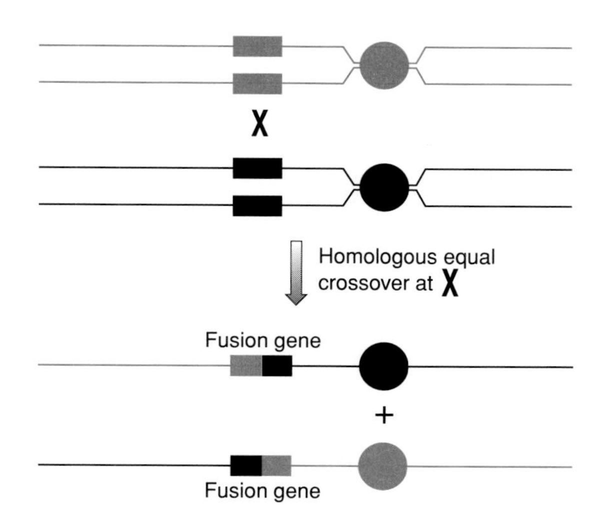

Figure 10.6: Homologous equal crossover can result in fusion genes.

The example shows how intragenic equal crossover occurring between alleles on nonsister chromatids can generate novel fusion genes composed of adjacent segments from the two alleles. *Note* that similar exchanges between genes on sister chromatids do not result in genetic novelty because the gene sequences on the interacting sister chromatids would be expected to be identical.

Unequal crossover is a form of recombination in which the crossover takes place *between nonallelic sequences* on nonsister chromatids of a pair of homologs (*Figure 10.7*). Often the sequences at which crossover takes place show very considerable sequence homology which presumably stabilizes mispairing of the chromosomes. Because crossover occurs between mispaired nonsister chromatids, the exchange results in a deletion on one of the participating chromatids and an insertion on the other. The analogous exchange between sister chromatids is called **unequal sister chromatid exchange** (see *Figure 10.7*). Both mechanisms occur predominantly at locations where the tandemly repeated units are moderate to large in size. In such cases, the very high degree of sequence homology between the different repeats can facilitate pairing of nonallelic repeats on nonsister chromatids or sister chromatids. If chromosome breakage and rejoining occurs while the chromatids are mispaired in this way, an insertion or deletion of an integral number of repeat units will result.

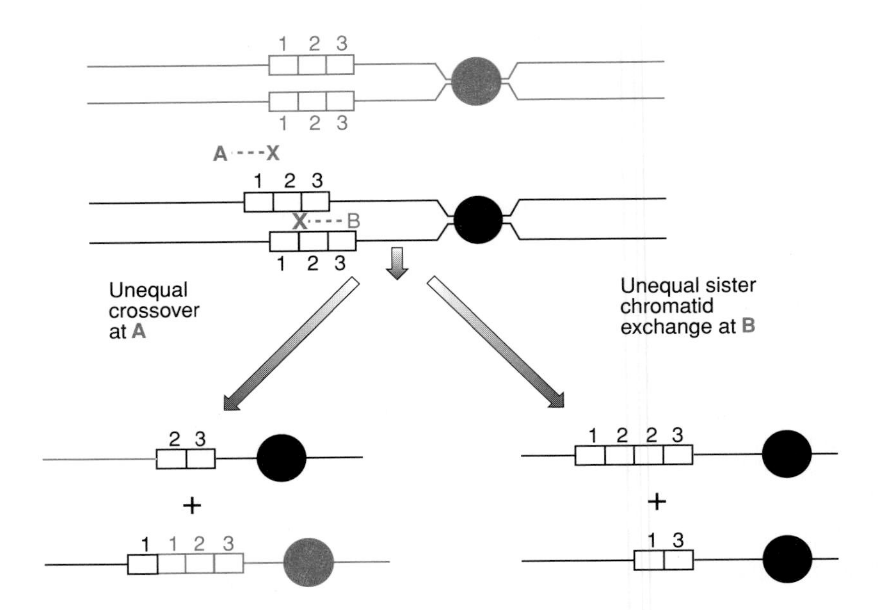

Figure 10.7: Unequal crossover and unequal sister chromatid exchange cause insertions and deletions.

The examples illustrate unequal pairing of chromatids within a tandemly repeated array. Unequal crossover involves unequal pairing of nonsister chromatids followed by chromatid breakage and rejoining. Unequal sister chromatid exchange involves unequal pairing of sister chromatids followed by chromatid breakage and rejoining. For the sake of simplicity, the breakages of the chromatids are shown to occur between repeats, but of course breaks can occur within repeats. *Note* that both types of exchange are *reciprocal* – one of the participating chromatids loses some DNA, while the other gains some.

Note that such exchanges are *reciprocal* – both participating chromatids are modified, in one case resulting in an insertion, and in the other case in a complementary deletion.

Unequal sister chromatid exchange is thought to be a major mechanism underlying VNTR polymorphism in the rDNA clusters. Unequal crossover is also expected to occur comparatively frequently in complex satellite DNA repeats and at tandemly repeated gene loci. In the latter case, unequal crossover is known to generate pathogenic deletions at some loci (see *Figure 10.15*). Such exchanges can also lead to *concerted evolution* (see page 219) by causing a particular variant to spread through an array of tandem repeats, resulting in *homogenization* of the repeat units (see *Figure 10.8*).

Occasionally, unequal crossover and unequal sister chromatid exchanges can occur at regions where there is little homology. This is likely to be the case when such mechanisms first generate a tandemly duplicated locus following mispairing of nonallelic repeats such as two *Alu* repeats or even smaller elements (*Figure 10.9*).

Gene conversion events may be relatively frequent in tandemly repetitive DNA

Gene conversion describes a *nonreciprocal* transfer of sequence information between a pair of nonallelic DNA sequences (*interlocus gene conversion*) or allelic sequences (*interallelic gene conversion*). One of the pair of interacting sequences, the **donor**, remains unchanged, but the other DNA sequence, the **acceptor**, is changed so that it gains some sequence copied from the donor sequence (*Figure 10.10*). The sequence exchange is a directional one because the acceptor sequence is modified by the donor sequence, but not the other way round. One possible mechanism envisages formation of a heteroduplex between a DNA strand from the donor gene and a complementary strand from the acceptor gene. Following heteroduplex formation, conversion of an acceptor gene segment may occur by **mismatch repair** – DNA repair enzymes recognize that the two strands of the heteroduplex are not perfectly matched and 'correct' the DNA sequence of the acceptor strand to make it perfectly complementary in the converted region to the sequence of the donor gene strand (see *Figure 10.10*).

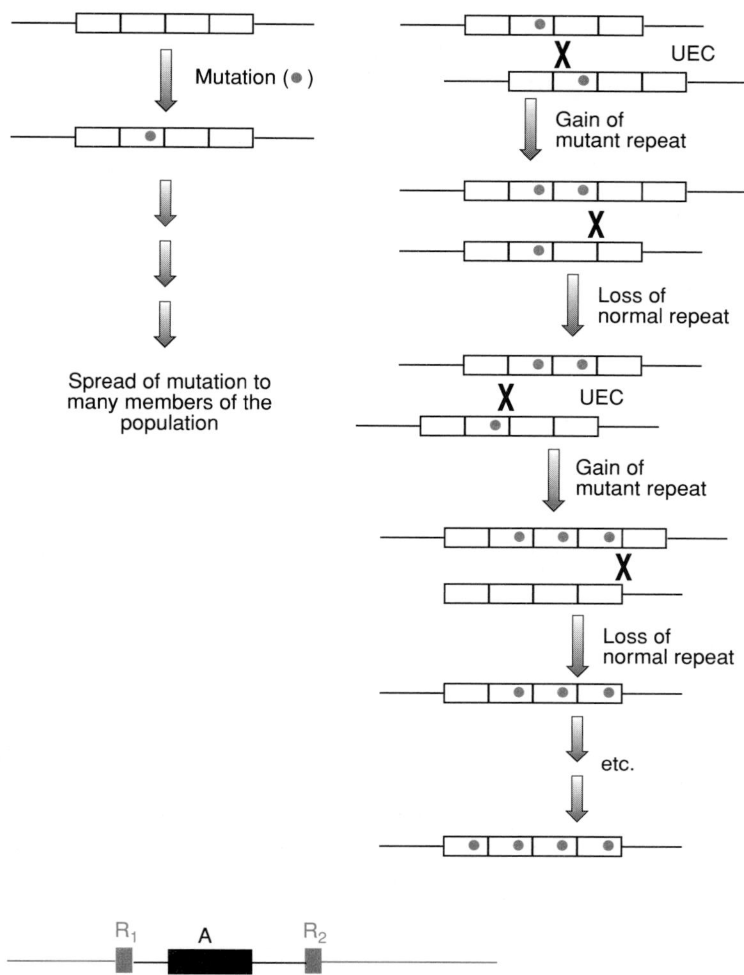

Figure 10.8: Unequal crossover in a tandem repeat array can result in sequence homogenization.

Note that the initial spread of the novel sequence variant to the same position in the chromosomes of other members of a sexual population can result by random genetic drift (see *Box 10.1*). Once the mutation has achieved a reasonable population frequency (left panel) it can spread to other positions within the array (right panel). This can occur by successive gain of mutant repeats as a result of unequal crossover (or unequal sister chromatid exchanges) and occasional loss of normal repeats. Eventually the mutant repeat can replace the original repeat sequence at all positions within the array, leading to sequence homogenization for the mutant repeat. Such sequence homogenization is thought to result in species-specific concerted evolution for repetitive DNA sequences (see page 219). UEC, unequal crossover.

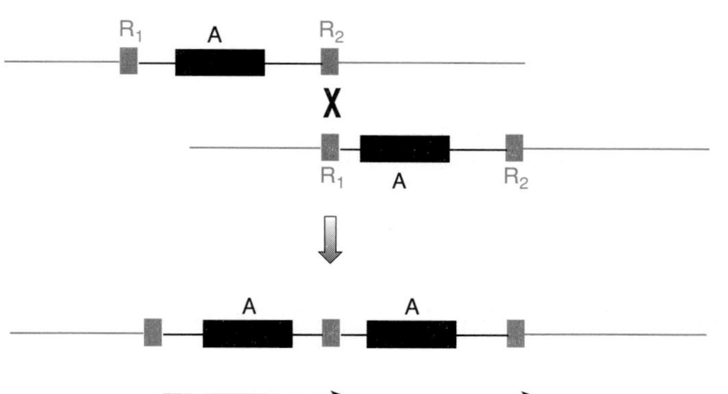

Figure 10.9: Tandem gene duplication can result from unequal crossover or unequal sister chromatid exchange, facilitated by short interspersed repeats.

The double arrow indicates the extent of the tandem gene duplication of a segment containing gene A and flanking sequences. Original mispairing of chromatids could be facilitated by a high degree of sequence homology between nonallelic short repeats (R_1, R_2). *Note* that the same mechanism can result in large scale deletions.

Gene conversion has been well-described in fungi where all four products of meiosis can be recovered and studied (*tetrad analysis*). In humans and mammals it is not possible to do this and so gene conversion cannot be demonstrated unambiguously in higher organisms (it can never be distinguished from double crossover events, for example, although double crossovers occurring in very close proximity would normally be expected to be extremely unlikely). Despite the difficulty in identifying gene conversion in complex organisms, there are numerous instances in mammalian genomes where an allele at one locus shows a pattern of mutations which strongly resembles those found in alleles at another locus of the same species. Such evidence suggests gene conversion-like exchanges between loci.

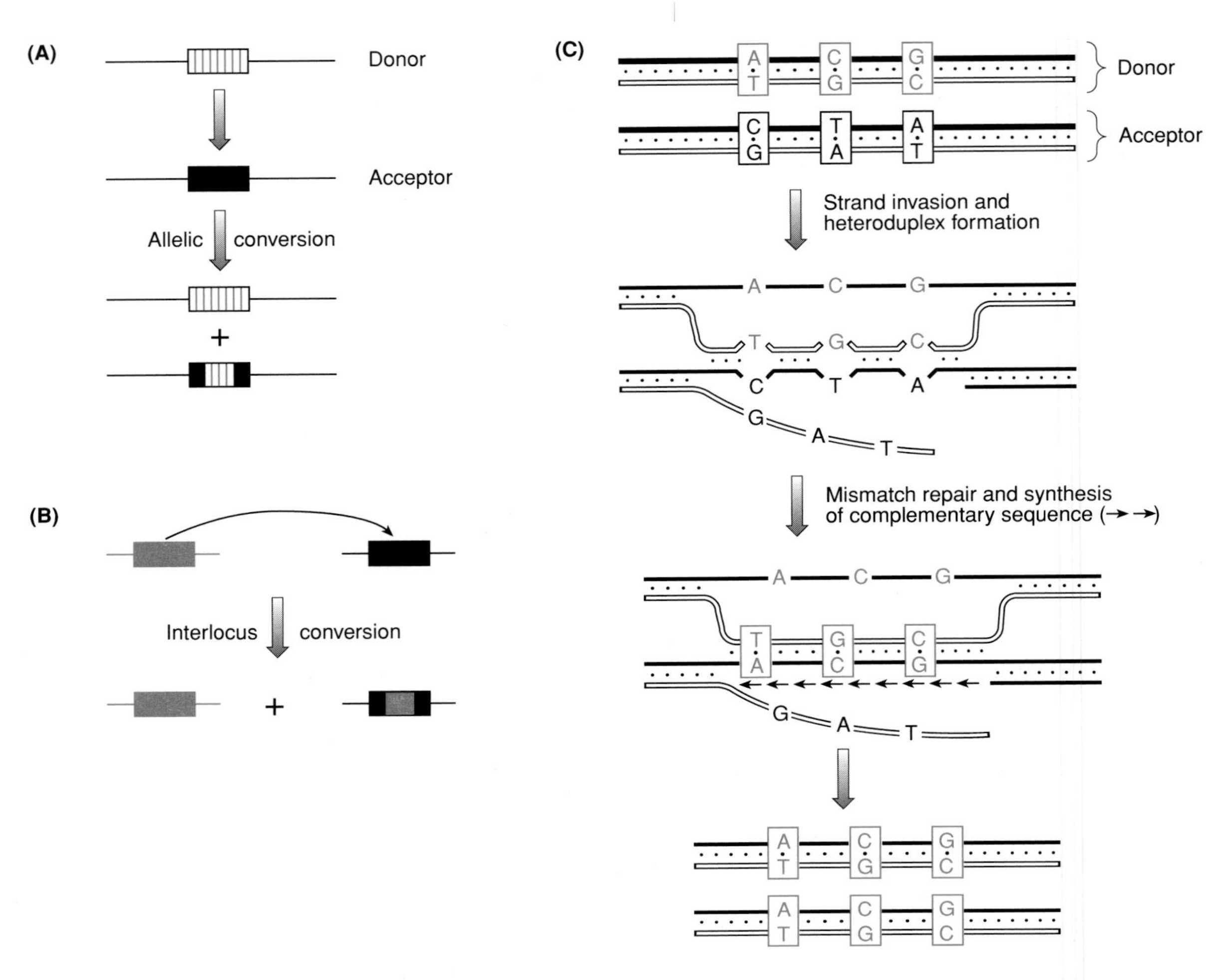

Figure 10.10: Gene conversion involves a nonreciprocal sequence exchange between allelic or nonallelic genes.

(**A**) Interallelic gene conversion. Note the nonreciprocal nature of the sequence exchange – the donor sequence is not altered but the acceptor sequence is altered by incorporating sequence copied from the donor sequence.
(**B**) Interlocus gene conversion. This is facilitated by a high degree of sequence homology between nonallelic sequences, as in the case of tandem repeats.
(**C**) Mismatch repair of a heteroduplex. This is one of several possible models to explain gene conversion. The model envisages invasion by one strand of the donor sequence (–) to form a heteroduplex with the complementary (+) strand of the acceptor sequence, thereby displacing the other strand of the acceptor. Mismatch repair enzymes recognize the mispaired bases in the heteroduplex and 'correct' the mismatches so that the (+) acceptor sequence is 'converted' to be perfectly complementary in sequence to the (–) donor strand. Subsequent replication of the (–) acceptor strand and sealing of nicks results in completion of the conversion.

Although simple comparisons of two sequences may be suggestive, the evidence for gene conversion is most compelling when a new mutant allele can be compared directly with its progenitor sequence. Certain highly mutable loci lend themselves to this type of analysis. In particular, some hypervariable minisatellite loci have high germline mutation rates (often 1% or more per gamete) and individual repeats often show nucleotide differences so that repeat subclasses can be recognized. Germline mutations can be studied by detecting and characterizing mutant minisatellite alleles in individual gametes. To do this, PCR analysis has been conducted on multiple dilute aliquots of DNA isolated from the sperm of an individual (**small pool PCR**), where each aliquot is calibrated to contain a few, perhaps 100, input molecules (Jeffreys *et al.*, 1994). The PCR products recovered from individual pools can then be typed to identify any new mutations that result in a novel allele whose length is sufficiently different as to be distinguishable from the progenitor allele.

Analyses of the patterns of germline mutation at three such loci have failed to identify exchanges of flanking markers and have shown that most mutations occurring at these loci are polar, involving the preferential gain of a few repeats at one end of a tandem repeat array. There is a bias towards gain of repeats and evidence was obtained for nonreciprocal sequence exchange between alleles, suggesting interallelic gene conversion (Jeffreys *et al.*, 1994). Evidence for interlocus gene conversion has also been obtained in human genes, notably the steroid 21-hydroxylase gene (see pages 267–269).

Pathogenic mutations

Pathogenic mutations are preferentially located at certain types of intragenic DNA sequence

Pathogenic mutations can occur at three types of DNA sequence at a gene locus.

- *The coding sequence of the gene.* This is where the great majority of recorded pathogenic mutations have been identified. Those due to nucleotide substitution are, in the vast majority of cases, nonsynonymous substitutions and mostly occur at first and second base positions of codons. However, very rarely, a synonymous codon substitution is not neutral as expected, but may cause disease by activating a cryptic splice site (see page 261). Because of its relatively high mutability, the CpG dinucleotide is often located at hotspots for pathogenic mutation in coding DNA (see Cooper and Youssoufian, 1988). Other hotspots include tandem repeats within coding DNA (see below).
- *Intragenic noncoding sequences.* This is restricted to sequences which are necessary for correct expression of the gene, such as important intronic elements, notably the highly conserved GT and AG dinucleotides at the ends of introns, but also conserved elements of the untranslated sequences. Often such mutations represent a small component (~10–15%) of the total pathogenic mutations at a gene locus (Cooper *et al.*, 1995). However, in some disorders pathogenic splicing mutations may be common. In the case of the collagen disorder osteogenesis imperfecta they constitute a very common pathological mutation which is second in frequency only to substitutions leading to replacement of the highly conserved, structurally important glycine residues. The collagen genes have small exons and a comparatively large number of introns (often more than 50 and as many as 106 in the case of the *COL7A1* gene; see page 157), making them exceptional targets for splicing mutations. Occasional pathogenic mutations have been recorded in the 5′ UTS (such as in the case of hemophilia B Leyden) and appear to exert their effect at the transcriptional level. Several examples are also known of pathogenic mutations in the 3′ UTS (see Cooper *et al.*, 1995).
- *Regulatory sequences outside exons.* Most mutations located in regulatory sequences have been identified in conserved elements located just upstream of the first exon, notably promoter elements. In addition, other more distantly located regulatory elements may be sites of pathological mutation. For example, deletions which eliminate the β-globin LCR (see *Figure 8.6*) but leave the β-globin gene and its promoter intact result in almost complete abolition of β-globin gene expression and contribute to β-thalassemia. Clearly, in some cases a gene may be regulated by the product of a distantly related gene. For

example, in the case of rare variants of α-thalassemia with mental retardation, the α-globin gene and its promoter may show no evidence of pathological mutation and the disease maps to an X-linked gene which encodes a transcription factor, one of whose target sequences is presumably the α-globin gene (Gibbons *et al.*, 1995).

Because of the large size of the human nuclear genome, most pathogenic mutations occur in nuclear DNA sequences. However, due to the large amount of nonfunctional DNA in the nuclear genome, most mutations in nuclear DNA are nonpathogenic. By comparison, the mitochondrial genome is a small target for mutation (about 1/200 000 of the size of the nuclear genome). The proportion of clinical disease due to pathogenic mutation in the mitochondrial genome might therefore be expected to be extremely low. However, unlike the nuclear genome the great bulk of the mitochondrial genome is composed of coding sequence, and mutation rates in mitochondrial genes are thought to be about 10 times higher than that in their nuclear counterparts. Accordingly, mutation in the mitochondrial genome is a significant contributor to human disease (see page 410).

Many different factors govern the expression of pathogenic mutations

The degree to which a pathogenic mutation results in an aberrant phenotype depends on several factors:

- *the mutation class and the way in which the expression of the mutant gene is altered.* This may depend on the location of the mutation within the gene (see *Table 10.4*). Most mutations will result in abolition or substantial reduction of gene expression, but some will lead to inappropriate expression (e.g. overexpression or *ectopic expression*, that is expression in tissues where the gene is not normally expressed).
- *The degree to which aspects of the aberrant phenotype are expressed in the heterozygote.* The presence of a single normal allele may be sufficient to maintain a

Table 10.4: Effect of location and class of mutation on gene function

Location and nature of mutation	Effect on gene function	Comments
Extragenic mutation	Normally none	Rare mutations may result in inactivation of distant regulatory elements required for normal gene expression (see *Figure 8.6*)
Multigene deletion	Abolition	Associated with contiguous gene syndromes (see *Figure 15.4*)
Whole gene deletion	Abolition	
Whole gene duplication	Can have effect due to altered gene dosage	Large duplications including the peripheral myelin protein 22 gene can cause Charcot–Marie–Tooth syndrome (see *Figure 15.6*)
Whole exon deletion	Abolition or modification	May cause shift in reading frame; protein often unstable
Within exon	Abolition	If loss/change of key amino acids, shift of the reading frame or introduction of premature stop codon
	Modification	If nonconservative substitutions, small in-frame insertions or other mutations at some locations
	None	If conservative/silent substitutions or mutation at nonessential sites
Whole intron deletion	None	
Splice site mutation	Abolition or modulation of expression	Conserved GT and AG signals are critically important for normal gene expression. Mutations may induce exon skipping
Promoter mutation	Abolition or modulation of expression	Deletion, insertion or substitution of nucleotides within promoter may alter expression. Complete deletion abolishes function
Mutation of termination codon	Modification	Additional amino acids are included at the end of the protein until another stop codon is reached
Mutation of poly(A) signal	Abolition or modulation of expression	Deletion, insertion or substitution of nucleotides within poly(A) site may alter expression. Complete deletion abolishes function
Elsewhere in introns/UTS	Usually none	

clinically normal phenotype (as in recessively inherited disorders), or a milder phenotype when compared with that of mutant homozygotes, as in dominantly inherited disorders where the mutation is a simple *loss of function* mutation (see page 403ff.).

- *The proportion and nature of cells in which the mutant gene is present.* Generally, mutations which are present in all the cells of an individual (inherited mutations) or in many of them (somatic mutations acquired very early in development) are likely to have a more profound effect than those present in a few cells (somatic mutations which arise at much later stages) or in cell types where the relevant gene is not expressed. Cancers, however, arise from unregulated division of cells produced from a single original mutant cell.
- *The parental origin of the mutation.* This is only known to be important in the case of a very few genes (see page 409 and *Figure 15.3*).

Most splicing mutations alter a conserved sequence required for normal splicing, but some occur in sequences not normally required for splicing

Many genes naturally undergo alternative forms of RNA splicing (page 167). In addition, mutations can sometimes produce an aberrant form of RNA splicing which is pathogenic. Sometimes this results in the sequences of whole exons being excluded from the mature RNA (**exon skipping** – see below). On other occasions, the abnormal splicing pattern may exclude part of a normal exon or result in new exonic sequences. Point mutations which alter a conserved sequence that is normally required for RNA splicing are comparatively common. Occasionally, however, aberrant splicing of a gene can be induced by mutation of other sequence elements which are not normally involved in splicing.

Mutations which alter important splice site signals

Often such mutations occur at the essentially invariant GT and AG dinucleotides located respectively at the start of an intron (*splice donor*) or at its end (*splice acceptor*). Flanking these important signals, however, are other conserved sequence elements (see *Figure 1.15*) which, if mutated, can also cause aberrant splicing. Mutations which alter such sequences can have different phenotypic consequences, depending on whether there is failure of splicing, or the use of an alternative illegitimate or natural splice site:

- *failure of splicing causing read-through of an intronic sequence.* This can occasionally result, for example, when the intron in question is quite small and the neighboring sequence lacks alternative legitimate splice sites or *cryptic splice sites* (sequences which resemble the consensus splice site sequences but which are not normally used by the splicing apparatus) (*Figure 10.11A*). The introduction of intronic sequence into a mature mRNA will, at the very least, introduce additional amino acids and may cause a frameshift.
- *Use of a cryptic splice site.* The splicing apparatus uses instead a nearby cryptic (illegitimate) splice site. Because individual splice donor and splice acceptor sequences often show some variation from the consensus sequences shown in *Figure 1.15*, cryptic splice sites may not be difficult to find (the β-globin gene has quite a variety of cryptic splice sites; see Cooper *et al.*, 1995). The use of an intronic cryptic splice site will introduce new amino acids, while using an exonic cryptic splice site will result in a deletion of coding DNA (*Figure 10.11A*). It should be noted that some apparently innocuous synonymous mutations in coding DNA may not be neutral as expected because they activate a cryptic splice site and can be pathogenic (Richard and Beckmann, 1995).

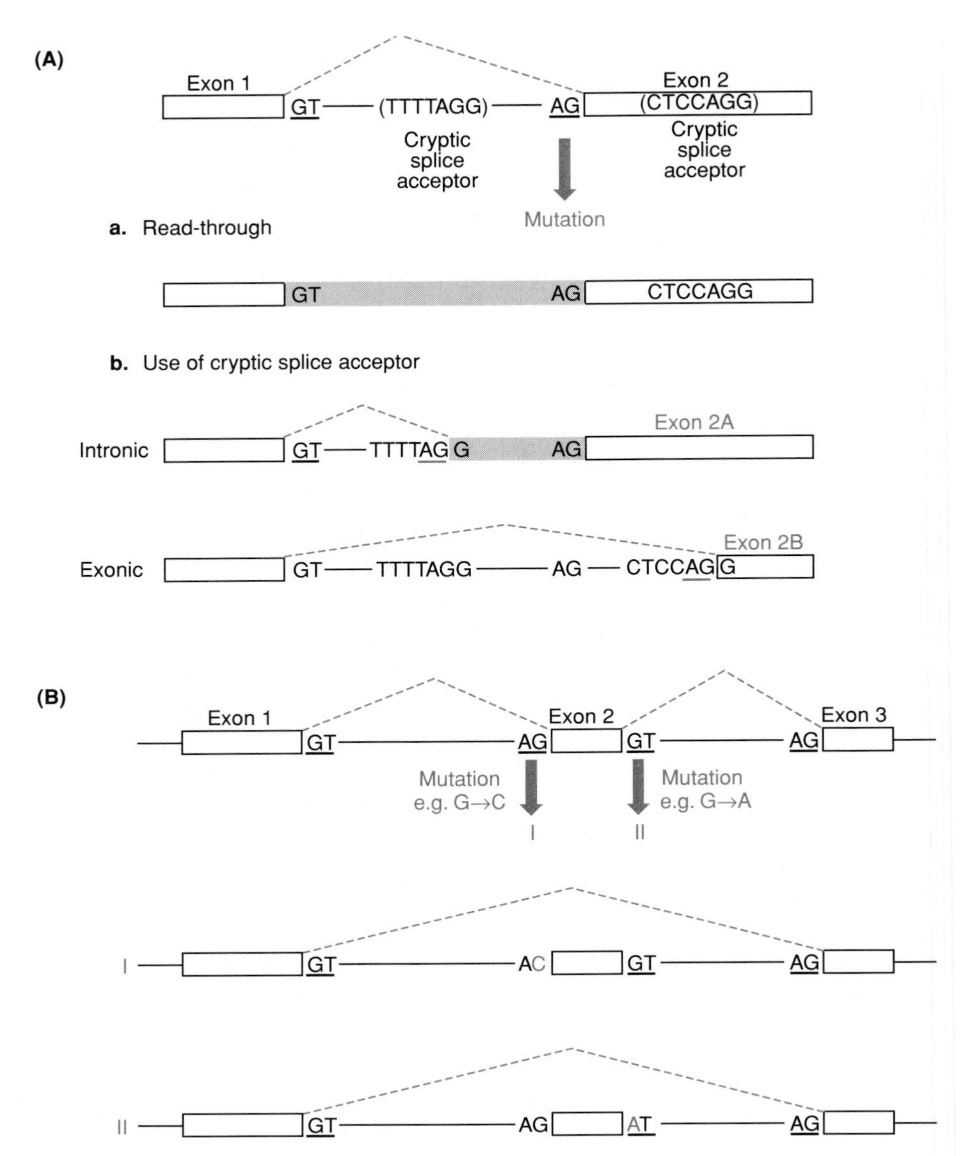

Figure 10.11: Mutations in conserved splice sites can cause altered exons or exon skipping.

(**A**) Altered exons as a result of read-through (a) or use of a cryptic splice site (b). In this illustration the gene has two exons and the variable presence of sequences similar to functional splice acceptor sites is considered. If there are no cryptic splice acceptor sites, and mutation destroys the splicing potential of the natural splice acceptor, splicing may not occur at all. As a result, there will be simple *read-through* of the intron and the gene will effectively be treated as a single large exon. If, however, there are cryptic splice sites, the splicing apparatus may be able to use them. Use of the cryptic splice site in the intron will result in a novel exon 2A which is increased in size by including a sequence normally present in the intron. If, on the other hand, a cryptic splice site in exon 2 is used, the resulting exon 2B will lack some of the normal exon 2 sequence.
(**B**) Exon skipping. The gene has three exons. Mutation at the splice acceptor site, GT, in intron 1 can result in skipping of the *downstream* exon 2 (I). In contrast, mutation at a splice donor site (e.g. the splice donor site, AG, of intron 2) can result in skipping of the *upstream* exon 2 (II).

- *Exon skipping*. The splicing apparatus uses an alternative legitimate splice site. Mutation of a splice donor sequence results in skipping of the upstream exon while mutation of the splice acceptor sequence results in skipping of the downstream exon (*Figure 10.11B*). Often, the exclusion of an exon has a profound effect on gene expression: it may result in a frameshift, an unstable RNA transcript, or a nonfunctional polypeptide because of a loss of a critical group of amino acids.

Mutations of sequences which are not normally important for RNA splicing

Occasionally, a mutation which results in aberrant RNA splicing does not occur at a sequence that is important for normal RNA splicing. Sometimes the mutation occurs at a cryptic splice site which may lie near a natural splice site. The mutation may then activate the cryptic splice site so that it can participate in RNA splicing, leading to the production of an altered exon (*Figure 10.12*). In other cases, mutations which occur within exons but not at cryptic splice sites can induce skipping of that exon (see below).

Nonsense mutations can have at least three different phenotypic consequences

Nonsense mutations are usually associated with severe phenotypes. However, in some cases, the effect on gene expression can be reduced by inducing skipping of the exon in which the mutation occurs. Evidence has been obtained for three different phenotypic consequences of nonsense mutations:

- *truncated protein.* Translation occurs to give a product that lacks some of the normal C-terminal sequence *truncated proteins.* Although demonstrated only rarely *in vivo* (see, for example, Lehrman *et al.*, 1987), the success of the artificial *protein truncation* test (see page 397) for screening mutations may suggest that this is not an infrequent phenotypic consequence. The effect on gene expression will depend on the stability of the polypeptide product, the extent of the truncation and the functional importance of the missing amino acids. Interference by the mutant polypeptide with the wild-type polypeptide may produce a severe phenotype.
- *Unstable mRNA.* There have been several reports where the mRNA which contains a premature termination codon is unstable and difficult to detect. In some cases, at least, the instability does not appear to be due to the failure of the mRNA to engage ribosomes; instead the defect in expression appears to

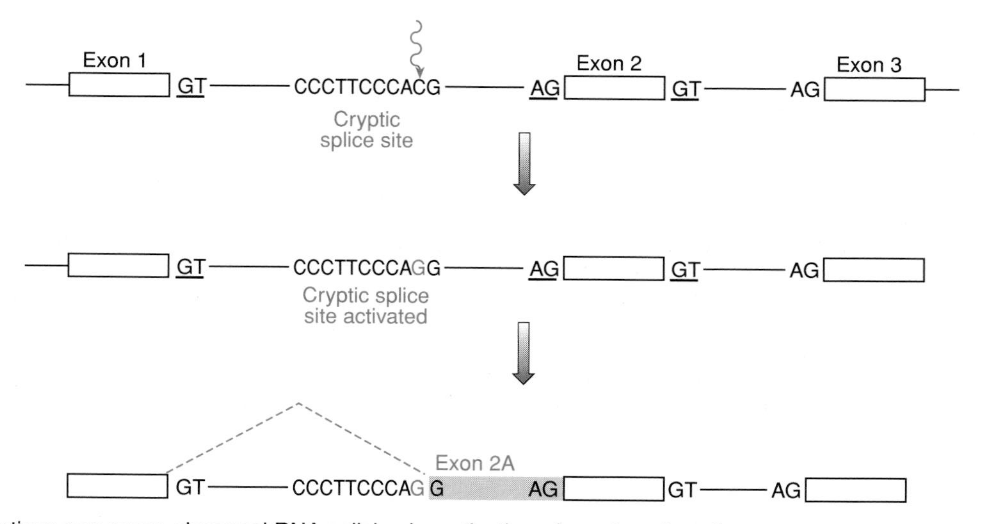

Figure 10.12: Mutations can cause abnormal RNA splicing by activation of cryptic splice sites.

A mutation can result in the alteration of a sequence which is not important for RNA splicing so as to create a new, alternative splice site. In the example illustrated, the mutation is envisaged to change a single nucleotide in intron 1. The nucleotide happens to occur within a *cryptic splice site* sequence that is closely related to the splice acceptor consensus sequence but, unlike the cryptic splice acceptor sites in *Figure 10.11*, shows a difference with respect to the conserved AG dinucleotide (see *Figure 1.15*). The mutation overcomes this difference and so can activate the cryptic splice site so that it competes with the natural splice acceptor site. If it is used by the splicing apparatus, a novel exon, exon 2A, results, which contains additional sequence which may or may not result in a frameshift.

occur prior to accumulation of the mRNA in the cytoplasm (Baserga and Benz, 1992).

- *Exon skipping.* Some nonsense mutations appear to induce skipping of constitutive exons *in vivo* (Dietz *et al.*, 1993). If the exon skipping does not result in a frameshift and the deleted amino acids are not essential for function, the phenotypic consequences of the mutation may be less severe than expected. Exon skipping is also known to be induced by other mutations which occur within exons, such as certain intra-exonic deletions. It is also known to occur naturally in the case of some exons, providing an alternative shortened transcript in addition to the normal one, a form of alternative RNA splicing.

The pathogenic potential of repeated sequences

The human genome, like other mammalian genomes, has a very high proportion of DNA sequences that are repeated. Tandem repeats in coding DNA include very short nucleotide repeats, moderately sized repeats and very large repeats that can include whole genes. Depending on the degree of sequence homology between the repeats, tandem repeats are liable to a variety of different genetic mechanisms causing sequence exchange between the repeats (*Table 10.5*). Often such sequence exchanges result in changes in the number of tandem repeats, either a reduction (deletion) or an increase (duplication/expansion), and occasionally they cause an alteration in the DNA sequence. Interspersed repeats can also cause pathogenic mutations by a different variety of mechanisms (see *Table 10.5*).

Table 10.5: Repeated DNA sequences often contribute to pathogenesis

Type of repeated DNA	Type of mutation	Mechanism and examples
Tandem repeats		
Very short repeats within genes	Deletion	Slipped strand mispairing (see *Figure 10.5*). Examples in *Figure 10.13*
	Frameshifting insertion	Slipped strand mispairing
	Triplet repeat expansion	Initially by slipped strand mispairing?; subsequently large-scale expansion by unknown mechanism
Moderate sized intragenic repeats	Intragenic deletion	UEC/UESCE[a] (see *Figure 10.7)*
Large tandem repeats containing whole genes	Partial or total gene deletion	UEC/UESCE[a] (*Figure 10.7*). Examples in *Figure 10.15*
	Alteration of gene sequence	Gene conversion (*Figure 10.10*). Examples in *Figures 10.15 and 10.16*
	Duplication causing gene dosage-related aberrant expression	UESCE[a] – 1.5 Mb duplication in Charcot–Marie–Tooth 1A (see *Figure 15.6*)
Interspersed repeats		
Short direct repeats	Deletion	Slipped strand mispairing or intrachromatid recombination?
Interspersed repeat elements (e.g. *Alu* repeats)	Deletion	UEC/UESCE[a]
	Duplication	UEC/UESCE[a]
Inverted repeats	Inversion	Intrachromatid exchange, e.g. Factor VIII (see *Figure 10.18*)
Active transposable elements	Intragenic insertion by retrotransposons	Retrotransposition (*Figures 8.7* and *8.11*). Examples, see page 271

[a]UEC, unequal crossover; UESCE, unequal sister chromatid exchange.

Slipped strand mispairing of short tandem repeats predisposes to pathogenic deletions and frameshifting insertions

Insertions and deletions in coding DNA are rare because they usually introduce a translational frameshift. However, occasionally, a series of tandem repeats of a small number of nucleotides occurs by chance in the coding sequence for a polypeptide. Such repeats, like classical microsatellite loci, are comparatively prone to mutation by slipped strand mispairing. As a result, the copy number of tandem repeats is liable to fluctuate, introducing a deletion or an insertion of one or more repeat units. If the mutation occurs in polypeptide-encoding DNA, a resulting deletion will often have a profound effect on gene expression. Frameshifting deletions will normally result in abolition of gene expression. Even if the deletion does not produce a frameshift, deletions of one or more amino acids can still be pathogenic (*Figure 10.13*). Small frameshifting insertions will also be expected to lead to loss of gene expression and often the insertion is a tandem repeat of sequences flanking it. However, nonframeshifting insertions would often not be expected to be pathogenic, unless the insertion occurs in a critically important region, destabilizing an essential structure or impeding gene function in some way. Note that extremely large triplet repeat expansions can lead to disease by mechanisms that are not understood at present (see next section).

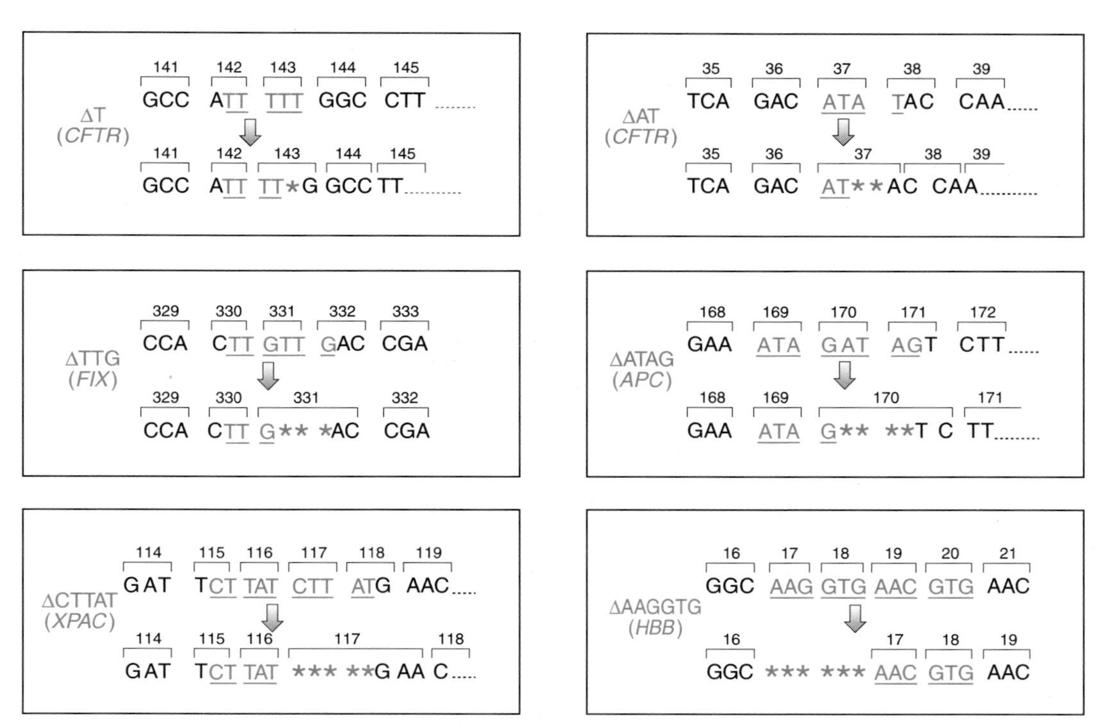

Figure 10.13: Short tandem repeats are deletion/insertion hotspots.

The six deletions illustrated are examples of pathogenic deletions occurring at tandemly repeated units of from 1 to 6 bp. The deletions of 3 and 6 bp do not cause frameshifts, and pathogenesis is thought to be due to removal of one or two amino acids that are critically important for polypeptide function. *Note* that in the case of the 6-bp deletion the original tandem repeat is not a perfect one. Genes (and associated diseases) are: *CFTR*, cystic fibrosis transmembrane regulator; *FIX*, factor IX (hemophilia B); *APC*, adenomatous polyposis coli; *XPAC*, xeroderma pigmentosa complementation group C; *HBB*, β-globin (β-thalassemia). Original references are listed in Appendix 3 of Cooper and Krawczak (1993). Though not illustrated here, small insertions are often tandem repeats of sequences flanking them (see Table 8.1 of Cooper and Krawczak, 1993).

Rapid large-scale expansion of intragenic triplet repeats can cause a variety of diseases

The discovery (Fu *et al.*, 1991) that human disease can be caused by large-scale expansion of highly unstable trinucleotide repeats was quite unexpected. Studies in other organisms had not revealed precedents for such a phenomenon, but the list of human examples is growing rapidly – at the time of writing it has reached 10. The pathological mechanisms by which expanded repeats cause disease are discussed in Chapter 15. Here we are concerned with the nature and mechanism of the DNA instability.

Tandem trinucleotide repeats are not infrequent in the human genome. Although there are 64 possible trinucleotide sequences, when allowance is made for cyclic permutations $(CAG)_n = (AGC)_n = (GCA)_n$ and reading from either strand [5′$(CAG)_n$ on one strand = 5′$(CTG)_n$ on the other], there are only 10 different trinucleotide repeats (*Notebox*). Most of these are known as usefully polymorphic microsatellite markers but, in addition, certain repeats of CAG/CTG and CCG/GGC show anomalous behavior.

AAC/GTT
AAG/CTT
AAT/ATT
ACC/GGT
ACG/CGT
ACT/AGT
AGG/CCT
ATC/GAT
CAG/CTG
CCG/CGG

The 10 possible trinucleotide repeats. Both DNA strands are shown; all other trinucleotide repeats are cyclic permutations of one or another of these 10.

In each case, repeats below a certain length are stable in mitosis and meiosis while, above a certain threshold length, the repeats become extremely unstable. These unstable repeats are virtually never transmitted unchanged from parent to child. Both expansions and contractions can occur, but there is a bias towards expansion. The average size change often depends on the sex of the transmitting parent, as well as the length of the repeat. The unstable expanding trinucleotide repeats fall into three classes (see *Table 10.6*):

- several genes contain $(CAG)_n$ repeats within the coding sequence, translated as polyglutamine tracts in the protein product. Typically, the stable and non-pathological alleles have 10–30 repeats, while unstable pathological alleles have modest expansions, often in the range of 40–100 repeats. Transcription and translation of the gene are not affected by the expansion.

Table 10.6: Unstable trinucleotide repeats in the human genome

Disease	MIM no.	Location of gene	Location of repeat	Repeat sequence	Normal length	Pre-mutation	Full mutation
Huntington disease	143100	4p16.3	Coding	$(CAG)_n$	9–35	?	37–100
Kennedy disease	313200	Xq21	Coding	$(CAG)_n$	17–24	—	40–55
Spino-cerebellar ataxia 1 (*SCA1*)	164400	6p23	Coding	$(CAG)_n$	19–36	?	43–81
Dentatorubral-pallidoluysian atrophy (*DRPLA*)	125370	12p	Coding	$(CAG)_n$	7–23	?	49– >75
Machado–Joseph disease (*MJD, SCA3*)	109150	14q32.1	Coding	$(CAG)_n$	12–36	?	67– >79
Fragile X site A (*FRAXA*)	309550	Xq27.3	5′ UTR	$(CGG)_n$	6–54	50–200	200– >1000
Fragile X site E (*FRAXE*)	309548	Xq28	?	$(CCG)_n$	6–25	?	>200
Fragile X site F (*FRAXF*)	600226	Xq28	?	$(GCC)_n$	6–29	?	>500
Fragile 16 site A (*FRA16A*)	136580	16q22	?	$(CCG)_n$	16–49	—	1000–2000
Myotonic dystrophy (*DM*)	160900	19q13	3′ UTR	$(CTG)_n$	5–35	37–50	50–4000

- Some $(CGG)_n$ repeats in noncoding sequences can expand massively from a normal copy number of 10–50 up to hundreds or thousands of repeats. By unknown means, the expanded repeats affect DNA methylation and chromatin structure, producing inducible chromosomal fragile sites (see page 408). Concurrently, expression of adjacent genes is inhibited.
- Uniquely, a $(CTG)_n$ repeat in the 3′ untranslated region of the myotonic dystrophy kinase gene (*DMK*) at 19q13 has 5–35 repeat units in normal people, but up to 2000 units in people with myotonic dystrophy (*DM*, MIM 160900). There is a perfect correlation between the repeat expansion and the disease, even though the repeat has no evident effect on transcription or the structure of the gene product.

Our understanding of the genetic mechanisms underlying expansion of triplet repeats is imperfect. Slipped strand mispairing (see *Figure 10.5*) could account for modest increases in sizes of tandem repeat arrays. Alleles with large numbers of repeats are likely to be more susceptible to unequal sister chromatid exchange, which could produce large changes of length. The fact that all the repeats are CG-rich, and therefore relatively resistant to denaturation, may be signficant. It has been suggested that long repeats could form abnormal DNA structures.

Intergenerational changes are normally reported as parent–child comparisons of blood lymphocyte DNA. There is little information about when in gametogenesis, fertilization or embryogenesis the changes arise. Limited studies of sperm show that highly expanded *DM* and *FRAXA* (fragile-X syndrome) repeats are not transmitted by affected males, although modest expansions can be. The largest expansions in Huntington disease (which, however, are small compared with large *FRAXA* or *DM* expansions) are seen in sperm, consistent with the observation that the severest cases inherit the disease from their father. At least for *FRAXA, DM* and Kennedy disease, the expanded repeats are mitotically unstable, so that a blood sample shows a smear of heterogeneous expanded repeats sizes. However, *in vitro*, even large repeats are stable. Thus, whatever the mechanism, it is not operative in all cells.

Strong evidence for a mechanism based on mispairing comes from the observation that only homogeneous repeats are unstable. In spinocerebellar ataxia type 1, 123/126 normal sized CAG repeats were interrupted by one or two CAT triplets, while 30/30 expanded alleles contained no interruption (Chung *et al.*, 1993; see *Figure 10.14*). It is probably generally true that interrupted trinucleotide repeats are stable. One problem with all these mispairing mechanisms is that they should result in contractions as well as expansions and this is not seen. Instead, after a certain threshold size, there appears to be a clear bias towards continued expansion of the size of the repeat unit array. Because understanding of trinucleotide repeats is progressing very rapidly at the time of writing, the reader is advised to consult a recent review for more information.

Tandemly repeated and clustered gene families may be prone to pathogenic unequal crossover and gene conversion-like events

Many human and mammalian gene clusters contain nonfunctional pseudogenes which may be closely related to functional gene members. Interlocus sequence exchanges between pseudogenes and functional genes can result in disease by removing or altering some or all of the sequence of a functional gene. For example, unequal crossover (or unequal sister chromatid exchange) between a functional gene and a related pseudogene can result in deletion of the functional gene or the formation of fusion genes containing a segment derived from the pseudogene. Alternatively, the pseudogene can act as a donor sequence in gene conversion

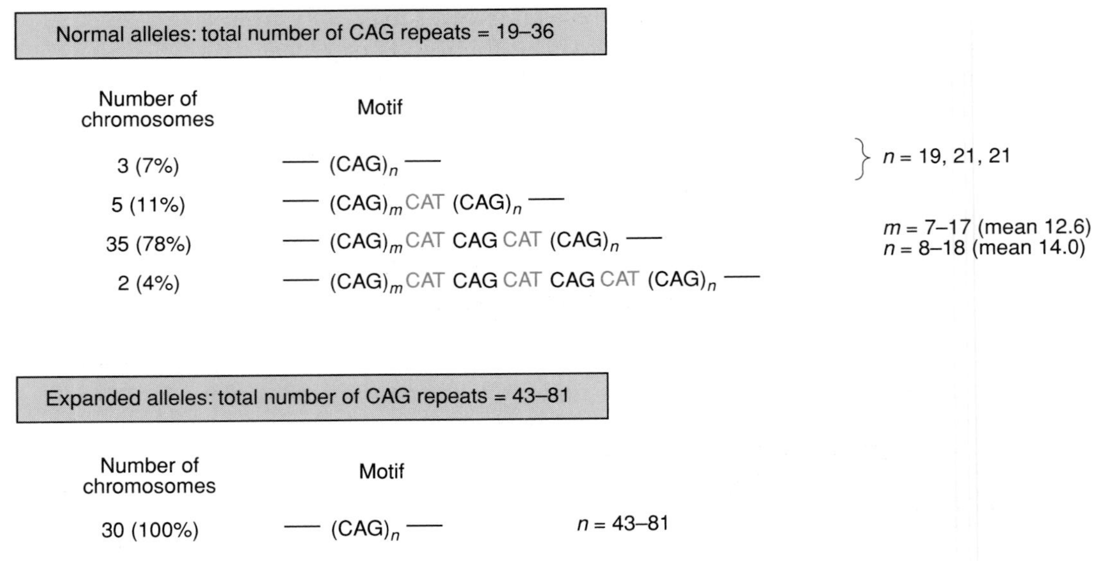

Figure 10.14: Uninterrupted triplet repeats are more prone to expansion.

Analysis of the *SCA1* spinocerebellar ataxia gene by Chung *et al.* (1993) showed that all the presumed stable alleles from normal subjects had interrupted repeats except the three with the shortest runs. However, all of the unstable expanded alleles found on disease chromosomes had uninterrupted repeats.

events and introduce deleterious mutations into the functional gene. The classical example is steroid 21-hydroxylase deficiency, where virtually 100% of pathological mutations arise as a result of sequence exchanges between the functional 21-hydroxylase gene, *CYP21B*, and a very closely related pseudogene, *CYP21A*. The two genes occur on tandemly repeated DNA segments approximately 30 kb long which also contain other duplicated genes, notably the complement C4 genes, *C4A* and *C4B* (see *Figure 10.15*). Large pathogenic deletions uniformly result in removal of about 30 kb of DNA, corresponding to one repeat unit length, and analysis of *de novo* 21-hydroxylase deficiency mutations has provided strong evidence for pathogenic deletions arising as a result of meiotic unequal crossover (Sinnott *et al.*, 1990). Virtually all of the 75% of pathological mutations which are point mutations are copied from deleterious mutations in the pseudogene, suggesting a gene conversion mechanism (see *Figures 10.15* and *10.16*). Analysis of one such mutation which arose *de novo* suggests that the conversion tract is a maximum of 390 bp (Collier *et al.*, 1993). Gene conversion events are also found in the duplicated C4 genes, both of which are normally expressed. A likely priming event for conversions in the *CYP21–C4* gene cluster is unequal pairing of chromatids so that a *CYP21A–C4A* unit pairs with a *CYP21B–C4B* unit (see *Figure 10.15*).

Interspersed repeats often predispose to large deletions and duplications

Short direct repeats

In several cases, the endpoints of deletions are marked by very short direct repeats. For example, the breakpoints in numerous pathological deletions of the mitochondrial genome occur at perfect or almost perfect short direct repeats. Of these, the most common is a deletion of 4977 bp which has been found in multiple patients with Kearns–Sayre syndrome, an encephalomyopathy characterized by external ophthalmoplegia, ptosis, ataxia and cataract. The deletion results in elimination of the intervening sequence between two perfect 13-bp repeats and loss of the sequence of one of the repeats (*Figure 10.17*). Recombination does not appear to occur in the mitochondrial genome and Schoffner *et al.* (1989) have postulated that

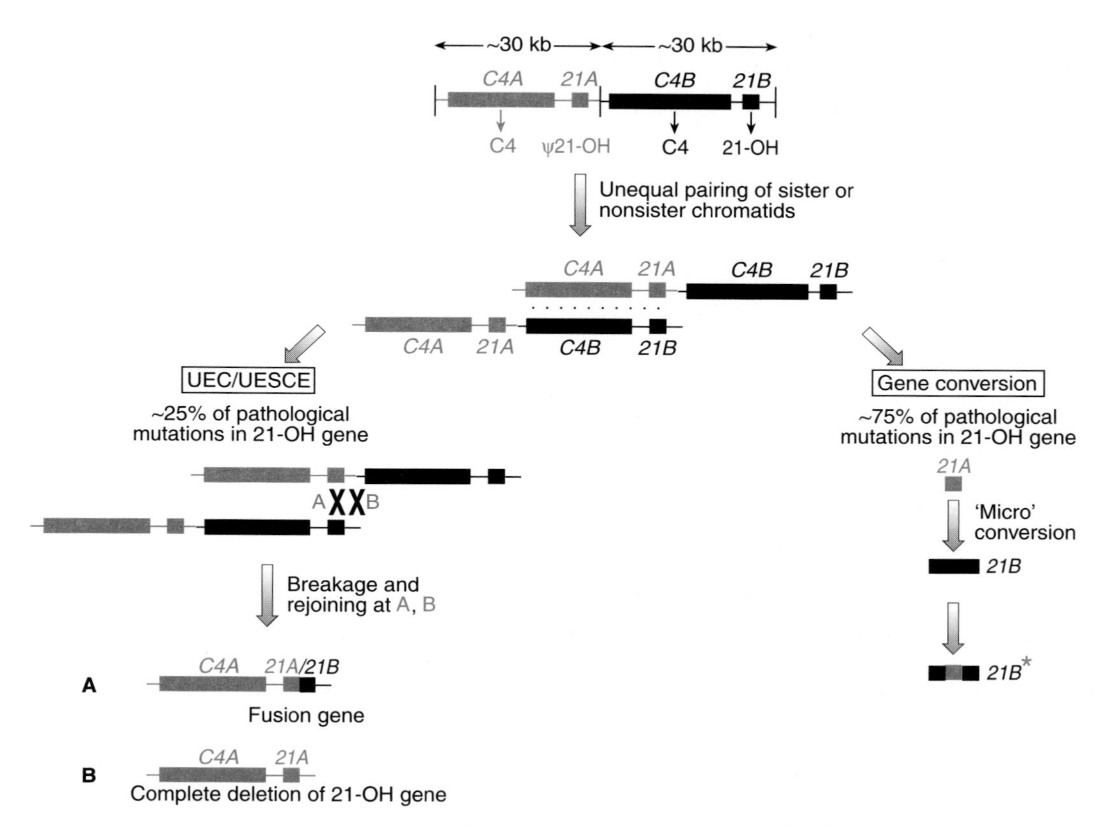

Figure 10.15: Virtually 100% of 21-hydroxylase gene mutations are due to sequence exchange with a closely related pseudogene.

The duplicated complement *C4* genes and steroid 21-hydroxylase genes are located on tandem 30-kb repeats which show about 97% sequence identity. Both the *C4A* and *C4B* genes are expressed to give complement C4 products; the *CYP21B* gene (21B) encodes a 21-hydroxylase product, but the *CYP21A* (21A) gene is a pseudogene. About 25% of pathological mutations at the 21-hydroxylase locus involve a 30-kb deletion resulting from unequal crossover (UEC) or unequal sister chromatid exchange (UESCE). The remaining mutations are point mutations where small-scale gene conversion of the *CYP21B* gene occurs – a small segment of the *CYP21A* gene containing deleterious mutations is copied and inserted into the *CYP21B* gene replacing a short segment of the original sequence (see *Figure 10.10C* for one possible mechanism). Possibly gene conversion events are, like UEC and UESCE, primed by unequal pairing of the tandem repeats on sister or nonsister chromatids.

Location of mutation	Normal 21-OH gene sequence (*CYP21B*)	Mutant 21-OH gene sequence	21-OH pseudogene sequence (*CYP21A*)
Intron 2	CCCACCTCC	CCCAGCTCC	CCCAGCTCC
Exon 3 (codons 110–112)	G**GA GAC TAC** TC **Gly Asp Tyr Ser**	G(.........)TC Val	G(.........)TC
Exon 4 (codon 172)	ATC ATC TGT Ile **Ile** Cys	ATC AAC TGT Ile Asn Cys	ATC AAC TGT
Exon 6 (codons 235–238)	ATC GTG GAG ATG **Ile Val** Glu **Met**	AAC GAG GAG AAG Asn Glu Glu Lys	AAC GAG GAGAAG
Exon 7 (codon 281)	CAC GTG CAC His **Val** His	CAC TTG CAC His Leu His	CAC TTG CAC
Exon 8 (codon 318)	CAC CAG GAG His **Gln** Glu	CTG TAG GAG Leu STOP	CTG TAG GAG
Exon 8 (codon 356)	CTG CGG CCC Leu **Arg** Pro	CTG TGG CCC Leu Trp Pro	CTG TGG CCC

Figure 10.16: Pathogenic point mutations in the steroid 21-hydroxylase gene originate by copying sequences from the 21-hydroxylase pseudogene.

The copying is thought to involve a gene conversion-like mechanism (see *Figures 10.15* and *10.10C*).

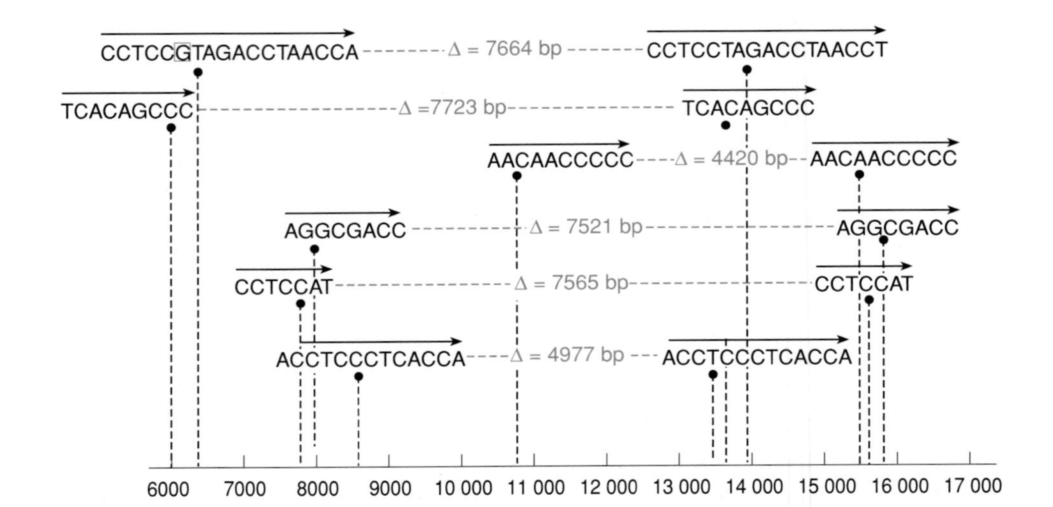

Figure 10.17: Short direct repeats mark the endpoints of many pathogenic deletions in the mitochondrial genome.

Note that as recombination does not occur within the mitochondrial genome, one likely mechanism to explain the deletions is slipped strand mispairing (see text).

such deletions arise by a replication slippage mechanism, similar to that occurring at short tandem repeats (see *Figure 10.5*). Partial duplications of the mitochondrial genome are also distinctive features of certain diseases, notably Kearns–Sayre syndrome. The ends of the duplicated sequences, like those of the common deletions, are often marked by short direct repeats, and the mechanisms of duplication and deletion appear to be closely related (see Poulton and Holt, 1994).

The Alu *repeat as a recombination hotspot*

Some large-scale deletions and insertions may be generated by pairing of non-allelic interspersed repeats, followed by breakage and rejoining of chromatid fragments. For example, the *Alu* repeat occurs approximately once every 4 kb and mispairing between such repeats has been suggested to be a frequent cause of deletions and duplications. Some large genes have many internal *Alu* sequences in their introns or untranslated sequences, making them liable to frequent internal deletions and duplications. For example, the 45-kb low density lipoprotein receptor gene has a relatively high density of *Alu* repeats (approximately one every 1.6 kb). A very high frequency of pathological deletions in this gene are likely to involve an *Alu* repeat, usually at both endpoints, and occasional pathogenic intragenic duplications also involve *Alu* repeats (see Hobbs *et al.*, 1990). Such observations have suggested a general role for *Alu* sequences in promoting recombination and recombination-like events. Initial gene duplications in the evolution of clustered multigene families may often have involved an unequal crossover event between *Alu* repeats or other dispersed repetitive elements. It should be noted, however, that some *Alu*-rich genes do not appear to be loci for frequent *Alu*-mediated recombination.

Pathogenic inversions can be produced by intrachromatid recombination between inverted repeats

Occasionally, clustered inverted repeats with a high degree of sequence identity may be located within or close to a gene. The high degree of sequence similarity between inverted repeats may predispose to pairing of the repeats by a mechanism that involves a chromatid bending back upon itself. Subsequent chromatid breakage at the mispaired repeats and rejoining can then result in an inversion, in much the same way as the natural mechanism used for the production of some immunoglobulin κ light chains (see *Figure 7.19*). The classic example of pathogenic inversions caused by such a mechanism occurs in more than 40% of cases of severe hemophilia

A. Intron 22 of the factor VIII (F8) gene contains a CpG island from which two internal genes, *F8A* and *F8B*, are transcribed in opposite directions: the *F8A* gene is transcribed in the opposite direction to the factor VIII gene, while the *F8B* gene is transcribed in the same direction as the factor VIII gene (see *Figure 10.18*). The *F8A* gene belongs to a gene family with two other closely related members several hundred kilobases upstream of the factor VIII gene, but in the opposite orientation. As a result, the region between the *F8A* gene and the other two members is susceptible to inversions – the *F8A* gene can pair with either of the other two members on the same chromatid, and subsequent chromatid breakage and rejoining in the region of the paired repeats results in an inversion which disrupts the factor VIII gene (Lakich *et al.*, 1993, see *Figure 10.18*).

DNA sequence transposition is not uncommon and can cause disease

As described in Chapter 8, pages 197–202, a proportion of moderately and highly repeated interspersed elements are capable of transposition via an RNA intermediate. Defective gene expression due to DNA transposition is comparatively rare and represents only a small component of molecular pathology. However, several examples have been recorded of genetic deficiency due to insertional inactivation by retrotransposons. For example, in one study, hemophilia A was found to arise in two out of 140 unrelated patients as a result of a *de novo* insertion of a LINE-1 (*Kpn*) repeat into an exon of the factor VIII gene (Kazazian *et al.*, 1988). Other instances are known of insertional inactivation by an actively transposing *Alu* element, as in a case of neurofibromatosis type 1 (Wallace *et al.*, 1991). Additionally, a number of other examples have been recorded of pathogenesis due to intragenic insertion of undefined DNA sequences.

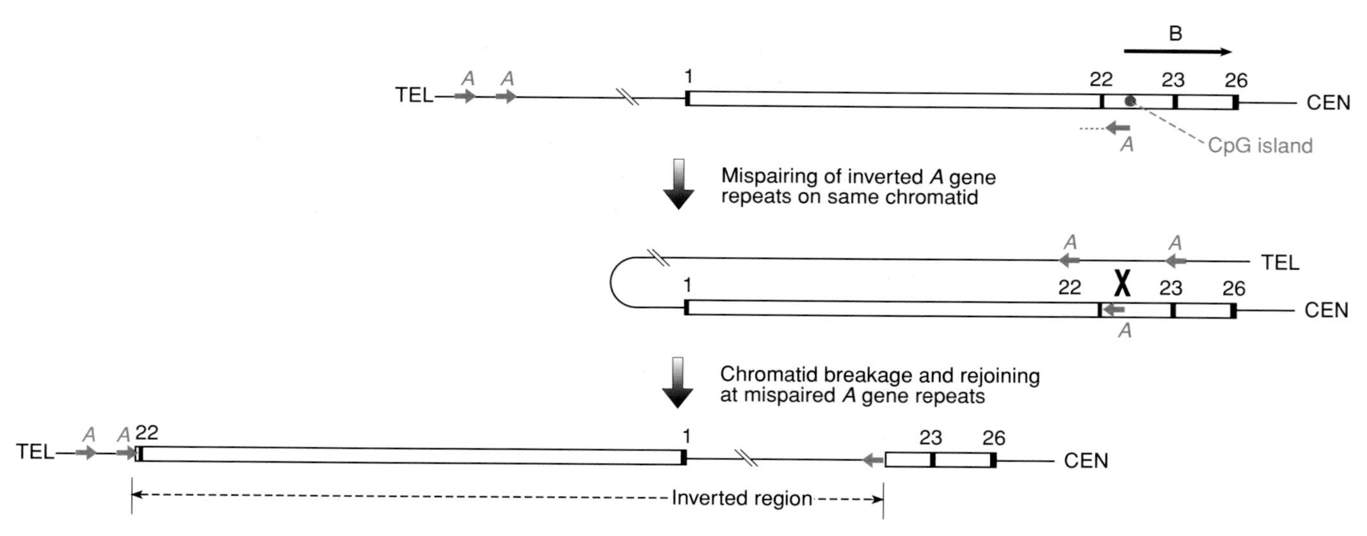

Figure 10.18: Inversions disrupting the factor VIII gene result from intrachromatid recombination between inverted repeats.

For the sake of clarity only exons 22, 23 and the first and last exons (1 and 26) of the factor VIII gene (*F8*; open box) are shown. Intron 22 of this gene contains a CpG island from which the (B) *F8B* gene internal is transcribed in the same direction as the factor VIII gene, and expression involves splicing of a novel exon within intron 22 on to exons 23–26 of the factor VIII gene. The internal *F8A* gene (A) is also transcribed from the intron 22 CpG island but in the opposite direction. Two other sequences closely related to *F8A* are found about 500 kb upstream of the factor VIII gene and are transcribed in the same direction as the factor VIII gene and the opposite direction to that of the *F8A* gene. The high degree of sequence identity between the three members of the *F8A* gene family means that pairing of the *F8A* gene with one of the other two members on the same chromatid can occur by looping back of the chromatid. Subsequent chromatid breakage and rejoining can result in an inversion of the region between the *F8A* gene and the other paired family member, resulting in disruption of the factor VIII gene (see Lakich *et al.*, 1993).

Further reading

Cooper DN, Krawczak M. (1993) *Human Gene Mutation*. BIOS Scientific Publishers, Oxford.

Crow JF. (1993) *Environ. Mol. Mutagen.*, **21**, 122–129.

References

Baserga S, Benz EJ Jr. (1992) *Proc. Natl Acad. Sci. USA*, **89**, 2935–2939.

Britten RJ. (1986) *Science*, **231**, 1393–1398.

Brown WM, Prager EM, Wang A, Wilson AC. (1982) *J. Mol. Evol.*, **18**, 225–239.

Cairns J. (1975) *Nature*, **255**, 197–200.

Chen X, Presser R, Simonetti S, Sadlock J, Jagiello G, Schan EA. (1995) *Am. J. Hum. Genet.* **57**, 239–247.

Chung MY, Ranum LPW, Duvick IA, Servadio A, Zoghbi HY, Orr HT. (1993) *Nature Genetics*, **5**, 254–258.

Collier PS, Tassabehji M, Sinnott PJ, Strachan T. (1993) *Nature Genetics*, **3**, 260–265; **4**, 101.

Collins DW, Jukes TH. (1994) *Genomics*, **20**, 386–396.

Cooper DN, Youssoufian H. (1988) *Hum. Genet.*, **78**, 151–155.

Cooper DN, Smith BA, Cooke HJ, Niemann S, Schmidtke J. (1985) *Hum. Genet.*, **69**, 201–205.

Cooper DN, Krawczak M, Antonorakis SE. (1995) In: *The Metabolic and Molecular Bases of Inherited Disease*, 7th Edn (eds CR Scriver, AL Beaudet, WS Sly, D Valle), pp. 259–291. McGraw-Hill, New York.

Dietz HC, Valle D, Francomano C, Kendzioz RJ Jr , Pyeritz RE, Cutting GR. (1993) *Science*, **259**, 680–683.

Dover GA. (1995) *Nature Genetics*, **10**, 254–256.

Fu YH *et al.* (1991) *Cell*, **67**, 1047–1058.

Gibbons RJ, Picketts DJ, Villard L, Higgs DR. (1995) *Cell*, **80**, 837–845.

Hauswirth W, Laipis P. (1985) In: *Achievements and Perspectives of Mitochondrial Research* (ed. E Quagliarello), pp. 49–59. Elsevier, Amsterdam.

Hayashi J-I, Takemitsu M, Goto Y-I, Nonaka I. (1994) *J. Cell Biol.*, **125**, 43–50.

Hobbs HH, Russell DW, Brown MS, Golding JL. (1990) *Annu. Rev. Genet.*, **24**, 133–170.

Holmquist GP, Filipski J. (1994) *Trends Ecol. Evol.*, **9**, 65–69.

Howell N, Halvorson S, Kubacka I, McCullough DA, Bindoff LA, Turnbull DM. (1992) *Hum. Genet.*, **90**, 117–120.

Imbert G, Kretz C, Johnson K, Mandel JL. (1993) *Nature Genetics*, **4**, 72–76.

Jeffreys A, Tamaki K, MacLeod A, Monckton DG, Neil DL, Armour JAL. (1994) *Nature Genetics*, **6**, 136–145.

Kazazian HH, Wong C, Youssoufian H, Scott AF, Phillips DG, Antonorakis SE. (1988) *Nature*, **332**, 164–166.

Koehler CM, Lindberg GL, Brown DR *et al.* (1991) *Genetics*, **129**, 247–255.

Lakich D, Kazazian HH Jr, Antonarakis SE, Gitschier J. (1993) *Nature Genetics*, **5**, 236–241.

Lehrman MA, Schneider WJ, Brown MS *et al.* (1987) *J. Biol. Chem.*, **262**, 401–410.

Levinson G, Gutman GA. (1987) *Mol. Biol. Evol.*, **4**, 203–221.

Li W-H, Grauer D. (1991) *Fundamentals of Molecular Evolution.* Sinauer Associates, Sunderland, MA.

Mahtani MM, Willard HF. (1993) *Hum. Mol. Genet.*, **2**, 431–437.

Poulton J. (1995) *Am. J. Hum. Genet.*, **57**, 224–226.

Poulton J, Holt IJ. (1994) *Nature Genetics*, **8**, 313–315.

Richard I, Beckmann JS. (1995) *Nature Genetics*, **10**, 259.

Schimmin LC, Chang BH-J, Li W-H. (1993) *Nature*, **362**, 745–747.

Schoffner JN. (1989) *Proc. Natl Acad. Sci. USA*, **86**, 7952–7956.

Sharp P and Matassi G. (1994) *Curr. Opin. Genet. Dev.*, **4**, 851–860

Sinnott PJ, Collier S, Costigan C, Dyer PA, Harris R, Strachan T. (1990) *Proc. Natl Acad. Sci. USA*, **87**, 2107–2111.

Tanaka M, Ozawa T. (1994) *Genomics*, **22**, 327–335.

Vogel F, Motulsky AG. (1986) *Human Genetics. Problems and Approaches.* Springer Verlag, Berlin.

Wada K, Aota S, Tsuchiya R, Ishibashi F, Gojobori T, Ikemura T. (1990) *Nucleic Acid Res.*, **18** (Suppl.), 2367–2400.

Wallace MR, Andersen LB, Saulino AM, Gregory PE, Glover TW, Collins FS. (1991) *Nature*, **353**, 864–868.

Weber JL, Wong C. (1993) *Hum. Mol. Genet.*, **2**, 1123–1128.

Whitfield LS, Lovell-Badge R, Goodfellow PN. (1993) *Nature*, **364**, 713–717.

Physical mapping

Chapter 11

A wide variety of physical mapping strategies have been used to analyze the DNA of complex eukaryotic genomes. In the following sections, an arbitrary distinction is made between two classes of physical mapping:

- **low resolution physical mapping** – the smallest map unit that can be resolved is typically one to several megabases of DNA;
- **high resolution physical mapping** – the resolution is typically very high, from hundreds of kilobases to a single nucleotide.

Since mammalian DNA has only a very small percentage of coding DNA (~3% in the case of the human genome), a variety of physical mapping methods have been developed for selectively studying transcribed sequences, the most important and, arguably, most interesting component of the genome.

Low resolution physical mapping

Somatic cell hybrid panels can permit chromosomal localization of any human DNA sequence

Under certain experimental conditions, cultured cells from different species can be induced to fuse together, thereby generating **somatic cell hybrids**. In human genetic mapping, hybrid cells are typically constructed by fusing human cells and rodent (usually mouse or hamster) cells. The initial fusion products are described as **heterokaryons** because the cells contain both a human and a rodent nucleus. Eventually, heterokaryons proceed to mitosis, resulting in dissolution of the two nuclear envelopes. Thereafter, the human and rodent chromosomes are brought together in a single nucleus. The hybrid cells are unstable initially; for reasons that remain unknown, most human chromosomes fail to replicate in subsequent rounds of cell division, and are lost. This gives rise eventually to a variety of more or less stable hybrid cell lines, each with the full set of rodent chromosomes plus a few human chromosomes (*Figure 11.1*). The loss of the human chromosomes occurs essentially at random but can be controlled by selection (see *Box 11.1*).

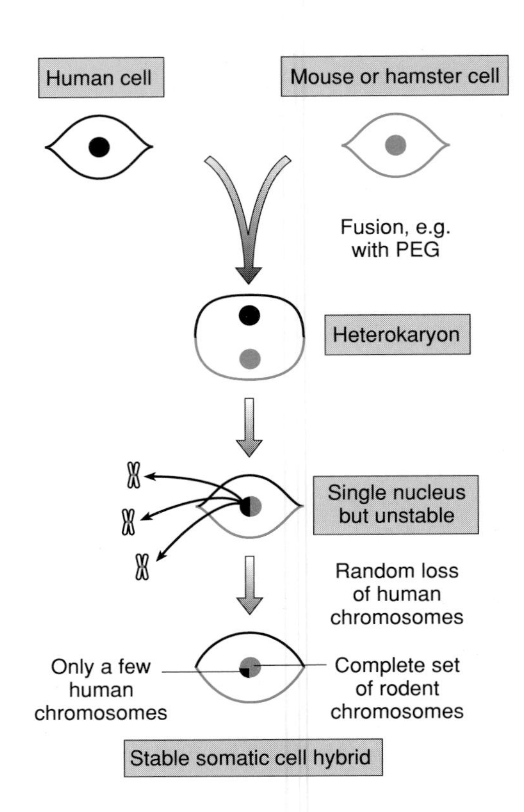

Figure 11.1: Fusion of cells from different species can result in stable somatic cell hybrids.

The example shows how stable human–rodent somatic cell hybrids can be generated following initial fusion using polyethylene glycol (PEG). For reasons that are not understood, human chromosomes are selectively lost from the initial fusion products. The loss occurs essentially at random so that eventually the stable products of a single fusion experiment will include a variety of cells with different complements of human chromosomes. They can be cloned to establish individual cell lines with a specific complement of human chromosomes. The identity of the human chromosomes can be established by PCR-based typing for chromosome-specific markers (see text).

Box 11.1: Selecting for the chromosomal content of hybirds

Hybrids can be selected for retention of a given human chromosome or chromosome fragment if it corrects an otherwise lethal abnormality in the rodent cell. Frequently used systems include:

- **HAT selection.** Somatic cell hybrids can be forced to retain human chromosome 17 using thymidine kinase-deficient (TK^-) rodent cells and growing the hybrids in HAT (hypoxanthine–aminopterin–thymidine) medium. TK^-cells are killed in HAT medium, but are rescued by the human *TK* gene on chromosome 17.
- **G418 selection.** Radiation hybrids (see below) can be selected for the presence of a particular human chromosome segment if it has been tagged by incorporation of a neomycin resistance (*neo*^{R}) gene. The neomycin analog G418 kills nonresistant cells. Neo^R is a typical example of a *dominant selectable marker*.

The human chromosomes in somatic cell hybrids can conveniently be identified by PCR screening with sets of chromosome-specific primers (Abbott and Povey, 1991). By collecting hybrid cell lines with different sets of human chromosomes, it is possible to generate a **hybrid cell panel** that can be used to map any human DNA sequence to a specific chromosome. To do this, the presence of the human sequence of interest in each of the hybrid cell lines is assayed. A PCR assay can be used (with primers specific for that sequence) or the relevant DNA sequence can be labeled and used as a hybridization probe (assuming that the correct human signal can be distinguished from any similar sequences that may be present in the rodent genome). Localization of a human DNA sequence to a single chromosome can be inferred by deduction: the chromosome must be common to all cell lines which are positive for the test and absent from all cell lines which are negative for the test, that is, all hybrids should be *concordant* (see *Table 11.1*)

Table 11.1: Mapping of a gene for microfibril-associated glycoprotein (MAGP) to human chromosome 1 using a panel of 16 somatic cell hybrids

	Human chromosome																						
MAGP/chromosome	1	2	3	4	5	6	7	8	9	10	11	12	13	14	15	16	17	18	19	20	21	22	X
Concordant hybrids																							
+/+	7	3	4	3	2	5	0	6	4	1	2	5	2	6	4	6	2	6	6	3	6	7	2
-/-	9	8	3	6	6	6	7	6	4	9	4	6	3	3	4	6	9	5	5	4	6	5	3
Discordant hybrids																							
+/-	0	3	2	2	5	1	5	1	4	6	2	2	5	1	3	1	5	1	0	4	0	0	0
-/+	0	2	7	3	3	3	2	4	3	1	6	4	6	5	6	3	1	5	5	6	4	4	2
Total discordant hybrids	0	5	9	5	8	4	7	5	7	7	8	6	11	6	9	4	6	6	5	10	4	4	2
Total informative hybrids[a]	16	16	16	14	16	15	14	17	15	17	14	17	16	15	17	16	17	17	16	17	16	16	7
Percentage discordant hybrids	0	31	56	36	50	27	50	29	47	41	57	35	69	40	53	25	35	35	31	59	25	25	29

[a]Chromosomes with rearrangements or present at a frequency of 0.1 or less were excluded.
The assignment to human chromosome 1 is indicated by complete concordance of the informative hybrids: all seven hybrids which possessed human chromosome 1 tested positive for MAGP and all nine hybrids which lacked human chromosome 1 tested negative for MAGP. There was discordance between the hybrids for all of the other human chromosomes. *Note* that the chromosome complement of individual hybrids, although generally stable, may occasionally undergo changes, and apparent discordance between hybrids may sometimes be found for a chromosome that contains the locus of interest. Reproduced from Faraco *et al.* (1995) with permission from Academic Press Inc.

Chromosomal localization of human DNA clones can be established directly and rapidly using panels of monochromosomal hybrids generated by microcell fusion

A disadvantage of traditional somatic cell hybrids is that the hybrid cells generally contain several human chromosomes rather than just a single human chromosome. In order to limit the amount of human genetic material transferred to a recipient rodent cell, the technique of **microcell fusion** (Fournier and Ruddle, 1977) can be applied. The first step is to subject the donor cells to prolonged mitotic arrest by continued exposure to an inhibitor of mitotic spindle formation, such as colcemid. As a result of this treatment, the chromosome content of a cell is partitioned into discrete subnuclear packets (*micronuclei*). The micronuclei can be physically isolated from the cells by centrifugation in the presence of cytochalasin B (a mitotic spindle inhibitor), resulting in the formation of *microcells*, particles consisting of a single micronucleus and a thin rim of cytoplasm surrounded by an intact plasma membrane. In much the same way as for normal donor cells, the microcells can be fused with recipient cells (*microcell-mediated chromosome transfer*). The resulting hybrids are known as *microcell hybrids*. Some contain a few donor chromosomes, but the simplest contain a single donor chromosome (**monochromosomal hybrids**) (see, for example, Warburton *et al.*, 1990).

The big advantage of monochromosomal hybrids is that they permit unambiguous evidence for the presence or absence of a defined human DNA on a specified chromosome. Monochromosomal human–rodent hybrid cell lines have now been established for each of the human chromosomes. This means that if the sequence of a particular human DNA is known, a PCR assay can be established and used to type a complete monochromosomal hybrid panel, thereby establishing the chromosomal location quickly. Alternatively, if no DNA sequence is available, a human fragment of interest can still be labeled and used as a hybridization probe against a Southern blot of genomic DNA samples from a monochromosomal hybrid panel.

Subchromosomal mapping is possible using hybrid cells containing human chromosome fragments

The above approaches have permitted assignments to whole chromosomes. In order to obtain subchromosomal localizations, specialized hybrids are required which contain only part of a human chromosome. The subchromosomal fragments may result from spontaneous chromosome breakage as a result of translocations or deletions, or they may be artificially induced.

Translocation and deletion hybrid panels

Somatic cell hybrids can be made by fusing rodent cells to human cells which are known to contain translocation chromosomes or chromosomes which contain a cytogenetically visible deletion, such as an interstitial deletion. The resulting hybrid cells can be screened for the presence of the abnormal chromosome. Thereafter, cell lines are made from hybrids which contain the abnormal chromosome but which *lack the normal homolog of the chromosome of interest* (**translocation hybrids** and **deletion hybrids**) (*Figure 11.2*). Clearly, for this approach to be useful in mapping a chromosome, several different natural breakpoints must be available for that chromosome, a condition which may not always be met.

Chromosome-mediated gene transfer

Because of the limited availability of spontaneous chromosome breakpoints, a more general mapping approach would involve artificially breaking human chromosomes and transferring the subchromosomal fragments into rodent cells. One of the first techniques to use this approach was **chromosome-mediated gene transfer** (**CMGT**), a procedure in which fragments of purified chromosomes are transferred into recipient cells in the presence of calcium phosphate (see Porteous, 1987). Mitotic chromosomes from a suitable cell type, such as a human fibroblast, are purified and allowed to co-precipitate with calcium phosphate on to the surface of a suitable recipient cell line in monolayer culture, such as mouse fibroblasts. Human chromosome fragments can then enter the mouse cells and integrate into the chromosomes, resulting in stable transformation. As a result, hybrids can be established which retain subchromosomal segments of human DNA (*transgenomes*) of a size that is useful for mapping (usually in the range of 1–50 Mb). However, the transgenomes are prone to frequent rearrangements, so that the use of CMGT as a mapping tool is very limited. Instead, it is more suited to functional assays of complex loci.

Irradiation fusion gene transfer

This is a procedure where chromosome fragments generated by lethal irradiation of donor cells are rescued by fusion with suitable recipient cells (see Walter and Goodfellow, 1993). In human genetics, this method has been popular for use on somatic cell hybrids containing a single human chromosome. Controlled X-ray irradiation cleaves all the chromosomes in the hybrid essentially at random into comparatively small pieces. In order to rescue fragments from the human chromosome, the lethally irradiated hybrid cell is fused with a rodent cell, and a selection system is applied to ensure the survival only of hybrid cells (*Figure 11.3*). Cells containing human sequences can then be identified conveniently by screening for the *Alu* repeat sequence (which occurs on average approximately once every 4 kb in human DNA – see page 199). Stable propagation of the human chromosome fragments in such cells is only possible if the fragments have inserted into the rodent chromosomes. Initially, therefore, there may be a comparatively large population of chromosome fragments before continued culture of the cells selects for those fragments which have stably integrated into the rodent chromosomes (breakages in

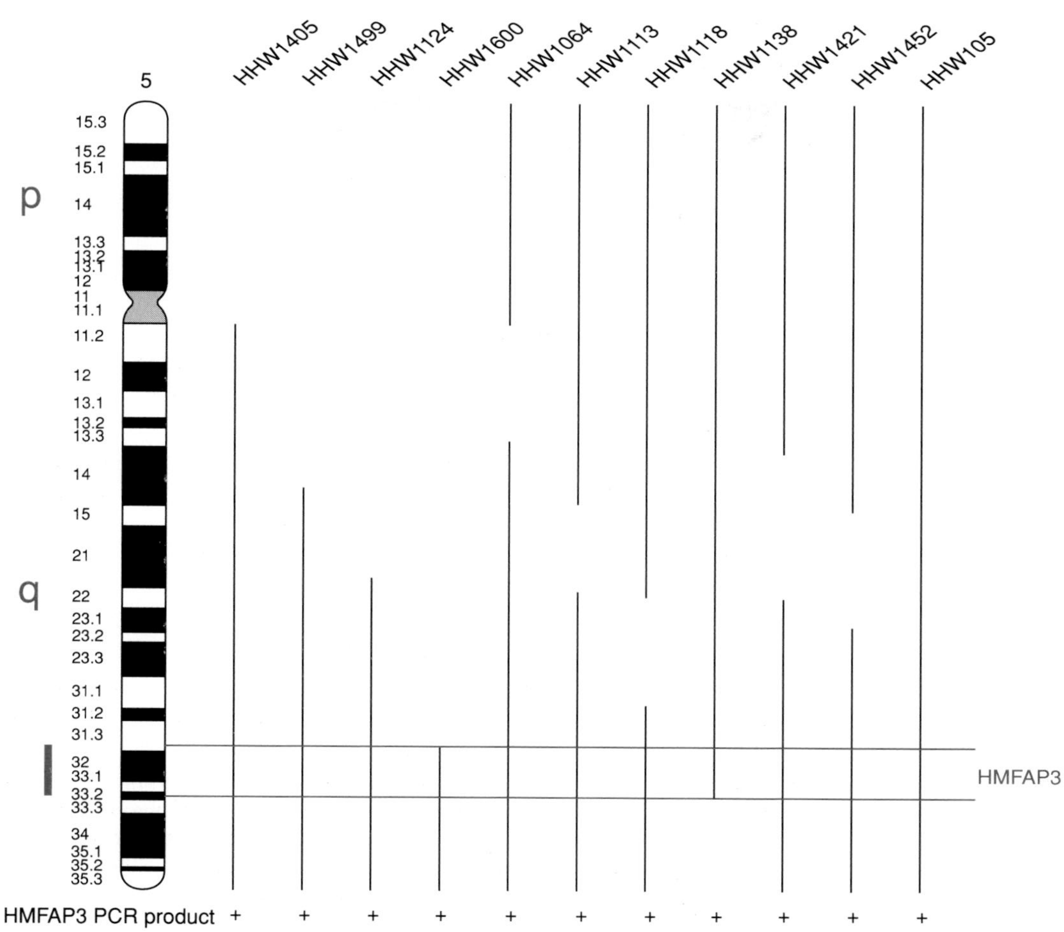

Figure 11.2: Subchromosomal localization can be achieved by mapping against a panel of hybrid cells containing translocation or deletion chromosomes.

The figure illustrates PCR-based mapping of the human microfibrillar protein MFAP3 using a panel of 5q translocation and deletion hybrids. Vertical black bars to the right indicate the extent of human chromosome 5 sequences which are retained in the hybrids. Hybrids HHW1405, 1499, 1124 and 1600 contain translocation chromosomes with 5q breakpoints and retention of the segment distal to the breakpoint. By contrast, translocation hybrid HHW1138 retains material proximal to the 5q breakpoint. Hybrids HHW1064, 1113, 1118, 1421 and 1452 have different interstitial deletions of 5q. The solid red vertical bar to the left indicates the inferred subchromosomal location as defined by breakpoints in hybrids HHW1600 and HHW1138 (red horizontal lines near bottom). Reproduced from Abrams *et al.* (1995) with permission from Academic Press Inc.

chromosomes are quite common, necessitating effective DNA repair systems). This results in a collection of *irradiation-reduced hybrids*, often called **radiation hybrids.**

When DNA samples from a panel of such radiation hybrids are screened by hybridization against a series of DNA clones, or by corresponding PCR assays, the patterns of cross-reactivity can be interpreted statistically to produce a linear map order for the DNA clones (Cox *et al.*, 1990). This is so because the nearer two DNA sequences are on a chromosome, the lower the probability of separating them by the chance occurrence of a breakpoint between them. The frequency of breakage between two markers can be defined by a value θ which is analogous to a recombination frequency in meiotic mapping, but in this case varies from 0 (the two markers are never separated) to 1.0 (the two markers are always broken apart and are therefore unlinked). Because θ underestimates the distance between markers that are far apart on the same chromosome, a more accurate estimate is provided by a mapping function, $D = -\ln(1-\theta)$, which assumes no interference and is analogous to the Haldane mapping function used in meiotic linkage analysis (see page 315). *D*

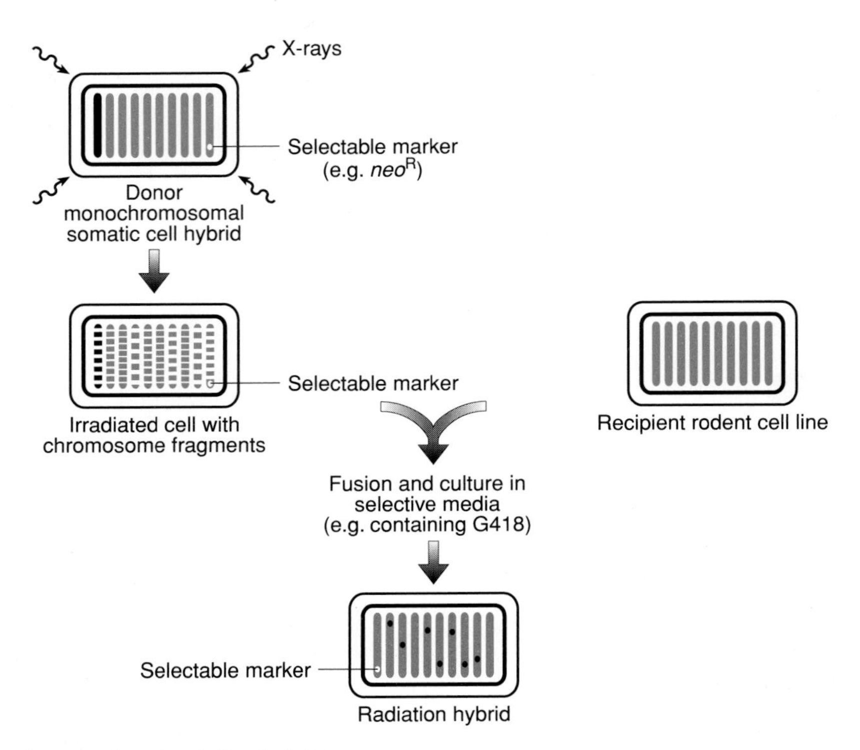

Figure 11.3: Construction of conventional radiation hybrids.

Conventional human radiation hybrids are constructed from an irradiated donor human–rodent hybrid cell line and a recipient rodent cell line. The donor cell line contains a single human chromosome (black) and the full complement of rodent chromosomes (red), one of which contains an integrated dominant selectable marker such as a neo^R neomycin resistance gene (see *Box 11.1*). Following irradiation of the donor cell line, all the chromosomes are fragmented to give several fragments per chromosome. Culture of the fusion products in a medium containing the neomycin analog G418 selects against growth of the original recipient rodent cells. The resulting radiation hybrids will contain the selectable marker plus several fragments of the human chromosome integrated randomly into rodent chromosomes.

is measured in *centiRays* (*cR*) and, because it is dependent on the dosage of radiation, it is referenced against the number of rads. For example, a distance of 1 cR_{8000} between two markers represents a 1% frequency of breakage between them after exposure to 8000 rad of X-rays (see *Figure 11.4*).

Typically about 100–200 hybrids are required to construct a map of a human chromosome. This type of approach is not practically feasible for whole genome mapping because of the huge number of hybrid cells that would be required. However, a variant form of radiation hybrid mapping involves the use of human fibroblast cells as the starting cells instead of a monochromosomal somatic cell hybrid. This has the attraction of offering construction of reasonably high resolution maps of the entire genome with a single panel of 100–200 hybrids (Walter *et al.*, 1994).

Chromosomal *in situ* hybridization has been revolutionized by fluorescence *in situ* hybridization techniques

Chromosomal *in situ* hybridization typically uses an air-dried microscope slide preparation of metaphase chromosomes, in which the chromosomal DNA has been denatured by exposure to formamide before hybridization with a suitable probe (see page 126). Isotopes such as ^{32}P are not suitable for probe labeling because their high energy causes scattering of the signal. As a result, the radioisotope of choice in traditional *in situ* hybridization experiments has been ^{3}H. However, the very weak emission has required long developing times (from weeks to months in many cases). Owing to the high background noise, the genuine signal is not easily

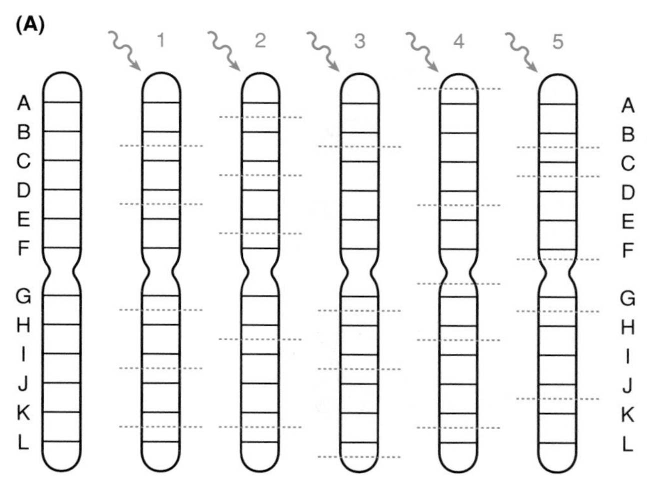

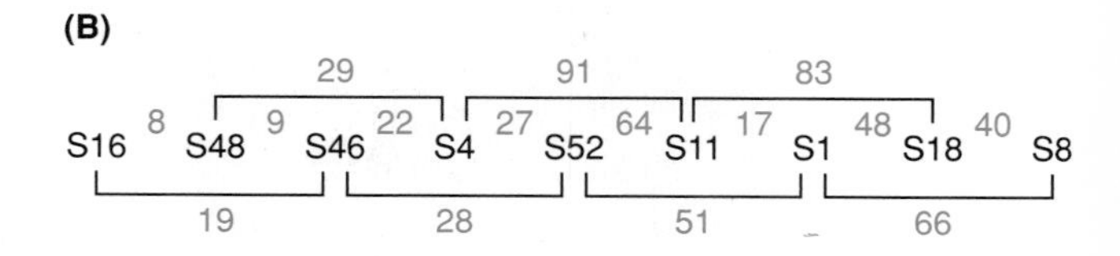

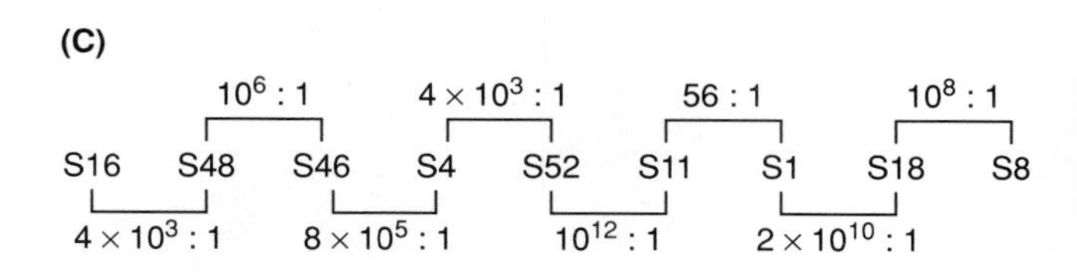

Figure 11.4: Constructing radiation hybrid maps.

(A) Breakpoints occur randomly. Five possible examples of breakpoints (dashed red lines) on the same type of chromosome are shown. Markers close together will tend to occur on the same fragment, e.g. A and B in all cases other than example 2. Thus, if a radiation hybrid contains marker A it will frequently also contain marker B, but rarely a distant marker such as L.
(B) Ordering of markers on human 21q. The order of markers *D21S16–D21S8* as inferred by Cox *et al.* (1990) from radiation hybrid mapping is shown. Figures on the top panel refer to distances between markers in centiRays$_{8000}$. For example, the S16–S48 interval is 8 cR$_{8000}$: at a radiation dose of 8000 rad, there is 8% frequency of breakage between them, and so a 92% chance they will occur together on one fragment.
(C) Odds ratios refer to the likelihood of the indicated order for pairs of markers compared with that with the markers inverted. For example, the calculated likelihood for the order S16–S48–S46–S4 is 10^6 times greater than for the order S16–S46–S48–S4.

identified and complex statistical analysis is required to differentiate genuine signal from background noise.

Standard fluorescence in situ *hybridization*

Recently, the sensitivity and resolution of *in situ* hybridization has been increased significantly by the development of **fluorescence *in situ* hybridization (FISH)** (see Trask, 1991; van Ommen *et al.*, 1995). In this technique, the DNA probe is labeled by incorporation of modified nucleotides, obtained by covalent binding of a *reporter molecule* (a molecule, such as biotin or digoxigenin which can be detected by specific binding to another molecule; see *Figure 5.6* for examples of modified nucleotides). Following hybridization and washing to remove excess probe, the chromosome preparation is incubated in a solution containing a fluorescently labeled *affinity molecule* which binds to the reporter on the hybridized probe. To increase the intensity of the hybridization signal, large DNA probes are preferred, usually cosmid clones containing around 40 kb of insert. Because such large sequences will contain many interspersed repetitive DNA sequences, it is necessary to use **chromosome *in situ* suppression hybridization** (Lichter *et al.*, 1990). Essentially, this is a form of *competition hybridization* (see *Box 5.3*): before the main hybridization, the probe is mixed with a large excess of unlabeled total genomic DNA and denatured, thereby saturating the repetitive elements in the probe, so that they no longer mask the signal generated by the unique sequences.

FISH has the advantage of providing rapid results which can be scored conveniently by eye using a fluorescence microscope. In metaphase spreads, positive signals show as double spots, corresponding to probe hybridized to both sister chromatids (*Figure 11.5*). Using sophisticated image processing equipment and reporter-binding molecules carrying different fluorophores, it is possible to map and order several DNA clones simultaneously. At present the maximum resolution of conventional FISH procedures on metaphase chromosomes is several megabases. The use

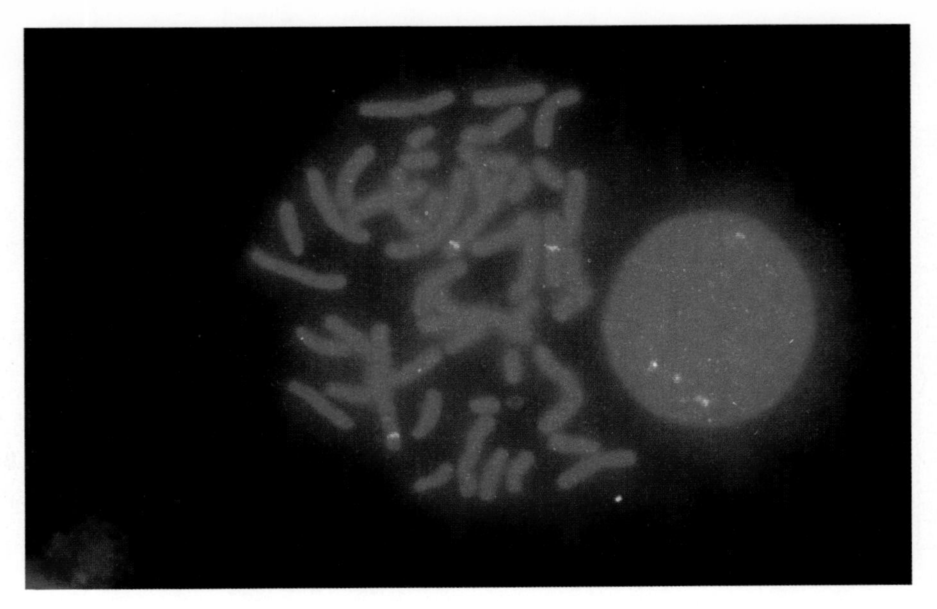

Figure 11.5: Fluorescence *in situ* hybridization.

An example of FISH using a YAC clone (CEPH 886e7) as a probe against metaphase chromosomes from a child with facial features similar to Cornelia de Lange syndrome. The YAC clone is known to map to the 3q26.3 region which is thought to contain a gene for Cornelia de Lange syndrome (MIM 122470). The FISH results show positive hybridization, as expected, to distal long arms of both chromosome 3s and also to a 3q26.3 segment that has inserted into chromosome 10. Trisomy of the 3q26.3 region produces a phenotype resembling mild Cornelia de Lange syndrome. Reproduced from Ireland *et al.* (1995) with permission from the BMJ Publishing Group.

of the more extended prometaphase chromosomes can permit 1 Mb resolution but, because of problems with chromatin folding, two differentially labeled probe signals may appear to be side-by-side, unless they are separated by distances greater than 2 Mb. Recently, however, new variations have been developed, permitting very high resolution (see page 284).

Chromosome painting

A special application of FISH has been the use of DNA probes where the starting DNA is composed of a large collection of different DNA fragments from a single type of chromosome. Such probes can be prepared by combining all human DNA inserts in a *chromosome-specific DNA library* (see page 284). The resulting hybridization signal represents the combined contributions of many loci spanning a whole chromosome and causes whole chromosomes to fluoresce (**chromosome painting**). A few differently colored fluorescence labels (*chromosome paints)* can be used in different ratios to provide numerous different colors for labeling chromosomes, thereby providing a *molecular karyotype* (Dauwerse *et al.*, 1992). The technique of chromosome painting has been extended recently by the ability to paint subchromosomal regions. This is possible using a mixed DNA probe corresponding to a particular subchromosomal region, as obtained from *chromosome microdissection DNA libraries* (see page 374).

Chromosome painting has found increasing applications in defining *de novo* rearrangements and *marker chromosomes* (see *Box 2.3*) in clinical and cancer cytogenetics (see *Figure 11.6* for an example). It is particularly helpful in cancer cytogenetics for two reasons. Chromosome preparations from tumors are often of poor quality, but information can often be obtained with the aid of chromosome painting. Additionally, complex chromosome rearrangements are particularly frequent in tumor samples and chromosome paints can be used in combination with standard probes to help recognize particular chromosome segments.

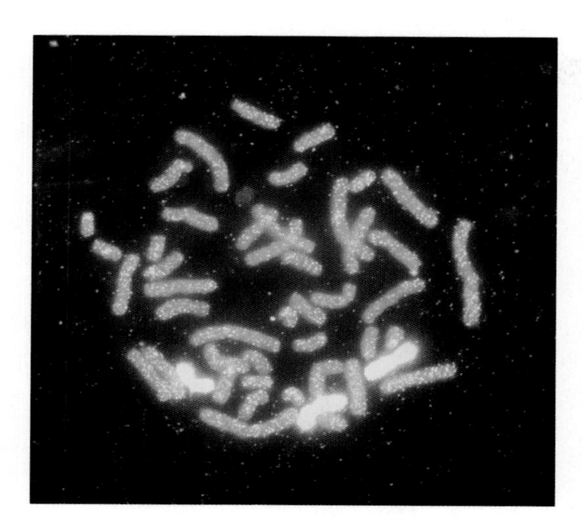

Figure 11.6: Chromosome painting can be used to define chromosome rearrangements.

The figure represents a chromosome painting analysis on a chromosome preparation from a boy with cleft lip, short stature, undescended testicles and retarded development. Conventional cytogenetic analysis revealed that 72% of the boy's blood cells contained an additional chromosome and suggested the possibility of Klinefelter's syndrome (47, XXY). The subsequent chromosome painting analysis shown here was performed with a chromosome 9 paint (i.e. a probe consisting of a mixture of insert DNAs from multiple loci spanning chromosome 9). The additional chromosome turned out to be an isochromosome for the short arm of chromosome 9. Figure kindly provided by Ian Cross, University of Newcastle upon Tyne, UK.

Flow cytometry permits sorting of individual chromosomes and the construction of chromosome-specific DNA libraries

Human chromosomes show considerable variation in size and DNA content (see *Table 7.3*). In addition, the base composition can vary considerably: chromosomes that are known to be gene-rich or gene-poor (see *Figure 7.5*) have a comparatively high % (G + C) or % (A + T) respectively. For example, chromosomes 21 and 22 are similar in size and in DNA content, but chromosome 21 is gene-poor and has a relatively high % (A + T), whereas chromosome 22 is gene-rich and has a relatively high % (G + C). As a result of the size differences and differences in base composition, human chromosomes can be separated by **flow sorting** (also called **flow cytometry**; see Bartholdi *et al.*, 1987). In this technique, chromosome preparations are stained with a DNA-binding dye which can fluoresce in a laser beam. The amount of fluorescence exhibited by a given chromosome is proportional to the amount of dye bound, which in turn is largely proportional to the amount of DNA, and hence the size of the chromosome. Additionally, certain dyes, such as chromomycin A_3, show preferential binding to GC-rich sequences, while others, such as Hoechst 33258, bind preferentially to AT-rich regions.

Because individual chromosomes can bind different amounts of fluorescent dyes, they can be fractionated in a **flow cytometer**. A stream of droplets containing stained chromosomes is passed through a finely focused laser beam at a rate of about 1000–2000 chromosomes per second, and the fluorescence of individual chromosomes present in individual droplets is monitored by a photomultiplier. The intensity of fluorescence of the different chromosomes can then be recorded. By using sort windows to define a range of fluorescence intensities corresponding to a particular type of chromosome, droplets containing the desired chromosome can be given an electric charge and deflected from the main stream on to a separate collecting grid (*Figure 11.7A*).

The resulting **flow karyogram** (see *Figure 11.7B*) may show good separation of some human chromosomes; others such as chromosomes 9–12 are not well separated. However, monochromosomal somatic cell hybrids (see page 277) can be used instead, and often result in good separation of the human chromosome. As a result, panels of chromosome-specific DNA representing each of the 24 human chromosomes have been obtained. These can be used to obtain a chromosomal localization for any human DNA fragment: the fragment is labeled and hybridized to dot-blots of DNA from the 24 chromosomes (*'flow-blots'*), or a PCR assay can be used, if

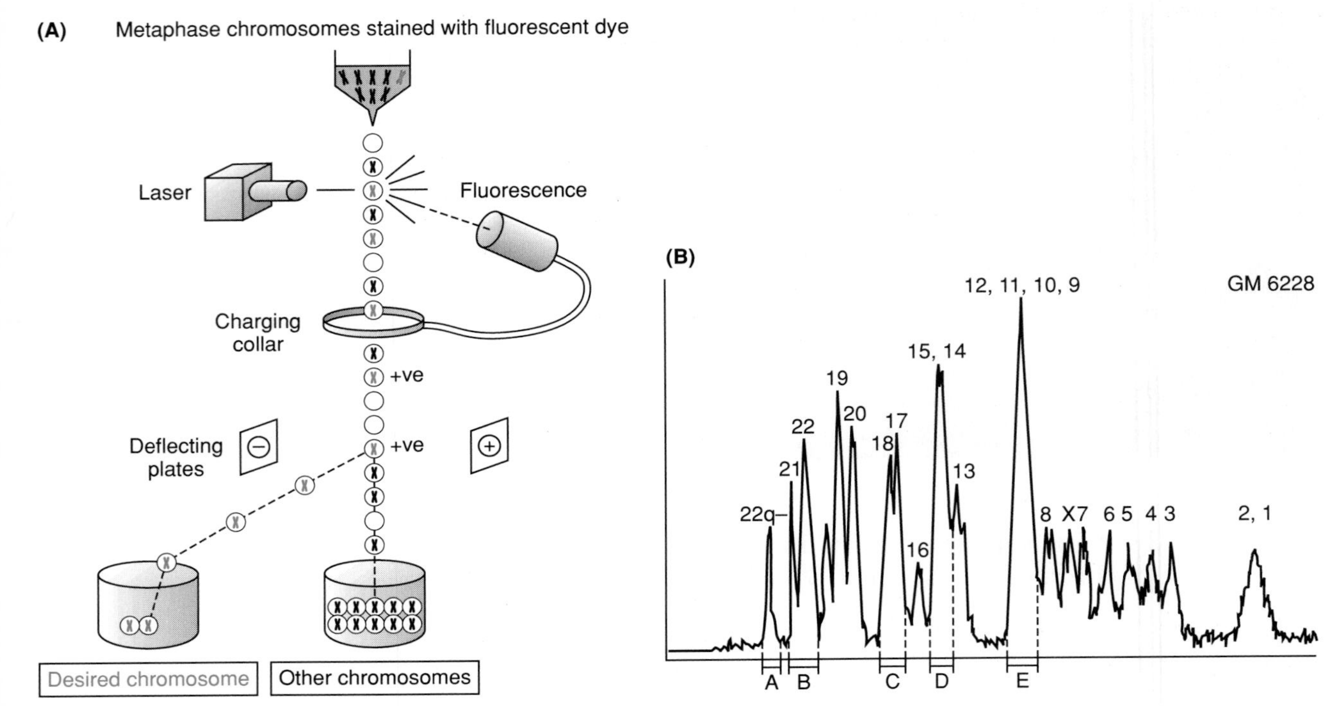

Figure 11.7: Fractionating chromosomes in a flow cytometer.

(A) Principle of chromosome fractionation. The stream of fluorescent dye-labeled chromosomes consists of very fine droplets, with each droplet containing at most one chromosome. As they pass through the laser beam they fluoresce and a photomultiplier tube records the intensity. Droplets containing a desired chromosome (as measured by the fluorescence intensity) can be given an electric charge using a charging collar, e.g. a positive (+ve) charge. After passing between electrically charged deflecting plates, the charged droplets can be deflected from the uncharged droplets, allowing sorting of a desired chromosome type (shown in red).
(B) A flow karyogram. The example is from the cell line GM6228 which has an unbalanced constitutional translocation t(11;22)(q23;q11). A–E: sorting windows used to collect specific chromosomes or sets of chromosomes. Reproduced from Cotter *et al.* (1989) with permission from Academic Press Inc.

appropriate. Additionally, purified DNA from a specific type of chromosome can be amplified by cell-based or PCR-based DNA cloning to generate **chromosome-specific DNA libraries** (Davies *et al.*, 1981). Chromosome libraries constructed using cosmid vectors have been particularly useful because YACs from a specific chromosome can be related quickly to cosmid clones from the same region. In addition, it has been possible in some cases to construct human chromosome-specific YAC libraries.

High resolution physical mapping: chromatin (or DNA) fiber FISH and restriction mapping

The physical mapping methods described above typically have a lower resolution limit of one to several megabases; they are complemented by molecular mapping methods which can map DNA in the range 1 bp to several megabases. DNA sequencing (page 303) provides the ultimate physical map by determining the linear order of single nucleotides, but mapping large DNA regions by sequencing is currently technically difficult (see below). Two major techniques provide map resolution of less than 1 Mb: restriction mapping and high resolution FISH on naturally extended chromosomes or artificially extended chromatin or DNA fibers.

Very high resolution FISH mapping can be achieved by hybridizing probes to extended chromosomes or artificially extended chromatin or DNA fibers

The linear length of DNA in an average sized human chromosome is about 5 cm, but it is compacted by various hierarchies of folding in metaphase chromosomes (see *Figure 2.3*). As a result, standard *in situ* hybridization against the highly condensed chromatin of metaphase and prometaphase chromosomes does not have a high mapping resolution, with typical upper limits of about one to several megabases of DNA. To obtain higher resolution, DNA probes are hybridized to the naturally extended interphase chromosomes or to artificially stretched chromatin or DNA fibers prepared by a variety of different methods.

In situ *hybridization using conventional interphase chromosomes*

The chromosomes of interphase nuclei are much more extended than metaphase or prometaphase chromosomes. As a result, FISH analyses of the chromosomes interphase nuclei can permit a high mapping resolution, and can help to determine the physical mapping order of some syntenic DNA clones. Because the extended chromosomes may loop back at certain regions, an accurate linear order for syntenic DNA clones requires a statistical analysis on the mapping results from many individual interphase nuclei (see *Figure 11.8*). Even so, the method is best suited to ordering sequences which are separated by intervals within the 50–500 kb range.

FISH mapping on artificially extended chromatin or DNA fibers

High resolution FISH can also be conducted using procedures which cause the DNA of chromosomes on a microscope slide to be extended prior to hybridization (**extended chromatin fiber FISH**; see Houseal and Klinger, 1994, for references). One such method, DIRVISH (direct visual hybridization), involves lysing cells with detergent at one end of a glass slide, tipping the slide, and allowing the DNA in solution to stream down the slide. Such preparations permit extremely high mapping resolutions: from over 700 kb to under 5 kb (see *Figure 11.9*). More recently, the principle of artificially stretching chromatin fibers has been extended to protein-free DNA: the target DNA is prepared from cells embedded in pulsed field gel electrophoresis blocks (see *Figure 11.11*). This **DNA FIber-FISH** method used unfixed linearized DNA fibers on a microscope slide and has a resolution of from 500 kb to a few kilobases (see Heiskanen *et al.*, 1995).

Long-range restriction mapping requires enzymes that cut DNA infrequently and fractionation of large restriction fragments by pulsed-field gel electrophoresis

Restriction mapping permits molecular mapping with resolutions which depend on the frequency of the recognition site. Most restriction enzymes which recognize a 4- or 6-bp sequence typically cut vertebrate DNA once every few hundred or few thousand base pairs. The recognition sequences for *rare-cutter* restriction nucleases are typically 6–8 bp long and contain one or more CpG dinucleotides which are rare in vertebrate DNA (see page 15). As a result, they generate fragments that are typically several hundred kilobases in size (see *Table 4.1* for some examples). Small restriction fragments can be size fractionated conveniently by agarose gel electrophoresis. However, the ability of conventional agarose gel electrophoresis to fractionate large DNA fragments is very limited. The method relies on a *sieving* effect: DNA molecules pass through pores in the agarose gel and small molecules are able to migrate more quickly through the uniform size pores (see *Figure 11.10A*). Above a certain size of DNA fragment, however, the sieving effect is no longer effective and the resolution of DNA fragments above 40 kb is extremely limited.

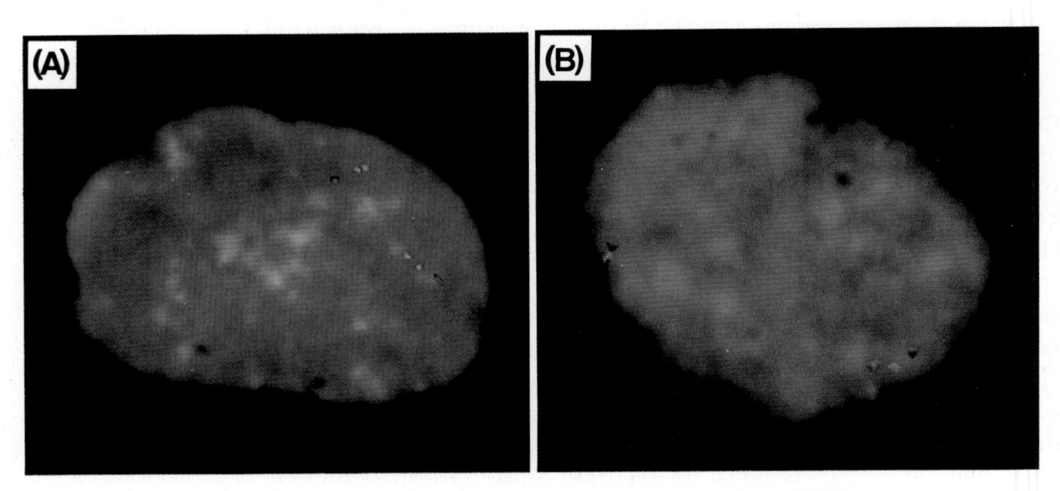

Figure 11.8: Determining the map order of syntenic DNA clones by three-color interphase FISH.

The examples illustrate YAC clones from 3p14 as follows.
(**A**) YACs 74B2 (red, R), 468B10 (orange, O) and 403B2 (green, G).
(**B**) YACs D20F4 (red, R), 168A8 (orange, O) and 258B7 (green, G).
In example A, the orange color appears between green and red on both chromosomes 3. The order 74B2–468B10–403B2 is presumed to be correct because the order ROG was observed on 33/37 occasions. In example B the observed RGO order was seen in 30/34 occasions, consistent with a correct order of D20F4–258B7–168A8. Alternative map orders are thought to be due to fold-back of the extended chromosomes. Reproduced from Wilke *et al.* (1994) with permission from Academic Press Inc.

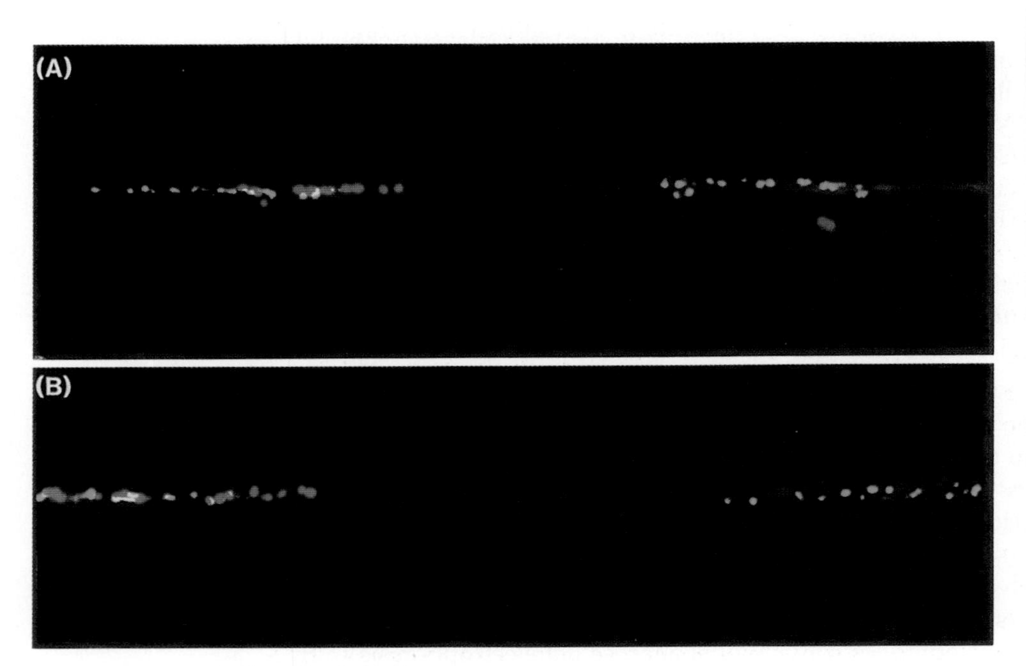

Figure 11.9: Extended chromatin fiber (ECF) FISH.

The example illustrates two-color FISH on artificially extended chromatin fibers. Signals are generated from three YACs which map to the 5q34–q35 region. YACs 786F6 and 935F5 known to map to opposite sides of a tumor translocation breakpoint were visualized using a green FITC (fluorescein isothiocyanate) labeling system. YAC 746B2, which maps to the same side of the tumor breakpoint as 786F6, was visualized using a red (rhodamine) labeling system. *Note* the partial overlap between 786F6 and 746B2, and the gap between them and 953F5, suggesting the order: 746B2/786F6–953F5. Reproduced from Haaf and Ward (1994) by permission of Oxford University Press.

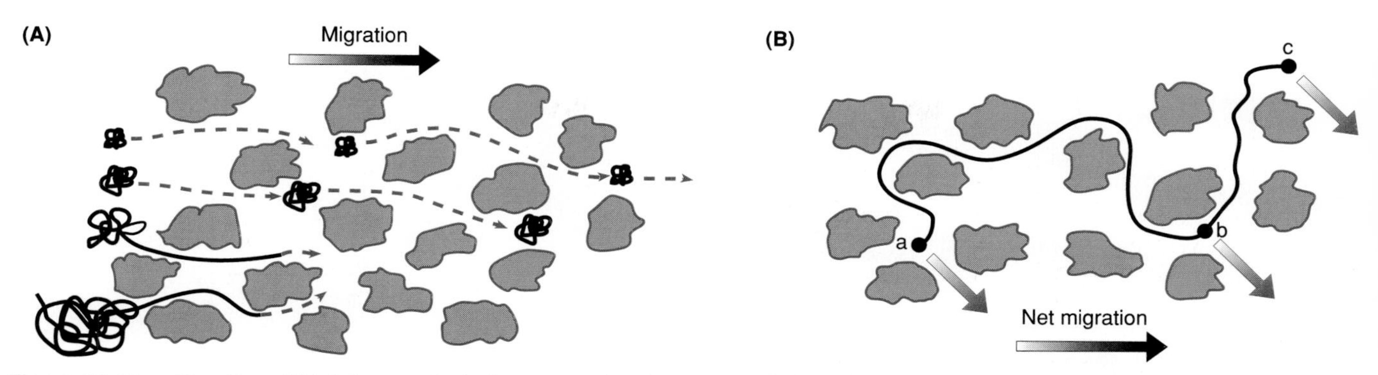

Figure 11.10: Migration of DNA fragments during conventional agarose gel electrophoresis and pulsed-field gel electrophoresis.

(A) Conventional agarose gel electrophoresis. Small DNA fragments can pass through the pores of the gel by a sieving effect: very small fragments migrate more rapidly than larger fragments. Above a certain size threshold, DNA fragments are so large that compact forms are too large to pass through the pores: the DNA fragment needs to adopt an extended conformation with a leading end migrating into the gel, and separation according to size is poor.
(B) Pulsed-field gel electrophoresis. Extended DNA fragments need to adopt a new conformation in response to an altered electric field. In the example, alternating electric fields are imagined to operate in two directions (top left → bottom right; bottom left → top right), resulting in net migration from left to right (see *Figure 1.11*). In response to the new electric field shown by red arrows, the DNA strand needs to re-orientate: the end marked c, which was previously the leading end (when the electric field was in the bottom left → top right orientation), may become the leading end again, possibly after initial migration of a loop at position b. The time taken to re-orientate and present the best conformation for advancing in the new direction is strictly size dependent, resulting in good separation of very large fragments.

Large-scale restriction mapping of human DNA can be done directly on the large inserts of YACs or by means of Southern blot hybridization of genomic DNA. Conventionally prepared genomic DNA is not a suitable target for large-scale restriction mapping, because the procedures involved in lysing the cells and purifying the DNA result in shear forces causing considerable fragmentation of the DNA. Instead, the DNA is isolated in such a way as to minimize artificial breakage of the large molecules prior to digesting with restriction endonucleases. To prepare high molecular weight DNA, samples of cells, for example white blood cells, are mixed with molten agarose and then transferred into wells in a block-former and allowed to cool. As a result, the cells become entrapped in solid agarose blocks (*Figure 11.11*). The agarose blocks are removed and incubated with hydrolytic enzymes which diffuse through the small pores in the agarose and digest cellular components, but leave the high molecular weight chromosomal DNA virtually intact. Individual blocks containing purified high molecular weight DNA can then be incubated in a buffer containing a rare-cutter restriction endonuclease.

In order to separate large restriction fragments, **pulsed-field gel electrophoresis (PFGE)**, a modified form of agarose gel electrophoresis, is used. Agarose blocks containing the large DNA fragments to be separated are placed in wells at one end of an agarose gel and the DNA migrates in the electric field. However, during a PFGE run, the relative orientation of the gel and the electric field is periodically altered, typically by setting a switch to deliver brief pulses of power, alternatively activating two differently oriented fields (*Figure 11.11*). Variants of the technique use a single electric field but with periodic reversals of the polarity (*field inversion gel electrophoresis*), or periodic rotation of the gel or electrodes. In common to each variation is the principle of a discontinuous electric field so that the DNA molecules are intermittently forced to change their conformation and direction of migration during their passage through the gel. The time taken for a DNA molecule to alter its conformation and re-orient itself in the direction of the new electric field is strictly size dependent; as a result, DNA fragments up to several megabases in size can be fractionated efficiently (see *Figure 11.10B*).

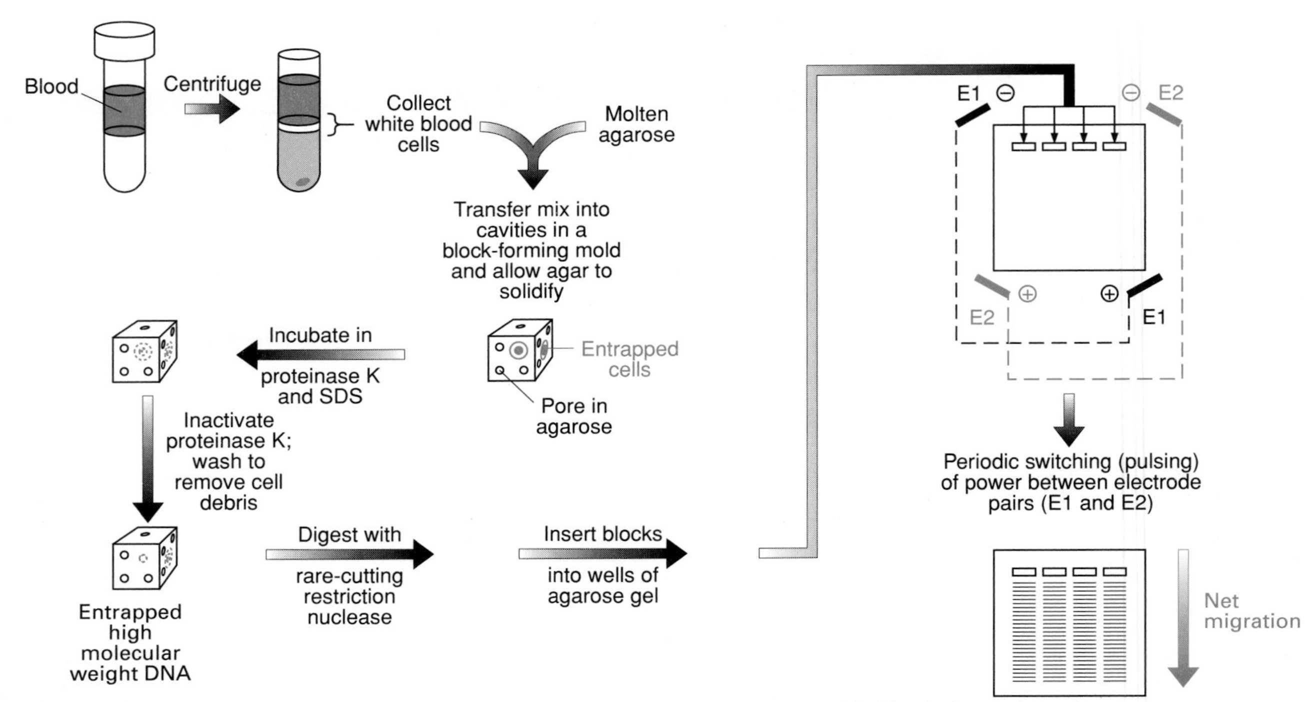

Figure 11.11: Fractionation of high molecular weight DNA from blood cells by pulsed-field gel electrophoresis.

Assembly of clone contigs

A primary goal of physical mapping is to assemble comprehensive series of DNA clones with overlapping inserts (clone contigs)

The construction of the ultimate physical map (the complete nucleotide sequence) requires considerable time and effort in the case of a very large DNA molecule such as that found in a chromosome. In order to provide a framework for this to be done efficiently, a series of cloned DNA fragments need to be assembled which collectively provide full representation of the sequence of interest. To ensure that there is complete representation, and no gaps, the series of clones should contain overlapping inserts forming a comprehensive **clone contig** (*Figure 11.12A*). In principle, contig assembly is facilitated by the way in which genomic DNA libraries are constructed: as part of the strategy for maximizing the representation of a library, the genomic DNA is deliberately subjected to *partial digestion* with a restriction endonuclease (see page 96). As a result, individual genomic DNA clones usually contain DNA sequences that partially overlap with that found in at least some other clones in the library (see *Figure 11.12B*). The cloning step means that the individual DNA fragments are sorted into different cells and so the original positional information of the fragments (how they were related to each other on the original chromosomes) is lost. However, such information can be retrieved by a variety of methods which can identify clones with overlapping inserts.

Chromosome walking means establishing clone contigs from fixed starting points

One widely used technique for identifying clones with overlapping inserts is to use a specific DNA probe from one clone to screen a DNA library. The positively

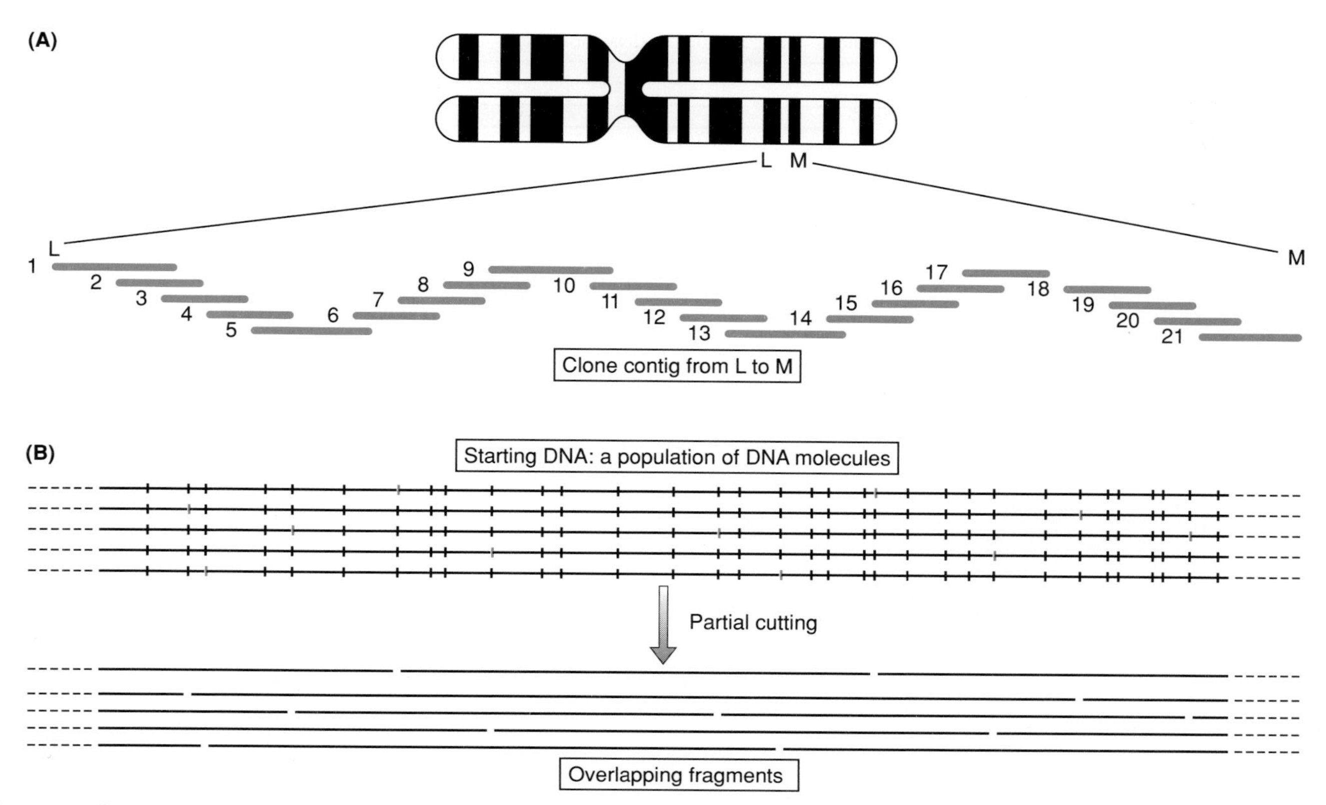

Figure 11.12: A clone contig consists of a linear series of DNA clones with overlapping inserts which originated by partial cutting of genomic DNA during DNA library construction.

(A) Clone contig. The contig consists of 21 clones arranged as a series of clones with partially overlapping inserts. All the genomic DNA between chromosomal sites L and M is represented in the contig.
(B) Generating overlapping DNA fragments by partial restriction cutting. During the construction of a genomic DNA library, the genomic DNA is *partially digested* with a restriction endonuclease. Of the available restriction sites in the chromosomal DNA (vertical bars in top panel), the chosen conditions will ensure that only a very small minority (red vertical bars) will be cleaved. Because the choice of which site is cut is essentially random, and because the starting DNA will contain numerous identical sequences, a series of overlapping fragments will be produced.

hybridizing clones should contain a DNA sequence that is closely related to the probe, including clones which contain sequences which partly overlap that found in the probe. This has often involved the preparation of a so-called *end probe* from the starting DNA clone: a fragment located at one end of the insert DNA and preferably present as single copy DNA, is purified and labeled. Positively hybridizing clones can then be purified and new, distal end probes can be prepared for further rounds of hybridization screening of the DNA library. In the case of genomic DNA libraries, this permits the assembly of a clone contig by bidirectional **chromosome walking** from a fixed starting point (*Figure 11.13*). In the case of mammalian genomes, chromosome walking has frequently involved screening cosmid libraries (*cosmid walking*). Recently, cosmid walking has been simplified enormously by avoiding the need to prepare end probes. Instead, the entire insert can be labeled and used as a probe under conditions which suppress the hybridization signal from DNA sequences in the probe which are highly repetitive in the human genome (see *Box 5.3*). Nevertheless, the signal from some other repetitive sequences may be difficult to suppress, causing occasional difficulties with this approach.

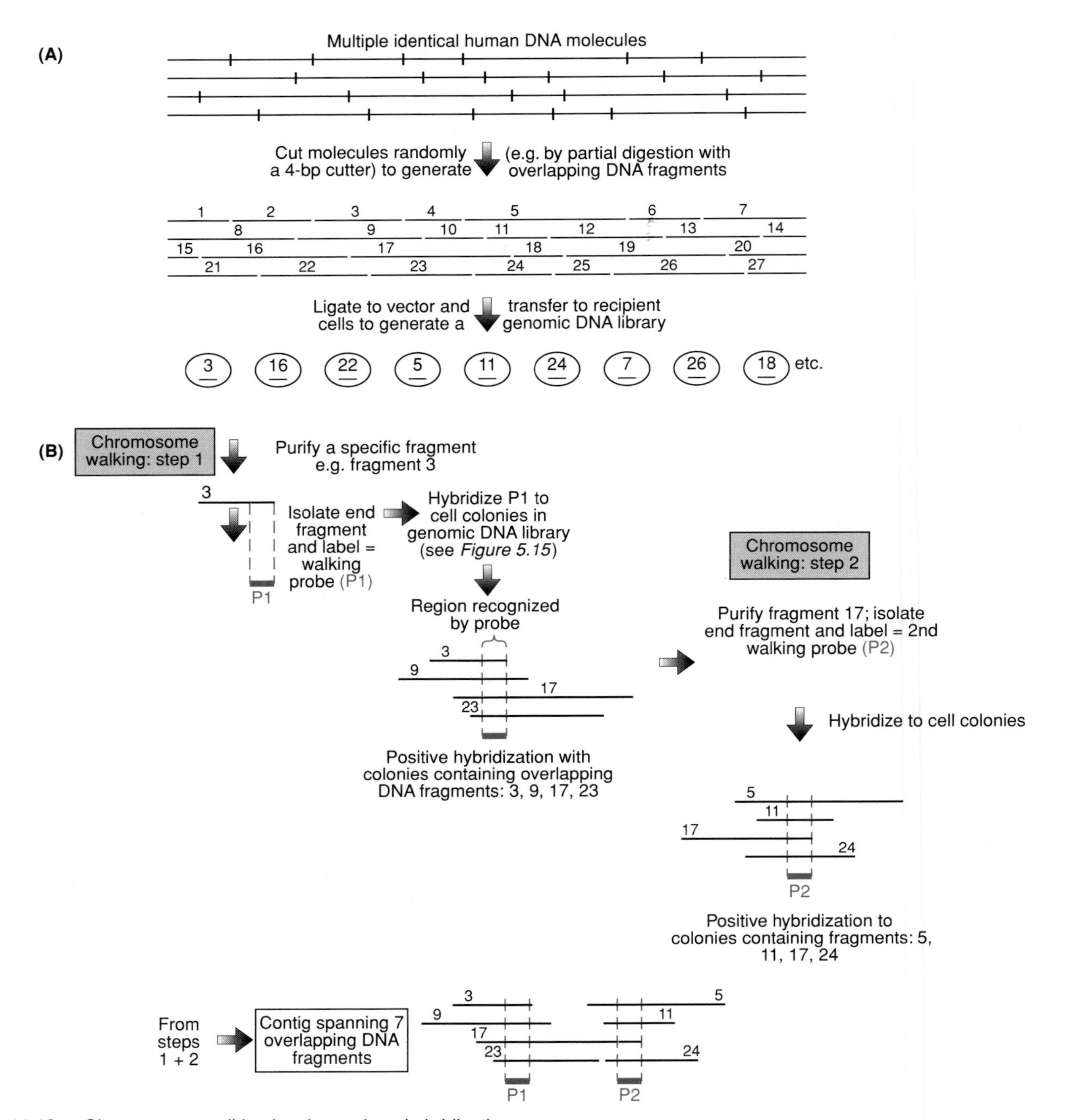

Figure 11.13: Chromosome walking by clone–clone hybridization.

The starting DNA is a population of DNA molecules from many cells. Random cutting by partial restriction endonuclease digestion of identical DNA molecules generates a series of overlapping fragments. For example, fragment 3 from the first molecule shows partial overlaps with fragments 9, 16, 17, 22 and 23 from the other three molecules. After cell-based cloning to make a library, the different fragments are sorted into different host cells. In this example, the starting point is cloned fragment 3 and a monodirectional chromosome walk (towards the right) is shown by using end fragments as hybridization probes for screening a genomic DNA library.
A bidirectional walk starting from fragment 3 would involve isolating a second end fragment from the other end of the insert. In the example shown this would be expected to hybridize to clones containing fragments 9, 16 and 22, and subsequent isolation of a second walking probe from, say, the distal end of fragment 22.
Note that, for cosmid clones, it is no longer necessary to isolate end probes: walking can be done by using whole cosmids as hybridization probes under conditions that suppress signals from repetitive DNA elements (whole clone–clone hybridization).

Chromosome walking with YACs often involves PCR-based library screening

Chromosome walking by whole clone–clone hybridization is not practically feasible with mammalian YACs: the large amount of repetitive DNA in the inserts means that blocking of the repetitive DNA signal during hybridization (see page 119) is technically difficult. Instead, techniques are used to recover short *end fragments* from individual YACs. The YAC DNA is cleaved with a restriction enzyme which is known to cleave the YAC vector sequence and which cuts frequently in human genomic DNA. Among the cleavage products there will be end fragments containing both the unknown terminal sequence from the insert DNA and the adjacent known vector sequence. Such sequences can then be amplified using various PCR-based *genomic walking methods*: a primer for a characterized sequence (in this case the sequence of the YAC vector adjacent to the cloning site) is used to permit access to an adjacent uncharacterized sequence (in this case the sequences at the ends of the insert). Frequently used methods include *inverse-PCR* and *bubble-linker PCR* (see pages 140–142 for the general principles, and Riley *et al.*, 1990, for one practical example).

YAC insert end fragments can be used as hybridization probes to screen colony filters from a YAC library. A widely used alternative, however, is to sequence an end fragment and then design oligonucleotide primers to permit a specific PCR assay for this sequence. The resulting PCR assay can be used to screen YAC libraries which have been distributed as hierarchical sets of agarose plugs each containing pools of YAC clones (see *Figure 11.14*). Positive results will identify YACs with overlapping inserts from which new end fragments can be rescued to continue the YAC walk. As, however, YAC clones often have *chimeric* inserts (composed of two or more fragments that are derived from noncontiguous portions of the genome), the chromosomal location of any new YACs identified in a YAC walk ought to be verified by FISH analyses.

Clone contigs can be assembled rapidly over large fractions of a genome by random clone fingerprinting

Chromosome walking is a highly directional and location-restricted procedure for generating clone contigs: a great deal of effort is invested in generating a contig from a fixed starting point on a specific chromosome, often as a prelude to *positional cloning* (see pages 372–387). In order to build clone contigs spanning large amounts of a chromosome, or even a whole genome, a more general, labor-efficient mapping approach is needed. To meet this need, a variety of **clone fingerprinting** techniques have been used to type clones at random and then integrate the information in order to identify clone contigs over large regions of a genome. For example, this approach has been used to great effect in assembling YAC contigs over significant amounts of the human genome (see page 343). Such clone fingerprinting techniques should be simple and rapid. As a result, DNA sequencing and restriction mapping are not techniques of choice: they are comparatively laborious. Instead, a variety of rapid clone fingerprinting techniques are preferred.

Repetitive DNA fingerprinting

Mammalian DNA has a very high component of highly repeated, interspersed DNA sequences such as the human *Alu* and LINE-1 (*Kpn*) repeats (occurring on average about once every 4 kb and 50 kb respectively – see pages 199–202). Because the spacing between such interspersed repetitive DNA elements will vary in different genomic regions, clones which contain overlapping DNA sequences may be identified by similar patterns following hybridization of repetitive DNA probes to

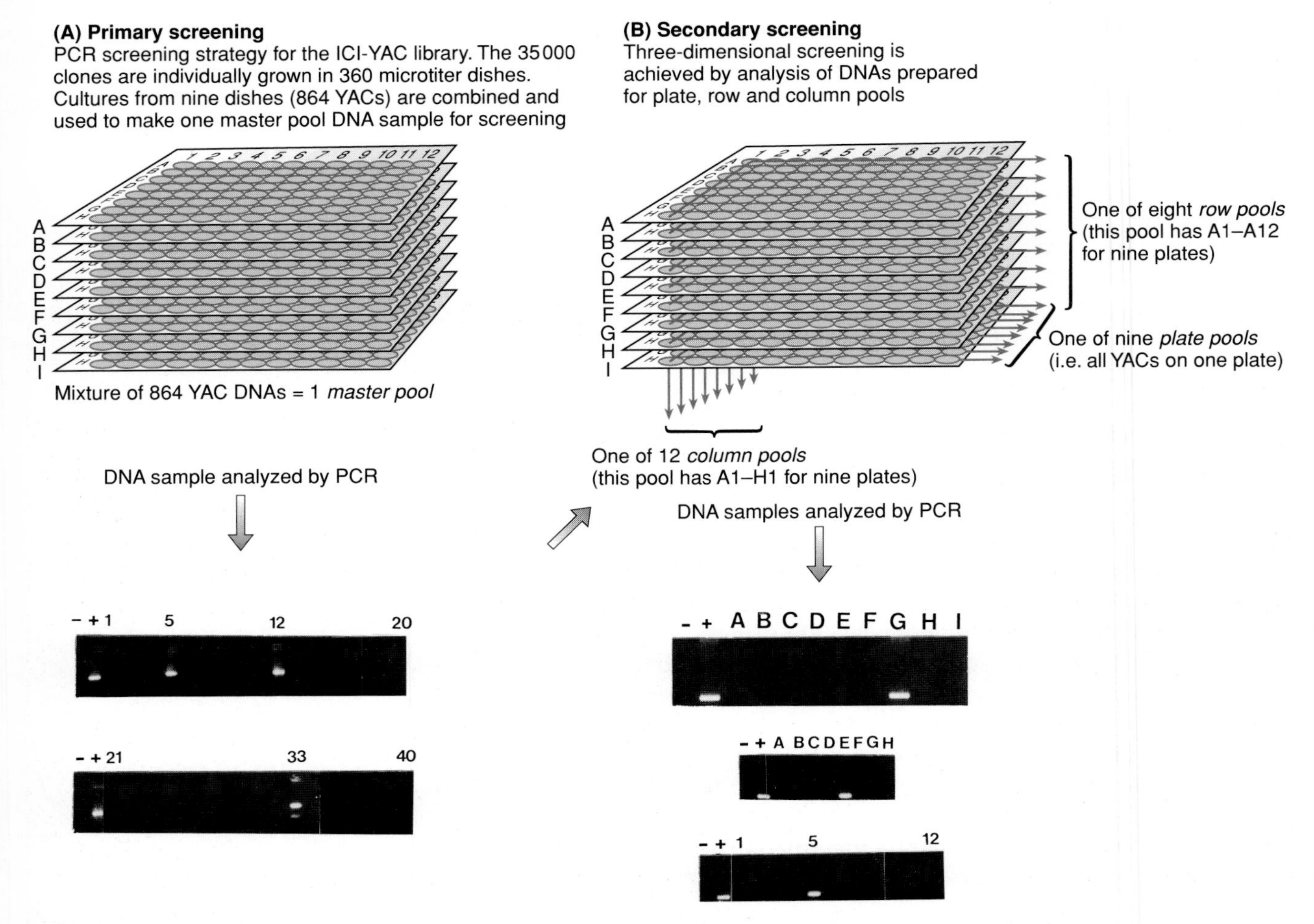

Figure 11.14: Screening of YAC libraries by PCR.

A YAC library can be screened for the presence of clones containing a specific sequence, provided there is a specific PCR assay for that sequence. Amongst other applications, this provides a convenient method for chromosome walking using YACs: a PCR assay for a sequence at the end of one YAC can be used to identify other YACs with overlapping sequences. The example illustrates screening of the human ICI YAC library generated by Anand *et al.* (1990). Approximately 35 000 individual clones were individually deposited into the 96 wells of 360 microtiter dishes. To facilitate screening, a total of 40 *master pools* were generated by combining all 864 clones in sets of 9 microtiter dishes (plates A–I). Modified from Jones *et al.* (1994) with permission from Academic Press Inc.
(A) *Primary screening* involves PCR assay of the 40 master pools. In this example, three master pools were positive when referenced against positive (+) and negative (–) controls: pools 5, 12, 23.
(B) *Secondary screening* identifies single YACs by assaying different subsets of the 864 YACs in a positive master pool, in this case master pool 12. Three-dimensional screening of each of nine *plate pools* (of 96 YACs each), eight *row pools* (of 106 YACs each) 12 *column pools* (of 72 YACs each) and identified a positive YAC in plate 12G (top panel), row E (middle panel), column 5 (bottom panel). The example here involved screening for YACs containing an anonymous X chromosome sequence. Photos were kindly provided by Dr Sandie Herrell, University of Newcastle upon Tyne.

Southern blots of the clone DNA digested with a suitable restriction endonuclease. For example, a whole genome clone fingerprinting approach has been applied to mapping the human genome largely on the basis of repetitive DNA fingerprinting of YACs (Bellanne-Chantelot *et al.*, 1992).

Sequence-tagged site content mapping

If a small amount of DNA sequence (e.g. a few hundred nucleotides) is available for a specific DNA clone, a PCR assay can be developed which is specific for that sequence. The DNA sequence is examined to identify short regions from which oligonucleotide primers can be designed for use in a standard PCR reaction. The site on the original genomic DNA from which the sequence is derived can then be thought of as being *tagged* by the ability to assay for that sequence. A site for which a specific PCR assay is available is therefore described as a **sequence-tagged site (STS)** (see *Box 11.2*).

If numerous individual STS sequences have been positioned on a chromosome or subchromosomal area, and YACs have been mapped to the same general area (e.g. by FISH), the ability to type rapidly for STSs allows a rapid method for identifying YACs with overlapping sequences (**STS content mapping** – see *Figure 11.15*).

Box 11.2: The importance of sequence-tagged sites (STSs)

Sequence-tagged sites are important mapping tools simply because the presence of that sequence can be assayed very conveniently by PCR. Most STSs are nonpolymorphic and can be assigned to a specific chromosomal location by typing a panel of somatic cell hybrids, and to subchromosomal localizations by typing panels of suitable hybrids (pages 278–280). Some STSs may be known to be derived from larger clones (e.g. gene clones) that have been physically mapped to a particular subchromosomal localization. In addition, microsatellite markers are a useful subset of STSs which are highly polymorphic and can be assigned a subchromosomal localization as a result of linkage analyses. STSs therefore represent a critical link between genetic and physical maps: both genetic and physical markers can be translated into the 'common language' of the STS. An example of how an STS is developed from a DNA sequence is shown below.

Rough sequence: 200 nucleotides.

Primers: chosen from underlined sequences, both 16 nucleotides long.

A (forward primer in bold, identical to the sense sequence from +50 to +66),

B (reverse primer, in bold, corresponding to antisense strand for the sequence from +175 to +190).

Note that the sequences from +1 to +50 and from +191 to +200 are extremely GC-rich with multiple runs of G and of C, which are not optimal sequences for designing PCR primers. Hence the decision to use internal primers.

STS: 141 nucleotides defined by primer ends: +50 to +190

```
   1                                                  40
5′ CCCAGCGGGC CCGCGGCGCA GGGGCCCGGC GGGGCCCTGG
   41               5′— Primer A → 3′                 80
   GGCCGCCCGG CAGTGAGCAT CAGATACAGA ACCTAGACGA
   80                                                120
   ACCTAGGACC AGTACCTACA AGGTACTCTA GATGATCTAT
   121                                               160
   ACTGAGGATC CTATTCAGAT CCTAGGTACC ACACTGATTA
   161                                               200
   AGGATACTAG CTATACGGAC ATGGCATTAC ACCCCCGGGG 3′
                   |||||| ||||||||||
              ←3′ TGCCTG TACCGTAATG 5′
                      Primer B
```

In a complex genome, such as the human genome, the chances of a 16 nucleotide long primer binding by chance to a related but different sequence other than the intended target is not insignificant. However, the chance of both primers binding to unintended related sequences which just happen to be *both in close proximity and also in a suitable orientation* is normally very low. The specificity of the reaction can be assayed simply by size-fractionating the amplification products on an agarose gel. If there is a single strong PCR product of the expected approximate size (141 bp in the above example), there is an excellent chance that the assay is specific for the intended target sequence.

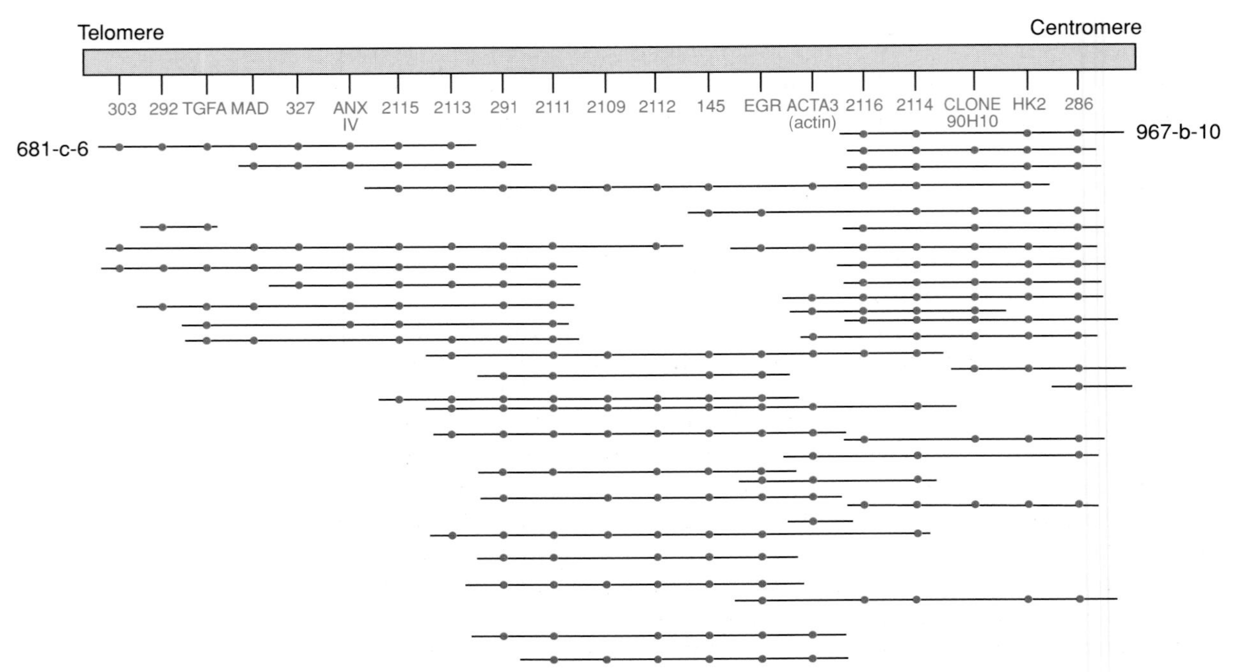

Figure 11.15: Assembling YAC contigs by STS content mapping.

The YAC contig shown was assembled for a region of human 2p12 which contains the limb girdle muscular dystrophy locus *LGMD2B*. Vertical lines indicate the positions of indicated STSs. STSs indicated by simple numbers represent microsatellite markers; others are known genes or an expressed sequence (clone 90H10). Horizontal lines indicate individual YACs, with filled circles denoting the presence of specific STSs. For the sake of clarity, YACs other than the most distal YAC (681-c-6) and the most proximal YAC (967-b-10) are not labeled. Data provided by Dr Rumaisa Bashir, University of Newcastle upon Tyne.

Hybridization using IRE-PCR products

Interspersed repeat element (IRE)-PCR is an exceptional PCR reaction in which only a single primer is used, corresponding to a common repetitive element in a genome of interest. It relies on the fact that, if the repetitive element is very common, two such elements will often be found in close proximity and in opposite orientations. On such occasions, a single primer can bind to the two elements and permit amplification of the sequence between them (see *Figure 6.12*). The *Alu* repeat is one such element, occurring on average about once every 4 kb in the human genome. **Inter *Alu*-PCR** (often described as ***Alu*-PCR**) can generate a variety of fragments from most human YACs. Individual or pooled *Alu*-PCR products which consist of single-copy DNA (except for the incorporated primers) can then be assembled from a YAC of interest and used as hybridization probes for screening other YACs or collections of inter *Alu*-PCR products derived from them.

Clone gridding and increasing automation have facilitated large-scale physical mapping efforts

Using current technologies, physical mapping of complex genomes is arduous and continues to be hampered by various practical difficulties:

- **chimeric DNA clones**. Cloning artifacts can mean that cells contain two non-contiguous pieces of DNA from the desired genome. This may occur as a result of *co-ligation* of two different restriction fragments (*Figure 4.2A*), but often results from *co-transformation* (*Figure 4.2B*) in clones with large inserts, and YAC clones are often chimeric.
- **Repetitive DNA**. Arranging clones in a physical order means relying on some clone-specific features. DNA sequences that are repeated can make the task of ordering clones a difficult one, and may cause problems in sequencing DNA.

- **Unclonable sequences.** Some DNA sequences, including classes of repetitive DNA that are not well represented in bacterial and yeast cells, may not easily be propagated in such cells. A variety of different genomic libraries may need to be screened (including those constructed with vectors containing a low copy number origin of replication such as BAC and P1 libraries; see pages 101–103).

A major requirement for efficient physical mapping of complex genomes is the need for **gridding** of DNA libraries. Following transformation of the host cells, individual bacterial or yeast colonies are picked into individual wells of microtiter dishes and stored in numerous multiwell dishes. As a result, each cell clone can be assigned an identifying grid co-ordinate, comprising dish number, row number and column number. When libraries need to be screened by hybridization with a labeled DNA probe, it is again most efficient to have previously prepared *high density gridded colony filters* in which the colonies have been spotted on to the appropriate membrane in a gridded fashion with a uniform number of rows and columns. This simplifies identification of hybridization-positive colonies. More importantly, it facilitates integration of the results from numerous independent research groups. This is possible because of increasing automation. Robots have been designed to pick and spot colonies, and numerous copies of **reference libraries** can be distributed, either as cell cultures or, more conveniently, as sets of colony filters and sent to researchers throughout the world. Hybridization data can then be sent back from individual laboratories to the distributing centers and the fixed geometry of the library co-ordinates simplifies the task of integrating the information received.

The Human Genome Project (see Chapter 13) is one of several projects to benefit from such technologies, and sets of hybridization filters containing gridded colonies for total genome YAC libraries, chromosome-specific cosmid libraries and other libraries have been distributed extensively. In addition, pools of yeast colonies have been distributed in agarose plugs to permit PCR typing for a sequence of interest. Again, this is done in a hierarchical co-ordinate manner to simplify the screening of libraries (see *Figure 11.14*).

Constructing transcript maps and identifying genes in cloned DNA

A wide variety of different methods can be used to identify genes in cloned DNA

A primary goal for physical mapping is identifying the locations of genes within a clone contig that has been localized to a specific chromosomal region. In principle, two major features permit the DNA of genes to be distinguished from DNA that does not have a coding function.

- **Expression**: all active genes are capable of making an RNA product, which, in the vast majority of cases, is mRNA. Mammalian genes usually contain exons and so the initial RNA transcript usually needs to undergo splicing.
- **Sequence conservation**: because genes execute important cellular functions, mutations which alter the sequence of the product will often be disadvantageous and are rapidly eliminated by natural selection (see *Box 10.1*). The sequence of coding DNA and important regulatory sequences is therefore more strongly conserved in evolution than that of noncoding DNA. In addition, premature termination codons in coding DNA are selected against. This means that genes often contain comparatively long **open reading frames**

(**ORFs**); in noncoding DNA the DNA triplets corresponding to termination codons are not selected against and ORFs are usually comparatively short. Some exons, however, may be quite small but can often be detected using computer programs to analyze the relevant DNA sequence (see page 303).

In addition, vertebrate genes are often associated with **CpG islands** (page 15; see Cross and Bird, 1995). These features have permitted a variety of different methods for identifying genes in cloned vertebrate DNA (Collins, 1991), of which the most commonly used are described below (see *Box 11.3*).

Box 11.3: Commonly used methods for identifying genes in cloned DNA

Method	Comments
Zoo-blotting	A DNA clone is hybridized at reduced hybridization stringency against a Southern blot of genomic DNA samples from a variety of animal species, a *zoo-blot*. Depends on coding DNA being more strongly conserved during evolution than noncoding DNA (see page 297 and *Figure 11.16*).
CpG island identification	Many vertebrate genes have associated *CpG islands*, hypomethylated GC-rich sequences usually having multiple rare-cutter restriction sites (see Cross and Bird, 1995). • *Identification by restriction mapping*. DNA clones are usually hybridized against Southern blots of genomic DNA cut with *Sac*II, *Eag*I or *Bss*HII to identify clustering of rare-cutter sites (see page 297 and *Figure 11.18*) • *Island-rescue PCR*. This is a way of isolating CpG island sequences from YACs by amplfying sequences between islands and neighboring *Alu* repeats (see page 30).
Hybridization to mRNA/cDNA	A genomic DNA clone can be hybridized against a Northern blot of mRNA from a panel of culture cell lines, or against appropriate cDNA libraries (see *Figures 5.15* and *11.6*).
Exon trapping	This is essentially an artificial RNA splicing assay (see page 299 and *Figure 11.19*). It relies on the observation that the vast majority of mammalian genes contain multiple exons which need to be spliced together at the RNA level.
cDNA selection or capture	These techniques involve repeated purification of a subset of genomic DNA clones which hybridize to a given cDNA population (see page 301 and *Figure 11.20*).
Computer analysis of DNA sequences	• *Homology searches*. Any DNA sequence obtained from a genomic DNA clone can be compared against all other available DNA sequences in sequence databases. Significant homology to known coding DNA or gene-associated sequences may indicate a gene (see page 303). • Exon prediction. A variety of computer programs have been developed to identify the locations of possible exons in nucleotide sequences (see page 303).

Genes can be identified easily by hybridizing DNA clones against Northern blots, cDNA libraries, zoo-blots and Southern blots of genomic DNA digested with rare-cutter restriction endonucleases

Of the various methods of identifying genes, a few are used as a first line of attack because they are relatively straightforward. In each case, the method involves using the genomic DNA clone as a hybridization probe. Inserts in cosmid clones and subclones are often useful hybridization probes. Larger clones such as YACs occasionally can be used as hybridization probes, but are generally less amenable to such approaches.

Hybridization to RNA/cDNA

A candidate DNA clone can be hybridized against a Northern blot containing a panel of mRNA or total RNA samples isolated from a variety of different tissues (brain, heart, lung, liver, kidney, etc.). Positive hybridization may indicate the presence of a gene within the cloned fragment and may suggest a suitable cDNA library for screening. This approach has been facilitated enormously by using whole cosmid clones as hybridization probes in competition hybridization (see *Box 5.3*). They may fail, however, for two reasons. First, significant expression of the gene may be restricted to a cell population or developmental stage which is not represented in the Northern blot panel or those cDNA libraries which are selected for screening. Additionally, there may be a problem with the intensity of the hybridization signal. For example, the proportion of exon sequence in the probe may be very low (the average exon size is only about 170 bp and a 40-kb insert in a cosmid may by chance contain only one exon). If this is the case and if the gene is not strongly expressed in the relevant tissue (so that it is not well represented in the RNA samples or cDNA libraries which are being screened), detection of positive hybridization signals may be difficult. Occasional transcribed repeats may also be problematic.

Zoo-blot hybridization

Coding DNA sequences are subject to considerable selection pressure to conserve biologically important sequences. By contrast, noncoding DNA sequences accumulate mutations comparatively rapidly and are not well conserved between species. A **zoo-blot** is a Southern blot of genomic DNA samples from a wide variety of different species. A genomic DNA clone which shows positive hybridization signals against the DNA of a variety of different species would be expected, therefore, to contain coding DNA sequences that have been strongly conserved during evolution. Some mammalian genes, often with a crucially important function in development, are so highly conserved that they will show significant hybridization signals with evolutionarily distant species such as yeast, *Drosophila* and *Caenorhabditis elegans*. Others may only show significant hybridization signals to mammals (see *Figure 11.16*).

CpG island identification

CpG islands (sometimes known as **HTF islands** – see Cross and Bird, 1995) are short (~1 kb long) hypomethylated GC-rich sequences which are often found at the 5′ ends of vertebrate genes (see *Figure 11.17*). In the human genome, an estimated 56% of genes are associated with such sequences (Antequera and Bird, 1993). They include all examples of housekeeping genes and genes that are widely expressed, and a significant portion, perhaps about 40% of genes which show tissue-specific or restricted expression patterns (see Larsen *et al.*, 1992).

The % (G + C) of CpG island sequences generally exceeds 60% and they have a high concentration of the CpG dinucleotide; 10–20 times greater than that of bulk vertebrate DNA. As a result, CpG islands often have restriction sites for a variety of rare-cutter restriction nucleases which cleave at GC-rich sequences containing one or two CpG dinucleotides. For example, each of the enzymes *Sac*II (CCGCGG) *Eag*I (CGGCCG) and *Bss*HII (GCGCGC) is expected to cut, on average, about 1.2 times within an island (Bird, 1987), but very rarely outside islands. Close clustering of such restriction sites in genomic DNA is often indicative of a CpG island, and can be identified by using genomic DNA clones as hybridization probes against Southern blots of suitably digested genomic DNA samples (see *Figure 11.18*). Note, however, that the method is not applicable to the substantial number of genes which have no associated CpG islands.

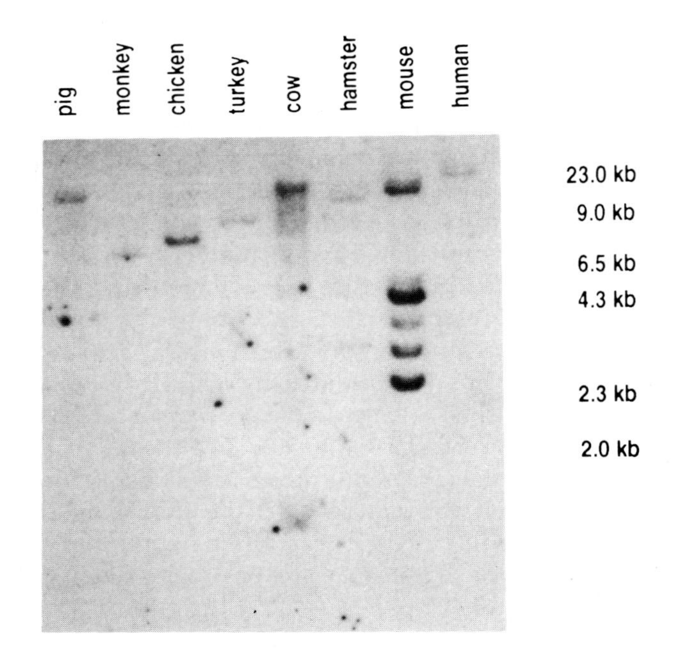

Figure 11.16: Zoo blot hybridization is an assay for DNA sequences that are highly conserved between species.

The example shows hybridization using a cDNA clone from the *NF2* (neurofibromatosis type 2) gene against a Southern blot of genomic DNA samples from the indicated species. Reproduced from Claudio *et al.* (1994) by permission of Oxford University Press.

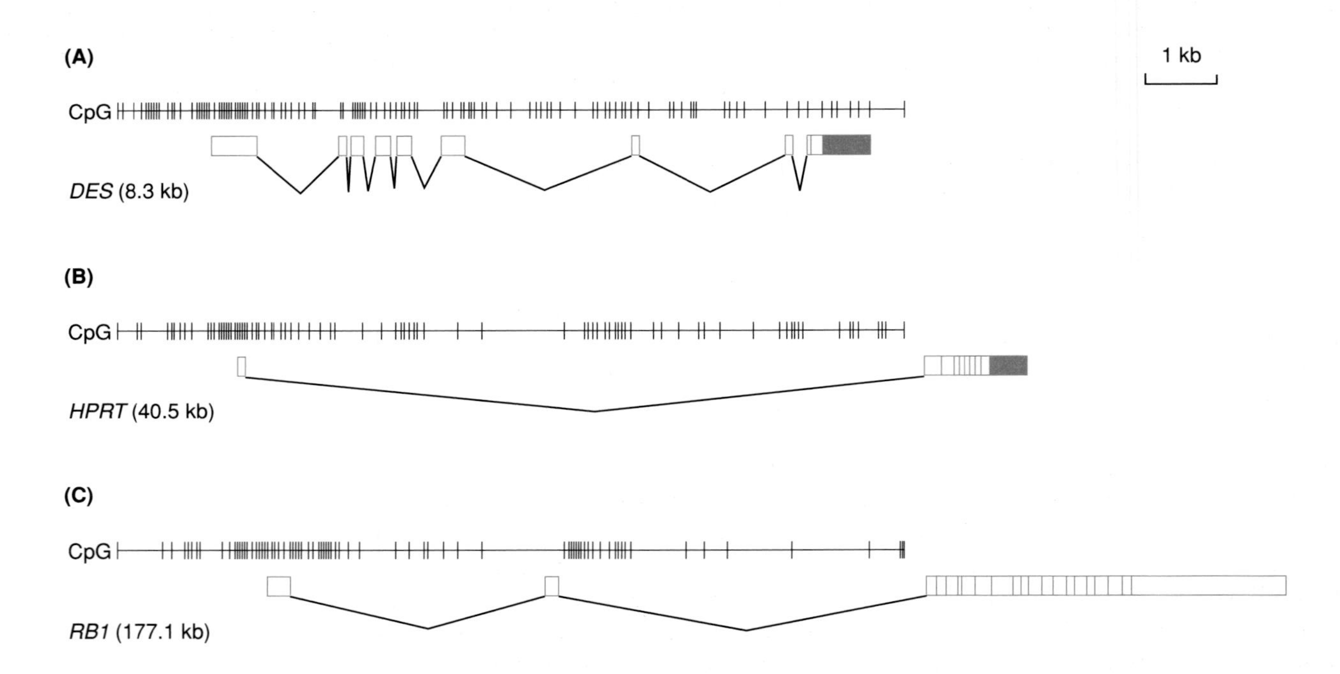

Figure 11.17: CpG island structure in three human genes.

Vertical bars represent the positions of the dinucleotide CpG in 10 kb DNA sequences representing: **(A)** the human desmin *(DES)* gene and the 5′ ends of **(B)** the human hypoxanthine phosphoribosyl transferase (*HPRT)* and **(C)** retinoblastoma (*RB1*) genes. Full gene lengths are indicated in brackets after the gene name. Boxes represent exons with splicing patterns shown only for the exons found in the first 10 kb. Open and tinted regions represent translated and untranslated sequences respectively.

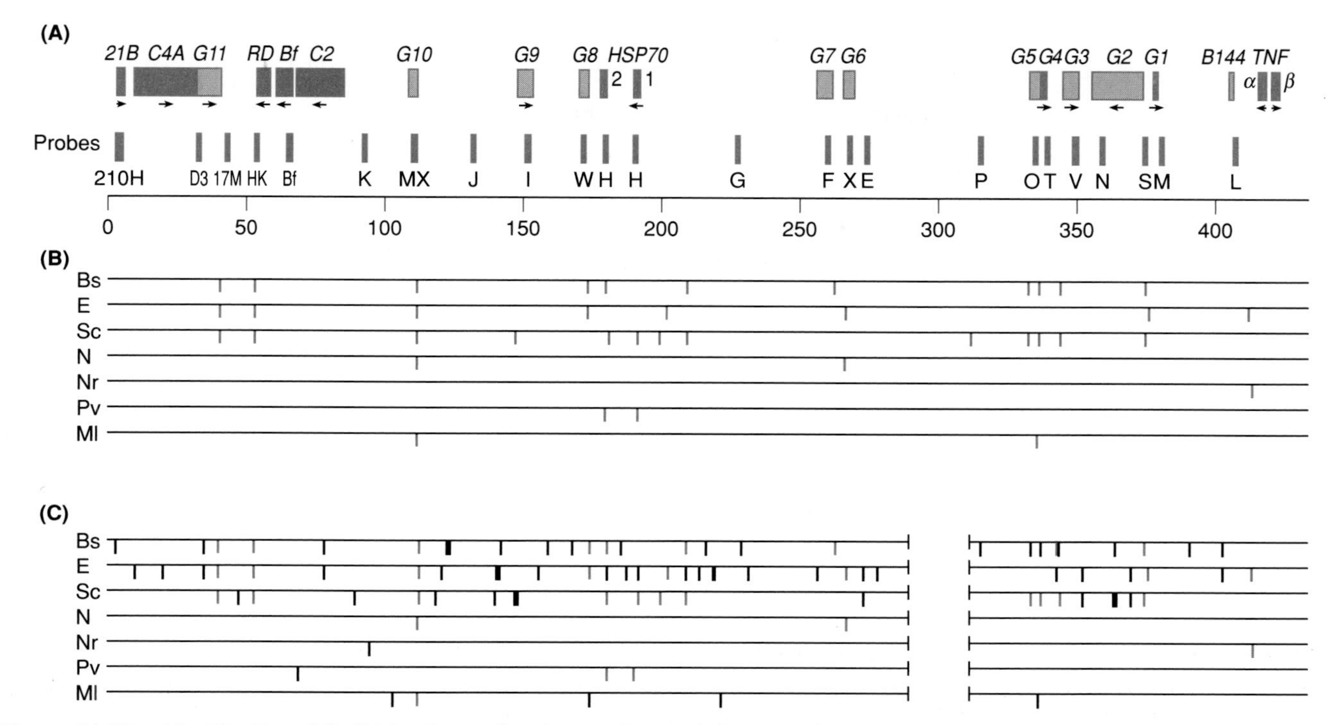

Figure 11.18: Identification of CpG island-associated genes by restriction mapping.

The figure shows restriction mapping in the class III region of the human leukocyte antigen (HLA) complex using the enzymes *Bss*HII (Bs), *Eag*I (E), *Sac*II (Sc), *Not*I (N), *Nru*I (Nr), *Pvu*I (Pv) and *Mlu*I (Ml). The restriction mapping was done by two methods. In one method, probes derived from the indicated regions **(A)** were hybridized to Southern blots of digested human genomic DNA to give the maps shown in **(B)**. The second method involved direct restriction mapping on isolated cosmid clone to give the maps shown in **(C)**. CpG islands were indicated by close clustering of rare-cutter sites (especially those for *Bss*HII, *Eag*I and *Sac*II) and were found in the genomic DNA associated with several genes, e.g. G11, RD, G10, etc. *Note* that many of the rare-cutter enzymes do not cleave at methylated sequences. As a result, there are numerous restriction sites in the cosmids because propagation of the human sequences in *E. coli* leads to removal of methyl groups originally present in genomic DNA. As CpG islands are hypomethylated in genomic DNA, however, they are susceptible to cleavage. Redrawn from Sargent *et al.* (1986) by permission of Oxford University Press.

Exon trapping/exon amplification techniques identify expressed sequences by their capacity for engaging in an artificial RNA splicing assay

The gene identification methods described above have their limitations: restricted expression patterns of some genes may make them difficult to identify; genomic DNA clones may give very weak hybridization signals if the percentage of exon sequence is very low; and, as we have seen (page 297), many genes which are expressed in a tissue-specific or restricted manner do not have associated CpG islands. An alternative is to identify a gene by the ability of its exons to engage in an artificial RNA splicing assay. RNA splicing involves fusion of exonic sequences at the RNA level and excision of intronic sequences (see page 17). Spliceosomes are able to accomplish this *in vivo* by recognizing certain sequences at exon/intron boundaries: a *splice donor* sequence at the junction between an exon and its downstream (3′) intron, and a *splice acceptor* sequence at the junction between an exon and its upstream (5′) intron (see *Figure 1.15*). A cosmid or other suitable genomic DNA clone containing an internal exon flanked by intronic sequences will therefore contain functional splice donor and acceptor sequences.

Exons can be identified in cloned genomic DNA by subcloning the DNA into a suitable expression vector and transfecting into an appropriate eukaryotic cell line in which the insert DNA is transcribed into RNA and the RNA transcript undergoes

RNA splicing. Such techniques are known as **exon trapping** (often called **exon amplification** if a PCR reaction is employed to recover the exons from a cDNA copy of the spliced RNA). For example, in the method of Church *et al.* (1994), the DNA is subcloned into a plasmid expression vector pSPL3 which contains an **artificial minigene** that can be expressed in a suitable host cell. The minigene consists of:

- a segment of the simian virus 40 (SV40) genome which contains an origin of replication plus a powerful promoter sequence;
- two splicing-competent exons separated by an intron which contains a multiple cloning site;
- an SV40 polyadenylation site.

The recombinant DNA is transfected into a strain of monkey cells, known as **COS cells**. COS cells were derived from monkey CV-1 cells by artificial manipulation, leading to integration of a segment of the SV40 genome containing a defective origin of SV40 replication. The integrated SV40 segment in COS cells allows any circular DNA which contains a functional SV40 origin of replication to replicate independently of the cellular DNA. Transcription from the SV40 promoter results in an RNA transcript which normally splices to include the two exons of the minigene. If the DNA cloned into the intervening intron contains a functional exon, however, the foreign exons can be spliced to the exons present in the vector's minigene. After making a cDNA copy using reverse transcriptase, PCR reactions using primers specific for vector exon sequences should distinguish between normal splicing and splicing involving exons in the insert DNA (see *Figure 11.19*).

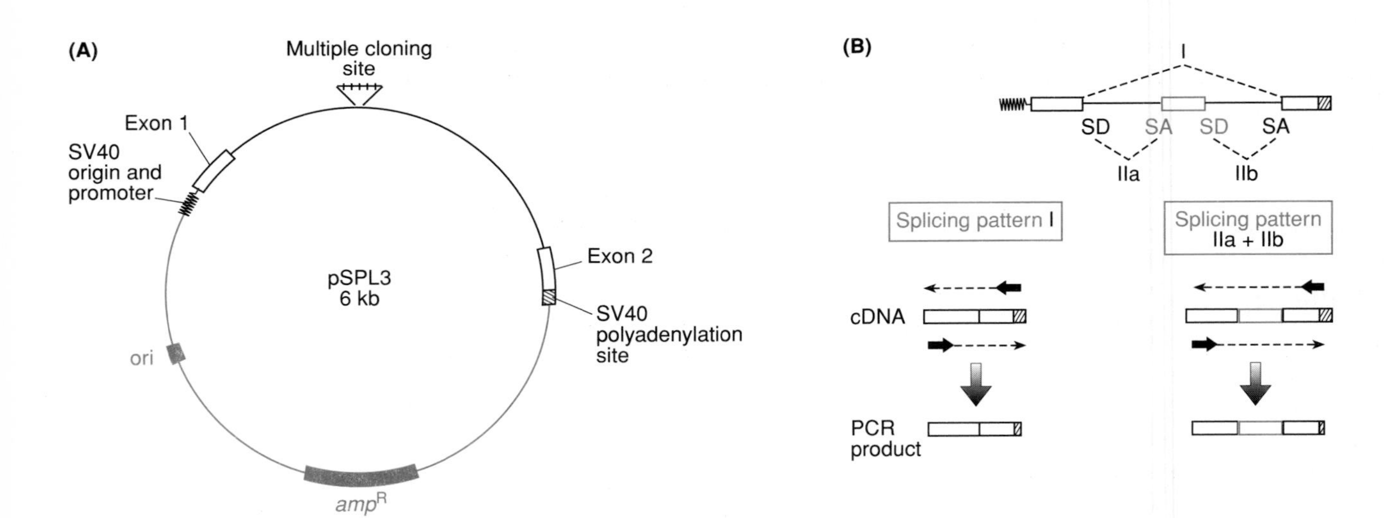

Figure 11.19: Exon trapping using the pSPL3 vector.

(A) The pSPL3 plasmid vector. This shuttle vector can be propagated in *E. coli* (using the ori origin of replication and selection for ampicillin resistance) and also in monkey COS cells (using the functional SV40 origin of replication) (Church *et al.*, 1994). The pSPL3 vector contains a minigene (in black): transcription occurs from the SV40 promoter and the RNA undergoes splicing under control of the host cell's RNA splicing machinery, resulting in fusion of the two vector exon sequences.
(B) Splicing patterns. The normal splicing pattern which is seen when only the vector exons are present is indicated by splicing pattern I. If a genomic DNA fragment cloned into pSPL3 contains an exon with functional splice donor (SD) and splice acceptor (SA) sequences, a different splicing pattern (IIa + IIb) may occur. The two splicing patterns can be distinguished at the cDNA level by using various vector-specific PCR primers and size-fractionation on gels can lead to recovery of the amplified exon from genomic DNA.

Genes in complex DNA clones, such as YACs, can be identified by forming heteroduplexes with cDNAs and by PCR-based methods of amplifying CpG island sequences

The techniques described on pages 297–300 have been popular when dealing with inserts from cosmid clones or subclones. Increasingly, however, large-scale physical mapping procedures rely on building YAC contigs. For any YAC, *cognate* cosmids (cosmid clones containing sequences from the same loci as those present in the YAC) can be identified by various procedures, including using (small) YACs as hybridization probes to screen chromosome-specific cosmid libraries (see page 284) or preparing cosmid libraries from the DNA of a purified YAC. An alternative is to try to identify expressed sequences directly from the YACs.

cDNA selection/capture

Direct cDNA selection techniques (also called *direct selection* and *cDNA selection*) involve forming genomic DNA/cDNA heteroduplexes by hybridizing a complex cloned DNA, such as the insert of a YAC, to a complex mixture of cDNAs, such as the inserts from all cDNA clones in a cDNA library (Lovett, 1994). The principle underlying the technique is that cognate cDNAs corresponding to genes found within the YAC will bind preferentially to the YAC DNA; several rounds of hybridization should lead to a huge enrichment of the desired cDNA sequences, enabling the identification of the corresponding genes. Considerable blocking of repetitive DNA sequences is required. Early approaches used immobilized YACs, but more modern approaches have used a solution hybridization reaction and biotin–streptavidin capture methods (see *Figure 11.20*). Like all expression-based systems, the method depends on appropriate levels of gene expression (the cognate cDNAs should not be too rare in the starting population). Additionally, genes containing very short exons may be missed because the heteroduplexes formed with cognate cDNAs may not be sufficiently stable. Another problem is that cDNAs may bind to pseudogenes which show a high degree of homology to the cognate functional genes.

Coincident sequence cloning is a related PCR-based technique for cloning those sequences which are coincident between any two DNA populations (such as a genomic DNA clone and the inserts from a cDNA library) by forming heteroduplexes between related (co-incident) sequences in the two populations. The most modern version, *end ligation coincident sequence cloning*, has, however, the advantage of an end ligation reaction which endows the method with great specificity (see also page 380).

Island rescue PCR

Island rescue PCR (IRP) can permit selective amplification of CpG island sequences from human YACs. The method depends on the high copy number of *Alu* repeats (so that there is a high chance of an *Alu* repeat in the vicinity of a CpG island) and the frequent occurrence of restriction sites for rare-cutters such as *Bss*HII in CpG islands. The YAC DNA is cut with a suitable rare-cutter restriction nuclease and fragments are ligated to a bubble-linker primer (see *Figure 6.12*) with a suitably complementary overhang. CpG island-*Alu* PCR is then possible using an *Alu*-specific primer and a bubble-linker-specific primer (Valdes *et al.*, 1994).

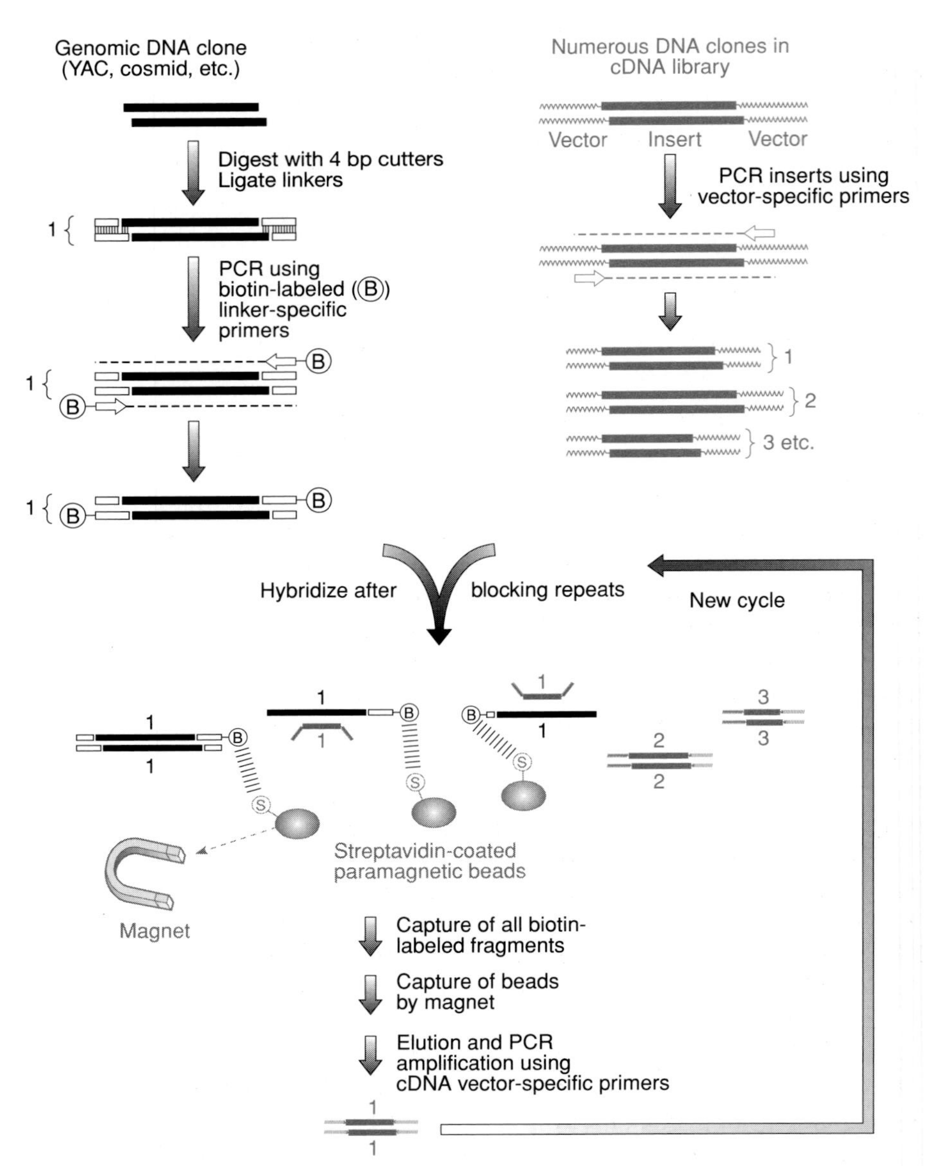

Figure 11.20 cDNA selection using magnetic bead capture.

The method relies on heteroduplex formation between single strands of a single genomic DNA clone (in black, numbered 1) and of a complex cDNA population, such as the inserts of a cDNA library (in red, numbered 1, 2, 3, etc.). The genomic DNA strands are labeled with a biotin group (attached to PCR primers which are incorporated during amplification). The hybridization reaction will favor heteroduplex formation involving those cDNA clones *cognate* with the genomic DNA clone. In this example, genomic DNA clone 1 and cDNA clone 1 are envisaged to be cognate, i.e. contain common sequences, allowing opposite sense strands to bond together, giving a heteroduplex. Hybridization products with a biotin group (including genomic DNA–cDNA heteroduplexes) will bind to streptavidin-coated paramagnetic beads and can be removed from other reaction components by a magnet. The separated beads can then be treated to elute the biotin-containing molecules and, by using PCR primers specific for the vector sequences flanking the cDNA, the bound cDNA can be amplified. This population is submitted to further hybridization cycles to enrich for the desired cDNA.

Transcript maps can be obtained by physically mapping randomly generated expressed sequence tags

The expressed component of complex genomes, such as mammalian genomes, may constitute only a few per cent of the total genome. As described in the preceding sections, one way of building transcript maps is to identify genes in defined clone contigs. An alternative, and more general, method is to obtain partial sequences of numerous randomly selected cDNA clones and then place these on physical maps.

Even a short sequence of, say 200 bp, from a cDNA clone permits sequence-specific primers to be designed so that a PCR assay can be developed that is specific for that sequence. This effectively means that an STS is available for an expressed sequence, a so-called **expressed sequence tag (EST)**. An individual EST can be mapped by a PCR assay to: a chromosome (using monochromosomal hybrids, page 277); a sub-chromosomal location (using radiation, translocation and deletion hybrids, pages 278–280); or to defined genomic clones (e.g. using panels of YACs from specific YAC contigs).

Genes can be identified in cloned DNA by computer analysis of the DNA sequence

Once a DNA clone has been sequenced (see following section), computer analyses can be used to determine whether the sequence is likely to represent part of a gene. Two major types of software are used:

- **homology searches.** The nucleotide sequence and the inferred amino acid sequences for all three reading frames are compared against all available DNA and protein sequences which have been recorded in electronic databases (see page 355 for examples of sequence databases). Any significant matching between the test sequence and the sequence of a known gene, cDNA or protein, whether of human or nonhuman origin, indicates a gene-associated sequence (either a functional gene, pseudogene or gene fragment (see page 517 for an example).
- **Exon prediction**. This approach is becoming increasingly important as large-scale DNA sequencing projects gather momentum. The programs are designed to scan a DNA sequence in order to identify the locations of likely exons by screening for conserved sequences found at exon/intron junctions and the splice branch site (see *Figure 1.15*), and the presence of comparatively long ORFs, etc. As yet, however, even the best such programs, such as the GRAIL software, have been only moderately successful in identifying exons when tested against genes whose exon organizations had previously been established (Lopez *et al.*, 1994). When the relevant gene is GC-rich, however, exon prediction can be quite accurate. A recent successful application has been the characterization of the adult polycystic kidney disease gene *PKD1*, which proved to be problematic because of the existence of several transcriptionally active copies of closely related *PKD1*-like sequences mapping centromeric to *PKD1* on chromosome 16p13.1. Following the initial identification of a partial cDNA clone, genomic clones were obtained from the *PKD1* locus and a total of 54 kb of DNA sequence was obtained upstream of the poly(A) signal. From this sequence the GRAIL2 program was able to identify a total of 46 exons (American PKD1 Consortium, 1995), a figure which was subsequently confirmed using more conventional methods.

DNA sequencing: towards the ultimate physical map

DNA sequencing usually involves enzymatic DNA synthesis in the presence of base-specific dideoxynucleotide chain terminators

Formerly, chemical DNA sequencing methods were often employed, using base-specific chemical modification and subsequent cleavage of the DNA. Currently,

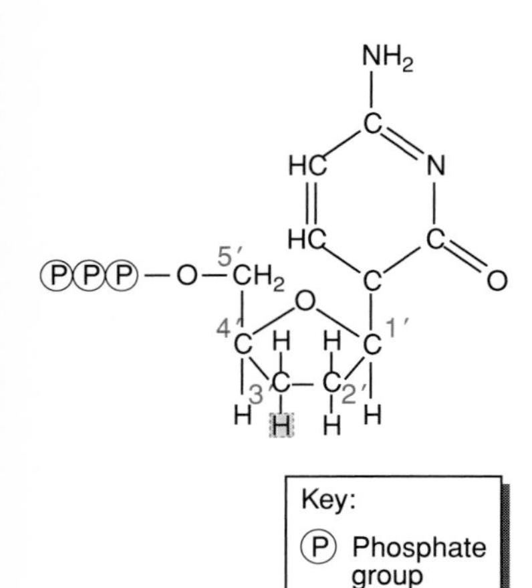

Figure 11.21: Structure of a dideoxynucleotide, 2′,3′ dideoxy CTP.

Note that the hydroxyl group which is attached to carbon 3′ in normal nucleotides (see *Figure 1.2*) is replaced by a hydrogen atom.

however, the vast majority of DNA sequencing is carried out using an enzymatic method: the DNA to be sequenced is provided as a single-stranded template for synthesis of new DNA strands by a DNA polymerase. In addition to the normal nucleotide precursors, DNA synthesis is carried out in the presence of base-specific **dideoxynucleotides (ddNTPs)**. The latter are analogs of the normal dNTPs but differ in that they lack a hydroxyl group at the 3′ carbon position as well as the 2′ carbon (*Figure 11.21*). A dideoxynucleotide can be incorporated into the growing DNA chain by forming a phosphodiester bond between its 5′ carbon atom and the 3′ carbon of the previously incorporated nucleotide. However, since ddNTPs lack a 3′ hydroxyl group, any ddNTP that is incorporated into a growing DNA chain cannot participate in phosphodiester bonding at its 3′ carbon atom, thereby causing abrupt termination of chain synthesis.

The template DNA for DNA sequencing must be single-stranded DNA, and a complementary **sequencing primer,** often about 20 nucleotides long, is required to bind specifically to a region of template DNA flanking the region whose sequence is to be determined. Four parallel base-specific reactions are conducted using a mix of all four dNTPs and also a small proportion of one of the four ddNTPs. By setting the concentration of the ddNTP to be very much lower than that of its normal dNTP analog, chain termination will occur randomly at one of the many positions containing the base in question. Each reaction is therefore a *partial reaction*: chain termination occurs randomly at only a very small minority of the possible bases *in any one DNA strand*. However, the DNA to be sequenced in a DNA sequencing reaction is a *population* of (usually) identical molecules. As a result, for each type of single-stranded DNA template, a particular base-specific reaction will generate *a collection of DNA fragments of different sizes, with a common 5′ end (defined by the sequencing primer) but variable 3′ ends (all of which terminate with the chosen ddNTP* (*Figure 11.22*).

Fragments that differ in size by even a single nucleotide can be separated on a **denaturing polyacrylamide gel** (see *Box 11.4*). The differently sized fragments can be detected by incorporating labeled groups into the reaction products, either by incorporating labeled nucleotides or by using a primer with a labeled group. The sequence can then be read off by reading from the bottom of the gel to the top, a direction that gives the 3′ → 5′ sequence of the complementary strand of the provided DNA template (see *Figure 11.22*).

DNA sequencing is increasingly being conducted using fluorescent labeling systems and automated detection systems

Traditional dideoxy sequencing methods have employed **radioisotope labeling**: the dNTP mix contains a proportion of radiolabeled nucleotides which are incorporated within the growing DNA chains. Following electrophoresis, typically the gel is dried and an autoradiographic film is placed in contact with the dried gel. After a suitable exposure time, the film is developed, giving a characteristic pattern of dark bands (*Figure 11.22*). ^{32}P-labeled nucleotides are not very suitable for this purpose: the high energy β radiation causes considerable scattering of the signal, leading to diffuse bands. Instead, ^{35}S- or ^{33}P-labeled nucleotides have been used.

Large-scale DNA sequencing efforts are dependent on improving efficiency by partial automation of the technologies involved. One major improvement in recent years has been the development of automated procedures for fluorescent DNA sequencing (Wilson *et al.*, 1990). These procedures generally use primers or dideoxynucleotides to which are attached **fluorophores** (chemical groups capable of fluorescing). During electrophoresis, a monitor detects and records the fluorescence signal as the DNA passes through a fixed point in the gel (*Figure 11.23A*). The

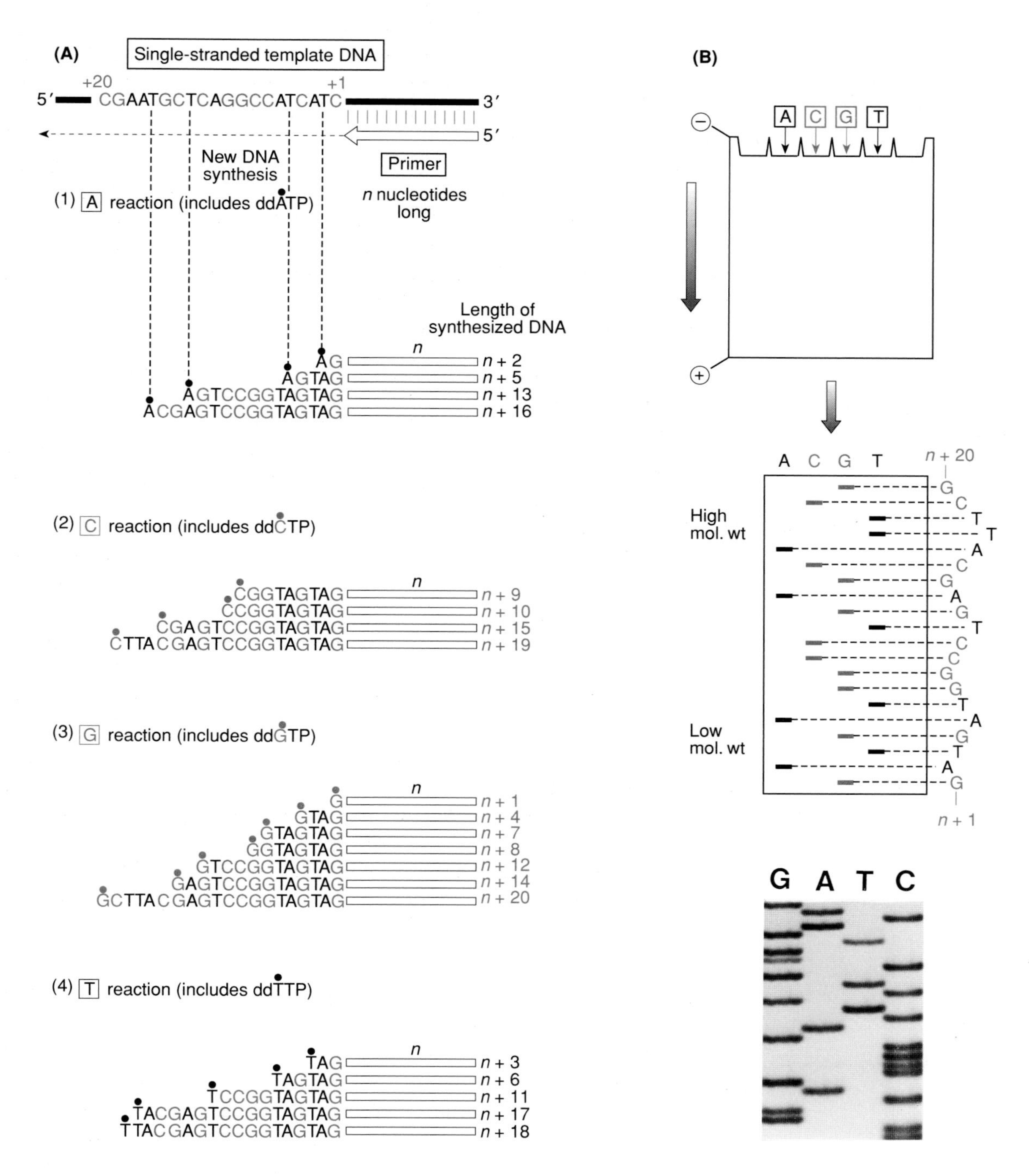

Figure 11.22: Dideoxy DNA sequencing relies on synthesizing new DNA strands from a single-stranded DNA template and random incorporation of a base-specific dideoxynucleotide to terminate chain synthesis.

(A) Principle of dideoxy sequencing. The sequencing primer binds specifically to a region 3′ of the desired DNA sequence and primes synthesis of a complementary DNA strand in the indicated direction. Four parallel base-specific reactions are carried out, each with all four dNTPs and with one ddNTP. Competition for incorporation into the growing DNA chain between a ddNTP and its normal dNTP analog results in a population of fragments of different lengths. The fragments will have a common 5′ end (defined by the sequencing primer) but variable 3′ ends, depending on where a dideoxynucleotide (shown with a filled circle above) has been inserted. For example, in the A-specific reaction chain, extension occurs until a ddA nucleotide (shown as A with a filled black circle above) is incorporated. This will lead to a population of DNA fragments of lengths $n+2$, $n+5$, $n+13$, $n+16$ nucleotides, etc.
(B) Conventional DNA sequencing. This generally involves using a radioactively labeled nucleotide and size-fractionation of the products of the four reactions in separate wells of a polyacrylamide gel. The dried gel is submitted to autoradiography, allowing the sequence of the complementary strand to be read (from bottom to top). The bottom panel illustrates a practical example, in this case a sequence within the gene for type II neurofibromatosis.

Box 11.4: Denaturing gels

Denaturing gels are used to size-fractionate single-stranded DNA or RNA molecules. The initial nucleic acid sample may be double-stranded and requires to be denatured (usually by heating the DNA in a solution which often contains a chemical denaturant such as formamide) prior to loading on the gel. The gel also contains a high concentration of chemical denaturant to ensure that any complementary strands remain single-stranded. The chemical denaturants used are strongly polar simple molecules, typically containing amino (NH_2) and carbonyl (C=O) groups. Such chemical groups effectively compete for hydrogen bond formation with the amino and carbonyl groups of bases, and so can disrupt the hydrogen bonding between base pairs (see *Figure 1.3*). Frequently used denaturants in gels include urea, formamide and formaldehyde.

$NH_2-C(=O)-NH_2$ Urea

$H-C(=O)-NH_2$ Formamide

$H-C(=O)-H$ Formaldehyde

use of different fluorophores in the four base-specific reactions means that, unlike conventional DNA sequencing, all four reactions can be loaded into a single lane. The output is in the form of intensity profiles for each of the differently colored fluorophores (*Figure 11.23B*), but the information is simultaneously stored electronically. This precludes transcription errors when an interpreted sequence is typed by hand into a computer file. Recent advances in technology mean that the accuracy of DNA sequencing using automated methods is acceptably high.

Single-stranded DNA clones for use in DNA sequencing are generally obtained using M13 or phagemid vectors

DNA templates provided in the form of isolated single-stranded DNA are preferred for conventional ddNTP sequencing: the sequences are clearer and easier to read. Single-stranded PCR products can be obtained by *asymmetric PCR* (page 143). Usually, however, single-stranded recombinant DNA clones are used as templates. They are obtained using vectors based on certain bacteriophages which naturally assume a single-stranded DNA form at some stage in their life cycle. Because the vector sequence is already known, it is often convenient to use a single vector-specific sequencing primer (*universal sequencing primer*) which is complementary to a sequence in the vector adjacent to the cloning site (*Figure 11.24*).

M13 vectors

M13 is one of a group of filamentous bacteriophages (including the fd and fl phages) which can infect certain strains of *E. coli*. The genomes of these phages consist of single-stranded circular DNA molecules about 6.4 kb long, which are very highly related to each other in DNA sequence and enclosed in a protein coat, forming a long filamentous structure. After adsorption to the bacterium, the phage genome enters the bacterial cell where it is converted from the single-stranded form to a double-stranded form, the *replicative form* (*RF*). The latter serves as a template for making numerous copies of the genome and, after a certain time, a phage-encoded product switches DNA synthesis towards production of single strands. The latter migrate to the cell membrane where they are enclosed in a protein coat, and hundreds of mature phage particles are extruded from the infected cell without cell lysis. M13 vectors are based upon the double-stranded RF form and have a multiple cloning site for generating double-stranded recombinant DNA circles. The latter can be transfected into suitable strains of *E. coli*. After a certain period, phage particles are harvested and stripped of their protein coats to release single-stranded recombinant DNA for direct use as templates in DNA sequencing reactions (*Figure 11.25A*).

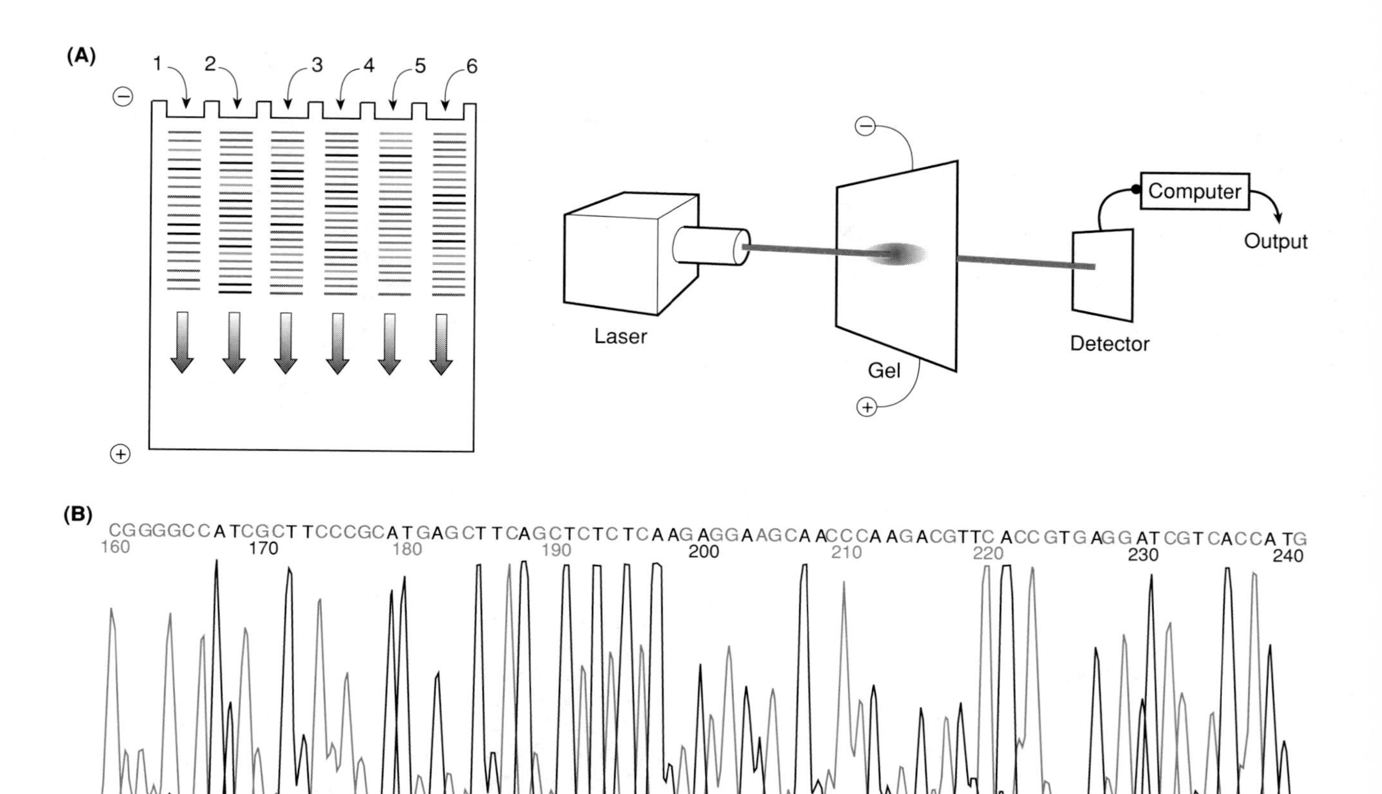

Figure 11.23: Automated DNA sequencing using fluorescent primers.

(A) Principles of automated DNA sequencing. Automated DNA sequencing involves loading all four reaction products into single lanes of the electrophoresis gel and capture of sequence data during the electrophoresis run. Four separate fluorescent dyes are used as labels for the base-specific reactions (the label can be incorporated by being attached to a base-specific ddNTP, or by being attached to the primer and having four sets of primers corresponding to the four reactions). During the electrophoresis run, a laser beam is focused at a specific constant position on the gel. As the individual DNA fragments migrate past this position, the laser causes the dyes to fluoresce. Maximum fluorescence occurs at different wavelengths for the four dyes, and the information is recorded electronically and the interpreted sequence is stored in a computer database.
(B) Example of DNA sequence output. This shows a typical output of sequence data as a succession of dye-specific (and therefore base-specific) intensity profiles. The example illustrated represents part of exon 1 of the neurofibromatosis type 2 (*NF2*) gene. Data provided by Susan Mason, University of Newcastle upon Tyne.

Phagemid vectors

Phagemid vectors are plasmids which have been artificially manipulated so as to contain a small segment of the genome of a filamentous phage, such as M13, fd or fl. The selected phage sequences contain all the *cis*-acting elements required for DNA replication and assembly into phage particles. They permit successful cloning of

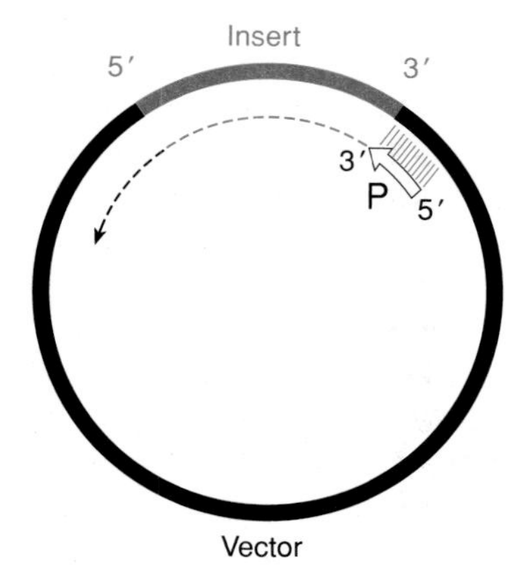

Figure 11.24: A universal sequencing primer can be used to sequence many different template DNAs.

DNA templates for DNA sequencing are often single-stranded recombinant DNA molecules. Different clones will often contain different inserts within the same vector molecule. As a result, a universal sequencing primer (P) can be designed to be complementary to a short vector sequence located next to the cloning site(s), allowing sequencing of different insert DNAs.

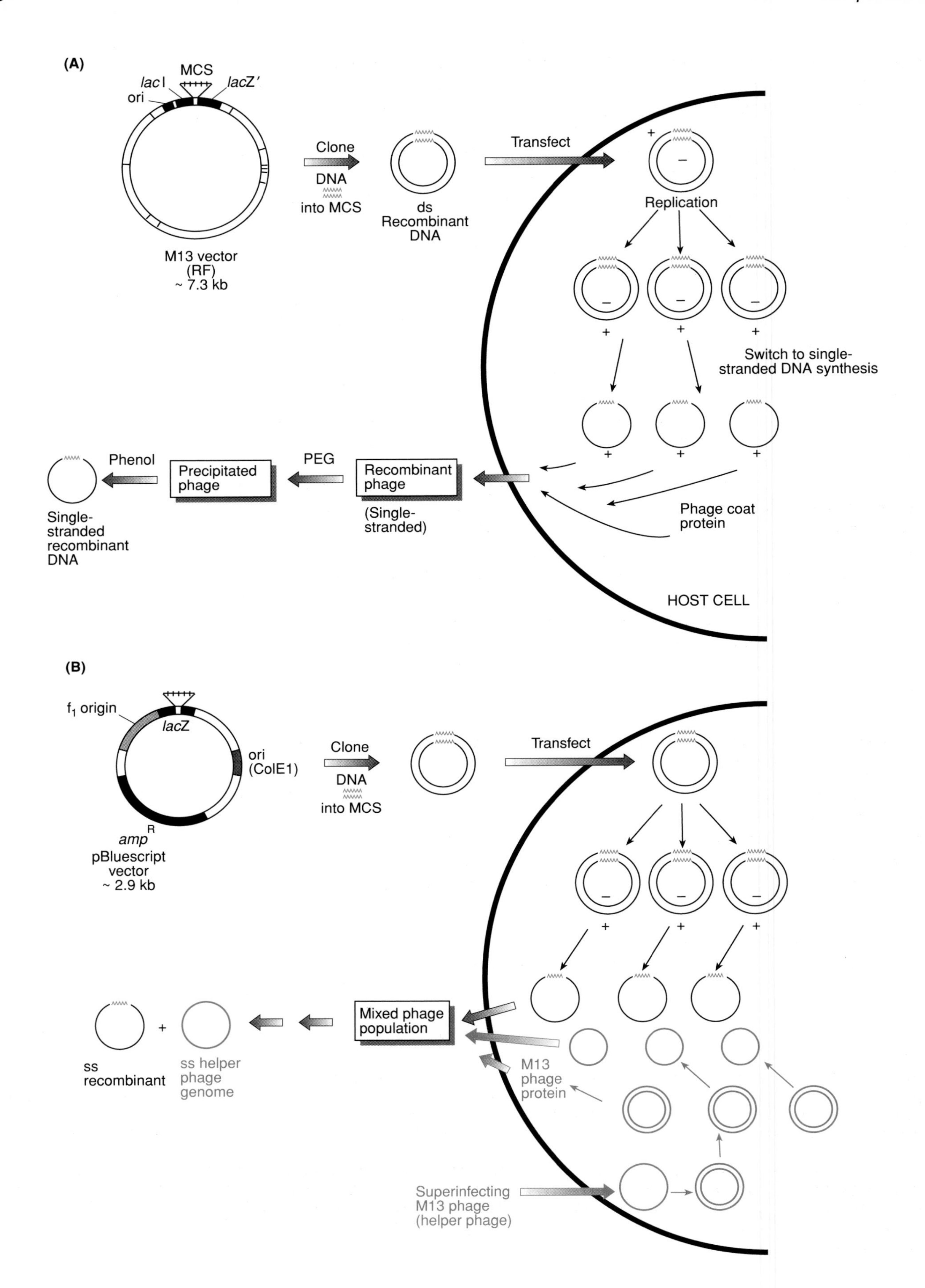
(A)
MCS
lacI
lacZ′
ori
M13 vector
(RF)
~ 7.3 kb
Clone
DNA
into MCS
ds
Recombinant
DNA
Transfect
Replication
Switch to single-
stranded DNA synthesis
Phenol
Precipitated
phage
PEG
Recombinant
phage
(Single-
stranded)
Single-
stranded
recombinant
DNA
Phage coat
protein
HOST CELL
(B)
f1 origin
lacZ
ori
(ColE1)
amp R
pBluescript
vector
~ 2.9 kb
Clone
DNA
into MCS
Transfect
Mixed phage
population
ss
recombinant
ss helper
phage
genome
M13
phage
protein
Superinfecting
M13 phage
(helper phage)

inserts several kilobases long (unlike M13 vectors in which such inserts tend to be unstable). Following transformation of a suitable *E. coli* strain with a recombinant phagemid, the bacterial cells are *superinfected* with a filamentous **helper phage,** such as fl, which is required to provide the coat protein. Phage particles secreted from the superinfected cells will be a mixture of helper phage and recombinant phagemids (*Figure 11.25B*). The mixed single-stranded DNA population can be used directly for DNA sequencing because the primer for initiating DNA strand synthesis is designed to bind specifically to a sequence of the phagemid vector adjacent to the cloning site. Commonly used phagemid vectors include the pEMBL series of plasmids and the pBluescript family (see *Figure 11.12B*).

Cycle sequencing is frequently used to establish DNA sequences from initially double-stranded DNA templates

Double-stranded DNA templates can be used in standard dideoxy sequencing by denaturing the DNA prior to binding the oligonucleotide primer. However, the quality of sequences from initially double-stranded DNA templates is often poor. **Cycle sequencing**, also called **linear amplification sequencing**, is a kind of PCR sequencing approach which overcomes this problem (see Kretz *et al.*, 1994). Like the standard PCR reaction, it uses a thermostable DNA polymerase and a temperature cycling format of denaturation, annealing and DNA synthesis. The difference is that cycle sequencing employs only one primer and includes a ddNTP chain terminator in the reaction. The use of only a single primer means that unlike the exponential increase in product during standard PCR reactions, the product accumulates linearly (see *Figure 11.26*). Because the product accumulates during the reaction, and because of the high temperature at which the sequencing reactions are carried out, and the multiple heat denaturation steps, small amounts of double-stranded plasmids, cosmids, λ DNA and PCR products may be sequenced reliably without a separate heat denaturation step.

Large-scale sequencing projects use a shotgun cloning approach for preparing DNA templates

The advantage of cloning the desired DNA for sequencing into a vector such as M13 is that only a single *universal primer* need be used for sequencing. However, the maximum amount of DNA sequence that can be read with consistently high accuracy from a single gel run is typically a few hundred nucleotides. Thus, only the first few hundred nucleotides of an insert can be sequenced using a vector-specific primer. As a result, different sequencing strategies may be adopted.

- **Primer-walking sequencing** is a strategy that can be employed to obtain the cumulative sequence of a single large DNA template. Once some initial sequence has been obtained, successive sequencing steps are carried out using

Figure 11.25: Producing single-stranded recombinant DNA using M13 and phagemid vectors.

(A) M13 vectors. M13 vectors are replicative (RF) forms of M13 derivatives containing a nonfunctional component of the *lacZ* β-galactosidase system which can be complemented in function by the presence of a complementary *lacZ* component in the *E. coli* JM series. The double-stranded M13 recombinant DNA enters the normal cycle of DNA replication to generate numerous copies of the genome, prior to a switch to production of single-stranded DNA (+ strand only). Mature recombinant phage exit from the cell without lysis.

(B) Phagemid vectors. The pBluescript series of plasmid vectors contain two origins of replication: a normal one from *Col*E1 and a second from phage f1 which, in the presence of a filamentous phage genome, will specify production of single-stranded DNA. Superinfection of transformed cells with M13 phage results in two types of phage-like particles released from the cells: the original superinfecting phage and the plasmid recombinants within a phage protein coat. Sequencing primers specific for the phagemid vector are used to obtain unambiguous sequences.

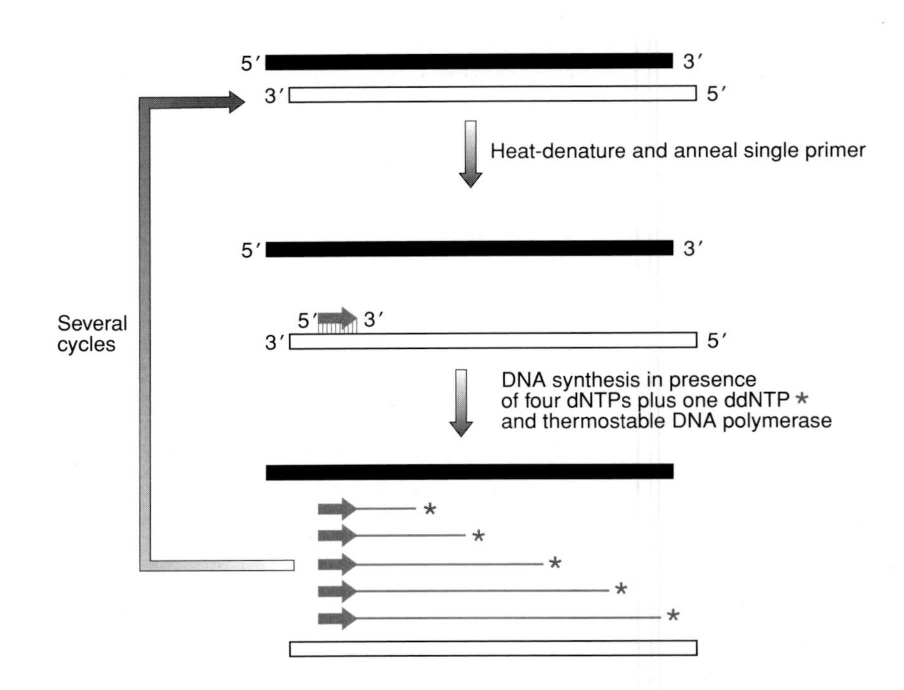

Figure 11.26: Cycle sequencing involves linear amplification using a single primer to initiate DNA synthesis.

Cycle sequencing using the dideoxynucleotide method involves setting up four parallel DNA sequencing reactions in which DNA synthesis occurs, using a mix of all four dNTPs plus one of the four ddNTPs. The reactions resemble PCR reactions because they involve the same thermocycling format as PCR, with three steps: heat denaturation, annealing of primer, and DNA synthesis using a heat-stable DNA polymerase. However, since only a single primer is used, the product accumulates in a linear fashion, rather than exponentially as in PCR.

insert-specific sequencing primers: the new sequence data are used to design a new sequencing primer for sequencing into the next region.

- **Shotgun sequencing** is a method for building up a complete sequence by randomly sequencing clones with overlapping inserts. Large-scale DNA sequencing projects, such as genome projects, usually involve establishing cosmid contigs and then sequencing the inserts of whole cosmids. Although primer-walking sequencing strategies could, in principle, be applied, they are not very efficient. Instead the sequence is typically obtained using a **shotgun cloning** method for preparing DNA templates. To do this, the insert DNA is subjected to partial digestion with a 4-bp cutter and the partially overlapping fragments are cloned at random into a suitable M13 or phagemid vector and sequenced. The sequence data obtained are fed into a computer programmed to detect overlaps between sequences and to assemble a composite sequence. Inevitably, this means some wastage of effort: the same sequence can be obtained over and over again if a particular region is, by chance, represented by many different clones.

Further reading

Dracopoli NC *et al.* (1995) (eds) *Current Protocols in Human Genetics.* John Wiley & Sons, Chichester.

References

Abbott C, Povey S. (1991) *Genomics*, **9**, 73–77.

Abrams WR, MA R-I, Kucich U *et al.* (1995) *Genomics*, **26**, 47–54.

American PKD1 Consortium (1995) *Hum. Mol. Genet.*, **4**, 475–482.

Anand R, Riley JH, Butler R, Smith JC, Markham AF. (1990) *Nucl. Acid, Res.*, **18**, 1951–1956.

Antequera F, Bird A. (1993) *Proc. Natl Acad. Sci. USA*, **90**, 11995–11999.

Bartholdi M, Meyne J, Albright K *et al.* (1987) *Methods Enzymol.*, **151**, 252–267.

Bellanné-Chantelot C, Lacroix B, Ougen P *et al.* (1992) *Cell*, **70**, 1059–1068.

Church DM, Stotler CJ, Rutter JL, Murrell JR, Trofatter JA, Buckler AJ. (1994) *Nature Genetics*, **6**, 98–105.

Claudio JO, Marineau C, Rouleau G. (1994) *Hum. Mol. Genet.*, **3**, 185–190.

Collins F. (1991) *Nature Genetics*, **1**, 3–6.

Cotter F, Nasipuri S, Lam G, Young BD. (1989) *Genomics*, **5**, 470–474.

Cox DR, Burmeister M, Price ER, Kim S, Myers RM. (1990) *Science*, **250**, 245–250.

Cross SH, Bird AP. (1995) *Curr. Opin. Genet. Dev.*, **5**, 309–314.

Dauwerse JG, Wiegant J, Raap AK, Breuning MH, van Ommen GJB. (1992) *Hum. Mol. Genet,*. **1**, 593–598.

Davies K, Young BD, Elles RG, Hill ME, Williamson R. (1981) *Nature*, **293**, 374.

Faraco J, Bashir M, Rosenbloom J, Francke U. (1995) *Genomics*, **25**, 630–637.

Fournier REK, Ruddle FH. (1977) *Proc. Natl Acad. Sci. USA*, **74**, 319–323.

Haaf T, Ward D. (1994) *Hum. Mol. Genet.*, **3**, 629-–33.

Heiskanen M, Hellsten E, Kallioniemi O-P, Makela TP, Alitalo K, Peltonen L, Palotie A. (1995) *Genomics*, **30**, 31–36.

Houseal TW, Klinger KW. (1994) *Hum. Mol. Genet.*, **3**, 1215–216.

Ireland M, English C, Cross I, Lindsay S, Strachan T. (1995) *J. Med. Genet.*, **32**, 837–838.

Kretz K, Callen W, Hedden V. (1994) *PCR Methods Applic.*, **3**, S107–S112.

Larsen F, Gundersen G, Lopez R, Prydz H. (1992) *Genomics*, **13**, 1095–1107.

Lichter P, Tang CJ, Call K *et al.* (1990) *Science*, **247**, 64–69.

Lopez R, Larsen F, Prydz H. (1994) *Genomics*, **24**, 133–136.

Lovett M. (1994) *Trends Genet.*, **10**, 352–357.

Porteous DJ. (1987) *Trends Genet.*, **3**, 177–182.

Riley J, Butler R, Ogilvie D *et al.* (1990) *Nucleic Acids Res.*, **18**, 2887–2890.

Sargent CA, Dunham I, Campbell RD. (1989) *EMBO J.*, **8**, 2305–2312.

Trask BJ. (1991) *Trends Genet.* **7**, 149–154.

Valdes JM, Tagle DA, Collins FS. (1994) *Proc. Natl Acad. Sci. USA*, **91**, 5377–5381.

van Ommen G-JB, Breuning MH, Raap AK. (1995) *Curr. Opin. Genet. Dev.*, **5**, 304–308.

Walter MA, Goodfellow PN. (1993) *Trends Genet.*, **9**, 352-356.

Walter MA, Spillett DJ, Thomas P, Weiseenbach J, Goodfellow PN. (1994) *Nature Genetics*, **7**, 22–28.

Warburton D, Gersen D, Yu M-T, Jackson C, Handelin B, Houseman D. (1990) *Genomics*, **6**, 358–366.

Wilke CM, Guo S-W, Hali BK *et al.* (1994) *Genomics*, **22**, 319–326.

Wilson RK, Chen C, Avdalovic N, Burns J, Hood L. (1990) *Genomics*, **6**, 626–634.

Genetic mapping

Chapter 12

Recombinants and nonrecombinants

In principle, genetic mapping in humans is exactly the same as genetic mapping in any other sexually reproducing diploid organism. The aim is to discover how often two loci are separated by meiotic recombination. Consider a person who is heterozygous at two loci and so types as A_1A_2 B_1B_2. Suppose the alleles A_1 and B_1 in this person came from one parent, and A_2 and B_2 from the other. Any of that person's children who inherit one of these parental combinations (A_1B_1 or A_2B_2) are **nonrecombinant**, whereas children who inherit A_1B_2 or A_2B_1 are **recombinant** (*Figure 12.1*). The proportion of children who are recombinant is the **recombination fraction** between the two loci A and B.

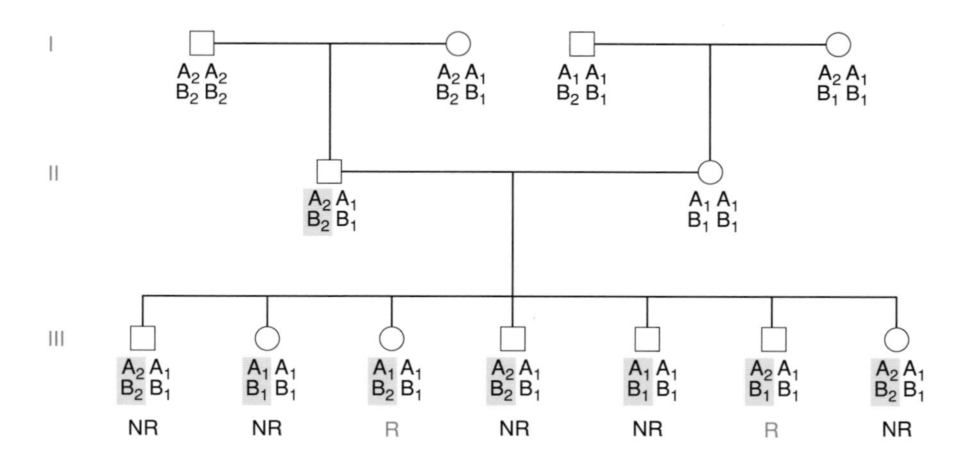

Figure 12.1: Recombinants and nonrecombinants.

Alleles at two loci (locus A, alleles A_1 and A_2; locus B, alleles B_1 and B_2) are segregating in this family. Where this can be deduced, the combination of alleles a person received from his or her father is boxed. Persons in generation III who received either A_1B_1 or A_2B_2 from their father are the product of nonrecombinant sperm; persons who received A_1B_2 or A_2B_1 are recombinant. The information shown does not enable us to classify any of the individuals in generations I and II as recombinant or nonrecombinant, nor does it identify recombinants arising from oogenesis in individual II_2.

Genetic and physical map distances

The recombination fraction is a measure of genetic distance

If the two loci are on different chromosomes they will segregate independently. At the end of meiosis I, whichever daughter cell receives allele A_1, there is a 50% chance that it will receive allele B_1 and a 50% chance it will receive B_2. Thus, on average, 50% of the children will be recombinant and 50% nonrecombinant. The recombination fraction is 0.5. If the loci are **syntenic**, that is if they lie on the same chromosome, then they might be expected always to segregate together, with no recombinants. However, this simple expectation ignores meiotic recombination. During prophase of meiosis I, pairs of homologous chromosomes synapse and exchange segments (*Figure 2.11*). A crossover, if it occurs between the positions of the two loci, will create two recombinant chromatids carrying A_1B_2 and A_2B_1. Bivalents in prophase I of meiosis, when crossovers are visible as chiasmata, consist of four chromatids. Only two chromatids are involved in any particular crossover. Thus one crossover creates two recombinant chromatids and leaves two nonrecombinant, giving 50% recombinants.

Recombination will rarely separate loci which lie very close together on a chromosome, because only a crossover located precisely in the small space between the two loci will create recombinants. Therefore sets of alleles on the same small chromosomal segment tend to be transmitted as a block through a pedigree. Such a block of alleles is known as a **haplotype**. Haplotypes mark recognizable chromosomal segments which can be tracked through pedigrees and through populations. When not broken up by recombination, haplotypes can be treated for mapping purposes as alleles at a single highly polymorphic locus.

The further apart two loci are on a chromosome, the more likely it is that a crossover will separate them. Thus the recombination fraction is a measure of the distance between two loci. Recombination fractions define **genetic distance**, which is not the same as **physical distance**. Two loci which show 1% recombination are defined as being 1 **centimorgan (cM)** apart on a genetic map. However, for distances above about 5 cM, human genetic map distances are not simple statements of the recombination fraction between pairs of loci. Loci which are 40 cM apart on a genetic map will show rather less than 40% recombination. This reflects the fact that recombination fractions never exceed 0.5, however far apart the loci are.

Recombination fractions do not exceed 0.5 however great the physical distance

A single recombination event produces two recombinant and two nonrecombinant chromatids. When loci are well separated there may be more than one crossover between them. Double crossovers can involve two, three or four chromatids, but *Figure 12.2* shows that the overall effect, averaged over all double crossovers, is to give 50% recombinants. Loci very far apart on the same chromosome might be separated by three, four or more crossovers. Again, the overall effect is to give 50% recombinants.

Mapping functions define the relationship between recombination fraction and genetic distance

The mathematical relationship between recombination fraction and genetic map distance is described by the **mapping function**. A mapping function is needed in multilocus mapping (page 324) to convert the raw data on the recombination

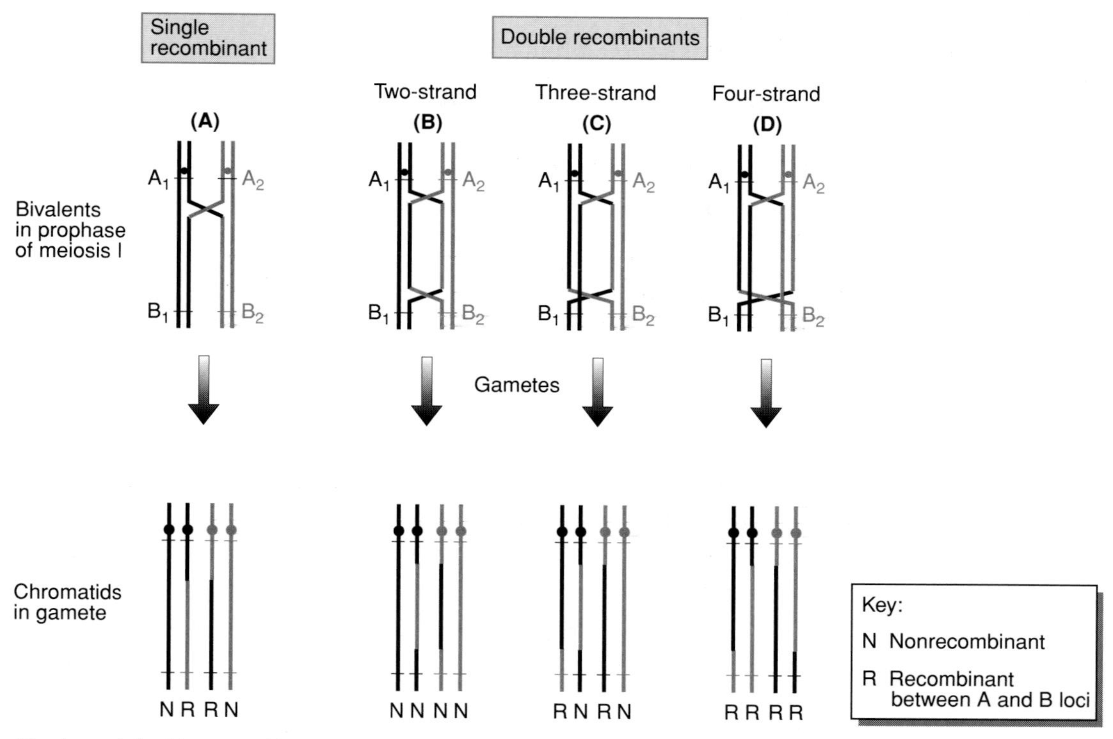

Figure 12.2: Single and double recombinants.

Each crossover involves two of the four chromatids of the two synapsed homologous chromosomes. The black chromosome carries alleles A_1 and B_1 at two loci, while the red chromosome carries alleles A_2 and B_2. Gametes in which the chromatid is the same color at the two loci are nonrecombinant for these loci, those where the chromatids are different colors are recombinant.
(A) A single crossover generates two recombinant and two nonrecombinant chromatids.
(B) A two-strand double crossover leaves flanking markers nonrecombinant on all four chromatids.
(C) A three-strand double crossover leaves flanking markers recombinant on two of the four strands.
(D) A four-strand double crossover generates 100% recombinants. The three types of double crossover occur in random proportions, so the average effect of a double crossover is to give 50% recombinants.

fraction into a genetic map. The simplest map function, Haldane's function, assumes that crossovers occur at random along a bivalent and have no influence on one another. This gives a map function

$$w = -\tfrac{1}{2}\ln(1 - 2\theta)$$

or

$$\theta = \tfrac{1}{2}[1 - \exp(-2w)],$$

where w is the map distance and θ the recombination fraction; as usual ln means logarithm to the base e, and exp means 'e to the power of'.

However, we know that the assumption that crossovers occur at random is not true. The presence of one chiasma inhibits formation of a second chiasma nearby. This phenomenon is called **interference**. Interference is well studied in *Drosophila* and yeast, and has recently been demonstrated formally in man (Schmitt *et al.*, 1994). A variety of mapping functions exist which allow for varying degrees of interference, and it is not clear which is the most appropriate for human mapping. A widely used mapping function is Kosambi's function:

$$w = \tfrac{1}{4}\ln[(1 + 2\theta)/(1 - 2\theta)]$$

or

$$\theta = \tfrac{1}{2}[\exp(4w) - 1]/[\exp(4w) + 1].$$

The interested reader should consult Ott's book (see Further reading) for a fuller discussion of mapping functions.

The relation between physical and genetic distances is not constant

Chiasma counts in human male meiosis show an average of 49 crossovers per cell (Morton *et al.*, 1982). Since each crossover gives 50% recombinants, the chiasma count implies a total male genetic map length of 2450 cM. The best estimate from linkage mapping, adding together all the chromosome lengths, is 2644 cM (Gyapay *et al.*, 1994). Chiasmata are more frequent in female meiosis (exemplifying Haldane's rule that the heterogametic sex has the lower chiasma count), and the total female map length in the study of Gyapay *et al.* (1994) was 4481 cM. Thus over the 3000 Mb genome, 1 male cM averages 1.13 Mb and 1 female cM averages 0.67 Mb; the sex-averaged figure is 1 cM = 0.9 Mb. These figures are useful rules of thumb but the actual correspondence varies widely for different chromosomal regions. The most extreme deviation is shown by the pseudoautosomal region at the tip of the short arms of the X and Y chromosomes (see *Figure 9.7*). Males have an *obligatory* crossover within this 2.6-Mb region, so that it is 50 cM long. Thus, for this region in males 1 Mb = 19 cM, whereas in females 1 Mb = 2.7 cM. In general, there is more recombination towards the telomeres of chromosomes and less towards the centromeres, especially in males (*Figure 12.3*). Uniquely, the Y chromosome, outside the pseudoautosomal region, has no genetic map because it is not subject to synapsis and crossing over in normal meiosis. The X chromosome of course undergoes normal recombination in females, and can be genetically mapped in female meioses.

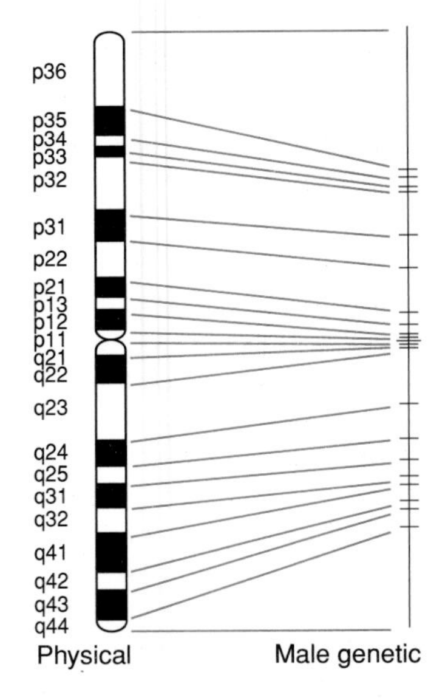

Figure 12.3: Comparison of physical map and male genetic map of chromosome 1.

The number of chiasmata within each chromosome band was noted on 232 bivalents from six men. On the genetic map, the length of a band (between adjacent cross-lines) is proportional to the fraction of all chiasmata falling in that band. Regions which are long on the genetic map but short on the physical map are hotspots for recombination. *Note* the paucity of crossovers near the centromere and the high frequency in the terminal bands 1p36 and 1q44. Data of Hultén *et al.* (1982).

Genetic markers

Human genetic mapping required the development of genetic markers

Since most people are interested in genes rather than the noncoding intergenic DNA, the best map would seem to be one showing the order and distances apart of all the genes. Maps of this sort are readily constructed in yeast, *Drosophila* and the mouse, but not in humans. Human families in which two diseases are segregating are extremely rare. Even if such families can be found, they may well have very few children or be unsuitable for genetic analysis in some other way. For this reason most human genetic mapping uses **markers**. There are two general approaches:

(i) **disease–marker mapping** is used for locating disease genes.

(ii) **Marker–marker mapping** is used to construct framework maps of markers. These aid high-resolution disease–marker mapping and help to relate genetic and physical maps by locating the same loci on maps of both types.

Genetic markers are mendelian characters which are sufficiently polymorphic to give a reasonable chance that a randomly selected person will be heterozygous. When a polymorphic marker is used, families can be selected for linkage analysis because they have an interesting disease or because they have a good structure for mapping, with a reasonable hope that family members will not all be homozygous (and hence uninformative) for the marker.

Requirements for gene mapping

Gene mappers could not set out to map a disease with a reasonable hope of success until markers were available which were spaced throughout the genome. Although in theory linkage can be detected between loci 40 cM apart, the amount of data required to do this is prohibitive. Ten meioses are sufficient to give evidence of linkage if there are no recombinants, but 85 meioses would be needed to give equally strong evidence of linkage if the recombination fraction was 0.3 (see *Figure 12.5* for a guide to these calculations). Obtaining enough family material to test much more than 50 meioses can be seriously difficult for a rare disease. Thus mapping requires markers spaced at intervals no greater than about 20 cM across the genome. Given the genome lengths calculated above, this means that we need a minimum of 200 markers. In fact much denser maps, down to 1 cM average spacing of markers, are needed to guide progress from initial mapping of a disease through to cloning the gene, and such maps are now becoming available (Gyapay *et al.*, 1994).

Any character which is mendelian and polymorphic can be used as a marker. It helps if the character can be scored easily and cheaply using readily available material, such as blood cells, but the most important thing is that it should be polymorphic. *Box 12.1* summarizes the development of human genetic markers, from blood groups and polymorphisms of serum proteins revealed by starch gel electrophoresis through to DNA microsatellites.

DNA polymorphisms as markers

DNA polymorphisms provided, for the first time, a set of markers which were sufficiently numerous and spaced across the entire genome. Their development allowed disease gene mapping to start in earnest. DNA markers have the additional advantages that they can all be typed by the same technique, and their chromosomal location can be found easily using FISH or other physical mapping methods (page 281). Thus DNA-based genetic markers can be cross-referenced to physical maps. This avoids the frustrating situation which arose when the long-sought cystic fibrosis gene (*CFTR*) was first mapped. Linkage was established to a protein polymorphism of the enzyme paraoxonase, but the chromosomal location of the paraoxonase gene was not known (Eiberg *et al.*, 1985).

The polymorphism information content measures how informative a marker is

Limitations of RFLPs as markers

The first generation of DNA markers were **restriction fragment length polymorphisms (RFLPs)**, owing their existence to **restriction site polymorphisms**. RFLPs were initially typed by hybridizing Southern blots of restriction digests with radiolabeled probes (see *Figure 5.11*). They suffer from two limitations. The original technology was quite laborious and required plenty of time, money and DNA. This

Box 12.1: The development of human genetic markers

Type of marker	No. of loci	Features
Blood groups 1910–1960	~ 20	May need fresh blood, rare antisera Genotype cannot always be inferred from phenotype because of dominance No easy physical localization
Electrophoretic mobility variants of serum proteins 1960–1975	~ 30	May need fresh serum, specialized assays No easy physical localization Often limited polymorphism
HLA tissue types 1970–	1 (haplotype)	One linked set Highly informative Can only test for linkage to 6p21.3
DNA RFLPs 1975–	$> 10^5$ (potentially)	Two allele markers, maximum PIC 0.375 Initially required Southern blotting, now PCR Easy physical localization
DNA VNTRs (minisatellites) 1985–	$> 10^4$ (potentially)	Many alleles, highly informative Type by Southern blotting Easy physical localization Tend to cluster near ends of chromosomes
DNA VNTRs (microsatellites) (di-, tri- and tetra-nucleotide repeats) 1989–	$> 10^4$ (potentially)	Many alleles, highly informative Can type by automated multiplex PCR Easy physical localization Distributed throughout genome

PIC, polymorphism information content; VNTR, variable number of tandem repeats

made a whole genome search using RFLPs an heroic, though by no means impossible, undertaking. Nowadays this is less of a problem because they can usually be typed by PCR. More fundamentally, their informativeness as markers is limited. For linkage analysis we need **informative meioses** (see *Box 12.2*).

The examples in *Box 12.2* show that a meiosis is not informative with a given marker if the subject is homozygous for the marker, and also in half of the cases where the subject is heterozygous but the spouse has the same heterozygous genotype. If there are marker alleles A_1, A_2, A_3... with gene frequencies p_1, p_2, p_3..., then the proportion of people who are homozygous is $p_1^2 + p_2^2 + p_3^2 + \ldots$ (page 76). The proportion of people who are heterozygous A_1A_2 is $2p_1p_2$, so the proportion of couples who are both heterozygous A_1A_2 is $4p_1^2p_2^2$. Half their children will also be A_1A_2 and therefore uninformative. Thus the **polymorphism information content (PIC)** of a marker is given by:

$$\mathrm{PIC} = 1 - \sum_{i=1}^{n} p_i^2 - \sum_{i=1}^{n} \sum_{j=i+1}^{n} 2p_i^2 p_j^2$$

where p_i is the frequency of the ith allele.

Box 12.2: Informative and uninformative meioses

A meiosis is **informative** for linkage when we can identify whether or not the gamete is recombinant. Consider the male meiosis which produced the paternal contribution to the child in the four pedigrees below:

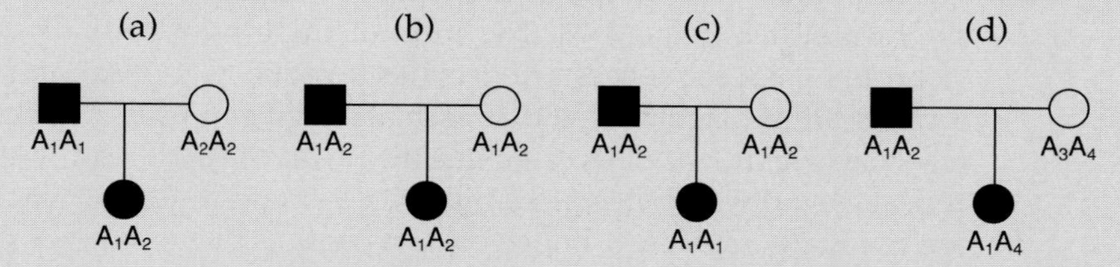

(a) This meiosis is uninformative: the marker alleles in the homozygous father cannot be distinguished.
(b) This meiosis is uninformative: the child could have inherited A_1 from father and A_2 from mother, or vice versa.
(c) This meiosis is informative: the child inherited A_1 from the father.
(d) This meiosis is informative: the child inherited A_1 from the father.
We assume that the father has a dominant condition that he inherited along with marker allele A_1.

The second term takes out homozygotes and the third term takes out half the matings of similar heterozygotes. *Table 12.1* gives examples.

Highly informative multiallelic markers

Restriction site polymorphisms have only two alleles: the site is present or it is absent. *Table 12.1* shows that restriction site polymorphism markers are useably informative if the rarer allele has a frequency of at least 0.2, but the PIC cannot exceed 0.375 for autosomal markers. Disease mapping using these polymorphisms was frustrating because all too often a key meiosis in a family turned out to be uninformative. **Minisatellite (VNTR) markers** were a great improvement. The VNTRs (page 196) gave a large number of alleles and high PIC values. With minisatellites, most meioses are informative. Still, the technical problems of Southern blotting and radioactive probes were an obstacle to easy mapping.

PCR changed all this and finally made mapping quick and easy. Primer sequences are available for many of the old two-allele restriction site polymorphisms. A sequence including the variable restriction site is amplified by PCR, the product is incubated with the appropriate restriction enzyme and then run out on a gel to see

Table 12.1: Polymorphism information content (PIC) of a marker

Location of marker A = autosomal, X = X-linked	Alleles (frequencies)	Heterozygosity	PIC
A	1 (1)	0	0
A	2 (0.5, 0.5)	0.50	0.375
A	2 (0.4, 0.6)	0.480	0.365
A	2 (0.3, 0.7)	0.420	0.332
A	2 (0.2, 0.8)	0.320	0.267
A	2 (0.1, 0.9)	0.180	0.164
A	4 (0.25, 0.25, 0.25, 0.25)	0.750	0.703
A	10 (all 0.1)	0.900	0.891
X	2 (0.5, 0.5)	0.500	0.500
X	4 (0.25, 0.25, 0.25, 0.25)	0.750	0.750

A marker with a PIC of 0 is never informative, one with a PIC of 1 is always informative. The PIC for autosomal markers is calculated by the formula shown above (page 318). For X-linked, but not autosomal, markers the PIC is equal to the heterozygosity.

if it has been cut (see *Figure 6.4*). These RFLPs can still be useful if strategically placed near a gene of interest. Classical minisatellites are difficult to handle by standard PCR protocols because large alleles may fail to amplify. Additionally, they tend to be clustered in subtelomeric regions of chromosomes. Thus, minisatellites are now little used for mapping. The standard tools for linkage analysis are now **microsatellites** (page 197). The bulk of these are $(CA)_n$ repeats. However, dinucleotide repeat sequences are peculiarly prone to replication slippage during PCR amplification (page 270) so that each allele gives a little ladder of 'stutter bands' on a gel (see *Figure 6.6*). This can make the results hard to read. Tri- and tetranucleotide repeats usually give clearer results with a single band from each allele, and so these are gradually replacing dinucleotide repeats as the markers of choice. Much effort is being devoted to producing compatible sets of microsatellite markers which can be amplified together in a multiplex PCR reaction, and have allele sizes which allow them to be run in the same gel lane without producing overlapping bands. With fluorescent labeling in several colors and automated gels, the goal of mapping before lunch looks attainable.

Two-point mapping

Lod scores are the statistical measure of the evidence for linkage

Having collected suitable families and typed them with an informative marker, how do we know when we have found linkage? Suppose we find two recombinants in 20 meioses. If the disease and marker are unlinked, we would expect 10 recombinants. A chi-squared (χ^2) test gives a value $[(2–10)^2/10 + (18–10)^2/10] = 12.8$ with 1 degree of freedom. This is a highly significant deviation from the null hypothesis of 50% recombinants. What this calculation ignores is the inherent improbability that two loci, chosen at random, should be linked. With 22 pairs of autosomes to choose from, it is not likely they would be located on the same chromosome (syntenic) and, even if they were, loci well separated on a chromosome are unlinked.

The prior probability of linkage

The likelihood that two loci should be linked (the prior probability of linkage) has been much argued over, but estimates of about one in 50 are widely accepted. Common sense tells us that if something is inherently improbable, we require strong evidence to convince us it is true. This common sense can be quantified in a Bayesian calculation (see *Box 12.3)*, showing that the conventional $p = 0.05$ threshold of significance requires 1000:1 odds from the linkage analysis.

Scoring recombinants

When families are large and recombinants can be counted, analysis is simple – but this happy state is all too rare. Families with interesting diseases are seldom large enough for results from one family alone to reach statistical significance. Thus we need to combine data from several families. Moreover, the imperfect structure of human families means that often recombinants cannot be identified unambiguously. *Figure 12.4* shows an example of both problems. If this is a rare disease, no researcher would be willing to discard the family. Some method is needed to extract the linkage information from a collection of such imperfect families. This is achieved by using computer-generated **lod scores**.

Box 12.3: Bayesian calculation of linkage threshold

Hypothesis: loci are	**linked** (recombination fraction = θ)	**not linked** (recombination fraction = 0.5)
Prior probability	1/50	49/50
Conditional probability: 1000:1 odds of linkage lod score $Z(\theta) = 3.0$	1000	1
Joint probability (prior × conditional)	20	~ 1

Because of the low prior probability that two randomly chosen loci should be linked, evidence giving 1000:1 odds in favor of linkage is required in order to give overall 20:1 odds in favor of linkage. This corresponds to the conventional $p = 0.05$ threshold of statistical significance. The calculation is an example of the use of Bayes' formula to combine probabilities (see *Box 16.1* and *Figure 16.14*). See text for description of the lod score.

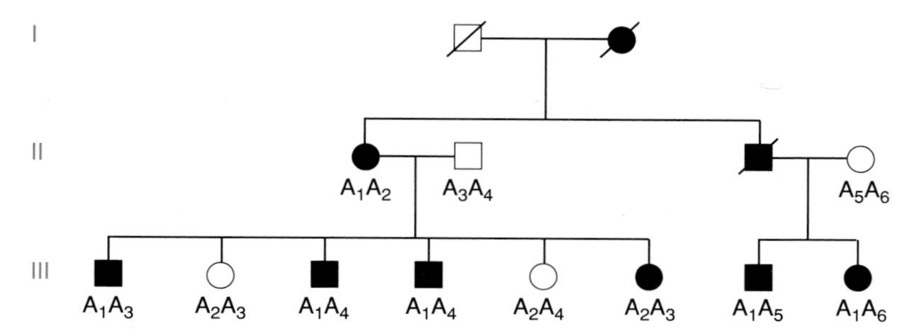

Figure 12.4: A family where recombinants cannot be identified unambiguously.

The autosomal dominant disease (filled symbols) is most likely segregating with marker allele A_1, and individual III_6 is the only recombinant. However, we cannot be certain that individual II_1 did not inherit the disease with marker allele A_2 from her mother, in which case there are many recombinants. III_7 and III_8 have also inherited marker allele A_1 along with the disease from their father, but we cannot be sure whether their father's allele A_1 is identical by descent to the allele A_1 in his sister II_1. There may be two copies of allele A_1 among the four grandparental marker alleles. The likelihood of this depends on the gene frequency of allele A_1. Thus, although this pedigree contains linkage information, extracting it is problematic.

The lod score, Z, introduced by Morton (1955), is the logarithm of the odds that the loci are linked (with recombination fraction θ) rather than unlinked (recombination fraction 0.5). Being a function of the recombination fraction, lod scores are calculated for a range of θ values and the maximum value $\hat{Z}$ estimated. The overall probability of linkage in a set of families is the product of the probabilities in each individual family, therefore lod scores (being logarithms) can be added up across families.

Calculating lod scores

Lod scores are calculated by looking at each meiosis in turn and comparing the likelihood of the observed genotypes on the alternative hypotheses of linkage (with recombination fraction θ) or no linkage. *Figure 12.5* gives two examples. In *Figure 12.5A* we could identify recombinants with certainty because the grandparental types enable us to define the mother's **phase**, that is the combination of alleles (haplotype) which she inherited from each parent. The meioses are **phase-known**. *Figure 12.5B* shows exactly the same family but with phase-unknown meioses because the

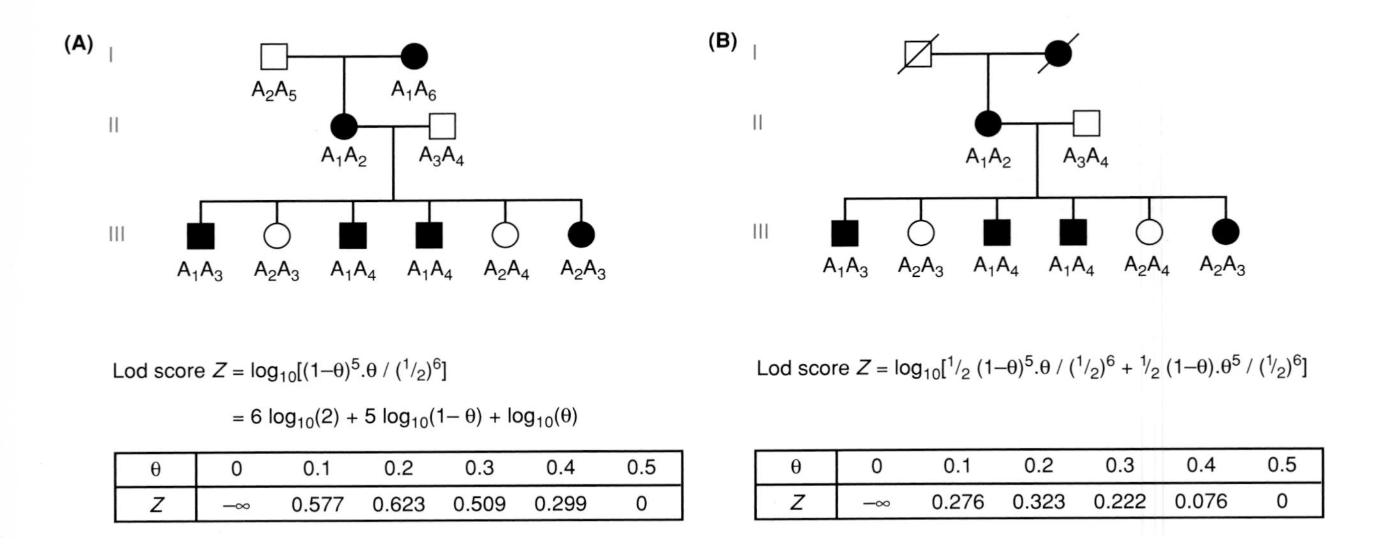

(A)

θ	0	0.1	0.2	0.3	0.4	0.5
Z	–∞	0.577	0.623	0.509	0.299	0

(B)

θ	0	0.1	0.2	0.3	0.4	0.5
Z	–∞	0.276	0.323	0.222	0.076	0

Figure 12.5: Calculation of lod scores

(A) All meioses are phase-known. We can identify III_1–III_5 unambiguously as nonrecombinant and III_6 as recombinant. For each meiosis, the likelihood of it being nonrecombinant, given that the loci are truly linked, is $1-\theta$ and the likelihood of it being recombinant is θ; if the loci are in fact unlinked the likelihood is 1/2. The overall likelihood of five nonrecombinants and one recombinant, given linkage, is $(1-\theta)^5.\theta$, and the likelihood given no linkage is $(1/2)^6$. The lod score is the logarithm of the ratio of these likelihoods.
(B) The same family, but phase-unknown. The mother, II_1, could have inherited either marker allele A_1 or A_2 with the disease; thus her phase is unknown. The lod score calculation must allow for either possible phase, with equal prior probability. The overall result is obtained by summing each lod score, weighted by its prior probability.

grandparents are not available. The lod score calculation must allow for either possible phase in the mother, with equal prior probability. The overall result is obtained by summing each lod score weighted by its prior probability.

Calculating the full lod score for the family in *Figure 12.4* is more difficult. To calculate the likelihood that III_7 and III_8 are recombinant or nonrecombinant, we must take likelihoods calculated for each possible genotype of I_1, I_2 and II_3, weighted by the probability of that genotype. For I_1 and I_2, the genotype probabilities depend on both the gene frequencies and the observed genotypes of II_1, III_7 and III_8. Genotype probabilities for II_3 are then calculated by simple mendelian rules. The branching tree of probabilities is calculable, but only by computer. Thus human linkage analysis, except in the very simplest cases, is entirely dependent on computer programs, all of which implement a procedure (the Elston–Stewart algorithm) for handling these branching trees of genotype probabilities given a pedigree structure and a table of gene frequencies. The best known programs are LIPED and MLINK (part of a package called Linkage). Both of these can be run on a personal computer. The general theory of linkage analysis is excellently covered in the book by Ott (see Further reading), while the book by Terwilliger and Ott (see Further reading) is full of practical advice indispensable to anybody undertaking human linkage analysis.

Lod scores of +3 and –2 are the criteria for linkage and exclusion

All lod scores are zero at $\theta = 0.5$ since they are then measuring the ratio of two identical probabilities, and $\log_{10}(1) = 0$. For smaller recombination fractions the critical thresholds for a single test are $Z = 3.0$ and $Z = -2.0$ (*Figure 12.6*). As briefly explained above (*Box 12.3*), $Z = 3.0$ or 1000:1 odds is the threshold for accepting linkage, with a 5% chance of error. Linkage can be rejected if $Z < -2.0$. Values of Z between –2 and +3 are inconclusive. See below for discussion of multiple tests; for a fuller discussion of these thresholds, the reader should consult Ott's book (see Further reading).

Positive lod scores

If there are no recombinants, the lod score will be maximum at θ = 0. If there are recombinants, Z will peak at the most likely recombination frequency (0.167 = 1/6 for the family in *Figure 12.5A*, but harder to predict for *Figure 12.5B*). Confidence intervals are hard to deduce analytically, but a widely accepted rule of thumb suggests they extend to recombination fractions at which the lod score is 1 unit below the peak value (the '**lod-1 rule**'). Thus, curve 2 in *Figure 12.6* gives acceptable evidence of linkage ($Z > 3$) with the most likely recombination fraction 0.23 and confidence intervals 0.17–0.32. The curve will be more sharply peaked the greater the amount of data, but in general peaks are quite shallow. It is important to remember that distances on human genetic maps are often very imprecise estimates.

Negative lod scores: exclusion mapping

Negative lod scores exclude linkage for the region where $Z < -2$. Curve 3 on *Figure 12.6* excludes the disease from 12 cM either side of the marker. While gene mappers hope for a positive lod score, exclusions are not without value. They tell us where the disease is not, and therefore narrow down the range of locations where it might be. If enough of the genome is excluded, only a few possible locations may remain. The EXCLUDE program (Edwards, 1987) turns negative linkage data into a diagram of remaining candidate locations (*Figure 12.7*). Note, however, that there are two fundamentally different types of exclusion data. If a marker gives negative lod scores in families, then typing an independent set of families for a nearby marker can greatly extend the region of exclusion; but typing the original families for a second closely linked marker gives very little extra information – we already knew the disease did not map to that position in those families. The EXCLUDE program assumes that the lod scores are independent, and so can exaggerate the true exclusion achieved in studies based on repeated typing of a fixed collection of families.

The threshold of significance in whole genome searches

In disease studies, families are typed for marker after marker until positive lods are obtained. The question arises whether the $Z = 3$ threshold is still appropriate. If 50

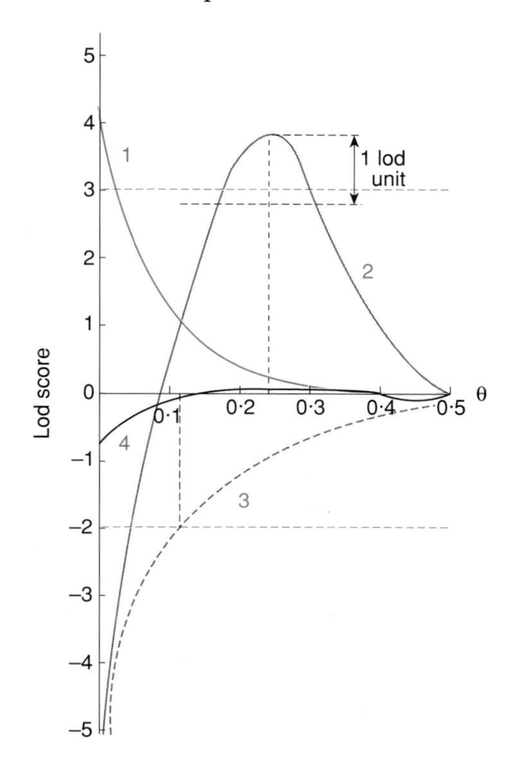

Figure 12.6: Lod score curves.

Graphs of lod score against recombination fraction from a hypothetical set of linkage experiments. Curve 1: evidence of linkage ($Z > 3$) with no recombinants. Curve 2: evidence of linkage ($Z > 3$) with the most likely recombination fraction being 0.23. Curve 3: linkage excluded ($Z < -2$) for recombination fractions below 0.12; inconclusive for larger recombination fractions. Curve 4: inconclusive at all recombination fractions.

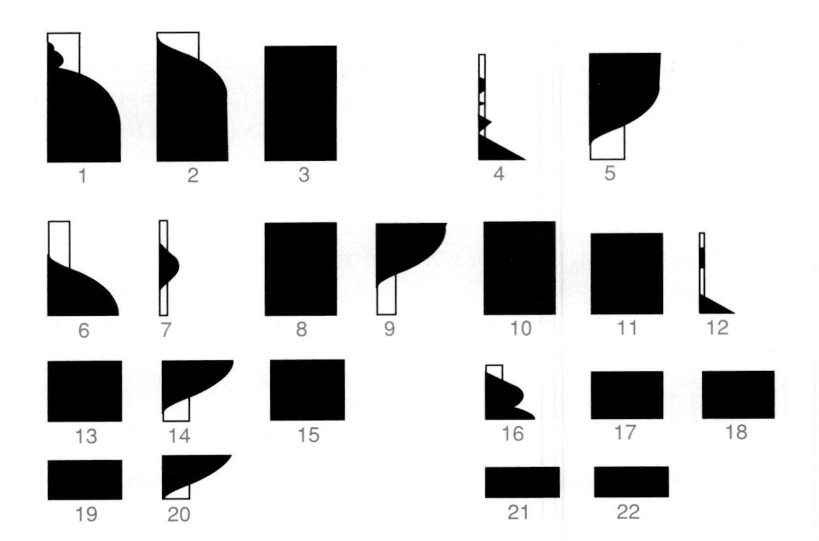

Figure 12.7: Exclusion map.

A disease gene is being sought by linkage analysis using a large number of markers. At this stage in the study all lod scores are negative. Each autosome is initially represented as a solid rectangle. Chromosomal regions excluded by negative linkage data are white, with the length of the black bar representing the probability that the disease may map to that position. Reproduced from Foy *et al.* (1990) with permission from The University of Chicago Press.

markers have been used, the chance of a spurious positive result is greater than if only one marker is used. A stringent procedure would multiply the p value by 50 before testing its significance. The threshold lod score for a study using n markers would be $3 + \log(n)$, that is a lod score of 4 for 10 markers, 5 for 100, etc. However, this may be over stringent. Linkage data are not independent. If a gene does not map to one location, then the prior probability that it maps to another location is raised. This is a difficult area, but it is generally agreed that the full correction is over stringent, at least for an extensive search through the whole genome. In practice, lod scores below 5, whether with one marker or many, should be regarded as provisional. Confirmatory evidence could include:

- more linkage data to raise the lod score;
- a patient with the disease who has a chromosomal abnormality at or encompassing that location;
- mapping of a plausibly homologous gene in the mouse to the corresponding mouse location.

To put the problem in perspective, many of the larger disease mapping studies are producing lods of 20–100, far beyond the reach of these worries.

Multipoint mapping

Multipoint linkage can locate a disease locus on a framework of markers

Linkage analysis can be more efficient if data for more than two loci are analyzed simultaneously. Multilocus analysis is particularly useful for establishing the chromosomal order of a set of linked loci. Experimental geneticists have long used three-point crosses for this purpose. The rarest recombinant class is that which requires a double recombination. In *Table 12.2*, the gene order A–C–B is immediately apparent. This procedure is more efficient than estimating the recombination fractions for intervals A–B, A–C and B–C separately in a series of two-point crosses.

A second advantage of multilocus mapping in humans is that it helps overcome problems caused by the limited informativeness of markers. Some meioses in a

Table 12.2: Gene ordering by three-point cross

Class of offspring	Position of recombination (x)	Number
ABC/abc abc/abc	Nonrecombinant	853
ABc/abc abC/abc	(A, B)–x–C	5
Abc/abc aBC/abc	A–x–(B,C)	47
AbC/abc aBc/abc	B–x–(A,C)	95

A cross has been set up between mice heterozygous at three linked loci (ABC/abc) and triple homozygotes abc/abc. The offspring are classified as shown. The rarest class of offspring will be those whose production requires two crossovers. Of the 1000 animals, 142 (95+47) are recombinant between A and B, 52 (47+5) between A and C, and 100 (95+5) between B and C. Only five animals are recombinant between A and C, but not between A and B, so these must have double crossovers, A–x–C–x–B. Therefore the map order is A–C–B. The genetic distances are approximately A–(5 cM)–C–(10 cM)–B.

family might be informative with marker A, and others uninformative for A but informative with the nearby marker B. Only simultaneous linkage analysis of the disease with markers A and B extracts the full information. This is less crucial now that we have highly informative microsatellite markers rather than two-allele RFLPs, but it remains an advantage of multipoint linkage. For disease–marker mapping, multipoint mapping has limited utility (see below), but for marker–marker mapping, where the aim is to create an ordered framework of markers, multipoint mapping is essential.

Multipoint mapping by computer

For disease–marker mapping the starting point is usually a 2-point lod score showing that the disease maps near one particular marker, plus a marker map locating that marker within a framework of markers. The marker map is taken as given, and the aim is to locate the disease gene in one of the intervals of the marker map. The LINKMAP program from the Linkage package is normally used for this. LINKMAP notches the disease locus across the marker framework, calculating the overall likelihood of the pedigree data at each position. The result (*Figure 12.8*) is a curve of likelihood against map location. The highest peak marks the most likely position of the disease locus. LINKMAP is also useful for exclusion mapping: if the curve stays below a lod score of –2 across the region, then it is excluded from that region. One point to check when considering LINKMAP curves is the likelihood measure used on the *y*-axis. The measure originally proposed was the **location score**, calculated as

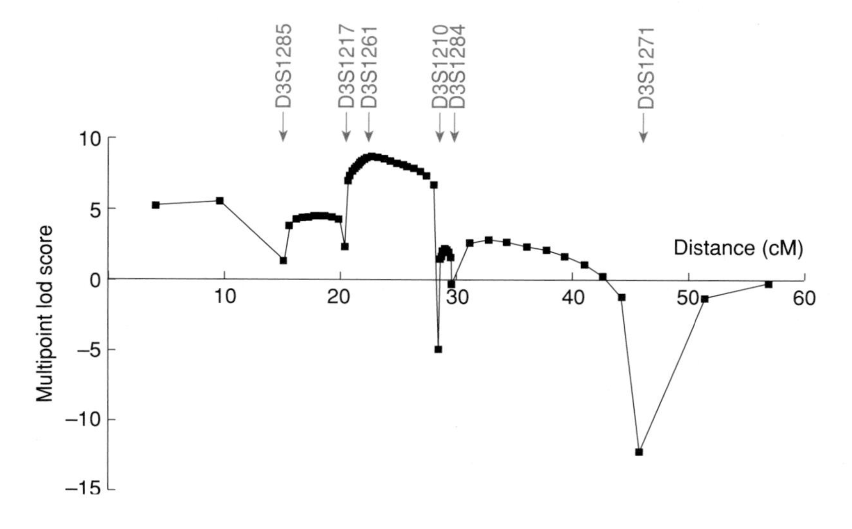

Figure 12.8: Multilocus mapping in man.

The horizontal axis is a map of markers and the vertical axis is the lod score. The LINKMAP program has moved the unmapped disease locus across the map, calculating the lod score at each position. Lod scores dip to strongly negative values near to the position of markers which show recombinants with the disease. The highest peak shows the most likely location. Odds in favor of this position are measured by the degree to which the highest peak overtops its rivals. Redrawn from Hughes *et al.* (1994) with permission from the author.

$-2\ln(LR)$ where ln means a natural logarithm (to the base e), and LR is the likelihood ratio for this location versus a location off the end of the map. The justification was that this measure should have χ^2 distribution, and so allow a p value to be calculated. Most recent publications use $\log_{10}(LR)$, the **multipoint lod score**, as in *Figure 12.8*. One unit of a multipoint lod score is 4.6 units of a location score.

Limitations of multipoint maps for locating disease genes

In our opinion, the use of multipoint linkage for disease–marker mapping is largely obsolete, except for exclusion mapping. Microsatellite markers make most meioses informative with most markers, and the real data are summarized by a table showing the position of each recombination on the map of markers. Little is gained by formal multipoint analysis. The apparently quantitative nature of LINKMAP output from analysis of this type of problem is largely spurious. Computational difficulties usually force the researcher to make simplifying assumptions: marker genotypes are often recoded so as to reduce the number of alleles considered, and usually the curve is computed in segments. These necessary simplifications mean the results are only semi-quantitative. Worse, peak heights depend crucially on the precise distances assigned to the marker framework, which in reality are seldom accurately known.

Multipoint linkage is essential for constructing marker–marker maps

The CEPH families

Disease–marker mapping suffers from the necessity of using whatever families can be found where the disease of interest is segregating. Such families will rarely have ideal structures. All too often the number of meioses is undesirably small, and missing persons mean that some meioses are phase-unknown. Marker–marker mapping avoids these problems. Most workers use a common collection of families assembled specifically for the purpose by the Centre pour l'Étude des Polymorphismes Humaines in Paris. These **CEPH families** have ideal structures for linkage: three generations, all four grandparents available and at least eight children. Immortalized cell lines from every individual ensure a permanent supply of DNA. Sample mix-ups and nonpaternity have long since been ruled out by typing with many markers.

The ideal structure of the CEPH families greatly eases the computational problems of large-scale linkage analysis – in disease mapping the programs spend most of their time trying to work out genotypes of people for whom there is inadequate information because of nonoptimal pedigree structures. Researchers who have generated markers can obtain DNA from the CEPH families in return for agreeing to submit their data to a common database. As this expands, increasingly it is becoming possible to localize every recombinant on every chromosome in every individual. Once this is achieved, a new marker can be mapped very easily. After assignment to a chromosome band by some physical localization technique, only a very small number of individuals with known recombinants in that region need be typed to place the new marker on the map. One minor drawback of using these standardized families is that there are only a finite number of recombinants, so that the end result will be to bin markers into a large but finite number of chromosomal segments, with no possibility of analysis to finer levels of discrimination.

Ordering loci on marker–marker maps

For constructing marker maps, multipoint linkage is essential. Ordering the loci is not a trivial problem. There are $n!/2$ possible orders for n markers. The latest marker maps (Gyapay *et al.*, 1994) average almost 100 markers per chromosome.

Something more intelligent than brute force computing must be used to work out the correct order. Physical mapping information can be immensely helpful here. Markers which can be typed by PCR can be used as sequence-tagged sites (STS; *Box 11.2*) and grouped into physically localized sets using somatic cell hybrids, radiation hybrids or YAC clones (page 291). Within sets, a hierarchical linkage analysis can identify the most unambiguously ordered subset of markers, then test a limited number of ways of slotting in the remaining markers by calculating the effect of inverting adjacent pairs. Markers are usually placed on published maps when the odds favoring the given order over the next best one tried are at least 1000:1 (cf. *Figure 11.4* for an equivalent physical map).

Distances on marker–marker maps

Distances on multipoint maps are at best approximate. One problem concerns the need to specify the ratio of male to female recombination rates. Over the whole genome this is known fairly precisely from chiasma counts, but the ratio is known to vary widely from one chromosomal region to another. The usual method is to estimate the ratio along with the distances as one extra parameter in a maximum likelihood estimation. The real variation in sex-specific recombination rates along a chromosome may be too complicated for the linkage program to model. Also, of course, the more parameters included, the more degrees of freedom there are, and so the less confidence one can place in the result. A second problem relates to the appropriate mapping function (page 315). Multipoint linkage programs require the mapping function to be specified, and the uncertainty as to which one is appropriate is another reason why the distance estimates should be regarded as indicative only. The inevitable laboratory typing errors also pose problems, because their effect is usually to introduce spurious recombinants. Error-checking routines test the extent to which the map can be shortened by omitting any single test result. Results which significantly lengthen the map (i.e. add recombinants) are suspect.

The value of marker framework maps

We have spent some time emphasizing the limitations of current marker maps, but this is only in order to caution the reader against the natural tendency to take the diagrams literally. With all their imperfections, the current generation of marker maps represent a major milestone in the progress of the Human Genome Project. The maps are strong on marker order, which is their primary use, and most of their weaknesses on marker distance will be overcome by integrating physical and genetic approaches. They are an incomparable tool for both genetic mapping of diseases and physical mapping in preparation for large-scale sequencing (*Figure 12.9*).

Problems with standard lod score analysis

Standard lod score analysis is a tremendously powerful method for scanning the genome in 20-Mb segments to locate a disease gene, but it has some drawbacks. These include:

- the need to specify a precise genetic model, detailing the mode of inheritance, gene frequencies and penetrance of each genotype. Such knowledge is not always available (see page 329).
- Limits on the ultimate resolution achievable. Once a marker is found for which all meioses are informative and nonrecombinant, linkage analysis comes to a halt. In typical collections of disease families, the target region thus identified

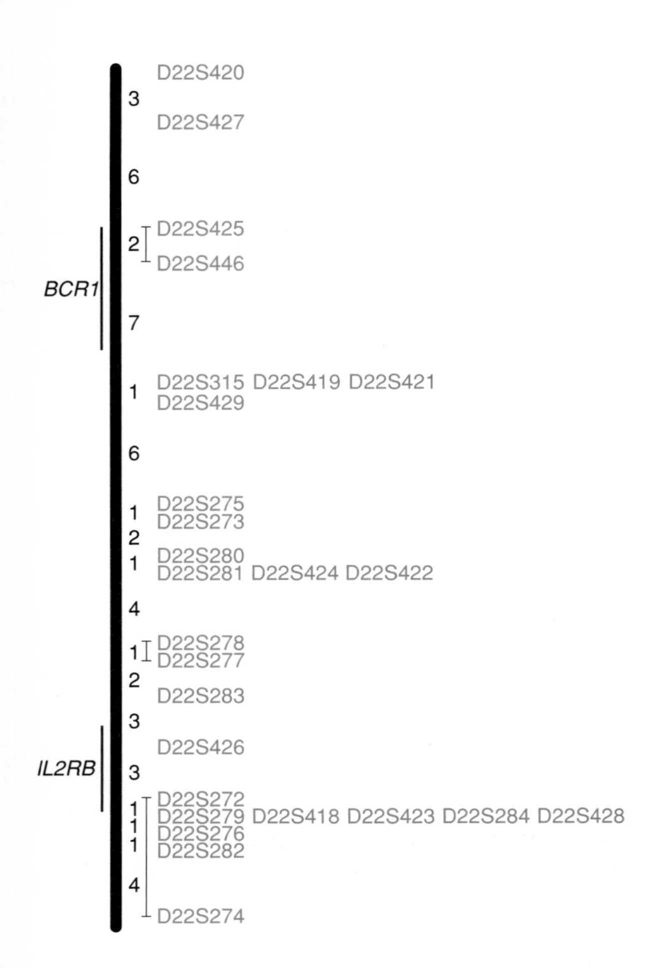

Figure 12.9: A high resolution genetic map of chromosome 22.

The loci shown are all microsatellite polymorphisms. Distances between loci are in cM. Approximate locations of two genes are shown to the left of the line. This map serves as a reference framework for mapping disease loci and for physical mapping. Redrawn from Gyapay *et al.* (1994) with permission from the author.

is likely to be 1 Mb or more, which is uncomfortably large for positional cloning of an unknown disease gene.

- Problems with locus heterogeneity. Where the same disease phenotype can be produced by mutations in two or more unlinked genes (page 65), linkage analysis can be extremely difficult. Even a dominant condition with large families can be problematic. It took years of collaborative work to show that tuberous sclerosis was caused by mutations at either of two loci, *TSC1* at 9q34 and *TSC2* at 16p13. With recessive conditions, the obstacles to direct linkage analysis may be almost insuperable, because the families are usually much smaller, so larger numbers of families need to be combined.
- Vulnerability to errors. False apparent recombinants can be generated by a number of errors (*Figure 12.10*).

Marker–marker linkage by typing sperm

One possible way to increase the resolution of marker–marker mapping is to type sperm instead of children. Humans have far too few children for optimal linkage analysis, but men produce untold millions of sperm, and modern PCR technology allows markers to be scored on single separated sperm from a doubly heterozygous man. Apart from technical problems, one drawback is that a single sperm cannot be resampled repeatedly to confirm interesting results, in the same way as a child can. Whole genome amplification (Zhang *et al.*, 1992; page 140) partially circumvents this problem. Individual spermatozoa are subjected to whole genome amplification followed by multiplex PCR amplification of markers from an aliquot. Further aliquots can be used to check any recombinants. Chakravarti (1994) discusses the first results

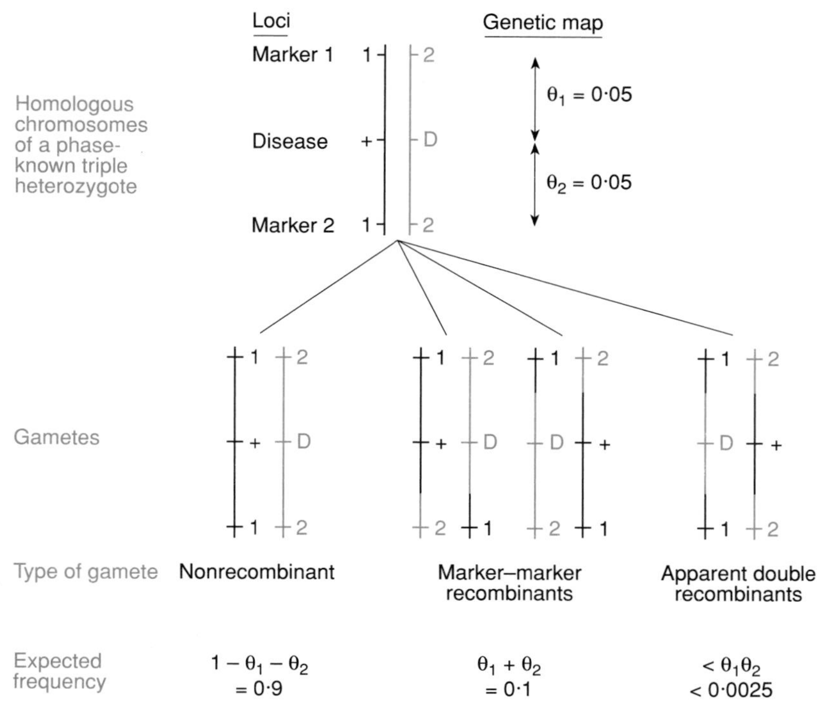

Figure 12.10: Apparent double recombinants suggest errors in the data.

Because of interference (page 315), the probability of a true double recombinant with markers 5 cM apart is exceedingly low, well below $0.05 \times 0.05 = 0.0025$. Apparent double recombinants usually signal an error in typing the markers, a clinical misdiagnosis or locus heterogeneity, so that the disease in this case does not map to locus D but elsewhere in the genome. Mutation in one of the genes or germinal mosaicism are rarer causes.

from this promising technique. Unfortunately sperm typing could not be used for disease–marker mapping, unless the disease mutations were already characterized.

Model-free linkage analysis

The need to specify a complete genetic model has been a serious problem for linkage analysis of imperfectly mendelian conditions. Behavioral geneticists in particular are acutely aware of the risk of generating spurious linkage through the use of a mis-specified model, as happened in schizophrenia (discussed by Byerley, 1989). One way around this problem has been the use of model-free methods of linkage analysis. These methods ignore unaffected people, but look for chromosomal segments that are shared by affected individuals.

It is important to distinguish segments **identical by descent (IBD)** from those **identical by state (IBS)**. If two sibs are both type 2-1 for a given marker, their alleles are certainly identical by state, but they may or may not be identical by descent. One or both parents could have been homozygous and the sibs could have inherited different chromosomes which happen to carry the same marker allele. Multiallele microsatellites are more efficient than two-allele markers for defining identity by descent, and multilocus multiallele haplotypes are better still.

Shared segment methods can be used within nuclear families (sib pair analysis, see below), within known extended families, or in populations that are descended from a small founder group. In the latter case, the affected people are probably unaware that they are related. Many generations and many meioses may separate them from their common ancestor, during which time repeated recombination may have reduced the shared chromosomal segment to a very small region. However, within this region there will be an association at the population level between the disease and a particular allele or haplotype (page 499).

In fact one can view family linkage and population association studies as the two ends of a continuum (*Figure 12.11*). The more generations separate two people from their common ancestor, the smaller the conserved ancestral segments. When the conserved segments are very small they are harder to find, so that more markers must be used and the risk of false positives is greater; but once found they map the disease with greater precision. Population associations can be used to map both mendelian and nonmendelian conditions. For mendelian conditions the optimal approach is to map the condition quickly but approximately by conventional family linkage studies, then to use allelic associations to refine the location. However, it is important to remember that no association will exist unless the affected people in the population are actually part of one big family. Diseases where mutations are frequent do not show allelic association, and can only be mapped within families.

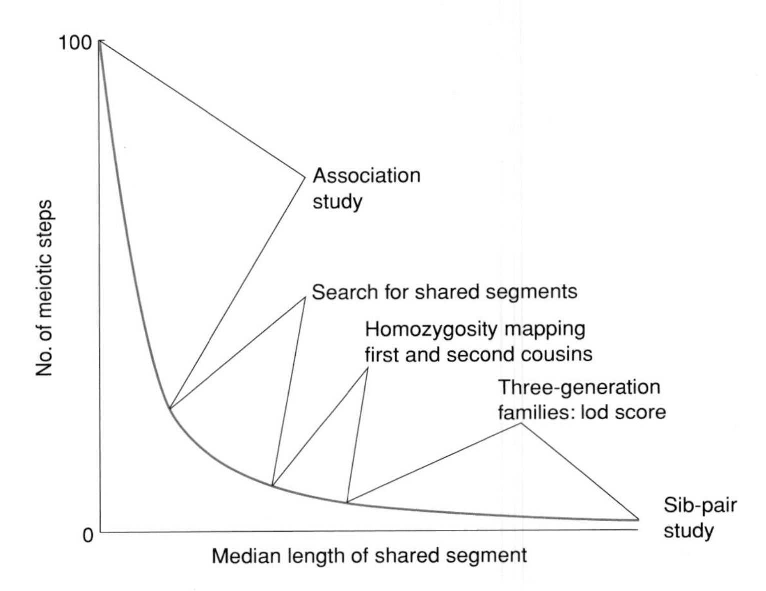

Figure 12.11: Family linkage and population association studies are two ends of a continuum.

The more meiotic steps separate two affected people, the smaller the ancestral chromosomal segment they will share. Adapted from Houwen *et al.* (1994).

Affected sib pairs allow model-free analysis in nuclear families

Picking a chromosomal segment at random, pairs of sibs are expected to share 0, 1 or 2 parental haplotypes with frequency 1/4, 1/2 and 1/4, respectively. However, if both sibs are affected by a genetic disease, then they will share the segment of chromosome carrying the disease locus. If the disease is dominant, they will share at least one parental haplotype, and if the disease is recessive they will share both haplotypes. This allows a simple form of linkage analysis (*Figure 12.12*). Affected sib pairs are typed for markers, and chromosomal regions sought where the sharing is above the level predicted by random segregation.

The degree of family clustering of a disease is expressed by the quantity λ_R, the risk to relative R of an affected proband compared with the population risk. Separate values can be calculated for each type of relative, for example λ_s for sibs. λ_s values can also be calculated to describe the degree of allele sharing at a given locus detected by sib pair analysis. Comparison of the locus-specific λ_s value with the overall λ_s for the disease gives an estimate of what proportion of the total family clustering is explained by the locus in question.

Although much less powerful than formal lod score analysis, sib pair analysis can be performed without making any assumptions about the genetics of the disease.

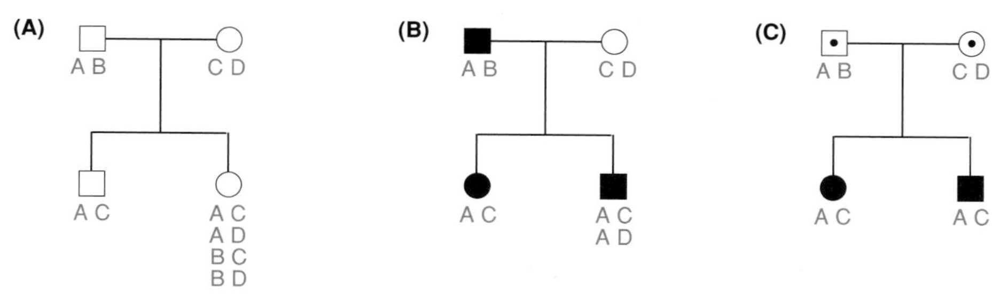

Figure 12.12: Sib pair analysis.

(A) By random segregation sib pairs share 0, 1 or 2 parental haplotypes 1/4, 1/2 and 1/4 of the time, respectively.
(B) Pairs of sibs who are both affected by a dominant condition share one or two parental haplotypes for the relevant chromosomal segment.
(C) Pairs of sibs who are both affected by a recessive condition share both parental haplotypes for the relevant chromosomal segment.

Thus it is used in nonmendelian diseases to seek susceptibility genes (page 496). For a susceptibility gene with low penetrance, we can infer the genotype of affected people (they must have the susceptibility allele), but not of unaffected people. It may be appropriate to use a halfway house between sib pair analysis and standard linkage analysis, **affected-only linkage analysis** (Weeks and Lange, 1988). An example of the systematic application to diabetes of sib pair analysis guided by λ_s values is given on page 498.

Homozygosity (autozygosity) mapping efficiently maps recessive conditions in extended families

Autozygosity is a term used to mean homozygosity for markers identical by descent. People with recessive diseases in consanguineous families are likely to be autozygous for markers linked to the disease locus. Suppose the parents are second cousins: they would be expected to share 1/32 of all their genes because of their common ancestry, and a child would be autozygous at only 1/64 of all loci. Autozygosity mapping becomes a powerful tool for linkage analysis if families can be found where two or more sibships, linked by inbreeding, each contain multiple affected people. Suitable families can be found in countries where inbreeding is common. The method has been applied successfully to locating genes for autosomal recessive hearing loss, which otherwise presents intractable problems because of extensive locus heterogeneity (Guilford *et al.*, 1994). It is model-free in that no assumptions need be made about penetrance or gene frequency, although in straightforward applications recessive inheritance is assumed.

Allelic association may be seen where most disease mutations in a population are derived from a few common ancestors

Linkage is a relationship between loci, but **association** is a relationship between alleles. **Allelic association** means that across the whole population, people who have a certain allele at one locus have a statistically more than random chance of having some particular allele at a second locus. Linkage is usually necessary, but never sufficient, for allelic association. For example, there is no association between Duchenne muscular dystrophy (DMD) and alleles of any marker, however closely linked. Because of strong natural selection, the half-life of a DMD mutation is only two generations, and unrelated boys with DMD normally carry different independent mutations. However, for some diseases, especially autosomal recessive and late-onset dominant diseases, the mutation rate is very low and a significant proportion of all the disease-bearing chromosomes in a population carry the same ancestral mutation. On the relevant chromosome segment apparently unrelated

Table 12.3: Allelic association in cystic fibrosis

Marker alleles	CF chromosomes	Normal chromosomes
X_1, K_1	3	49
X_1, K_2	147	19
X_2, K_1	8	70
X_2, K_2	8	25

Data from typing for the RFLP markers XV2.c (alleles X_1 and X_2) and KM19 (alleles K_1 and K_2) in 114 British families with a cystic fibrosis (CF) child. Chromosomes carrying the CF disease mutation tend also to carry allele X_1 of XV2.c and allele K_2 of KM19. Data derived from Ivinson *et al.* (1989).

affected people often carry marker alleles identical by descent. Cystic fibrosis provides a classic example (*Table 12.3*).

Only tightly linked markers will be shared. For recombination fraction θ, a fraction θ of chromosomes will lose the association each generation, and a fraction $(1-\theta)$ will retain it. After n generations, a proportion $(1-\theta)^n$ of chromosomes will retain the association. For 1% recombination, the half-life of the association is 68 generations, or about 1500 years. Note, however, that allelic association might be produced even in the absence of linkage by natural selection favoring a particular combination of alleles at two loci, or more importantly by stratification (genetic substructure) within the population (see page 499).

Allelic association offers the possibility of taking linkage analysis to resolutions unattainable in family studies. This approach was used with cystic fibrosis and with Huntington disease. In both cases, the result was helpful but not simple to interpret. There is no simple gradient of allelic association up which the researcher can walk (*Figure 12.13*). The figure shows cases of a strong association with a more distant marker and a weak association with a closer marker. The marker D4S95, closely linked to Huntington disease, detects RFLPs with three enzymes, *Taq*I, *Mbo*I and *Acc*I. Results confirmed in several independent studies show a

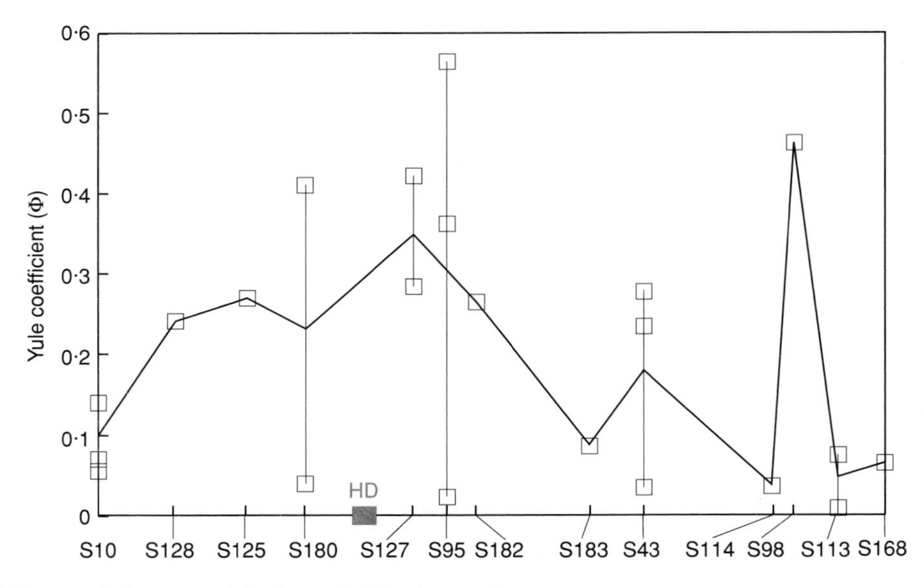

Figure 12.13: Allelic association around the locus for Huntington disease.

S10, S128 etc. are DNA markers, shown in their map positions relative to the HD locus. The total distance represented is 2500 kb. For some loci, several different RFLPs exist, which sometimes show very different allelic association, for example marker S95 (see text). Linkage disequilibrium is measured by the Yule coefficient. For two loci A and B, the Yule coefficient is $(p_{1,1}-p_{1,2}) / (p_{1,1}+p_{1,2}-2p_{1,1}p_{1,2})$, where $p_{1,1}$ and $p_{1,2}$ are the frequency of allele A_1 on chromosomes carrying allele B_1 and B_2, respectively. From Krawczak and Schmidtke (1994).

strong association with a particular *Acc*I and a particular *Mbo*I allele, but no association with either *Taq*I allele. The reasons for these discontinuities are not clear. Probably the associations reflect a complex history, with a combination of several independent mutations, recombination in one of a small founder population of disease chromosomes, and maybe an origin of some marker polymorphisms more recently than some disease mutations.

Allelic association can be remarkably powerful for mapping the private diseases of genetically isolated populations. A bold application of this principle enabled Houwen *et al.* (1994) to map the rare recessive condition, benign recurrent intrahepatic cholestasis, using only four individuals (two sibs and two supposedly unrelated people) from an isolated Dutch village. The ideal population for association studies is genetically homogeneous, different from its neighbors, and descended from a small number of original founders. The Finns are a classic case. Linguistically and genetically they are very different from their Nordic and Slav neighbors. Several recessive diseases that are common elsewhere are rare in Finland (e.g. cystic fibrosis), and several otherwise rare diseases are relatively frequent in Finland. Despite the great promise of allelic association, it is important to remember that it applies only to diseases and populations where most affected people are descended from a common ancestor who was a carrier.

Further reading

Ott, J. (1991) *Analysis of Human Genetic Linkage,* revised Edn. Johns Hopkins University Press, Baltimore, MD.

Terwilliger J, Ott J. (1994) *Handbook for Human Genetic Linkage.* Johns Hopkins University Press, Baltimore, MD.

References

Byerley WF. (1989) *Nature,* **340**, 340–341.

Chakravarti A. (1994) *Am. J Hum. Genet.,* **55,** 421–422.

Edwards JH. (1987) *J. Med. Genet.,* **24**, 539–543.

Eiberg H, Mohr J, Schmiegelow K, Nielsen LS, Williamson R. (1985) *Clin. Genet.,* **28,** 265–271.

Foy C, Newton VE, Wellesley D, Harris R, Read AP. (1990) *Am. J. Hum. Genet.,* **46,** 1017–1023.

Guilford P, Ben Arab S, Blanchard S, Levillers J, Weissenbach J, Belkahia A, Petit C. (1994) *Nature Genetics,* **6**, 24–28.

Gyapay G, Morisette J, Vignal A, Dib C, Fizames C, Millasseau P, Marc S, Bernardi G, Lathrop M, Weissenbach J. (1994) *Nature Genetics,* **7**, 246–339.

Houwen RHJ, Baharloo S, Blankenship K, Raeymaekers P, Juyn J, Sandkuijl LA, Freimer NB. (1994) *Nature Genetics,* **8**, 380–386.

Hughes A, Newton VE, Liu XZ, Read AP. (1994) *Nature Genetics,* **7**, 509–512.

Hultén MA, Palmer RW, Laurie DA. (1982) *Ann. Hum. Genet.,* **46**, 167–175.

Ivinson AJ, Read AP, Harris R, Super M, Schwarz M, Clayton Smith J, Elles R. (1989) *J. Med. Genet.*, **26**, 426–430.

Krawczak M, Schmidtke J. (1994) *DNA Fingerprinting*, p. 65. BIOS Scientific Publishers, Oxford.

Morton NE. (1955) *Am. J. Hum. Genet.*, **7**, 277–318.

Morton NE, Lindsten J, Iselius L, Yee S. (1982) *Hum. Genet.*, **62**, 266–270.

Schmitt K, Lazzeroni LC, Foote S, Vollrath D, Fisher EMC, Goradia TM, Lange K, Page DC, Arnheim N. (1994) *Am. J. Hum. Genet.*, **55**, 423–430.

Weeks DE , Lange K. (1988) *Am. J. Hum. Genet.*, **42**, 315–326.

Zhang L, Cui X, Schmitt K, Hubert R, Navidi W, Arnheim N. (1992) *Proc. Natl Acad. Sci. USA*, **89**, 5847–5851.

The Human Genome Project

Chapter 13

History, organization and goals of the Human Genome Project

The Human Genome Project is an international project whose ultimate aim is to obtain a complete description of the human genome by DNA sequencing. As the human mitochondrial genome has already been completely sequenced (see page 147), the genome under investigation is the nuclear genome. Because of the scale of the effort involved, it represents biology's first 'big science' project.

The Human Genome Project was conceived out of the need for a large-scale project to develop new mutation detection methods

A workshop held in Alta, Utah, in December 1984 was a major catalyst in the development of the Human Genome Project. Sponsored partially by the US Department of Energy (DOE), the workshop was intended to evaluate the current state of mutation detection and characterization and to project future directions for technologies to address current technical limitations. The growing roles of novel DNA technologies were discussed, notably the emerging gene cloning and sequencing technologies. Although such technologies had been in operation for about a decade, the efforts of individual laboratories to try and clone and characterize one gene at a time were considered to be wasteful of scientists' time and research resources. Because of the perceived technical obstacles, a principal conclusion was that methods were incapable of measuring mutations with sufficient sensitivity, unless an enormously large, complex and expensive program was undertaken. A subsequent report on *Technologies for Detecting Heritable Mutations in Human Beings* sparked the idea for a dedicated human genome project by the DOE, and in March 1986 it sponsored an international meeting in Santa Fe, New Mexico, to assess the desirability and feasibility of ordering and sequencing DNA clones representing the entire human genome. Virtually all participants concluded that such a project was feasible and would be an oustanding achievement in biology.

After extensive discussions with the US scientific community, the DOE responded to the Santa Fe meeting by issuing a Report on the Human Genome Initiative in the spring of 1987. Three major objectives were to be implemented: generation of refined physical maps of human chromosomes; development of support technologies and facilities for human genome research; and expansion of communication networks and of computational and database capacities. As implementation of this program began with a small number of pilot projects, other US organizations initiated their own studies of policy and strategy. In 1988, two additional widely circulated reports from the US Office of Technology Assessment and National Research Council appeared, and the US National Institutes of Health (NIH) set up an Office of Human Genome Research (later re-named the National Center for Human Genome Research) to co-ordinate NIH genome activities in co-operation with other US organizations. In the same year, the US congress officially gave approval to a 15-year US human genome project commencing in 1991. The required funding was estimated to be about $3 billion.

Organization of the Human Genome Project

The US Human Genome Project remains the major contributor to international research in this area. The co-ordination of this project has been entrusted to both the DOE (Offices of Energy Research and of Health and Environmental Research and Human Genome Program) and the US Department of Health and Human Services [Public Health Service, National Institutes of Health (NIH) and National Center for Human Genome Research]. The NIH clearly had a natural interest in the Human Genome Project, having been a major supporter of research into genetics and molecular biology. In addition, the DOE was considered to have an important role to play, following on from its long-standing program of genetic research directed at improving the ability to assess the effects of radiation and energy-related chemicals on human health. Recognizing their complementary activities, the NIH and DOE agreed to co-ordinate their individual genome activities. The NIH activity is largely channeled through the National Center for Human Genome Research, while DOE's genome activities are represented mainly by multidisciplinary programs underway at the Lawrence Berkeley National Laboratory, the Los Alamos National Laboratory and the Lawrence Livermore National Laboratory.

The Human Genome Project is now a truly international project and major programs have been established in the UK, France and Japan. Following the US lead, national human genome programs have also been established in Europe by Denmark, France, Germany, Italy, The Netherlands, the UK and the USSR (CIS), and also transnationally by the European Community. In addition, the Nordic countries have proposed a co-operative genome initiative which would allow them to contribute jointly to the human genome mapping effort. Outside Europe and the US, national genome projects have been established in Australia, Canada, Japan, Korea and New Zealand. The **Human Genome Organization (HUGO)** was established to co-ordinate the different national efforts, facilitate exchange of research resources, encourage public debate and advise on the implications of human genome research (McKusick, 1989). Conceived in 1988, HUGO is now administered through three centers: HUGO Europe (London), HUGO Americas (Bethesda, USA) and HUGO Pacific (Tokyo).

The complete nucleotide sequence of the human genome is only one of several goals of the Human Genome Project

The major rationale of the Human Genome Project is to acquire fundamental information concerning our genetic make-up which will further our basic scientific understanding of human genetics and of the role of various genes in health and disease. The major scientific thrust of the Human Genome Project concerns constructing high resolution genetic and physical maps as a prelude to the ultimate physical map, the complete sequence of the human genome. Much of this work is being carried out in a few major genome mapping centers. There is also extensive interaction with research focusing on mapping disease genes, which is currently conducted in numerous laboratories throughout the world (see *Figure 13.1*). The major 15-year goals also include a commitment to map and sequence the genomes of a variety of model organisms and to develop ancillary technologies including data analysis (see *Table 13.1*). Significant funding is also being provided for research on ethical, legal and social considerations, and support for technology transfer to the medical community.

At the outset of the Human Genome Project, it was clearly recognized that the DNA sequencing technology then available could not deliver the desired major goal (the complete sequence of the human genome in 15 years) within the budgetary constraints that were imposed. Improvements in DNA sequencing technology have been steady rather than dramatic, so that DNA sequencing costs have fallen progressively but without a huge increase in sequencing efficiency. At the time of writing, the largest regions of continuous human DNA sequence to be published were, with the single exception of a sequence in the *TCRB* gene cluster, less than 0.2 Mb long, although considerably longer DNA sequences have been sequenced in model organisms (see page 351). As a result of the perceived limitations in sequencing technology, the initial emphasis of the project has been on building high

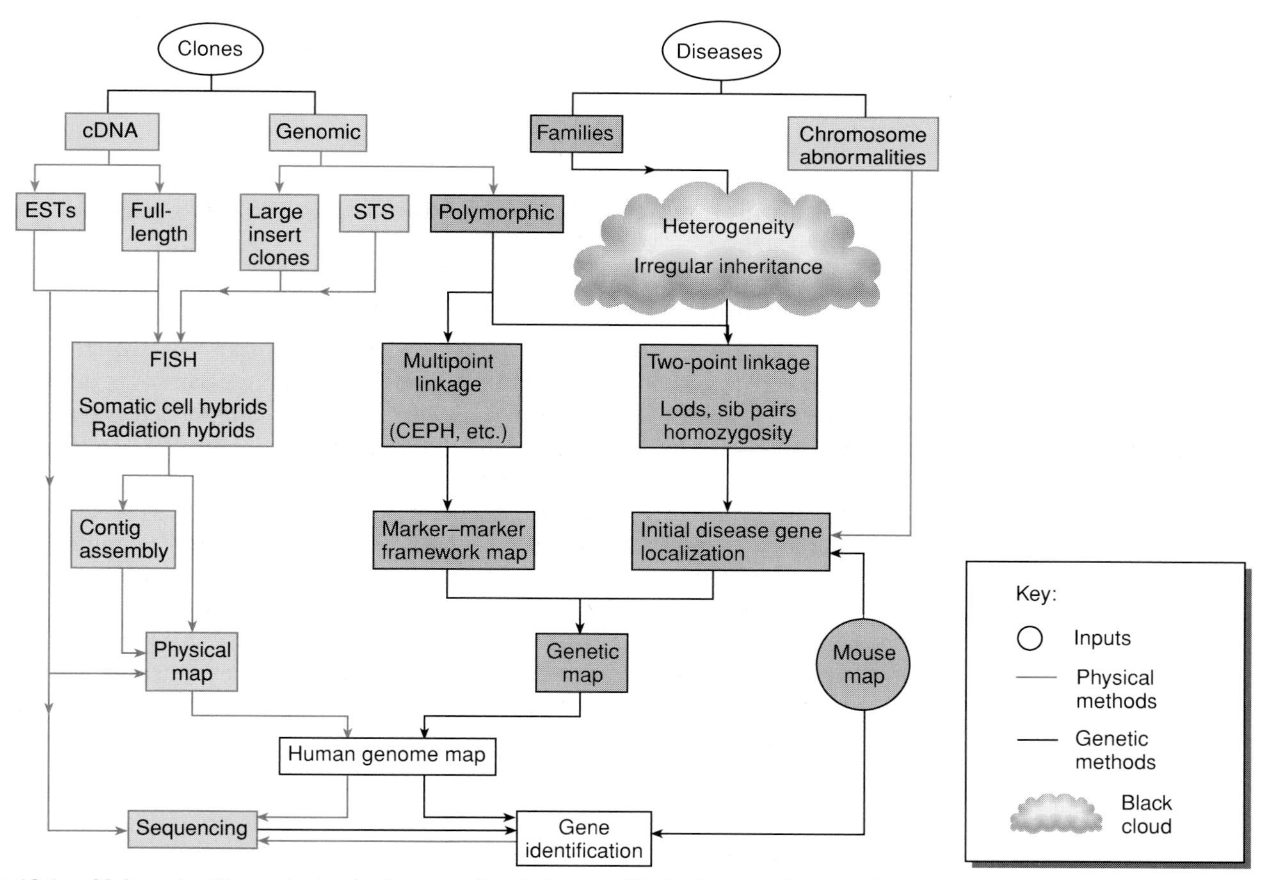

Figure 13.1: Major scientific strategies and approaches being used in the Human Genome Project.

The major scientific thrust of the Human Genome Project begins with the isolation of human genomic and cDNA clones (by cell-based cloning or PCR-based cloning). These are then used to construct high resolution genetic and physical maps prior to obtaining the ultimate physical map, the complete nucleotide sequence of the 3000 Mb nuclear genome. Inevitably, the project interacts with research on mapping and identifying human disease genes. In addition, ancillary projects include studying genetic variation (the Human Genome Diversity Project; see page 352); genome projects for model organisms (page 348) and research on ethical, legal and social implications. The data produced are being channeled into mapping and sequence databases (page 353) permitting rapid electronic access and data analysis. EST, expressed sequence tag (see page 346); STS, sequence tagged site (see page 293).

Table 13.1: Fifteen-year goals of the US Human Genome Project

- Construction of a high resolution genetic map of the human genome.
- Production of a variety of physical maps of all human chromosomes, and of the DNA of selected model organisms, with emphasis on maps that make the DNA accessible to investigators for further analysis.
- Determination of the complete sequence of human DNA and of the DNA of selected model organisms.
- Development of capabilites for collecting, storing, distributing and analyzing the data produced.
- Creation of appropriate technologies necessary to achieve these objectives.

resolution genetic and physical maps of the human genome rather than large-scale sequencing. The latter is not expected to be undertaken until there are significant clone contigs spanning individual chromosomes. Meanwhile, the progress of large-scale sequencing projects for certain model organisms (see below) has been monitored carefully in order to benefit from their experience prior to launching in to full-scale sequencing of the human genome. As a result of this experience, there is now some optimism that complete sequencing of the human genome may well be achieved ahead of the initial target date of 2005.

Human genetic maps

The first human genetic map was published in 1987 and was based on RFLP markers

Classical genetic maps for experimental organisms such as *Drosophila* and mouse have been available for decades, and have been refined continuously. They are constructed by crossing different mutants in order to determine whether the two gene loci are linked or not. For much of this period, human geneticists were envious spectators, because the idea of constructing a human genetic map was generally considered unattainable. Unlike the experimental organisms, the human genetic map was never going to be based on genes because the frequency of mating between two individuals suffering from different genetic disorders is vanishingly small. All this left was the possibility of a *genetic map based on polymorphic markers* which were not necessarily related to disease or to genes. As long as the markers showed mendelian segregation and were polymorphic enough so that recombinants could be scored in a reasonable percentage of meioses, a human genetic map could be obtained. The problem here was that, until recently, suitably polymorphic markers were just not available.

Classical human genetic markers consisted of protein polymorphisms, notably blood group and serum protein markers, which are both rare and not very informative (see *Box 12.1*). By 1981, only very partial human linkage maps could be obtained, and then only in the case of a few chromosomes. The identification of abundant DNA-based polymorphisms, however, called for a radical revision of thinking, and the early 1980s saw serious discussion of the possibility of constructing a complete human genetic map for the first time. Botstein *et al.* (1980) argued that it was feasible to construct a complete linkage map of the human genome using *RFLPs* (see page 122 and *Box 12.1*). The desirability of a complete linkage map of the human genome was clear. In addition to providing a framework for studying the nature of recombination in humans, there were several important areas which would benefit:

- **gene localization** – any gene for which a polymorphism could be typed could immediately be placed on the genetic map, and its chromosomal localization obtained. From the medical viewpoint, this had great appeal: linkage analyses could be employed in families segregating a disease gene, enabling the genes for many inherited conditions to be mapped for the first time.
- **Gene cloning** – gene localization would provide starting points for efforts to clone genes (see *positional cloning*, pages 372–387).
- **Diagnosis** – prenatal or pre-symptomatic diagnosis of inherited disease genes by linkage analysis would be facilitated enormously by having many DNA markers in the vicinity of the disease locus (Chapter 16).

Almost inevitably, the realization that a comprehensive human genetic map was now attainable sparked serious efforts to construct one. In 1987 the first such map was published based on the use of 403 polymorphic loci, including 393 RFLP markers (Donis-Keller *et al.*, 1987). This was a massive undertaking: the study required typing each member of 21 three-generation families for each polymorphic locus and analyzing the results by computer-based linkage analyses. The results provided 23 **linkage groups** (linear arrays of linked markers) corresponding to the 22 autosomes and the X chromosome (a genetic map can be constructed for the major pseudo-autosomal region on the Y chromosome but not for the rest of the Y chromosome, which does not engage in recombination, see pages 211 and 218). Important though this achievement was, there remained some serious drawbacks by comparison with the features that one would wish to see in an idealized genetic map (see *Table 13.2*). The average spacing between the markers (>10 cM) was still considerable. More significantly, a major limitation was the reliance on RFLP markers, which are not very informative markers and are difficult to type (see *Box 12.1*).

High resolution human genetic maps have been obtained recently largely through the use of microsatellite markers

The next breakthrough in constructing a human genetic map came as a result of the use of markers that were very much more polymorphic than RFLP markers. Hypervariable minisatellite polymorphisms could be considered as an alternative, but only in a limited way. This is so because they are not distributed widely throughout the genome; they have a marked tendency to occur near the telomeres. *Microsatellite markers* (also described as **short tandem repeat polymorphisms,** or **STRPs**) have the advantage of being abundant, dispersed throughout the genome,

Table 13.2: Towards an ideal genetic map

Marker density	The average spacing between polymorphic markers should be considerably less than 1 cM, so that virtually all regions of the genome will contain a marker in their immediate vicinity.
Marker typing	Each marker should be easy to type, and the information and facilities to enable the typing should be widely available.
Marker utility	The heterozygosity of each marker (see text) should be high, preferably approaching 1.0. This would mean that each marker would be informative in a large percentage of different meioses.
Marker order	The utility of the genetic map is clearly dependent on confidence that the sequential order of markers in a linkage group is correct. Genetic linkage analyses cannot provide absolute certainty concerning marker order. Instead, a statistical approach is used which gives an odds ratio in favor of a certain order. The confidence in marker order depends on finding as many recombinants as possible. For a high resolution map, this means scoring large numbers of informative meioses.
Reference families	A single set of reference families should be used; samples from each member of the reference families used to construct the map should be widely available, so that different laboratories could place new markers on the map and continue to improve it.
Map utility	New markers should be integrated into a single map whose resolution would continue to improve; different genetic maps using different sets of markers should be avoided.

highly informative and easy to type (see *Box 12.1*). By focusing on this type of marker, researchers at the Généthon laboratory in France were able quickly to provide a second-generation linkage map of the human genome (Weissenbach *et al.*, 1992). This *tour de force* involved selecting suitably polymorphic CA/TG repeats, mapping them to specific chromosomes by typing panels of human–rodent *somatic cell hybrids* (see page 275) and performing statistical linkage analyses on markers from individual chromosomes. A total of 813 markers, of which 605 showed a heterozygosity above 0.7, were organized into 23 linkage groups.

The major advance by the Généthon laboratory was made possible by large-scale funding from the French muscular dystrophy organization *Association Française Myopathique (AFM)* which enabled a factory approach to research, using large-scale automation. Subsequently, maps have been produced with ever increasing numbers of genetic markers, especially microsatellite markers, and ever increasing resolution. A variety of different genetic maps (using different marker sets) have been constructed (see Murray *et al.*, 1994, for references). The problem with this is twofold. First, not all genetic maps have used the same sets of reference families. Secondly, different genetic markers have been used to construct different genetic maps and the linkage relationship between markers in different genetic maps may not be known. In recognition of these limitations, the international research community have increasingly recognized the importance of collaboration:

- *reference families.* The most widely distributed set of reference families is that deriving from the *CEPH (Centre d'Études du Polymorphisme Humaine)* collaboration of more than 100 independent laboratories (Dausset *et al.*, 1990).
- *Integration of maps.* Studying the linkage relationship between markers used in different genetic maps provides an integrated map. The most recently published genetic map is such an example, and has about 4000 STRPs (mostly derived from the Généthon laboratory) plus 1800 other markers, mostly RFLPs, which together provided an average marker density of one per 0.7 cM (Murray *et al.*, 1994).

The resolution of the map published by Murray *et al.* (1994) was considerably higher than the average 2–5 cM resolution hoped for by the end of the first 5 years of the project. However, further advances in map resolution will be limited because of the poor returns involved in identifying novel microsatellite markers; recently discovered microsatellite markers often turn out to be identical to previously characterized markers. Additionally, there is a widespread perception that the current marker resolution is sufficient to provide the framework for constructing a comprehensive physical map of the human genome. As a result, the major effort has recently switched to the construction of high resolution physical maps of the human genome.

Physical maps of the human genome

A variety of different physical maps of the human genome are being constructed

Like the genetic map, a physical map of the human genome will consist of 24 component maps corresponding to the 24 different types of chromosome. The different genetic maps of the human genome that have been assembled so far all represent the same concept: sets of linked polymorphic markers (linkage groups) corresponding to different chromosomes. However, unlike this uniformity, a variety of different types of physical map are possible (see *Table 13.3, Figure 13.2*). In a sense, the first physical map of the human genome was obtained when cytogenetic banding techniques were used not only to distinguish the different chromosomes, but also to provide discrimination of different subchromosomal regions (see the human karyogram in *Figure 7.4*). Although the resolution of this map is coarse (an average sized chromosome band in a 550-band preparation contains ~6 Mb of DNA), it has been very useful as a framework for ordering the locations of human DNA sequences by chromosome *in situ* hybridization techniques (see pages 126 and 280–282).

Table 13.3: Different types of physical map can be used to map the human nuclear genome

Type of map	Examples/methodology	Resolution
Cytogenetic	Chromosome banding maps	An average band has several Mb of DNA
Chromosome breakpoint maps	Somatic cell hybrid panels containing human chromosome fragments derived from natural translocation or deletion chromosomes	Distance between adjacent chromosomal breakpoints on a chromosome is usually several Mb
	Monochromosomal radiation hybrid (RH) maps	Distance between breakpoints is often many Mb
	Whole genome RH maps	Resolution can be as high as 0.5 Mb
Restriction map	Rare-cutter restriction maps, e.g. *Not*I maps	Several hundred kb
Clone contig map	Overlapping YAC clones	Average YAC insert has several hundred kb of DNA
	Overlapping cosmid clones	Average cosmid insert is 40 kb
Sequence-tagged site (STS) map	Requires prior sequence information from ordered clones so that STSs can be ordered	A desired goal is an average spacing of 100 kb
Expressed sequence tag (EST) map	Requires cDNA sequencing then mapping cDNAs back to other physical maps	Highest possible average spacing is ~ 40 kb
DNA sequence map	Complete nucleotide sequence of chromosomal DNA	1 bp

Other maps have been obtained by mapping natural chromosome breakpoints (using panels of somatic cell hybrids containing fragments derived from translocation and deletion chromosomes; see page 278), or by mapping artificial breakpoints using radiation hybrids (RHs; see pages 278–280). Again, however, the resolution achieved by such hybrid cell panels (the average distance between neighboring breakpoints) can be quite limited for large parts of the genome. As a result, higher resolution physical maps are desirable. Clearly, the physical map which will provide the highest possible resolution, that of single base pairs, is the ultimate map: the complete nucleotide sequence of the genome. As this will not be achieved for some time, attention has been focused on constructing physical maps of intermediate resolution. Comprehensive RH maps and rare-cutter restriction maps have been achieved thus far for only a few human chromosomes. One example is chromosome 21 where a *Not*I restriction map has been published for the entire long arm (Ichikawa *et al.*, 1993). In addition, much of the current mapping effort is aimed at mapping of coding DNA sequences, thereby producing comprehensive *transcription maps* (see page 295).

Considerable progress has been made towards producing complete clone contig maps of several human chromosomes

A major intermediate goal of the Human Genome Project is to construct a complete **clone contig map** of the DNA of each of the 24 different types of human chromosome. This means relating different DNA clones to define a series of partially overlapping DNA molecules covering the entire length of a chromosome (page 288). A complete contig map of a chromosome would therefore comprise all the DNA, without any gaps (*contig* originated as a shortened form of the word *contiguous*).

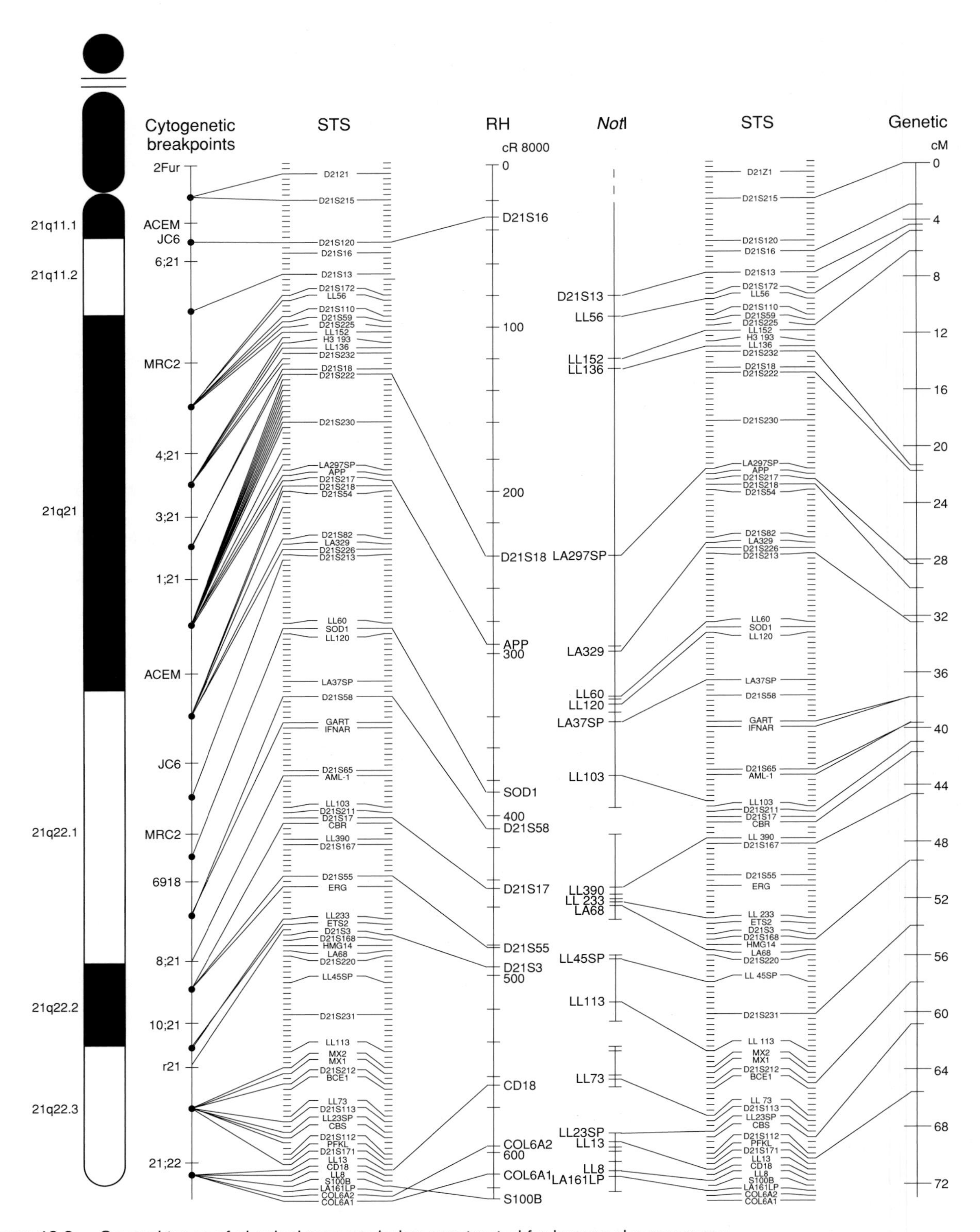

Figure 13.2: Several types of physical map are being constructed for human chromosomes.

The figure shows integration of several physical maps for the long arm of human chromosome 21. Next to the standard cytogenetic map on the left are the positions of chromosome 21 breakpoints observed largely from studying chromosome 21 translocations (6;21, 4;21, 3;21, 2;21, etc.). The STS map is shown twice to facilitate comparisons with the other maps, which also include the genetic linkage map of Chumakov *et al.* (1992), the PFGE-based *Not*I restriction map (Ichikawa *et al.,* 1993) and the radiation hybrid map (measured in centiRays after exposure to 8000 rad). Reproduced with permission from *Nature*, vol. 359, page 385. Copyright 1992 Macmillan Magazines Limited.

Identification of overlaps between the DNA segments of different clones can be done by a variety of different procedures (see pages 288–294). Because of their large inserts, yeast artificial chromosome (YAC) clones have been particularly useful in generating such contigs. However, in some parts of the genome which are of special interest (e.g. containing important genes), higher resolution contig maps have been constructed using overlapping cosmid, bacteriophage λ, P1, PAC and BAC clones. Ultimately, high resolution maps based on cosmid contigs will provide a suitable framework for sequencing whole chromosomes.

Cosmid mapping exercises have been heavily dependent on the construction of chromosome-specific cosmid libraries, and libraries of varying quality are available for essentially all human chromosomes. Such libraries have been widely used as *reference libraries* by the scientific community. Gridding of clones from the libraries in microtiter dishes has facilitated making replica copies of the libraries for distribution to various major scientific centers. More conveniently, it has been possible to make *high density gridded colony filters* in which copies of the clones are arrayed on hybridization filters for distribution to thousands of laboratories throughout the world. The filters can then be screened by hybridizing with a labeled probe of interest (see *Figure 13.3*), and the resulting hybridization data can be passed back to the distributing centers which, because of the fixed geometry of the clone positions, are able to integrate the information provided by numerous laboratories.

Although chromosome-specific YAC libraries have been constructed in some cases, most of the effort in constructing contig maps of individual human chromosomes has involved isolating and characterizing YAC clones from total genome YAC libraries. In order to construct a YAC clone contig map for a specific chromosome, *STS* (sequence tagged site) *content mapping* (see page 293) has been particularly valuable and has involved using only those STS markers (both polymorphic and nonpolymorphic) which have been shown to map to the chromosome under investigation.

Significant contig maps for individual human chromosomes were first obtained for chromosome 21 (Chumakov *et al.*, 1992) and the Y chromosome (Foote *et al.*, 1992). In the former case, the reported clone contig spanned about 45 Mb of the long arm; in the latter case the contig spanned about 29 Mb, corresponding to the euchromatic portion of the Y chromosome (the short arm, centromere and proximal long arm). The remainder of the Y chromosome is heterochromatic and consists of about 30 kb of highly repetitive DNA, based on individual repeats whose cumulative sequence length is only about 10 kb. At the time of writing, reasonably comprehensive YAC contig maps have been published for chromosomes 3, 12 and 22 and also integrated maps of chromosomes 16 and 19 which include high resolution cosmid contigs (see Little, 1995). In addition, physical maps for chromosomes 7 and 11 are all well advanced.

Physical maps of the entire human genome are being constructed using random clone fingerprinting methods

Because of the reliance on whole genome YAC libraries, workers at the CEPH laboratory in Paris took, at a very early stage, an alternative bold approach. Rather than proceeding with one chromosome at a time, they sought to assemble contig maps of all 24 chromosomes in parallel by *random clone fingerprinting* (page 291). This quickly led to the major advance of **a first generation physical map** of the human genome (Cohen *et al.*, 1993). This map was constructed by exhaustive screening of the CEPH YAC library which contains 33 000 YACs with an average insert size of

Figure 13.3: Gridded clone hybridization filters have facilitated physical mapping of the human genome.

The figure illustrates an autoradiograph of a membrane containing human YAC clones (i.e. total DNA from individual yeast clones containing human YACs). The membrane contains a total of 17 664 clones which had been gridded in arrays of a unit grid of 6×6 clones. The hybridization signals include weak signals from all clones by using a ^{35}S-labeled probe of total yeast DNA plus strongly hybridizing signals obtained with a ^{32}P-labeled unique sex chromosome probe (*DXYS646*). Original photo from Dr Mark Ross, Sanger Centre, Cambridge. Reproduced from Ross (1995) with permission from John Wiley & Sons Inc.

0.9 Mb, representing 10 haploid *genome equivalents* (see page 93). Overlaps between YAC clones were identified using three methods:

- *repetitive DNA fingerprinting* (page 291). Clone fingerprints of all 33 000 YACs were produced by detection of restriction fragments containing medium repeat sequences (THE and LINE-1 elements – see pages 198 and 201).
- *STS content mapping* (page 293). The STSs were chosen to be highly polymorphic microsatellite markers (STRPs) which had been placed on the human genetic map. A total of 2100 such markers were used to screen partially or totally the 33 000 YACs.
- Alu-*PCR probe hybridization*. The *Alu*-PCR reaction (see page 294) was used to derive clone-specific DNA fragments for hybridization assays to test the relatedness of different clones or clone pools.

The physical map published by Cohen *et al.* (1993) was recognized to be far from complete, with poor coverage of some chromosomes, notably chromosome 19. What this oustanding achievement did represent, however, was a framework for the scientific community to build upon, in order to produce future detailed maps of all the chromosomes. As the detailed mapping information was made widely available by electronic access through the *Internet* (see *Box 13.2*), numerous researchers throughout the world were quickly able to relate and use this information with respect to specific chromosomes, or often subchromosomal regions, that were of interest. Very recently, the same researchers have published an updated YAC contig map covering perhaps 75% of the human genome and consisting of 225 contigs with an average size of 10 Mb (Chumakov *et al.*, 1995).

The utility of the YAC contig maps has, however, been limited by some problems with their construction. A major disadvantage of YAC libraries is the high frequency of **chimeric clones:** in close to 50% of individual YACs, the DNA insert represents artifactually linked segments from noncontiguous portions of the genome (often as a result of recombination between different YACs – see Larionov *et al.*, 1994). A YAC that contains DNA segments from different chromosomes can be identified readily, but YACs that contain noncontiguous fragments from the same chromosome are harder to detect. Additionally, it has become clear that the methods used to identify overlaps between different clones are far from foolproof. As a result, considerable effort has been required subsequently to verify the map by, for example, using FISH mapping (see pages 280–286) to check the chromosomal locations of individual YACs. Because of the difficulty in establishing comprehensive contig maps, the goal of an STS map with an average spacing of 100 kb will not be met as envisaged at the end of the first 5 years of the project. Instead, a recent assessment of mapping progress now envisages that this will not be completed until the end of 1998 (Cox *et al.*, 1994).

EST projects are providing a rapid route to identifying and mapping the great majority of genes in the human genome

Whole genome sequencing or coding DNA sequencing?

From the outset of the Human Genome Project there has been much debate over whether to go for an all-out assault of indiscriminate sequencing of all 3 billion bases of the human genome, or whether to focus initially just on the coding DNA sequences. The average coding DNA of a human gene is about 1.5 kb, but human genes occur on average roughly once every 40–50 kb of DNA. As a result, coding DNA accounts for a mere 3% of the human genome (unlike *Saccharomyces cerevisiae* and *Caenorhabditis elegans* where the gene density is much higher – see pages 350–351).

Support for priority sequencing of the coding DNA is largely dependent on two arguments: (i) the coding DNA component contains the vast information content of the genome and so is by far the most interesting and medically relevant part; and (ii) it is such a small percentage of the genome that it can be achieved very quickly, and very cheaply, by comparison with efforts to sequence the entire genome. Against this view, supporters of the total genome sequencing approach have emphasized the difficulty of finding every mRNA expressed in all tissues, cell types and developmental stages, and the neglect of control sequences flanking the coding sequences and located within introns.

Identifying coding DNA

Identification and isolation of human coding DNA clones can be done by a variety of different routes.

- *Screening human cDNA libraries.* Because the expression of many genes is largely confined to particular tissues or organs, and in some cases to particular stages of embryonic or fetal development, this raises the question of the source of the cDNA libraries to be analyzed.
- *CpG island library screening.* A library of human CpG island sequences has been constructed recently (Cross *et al.*, 1994). Isolated clones can be used as tags to

identify a considerable fraction of human genes (>50% of human genes have been estimated to have CpG islands, see Antequera and Bird, 1993).

- *Exon trapping and cDNA selection.* This particularly applies to screening YAC and cosmid clones and other large insert genomic clones (see pages 299–302).

In addition, screening of human genomic DNA clones by hybridization to puffer fish genomic DNA libraries can facilitate identification of coding DNA sequences in the human clones, assuming that the gene in question is reasonably well conserved in evolution. This approach exploits the observation that the puffer fish has a very compact genome with much the same number of genes as the human genome, but less than one seventh of the noncoding DNA (see page 349).

Of the above approaches, large-scale cDNA sequencing has predominated: by sequencing a fragment of a cDNA clone one generates an **expressed sequence tag (EST)**, an STS which is specific for a coding DNA (Adams *et al.*, 1991). Often this has involved large-scale sequencing of short segments (typically 200–300 nucleotides) at the ends of individual cDNA clones. Because the 3′-untranslated sequence (3′-UTS) of human genes usually falls within the 400–800 nucleotide range, the 3′ EST sequences are usually derived from the 3′-UTS. Like all other STSs, ESTs have the virtue of being very simply assayed by PCR (see *Box 11.2*). Screening genomic DNA is usually possible (the 3′-UTSs in particular are rarely split by introns at the level of genomic DNA) and the relative lack of conservation of 3′-UTS sequences during evolution means that orthologous rodent sequences are not usually co-amplified when screening somatic cell hybrids (see below).

Identifying and characterizing ESTs

The recent focus of the Human Genome Project has undoubtedly shifted towards characterization of the coding DNA component and the development of a **human gene map** (placing of all 65 000–80 000 human genes on the physical map). This has happened because of the sheer speed with which large-scale automated sequencing approaches (see page 304) can generate thousands of ESTs, and the ability to map the ESTs quickly. A considerable variety of **EST projects** have been spawned, sequencing short fragments (typically 200–300 bp) of cDNA clones from brain, skeletal muscle, lymphocytes, liver, heart, testis and other tissues (see Boguski and Schuler, 1995; Sikela and Auffray, 1993). In many cases, the screening and usage of the cDNA libraries has been facilitated by using gridded arrays of clones in much the same way as for genomic DNA clones (see *Figure 13.3*). The **IMAGE consortium** (*integrated molecular analysis of gene expression*) has particularly fostered this development, and master arrays of clones representing unique genes are being prepared.

Until very recently, the EST approach has been dominated by the *Institute of Genomic Research* (*TIGR*) at Gaithersburg, Maryland (USA). Backed by major industrial funding, TIGR adopted an aggressive factory-style sequencing approach (see *Figure 13.4*) which has been very successful. At the time of writing, TIGR was rumored to have generated several hundred thousand ESTs which are believed to represent sequences from the majority of human genes (traditional EST approaches introduce an element of redundancy: an average gene is represented by several different ESTs). Although some of the TIGR sequences have been deposited in the publicly accessible electronic database dbEST (see *Table 13.5* and Boguski and Schuler, 1995), much of these data had not been made publicly available (see *Box 13.3*). Very recently, however, a considerable amount of the TIGR data has been published (Adams *et al.*, 1995). Recently, a collaboration between Merck Pharmaceuticals and the University of Washington has made a commitment to characterize up to 400 000 human ESTs, representing 3′ and 5′ sequences, and to make this freely available by April 1996. By September 1995 this consortium had placed a total of 170 000 human ESTs in dbEST (see Goodfellow, 1995).

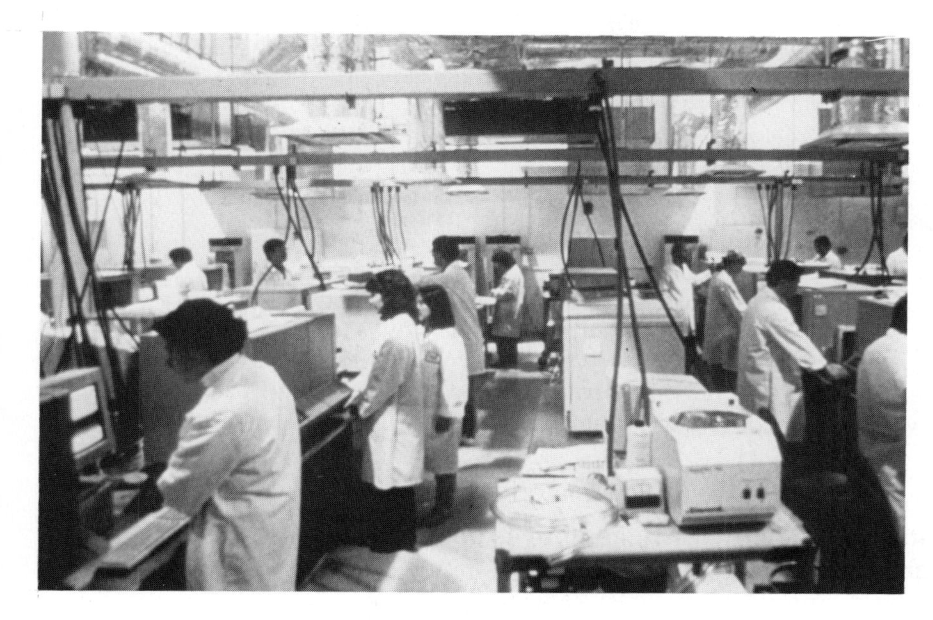

Figure 13.4: Factory-style DNA sequencing at the Institute for Genomic Research (TIGR).

By opting for large-scale automated DNA sequencing, the Institute for Genome Research at Gaithersburg, USA has made rapid progress in a variety of sequencing projects including a human EST program and the recently completed *Haemophilus influenzae* genome project (see page 348). Reproduced from Adams *et al.* (1994) with permission from the author.

Not all genes will be represented by ESTs, however, because some genes will be very poorly expressed or not expressed at all in the cDNA libraries analyzed, including genes which show highly restricted tissue or cell expression patterns and genes which are expressed at certain developmental stages. It has been expected that the EST program should be able to identify sequences from about 80% of all the human genes. The remainder will be identified as coding DNA sequences in genomic DNA clones (see *Box 11.3*).

Mapping ESTs and characterizing full-length transcripts will soon provide a comprehensive human gene map

Most ESTs are nonpolymorphic but can be assigned to chromosomal and subchromosomal locations by physical mapping techniques. Chromosomal localizations can be obtained rapidly using a PCR assay to screen for the presence of a particular EST in a panel of somatic cell hybrids (see page 275). Subchromosomal localization of human cDNA clones by FISH (pages 280–282) has also been possible (see, for example, Korenberg *et al.*, 1995).

One recent approach to mapping ESTs taken by a consortium of British, French and American laboratories is to map human ESTs by PCR assays of panels of *whole genome radiation hybrids* (see page 289, Boguski and Schuler, 1995). The Stanford whole-genome RH panel consists of 83 hybrids and results in maps of 0.5 Mb resolution, while the Genebridge (Généthon– Cambridge) RH panel produces lower resolution maps but allows markers to be linked more readily. About 30 000–40 000 ESTs are expected to be placed on RH maps by the summer of 1996. Another very important development has been to place ESTs on individual YACs by PCR assays of whole genome YAC libraries (Berry *et al.*, 1995; see Boguski and Schuler, 1995). In addition, to cDNA/EST mapping, construction of comprehensive transcription maps of human chromosomes may also involve mapping of exons trapped from cosmid and YAC clones (for a recent example, see Yaspo *et al.*, 1995).

Because ESTs are comparatively short sequences (~200–300 bp, while an average full-length human cDNA is ~2 kb), the same gene can be tagged by several different

ESTs. A full coding sequence could be obtained by sequencing the corresponding full-length cDNA. As each gene can be tagged by several different ESTs, some of which may overlap each other, another alternative involves screening ESTs against each other, a form of *EST walking* in which overlapping sequences are identified and integrated to form a full-length sequence (see Adams *et al.*, 1995).

Ancillary projects: genome maps of model organisms and the Human Genome Diversity Project

Mapping the human genome is not the only scientific project supported by the Human Genome Project. Support has also been given to projects which aim to sequence the DNA of model organisms and to attempts to identify and characterize genetic variation in different human populations (the *Human Genome Diversity Project*).

A variety of model organism genome projects are expected to aid understanding of human gene function and to serve as pilot projects for sequencing of the human genome

At the outset of the Human Genome Project, it was recognized that comprehensive maps of certain model organisms would be highly desirable. Such organisms include a variety of species, some of which have been particularly amenable to genetic analysis (see *Box 13.1*).

The need to study such organisms is evident from numerous examples where information derived from studies of the biology of model organisms has been essential to interpret data obtained in studies of humans and in understanding human biology. Research involving such models will continue to provide a basis for analyzing normal gene regulation, genetic diseases and evolutionary processes. In addition, large-scale sequencing of selected model organisms, notably the roundworm *C. elegans* and the yeast *S. cerevisiae,* are being viewed as pilot projects for sequencing the human genome (see below).

Rapid progress has been made in mapping the genomes of some model organisms

Thus far, considerable progress has been made in several of the model organism genome projects and large-scale sequence comparison studies have been initiated for selected homologous regions in different genomes (see Jones, 1995).

Bacterial genome projects

The *E. coli* and *Bacillus subtilis* genome projects are being stewarded respectively by research groups in the USA, and Japan, and by a consortium of European laboratories. The complete sequences are expected in a few years. Recently, however, a collaboration between TIGR at Gaithersburg, USA, and a team at the Johns Hopkins University has produced the complete sequence of the 1.8 Mb *Haemophilus influenzae* genome (Fleischmann *et al.*, 1995). A total of 1749 genes have been identified, giving an average gene density of about 1 per kilobase. At about the same time, the complete nucleotide sequence of the 580-kb *Mycoplasma genitalium* genome was also established by a team at TIGR.

Box 13.1: Model organisms for which genome projects are considered particularly relevant to the Human Genome Project

Escherichia coli (4.6 Mb). This bacterium has been well studied, with intensive investigations of gene structure, gene regulation and gene function.

Saccharomyces cerevisiae (14 Mb). This budding yeast is one of the most extensively studied eukaryotes and a large amount of information is known about its gene structure, gene regulation and gene function. It has been very amenable to genetic analyses, partly because of the high frequency of nonhomologous recombination.

Caenorhabditis elegans (100 Mb). This 1-mm long roundworm consists of 959 somatic cells. It is a model organism for developmental biology studies: the exact lineage of every one of its cells is known – information which is unknown in all other multicellular organisms. There is also a complete wiring diagram of the nervous system: all 302 neurons and the connections between them are known.

Drosophila melanogaster (165 Mb). The fruit fly has been the subject of extensive genetic analyses and dissection of gene structure, regulation and function. Crucially important gene functions and developmental processes that are highly conserved between species will continue to be major areas that are particularly relevant to understanding human biology.

Fugu rubripes rubripes (400 Mb). The puffer fish provides a compact model vertebrate genome (Brenner *et al.*, 1993): repetitive DNA sequences comprise only about 7.4% of the genome, and introns are generally very small. As a result, the exon content is very high and genes are much shorter than their human homologs (Baxendale *et al.*, 1995; see *Figure 19.6*). The strong conservation of coding sequences during evolution means that the puffer fish genome could provide a rapid way of identifying human coding DNA sequences (which account for only about 3% of the human genome).

The mouse (3000 Mb). This is the mammalian species with the most highly developed genetics. Its small size and short generation time have allowed large-scale mutagenesis programs and extensive genetic crosses. There is considerable subchromosomal conservation of linked genes between mouse and humans (see *Figure 14.13*). The ability to construct mice with pre-determined genetic modifications to the germline (by transgenic technology and gene targeting in embryonic stem cells) has been a powerful tool in studying gene expression and function and in creating mouse models of human disease (see Chapter 19).

The rat ***(Rattus norvegicus)*** (3000 Mb). Rats, being considerably larger than mice, have for many years been the mammal of choice for physiological, neurological, pharmacological and biochemical analysis. They may also provide genetic model systems for complex human vascular and neurological disorders, such as hypertension and epilepsy (for various reasons, there are no mouse models for such diseases). Genetic analysis in laboratory rats, however, is much less advanced than in mice, partly because of the relatively high cost of rat breeding programs and the current inability to modify the rat germline by gene targeting. Recently, however, a genetic linkage map has been constructed (Jacob *et al.*, 1995), and a genome project is actively under consideration.

The publication of the complete sequence of the *H. influenzae* genome meant that a milestone in biology had been reached: *for the first time, the genetic blueprint of a free-living organism had been determined* (up until then the largest genomes to have been sequenced came from viruses). Extensive use of state-of-the-art technology permitted very rapid progress: about 99% of the entire genome was sequenced in only 6 months using a random sequencing strategy similar to the ones being used for complex genomes (see *Figure 13.5*). The bacterium involved was a benign laboratory strain, but the wild-type causes ear infections and meningitis. The new *H. influenzae* database can now be put to various uses ranging from understanding evolutionary biology to working out the functions of specific genes (*Figure 13.6*).

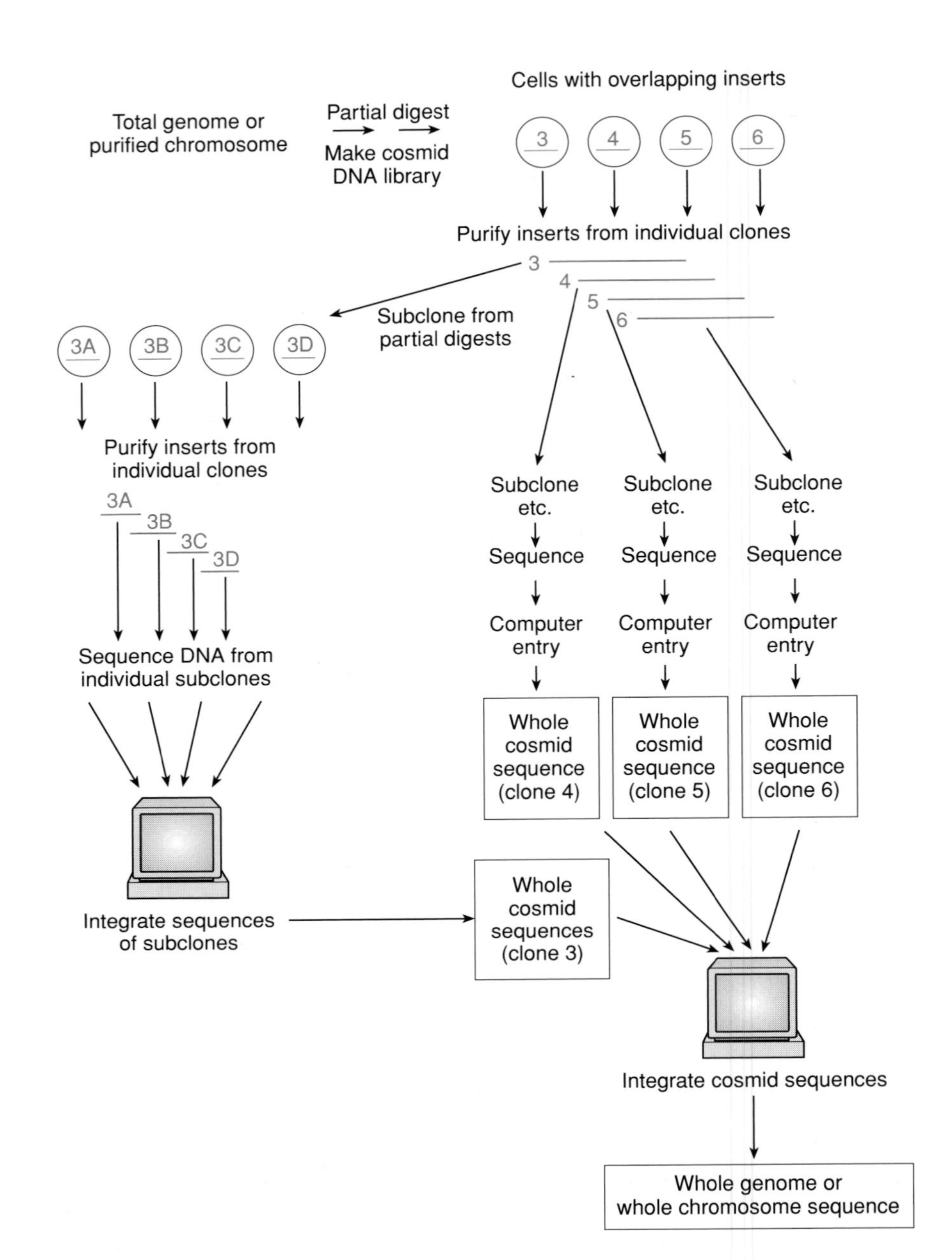

Figure 13.5: Random sequencing of subclones and computer integration of sequence data is being used in large genome sequencing projects.

The random sequencing strategy shown is typical of the ones used in large whole genome sequencing projects, where subclones from overlapping cosmids are sequenced at random and computer-based sequence alignment programs help to integrate the data. A basically similar random clone sequencing approach was used for sequencing the comparatively small *H. influenzae* genome, but reliance was placed on sequencing smaller insert clones instead of cosmids (see Fleischmann *et al.*, 1995).

The Saccharomyces cerevisiae *genome project*

Individual chromosomes of the budding yeast *S. cerevisiae* are being sequenced by European and American consortia. By mid-1995, about 10 Mb of the 14-Mb genome had been obtained, and the complete DNA sequences of eight of the 16 chromosomes had been published (see Oliver, 1995). The sequence of the entire genome is expected by early 1996. The available data suggest that yeast genes are closely clustered: on average, there is one gene every 2 kb.

The Caenorhabditis elegans *genome project*

The *C. elegans* genome project is being viewed as the major pilot model for large-scale sequencing of the human genome. Detailed physical and genetic maps of the genome have already been constructed, and a project to sequence the entire 100-Mb genome has been undertaken as a collaboration between researchers in the Sanger Centre (Cambridge, UK) and Washington University School of Medicine. Recently, the longest single segment of DNA ever to be sequenced was established by this

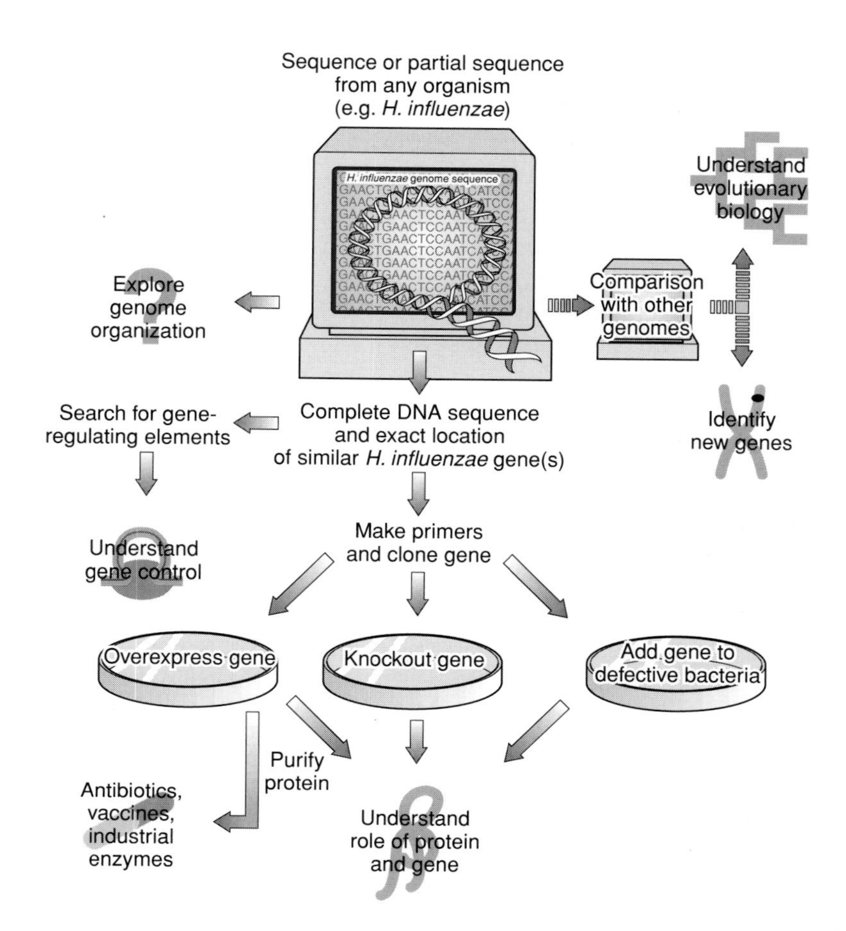

Figure 13.6: The complete sequence of a cellular genome opens new avenues of research and therapeutics.

The completed sequence of the bacterium *H. influenzae* obtained by Fleischmann *et al.* (1995) can be put to a variety of different uses, from understanding evolutionary biology to working out the functions of individual genes. Adapted with permission from Nowak (1995). Copyright 1995, American Association for the Advancement of Science.

consortium: 2.2 Mb of DNA from chromosome III of *C. elegans* (Wilson *et al.*, 1994). By mid-1995, a total of about 20 Mb of the *C. elegans* genome had been sequenced. The mapping data is being entered into the **ACEDB database** (a *Caenorhabditis elegans* database), and it is intended to use this database format to input physical mapping data for the Human Genome Project (see below). Analyses indicate that, on average, there is one gene every 5 kb, and comparison with published sequences from elsewhere reveals that about one gene in three shows similarities to previously known genes. In addition, a surprisingly high number of the genes (~25%) appear to occur as part of operons where individual genes are transcribed as part of large multigenic RNA transcripts (Zorio *et al.*, 1994). This large sequencing advance was made possible by automation of sequencing methods and, increasingly, every aspect of the technology is being automated, including robotic picking of M13 plaques, and preparation of template DNA for sequencing. Based on current progress, the full sequence is expected by 1998.

The Drosophila melanogaster *genome project*

The *D. melanogaster* genome project is being conducted largely as a collaboration between the University of California at Berkeley laboratory and a consortium of European laboratories (see Hartl and Palazzolo, 1993). Currently, only about 2 Mb of DNA have been sequenced, and the available information suggests a gene density of about 1 per 10–15 kb.

The mouse genome project

The most recent mouse genetic map has over 4000 STRP markers with an average spacing between markers of 0.35 cM, corresponding to a physical map distance of about 750 kb (Dietrich *et al.*, 1994). Because of various features (see *Box 13.1*), the mouse provides the model genome which is most relevant to the Human Genome Project. Of special importance are large chromosomal segments which show conservation of linkage (synteny) and of marker order between mice and humans (and other mammals; see *Figure 14.13*). Hence if a region of the mouse genome is mapped to high resolution, the information can be used to make predictions about the orthologous region of the human genome (and vice versa). This is particularly relevant to medical research because orthologous mouse and human mutants often show similar phenotypes so that positional cloning of a disease gene in one species may have considerable relevance to the other species.

Recently, there has been a commitment to develop very high resolution mouse genetic maps with average spacing of about 0.1 cM (200 kb) by large-scale **backcross mapping** (see *Box 14.4*). This involves a comprehensive backcross between two strains of mice, *Mus musculus* and *Mus spretus*, involving, for example, 1000 animals in the European mouse backcross effort (Breen *et al.*, 1994). A critical feature is that just about any type of DNA probe (i.e. cDNA, random single copy DNA fragments, microsatellites, etc.) will show variation between these species and so can be mapped on the backcross. Clearly, the mapping of coding DNA sequences will be of paramount interest to human geneticists; mapping of microsatellite markers is less important, because individual microsatellites are usually not conserved between species, except in the case of a proportion of those embedded within genes.

Other genome projects

Several other animal and plant genome projects are currently underway or are being actively considered. In some cases, the motivation is largely commercial, as in the case of mapping the genomes of various livestock animals (cow, sheep, pig, etc.) and cereal plants (wheat, rice, etc.). Because of its amenability to genetic analyses, the plant *Arabidopsis thaliana* has also been the subject of much attention, and a genome project is currently under active consideration. Two other contenders that may be particularly relevant to the Human Genome Project are possible projects to map the genomes of the puffer fish and the rat (see *Box 13.1*).

The Human Genome Diversity Project is an attempt to collect and study DNA sequence data from a variety of human populations

At the outset, the Human Genome Project was conceived as a project to obtain the nucleotide sequence of a collection of cloned human DNA fragments collectively amounting to one haploid genome. What it did not consider was the genetic diversity of humans. In order to address this issue, a small group of human geneticists proposed a global effort to survey genetic diversity among the world's peoples (Cavalli-Sforza *et al.*, 1991). The idea of a **Human Genome Diversity Project** quickly caught the imagination of the scientific community, was endorsed by HUGO, and was supported by funding agencies. Because certain marginalized human populations are threatened with the possibility of assimilation in the near future, the project had a certain emotional appeal as a means of conserving our genetic heritage, and could be considered to be a cultural obligation of the Human Genome Project.

The Human Genome Diversity Project was considered feasible because of the need to sample individuals only once: lymphocytes in a single blood sample taken from an individual can be immortalized (e.g. by treatment with Epstein–Barr virus; EBV), permitting a permanently renewable resource for studying the structure of the constituent DNA. Additionally, the great sensitivity of the PCR reaction means that DNA sequences can be amplified from easily accessible sources, such as saliva or hair roots, an important back-up in the case of individuals unwilling to donate a blood sample. The collection of DNA samples from, say, 25 anonymous individuals in each of 500 (out of the estimated total of 7000) ethnic groups world-wide was, therefore, a distinctly achievable aim. Key areas of research envisaged as benefitting from such a project include:

- *human origins, prehistoric population movements and social structure.*
- *Adaptation and disease.* As susceptibility to diseases can vary from one population to another, it is important to assess whether such variation is the result of adaptation to local conditions, or of random changes in the genetic constitutions. Important diseases of prime interest in this regard include diabetes, hypertension, thalassemia and sickle cell anemia.
- *Forensic anthropology.* The accuracy of DNA fingerprinting, a widely used tool in forensic science (see page 446), is dependent, in part, on knowing how the DNA markers detected in fingerprinting vary from one population to the next.

At the time of writing, the scale of The Human Genome Diversity Project had not been determined, although some preliminary estimates envisaged an initial 5-year project costing about $30 million. The collection will focus on populations that have been geographically isolated or have a distinct culture and language, which would be expected to provide more meaningful information than collections from ethnically heterogeneous urban populations. As well as collecting samples, the plan is to compare DNA sequences from the samples at a few dozen carefully chosen sites along the genome. In addition to the research areas described above, the project should give insights into the extent to which individuals differ from the reference human sequence charted in the Human Genome Project. The diversity project has, however, been controversial (see *Box 13.3*).

Data storage and access in the Human Genome Project

The avalanche of mapping information from the genome projects is being channeled into easily accessible electronic databases

To be useful, the enormous amount of information being generated in the genome projects must be easily accessible and in a form that is easily analyzed. Electronic databases are needed to serve this function (see Database issue of *Nucleic Acids Research,* 1994). Such databases can be accessed from remote computer stations throughout the world using the **Internet,** an international network that links computers throughout the world (see *Box 13.2*). In this way, distant users can input raw data which is processed by the database managers, and also analyze data already stored in the databases.

Sequence databases

Important nucleotide and protein sequence databases were established long before the genome projects. They record published nucleotide and protein sequences from all species and, increasingly, unpublished sequence data, because of the requirement of many scientific journals for authors to deposit the sequence information in approved databases prior to publication. The major sequence databases are located at dedicated centers in the USA (National Center of Biotechnology Information), UK (European Bioinformatics Institute) and Japan (National Institute of Genetics) (see *Table 13.4*) with co-ordination between the centers. By November 1995, GenBank listed a total of 385 Mb of nucleotide sequences, derived from a large number of

Box 13.2: A guide to the Internet

General aspects

The **Internet**, or Net as it is affectionately known, describes a worldwide system that links innumerable individual computers and local networks in universities, research laboratories, and commercial and government organizations of every type and size. It allows almost instant communication between computers anywhere in the world.

Accessing the Internet requires appropriate hardware, software and permissions. The basic hardware is a network card inside your computer and, if you are using a telephone line, a **modem**. Software includes basic programs to enable your computer to communicate across a network, and high-level tools such as a Web browser (see below).

Permissions for network access and e-mail (see below) are usually automatic within institutions; private users need to pay to join an access service (advertized in computer magazines). Once everything is set up, accessing the net is extremely simple, but setting up requires computer expertise and local knowledge. Consult your institution support service or local vendor.

The main uses of the Net can be divided into private, semi-private and public communications.

- Private communications include the various ways in which individuals can transfer data to or from a computer to which they have been granted access through a username and a password. Because of security risks, sensitive private communications are carried on private networks separate from the Internet. The main form of private communication on the Internet is a form of electronic mail known as **e-mail**, in which a message is sent that can be read only by the person addressed. This is achieved by using the person's e-mail address, which directs the message to the recipient's local computer, where access to it requires the correct username and password. Any computer data, including text, data, images, etc., can be sent by e-mail. E-mail has many advantages over conventional mail, phone or fax, but has the limitation that there are no public directories containing all e-mail addresses.
- Semi-private net applications center around moderated discussion groups, where registered members receive and can submit data, opinions, comments, etc.
- Different public applications exist. **Bulletin boards** are unmoderated discussion groups, which anyone can read and to which anyone can contribute. They exist on every conceivable topic. Another major application is the world-wide web (see below).

The world-wide web (WWW)

Probably the most revolutionary application of the Internet is the world-wide web (WWW). The Web, as it is affectionately known, is not a separate network, but a way of using the Net as a free public notice board or library or unlimited size and with unlimited access. Individuals or organizations choose to make available to Net users material which is located on their own computer. Net users find and view the material using a **web browser** program such as MOSAIC (a public domain program produced by the US National Center for Supercomputer Applications) or NETSCAPE NAVIGATOR (a commercial product currently available free of charge to academic users). The clever design of these programs allows the user to move effortlessly from site to site, collecting and collating material of every sort, including text, data, images, video clips and sound.

The key to the success of the Web is a text format known as hypertext. **Hypertext** is basically the same as regular text – it can be sorted, read, searched, or edited – with an important exception: *hypertext contains links within the text to other documents.* Clicking with a mouse on to a hypertext link in one document then allows the user to dial up another computer and access another hypertext document with its own hyperlinks. As a result, the hyperlinks can create a complex *web* of connections allowing the user to browse through innumerable documents stored in databases throughout the world. Various indexing programs (*'search engines'*) allow users to search for Web documents on a specific subject, such as the *yahoo* database maintained at Stanford University (http://www.yahoo.com).

In order to use the Web it is necessary to call up a document by its **uniform resource locator** (**URL**), an identifier which takes the form http://filename, so called because the standard language used to communicate within the Web is called the **hypertext transmission protocol** (**http**). For example, the URL of the WWW server at the European Bioinformatics Institute is http://www.ebi.ac.uk/, and if accessed will call up the *home page* (the default document seen when connecting to a Web server for the first time) seen in *Figure 13.7*.

Once accessed, the material can be stored and printed as desired. An essential feature of the Web, which exhilarates some people and worries others, is that access is completely open. There are no controls and no censorship, except within local institutions. Given the modest technology required, anybody can post anything on the Web. Thus information on the Web is very different from information in a peer-reviewed scientific journal, such as those we cite in chapter references. It may be good, poor, bad or even maliciously misleading. *Caveat emptor!* However, geneticists are exceedingly well served by many high quality databases, including those at the sites listed in *Tables 13.4* and *13.5*. Gaining access to these should be a high priority for any serious student of human genetics. For an introduction, see Harper (1995).

Table 13.4: Major nucleic acid and protein sequence databases

Database	Description and electronic addresses
GenBank	DNA and protein sequences. One of many databases distributed by the US National Center for Biotechnology Information (NCBI), NIH. *WWW-URL: http://www.ncbi.nlm.nih.gov; e-mail: ncbi.nlm.nih.gov*
EMBL	DNA sequences. Formerly at Heidelberg, but now distributed by the European Bioinformatics Institute (EBI) at Cambridge, UK, together with 30 other molecular biology databases. *WWW-URL: http://www.ebi.ac.uk; e-mail: ebi.ac.uk*
DDBJ	DNA database of Japan (Mishima). *WWW-URL:http://www.nig.oc.jp/home.html; e-mail: dd.bj@ddbj.nig.ac.jp*
PIR (protein information resource) – international	Protein sequences (single database distributed as a collaboration by: the US National Biomedical Research Foundation, Washington; the Martinsried Institute for Protein Sequences, Germany; and the Japan International Protein Information Database, Tokyo). *e-mail: PIRMAIL @ nbrf.georgetown.edu (USA); MIPS @ ehpmic.mips.biochem.mpg.de (Europe); EX5292 @JPN SUT30.BITNET (Japan)* Also accessible via the US NCBI (see above)
SWISS-PROT	Protein sequences. Maintained collaboratively by the University of Geneva and the EMBL data library. Distributed by several centers, such as the NCBI and EBI. Access via NCBI or EBI (see above)

The EMBL Outstation **European Bioinformatics Institute**

This is the the world-wide web (WWW) server of the European Bioinformatics Institute (EBI)which is located at Hinxton Hall near Cambridge in the UK. The EBI is an Outstation of EMBL

EMBL Nucleotide Sequence Database

SWISS-PROT Protein Sequence Database

EBI Databases

Data Submissions

Database query/retrieval

Sequence Similarity Searches

Documentation and Software

Network Navigation

Figure 13.7: The home page of the world-wide web server of the European Bioinformatics Institute.

species of which human entries predominated, accounting for about 30% of the total. Much of this data is derived from partial cDNA sequences which are also being stored in specialized databases such as *dbEST* (see below).

Table 13.5: Major databases of relevance to the Human Genome Project

Database	Description and electronic access
GDB (genome database)	The major human mapping database. Housed at Johns Hopkins University, School of Medicine. Permits interaction with the OMIM database (see below) *WWW-URL: http://gdbwww.gdb.org/; FTP: ftp.gdb.org; e-mail: help@gdb.org*
dbEST	Collection of partial sequences from cDNA clones (ESTs) *WWW-URL:http://www.ncbi.nih.gov/dbEST/index.html; FTP: ftp.gdb.org; e-mail: help@gdb.org*
OMIM (on-line Mendelian Inheritance in Man)	Electronic catalog of Victor McKusick's *Mendelian Inheritance in Man*, listing all known inherited human disorders. Housed at the Johns Hopkins University, and interactive with GDB (see above). Access via GDB
MGD (Mouse Genome Database)	The major mouse genome database. Based at the Jackson Laboratories, Bar Harbor, US *WWW-URL: http: //www.informatics.jax.org/;FTP: ftp.informatics.jax.org;e-mail: mgi-help@informatics.jax.org.*
Généthon	Human genetic maps based on linked $(CA)_n$ repeat markers, based in the Généthon laboratory near Paris *WWW-URL: http://www.genethon.fr/; FTP: ftp.genethon.fr*
CHLC (Co-operative Human Linkage Center)	Collaborative human genetic map database containing genotypes, marker data and linkage server, co-ordinated by Jeff Murray (University of Iowa) *WWW-URL: http://www.chlc.org/; FTP: ftp.chlc.org; e-mail:infoserver@chlc.org.*
CEPH	YAC-based physical maps of the human genome, based at the CEPH laboratory in Paris *WWW-URL:http://www.cephb.fr/bio/ceph-genethon-map.html* *FTP: ceph-genethon-map.genethon.fr; e-mail: ceph-genethon-map@cephb.fr*
Whitehead Institute	YAC-based physical maps of the human genome, based at the Whitehead Institute for Biomedical Research, Massachusett, USA *WWW-URL:http://www-genome.wi.mit.edu/*

Currently, the rate of sequencing is such that the number of entries in the major databases is doubling in size every 18 months, even though it is widely recognized that the rate of DNA sequencing is presently being held back by technological limitations. It is to be expected that the significant investment into research on technological developments will result in more rapid generation of DNA sequences. As the amount of data rises exponentially, so databases are having to be modified to maintain simple and efficient data. In addition to data input, sequence databases also offer suites of computer programs for users to analyze sequences (see page 515).

Mapping databases

With the advent of the genome projects, numerous electronic databases have been established to record the genetic and physical mapping information that is being generated. The central data resource for the human gene mapping effort, the **Genome Data Base (GDB),** was established at Johns Hopkins University School of Medicine in Baltimore, USA. GDB is regularly updated (see Fasman *et al.*, 1994). It collects, organizes, stores and distributes human genome mapping information, and also serves as a repository for genetic disease information. It also offers interaction with other databases, such as **OMIM** (see *Table 13.5*).

At the time of writing, GDB contains data in text form with a series of interacting manager programs, including *locus* (genes, D-segments, fragile sites, breakpoints, etc.), *polymorphism* (genes, D-segments), *probe* (PCR, ASO, clones) and *citation* managers, etc. The nomenclature used is decided by the HUGO nomenclature committee. Genes and pseudogenes are allocated symbols of usually two to six characters with a final P indicating a pseudogene. For anonymous DNA sequences, the convention is to use D (= DNA) followed by 1–22, X or Y to denote the chromosomal location, then S for a unique segment, Z for a chromosome-specific repetitive DNA family or F for a multilocus DNA family, and finally a serial number. The letter E following an anonymous clone number indicates that the clone is known to be expressed (see *Table 13.6*). As large-scale sequencing of the human genome begins, GDB will increasingly accommodate graphical displays of the type being currently used in the *C. elegans* database, ACEDB (see *Figures 13.8* and *13.9*).

Table 13.6: Human gene mapping nomenclature

Symbol	Interpretation
CRYB1	Gene for crystallin, β polypeptide 1
GAPD	Gene for glyceraldehyde-3-phosphate dehydrogenase
GAPDL7	GAPD-like gene 7, functional status unknown
GAPDP1	GAPD pseudogene 1
AK1	Gene for adenylate kinase, locus 1
AK2	Gene for adenylate kinase, locus 2
*PGK1*2*	Second allele at *PGK1* locus
B3P42	Breakpoint number 42 on chromosome 3
DYS29	Unique DNA segment number 29 on the Y chromosome
D3S2550E	Unique DNA segment number 2550 on chromosome 3, known to be expressed
D11Z3	Chromosome 11-specific repetitive DNA family number 3
DXYS6X	DNA segment found on the X chromosome, with a known homolog on the Y chromosome, and representing the 6th XY homolog pair to be classified
DXYS44Y	DNA segment found on the Y chromosome, with a known homolog on the X chromosome, 44th XY homolog pair
D12F3S1	DNA segment on chromosome 12, first member of multilocus family 3
DXF3S2	DNA segment on chromosome X, second member of multilocus family 3
FRA16A	Fragile site A on chromosome 16

A variety of other databases exist for the model organisms too, such as the **Mouse Genome Database,** MGD, which is housed at the Jackson Laboratories at Bar Harbor, USA. MGD stores data on mouse loci and probes and is integrated with the **Encyclopedia of the Mouse Genome,** an application that generates a graphical display of mouse genetic linkage maps using MGD locus and homology information. Such databases are useful to human gene mappers, because they provide human–mouse comparative mapping data (*Figure 13.10*). Clearly, data from the research community is fed into such dedicated databases. Recently, however, some of the major laboratories such as the CHLC genetic mapping collaboration, and the Genethon/CEPH laboratories, that are generating mapping data have established their own **file transfer protocol (ftp) servers** in order to provide the scientific community rapidly with ready access to recently produced data through Internet routes such as the World-Wide Web (see *Box 13.2*).

Very recently, there has been a rapid rise in the rate of data accumulation in the genome projects. By 31 December 1995, the GDB database included a total of 5895 human genes, 76 779 D segments and 18 716 polymorphisms. Most of the present GDB entries, however, have had a very short history. For example, the combined total of human genes plus markers that were entered into the GDB database in 1994 alone (> 40 000) was double that listed in the database at the beginning of the year (see *Figure 13.11*).

The Human Genome Project: friend or foe?

When the sequence is known, what will it mean and was it worth the effort after all?

If the sequence of all nucleotides in the human genome were printed using the typeface and line spacing in the present book, one page would contain 4 kb of sequence,

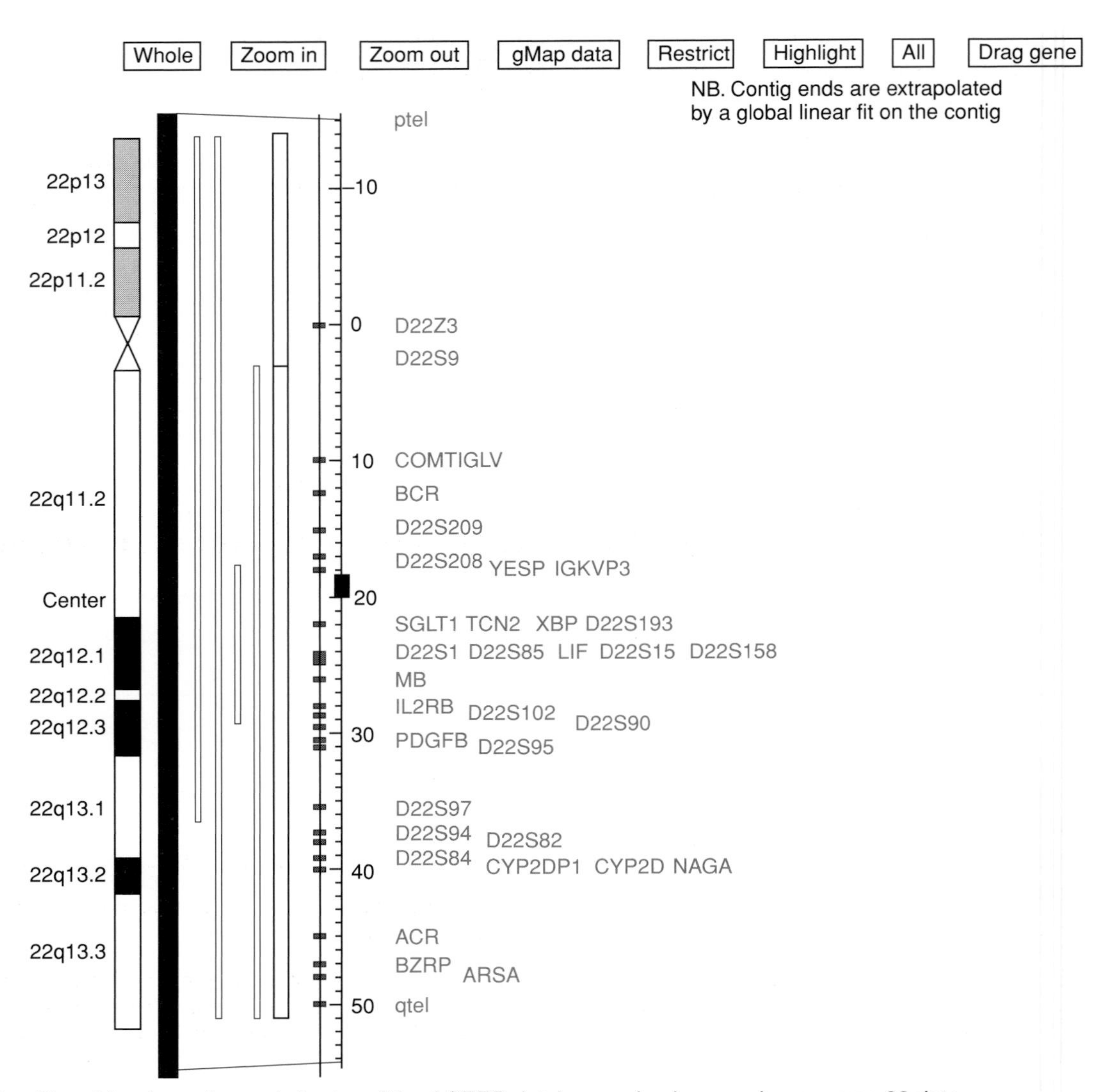

Figure 13.8: The gMap (genetic map) display of the ACEDB database using human chromosome 22 data.

Boxes running vertically next to the chromosome ideogram represent somatic cell hybrids. Reproduced from Bishop (1994) with permission from Academic Press Inc.

an average chromosome sequence would run into about 60 books of the same size as the present one, and the entire sequence would require 1200 such books. An average gene would be spaced over perhaps about three pages and be separated from its neighbors by about seven pages of noncoding sequence. Within an average 3-page gene sequence, the coding DNA would be arranged as a series of segments roughly two lines long and collectively occupy only about one third of a page. Could such a collection of unbelievably dull reading ever be of any value other than as a lending library for insomniacs?

This analogy has its serious point: what is the value of so much noncoding DNA in the human genome and is the sequence worth knowing? As is already known, there will be many very short regulatory sequences usually located very close to the coding DNA, but such sequences, together with the sequences that are necessary for chromosome function (see page 45) will still represent a tiny fraction of the total noncoding DNA in the human genome. Some have envisaged that this noncoding

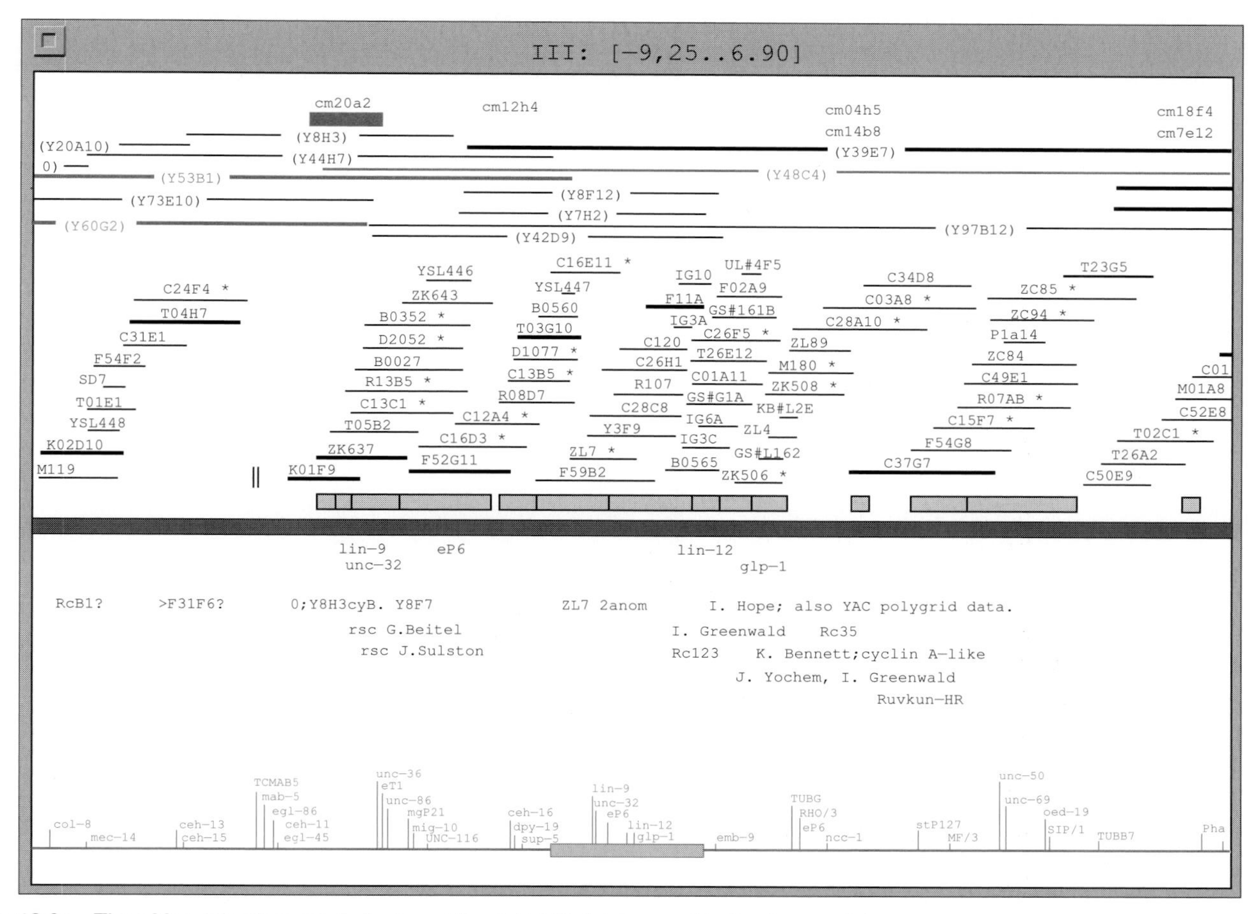

Figure 13.9: The pMap (physical map) display of the ACEDB database with *C. elegans* data.

The region around the *C. elegans* gene *lin-9* on chromosome III is shown. At the top of the display are a series of names which indicate cloned DNA probes. One of these has been picked and is highlighted (red box). Horizontal lines below this indicate the extent of the first YACs and, below this, cosmids. The three YACs to which the selected probe hybridizes are highlighted in red. Reproduced from Bishop (1994) with permission from Academic Press Inc.

DNA may still carry an important message, although clearly not one that is very well conserved during evolution. Analysis of available noncoding DNA sequences has revealed a nonrandom pattern, but the significance of this finding has been controversial, and most researchers feel that noncoding DNA is generally functionless, a collection of junk sequences that have been inserted into the genome during evolution. After all, the puffer fish carries no fewer genes than other fish, but its genome is only one eighth the size of many vertebrate genomes (*Box 13.1*).

Notwithstanding the general pessimism concerning the value of noncoding DNA, many researchers remain convinced that at least the *noncoding DNA within genes* must be doing something valuable: cells have put up with introns for hundreds of millions of years, patiently cutting out such sequences every time a gene is transcribed and splicing the coding parts together. It is not as if, once introns have been introduced, the cell cannot remove them from the coding DNA: functional processed genes can be formed by the mechansim shown in *Figure 8.7*, such as the *PDHA2* gene (see pages 216–217). Inevitably, however, most of the analysis of the human genome sequence will be initially confined to the coding DNA sequences, the most interesting and medically significant part.

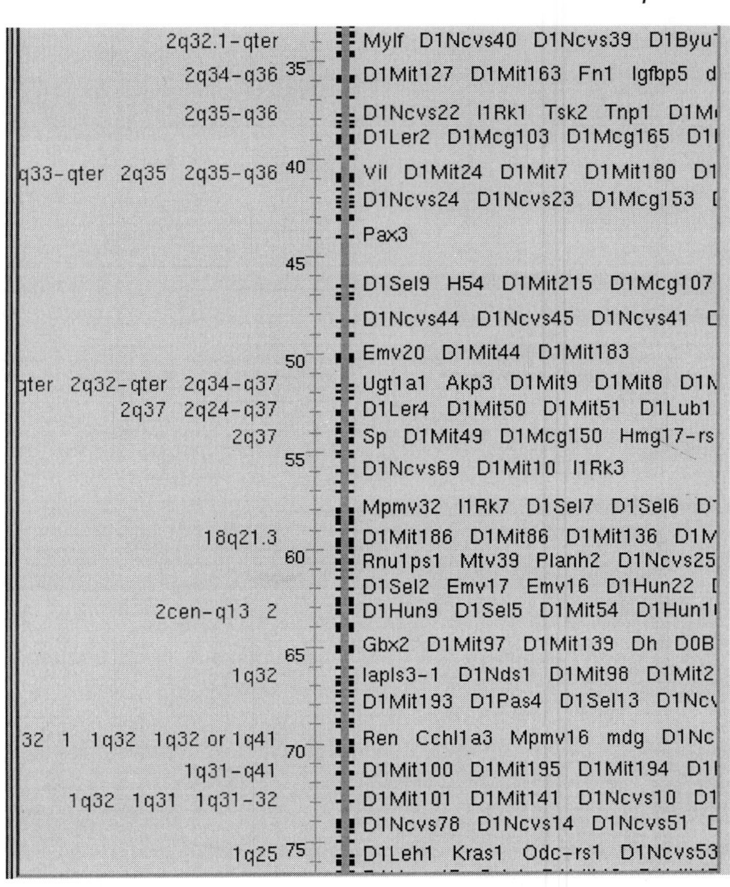

Figure 13.10: The Encyclopedia of the Mouse Genome permits graphical display of comparative man–mouse homologies.

The figure illustrates cross-homology between distal human chromosome 2 and a segment of mouse chromosome 1. Such cross-homology was helpful in the identification of the gene for Waardenburg syndrome type 1 (see page 389).

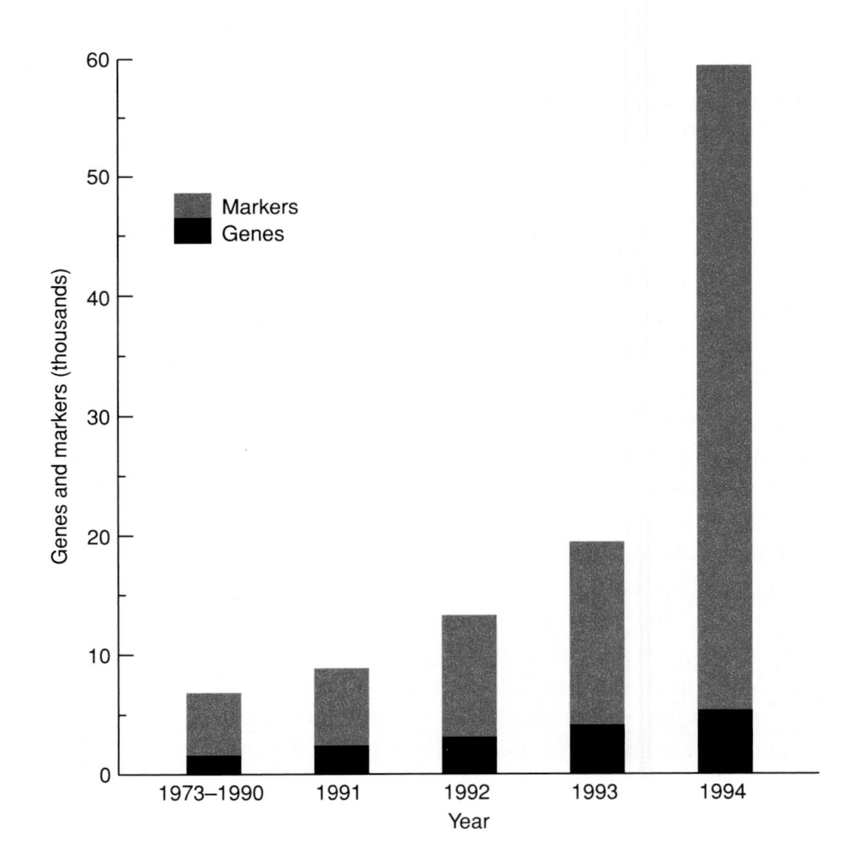

Figure 13.11: The rapid growth of mapping data in the Human Genome Project.

The figure illustrates the cumulative growth in the numbers of human genes and anonymous DNA markers that have been cataloged at the indicated time points. *Note* that progress in this area has been so rapid that about 40 000 genes and markers were first reported in 1994, twice the total number that had been reported before 1994.

The medical and scientific benefits of the Human Genome Project are expected to be enormous

For many human biologists and geneticists, the Human Genome Project represents an exciting, historic mission. Since its outset, the project has been especially justified by the expected medical benefits of knowing the structure of each human gene. Inevitably, this information will provide more comprehensive prenatal and pre-symptomatic diagnosis of disorders in individuals judged to be at risk of carrying a disease gene. Additionally, the information on gene structure will be used to explore how individual genes function and how they are regulated. Such information will provide sorely needed explanations for biological processes in humans. It would also be expected to provide a framework for developing new therapies for diseases, in addition to simple gene therapy approaches. More importantly, as mutation screening techniques develop, an expected benefit would be to alter radically the current approach to medical care, from one of treating advanced disease to preventing disease based on the identification of individual risk (see Guyer and Collins, 1993).

Exciting though such possibilities are, however, there may be unexpected difficulties in understanding precisely and comprehensively how some genes function and are regulated (cautionary precedents are the lack of progress in predicting protein structure from the amino acid sequence, and the lack of understanding of the precise ways in which the regulation of globin gene expression is co-ordinated, decades after the relevant sequences have been obtained). In addition, the single gene disorders which should be the easiest targets for developing novel therapies are very rare; the most common disorders are multifactorial. So although the data collected in the Human Genome Project will inevitably be of medical value, some of the most important medical applications may take some time to be developed.

The Human Genome Project has increasingly attracted criticism

The Human Genome Project has been controversial from its outset. Some of the criticism concerns scientific strategy and the ways in which it is being executed. As described on page 345, the goal of total genome sequencing as opposed to prioritizing the sequencing of coding DNA has been controversial. The way in which the Human Genome Diversity Project, in particular, has been conducted has attracted considerable criticism: sample collection from individuals has often been done insensitively, without proper consultation or explanation of the possible future use of the samples. As a result, there has been furious opposition to the diversity project by several of the endangered indigenous groups who were selected, without their consultation, as target populations. Because lymphocytes in blood samples can be immortalized by procedures such as treatment with EBV, this leaves open the possiblity of commerical exploitation of a valuable cell line (see *Box 13.3*). In such an environment of distrust, the project's aim of collecting and storing samples from endangered human populations can easily be seen as a curious priority, when virtually nothing is being done to address the causes of why the populations are endangered in the first place.

As the Human Genome Project enters a phase where high throughput, large-scale automation is required to drive progress, commercial concerns are increasingly taking a stake in the research. Because of their suitability for such work and the large financial investments to support their research, they can virtually monopolize certain research areas, such as the human cDNA sequencing effort. By subsequently seeking patents to protect their financial investment, they have then raised an important issue: *who owns the human genome?* (see *Box 13.3*).

Box 13.3: Patenting human cells and genes is controversial

Human cells. The unwelcome commercial exploitation of human cells has been acutely illustrated by the case of John Moore, an American leukemia patient. Without seeking his consent, without even informing him, the University of California doctor who treated him cultured white blood cells from his cancerous spleen into an 'immortal' cell line. Unusually, the resulting cell line was capable of producing blood proteins of great value in treating immunosuppressive disease. The University of California subsequently applied for and was granted a patent on its 'invention'. Thereafter, the cell line was sold to a biotechnology company for $1.7 million, again without John Moore's knowledge, and subsequently made a fortune for its new owners. Similarly, cell lines derived from samples from isolated human populations have been patented, because of their unusual qualities. A recent example is provided by the Hagahai people of Papua New Guinea. After samples had been collected for establishing cell lines at the US NIH, incidental studies of the cells showed evidence of the human T-lymphotrophic virus-1 (HTLV-1) virus, indicating that several of the individuals who donated the samples were carriers of HTLV-1. As HTLV-1 was considered a virus which could help combat leukemia, the US Department of Commerce patented the cell lines.

Human genes. In 1991, the US NIH applied for a patent for more than 7000 fragments of brain cDNA clones whose sequences had been established as part of an *EST mapping* exercise led by Dr Craig Venter. This attempt met with widespread opposition from the scientific community, especially since nothing was known about the functions of the expressed sequences. Under pressure, the US Patents Office rejected the applications. For the first time, the question of the ownership of the human genome had been raised and the idea of commercial monopoly of what is quite simply our genetic heritage appeared alarming and offensive to many.

As commercial involvement in the Human Genome Project increases rapidly, the issue continues to be controversial. A recent example again concerns Dr Venter who, after filing the previous NIH patent applications, subsequently left NIH to set up a new institute, the Institute of Genome Research at Gaithersburg, USA, funded by a commercial concern. Adopting a factory-style approach to EST sequencing (in terms of scale of automation and number of technicians) they were able to out-perform any research laboratory. As a result, they quickly were able to compile the world's largest human gene databank, currently believed to contain hundreds of thousands of cDNA sequences (individual genes can be represented by several cDNA fragments).

In April 1994 the drug company SmithKline Beecham invested £80 million for an exclusive stake in Venter's database and announced to scientists that they could have access to it only if they agreed to concede first rights to any patentable discovery. Again the prospect of a corporation trying to monopolize control of a large part of the expressed human genome alarmed the scientific community. Many felt that a case could be made for patenting, after identifying the function of a gene, but not before. Subsequently, however, another large drugs company, Merck, have funded a similar large-scale EST sequencing project, with the promise of making the resulting data freely available to the scientific community (see pages 346–347).

Without proper safeguards, the Human Genome Project could lead to discrimination against carriers of disease genes and to a resurgence of eugenics

Any major scientific advance carries with it the fear of exploitation. The Human Genome Project is no exception, and perceived benefits of the project can also have a downside. For example, when we know all the human genes and can detect large numbers of disease-associated mutations, there will be enormous benefit in targeting prevention of disease to those individuals who can be shown to carry disease genes. However, the same information can also be used to discriminate against such individuals by insurance companies. For example, there is the very real prospect of insurance companies insisting on large-scale genetic screening tests for the presence of genes that confer susceptibility to common disorders such as diabetes, cardiovascular disease, cancers and various mental disorders. Perfectly healthy individuals who happen to be identified as carrying such disease-associated alleles may then be refused life or medical insurance. Clearly such discrimination is practised on a small scale at the moment; what is alarming to many is the prospect of discrimination against a very large percentage of the individuals in our society. It is also important to preserve people's right not to know. A fundamental ethical principle in all genetic counseling and genetic testing is that genetic information should be generated only in response to an explicit request from a fully informed adult patient.

Another troublesome area is the question of biological determinism and whether comprehensive knowledge of human genes could foster a revival of *eugenics,* the application of selective breeding or other genetic techniques to 'improve' human qualities (Garver and Garver, 1994). In the past, negative eugenic movements in the US and Germany severely discriminated against individuals who were adjudged to be inferior in some way, notably by forcing them to be sterilized. The possibility also exists of a preoccupation with *genetic enhancement* to positively select for heritable qualities that are judged to be desirable (see page 585). In recognition of the above problems, the US Human Genome Project has devoted considerable resources to support research into the ethical, legal and social impact of the project.

Further reading

Engel LW. (1993) The Human Genome Project. History, goals and progress to date. *Arch. Pathol. Lab. Med.,* **117,** 459–465.

Glister P. (1994) *The Internet Navigator,* 2nd Edn. John Wiley & Sons, New York.

Guyer MS, Collins FS. (1993) The Human Genome Project and the future of medicine. *Am. J. Dis. Child.,* **147,** 1145–1152.

Wilkie T. (1993) *Perilous Knowledge: the Human Genome Project and its Implications.* Faber and Faber, New York.

Understanding our genetic inheritance. The US Human Genome Project: the first five years FY 1991–1995. (1990) US Department of Health and Human Services and US Department of Energy.

Genetic Information and Health Insurance. (1993) US NIH–DOE Working Group on Ethical and Social Implications of Human Genome Research.

The Genome Directory. (1995) A supplement published by the journal. *Nature,* **377,** 1S–379S.

Genome Information on the World-Wide Web. *Genome Digest*, October 1995, pp. 10–13.

Electronic information on the Human Genome Project can be found at many locations. Useful Web sites are found at the US National Center for Human Genome Research (NCHGR – *http://gaea.nchgr.nih.gov/*) and the US National Center for Genome Resources (NCGR – *http://www.ncgr.org/*).

References

Adams MD, Kelley JM, Gocayne JD *et al.* (1991) *Science,* **252,** 1651–1656.

Adams MD, Kerlavage AR, Kelley JM, Gocayne JD, Fields C, Fraser CM, Venter JC. (1994) *Nature,* **368,** 474–475.

Adams MD, Kerlavage AR, Fleischmann RD *et al.* (1995) *Nature,* **377** (Suppl.), 3–174.

Adams MD *et al.* (1995) *Nature,* **377**, 3S–174S.

Antequera F, Bird A. (1993) *Proc. Natl Acad. Sci. USA,* **90**, 11995–11999.

Ashworth LK, Batzer MA, Brandriffi B *et al.* (1995) *Nature Genetics,* **11**, 422–427.

Baxendale S, Abdulla S, Elgar G *et al.* (1995) *Nature Genetics,* **10**, 67–76.

Bentley DR, Dunham I. (1995) *Curr. Opin. Genet. Dev.,* **5**, 328–334.

Berry A, Stevens TJ, Walter NAR *et al.* (1995) *Nature Genetics,* **10**, 415–423.

Bishop MJ. (1994) (ed.) *Human Genome Computing*. Academic Press, New York.

Boguski M, Schuler GD. (1995) *Nature Genetics,* **10**, 369–371.

Botstein D, White RL, Skolnick M, Davis RW. (1980) *Am. J. Hum. Genet.,* **32**, 314–331.

Breen M, Deakin L, Macdonald B *et al.* (1994) *Hum. Mol. Genet.,* **3**, 621–627.

Brenner S, Elgar G, Sandford R, Macrae A, Venkatesh B, Aparicio S. (1993) *Nature,* **366**, 265–268.

Cavalli-Sforza LL, Wilson AC, Cantor CR, Cook-Deegan RM, King MC. (1991) *Genomics,* **11**, 490–491.

Chumakov IM, Rigault P, Guillou S *et al.* (1992) *Nature,* **359**, 380–386.

Chumakov IM, Rigault P, Le Gall I *et al.* (1995) *Nature,* **377** (Suppl.), 175–297.

Chumakov I *et al.* (1995) *Nature,* **377**, 175S–298S.

Cohen D, Chumakov I, Weissenbach J. (1993) *Nature,* **366**, 698–701.

Collins F, Galas D. (1993) *Science,* **262**, 43–46.

Cooper DN. (1995) In: *Functional Analysis of the Human Genome* (eds F Farzaneh, DN Cooper), pp. 43–68. BIOS Scientific Publishers, Oxford.

Cox DR, Green ED, Lander ES, Cohen D, Myers R. (1994) *Science,* **265**, 2031–2032.

Cross SH, Charlton JA, Nan X, Bird AP. (1994) *Nature Genetics,* **6**, 236–244.

Cuticchia AJ, Chipperfield MA, Porter CJ, Kearns W, Pearson PL. (1993) *Science,* **262**, 47–48.

Database issue of Nucleic Acid Research. (1994) *Nucleic Acids Res.,* **17**, 3441–3665.

Dausset J, Cann H, Cohen D, Lathrop M, Lalovel JM, White R. *et al.* (1990) *Genomics,* **6**, 575–577.

Dietrich WF, Miller JC, Steen RG *et al.* (1994) *Nature Genetics,* **7**, 220–245.

Donis-Keller H, Green P, Helms C *et al.* (1987) *Cell,* **51**, 319–337.

Fasman KH, Cuticchia AJ, Kingsbury DT. (1994) *Nucleic Acids Res.,* **22**, 3462–3469.

Fleischmann D, Adams MD, White O *et al.* (1995) *Science,* **269**, 496–512.

Foote S, Vollrath D, Hilton A, Page DC. (1992) *Science,* **258**, 60–66.

Garver KL, Garver B. (1994) *Am. J. Hum. Genet.,* **54**, 148–158.

Goodfellow P. (1995) *Nature,* **377**, 285–286.

Guyer MS, Collins FS. (1993) *Am. J. Dis. Child.,* **147**, 1145–1152.

Harper R. (1995) *Trends Genet.,* **11**, 223–228.

Hartl DL, Palazzolo MJ. (1993) In: *Genome Research in Molecular Medicine and Virology* (ed. KW Adolph), pp. 115–129. Academic Press, Orlando, FL.

Ichikawa H, Hosoda F, Arai Y, Shimizu K, Ohira M, Ohki M. (1993) *Nature Genetics,* **4**, 361–365.

Jacob A, Brown DM, Bunker RK *et al.* (1995) *Nature Genetics,* **9**, 63–69.

Jones SJM. (1995) *Curr. Opin. Genet. Dev.,* **5**, 349–353.

Korenberg JR, Chen X-N, Adams MD, Venter JC. (1995) *Genomics,* **29**, 364–370.

Larionov V, Kouprina N, Nikolaishvili N, Resnick MA. (1994) *Nucl. Acid Res.,* **20**, 4154–4162.

Little P. (1995) *Nature,* **377**, 286–287.

McKusick V. (1989) *Genomics,* **5**, 385–387.

Morishima Y *et al.* (1995) *Genomics,* **28**, 273–279.

Murray JC, Buetow KH, Weber JL *et al.* (1994) *Science,* **265**, 2049–2054.

Nowak R. (1995) *Science,* **269**, 468–470.

Oliver S. (1995) *Nature Genetics,* **10**, 253–254.

Ross MT. (1995) In: *Current Protocols in Human Genetics* (eds NC Dracopoli *et al.*). John Wiley & Sons, New York.

Sikela JM, Auffray C. (1993) *Nature Genetics,* **3**, 189–191.

Weissenbach J, Gyapay G, Dib C, Vignal A, Morissette J, Millasseau P, Vaysseix G, Lathrop M. (1992) *Nature,* **359**, 794–801.

Wilson J, Ainscough R, Anderson K *et al.* (1994) *Nature,* **368**, 32–38.

Yaspo M-L, Gellen L, Mott R, Korn B, Nizetic D, Pouska AM, Lehrach H. (1995) *Hum. Mol. Genet.,* **4**, 1291–1304.

Zorio DA, Cheng NN, Blumenthal T, Spieth J. (1994) *Nature,* **372**, 270–272.

Identifying human disease genes

Chapter 14

Principles and strategies in identifying disease genes

Before 1980, only a very few human genes had been identified as disease loci. Such early successes were very largely the result of exceptional characteristics: the biochemical basis of the disease had previously been established and purification of the gene product could be achieved without too much difficulty. Such advantages do not apply, however, to the great majority of diseases resulting from mutation in human genes. In the 1980s, the application of recombinant DNA technology offered new approaches to mapping and identifying the genes underlying inherited single gene disorders and somatic cancers, and the number of disease genes identified started to increase rapidly. With the subsequent advent of rapid PCR-based linkage studies and PCR mutation screening technologies, the identification of novel disease genes became commonplace and is currently occurring on a weekly basis.

It is important to note that not all 65 000–80 000 human genes will be identified as disease genes. Some genes are indispensable to embryonic function so that deleterious mutations result in embryonic lethality and go unrecorded in humans. In other cases, abolition of gene function may have no effect on the phenotype because of *genetic redundancy,* that is other nonallelic genes also supply the same function. *Mendelian Inheritance in Man,* the catalog of inherited human disorders, currently lists only about 5000 mendelian traits. As we shall see in Chapter 15, there is not a one-to-one correspondence between genes and clinical syndromes. Sometimes different mutations in one gene cause different phenotypes, and frequently the same disease phenotype can be caused by mutations in different genes.

The more common inherited disorders are the ones that are the most difficult to study by molecular genetics: a combination of different genes is often involved (oligogenic or polygenic disorders, see Chapter 18) as well as different environmental triggers. Similarly, the more common cancers involve cellular events in which multiple genes are involved. Not unexpectedly, therefore, the human disease genes that have been isolated to date are very largely those responsible for inherited single gene disorders, or somatic cancers where there is a single major susceptibility gene. This chapter will focus on attempts to identify these types of genes; the strategies used to identify genes involved in common diseases is considered in Chapter 18.

Four general strategies can be described for indentifying disease genes

The choice of strategy for identifying a disease gene depends on what resources (animal models, chromosomal abnormalities, clone libraries, etc.) are available, and on how much is known about the pathogenesis of the disease. Several of the possible strategies aim initially to identify a number of **candidate genes**, which then have to be tested individually for evidence that implicates them as the disease locus. Several different methods have been used to identify candidate genes (see *Box 14.1* and *Figure 14.1*), but mapping the disease to a specific subchromosomal localization is generally the most fruitful first step.

Box 14.1: General strategies for identifying human disease genes

Functional cloning (see page 369)
Some information about the function of the gene is exploited to isolate a gene clone. If the gene product is known, partial purification of the product can permit various strategies for identifying the underlying gene. Alternatively, a functional assay can be used to screen for the gene. This approach has been useful in only a few cases.

Positional cloning (see page 372 and *Table 14.1*)
This means isolating the gene knowing only its subchromosomal location, without using any information about the pathogenesis or the biochemical function. The general approach is to try to construct physical and genetic maps of the region, refine the subchromosomal localization, and then identify genes in the region to investigate as disease gene candidates. Positional cloning remains arduous, and is increasingly becoming unnecessary as information accumulates which allows a positional candidate gene approach (see below).

Position-independent candidate gene approaches (see page 387)
A candidate gene for a human disorder may be suggested without any knowledge of the chromosomal location. This can happen if the phenotype resembles another phenotype in animals or humans for which the gene is known, or if the molecular pathogenesis suggests that the gene may be a member of a known gene family. Such approaches have only rarely been successful, and have been overtaken by positional candidate gene approaches (see below).

Positional candidate gene approaches (see page 387 and *Table 14.2*)
Once a disease has been mapped, it is increasingly becoming possible to use database searches to identify candidate genes. With more and more human genes being mapped to specific subchromosomal regions, positional candidate gene approaches are set to dominate the field.

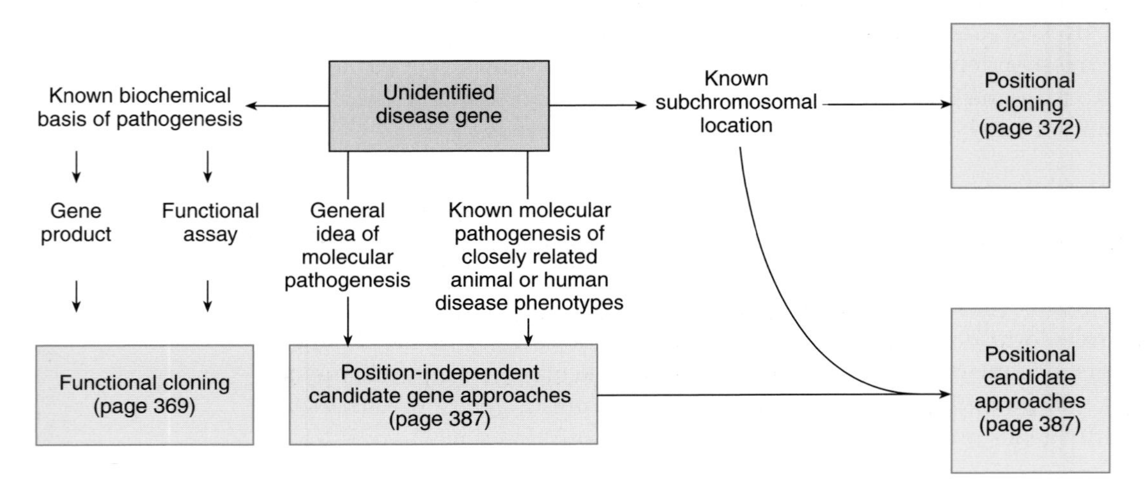

Figure 14.1: Four ways to identify a human disease gene.

Note that the two candidate gene methods start with the hypothesis of a candidate gene, but that ultimately all four methods identify candidate disease genes which then have to be tested for evidence of association with disease.

Positional cloning can be exceedingly laborious, and is usually not seriously attempted until genetic or physical mapping has located the disease gene to within about 1–2 Mb of DNA. For some disorders, such fine mapping is not practicable. If a disorder is rare, it may be very difficult to identify crossovers that narrow the location of the disease gene to a small interval, and the chance of identifying a patient with a suitable disease-associated cytogenetic abnormality which could refine the map location, may be virtually zero. In such cases candidate gene approaches are the only way forward.

Candidate disease genes must be tested for evidence of disease association

Each of the approaches listed in *Box 14.1* generates candidate disease genes which then have to be tested individually to determine the likelihood that they are associated with the disease. A good candidate gene should have an expression pattern consistent with the disease phenotype, although usually only a subset of expressing tissues are involved in the pathology (see page 419). Demonstrating that a candidate gene is likely to be the disease locus can be done by various means.

- *Mutation screening*. Screening for patient-specific point mutations in the candidate gene is by far the most popular method, because it is generally applicable and comparatively rapid. Identifying mutations in multiple affected individuals strongly suggests that the correct gene has been identified, but formal proof requires additional evidence.
- *Restoration of normal phenotype* in vitro. For some disorders, particularly those where mutations cause loss of function, the phenotype is reversible. If a cell line that displays the mutant phenotype can be cultured from the cells of a patient, transfection of a cloned normal allele into the cultured disease cells may result in restoration of the normal phenotype by complementing the genetic deficiency.
- *Production of a mouse model of the disease*. Once a putative disease gene is identified, a transgenic mouse model of the disease can be constructed. The orthologous mouse gene is identified and characterized and, if the human phenotype is known to result from loss of function, *gene targeting* can be used to introduce a deleterious germline mutation in the mouse ortholog. The mutant mice are expected to produce a phenotype resembling the human disease, this expectation may not always be met (see page 546). If the human disease phenotype results from a gain of function, then attempts to make a transgenic mouse model normally involve introduction of a disease allele into the mouse germline (see page 544).

Functional cloning

Functional cloning means using information about the function of an unidentified disease gene in order to identify the gene. Essentially two approaches are used: those which depend on the availability of purified gene product, and those for which a functional gene assay is required.

Partial purification of a gene product permits production of gene-specific oligonucleotides or specific antibodies which can be used to identify the gene

If the biochemical basis of an inherited disease is known, it may be possible to purify and partially characterize some of the gene product. If this is possible, two approaches can be taken to identify the disease gene.

Gene-specific oligonucleotides

This approach relies on the ability to isolate sufficient protein product to permit amino acid sequencing. Specific peptide bonds in the protein product can be cleaved using suitable proteolytic enzymes such as *trypsin* (cuts at the carboxyl end of lysine or arginine residues) or reagents such as *cyanogen bromide* (cuts at the carboxyl end of methionine residues). The amino acid sequence of the resulting peptides can be determined by chemical sequencing. This involves a repeated series of chemical reactions in an automated **amino acid sequencer**. In each cycle, an individual peptide is exposed to a chemical that covalently bonds to the N-terminal amino acid and subsequently cleaves this amino acid, permitting its identification by chromatographic methods. Overlapping peptides can be identified by sequence overlaps, enabling longer sequences to be established.

The resulting amino acid sequence is inspected to identify regions containing amino acids with minimal codon degeneracy (e.g. methionines and tryptophans are uniquely encoded by AUG and UGG codons, respectively). Once suitable regions have been identified, combinations of oligonucleotides are synthesized to correspond to all possible codon permutations. The resulting mix of *partially degenerate oligonucleotides* is labeled and used as a probe to screen cDNA libraries. As only one of the oligonucleotides in the mix will correspond to the authentic sequence, it is important to keep the number of different oligonucleotides low so as to increase the chance of identifying the correct target. Once a suitable cDNA clone is isolated, it can be used to screen a genomic DNA library in order to isolate genomic DNA clones for full characterization of the gene.

Identification of the hemophilia A gene illustrates this approach. Biochemical analysis of serum samples from individuals with hemophilia A had previously established a genetic deficiency of a blood clotting factor, factor VIII. Purification of factor VIII from plasma is not straightforward, partly because it is present in very low quantities. One approach involved isolating small quantities of clotting factor VIII from large volumes of pig blood by standard protein purification techniques (see *Table 19.1*). The purified product allowed the production of gene-specific oligonucleotide probes for library screening (see *Figure 14.2*).

Library screening by hybridization can be tedious when a complex mixture of oligonucleotides is used, because the results are greatly influenced by the hybridization conditions. A more rapid alternative is to use partially degenerate oligonucleotides as PCR primers. One early strategy was to use two such sets corresponding to amino acid sequences from different regions of the protein as primers. By using total cDNA from a suitable source as a template, a specific cDNA could be generated spanning the codons from the two different regions (see *Figure 6.9*). However, this approach demands considerable prior information about the sequence of the protein. A more convenient alternative is to prepare cDNA and ligate it to vector DNA molecules. PCR can then be conducted using one vector-specific primer and one primer composed of a panel of partially degenerate oligonucleotides.

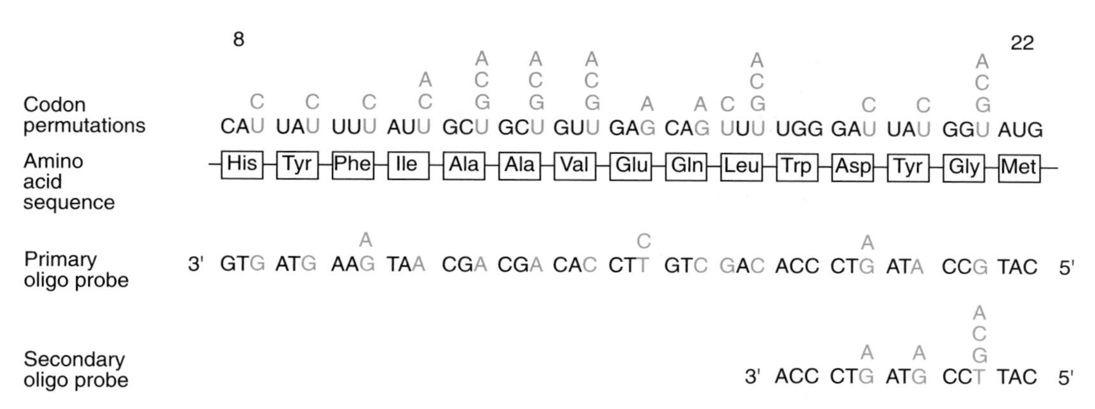

Figure 14.2: The factor VIII gene, the locus for hemophilia A, was cloned by product-directed oligonucleotide screening of DNA libraries.

The figure illustrates one way in which factor VIII DNA clones were obtained, following cleavage of purified porcine factor VIII protein into peptides and amino acid sequencing. The resulting sequences were inspected to identify regions with low codon redundancy. The top panel shows a sequence of 15 amino acids from His8 to Met22 in one of the peptides, with the possible codon permutations above (with variable nucleotides in red). This sequence was selected because of the generally low codon redundancy: two amino acids, Trp and Met, are specified by a single codon and another seven can be specified by just two alternative codons. A partially degenerate 45-bp antisense oligonucleotide probe was prepared and used as a primary hybridization probe to screen a porcine genomic DNA library, and thereafter secondary screening used a 15-bp antisense degenerate oligonucleotide probe corresponding to the sequence from Trp18 to Met22. The porcine factor VIII genomic clone was then used to screen human DNA libraries to identify the human gene (see Gitschier *et al.*, 1984).

Specific antibodies

If even small amounts of the normal protein product can be isolated, specific antibodies can be raised. The protein, or a peptide derived from it, is conjugated to a powerful immunogenic **hapten** (such as keyhole limpet hemocyanin) and the compound molecule injected into a rabbit or mouse. The hapten activates B lymphocytes, and the protein or peptide of interest activates helper T lymphocytes, leading to production of antibodies. Mouse or rabbit antibodies which are specific for the desired protein or peptide can then be used in various ways to identify a corresponding cDNA.

An early approach was to enrich for mRNA encoding the product in a cell-free *in vitro protein synthesis system*. This was the way in which the gene for phenylketonuria was identified. Phenylketonuria was known to be caused by a lack of the enzyme phenylalanine hydroxylase. The enzyme was purified from liver, where it was known to be expressed. Specific antibodies were raised and employed to immunoprecipitate polysomes containing phenylalanine hydroxylase mRNA (Robson *et al.*, 1982). The purified mRNA was then converted to cDNA and a specific cDNA clone was isolated.

This type of approach has been superseded by **antibody screening** of cDNA libraries. Antibody screening requires a cDNA library constructed by cloning cDNA from a relevant tissue into an expression vector. Inserts within the recombinant DNA clones are expected to be expressed within the host cell to produce foreign polypeptides. Specific antibodies can then be used to screen colony filters from such **expression cDNA libraries** to identify clones encoding the product of interest (see *Figure 19.1*).

Some disease genes have been identified by using a functional complementation assay.

In some cases, a **functional complementation** assay is possible even when the gene product is unknown. Yeast mutants have been used to identify human genes which perform specific biochemical functions. Defined yeast mutants are relatively abundant, and genetic analysis in yeast is particularly sophisticated because of the ease of performing homologous recombination. For proteins that have been very highly conserved during evolution, the human protein may be able to complement a yeast mutant defective in the corresponding protein. The procedure involves transforming the mutant yeast cells with different human DNA fragments in an effort to rescue the normal phenotype. This approach has been successful in identifying the human genes that specify various enzymes of purine and pyrimidine biosynthesis, and also some crucially important transcription factors, etc.

Human disease genes have also been identified by functional complementation cloning, particularly those involved in DNA repair. A variety of mammalian cell lines have been generated that are deficient in DNA repair. They show abnormal responses following exposure to UV irradiation or chemical mutagens. These mutant cells, or alternatively cells derived from patients with a DNA repair deficiency, can be transformed by fragments of normal human DNA or human chromosomes in order to produce a repair-competent phenotype. This was the way in which cDNA clones for the Fanconi's anemia group C (*FACC*) gene were first obtained (Strathdee *et al.*, 1992).

Positional cloning

Positional cloning means cloning a gene knowing nothing except its subchromosomal location (Collins, 1992, 1995). Often the initial localization defines a relatively large candidate region of 10 Mb or more. Such initial locations may come from varous sources; the most common are:

- *linkage analysis* identifies a marker showing a lod score greater than 3 in family studies (Chapter 12). This is the most common means of localization for inherited diseases.
- *Loss of heterozygosity screening* (*Figure 17.11*) identifies a chromosomal region which is commonly deleted in tumors. Tumor suppressor genes important in cancer are often first located this way (see page 467 for details).
- *Chromosomal abnormalities* can implicate a general region in a disease. For example, translocations and deletions identified Xp21 as the location of the Duchenne muscular dystrophy (DMD) gene (page 380).

Getting from such an initial localization to an identified gene is a major achievement. The first successful positional cloning efforts were published in 1986 and marked a triumphant new era for human molecular genetics. One after another, genes for important disorders such as DMD, cystic fibrosis (CF), Huntington's disease (HD), adult polycystic kidney disease, colorectal cancer, breast cancer, etc., were isolated. However, positional cloning can be exceedingly arduous, and by 1995 only about 50 inherited disease genes had been identified by this approach (Collins, 1995). Moreover, almost every one of these successes had depended on something other than pure positional cloning – usually the discovery of a large-scale mutation in one or more patients, or the existence of linkage disequilibrium. Now, with the ever increasing number of mapped human genes, and with the expectation that the great majority of all human genes will be placed on the physical map by about 1998 (see page 347), positional candidate approaches are set to overtake pure positional cloning (see *Figure 14.3*).

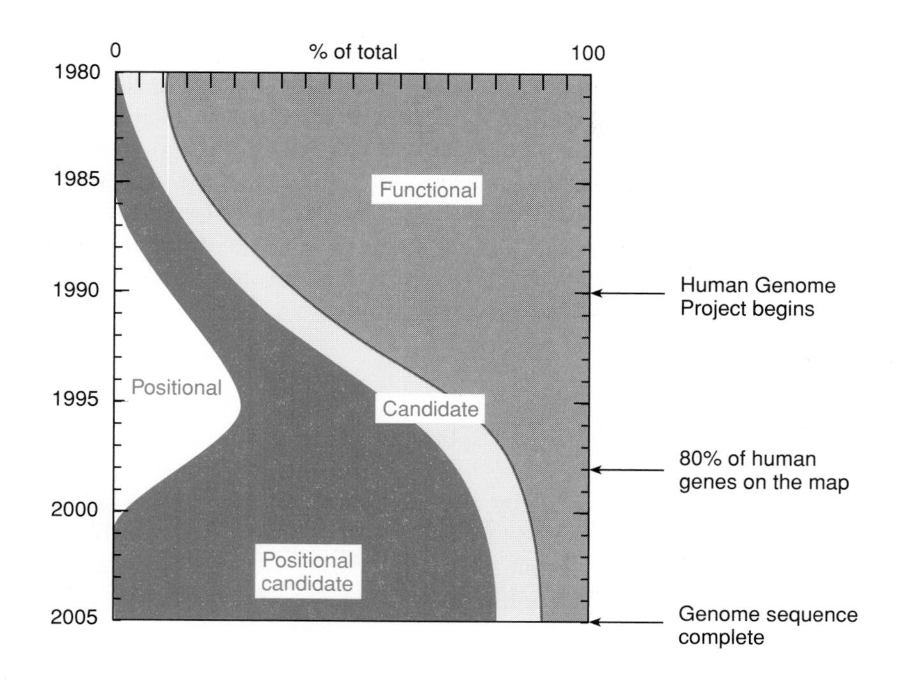

Figure 14.3: Positional candidate gene approaches are set to dominate other methods of identifying human disease genes.

Modified from Collins (1995) with permission from Nature America Inc.

Narrowing down the candidate region requires resources of clones and markers

A first requirement for narrowing down the candidate region is a series of clones and markers from within this region. Polymorphic markers are needed to generate a high resolution genetic map of the region, and clones are required both as a source of such markers and to start physical mapping. With the progress of the Human Genome Project, many of the resources required have already been generated on a genome-wide scale. In many chromosomal regions, existing high resolution genetic maps allow a disease gene to be placed between markers only 1 cM apart, if sufficient family material is available. The current large-scale efforts to construct detailed physical maps of human chromosomes mean that contigs of ordered clones (usually YACs) may well already have been constructed across the critical region. If not available, they must be constructed. It may be advantageous to use large-insert cloning vectors other than YACs (P1s, PACs or BACs, see pages 101–103) if the region is unstable or hard to clone in YACs. The large-insert clones can then be subcloned into cosmids for further study. Sometimes large-scale disease-associated mutations can be identified using these resources, and such mutations can help pinpoint the location of the disease gene.

Nowadays, such clones are found by screening publicly available large-insert libraries which are stored in a few major reference centers, and it is only the ordering of a modest number of clones into a contig that may have to be done in-house. In the late 1980s, when genetic, and especially physical, maps of human chromosomes were still at a primitive stage, and available libraries had small inserts, extra starting points for contig assembly were needed within candidate regions. Various ingenious methods were developed to rescue DNA clones from regions known to include a disease gene.

Chromosome microdissection cloning

Using micromanipulation techniques, a small chromosomal segment can be physically cut out from an individual chromosome from a cytogenetic preparation on a microscope slide (see Edstrom *et al.*, 1987). Extremely fine needles, or a laser beam, are used to perform the microdissection in a series of cells (*Figure 14.4*). The excised fragments, typically representing a single chromosomal band, are collected, pooled and the DNA extracted for use in constructing DNA libraries (*microdissection libraries;* see Ludecke *et al.*, 1989). This technical *tour de force* has been useful in generating DNA clones from several previously poorly characterized disease-associated subchromosomal regions (for an example, see Hampton *et al.*, 1991). However, microdissection libraries are difficult to construct; contamination by extraneous DNA has been a problem and the clone complexity (the number of different DNA sequences) has often been poor. This approach has very largely been superseded with the advent of large-scale YAC mapping.

Subtraction cloning

Subtraction cloning can be used to select clones of the DNA that is deleted in an individual with a chromosomal deletion. Subtraction cloning uses two DNA samples, a *test DNA* and a *driver DNA*. The test DNA is mixed with a large excess of driver DNA, denatured and re-annealed. By one means or another, double helices are selected in which both strands are test DNA. These preferentially represent sequences in the test DNA which are absent from the driver DNA. Its most celebrated application was in cloning the dystrophin gene (see page 381 and *Figure 14.9*). The test DNA came from a normal individual, and the driver DNA from a patient who had a deletion including the dystrophin gene. The test clones remaining after subtraction are enriched for DNA derived from the region missing in the affected patient.

In practice, subtraction cloning is a very difficult technique, which seldom succeeded with genomic DNA, although it worked better with cDNA. A modern version is **representational difference analysis** (**RDA**; see Lisitsyn, 1995). RDA uses several means, including selective PCR, to enrich sequences present in the test but not the driver DNA, and is showing some promise as a method for isolating regions

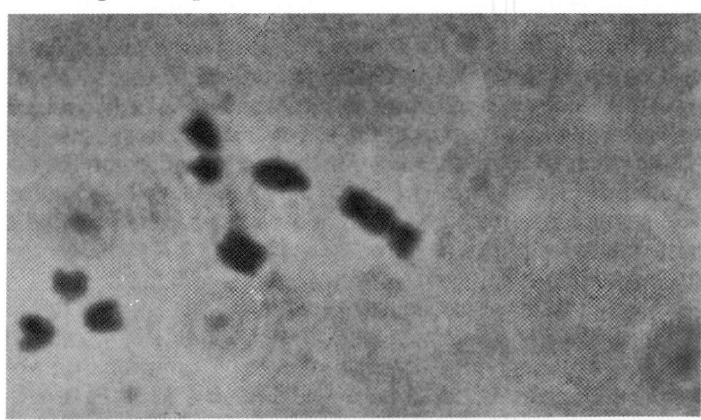

Figure 14.4: Chromosome microdissection permits excision of small subchromosomal segments from individual chromosomes for preparation of DNA libraries.

The upper photograph shows unstained metaphase chromosomes from a human–mouse somatic cell hybrid which contains a human chromosome 7 (center of photograph). The lower photograph shows the same preparation following microdissection of a small segment of 7q, with the dissected fragment lying close to the chromosome. This type of approach was applied in an effort to retrieve DNA from the cystic fibrosis region on 7q. Reproduced from Edstrom *et al.* (1987) with permission from Academic Press.

amplified or deleted in cancer cells (see Schutte *et al.*, 1995, for a recent example). Genes in such regions are positional candidate tumor suppressor or oncogenes (see Chapter 17).

Genetic analyses are the principal means of narrowing down the location of the disease gene

Crossover analysis

An initial genome-wide search for linkage will use markers spaced at relatively wide intervals, and so define only a broad candidate region. Once a candidate region is identified, it must be saturated with polymorphic markers. Increasingly, suitable markers which are known to map to that region will be available, but if not then YACs and cosmids must be isolated from the candidate region and screened for microsatellites, which are then tested for polymorphism and mapped in CEPH families. Multipoint mapping (*Figure 12.8*) is used to define the marker map but, for placing the disease locus within the marker framework, it is preferable to identify crossovers by analyzing haplotypes. Haplotypes which segregate with disease (*disease haplotypes)* are defined in individual pedigrees and recombinants are identified. This allows *proximal* and *distal* flanking markers to be defined (see *Figure 14.5*). Proximal markers are closer to the centromere than the disease locus, and distal markers lie telomeric of the disease locus. Mapping continues until the positions of the closest recombinations on either side of the disease locus have been located

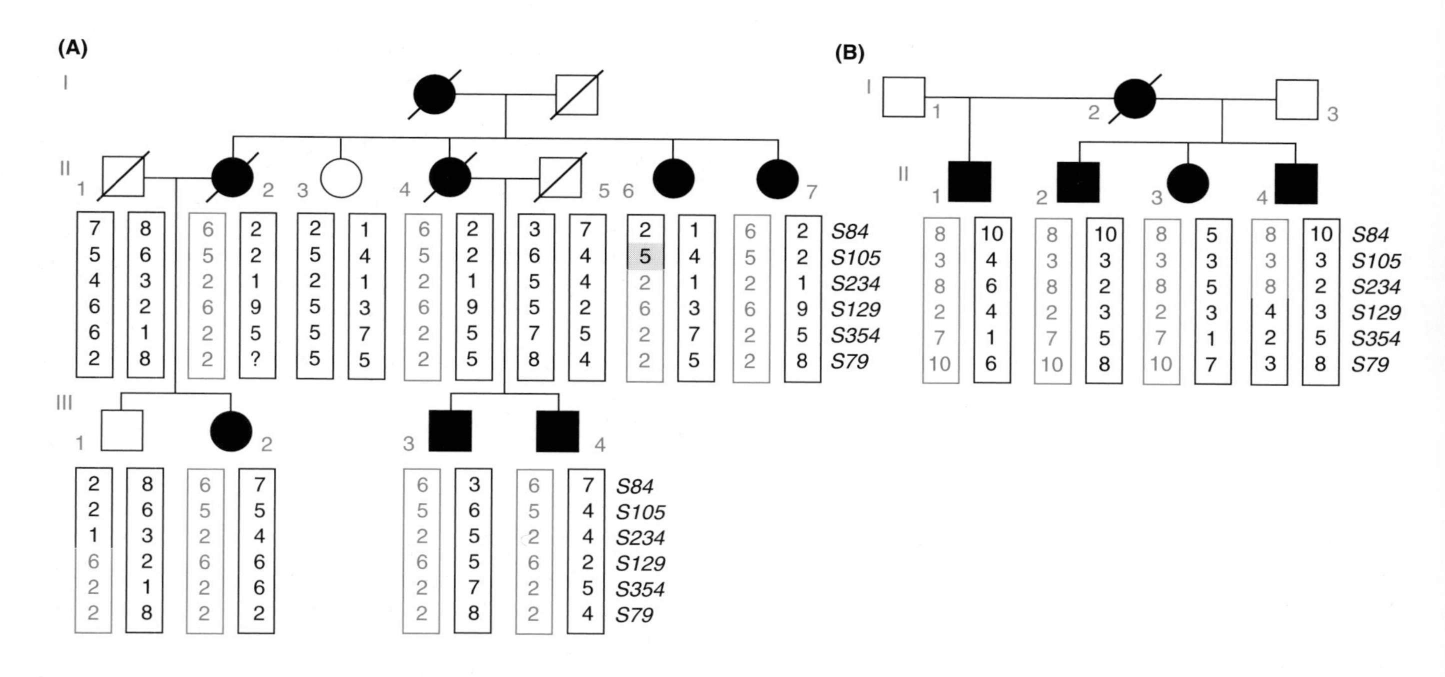

Figure 14.5: Crossover analysis seeks to map a gene by defining flanking proximal and distal recombinants.

The figure shows haplotype analysis in two pedigrees with a dominantly inherited skin disorder, Darier's disease, which had previously been shown to be linked to markers on 12q. Genotypes in II_1, II_2, II_4 and II_5 in pedigree **(A)** are inferred.
(A) In this family the disease gene segregates with the marker haplotype 6-5-2-6-2-2 between *D12S84* (proximal) and *D12S79* (distal). A crossover in II_6 indicates that the disease gene must map distal to *D12S84* (the positioning of *D12S105* is ambiguous because of presumed homozygosity for allele 5 in I_1 – compare the genotypes for II_3 and II_6).
(B) In this family the disease gene can be deduced to be proximal to *D12S129*. The combined data indicate that the Darier's disease gene must map between the proximal marker *D12S84* and the distal marker *D12S129*. Reproduced from Carter *et al.* (1994) with permission from Academic Press.

between pairs of closely spaced markers. This defines a candidate region which cannot be narrowed except by finding more families containing new recombinants.

When a single recombinant defines the region which is to be subjected to a major cloning exercise, it is important to consider possible sources of error. Meticulous clinical diagnoses are imperative. Key recombinations are more reliable if they occur in unambiguously affected people; some unaffected individuals may carry a *nonpenetrant disease gene* which can lead to them being misinterpreted as recombinant when in fact they are nonrecombinant. Sometimes, despite good positive lod scores, there appear to be recombinants with every marker tried – this is usually an indication that somebody has been diagnosed wrongly (labeled as affected when unaffected, or vice versa) or else that the disease gene in one or more of the families under investigation does not map to the candidate region. Alternatively, perhaps the markers are wrongly ordered on the genetic map.

Genetic mapping gives only a crude estimate of the size of the critical region, because genetic distances between markers are usually known with much less certainty than the map order (see page 327). However, except in the grossest cases, size estimates in centiMorgans are not relevant to positional cloning. What matters is the size in megabases of the critical region, to estimate how many YACs or other clones are required to build a contig across the region.

Linkage disequilibrium as a tool for high resolution genetic mapping

Linkage disequilibrium describes an association at the population level of a particular marker allele with a disease. As we saw in Chapter 12 (pages 331–333), linkage disequilibrium is seen only when many of the apparently unrelated affected people in the population in fact derive their disease chromosome from a shared ancestor. Detecting disequilibrium is not as simple as it might appear (see page 502), but when genuinely present it can be very helpful for positional cloning because it is an inherently short range phenomenon. Disequilibrium because of shared ancestors is usually seen only with markers within 1 Mb or less of a disease locus, and sometimes it is restricted to very close markers. Positional cloning of both the HD and CF genes was helped greatly by the observation of disequilibrium. However, the extent of disequilibrium between a disease and a marker allele is not determined solely by the distance between them. It depends also on unknowable details of the history, both of the population and of the mutation being studied. *Figure 12.13* illustrates this irregularity for the case of HD. Thus, although a useful guide, disequilibrium rarely pinpoints the location of a gene to a single clone of a contig. The example of CF is discussed below (pages 384–385).

Large-scale mutations in affected patients provide important clues to the physical location of the disease gene

Researchers are constantly on the alert for special patients or observations which will short-cut the labor of pure positional cloning. Discovering disease-associated cytogenetic abnormalities (translocations, interstitial deletions, inversions, etc.) has often provided crucial clues for identifying a disease gene (see *Box 14.2* and below).

Chromosomal translocations and inversions

If a person with an apparently balanced translocation or inversion is phenotypically abnormal, there are three possible explanations:

(i) the finding is coincidental;

(ii) the rearrangement is not in fact balanced – there is an unnoticed loss or gain of material;

Box 14.2: Pointers to the presence of large-scale mutations

Alert clinicians play a crucial role in identifying patients who may harbor a large-scale mutation. Possibilities include the following.

A cytogenetic abnormality in a patient with the standard clinical presentation
If a disease gene has been mapped to a certain subchromosomal location and then a patient with that disease is found who has a chromosome abnormality affecting that same location, the chromosome abnormality most probably caused the disease. In the case of balanced translocations and inversions, breakpoints are often located within the disease gene, or very close to it, and cloning the breakpoints often provides the quickest route to identifying the disease gene. Cytogenetic abnormalities where there is loss of DNA, notably interstitial deletions, may also be valuable; the breakpoints may be located some distance from the disease gene but, if the segment that is lost is small, mapping the breakpoints may enable the gene to be mapped to a small interval.

Most such patients will have *de novo* mutations, and the chance of finding them is highest for diseases where there is a high rate of new mutations, notably severe early onset dominant disorders (where most affected individuals do not reproduce, and are new mutations) and severe X-linked recessive disorders (where affected males do not reproduce and one third are new mutations; the other mutations originate in and are transmitted by carrier females. It is a difficult question whether performing chromosome analysis on all patients with *de novo* mutations is a worthwhile expenditure of research effort. However, reports of patients with chromosomal abnormalities are carefully scanned by researchers who hope to discover cases showing features of the disease they are investigating.

Additional mental retardation
Rare patients may have the expected phenotype, but in addition display severe mental retardation. Large chromosomal deletions almost always cause severe mental retardation, reflecting the involvement of a high proportion of our genes in fetal brain development. Thus, although the finding may be coincidental, such cases can be caused by deletions that eliminate the disease gene plus additional neighboring genes. They warrant cytogenetic and molecular analysis.

Contiguous gene syndromes
Very rarely a patient may appear to suffer from several different genetic disorders simultaneously. This may be due to the rare misfortune of acquiring two independent pathogenic mutations. In other cases, however, the unusual phenotype results from simultaneous deletion of neighboring genes. Contiguous gene syndromes are particularly well defined for X-linked diseases (see page 410).

(iii) one of the chromosome breakpoints causes the disease.

A chromosomal break can cause a loss-of-function phenotype if it disrupts the coding sequence of a gene or a nearby regulatory region. Alternatively, it could cause a gain of function, for example by splicing regulatory sequences from one gene to distal coding sequences from another gene, causing inappropriate expression (this is rare in inherited disease but common in tumorigenesis, see Chapter 17). In either case, the breakpoint provides a valuable clue to the exact physical location of the disease gene. The clue is valuable but not infallible: sometimes breakpoints can alter expression of a gene located hundreds of kilobases away by affecting the structure of large-scale chromatin domains (see page 174).

The precise location of a chromosome breakpoint can be defined in two main ways. FISH (page 280) can be used to identify DNA clones which span the breakpoint. Alternatively, different DNA clones from the relevant region can be used in turn to see if any can identify patient-specific restriction fragments, by hybridizing each clone to the patient's genomic DNA which has been digested with a rare-cutter restriction endonuclease and subjected to PFGE (page 286).

Chromosomal deletions and duplications

Chromosomal deletions are likely to cause loss-of-function abnormalities as a result of elimination of some or all of the gene sequence at the disease locus. Duplications are more rarely associated with abnormal phenotypes; when they are, the cause is usually disruption of a gene at the point of insertion or, more rarely, a dosage effect (see *Figure 15.6*).

Cytogenetically visible deletions or duplications involve many megabases of DNA. In the past, subtraction cloning was attempted using such deletions (see above, page 381). Presently, their main value is as a pointer to broad subchromosomal localization. Small-scale deletions (*microdeletions*) are more valuable for positional cloning. Deletions too small to be visible by normal cytogenetic analysis may, nevertheless, be large in molecular terms. Microdeletions of tens or hundreds of kilobases of DNA are not uncommon in some disorders (see pages 267–270). Often they are generated by interaction between repetitive sequences at or near the disease locus (see page 264ff.). Microdeletions can be identified by several methods.

Noninheritance of marker alleles. A microdeletion may eliminate a marker DNA locus. Individuals carrying such a deletion will be *hemizygous*, but may be mistakenly interpreted as being homozygous. Transmission of the disease chromosome is often accompanied by what appears to be nonmendelian segregation (see *Figure 14.6*).

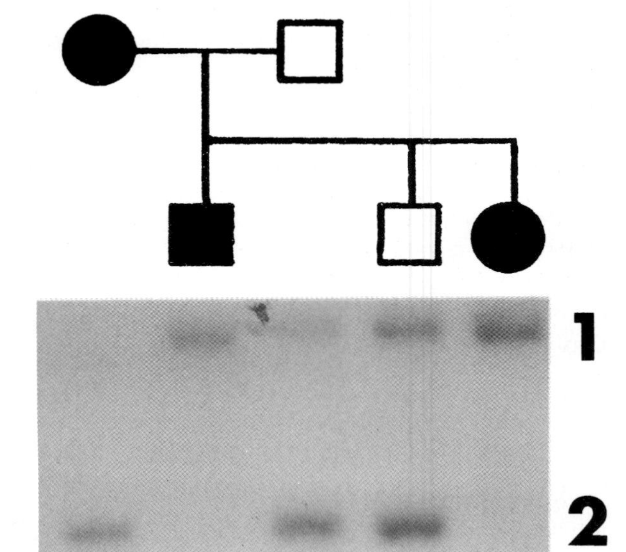

Figure 14.6: Deletions at a disease locus may result in non-inheritance of closely linked markers.

The pedigree represents a family with type 2 neurofibromatosis. Family members were typed for a *Taq*I RFLP at the closely linked neurofilament heavy chain (*NEFH*) locus, and reproducibly gave the indicated typing results. The apparent nonmendelian segregation was suspected to be due to a large deletion encompassing both the *NEFH* gene and the unidentified *NF2* gene. Subsequent FISH and PFGE analyses confirmed the existence of the deletion (see *Figure 14.7*). Reproduced from Watson *et al.* (1993) by permission of Oxford University Press.

Hybridization-based restriction mapping. This is a favorite way of screening for large-scale mutations. Genomic DNA samples from a panel of patients are digested with rare-cutter restriction endonucleases, and the fragments size-fractionated by PFGE. DNA probes which map in the vicinity of the disease locus can then be tested in turn to see if they can detect abnormal patient-specific hybridization bands. If a probe detects abnormal size bands in a patient's DNA digested separately with two or more different enzymes, a large-scale mutation is indicated (see *Figure 14.7A*). This method can detect not only deletions but also disease-associated duplications and insertions.

FISH mapping. Once suspected, a microdeletion can also be confirmed by this method. A DNA clone from the deleted interval is used as a hybridization probe against a metaphase chromosome preparation from the patient. For autosomal microdeletions, the probe should hybridize to only one of the two homologs (*Figure 14.7B*).

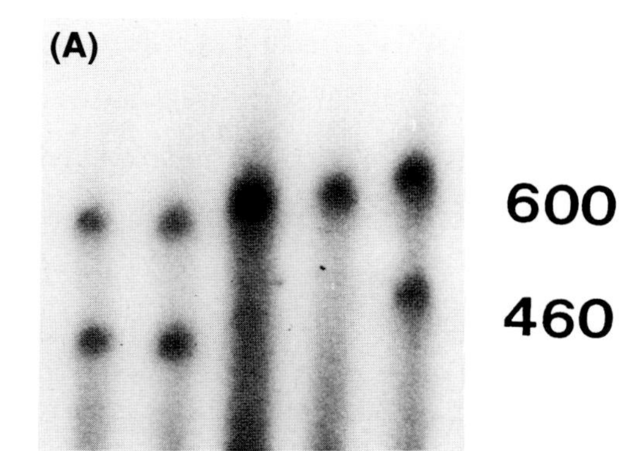

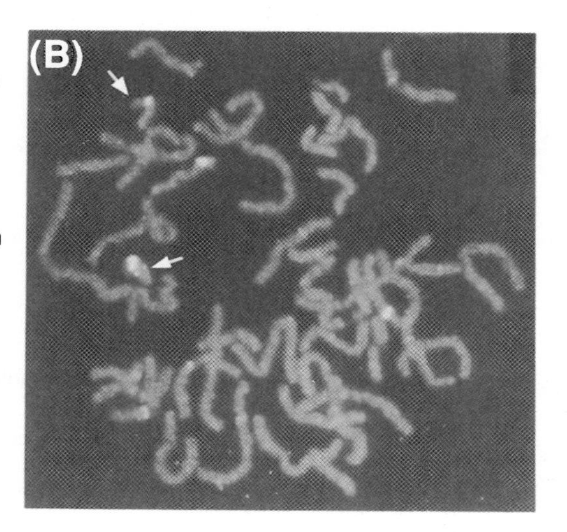

Figure 14.7: Microdeletions can be identified by restriction mapping using PFGE and by FISH.

The panels illustrate two sets of analyses to test whether affected members in the pedigree in *Figure 14.6* carried a small deletion of chromosome 22 in the vicinity of the *NF2* locus.
(A) PFGE analysis. A genomic DNA clone from the *NF2* gene region was hybridized to a Southern blot of genomic DNA from indicated family members, which had been digested with *NotI* and size-fractionated by PFGE. The 600-kb *NotI* fragment represents wild-type alleles. *Note* the additional, approximately 460-kb band found only in the affected individuals, resulting from a deletion which simultaneously eliminated the *NF2* gene.
(B) FISH analysis. Metaphase chromosome preparations from an affected individual in *Figure 14.6* were hybridized with two DNA probes. A probe that hybridizes to repeat sequences found at the centromeres of chromosomes 14 and 22 produces a strong signal on the two homologs for each chromosome. A cosmid from the *NF2* gene region, however, hybridizes to only one of the two chromosome 22 homologs. This is a single copy probe so it gives a fainter signal (a dot from each chromatid) than the repeat sequence probe. Reproduced from Watson *et al.* (1993) by permission of Oxford University Press.

Identifying genes within the critical region

Once an ordered contig of clones is in place across the critical region, the usual next step in positional cloning is to construct a **transcript map** of all expressed sequences in the region. The general methods for identifying unknown transcribed sequences from within a contig of genomic clones have been discussed in detail in Chapter 11, and are summarized briefly in *Box 14.3*. Sometimes the pathology of the disease under study suggests a particular investigation. For example, if the gene that is being sought is known to be expressed in muscle, individual clones within the clone contig can be used as hybridization probes to screen a muscle cDNA library to identify cognate cDNAs. Once suitably promising candidate genes have been identified by any of these methods, they must be tested for evidence that they are involved in the disease.

Box 14.3: Methods for identifying expressed sequences within genomic clones from a disease gene region

cDNA library screening, using as probes genomic clones from the candidate region.

cDNA selection, for ultra-sensitive detection of cDNAs derived from the candidate region (*Figure 11.20*).

Zoo blotting, to seek evolutionarily conserved sequences (*Figure 11.16*).

CpG island identification, to seek regions of undermethylated DNA which often lie close to genes (*Figure 11.18*).

Exon trapping, to find genomic sequences flanked by functional splice signals (*Figure 11.19*).

Sequence analysis, to detect genomic sequences having characteristics of exons (page 303).

Coincidence sequence cloning is a PCR-based method which involves selecting for sequences that are *co-incident* between two complex populations of DNA sequences (Brookes *et al.*, 1994). One application involves using YAC clones or other genomic DNA clones from a disease gene region as one DNA source and a second DNA source consisting of cDNA from cells in which the sought-after gene is thought to be expressed. In principle, this method is the PCR equivalent of the hybridization-based cDNA selection technique: the objective is to identify expressed sequences in the genomic DNA clones because of their coincidence with transcripts found in the cDNA (see Francis *et al.*, 1995, for one recent example).

Three classical examples of successful positional cloning

To illustrate how positional cloning has been applied to isolating disease genes in practice (see *Table 14.1*), we describe briefly how the genes causing DMD, neurofibromatosis type I (NF1) and CF were identified.

Table 14.1: Examples of inherited disease genes identified by positional cloning

Year	Disease	Cytogenetic rearrangement
1986	Duchenne muscular dystrophy	+
	Retinoblastoma	+
1989	Cystic fibrosis	−
1990	Neurofibromatosis type 1	+
	Wilms' tumor	+
1991	Aniridia	+
	Familial polyposis coli	+
	Fragile-X syndrome	+
	Myotonic dystrophy	−
1993	Huntington's disease	−
	Tuberous sclerosis	+
	von Hippel–Lindau disease	−
1994	Achondroplasia	−
	Early-onset breast/ovarian cancer	−
	Polycystic kidney disease	+
1995	Spinal muscular atrophy	−

Two different chromosomal abnormalities were used to clone the dystrophin gene

In the 1980s, several groups competed to clone the gene responsible for DMD (MIM 310200), and different approaches were used. The pioneering work of these groups, overcoming formidable technical difficulties to clone an unprecedented gene, was probably the major inspiration for most subsequent positional cloning efforts. This work has been well reviewed by Worton and Thompson (1988).

Localization of the DMD *gene by linkage analysis and through X–autosome translocations*

The gene for DMD was mapped to Xp21 by conventional linkage analysis as long ago as 1982 (the first disease to be so mapped – see page 440). Additional confirmation of this localization came from studies of rare affected females. These women, about 20 of whom have been described world-wide, occur sporadically in families with no history of DMD, and there is no evidence that they have inherited a conventional *DMD* mutation from either parent. Instead, they all carry balanced X–autosome translocations. In each case, the X chromosome breakpoint has disrupted the *DMD* gene and the pathogenesis results from an unusual mechanism. Although X inactivation is random, those cells in which the autosomal genes on the translocation chromosome were subject to inactivation imposed by the *XIST* gene are less viable and die, largely because of the reduced dosage of the autosomal genes. Thus, the cells that survive and give rise to the individual do not produce any dystrophin (see *Figure 14.8*). Although each woman has a different autosomal breakpoint, and many different autosomes are involved, the X chromosome breakpoint is always at Xp21, confirming the localization of the *DMD* locus to Xp21.

Isolation of the DMD *gene by subtraction cloning*

Kunkel's group in Boston used DNA from a boy ('BB' – see page 410) who had DMD and a cytogenetically visible Xp21 deletion. The technically very difficult procedure of subtraction cloning (see pages 374–375 and *Figure 14.9*) was used to isolate clones corresponding to sequences deleted in the patient.

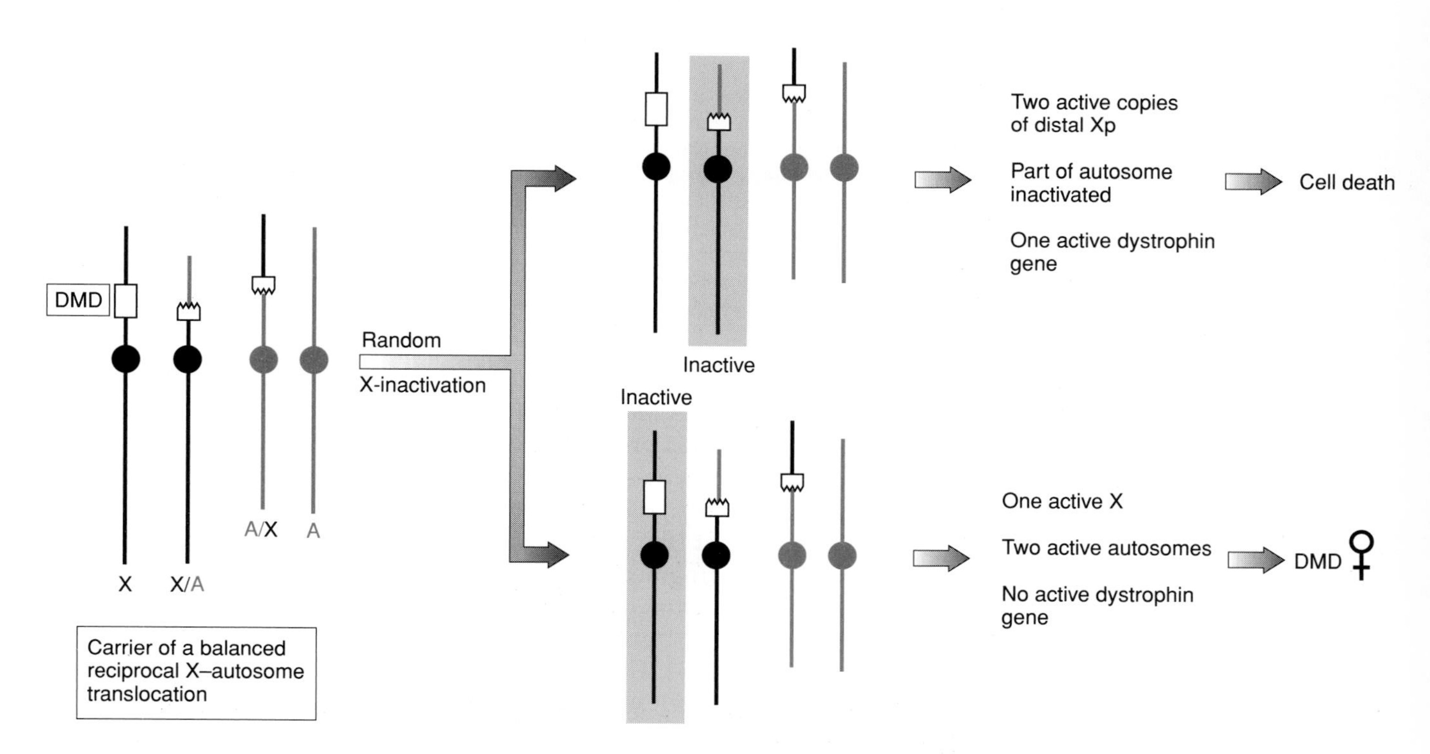

Figure 14.8: Nonrandom X inactivation occurs in female DMD patients with Xp21–autosome translocations.

The translocation is balanced, but the X chromosome breakpoint disrupts the dystrophin gene. X inactivation is random, but cells which inactivate the translocated X die because of lethal genetic imbalance. The embryo develops entirely from cells where the normal X is inactivated, leading to a woman with no functional dystrophin gene. The resulting failure to produce any dystrophin causes DMD.

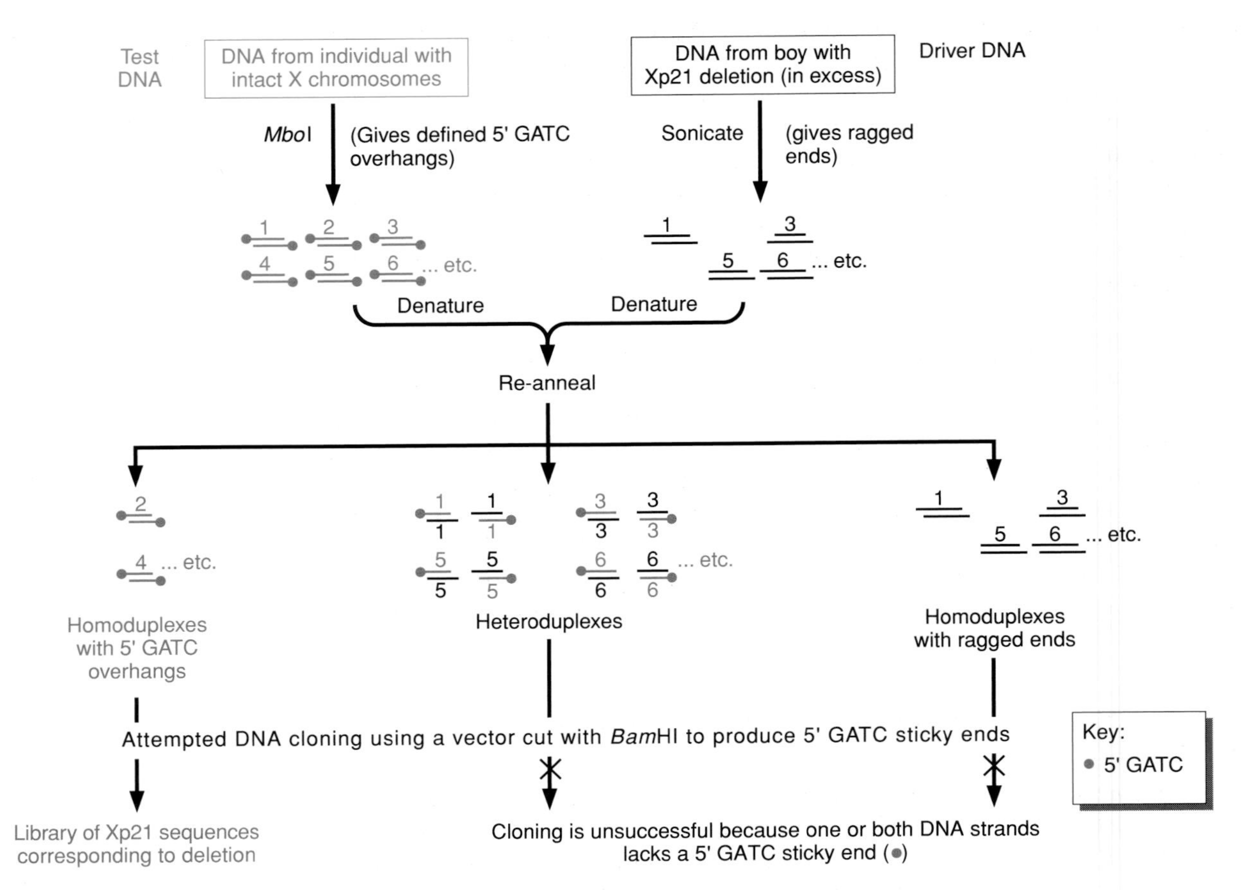

Figure 14.9: Subtraction cloning of Xp21 DNA sequences permitted identification of the *DMD* gene.

Denatured sonicated DNA from a DMD patient with a large deletion at Xp21 (the *driver DNA*) was mixed with *MboI*-digested denatured DNA from an individual with intact X chromosomes (the *test DNA*) and allowed to re-anneal. By having the driver DNA in vast excess, the single-stranded sequences in the test DNA are very much more likely to find and hybridize with a complementary strand within the driver DNA, than their original complementary strands. Because of the deletion, some Xp21 sequences (e.g. numbers 2 and 4) are not found in the driver DNA. The corresponding sequences in the test DNA cannot form heteroduplexes, and instead form homoduplexes. Such sequences will be the only population of renatured DNA in which both strands have sticky ends (5′ GATC overhangs), and so can be cloned selectively by ligation to a vector with complementary ends (see *Figure 4.4*). In practice, the sequence complexity of human genomic DNA means that the hybridization reaction normally proceeds very slowly, and a specialized technique (PERT – phenol-enhanced reassociation technique) was required to increase the reassociation rate (see Kunkel *et al.*, 1985).

To isolate the *DMD* gene, genomic DNA from a normal individual was digested with the restriction nuclease *Mbo*I, which cleaves DNA immediately 5′ of its recognition sequence GATC. This produced fragments (the *test DNA*) which were double-stranded, except for single-stranded GATC overhangs at the 5′ ends. The *Mbo*I fragments were denatured and combined with a 200-fold excess of denatured DNA from the deletion patient (the *driver DNA*), which had been sonicated to produce ragged ends. Annealed fragments were then cloned into a vector which had been cut with *Bam*HI. This vector could accept only fragments with overhanging 5′ GATC ends on *both* DNA strands (see *Figure 4.4*), and so duplexes containing the driver DNA were not cloned. Thus, Xp21 sequences from the test DNA were preferentially cloned, because they were unable to hybridize to the driver DNA where their counterparts were deleted.

Individual DNA clones in the resulting **subtraction library** were then used as probes in Southern blot hybridization against DNA samples from normal people and DMD patients. One clone, pERT87-8, detected deletions in DNA from about 7%

of cytogenetically normal DMD patients, and detected polymorphisms which proved to be tightly linked to DMD in family studies. These results showed that pERT87-8 was located much closer to the *DMD* gene than any previously isolated clones (in fact it was within the gene, in intron 13). Other nearby genomic probes were isolated, and then muscle cDNA libraries screened. Given the low abundance of dystrophin mRNA and, as we now know, the small size and widely scattered location of the exons, finding cDNA clones was far from easy, but eventually clones were identified, and subsequently the whole remarkable dystrophin gene (*Figure 7.11*) characterized.

Isolation of the DMD *gene by cloning a translocation breakpoint*

While Kunkel's group was working on subtraction cloning, Worton's group in Toronto was successful with a different approach. One of the affected women described above had an X;21 translocation with a breakpoint in the short arm of chromosome 21. Knowing that 21p is occupied by arrays of repeated rRNA genes (page 188), Worton's group prepared a genomic library and set out to find clones containing both rDNA and X chromosome sequences. This led to isolation of XJ (X junction) clones which, in a similar way to Kunkel's pERT87-8 probe, detected deletions and polymorphisms. XJ turned out to be located in intron 17 of the dystrophin gene.

The gene for NF1 was identified by mapping disease-associated translocations

In 1987, linkage analyses mapped the gene for NF1 (MIM 162200) to 17q11.2. Thereafter, two NF1 patients were identified with balanced translocations: t(1;17)(p34.3:q11.2) and t(17;22)(q11.2:q11.2). Because both patients had breakpoints on 17q11.2, it was considered likely that these breakpoints interrupted the *NF1* gene, causing loss of gene function.

To locate the breakpoints, individual DNA clones from a chromosome 17-specific DNA library were used as hybridization probes against total genomic DNA from the translocation patients and normal individuals, which had been digested with rare-cutter restriction endonucleases and size-fractionated by PFGE. One such clone, 17L1A, identified anomalous sized restriction fragments in the DNA from the t(1;17) translocation patient, suggesting that it mapped in the immediate vicinity of the translocation (Fountain *et al.*, 1989; see *Figure 14.10*).

Exhaustive cloning and mapping of the region identified by 17L1A led to the identification of a number of candidate genes, one of which was subsequently identified as the *NF1* gene by the demonstration of patient-specific mutations. It then became evident that both translocation breakpoints lay within the *NF1* gene, which is now known to span more than 300 kb of DNA. Three other genes were found to be in close proximity to the translocations, and turned out to be examples of genes occurring within another gene (see *Figure 7.7*).

The CF gene was cloned without the aid of large-scale mutations, but with the help of linkage disequilibrium

In 1985, studies of affected sib pairs (see *Figure 12.12*) showed that the gene for CF (MIM 219700) was linked to a protein polymorphism of the enzyme paraoxonase. At that time, the chromosomal location of the paraoxonase gene was not known (this illustrates one of the big advantages of DNA over protein polymorphisms, see *Box 12.1*). A rapid mapping effort located the paraoxonase gene to chromosome 7, and a variety of DNA markers were used to show that CF mapped to 7q31–q32. The

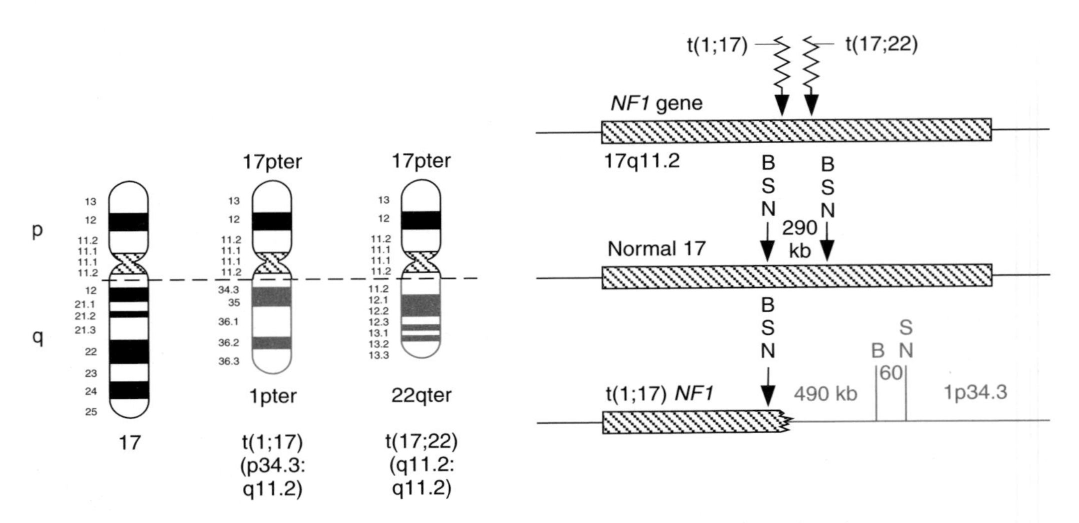

Figure 14.10: Positional cloning of the *NF1* gene was facilitated by disease-associated translocations.

The translocation breakpoints in the NF1 patients occur within the *NF1* gene (right panel) in a region flanked by closely clustered cleavage sites for the rare-cutter restriction endonucleases *Bss*H II (B), *Sac*II (S) and *Not*I (N). When this DNA region is cleaved by any one of these enzymes, a fragment of about 290 kb is normally produced. However, the translocation breakpoint results in novel restriction fragment sizes. For example, the t(1;17) translocation results in elimination of the distal *Bss*H II/*Sac*II/*Not*I cluster and the introduction of novel restriction sites from the chromosome 1 segment, giving a 490-kb *Bss*H II fragment and 550-kb *Sac*II and *Not*I fragments. Individual chromosome 17 DNA clones were labeled and hybridized in turn against PFGE Southern blots of appropriately digested DNA from the translocation patient until one of them (17L1A) was found to detect patient-specific fragments (see Fountain *et al.*, 1989).

MET oncogene was established as a proximal flanking marker and an anonymous clone D7S8 as a distal marker.

Despite an intensive world-wide search, no cases have been discovered of CF patients with translocation, inversion or deletion breakpoints at 7q31–q32, nor did any microdeletions emerge during the progress of the research. Since no large-scale mutations were ever found, identifying the CF gene proved an exceedingly arduous task. The gene was isolated only after extensive genetic mapping and exhaustive molecular characterization of the candidate region. A huge effort was devoted to screening genomic DNA libraries and to performing *chromosome walking* (see *Figure 11.13*) to generate clones mapping between the flanking markers D7S8 and *MET*. *Microdissection cloning* was also used in an effort to generate additional DNA clones from the candidate region (*Figure 14.4*). All this work was conducted before human YAC libraries were available, and so the chromosome walking used cosmid and phage λ clones. Thus, individual steps were only about 10–20 kb, and frustrated researchers talked about chromosome crawling.

To try to move along the chromosome in steps larger than 10–20 kb, an ingenious technique called **chromosome jumping** was used. The aim of this method is to hop from one genomic clone to another located several hundred kilobases away in the same subchromosomal region. Chromosome jumping uses libraries of special genomic DNA clones whose inserts are composed of two ligated but originally non-contiguous pieces of DNA, brought together by artificial circularization of large DNA restriction fragments (see *Figure 14.11*). Successful chromosome jumps in the candidate region provided new start points for further chromosome walking, until eventually most of the region was encompassed by DNA clones (see *Figure 14.12*).

Linkage disequilibrium (see page 331) provided valuable clues about the location of the CF gene. The mutation rate is very low and CF is maintained in the population

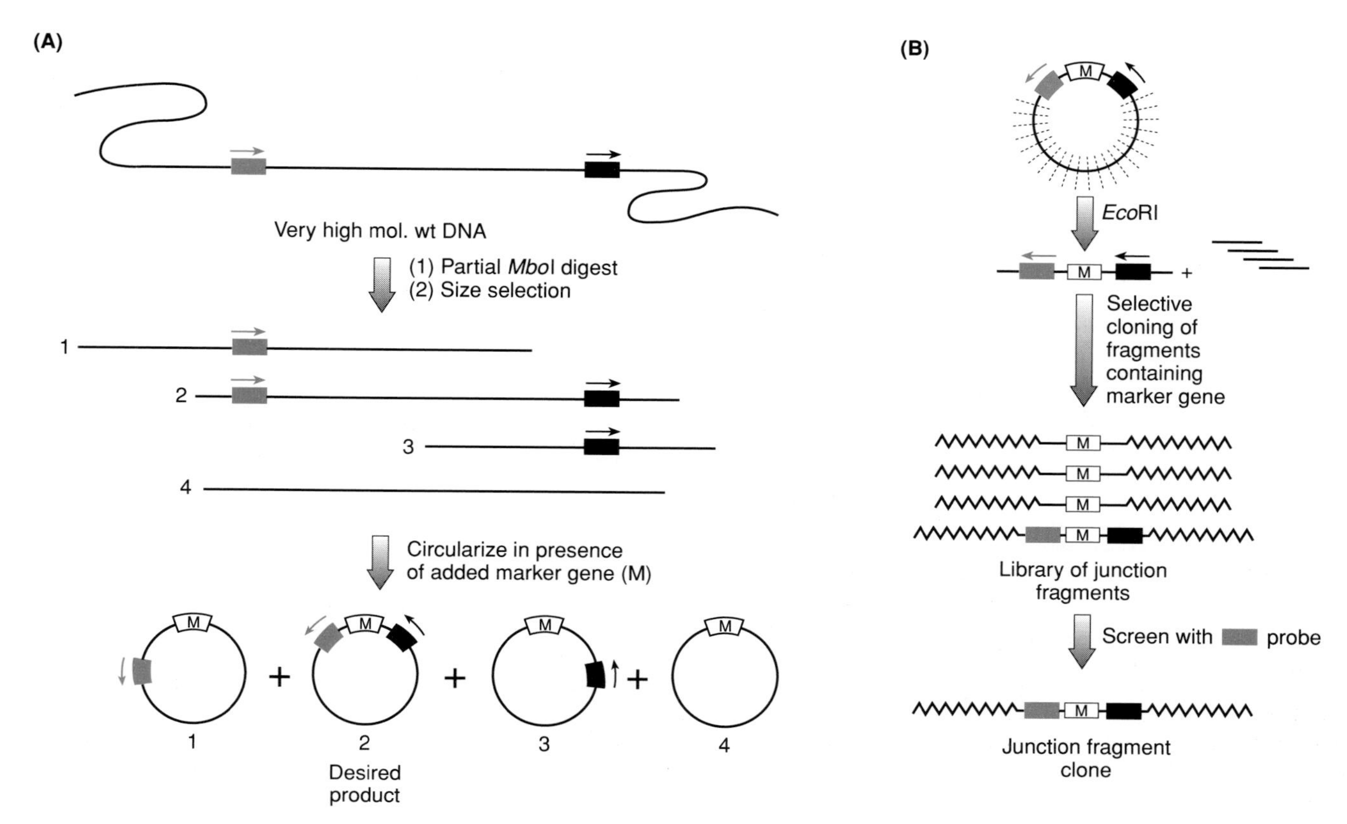

Figure 14.11: Construction of chromosome jumping libraries.

The filled red box represents a starting DNA sequence which, in the final small junction fragment clone, is present along with a marker gene M, (an amber suppressor gene, see page 95) and another segment of DNA (filled black box) that was located initially far away on the chromosome. Very high molecular weight DNA is subject to partial cutting with *Mbo*I to produce very large fragments. By diluting the DNA to very low concentrations before ligation, *intramolecular ligation* (cyclization) is favored (see *Figure 4.4*). The marker molecule is, however, present in excess and so can be included in the circles. Digestion with *Eco*RI cleaves the large circles many times, but not the marker gene (deliberately chosen to lack an *Eco*RI site). Cloning into a suitable λ vector containing an amber mutation selects for sequences containing the marker gene. Screening of the resulting library with the starting clone can then identify sequences initially located hundreds of kilobases away.

largely by *heterozygote advantage* (page 78). CF disease chromosomes in apparently unrelated people often derive from a distant common ancestor. The original flanking markers D7S8 and *MET* show only extremely weak disequilibrium, but some of the new markers such as *KM19* and *XV2.c* generated from clones within the candidate region showed strong disequilibrium (*Table 12.3*). Linkage disequilibrium data can be hard to interpret (see *Figure 12.13*) but, in the case of CF, a gradient of steadily increasing disequilibrium pointed quite effectively to the location of the 5′ end of the gene.

Eventual isolation of the *CFTR* gene, unaided by large-scale mutations, required extensive screening of cDNA libraries for the elusive cDNA. A further small difficulty was encountered in producing convincing evidence that the gene eventually cloned was indeed the site of mutations causing CF. Because of the powerful linkage disequilibrium, it was expected that most CF mutant chromosomes in the population would share a great deal of ancestral sequence. Therefore, showing that a particular sequence change (the ΔF508 mutation, page 437) was present on 70% of CF chromosomes did not prove in a wholly convincing manner that *CFTR* was the CF gene, still less that ΔF508 caused CF.

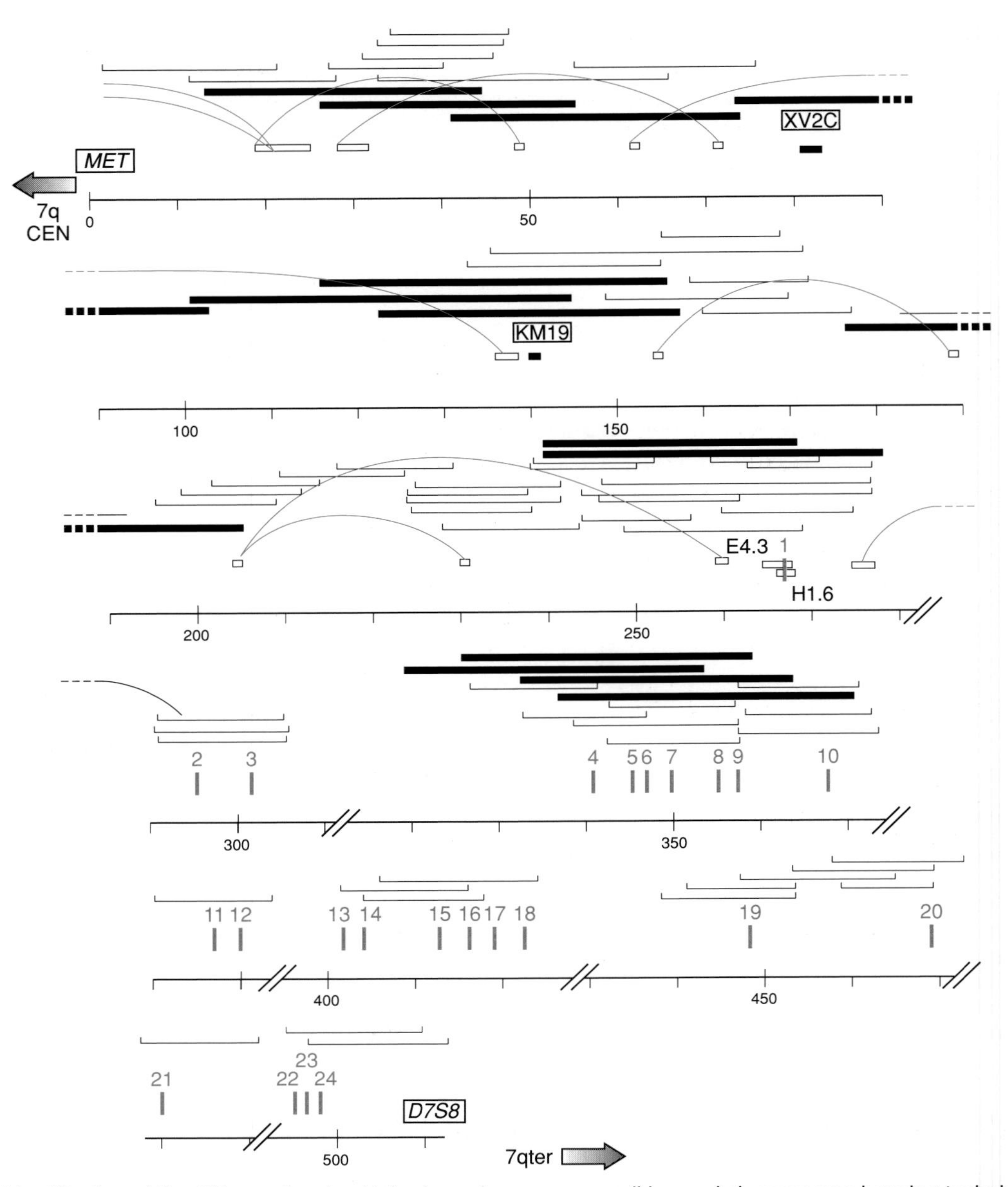

Figure 14.12: Identification of the CF gene involved laborious chromosome walking and chromosome jumping techniques.

Starting from the flanking markers *MET* (proximal) and *D7S8* (distal), an intervening region of about 500 kb was intensively mapped. Chromosome walking was used to identify overlapping λ and cosmid clones (short thin and long thick horizontal lines, respectively, above the restriction map). Chromosome jumping steps (red arcs) facilitated this process. After several false starts, the overlapping E4.3 and H1.6 clones, which contained evolutionarily conserved sequences (as detected by *zoo-blotting* – see *Figure 11.16*), were used to isolate a cognate cDNA clone. The cDNA clone was then used to map back to λ genomic clones and the gene was shown to contain 24 exons (see *Figure 19.3*). Gaps remained, however (e.g. between exons III and IV) and the full structure of the gene subsequently was obtained and shown to comprise 27 exons. Verification of the gene's involvement in CF was obtained by demonstrating patient-specific mutations (see text). Reproduced with permission from Rommens *et al.* (1989). Copyright 1989 American Association for the Advancement of Science.

ΔF508 could have been simply a neutral variant inherited along with CF on the ancestral disease chromosome, especially since the sequence change left the reading frame intact. The ΔF508 mutation is present in 3–4% of phenotypically normal individuals (we would now identify them as CF carriers). The fact that ΔF508 homozygotes were always severely affected was persuasive but not totally conclusive – ΔF508 could have been in more or less complete linkage disequilibrium with the real CF mutation. Biochemical and pathological knowledge was important, in showing that the *CFTR* gene encoded an ion channel, and that the pathogenesis of CF was ultimately caused by defective regulation of chloride ion transport across apical membranes. The subsequent identification of minority disease alleles, where the expected effect on gene expression was more obviously deleterious (e.g. G542X), provided further confirmation that the true disease locus had been identified.

Candidate gene approaches

Ultimately all approaches used to identify disease genes generate *candidate genes,* which then have to be tested individually to see if there is compelling evidence that they are associated with the disease in question (see page 393). However, two of the four major routes to identify a disease gene *start with the assumption of a candidate gene,* and these will be discussed in this section. A particular gene is hypothesized to be the locus for the disease, and the hypothesis is then tested by checking for evidence that the candidate gene is associated with the disease. Such *candidate gene approaches* may be based on particular properties of the product of the candidate gene which are consistent with its involvement in the pathogenesis. Clearly, candidate genes may be suggested for some disorders without knowing the subchromosomal location of the disease gene and/or candidate gene (**position-independent candidate gene approaches**). However, confidence in a particular candidate disease gene is increased substantially if it can be shown to map to the same subchromosomal region as the disease gene. Such **positional candidate gene approaches** have already been very valuable (*Table 14.2* lists some recent successes), but in the next few years they will completely dominate disease gene identification, simply because of the rate at which subchromosomal locations are being established for both disease genes and human genes in general (see Chapter 13).

Table 14.2: Examples of inherited disease genes identified by positional candidate methods

Disease	Affected protein
Alzheimer's disease	β-Amyloid protein precursor, apoE
Charcot–Marie–Tooth disease type 1A	Myelin protein zero (P_0)
Charcot–Marie–Tooth disease type 1B	Peripheral myelin protein 22
Familial melanoma	p16
Hereditary nonpolyposis colon cancer	*hMSH2, hMLH1, hPMS1, hPMS2*
Malignant hypothermia	Ryanodine receptor
Marfan syndrome	Fibrillin
Multiple endocrine neoplasia type 2A	Receptor tyrosine kinase *RET*
Retinitis pigmentosa	Peripherin, rhodopsin
Waardenburg syndrome type 1	Paired box gene *PAX3*

Candidate disease genes may be suggested by the pathogenesis, and by structural or functional relatedness to other genes known to be loci for similar disease phenotypes

Appropriate expression pattern or function

For some disorders, observations on the pathogenesis may immediately suggest candidate genes with an appropriate expression pattern or function. For example, neural tube defects are likely to involve genes that are expressed shortly before or during neurulation in the developing embryo (the 3rd–4th weeks of human embryonic development). A gene which shows expression in the neural tube at these stages is therefore a candidate. This can be demonstrated by *in situ* hybridization against mRNA in sections of embryos (see *Figure 5.16*), most conveniently by using mouse embryonic sections at the equivalent developmental stages (7.5–9.5 embryonic days in this case).

As another example, consider an inherited skin disorder in which the disease tissue is characterized by loss of adhesion between keratinocytes. This suggests the involvement of *desmosomes* (complex structures which act as intercellular rivets, notably in epithelial cells). Since the genes for several of the desmosome proteins have been characterized, mutation screening could be used to test individual genes for association with disease. However, screening for mutations in multiple candidate genes is time-consuming, and it would be useful to establish first of all a subchromosomal location for the disease. This would eliminate most of the possible candidates, so that attention could be focused on the one or two desmosomal protein genes mapping to the correct chromosome or subchromosomal location.

Homology to a gene implicated in an animal model of the disease

If an animal phenotype shows a striking similarity to a human disorder, then it might result from mutations in the animal ortholog of the human disease gene. Then, if the human gene is unknown but the animal gene is known, knowledge of the animal ortholog can be used to help identify and characterize the human gene. This can then be tested for its involvement in the disease. Such an animal model may have originated spontaneously, or have been created artificially by X-ray or chemical mutagenesis or even by gene targeting (see pages 536–539). Though valuable, this approach is less general than might be hoped, because mutations in orthologous genes in humans and mice (or other animals) not infrequently produce considerably different phenotypes (see page 546).

Homology, or functional relatedness, to a gene implicated in a similar human disease phenotype

Mutations at more than one locus may produce the same clinical result (*locus heterogeneity*, see page 419). In some cases, with hindsight, the phenotypes caused by mutations in different genes can be distinguished, in other cases not. If a gene is identified as the locus for one such disease, then genes closely related to it in sequence or function may be candidates for closely similar diseases. For example, the identified disease gene may be a member of a multigene family, other members of which would be candidates for similar diseases (see the example of fibrillin genes, opposite). Candidate genes may also be suggested on the basis of a close functional relationship to a gene known to be the locus for a disease with a similar phenotype. The genes could be related by encoding a receptor and its ligand, or other interacting components in the same metabolic or developmental pathway.

The strongest candidate genes are those which map to the same sub-chromosomal location as the disease gene

As we have seen, positional cloning can be exceedingly laborious, whereas testing known genes for patient-specific mutations is relatively easy. Thus, known genes located within the region to which a disease has been mapped are always worth considering as candidates. With more and more human genes being assigned to subchromosomal locations, there is a high chance that database searches will reveal one or more possible candidate genes in the appropriate location. Molecular pathology is complicated (see Chapter 15), and predictions of the biochemical function of an unknown disease gene are often proved wrong once the gene is isolated. Thus, the most compelling single attribute of a candidate gene is that it is located in the same subchromosomal region as the relevant disease gene.

Rhodopsin – a positional candidate for retinitis pigmentosa
The case of rhodopsin illustrates the value of knowing the location of a gene. Isolation and sequencing of the gene for human rhodopsin was reported in 1984, and it was mapped subsequently to 3q21–qter in 1986. Rhodopsin is a pigment of the rods of the retina, and so could always have been considered a good candidate for disorders involving hereditary retinal degeneration. Among the latter are the various forms of retinitis pigmentosa (RP), which are marked by progressive visual loss resulting from clumping of the retinal pigment. Although rhodopsin was a possible candidate gene for some forms of RP, it was only one of many proteins that were known to be involved in phototransduction. However, in 1989, linkage analyses in a large Irish RP family mapped their disease gene to 3q in the neighborhood of rhodopsin. Rhodopsin was now a serious candidate gene, and patient-specific mutations in the rhodopsin (*RHO*) gene were identified within a year (Dryja *et al.*, 1990).

Fibrillin genes – a family of positional candidates
Sometimes one positional assignment can lead to another, as in the case of Marfan syndrome (MFS, MIM 154700). The phenotype of MFS suggested some abnormality in a connective tissue component. Linkage analysis mapped the MFS gene to 15q, and subsequently the gene for the connective tissue protein fibrillin was localized to 15q21.1 by *in situ* hybridization. Fibrillin was then a positional candidate, with additional support from the pathology. A patient-specific missense mutation was soon demonstrated in the fibrillin (*FBN1*) gene (Dietz *et al* ., 1991). A second fibrillin gene was shown to map to 5q, which therefore became a candidate location for other Marfan-like phenotypes. A related condition, congenital contractural arachnodactyly (MIM 121050) was mapped to 5q and shown to be caused by mutations in the *FBN2* gene (Putnam *et al.*, 1995).

Human–mouse homology can aid in the identification of disease genes

A human disorder and its counterpart in an animal may often show sufficient phenotypic overlap to be recognizable as likely **disease homologs**. Human–mouse homologies provide particularly valuable clues towards identifying human disease genes for several reasons:

- *the detailed mouse genetic map*. Accurate placing of mutants on the mouse genetic map is possible by *backcross mapping* (see *Box 14.4*).
- *The considerable conservation of synteny*. This often extends to large segments of human and mouse chromosomes (see *Figure 14.13*). Once a subchromosomal location for a gene of interest is known in mouse or humans, it is often possible to predict the likely subchromosomal location of that gene in the other species. Sometimes, however, synteny is poorly conserved, making prediction difficult.
- *The considerable conservation of coding DNA*. As detailed in Chapter 9 (pages 230–231), exon sequences are usually well conserved between orthologous human and mouse genes. This means that once a human or mouse gene is isolated, a suitable DNA probe can be prepared to screen DNA libraries from the other species in order to identify and characterize the orthologous gene.

An example: Waardenburg syndrome and the* Splotch *mouse
Waardenburg syndrome type 1 (WS1, MIM 193500) illustrates the value of human–mouse comparisons. A pedigree of this autosomal dominant but variable

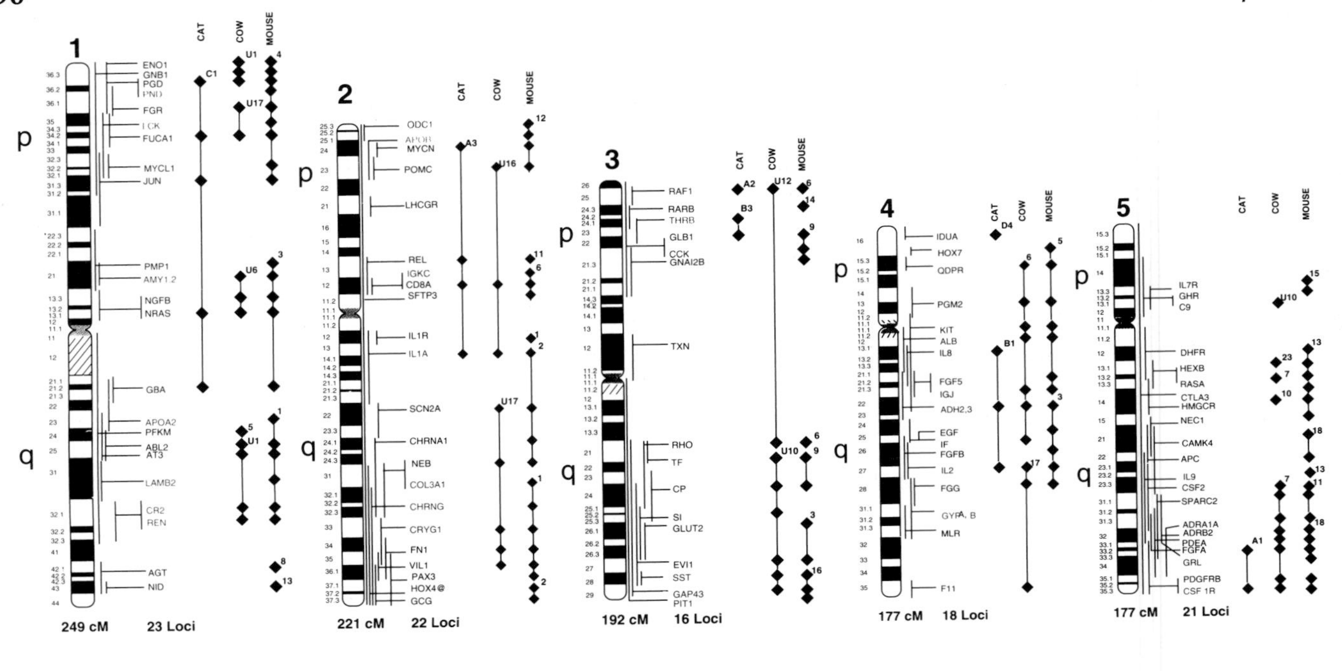
1
CAT
COW
MOUSE
p
q
ENO1
GNB1
PGD
PND
FGR
LCK
FUCA1
MYCL1
JUN
PMP1
AMY1,2
NGFB
NRAS
GBA
APOA2
PFKM
ABL2
AT3
LAMB2
CR2
REN
AGT
NID
249 cM 23 Loci
2
ODC1
APOB
MYCN
POMC
LHCGR
REL
IGKC
CD8A
SFTP3
IL1R
IL1A
SCN2A
CHRNA1
NEB
COL3A1
CHRNG
CRYG1
FN1
VIL1
PAX3
HOX4@
GCG
221 cM 22 Loci
3
RAF1
RARB
THRB
GLB1
CCK
GNAI2B
TXN
RHO
TF
CP
SI
GLUT2
EVI1
SST
GAP43
PIT1
192 cM 16 Loci
4
IDUA
HOX7
QDPR
PGM2
KIT
ALB
IL8
FGF5
IGJ
ADH2,3
EGF
IF
FGFB
IL2
FGG
GYPA, B
MLR
F11
177 cM 18 Loci
5
IL7R
GHR
C9
DHFR
HEXB
RASA
CTLA3
HMGCR
NEC1
CAMK4
APC
IL9
CSF2
SPARC2
ADRA1A
ADRB2
PDEA
FGFA
GRL
PDGFRB
CSF 1R
177 cM 21 Loci

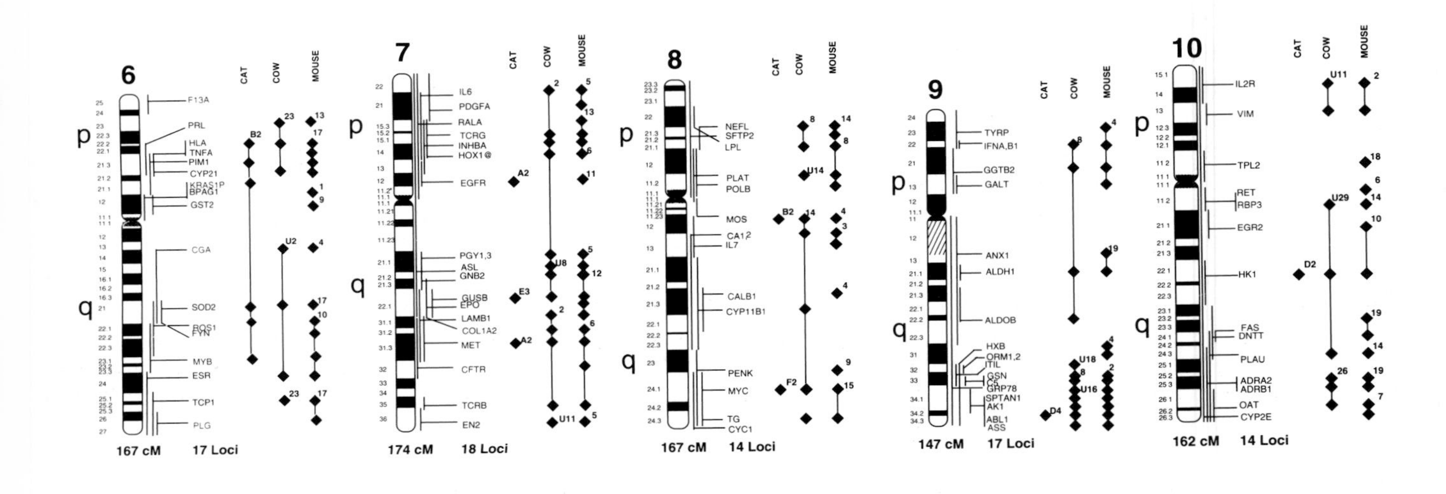
6
CAT
COW
MOUSE
p
q
F13A
PRL
HLA
TNFA
PIM1
CYP21
KRAS1P
BPAG1
GST2
CGA
SOD2
ROS1
FYN
MYB
ESR
TCP1
PLG
167 cM 17 Loci
7
IL6
PDGFA
RALA
TCRG
INHBA
HOX1@
EGFR
PGY1,3
ASL
GNB2
GUSB
EPO
LAMB1
COL1A2
MET
CFTR
TCRB
EN2
174 cM 18 Loci
8
NEFL
SFTP2
LPL
PLAT
POLB
MOS
CA1,2
IL7
CALB1
CYP11B1
PENK
MYC
TG
CYC1
167 cM 14 Loci
9
TYRP
IFNA,B1
GGTB2
GALT
ANX1
ALDH1
ALDOB
HXB
ORM1,2
ITIL
GSN
C5
GRP78
SPTAN1
AK1
ABL1
ASS
147 cM 17 Loci
10
IL2R
VIM
TPL2
RET
RBP3
EGR2
HK1
FAS
DNTT
PLAU
ADRA2
ADRB1
OAT
CYP2E
162 cM 14 Loci

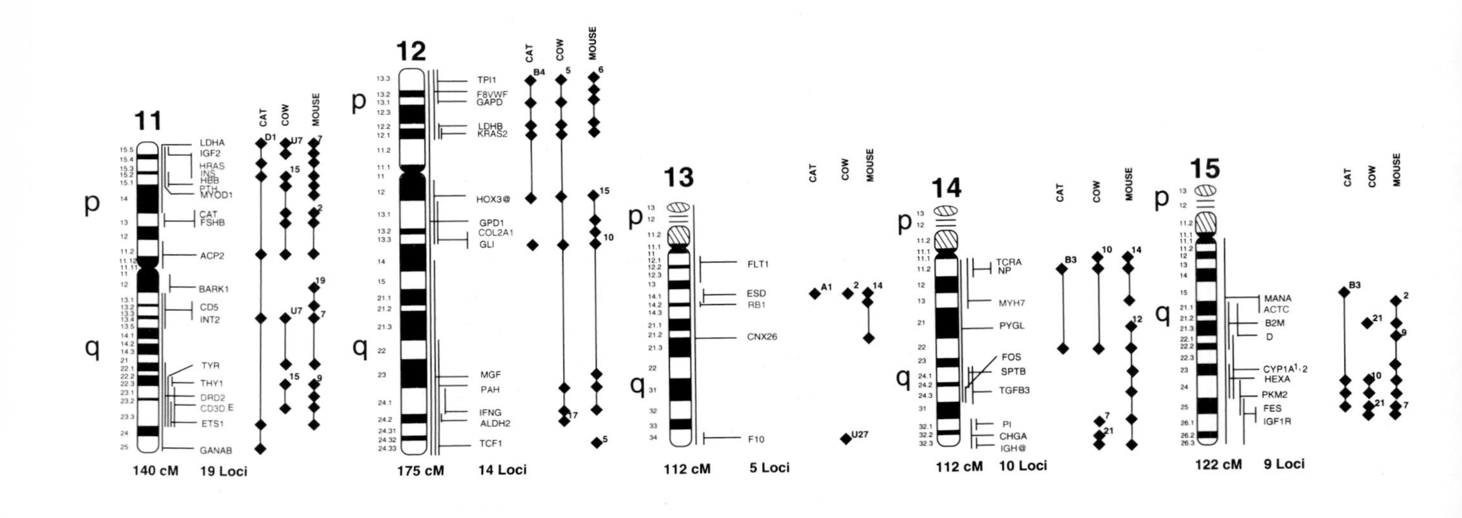
11
CAT
COW
MOUSE
p
q
LDHA
IGF2
HRAS
INS
HBB
PTH
MYOD1
CAT
FSHB
ACP2
BARK1
CD5
INT2
TYR
THY1
DRD2
CD3D,E
ETS1
GANAB
140 cM 19 Loci
12
TPI1
F8VWF
GAPD
LDHB
KRAS2
HOX3@
GPD1
COL2A1
GLI
MGF
PAH
IFNG
ALDH2
TCF1
175 cM 14 Loci
13
FLT1
ESD
RB1
CNX26
F10
112 cM 5 Loci
14
TCRA
NP
MYH7
PYGL
FOS
SPTB
TGFB3
PI
CHGA
IGH@
112 cM 10 Loci
15
MANA
ACTC
B2M
D
CYP1A1,2
HEXA
PKM2
FES
IGF1R
122 cM 9 Loci

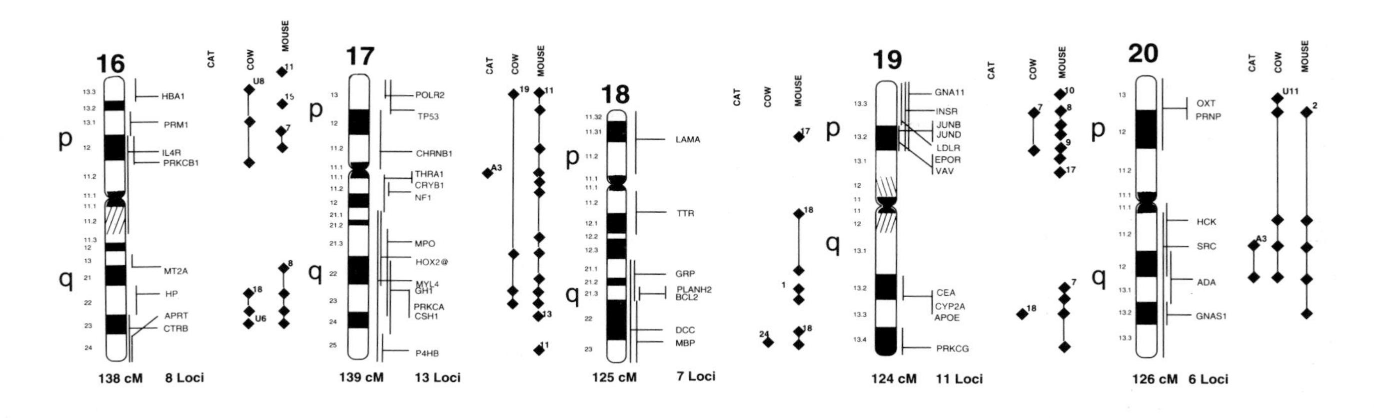

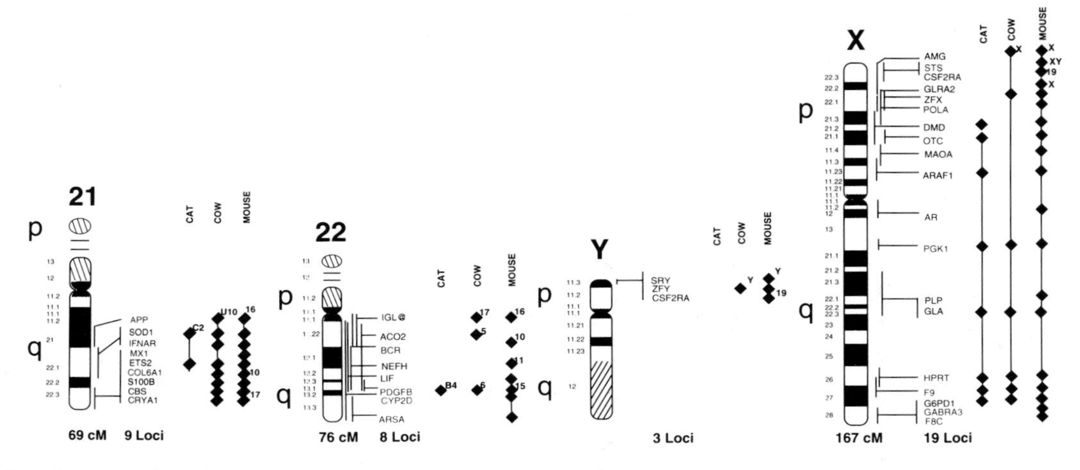

Figure 14.13: Conservation of synteny for mammalian genes is confined to subchromosomal segments.

Gene loci in cat, cow and mouse are shown as diamonds next to the orthologous human genes. Vertical lines linking two or more diamonds indicate that synteny is conserved for these loci and human orthlogs. *Note* the very considerable conservation of synteny for some chromosomes, notably the X chromosomes and also human chromosome 17 and mouse chromosome 11. Redrawn from O'Brien *et al.* (1993) with permission from Nature America Inc.

condition was shown in *Figure 3.4C*. The characteristic pigmentary abnormalities and hearing loss of WS1 are caused by absence of melanocytes from the affected parts (including the inner ear, where melanocytes are required in the stria vascularis of the cochlea in order for normal hearing to develop). Linkage analysis, aided by the description of a chromosomal abnormality in an affected patient, localized the gene for WS1 to the distal part of 2q. At this point, a likely mouse homolog emerged. The *Splotch (Sp)* mouse mutant has pigmentary abnormalities caused by patchy absence of melanocytes, and the *Sp* gene maps to a linkage group on mouse chromosome 1 that shows extensive conservation of synteny with distal human 2q (*Figure 13.10*).

Consideration of the pathogenesis provided further evidence that *WS1* and *Sp* are orthologous genes. The root cause of the phenotype lies in the embryonic neural crest, because melanocytes originate in the neural crest and migrate out to their final locations during embryonic development. Although heterozygous *Sp* mice resemble WS1 patients, homozygous *Sp* mice have neural tube defects and have been studied for many years as a model for human neural tube defects.

A positional candidate gene emerged when the murine *Pax-3* gene was mapped to the vicinity of the *Sp* locus. *Pax-3* is one of a family of genes (*PAX* genes) that encode

Box 14.4: Mapping mouse genes

Several methods are available for easy and rapid mapping of phenotypes or DNA clones in mice. Together with the ability to construct transgenic mice (Chapter 19), they make the mouse especially useful for comparisons with humans. Methods include:

Interspecific crosses (*Mus musculus* × *Mus spretus* or *Mus castaneus*)
The species have different alleles at many polymorphic loci, making it easy to recognize the origin of a marker allele. This is exploited in two ways:

- *constructing marker framework maps*. Several laboratories have generated large sets of F_2 backcrossed mice. Any marker or cloned gene can be assigned rapidly to a small chromosomal segment defined by two recombination breakpoints in the collection of backcrossed mice. For example, the collaborative European backcross was produced from a *M. spretus* × *musculus* (C57BL) cross. Five hundred F_2 mice were produced by backcrossing with *spretus*, and 500 by backcrossing with C57BL. All microsatellites in the framework map are scored in every mouse.
- *Mapping a new phenotype*. A cross must be set up specifically to do this but, unlike with humans, any number of F_2 mice can be bred to map to the desired resolution. *Musculus* × *castaneus* crosses are easier to breed than *musculus* × *spretus*.

Recombinant inbred strains
These are obtained by systematic inbreeding of the progeny of a cross, for example the widely used BXD strains are a set of 26 lines derived by over 60 generations of inbreeding from the progeny of a C57BL/6J × DBA/2J cross. They provide unlimited supplies of a panel of chromosomes with fixed recombination points. DNA is available as a public resource, and the strains function rather like the CEPH families do for humans (see page 326). Recombinant inbred strains are particularly suited to mapping quantitative traits (see Chapter 18), which can be defined in each parent strain and averaged over a number of animals of each recombinant type. Compared with mice from interspecific crosses, it may be harder to find a marker in a given region which distinguishes the two original strains, and the resolution is lower because of the smaller numbers.

Congenic strains
These are identical except at a specific locus. They are produced by repeated backcrossing, and can be used to explore the effect of changing just one genetic factor on a constant background.

Silver (1995) gives an overview of mouse genetics (see Further reading), and Copeland and Jenkins (1991) describe the uses of interspecific crosses.

transcription factors containing the paired box DNA-binding motif, and it is expressed in mouse embryos in the developing nervous system, including the neural crest. The sequence of *Pax-3* was almost identical to the limited sequence which had previously been published for an unmapped human genomic clone, *HuP2*. Such observations prompted mutation screening of *Pax-3* and *HuP2* and led to identification of mutations in *Splotch* mice and humans with WS1 (reviewed by Strachan and Read, 1994). As the underlying genes, *Pax-3* and *HuP2* were clearly orthologs, the *HuP2* gene was subsequently re-named *PAX3*.

Limitations of human–mouse homologies

Though enormously valuable as a guide to human–mouse homologies, conservation of synteny is not always sufficient to allow identification of positional candidates. This is illustrated by the *mi/MITF* locus, mutations in which cause another variant of Waardenburg syndrome, WS type 2 (MIM 193510) in man. Although the *mi* locus on mouse chromosome 6 has long been recognized as a likely candidate homolog of some form of WS, attempts to predict the location of the human homolog failed because synteny is very poorly conserved in this chromosomal region. Genes mapping close to *mi* have human homologs mapping to 3p25, 3q21–q24 and 10q11.2. Each of these locations was tested for linkage to WS2, with negative results. Not until the human homolog, *MITF*, had been cloned and mapped by FISH to 3p14, and WS2 in humans independently mapped by linkage to the same location, was there sufficient evidence to begin a successful search for mutations in man (Tassabehji *et al.*, 1994).

Mutation screening: confirming a candidate gene

Principles of mutation screening

However promising a candidate gene appears to be for a disease, it must be shown to be mutated in affected people. **Mutation screening** entails testing DNA samples from a sizeable panel of patients and control individuals. The first step is to design pairs of specific primers for use in amplifying portions of the coding DNA, either from a genomic DNA sample (if the exon/intron boundaries are known), or from cDNA generated by RT-PCR from mRNA of patients (*Figure 6.7*). The products of individual amplification reactions are then subjected to one or more rapid mutation screening procedures which are designed to detect point mutations (see below). Any sequence variants detected must be characterized by DNA sequencing. The point of the exercise is to demonstrate a variety of patient-specific mutations which are expected to have a deleterious effect on gene expression.

Mutation screening is often straightforward for diseases where a good proportion of patients carry independent mutations (typically severe early onset dominant or X-linked recessive disorders). A panel of DNA samples from patients will usually show a variety of different mutations if the correct gene is tested. When the disease phenotype results from loss of function (page 404), one might expect to identify a variety of mutations with an obviously deleterious effect on gene expression (e.g. nonsense mutations, frameshift mutations, etc.). *Figure 15.1* shows an example. If the identified mutations are absent from control samples, then the conclusion that the gene being tested really is the locus for the disease becomes almost inescapable. However, other circumstances may make the identification of mutations and the interpretation of mutation screening more difficult.

- *Unsuspected locus heterogeneity.* Often mutations in several different genes can give almost identical phenotypes, so that a panel of unselected patient samples may have pathogenic mutations in different genes. Identifying a candidate gene for one of the rarer disease loci can be difficult, as most samples will show no mutation in that gene. Ideally, one would use only samples from families with demonstrated linkage to the candidate region, but this may be impracticable – family sizes for recessive and some dominant disorders are often too small for independent linkage analyses, and in some severe dominant disorders most patients present as sporadic cases without a family history.
- *Mutational homogeneity.* This problem was discussed above (page 386) in connection with CF. Most apparently unrelated patients carry the same mutation, ΔF508.
- *Mutations are not unambiguously pathogenic.* It may be difficult to identify missense mutations as being pathogenic as opposed to being neutral variants with no major effect on gene expression. If the genome-wide average heterozygosity of 0.0037 (page 241) was applied to coding sequences, then screening

a 3-kb gene for mutations in a panel of 100 patients would reveal over 1000 sequence changes. Because of selection pressure to conserve gene function, coding sequences are much more highly conserved but, even so, such a sample will inevitably contain a few nonpathogenic sequence variants. If each variant is seen in one person in 10 000 in the population, it will almost certainly not be found in the panel of, say, 100 normal controls tested at the same time. Without functional testing, many variants cannot be classified definitely as pathogenic or neutral. It is likely, for example, that some of the 500 reported CF mutations are in fact rare neutral variants.

- *Mutations may be hard to find.* Large genes are more difficult to screen for mutations, and sometimes mutations seem very hard to find. Current examples include the *NF1* gene and the *PKD1* gene for adult polycystic kidney disease. Mutations in the *F8* gene causing severe hemophilia A seemed to be hard to find, until it was discovered that most of the missing mutations were large inversions which disrupted the gene (see *Figure 10.18*) but were not detected by the PCR methods normally used.

There are numerous methods for screening a panel of patients for mutations in a given gene

Mutation testing methods can be divided into two groups:

(i) *mutation detection* methods test a DNA sample for the presence or absence of one specific mutation;

(ii) *mutation screening* methods screen a sample for any deviation from the standard sequence.

The first set of methods is very useful in diagnostic work, but not for identifying candidate genes, where the mutation cannot be specified in advance (a possible exception is testing for expanded trinucleotide repeats, see below). They are described in Chapter 16. Some of the more advanced proposals for general screening methods ('DNA chips', etc., see page 430) could, in principle, be used for identifying candidate genes, but setting up the test requires heavy initial expenditure, so that again they will find their application in large-scale diagnostic work. For identifying mutations in candidate genes, 100% sensitivity is not necessary. If six different plausible mutations in a candidate gene can be picked up in a panel of 20 patients (and are absent in 100 controls), that is quite enough to establish that the right gene has been cloned. More important than 100% sensitivity is that the method should be flexible enough to be applied to a series of candidates before the correct gene is found. The relative merits of many methods have been discussed by Grompe (1993) and Prosser (1993).

Methods for general mutation screening, other than simple sequencing, usually test for differences between the sequence under test and some standard sequence. Most, but not all, rely on detecting mismatched bases formed when complementary strands of a mutant and a wild-type allele are allowed to hybridize to form a heteroduplex (**heteroduplex analysis**). This is likely to occur naturally when PCR product from a heterozygous person is denatured then cooled to allow single mutant strands to base-pair with complementary strands from the wild-type allele. For detecting homozygous mutations, or X-linked mutations in males, however, it would be necessary to add some wild-type DNA.

Heteroduplex mobility in polyacrylamide gels (see *Figure 14.14*)

The electrophoretic mobility of heteroduplexes in polyacrylamide gels is less than that of the homoduplexes, and they can be detected as extra slow moving bands.

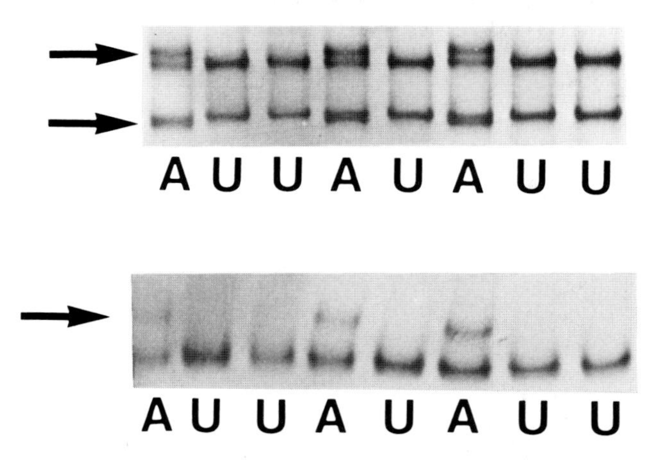

Figure 14.14: Heteroduplex and single strand conformational polymorphism analysis permit rapid point mutation screening.

Exon 2 of the *NF2* gene was PCR amplified from genomic DNA of eight members of a family in which a C→T base change mutation was segregating. Affected individuals are constitutionally heterozygous for this mutation. After denaturation, the samples were loaded on to a nondenaturing polyacrylamide gel. Some of each product re-annealed to give double-stranded DNA. This runs faster in the gel and gives the bands seen in the lower panel. The single-stranded DNA runs more slowly in the same gel (upper panel). Unaffected individuals (U) in the upper panel have two bands representing the different mobilities of the two complementary strands. However, the affected (A) individuals (tracks 1, 4 and 6) have two additional bands (arrows) as a result of the presence of a mutant allele. In the lower panel, upper heteroduplex bands (arrow) can also be seen in the DNA from the affected individuals. Original photos kindly provided by Dr David Bourn, University of Newcastle upon Tyne.

Resolution is better in specially formulated gel matrices (Hydro-link™, MDE™, etc.). If fragments of under 200 bp are tested, insertions, deletions and most single base substitutions are detectable (Keen *et al.*, 1991). Heteroduplex analysis is mostly used combined with single-strand conformation analysis (see below) on a single gel.

Single-strand conformational polymorphism analysis (SSCP or SSCA) (see *Figure 14.14*)

Single-stranded DNA has a tendency to fold up and form complex structures stablilized by weak intramolecular bonds, notably base-pairing hydrogen bonds. The electrophoretic mobilities of such structures on nondenaturing gels will depend not only on their chain lengths but also on their conformations, which are dictated by the DNA sequence. For SSCP, amplified DNA samples are denatured and loaded on a nondenaturing polyacrylamide gel. Primers can be radiolabeled or the unlabeled products detected by silver staining. Control samples must be run so that differences from the wild-type pattern can be noticed. SSCP is simple and adequately sensitive, but is inefficient for fragments longer than 200 bp, and does not reveal the nature or position of any mutation detected (Sheffield *et al.*, 1993). The precise pattern of bands seen is very dependent on details of the conditions. An elaboration of SSCP, *dideoxy fingerprinting*, analyzes each band in a sequencing ladder by SSCP, which is claimed to give 100% sensitivity (Sarkar *et al.*, 1992).

Chemical cleavage of mismatches (CCM) (see *Figure 14.15*)

A radiolabeled probe is hybridized to the test DNA, and mismatches detected by a series of chemical reactions that cleave one strand of the DNA at the site of the mismatch. CCM is one of the most sensitive methods of mutation detection, it can be applied to kilobase-length samples (whereas heteroduplex analysis or SSCP work badly for fragments over 200 bp long), and it reveals the position of the mismatch (by the size of the cleaved fragment). Its disadvantages are that the chemicals, especially osmium tetroxide, are very toxic, and some practice is required before it works well.

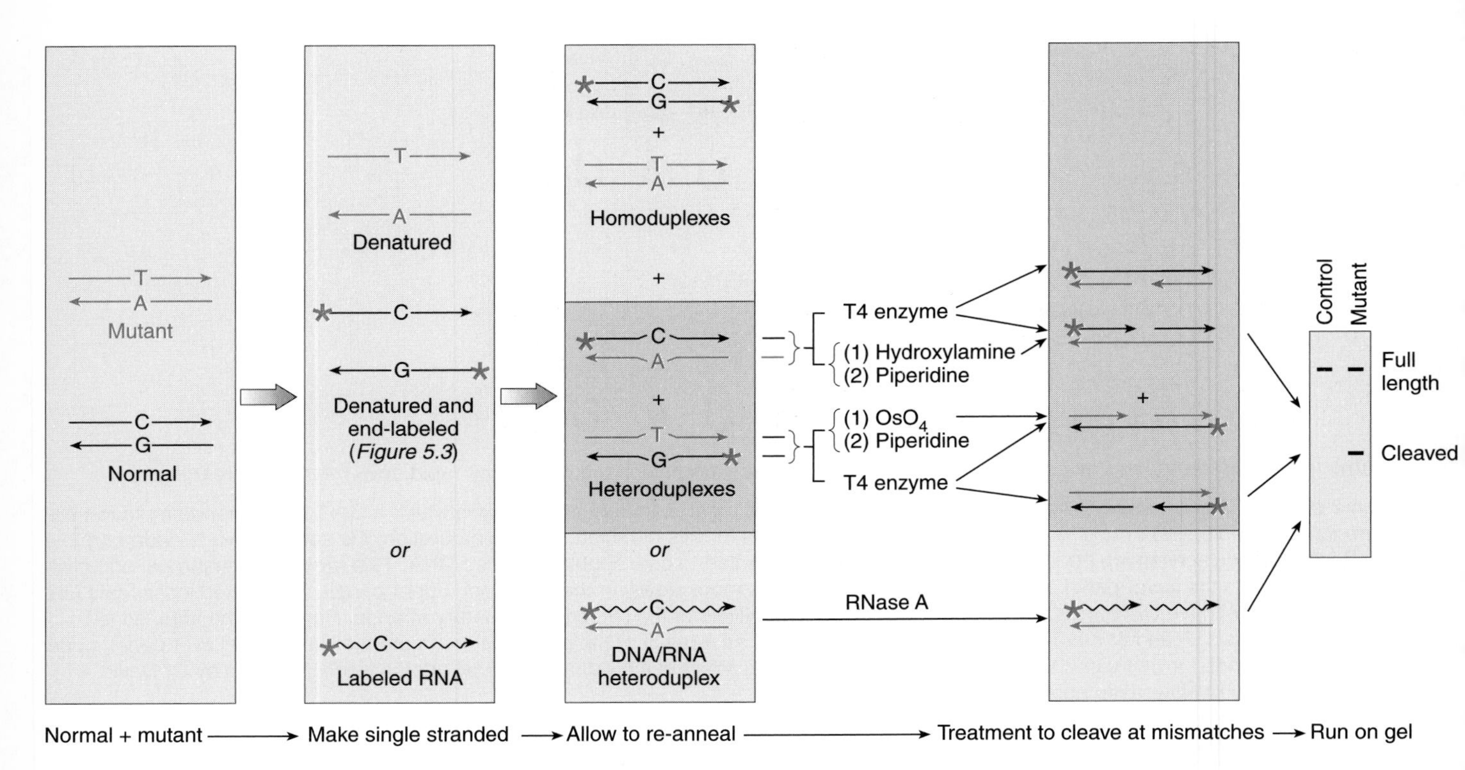

Figure 14.15: Cleavage of mismatches in heteroduplex.

The test sample is mixed with labeled wild-type sequence, denatured and allowed to renature. If the test sequence differs from the wild-type, heteroduplexes are formed which can be cleaved by chemical or enzymic treatment (see text). Cleavage is revealed by observing short labeled fragments on a gel. *Note* that only cleavage of the labeled probe is detected, not cleavage of the mutant DNA (shown in red). Thus, in this example, no cleavage product would be seen with OsO_4, which cleaves a strand containing a mismatched T.

Enzymatic cleavage of mismatches (see *Figure 14.15*)

Certain bacteriophage enzymes involved in DNA repair or recombination cleave mismatches, allowing an assay identical in principle to CCM, but without the unpleasant chemicals. T4 phage resolvase and endonuclease VII have been used for this purpose. This is a promising recent technique which has not yet been tested on a large scale (reviewed by Dean, 1995). An earlier version of the method used RNaseA to cleave mismatches in heteroduplexes between the test DNA and an RNA wild-type sequence.

Denaturing gradient gel electrophoresis (DGGE)

In this method (Cariello and Skopek, 1993), DNA duplexes are forced to migrate through an electrophoretic gel in which there is a gradient of increasing amounts of a denaturant (usually a chemical denaturant, but *temperature gradient gel electrophoresis* is also sometimes used). Migration continues until the DNA duplexes reach a position in the gel where the strands melt and separate, after which the denatured DNA does not migrate much further. A single base pair difference between a normal and a mutant DNA duplex is sufficient to cause them to migrate to different positions in the gel. DGGE is potentially highly sensitive, but it requires very careful design of primers, so that the sequence amplified has the right profile of melting domains. Sensitivity is improved by adding a tail of poly(G:C) (a *GC clamp*) to each primer. Because DGGE requires expensive GC-clamped primers and careful fine tuning for each individual gene, it is probably better suited to diagnostic work than to checking a series of candidate genes.

Protein truncation test (PTT) (see *Figure 14.16*)
The PTT is a specific test for frameshifts, splice sites or nonsense mutations which truncate a protein product. The procedure is to make cDNA by RT-PCR, using a special primer which carries at the 5′ end a T7 promoter followed by a eukaryotic translation initiator sequence. The cDNA is put into a coupled transcription–translation system, which uses the T7 promoter to make mRNA and the translation initiator to translate it. The protein product is then run out on a gel. If the product is full length, no truncating mutation is present in the sequence cloned. Truncating mutations result in shorter products, the size of which reveals the position of the mutation (van der Luijt *et al.*, 1994).

Clearly, the strength and weakness of the PTT is that it detects only certain classes of mutation. In some genes, such as the dystrophin, *APC* and *BRCA1* genes, missense

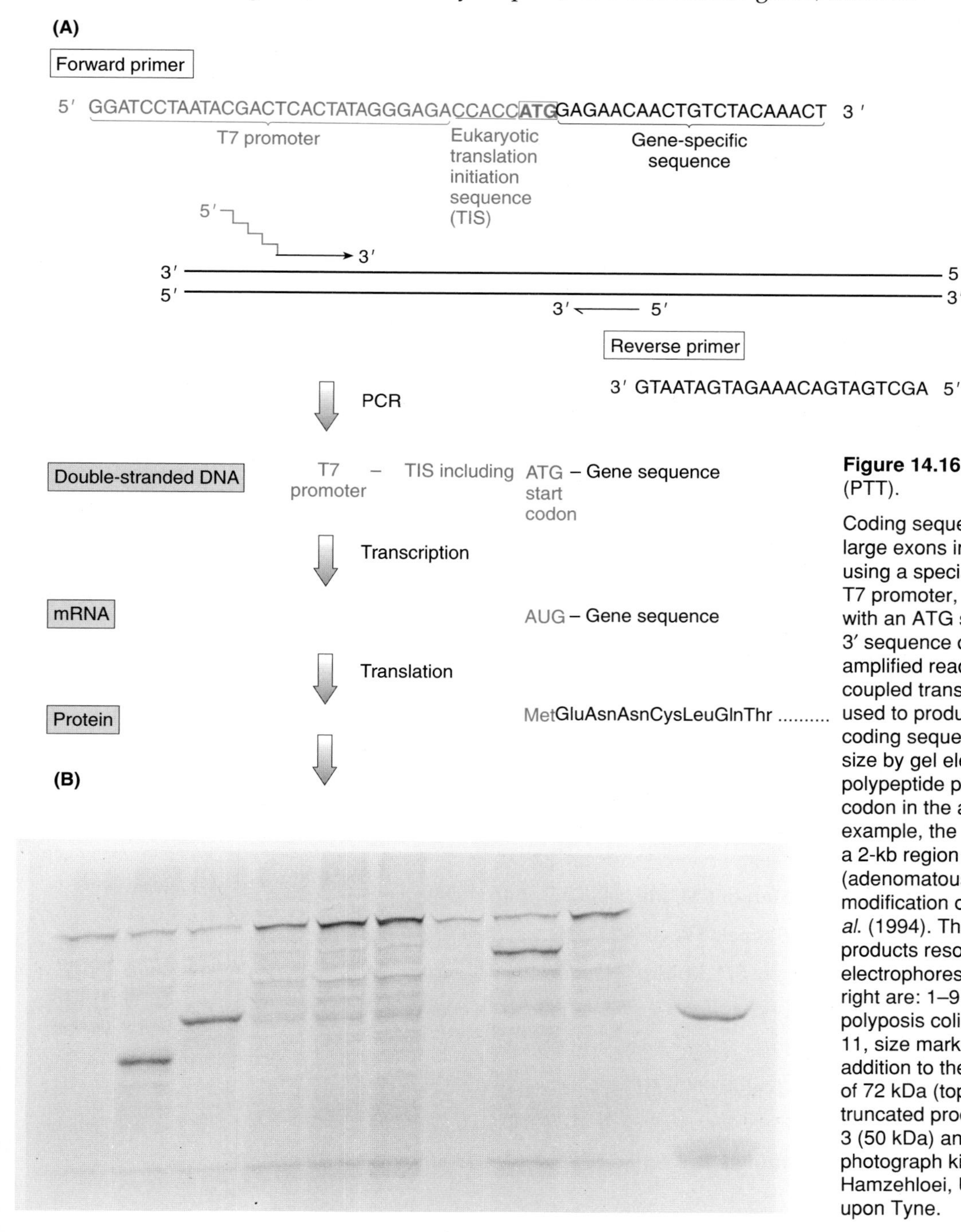

Figure 14.16: The protein truncation test (PTT).

Coding sequence without introns (cDNA or large exons in genomic DNA) is PCR amplified using a special forward primer that includes a T7 promoter, a eukaryotic translation initiator with an ATG start codon, and a gene-specific 3′ sequence designed so that the sequence amplified reads in-frame from the ATG. A coupled transcription–translation system is used to produce polypeptide from the amplified coding sequence, which is then checked for size by gel electrophoresis. A truncated polypeptide points to the presence of a stop codon in the amplified coding sequence. In this example, the primers were designed to amplify a 2-kb region in exon 15 of the APC (adenomatous polyposis coli) gene, using a modification of the method of van der Luijt *et al.* (1994). The bottom panel shows translation products resolved by SDS–polyacrylamide gel electrophoresis (PAGE). Lanes from left to right are: 1–9, unrelated familial adenomatous polyposis coli patients; 10, negative control; 11, size markers (46 kDa and 30 kDa). In addition to the wild-type translation product of 72 kDa (top band), *note* the presence of truncated products in lanes 2 (41 kDa), 3 (50 kDa) and 8 (64 kDa). Original photograph kindly provided by Tayebeh Hamzehloei, University of Newcastle-upon Tyne.

mutations are seldom found, and are probably usually nonpathogenic. For such genes, the PTT has several advantages. It conveniently ignores the nonpathogenic silent or missense base substitutions, and (like heteroduplex cleavage methods, but unlike SSCP) it reveals the approximate location of any mutation. Large exons, such as exon 15 of the *APC* gene, can be tested using genomic DNA rather than by RT-PCR (see van der Luijt *et al.*, 1995, and *Figure 14.16*). Several variants have been developed to give cleaner results, usually by incorporating an immunoprecipitation step.

Triplet repeat expansions

One test for a special class of mutations is nevertheless sometimes used in screening candidate genes. Expanded trinucleotide repeats are known to cause several inherited disorders, mostly of neurological origin (see pages 266 and 435). Often such disorders show *anticipation:* the disease presentation occurs at an earlier age and with increased severity in successive generations. If the disease under investigation shows any of these features, it may be worth screening for triplet repeat expansions.

There are only 10 possible different triplet repeats (page 266), so that DNA clones from the relevant subchromosomal location can be probed fairly easily with oligonucleotide probes corresponding to each repeat. Positive hybridization is followed by attempts to characterize the flanking regions, prior to investigating patient DNA samples for evidence of expansion of the repeats. Alternatively, the *repeat expansion detection method* (Schalling *et al.*, 1993) may permit detection, although not localization, of expanded repeats in unfractionated genomic DNA of affected patients.

Further reading

McKusick VA. (1994) *Mendelian Inheritance in Man,* 11th Edn. Johns Hopkins University Press, Baltimore.

Monaco AP. (1994) Isolation of genes from cloned DNA. *Curr. Opin. Genet. Dev.* **4**, 360–365.

Nelson DL. (1995) Positional cloning reaches maturity. *Curr. Opin. Genet. Dev.*, **5,** 298–303.

Silver LM. (1995) *Mouse Genetics: Concepts and Applications.* Oxford University Press, Oxford.

References

Brookes AJ, Slorach EM, Morrison KE *et al.* (1994) *Hum. Mol. Genet.*, **3**, 2011–2017.

Cariello NF, Skopek TR. (1993) *Mutat. Res.*, **288**, 103–112.

Carter SA, Bryce SD, Munro CS *et al.* (1994) *Genomics*, **24**, 378–382.

Collins FS. (1992) *Nature Genetics*, **1**, 3–6.

Collins FS. (1995) *Nature Genetics*, **9**, 347–350.

Copeland N, Jenkins NA. (1991) *Trends Genet.*, **7**, 113–118

Dean M. (1995) *Nature Genetics*, **9**, 103–104.

Dietz HC, Cutting GR, Pyeritz RE *et al.* (1991) *Nature*, **352**, 337–339.

Dietz HC, Cutting GR, Pyeritz RE *et al.* (1991) *Nature,* **352**, 337–339.

Dryja TP, McGee TL, Reichel E *et al.* (1990) *Nature,* **343**, 364–366.

Edstrom J-E, Kaiser R, Rohme D. (1987) *Methods Enzymol.,* **151**, 503–516.

Fountain JW, Wallace MR, Bruce MA *et al.* (1989) *Science,* **244**, 1085–1087.

Francis KJ, Nesbit MA, Theodosiou AM *et al.* (1995) *Genomics,* **27**, 366–369.

Gitschier J, Wood WI, Goralka TM *et al.* (1984) *Nature,* **312**, 326–330

Grompe M. (1993) *Nature Genetics,* **5**, 111–116.

Hampton G, Leuteritz G, Ludecke HJ *et al.* (1991) *Genomics,* **11**, 247–251.

Keen J, Lester D, Inglehearn C, Curtis A, Bhattacharya S. (1991) *Trends Genet.,* **7**, 5.

Kunkel LM, Monaco AP, Middlesworth W, Ochs HD, Latt SA. (1985) *Proc. Natl Acad. Sci. USA,* **82**, 4778–4782.

Lisitsyn NA. (1995) *Trends Genet.,* **11**, 303–307.

Ludecke HJ, Senger G, Claussen V, Horsthemke B. (1989) *Nature,* **338**, 348–350.

O'Brien SJ, Womack JE, Lyons LA, Moore KJ, Jenkins NA, Copeland NG. (1993) *Nature Genetics,* **3**, 103–112.

Prosser J. (1993) *Trends Biotechnol.,* **11**, 238–246.

Putnam EA, Zhang H, Ramirez F, Milewicz DM. (1995) *Nature Genetics,* **11**, 456–458.

Robson, KJH, Chandra, T, MacGillivray, RTA, Woo, SLC. (1982) *Proc. Natl. Acad. Sci. USA,* **79**, 4701–4705.

Roest PAM, Roberts RG, Sugino S, van Ommen GJB, den Dunnen JT. (1993) *Hum. Molec. Genet.,* **2**, 1719–1721.

Rommens JM, Iannuzzi MC, Kerem B-S *et al.* (1989) *Science,* **245**, 1059–1065.

Sarkar G, Yoon HS, Sommer SS. (1992) *Genomics,* **13**, 441–443.

Schalling M, Hudson TJ, Buetow KH, Housman DE. (1993) *Nature Genetics,* **4**, 135–139.

Schutte M, da Costa LT, Hahn SA *et al.* (1995) *Proc. Natl Acad. Sci. USA,* **92**, 5950–5954.

Sheffield VC, Beck JS, Kwitek AE, Sandstrom DW, Stone EM. (1993) *Genomics,* **16**, 325–332.

Strachan T, Read AP. (1994) *Curr. Opin. Genet. Dev.,* **4**, 427–438.

Strathdee CA, Gavish H, Shannan WR, Buchwald M. (1992) *Nature,* **356**, 763–767.

Tassabehji M, Newton VE, Read AP. (1994). *Nature Genetics,* **8**, 251–255.

van der Luijt R, Khan PM, Vasen H, van Leeuwen C, Tops C, Roest P, den Dunnen J, Fodde R. (1994) *Genomics,* **20**, 1–4.

Wang M, Tsipouras P, Godfrey M. (1995) *Am. J. Hum. Genet.,* **57**, A231.

Watson CJ, Gaunt L, Evans G, Patel K, Harris R, Strachan T. (1993) *Hum. Molec. Genet.,* 701–704.

Worton RG, Thompson MW. (1988) *Annu. Rev. Genet.,* **22,** 601–629.

Molecular Pathology

Chapter 15

Molecular pathology seeks to explain why a given genetic change should result in a particular clinical phenotype. We have already reviewed the nature and mechanisms of mutations in Chapter 10 (briefly summarized in *Box 15.1*); this chapter is concerned with their effects. Molecular pathology requires us to work out the effect of a mutation on the quantity or function of the gene product and to explain why the change is or is not pathogenic for any particular cell or stage of development.

Not surprisingly, given the complexity of genetic interactions, molecular pathology is at present a very imperfect science. Clinical symptoms are often the end result of a long chain of causation, and all too often they are not predictable or even readily comprehensible with our present state of knowledge. Nevertheless, as the emphasis of the Human Genome Project moves from cataloging genes to understanding their function, the study of molecular pathology has moved to center stage, and knowledge is progressing. For a growing number of well-studied genes, it is now becoming possible to predict the phenotypic effect of a given DNA sequence change.

One of the major advantages of studying humans rather than laboratory organisms is that the healthcare systems world-wide act as a gigantic and continuous mutation screen. As a consequence, for most inherited diseases where the gene responsible has been identified, many different mutations are known. We cannot do experiments on humans or breed them to order, but humans provide unique opportunities to observe the phenotypic effects of many different changes in a given gene.

Box 15.1: The main classes of mutation

Deletions ranging from 1 bp to megabases

Insertions including duplications

Single base substitutions:

- **Missense mutations** replace one amino acid with another in the gene product.
- **Nonsense mutations** replace an amino acid codon with a stop codon.
- **Splice site mutations** create or destroy signals for exon/intron splicing.

Frameshifts can be produced by deletions, insertions or splicing errors.

See Chapter 10 for a full discussion of classes of mutation and the mechanisms producing them.

This generates hypotheses, which must then be tested in animals. Thus, investigations of naturally occurring human mutations are complemented by studies of specific mutations in transgenic animals (see page 530).

Nomenclature and databases of mutations

The preferred nomenclature of genes and alleles is laid down by the Nomenclature Committees at the various Human Gene Mapping meetings (McAlpine *et al.*, 1992). Allelic variants are collected in the databases such as OMIM (see page 61) and numbered serially, using the locus name followed by an asterisk and a serial number, for example *HBB*0403*. A valuable summary of genetic nomenclature for many different organisms, including man, was published as a supplement to *Trends in Genetics* in March 1995 (see Further reading).

For describing mutations (rather than simply listing them), a nomenclature is needed which identifies the sequence change. *Table 15.1* summarizes the recommended conventions. Details of mutations are stored in OMIM as noted above, and more usefully compiled by Cooper and Krawczak (1995).

Table 15.1: Nomenclature for describing mutations

Amino acid substitution

Use the one-letter codes: A, alanine; C, cysteine; D, aspartic acid; E, glutamic acid; F, phenylalanine; G, glycine; H, histidine; I, isoleucine; K, lysine; L, leucine; M, methionine; N, asparagine; P, proline; Q, glutamine; R, arginine; S, serine; T, threonine; V, valine; W, tryptophan; Y, tyrosine; X means a stop codon.

R117H – amino acid 117 changed from arginine to histidine
G542X – glycine 542 replaced by stop codon
(these are often written as Arg117 →His and Gly542 →Stop, respectively).

Nucleotide substitution

Nucleotide notations are often prefixed with **nt** (nucleotide) or **np** (nucleotide pair). The numbering normally refers to the sense strand of the cDNA (except for mitochondrial mutations). Since 5′ ends of many cDNAs are not well defined, it may be necessary to number with reference to a specific published sequence. Nucleotides within introns (which will not be listed or numbered in the cDNA sequence) can be specified by their position relative to the nearest exon.

1162(G →A) or **np1162(G →A)** – replace guanine 1162 (in the cDNA) by adenine.
621+1(G →T) – replace guanine with thymine at the first position of the intron immediately after position 621 (the last nucleotide in an exon).
1781–5(C →A) – replace cytosine in the intron five bases 5′ of position 1781 with adenine.

Deletions and insertions

Delta-F508 or **ΔF508** – delete the codon for phenylalanine 508.
nt6232(del5) – delete five nucleotides, the first of which is at position 6232.
nt409(insC) – insert cytidine after nucleotide position 409.

See Beaudet and Tsui (1993). Hemoglobinopathies have traditionally been described using a slightly different system (see *Table 16.4*).

Classification of mutations: loss of function versus gain of function

The convenient classification of alleles into **A** and **a** hides a vast diversity of DNA sequence changes. About 500 different cystic fibrosis mutant alleles have been described, and a similar number of different mutations in the β-globin gene. There is no reason why these should all fit into a few tidy categories. In principle, however, mutation of a gene might cause a phenotypic change in either of two ways:

- the product may have reduced or no function (**loss-of-function mutation**);
- the product may function in an abnormal way (**gain-of-function mutation**).

With some genes, mutations of both types are known, and they produce very different phenotypes (page 422). In general, loss-of-function mutations produce recessive phenotypes. This is because, for most gene products, the precise quantity is not crucial, and we can get by on half the normal amount. Thus, most inborn errors of metabolism are recessive. Occasionally, however, 50% of the normal level is not sufficient for normal function and one sees dominant **dosage effects** (see *Table 15.3*). Gain-of-function mutations also produce dominant phenotypes. Few mutations involve the product doing something completely novel. More often, we see an escape from normal controls – the mutant gene causes an abnormal phenotype because it functions in the wrong cell, at the wrong time or to an excessive extent (see page 412). A **dominant-negative** effect occurs when the mutant product not only loses its own function but also prevents the product of the normal allele from functioning in a heterozygous person. Dominant-negative effects are seen particularly with proteins that work as dimers or multimers (see below, page 414).

Inevitably, some mutations cannot be classified easily as either loss or gain of function. Sometimes the problem is mainly semantic – has a permanently open ion channel lost the function of closing or gained the function of inappropriate opening? At other times the difficulty is more fundamental, as with dominant-negative effects. Nevertheless, the distinction between loss of function and gain of function provides a useful first approach to classifying mutations.

Changes can be quantitative or qualitative

The hemoglobinopathies (page 423) illustrate the difference between quantitative and qualitative alterations in a gene product. In the thalassemias, the problem is an inadequate quantity of globin, whilst in other hemoglobinopathies the problem is an abnormal globin. At one time it was thought that this distinction might reflect two fundamentally different classes of mutation: qualitative changes (structural abnormalities) caused by mutations in coding sequences, and quantitative changes caused by mutations in control sequences. However, it has become apparent that many thalassemias are caused by structural abnormalities which make the globin mRNA or protein unstable and, in general, one must look at each individual mutation to determine its likely pathogenic effect.

Loss of function is likely when point mutations in a gene produce the same pathological change as deletions

Purely genetic evidence, without biochemical studies, can often suggest whether a phenotype is caused by the loss or gain of function. When a clinical phenotype results from the loss of function of a gene, we would expect any change that inactivates the gene product to produce the same clinical result. We should be able to find point mutations that have the same effect as mutations that delete or disrupt the gene. Waardenburg syndrome type 1 (MIM 193500) provides an example. As *Figure 15.1* shows, causative mutations in the *PAX3* gene include amino acid substitutions, frameshifts, splicing mutations and, in some patients, complete deletion of the *PAX3* sequence. These observations demonstrate that Waardenburg syndrome is caused by loss of function of *PAX3*.

Gain of function is likely when only a specific mutation in a gene produces a given pathology

Gain of function is likely to require a much more specific change than loss of function. The mutational spectrum in gain-of-function conditions should be correspondingly more restricted, and deletion or disruption of the gene should not produce the

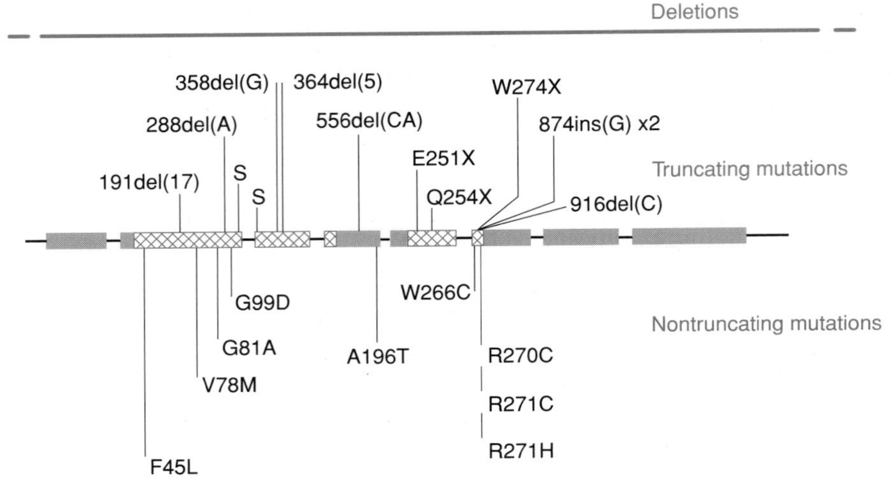

Figure 15.1: Loss-of-function mutations in the *PAX3* gene.

The *PAX3* gene has eight exons. Sequences known to be crucial to the function of the gene product are cross-hatched. All the mutations shown were found in patients with Waardenburg syndrome. S denotes a splice site mutation. The mutations written above the gene diagram would be expected to lead to a truncated gene product because of premature termination after a frameshift or splicing mutation; those written below the diagram would be expected to produce a full-length protein but with amino acid substitutions. Some patients with Waardenburg syndrome have the entire gene deleted. Since all these different changes produce the same clinical result, it is likely that the cause is loss of function of the gene. Data from Tassabehji *et al.* (1995).

condition. Huntington disease (HD, MIM 143100) is a likely example. As we saw in Chapter 10 (page 266), the only mutation in the *HD* gene associated with HD is expansion of the trinucleotide repeat. No missense, nonsense, frameshift or splice mutations have been found. One reported case confirms that loss of function of one copy of the *HD* gene does not cause HD: Ambrose *et al.* (1994) found a phenotypically normal individual with a translocation breakpoint disrupting the *HD* gene. Patients with chromosomal deletions including the *HD* gene also fail to manifest HD, although these patients (unlike the translocation case) have other abnormalities that might prevent HD being noted even if it was present. A final point favoring gain of function in HD is the fact that homozygotes are clinically identical to heterozygotes. Proof that HD is a gain-of-function disease however will require biochemical demonstration of the gained function.

Loss-of-function mutations

Loss of function may be partial or total. Alleles which produce no product are often termed **null alleles**. Sometimes the gene product can be detected immunologically even though it is nonfunctional, indicating that the protein can still be synthesized and exported. Such mutants are often termed **CRM$^+$** (positive for cross-reacting material).

Mutations leading to deficiency of a protein are not necessarily located in the gene which encodes it

Agammaglobulinemia (lack of immunoglobulins, leading to clinical immunodeficiency) is often mendelian. It is natural to assume the cause would be mutations in the immunoglobulin genes. However, the immunoglobulin loci are located on chromosomes 2, 14 and 22 (*Table 7.13*), and agammaglobulinemias do not map to these locations. Many forms are X-linked. Remembering the many steps needed to turn a newly synthesized polypeptide into a correctly functioning protein (pages 24–26), this lack of one-to-one correspondence between the mutation and the protein structural gene should not come as any great surprise. Failures in immunoglobulin gene processing, in B-cell maturation, or in more fundamental processes of the immune system will all produce immunodeficiency.

One gene defect can sometimes produce multiple enzyme defects. I-cell disease or mucolipidosis II (MIM 252500) is marked by deficiencies of multiple lysosomal enzymes. The primary defect is not in the structural gene for any of these enzymes, but in an *N*-acetylglucosamine-1-phosphotransferase which phosphorylates mannose residues on the glycosylated enzyme molecules. The phosphomannose is a signal that targets the enzymes to lysosomes; in its absence the lysosomes lack a whole series of enzymes. The overall message is that one should not be naive when speculating about the gene defect underlying a clinical syndrome.

Many different changes to a gene can destroy its function

Table 15.2 lists the main ways in which changes to a gene can reduce or destroy its function. In general, it is not easy to predict whether the result of a particular mutation will be neutral, a partial loss of function or total loss. Predictions can be made with reasonable confidence when the gene is disrupted or deleted. Otherwise prediction is possible only when the effects (biochemical and clinical) of many other mutations in the same gene are known. An added complication comes from the fact that most people with recessive conditions whose parents are unrelated are **compound heterozygotes**, with two different mutations. If the mutations both cause simple loss of function, but to differing degrees, the less severe allele dictates the overall loss of function, and hence the clinical effect.

Table 15.2: Fifteen ways to reduce or abolish the function of a gene product

Change	Example
Delete the entire gene	Most α-thalassemia mutations (page 423)
Delete part of the gene	60% of Duchenne muscular dystrophy mutations *(Figure 15.2)*
Disrupt the gene structure:	
(i) by a translocation	X–autosome translocations in women with Duchenne muscular dystrophy (*Figure 14.8*)
(ii) by an inversion	Inversion in *F8* gene (*Figure 10.18*)
Insert a sequence into the gene	Insertion of LINE-1 repetitive sequence (see page 271) into *F8* gene in hemophilia A
Inhibit or prevent transcription	Fragile-X full mutation (*Table 10.6*)
Promoter mutation reducing mRNA levels	β-Globin –29 (A→G) mutation (*Table 16.4*)
Decrease mRNA stability	Hb-Constant Spring (page 423)
Inactivate donor splice site (*Figure 10.11*) (causing read-through into intron)	*PAX3* 451+1(G→T) mutation (*Figure 15.1*)
Inactivate donor or acceptor splice site (*Figure 10.11*) (causing exon to be skipped)	*PAX3* 452-2(A→G) mutation (*Figure 15.1*)
Activate cryptic splice site (*Figure 10.12*) (causes gain/loss of coding sequence)	β-Globin intron 1–110(G→A) mutation causing Mediterranean β-thalassemia (*Table 16.4*)
Introduce a frameshift in translation	*PAX3* 874ins(G) mutation (*Figure 15.1*)[a]
Convert a codon into a stop codon	*PAX3* Q254X mutation (*Figure 15.1*)
Replace an essential amino acid	*PAX3* R271C mutation (*Figure 15.1*)
Prevent post-transcriptional processing	Cleavage-resistant pro-insulin in familial hyperproinsulinemia
Prevent correct cellular localization of product	ΔF508 mutation in cystic fibrosis

See *Table 10.4* for a classification of mutations by their nature and location in the gene. [a]The *PAX3* 874ins(G) mutation introduces a seventh G into a run of six Gs; it has arisen independently more than once and illustrates the relatively high frequency of slipped-strand mispairing (page 265).

Deletions and insertions

The effects of small deletions or insertions depend partly on whether or not they generate frameshifts (i.e. whether the number of nucleotides added or removed is a multiple of 3). For example, two-thirds of mutations in the dystrophin gene are deletions of one or more exons, but some of these deletions cause the lethal condition Duchenne muscular dystrophy (DMD), whereas others cause the milder condition Becker muscular dystrophy (BMD). The explanation does not lie in the size of the deletions (some BMD deletions are larger than some causing DMD), nor in their position within the dystrophin gene (BMD and DMD deletions overlap), but in whether or not they create a frameshift (Koenig *et al.*, 1989). The dystrophin gene consists of 79 small exons (*Figure 7.11*). When exons are classified according to the reading frame position of their boundaries, it emerges that almost all deletions causing DMD involve frameshifts, whereas almost all BMD deletions are frame neutral (*Figure 15.2*). Supporting this interpretation, Western blots show no detectable dystrophin in muscle biopsies of DMD patients, but dystrophin of reduced molecular weight in BMD patients.

Splicing variants

Abolition of a splice site or activation of a cryptic splice site (*Figure 10.11*) normally abolishes gene function. Exons are skipped, or intronic sequence retained in the mature mRNA. Such changes often introduce frameshifts which exacerbate their effects. Occasionally, function is rescued by alternative splicing. Alternative splicing sometimes rescues other types of mutation as well. This may be the explanation for the 1.5% of patients with dystrophin gene deletions whose disease severity (DMD or BMD) does not fit the frameshift hypothesis described above. There is also evidence (Dietz and Kendzior, 1994) that nonsense mutations can trigger skipping of the exon involved, although the mechanism of this effect is totally obscure.

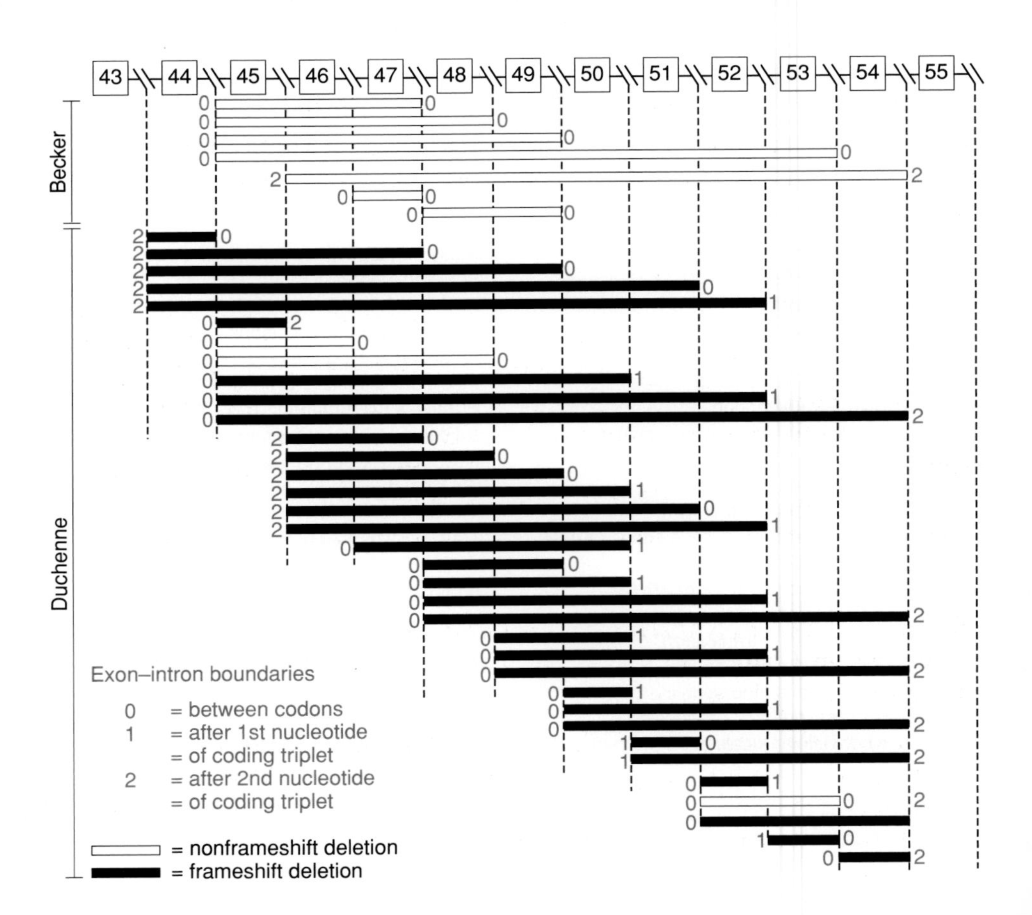

Figure 15.2: Deletions in the central part of the dystrophin gene are associated with Becker and Duchenne muscular dystrophy.

Numbered boxes represent exons 43–55. See *Figure 9.18* and *Box 9.2* (page 225) for more detail on exon/intron boundaries, and *Figure 16.7* to see how these deletions are characterized in the laboratory.

Amino acid substitutions

Without biochemical experiments, predicting the effect of a missense mutation is difficult. Substitutions are more likely to cause loss of function if they:

- involve a nonconservative amino acid change (see *Box 10.2*, page 246).
- Affect a part of the protein known to be functionally important. For example, the substitutions in *Figure 15.1*, all of which cause loss of function, are concentrated in the *paired domain* and *homeodomain* of the PAX3 protein, whereas the truncating mutations are distributed more widely across the gene.
- Involve an amino acid which is conserved across species or between members of a gene family (page 185).

Effects on transcription, RNA processing and translation

When thinking about the possible effect of a mutation in a coding sequence, it is important to consider all stages of gene expression. A change causing an apparently innocuous amino acid substitution or even a synonymous codon substitution may abolish a splice site or activate a cryptic site (see *Figure 10.11*). The amino acid substitution A196T in the *PAX3* gene (*Figure 15.1*) was found in a patient with Waardenburg syndrome, indicating that it abolished the function of the PAX3 protein. This was unexpected because A196T lies outside the known functionally important parts of the PAX3 protein and is not a highly conserved residue. Inspection of the genomic DNA sequence, however, shows that the nucleotide changed is the last nucleotide of an exon. 3′ Nucleotides of exons are usually G (*Figure 1.15*), and a G $\rightarrow$ C substitution at this position (as in A196T) is known to affect splicing efficiency (Krawczak *et al.*, 1992). Another example is the β-globin mutation D26K in hemoglobin E. This causes β-thalassemia, an unexpected clinical effect. At the DNA level, codon 26 is close to a donor splice site in codon 30 (*Figure 1.19*), and the G $\rightarrow$ A change reduces splicing efficiency.

Mutation of a gene is not the only way to abolish its function

Genes can lose (or gain) function not only by mutation, deletion or duplication, but also by inappropriate operation of any of the mechanisms controlling normal gene expression in cells (page 161).

Altered chromatin structure

Specific gene expression is partly controlled by long-range chromatin structures that allow or prevent transcription of genes within an extended domain (page 190). Chromosomal rearrangements, such as translocations or deletions, can alter the local chromatin structure in ways which are not currently well understood. We described three examples of loss of function caused by such effects (aniridia, campomelic dysplasia and fascio-scapular-humeral muscular dystrophy) in Chapter 7, page 174. Some of the diseases caused by expanded trinucleotide repeats (*Table 10.6*) also probably result from inactivation of a gene by alteration of a chromatin domain.

Molecular pathology of triplet repeat diseases

As we saw in Chapter 10 (page 266), pathogenic unstable trinucleotide repeats appear to fall into three groups.

- Moderately expanded $(CAG)_n$ repeats within coding regions encode expanded polyglutamine tracts in proteins, and these cause an unidentified gain of function (see the discussion of Huntington disease above, page 404). Possibly they cause proteins to oligomerize (Stott *et al.*, 1995).

- The massive expansions of $(CGG)_n$ in the fragile X syndromes FRAXA and FRAXE cause loss of function of nearby genes. The immediate cause of FRAXA fragile X syndrome is silencing of the *FMR1* gene by methylation of its promoter. Large runs of (CGG) in the 5′ untranslated region can also inhibit translation of mRNA (Feng *et al.*, 1995). Rare patients with typical fragile X syndrome have a normal sized CGG repeat but have intragenic deletions or point mutations in the *FMR1* gene, proving that the phenotype is caused by loss of function of that gene (see de Lugenbeel *et al.*, 1995). How expansion of the CGG repeat causes methylation of the promoter is not clear. Probably expansion alters both the replication timing and methylation status of a chromatin domain. The trinucleotide repeat is itself a target for methylation. The altered chromatin structure can be visualized under special culture conditions as the fragile site. Other folate-inducible chromosomal fragile sites such as *FRAXF*, *FRA16A* and *FRA11B* (Jones *et al.*, 1995) have similar expanded repeats. Such expansions are pathogenic only if important genes are located within the affected domain.
- The molecular pathology of myotonic dystrophy is not understood at present. A $(CTG)_n$ repeat in the 3′ untranslated part of a protein kinase gene (*DMPK*) undergoes massive expansion (*Table 10.6*). No other mutation has been found in any patient with myotonic dystrophy, which suggests a specific gain of function, but the nature of the gain is unknown. Although changes have been reported in transcription and mRNA processing of *DMPK* genes carrying an expanded repeat, the changes reported are small and inconsistent. Maybe the expansion affects adjacent genes. The region of chromosome 19 containing the *DMPK* gene is very gene-rich, and the expanded repeat lies within a CpG island probably controlling expression of nearby genes (Boucher *et al.*, 1995).

Disturbances of methylation and/or imprinting

Imprinted genes (page 70) are expressed from only one of the two homologous chromosomes, probably as a result of parent-specific methylation (page 170). Imprinting is a normal developmental process for a small number of genes, but molecular events involving imprinting can also lead to unusual molecular pathology, best illustrated by Prader–Willi and Angelman syndromes.

- Prader–Willi syndrome (PWS; MIM 176270) is characterized by hypotonia in infancy (floppy baby), moderate mental retardation, hypogenitalism in boys and, later in childhood, gross obesity caused by uncontrollable eating.
- Angelman syndrome (AS; MIM 105830) features severe mental retardation, growth retardation, jerky movements, inappropriate laughter and fair coloration.

Both conditions are usually associated with deletions of 15q12 that appear identical under the microscope. Molecular studies show that the deletions usually overlap, but rare small deletions define separate critical regions for PWS and AS. Several candidate genes have been cloned from these regions. Studies of the parental origin of affected chromosomes and of nondeletion cases have revealed a fascinating picture (*Figure 15.3*). 15q12 contains two adjacent but oppositely imprinted domains. Maternal and paternal imprints can be recognized by their different methylation patterns. Lack of expression of the paternally expressed domain causes PWS, and lack of expression of the maternally expressed domain causes AS. Chromosomal deletions, uniparental disomy and errors in the imprinting process can all cause lack of expression. In AS, but not PWS, familial cases occur, who have no obvious deletion or pattern of inappropriate methylation. They might have point mutations in the (unknown) AS gene, or maybe some as yet undefined disturbance of imprinting.

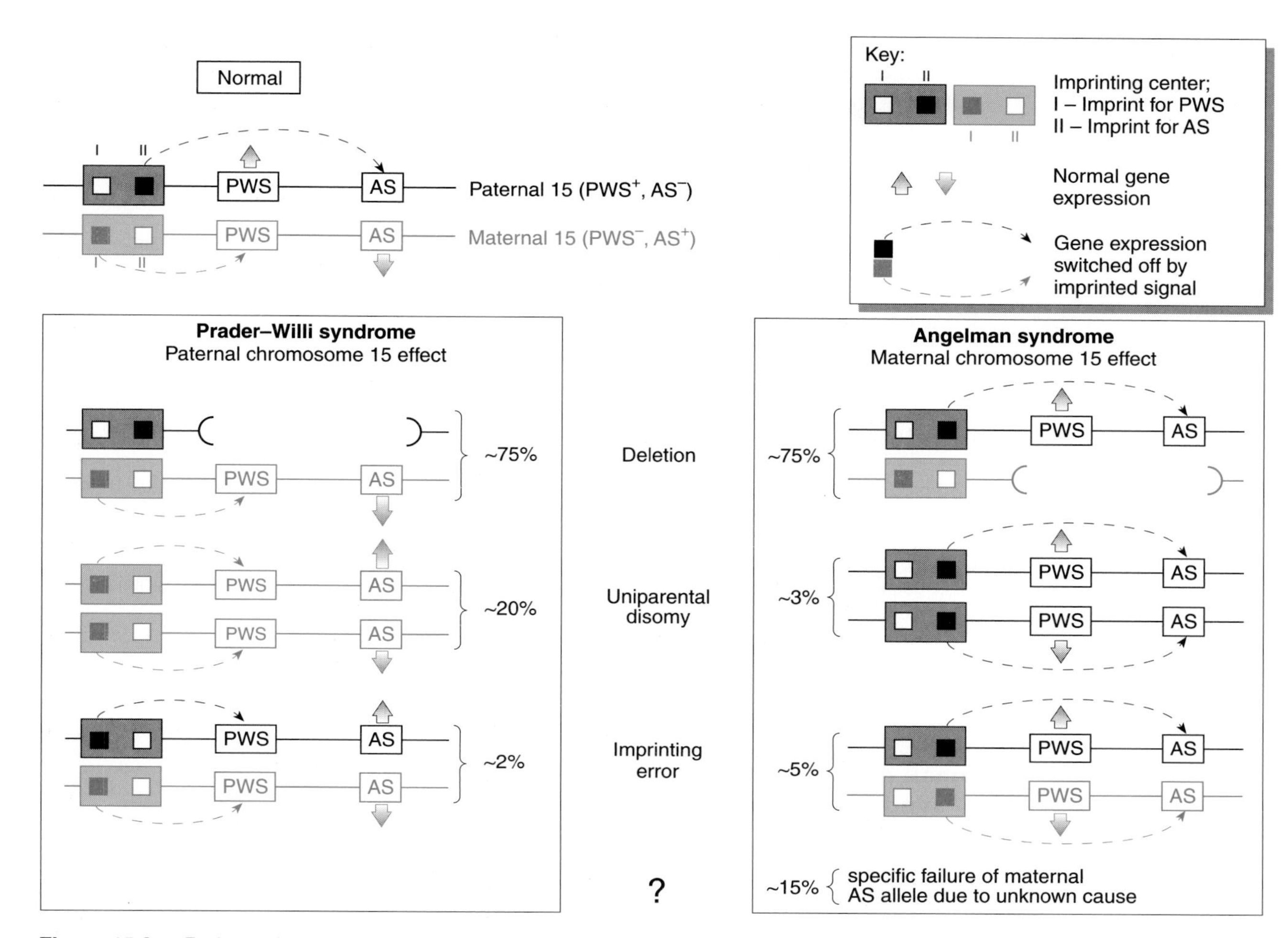

Figure 15.3: Pathogenic mechanisms in Prader–Willi (PWS) and Angelman (AS) syndromes.

PWS and AS are caused by lack of expression of their respective genes. These lie close together on chromosome 15. Deletions (usually encompassing both genes) are the commonest cause of either syndrome (top lines in boxes). However, only deletion or nonexpression of the paternal copy of PWS (black) or the maternal copy of AS (red) has any effect, because the other alleles are normally silenced by an imprinting center located some distance away (dashed arrows). If one homolog is missing becuase of uniparental disomy (middle lines in boxes), the effect is the same as deletion. Rarely, an imprinting error mimics the effect of uniparental disomy (bottom lines in boxes). Some cases of AS appear to result from specific nonexpression of the maternal AS gene, without the methylation pattern associated with a paternal imprint. See Lalande (1994) for further details.

Imprinting is controlled by a region some distance centromeric of the PWS and AS candidate regions, a long range control of chromatin domain structure reminiscent of the globin locus control region (pages 174 and 190).

An apparently similar pair of adjacent but oppositely imprinted domains at 11p15.5 includes the *IGF2* and *H19* genes (Chapter 7, *Table 7.11*), and disturbance of the normal pattern of expression probably causes Beckwith–Wiedemann syndrome (MIM 130650).

Loss-of-function mutations can produce dominant phenotypes through haploinsufficiency or dominant-negative effects

Dosage-sensitive loci

Loss-of-function mutations usually produce recessive phenotypes. For a limited number of genes, however, a 50% reduction in the dosage of the gene product leads to phenotypic changes. **Haploinsufficiency** is the term used to describe this situation. Homozygous loss usually results in much more severe effects but, as we saw in

Chapter 3 (page 61), it is still appropriate to describe the heterozygous phenotype as dominant. Abnormally high dosage at dosage-sensitive loci may also produce phenotypic effects. Dosage effects are discussed below (see page 416). *Table 15.3* lists some loci where dosage effects appear to operate.

Table 15.3: Phenotypes probably caused by single gene dosage effects

Gene product	Dosage w.r.t. normal	Phenotype resulting	MIM no. or ref.
C1 esterase inhibitor	50%	Angioneurotic edema	106100
Fibrillin	50%	Marfan syndrome	154700
PAX3 protein	50%	Waardenburg syndrome type 1	193500
PAX6 protein	50%	Aniridia	106210
LDL receptor	50%	Hypercholesterolemia	143890
PMP22 myelin protein	50%	Tomaculous neuropathy	162500
	150%	Charcot–Marie–Tooth neuropathy 1A	118220
DSS product	200%	46,XY females	Bardoni *et al.* (1994)

It can be difficult to prove that a phenotype depends on a pure gene dosage effect, and not all the examples in this table may stand the test of time.
DSS, dosage-sensitive sex reversal; LDL, low density lipoprotein; PMP, peripheral myelin protein.

Dominant-negative effects

Dominant-negative effects (see page 403) are discussed below (see page 414 and *Figure 15.7*).

Small deletions on the X chromosome produce contiguous gene syndromes in males

The X chromosome contains 140 Mb of DNA (*Table 7.3*) and maybe 2000 genes. Deletions at or below the limit of microscopic visibility (<5 Mb) may remove strings of adjacent genes and, in males, produce well-defined **contiguous gene syndromes** (*Figure 15.4*). Affected males show superimposed features of several different X-linked mendelian diseases. A classic case was the boy 'BB' who suffered from DMD, chronic granulomatous disease and retinitis pigmentosa, together with mental retardation (Francke *et al.*, 1985). He had a chromosomal deletion in Xp21 that removed a series of contiguous genes (and provided investigators with the means to clone the dystrophin and cytochrome b_{245} genes; see *Figure 14.9*). Deletions of the tip of Xp are seen in another set of contiguous gene syndromes. Successively larger deletions remove more genes and add more diseases to the syndrome (Ballabio and Andria, 1992). Microdeletions are relatively frequent in some parts of the X chromosome (e.g. Xp21, proximal Xq) but rare or unknown in others (e.g. Xp22.1–22.2, Xq28). No doubt deletion of certain individual genes, and visible deletions in gene-rich regions, would be lethal.

Comparably sized deletions on autosomes rarely produce clear-cut contiguous gene syndromes, because in a heterozygous deletion carrier only the few dosage-sensitive genes would affect the phenotype (*Figure 15.4*). However, such microdeletions are proving to underlie an increasing number of well-known syndromes. They are discussed below (page 418 and *Table 15.7*).

Mitochondrial diseases can be caused by loss-of-function mutations in mtDNA, but the relationship between genotype and phenotype is complex

Mitochondrial diseases (page 65) can be caused by point mutations, deletions and duplications which abolish the function of genes in the densely packed mitochondrial

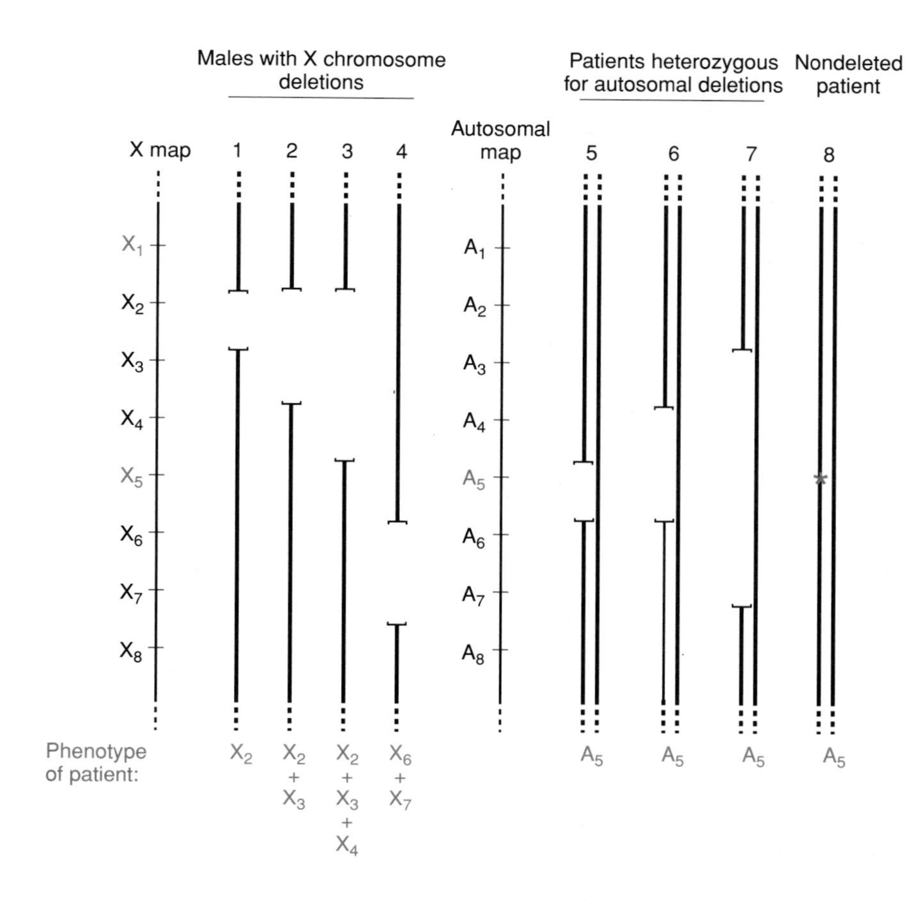

Figure 15.4: X-linked and autosomal microdeletion syndromes.

On the X chromosome, deletion of genes X_1 or X_5 is lethal in males. Patients 1–3 show a nested series of contiguous gene syndromes, patient 4 a nonoverlapping contiguous gene syndrome. On the autosome, only gene A_5 is dosage sensitive. Patients 5–7, with different sized deletions, all show the same phenotype as patient 8, who is heterozygous for a loss-of-function point mutation in the A_5 gene.

genome (*Figure 7.2*). Cells typically contain thousands of mtDNA molecules. An important aspect of molecular pathology is whether every mtDNA molecule carries the causative mutation (*homoplasmy*), or whether the cell contains a mixed population of normal and mutant mitochondria (*heteroplasmy*; see *Box 10.3*)

In some cases (for example the nucleotide 11 778 mutation responsible for 60–70% of Leber's hereditary optic atrophy), some patients are heteroplasmic while others appear homoplasmic for the same mutation. The phenotype may depend on the proportion of abnormal mtDNA in some critical tissue. This proportion can be very different in mother and child because of the random segregation of mtDNA molecules at cell division.

Mitochondrial mutations appear to evolve within individuals. The same individual can carry both deletions and duplications, and the proportion can change with time (Poulton, 1995). Variants tend to be more frequent in older people, and an accumulation of mitochondria deficient in energy production may be a factor in aging. Some mitochondrial diseases appear to be of a quantitative nature: small mutational changes accumulate which reduce the energy-generating capacity of the mitochondrion and, at some threshold deficit, clinical symptoms appear. External agents may precipitate disease in a susceptible individual – for example, a mitochondrial 12S RNA sequence variant appears to make individuals susceptible to hearing loss after treatment with aminoglycoside antibiotics. The same variant can interact with an unidentified autosomal locus to produce congenital hearing loss even in people who have never been given aminoglycoside antibiotics (Prezant *et al.*, 1993).

Not all mitochondrial abnormalities are caused by mutations in the mitochondrial genome. Many mitochondrial components are encoded by nuclear genes (see *Box 7.1*) and nuclear genes control mitochondrial processes.

Gain-of-function mutations

Gain-of-function mutations produce dominant phenotypes by a variety of mechanisms (*Table 15.4*; see Wilkie, 1994). Gains of truly novel functions are very unusual except in cancer (see below); in inherited diseases, gain of function usually means that the gene is expressed at the wrong time in development, in the wrong tissue, in response to the wrong signals, or at an inappropriately high level.

Table 15.4: Mechanisms of gain-of-function mutations

Malfunction	Gene	Disease	MIM no.
Acquire new substrate	*PI* (Pittsburgh allele)	α_1- Antitrypsin deficiency	107400
Overexpression	*PMP22*	Charcot–Marie–Tooth disease	118200
Receptor permanently 'on'	*GNAS1*	McCune–Albright	174800
Ion channel inappropriately open	*SCN4A*	Paramyotonia congenita	168300
Structurally abnormal multimers	*COL2A1*	Osteogenesis imperfecta	Various
Chimeric gene	*BCR–ABL*	Chronic myeloid leukemia	151410
Unknown	*HD*	Huntington disease	143100

Acquisition of a novel function is rare in inherited disease but common in cancer

A rare case of an inherited point mutation conferring a truly novel function on a protein is the Pittsburg allele at the *PI* locus (*Figure 15.5;* Owen *et al.*, 1983). Generation of novel genes by an exon-shuffling mechanism is commonly seen in cancer cells. Many tumor-specific chromosomal rearrangements produce chimeric genes which combine functions of two different genes in a way that leads to uncontrolled cell proliferation (see *Table 17.3*).

Overexpression of a normal gene product may be pathogenic

Large increases in gene dosage (gene amplification) are a common mechanism by which proto-oncogenes are activated in cancer cells (see page 462) but, in inherited disease, increased copy number of a gene is not a common pathogenic mechanism. Most genes are not dosage sensitive. For example, people have varying numbers of green color vision pigment genes at Xq28, and of human leukocyte antigen (HLA)-DRβ genes at 6p21, but these are normal nonpathological polymorphisms. One example of an inherited disease caused by increased gene dosage is Charcot–Marie–Tooth disease 1A (CMT1A; also known as hereditary motor and sensory neuropathy 1A, HMSN1; MIM 118220). As *Figure 15.6* shows, overexpression of the *PMP22* gene by increased copy number or activating mutations causes CMT1A, whereas underexpression causes a different phenotype, hereditary neuropathy with pressure palsies or tomaculous neuropathy (MIM 162500). The clinical effects are thought to be due to an abnormal balance of components in the myelin of peripheral nerves.

Qualitative changes in a gene product can cause gain of function

Although gains of truly novel functions are very rare in inherited disease, mutations modifying the level, timing or tissue specificity of activity of a protein can produce dominant gain-of-function phenotypes.

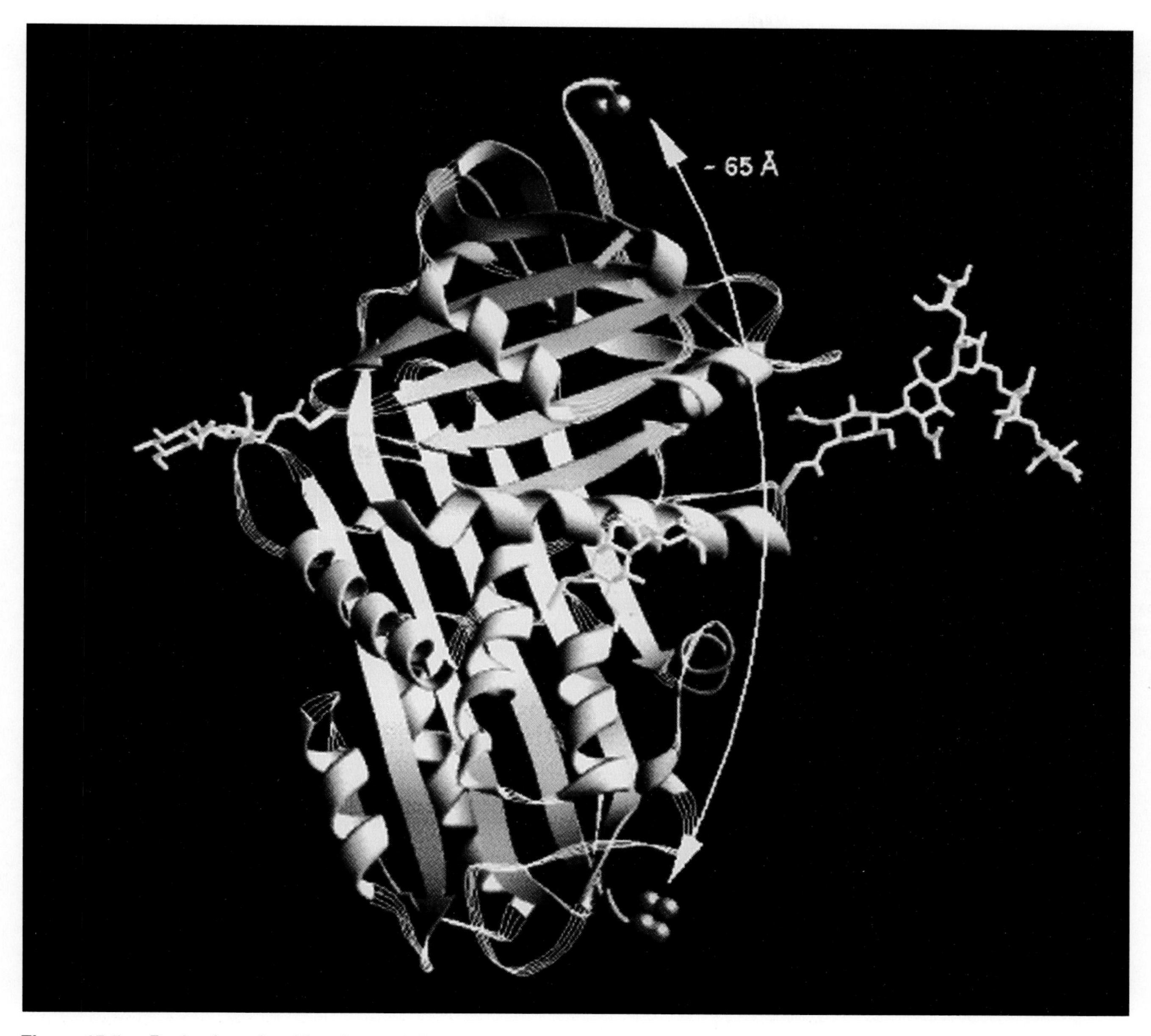

Figure 15.5: Basis of a gain-of-function mutation.

The α_1-antitrypsin molecule inhibits elastase. Methionine 358 in the reactive center acts as a 'bait' for elastase; when the peptide link between Met358 and Ser359 is cleaved, the two residues spring 65 Å apart, as shown here. Elastase is trapped and inactivated. The Pittsburgh variant has a missense mutation M358R which replaces the methionine bait with arginine. This destroys affinity for elastase but creates a bait for thrombin. As a novel constitutively active antithrombin, the Pittsburgh variant produces a lethal bleeding disorder. Image from University of Geneva ExPASy molecular biology World-Wide Web server.

Inappropriately open ion channels are the cause of the muscle disease paramyotonia congenita (MIM 168300). The *SCN4A* gene encodes the α-subunit of the adult muscle sodium ion channel. Three mutations G1306V, T1313M and V1589M all affect highly conserved residues which help close the channel. The mutations do not leave the channel permanently open (probably that would be lethal), but they make it sluggish at low temperatures, so that in the cold it stays open too long. People heterozygous for any of these mutations cannot make their muscles work when they are cold. A different set of mutations in the same *SCN4A* gene lead to another mendelian condition, hyperkalemic periodic paralysis [MIM 170500; see McClatchey *et al.* (1992) for a discussion of the molecular pathology].

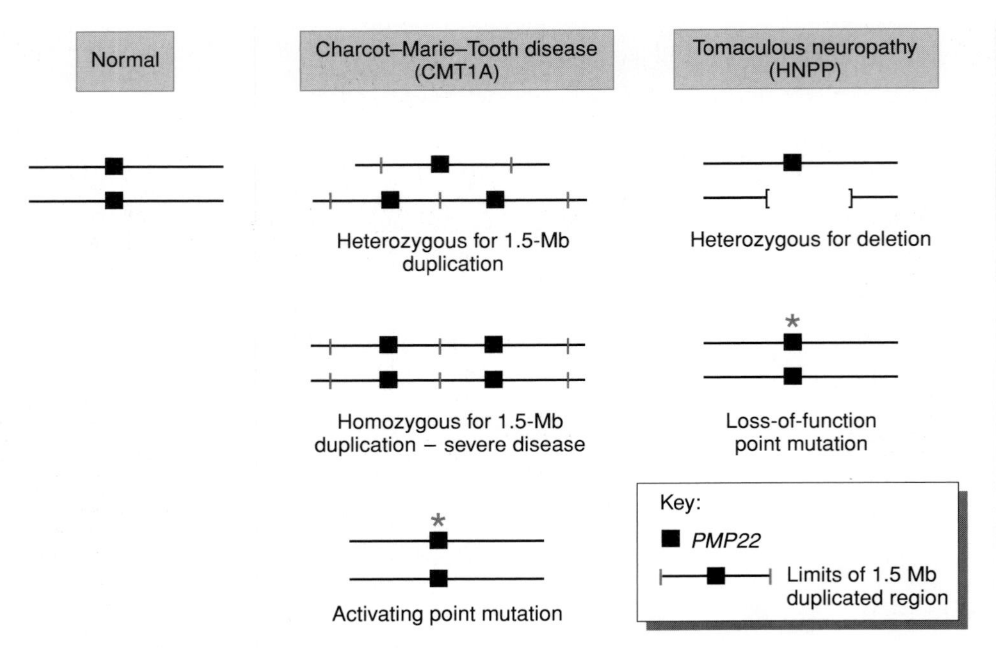

Figure 15.6: Gene dosage effects with the *PMP22* gene.

Most patients with Charcot–Marie–Tooth disease are heterozygous for a 1.5-Mb duplication at 17p11.2, including the gene for peripheral myelin protein, *PMP22* (black square). A patient homozygous for the duplication had very severe disease. Some patients have only two copies of the *PMP22* gene, but one copy carries an activating mutation. Deletion or loss-of-function mutation of the *PMP22* gene is seen in patients with tomaculous neuropathy (Patel and Lupski, 1994).

In McCune–Albright syndrome or polyostotic fibrous dysplasia (PFD; MIM 174800), there is an abnormality in one of the G-proteins that couple cell surface receptors to the adenylyl cyclase intracellular signal transduction system (Weinstein *et al.*, 1991). Mutation of the *GNAS1* gene results in an overactive G-protein subunit and an inappropriate response to extracellular signals. This results in a variety of endocrine abnormalities such as precocious puberty. PFD is known only as a somatic condition in mosaics – probably constitutional mutations would be lethal. Very similar mechanisms, involving uncontrolled cell proliferation, are common in cancer (Chapter 17). Loss-of-function mutations of the same gene often underlie a different disease, Albright's hereditary osteodystrophy (*Table 15.10*).

Mutations in components of multimeric protein complexes can have dominant-negative effects

Collagen mutations

As mentioned above, mutations affecting collagens often show classic dominant-negative effects. Collagens, the typical structural proteins of connective tissue, form a large family of related proteins. Collagens are built of triple helices of polypeptide chains. Some are homotrimers, others heterotrimers. In most but not all collagens, the triple helical molecules are assembled into close-packed cross-linked arrays to form rigid fibrils. A mutant collagen molecule that packs wrongly causes more problems than one that is simply not recognized as a collagen. Gene deletions or major structural abnormalities cause at worst a 50% reduction in the yield of functional molecules. However, a single chain which packs wrongly into the triple helix can wreck the structure (*Figure 15.7*). The term '**protein suicide**' has been used to describe this dominant-negative effect. The clinical consequences of collagen mutations are discussed below (page 420).

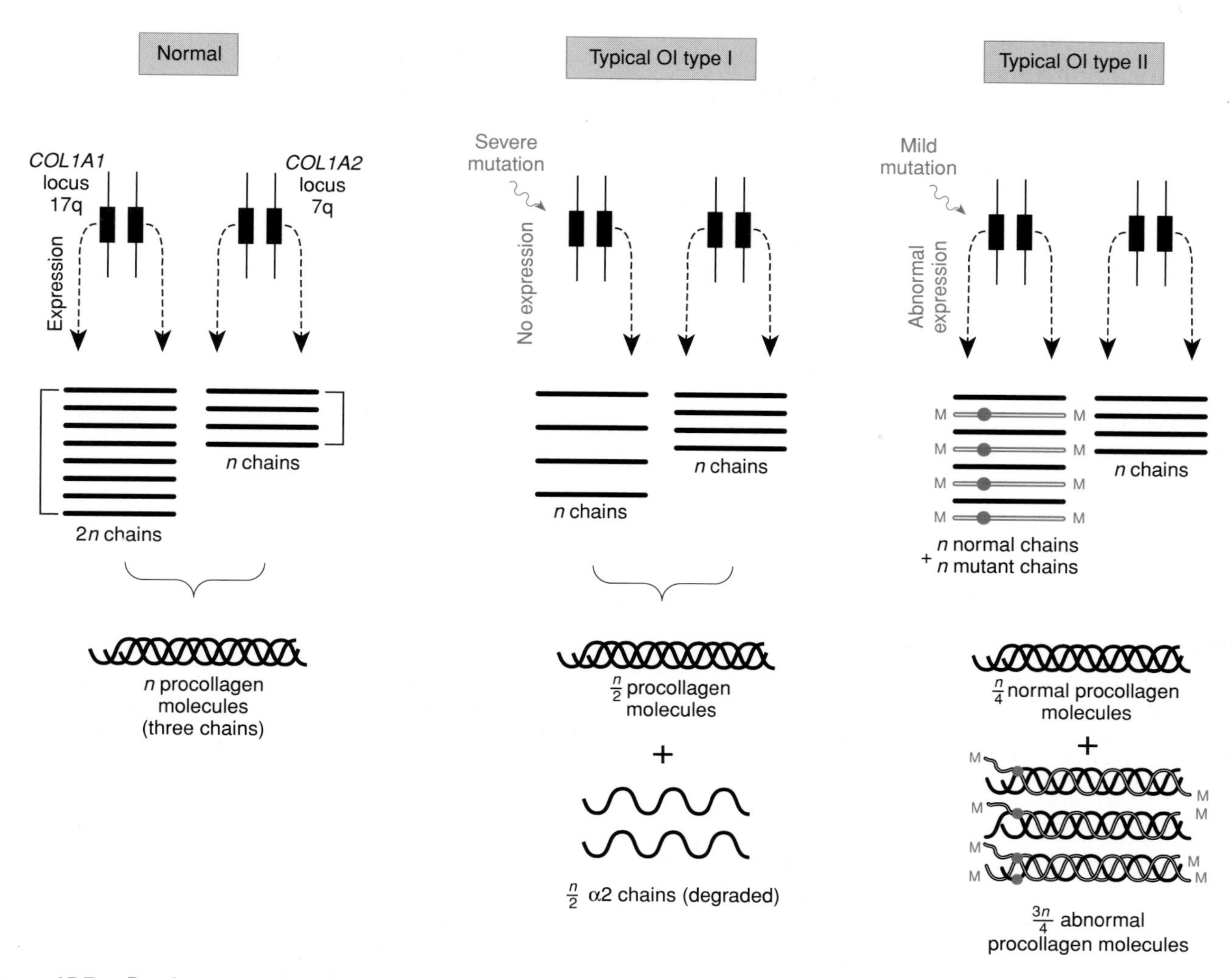

Figure 15.7: Dominant-negative effects of collagen gene mutations.

Collagen fibrils are built of arrays of triple-helical procollagen units. The type I procollagen comprises two chains encoded by the *COL1A1* gene and one encoded by *COL1A2*. Null mutations in either gene have a less severe effect than mutations encoding polypeptides which cause the triple helix to be nonfunctional. See *Table 15.8* for details.

Other dominant-negative mutations

Many nonstructural proteins also show dominant negative effects. We shall mention two examples. Helix–loop–helix and leucine zipper transcription factors bind DNA as dimers (*Figure 7.10*). Mutants defective in dimerization often cause recessive phenotypes, but mutants which are able to sequester functioning molecules into inactive dimers give dominant phenotypes (Hemesath *et al.*, 1994). Cell surface tyrosine kinase receptors provide a second example. These respond to binding of their ligand by dimerizing, which triggers autophosphorylation and transmission of the signal into the interior of the cell. One cause of Hirschsprung's disease is a dominant-negative loss-of-function mutation in the tyrosine kinase encoded by the *RET* gene. Mutant receptor molecules form nonfunctional dimers (Pasini *et al.*, 1995).

Gene dosage effects

Mutations, as we have seen, can be classified into gain of function and loss of function categories. Loss of function at some loci will become evident at, say, 50% of the normal level of product, whilst in others effects first emerge when the level of product drops to 10% or 5% of normal. The amount of product needed for normal function may differ in different cells, so that a range of phenotypes may be seen, depending on the level of functional gene product. This is well illustrated by the effect in males of loss-of-function mutations in the X-linked gene for the enzyme HPRT (hypoxanthine guanine phosphoribosyl transferase; *Table 15.5*).

Table 15.5: A dosage-sensitive phenotype: effects of decreasing levels of hypoxanthine adenine phosphoribosyl transferase on phenotype (Sege-Peterson *et al.*, 1993)

HPRT activity (% of normal)	Phenotype
>60	Normal
8–60	Neurologically normal, hyperuricemia (gout)
1.6–8	Neurological problem (choreoathetosis)
1.4–1.6	Lesch–Nyhan syndrome (self-mutilation, choreoathetosis) but intelligent
<1.4	Classical Lesch–Nyhan syndrome (MIM 308000: self-mutilation, choreoathetosis and mental retardation)

Loci where a 50% reduction has phenotypic effects are sensitive to gene dosage. Presence of only one normal and one nonfunctional copy of the gene results in a phenotypic effect. Loss-of-function mutations at such loci are dominant, at least if they cause total loss of function. Some loci are also sensitive to moderately increased gene dosage.

Chromosomal aneuploidies demonstrate the importance of correct gene dosage

Having the wrong number of chromosomes has serious, usually lethal, consequences (*Table 15.6*). Even if each individual chromosome is in itself normal, chromosomes have to be present in the correct numbers. The extra chromosome 21 in a man with Down syndrome is a perfectly normal chromosome, inherited from a normal parent. The syndrome must be caused by the 50% greater dosage of

Table 15.6: Consequences of numerical chromosome abnormalities

Polyploidy		
Triploidy (69,XXX, XXY or XYY) 2% of all conceptions; almost never liveborn; do not survive		
Aneuploidy		
Autosomes	*nullisomy* (missing a pair of homologs)	
		Pre-implantation lethal
	monosomy (one chromosome missing)	
		Embryonic lethal
	trisomy (one extra chromosome)	
		Usually embryonic or fetal lethal
		Trisomy 13 (Patau syndrome) and trisomy 18 (Edward syndrome) may survive to term
		Trisomy 21 (Down syndrome) may survive to age 40
Sex chromosomes		
	XXX, XXY, XYY	Relatively minor problems, normal lifespan
	45,X	99% abort spontaneously; survivors are of normal intelligence but infertile and show minor physical signs

See pages 52–53 for nomenclature and mechanisms of origin of chromosome abnormalities.

chromosome 21. Monosomies have even more catastrophic consequences than trisomies. Having the wrong number of X chromosomes has only minor effects, compared with autosomal aneuploidy, because X inactivation (page 172) ensures dosage compensation. Unbalanced chromosome structural abnormalities (page 54) have phenotypic effects very roughly proportional to the amount of extra or missing material. Chromosome abnormalities clearly demonstrate the crucial importance of correct gene dosage, but do not explain why gene dosage should be so critical.

Dosage effects may be seen where a gene product interacts with other proteins or DNA sequences

One might reasonably ask why there should be dosage sensitivity for any gene product. Why has natural selection not managed things better? If a gene is expressed so that two copies produce a barely sufficient amount of product, selection for variants with higher levels of expression should lead to the evolution of a more robust organism with no obvious price to be paid. The answer is that, in most cases, this has indeed happened – relatively few genes are dosage sensitive. However, certain functions are inherently dosage sensitive (see Fisher and Scambler, 1994). These include:

- gene products which are part of a quantitative signaling system whose function depends on partial or variable occupancy of a receptor, DNA-binding site, etc;
- gene products which compete with each other to determine a developmental or metabolic switch;
- gene products which co-operate with each other in interactions with fixed stoichiometry (such as the α- and β-globins and many structural proteins).

We see that it is gene interactions which lie at the root of dosage sensitivity. In each case the gene product is titrated against something else in the cell, and so changes in gene dosage cause malfunctions. What matters is not the correct absolute level of a gene product, but the correct *relative levels* of interacting products. Genes whose products act essentially alone, such as many soluble enzymes of metabolism, seldom show dosage effects. Because of the interactions, pathogenic effects caused by gene dosage are subject to modification by changes elsewhere in the genome, so that the phenotypes often show variable expression, even within families.

'Chromosomal' phenotypes

An experienced physician can often guess that a patient has a chromosomal abnormality, without being able to guess which chromosome is involved. Nearly always there are multisystem developmental abnormalities and mental retardation, usually growth retardation, fits and a dysmorphic facial appearance. Even specific features such as low-set ears are strikingly frequent. Since embryonic development and brain function are the two human processes which depend on the greatest numbers of genes (which is why fetal brain cDNA libraries are so often used for screening), it seems plausible that these functions should be the most vulnerable to random disturbances of gene dosage. Nevertheless, it is curious that the phenotypes have so much in common, regardless of the particular chromosomal region involved. It is an interesting question whether this simply reflects the involvement of a number of individual dosage-sensitive loci, or whether some other mechanism might operate at the chromosomal level.

The problem of triploidy

If relative, rather than absolute gene dosage is the critical factor, then it is not obvious why triploidy (page 51) is lethal in humans and other animals. With three copies of every autosome, the dosage of autosomal genes is balanced and should not cause problems. Triploids are always sterile (because triplets of chromosomes cannot pair and segregate correctly in meiosis), but many triploid plants are in all other respects healthy and vigorous. The lethality in animals may be explained by X–autosome dosage imbalance, together perhaps with incorrect dosage of imprinted genes (page 70).

Chromosomal microdeletions

An increasing number of clinical syndromes are proving to be caused by *microdeletions* or occasionally microduplications (*Table 15.7*). Typically, microdeletion syndromes involve complex but recognizable phenotypes, and usually they occur sporadically, with no previous family history. Once the cause is recognized, further cases can be diagnosed easily by FISH (page 280) using a probe from the deleted region. As discussed above (page 410; *Figure 15.4*), autosomal microdeletions rarely produce well-defined contiguous gene syndromes like X chromosome microdeletions, because only dosage-sensitive genes contribute to the phenotype (except for rare cases where the deletion uncovers a recessive point mutation on the homolog). However, in a few cases, the effects of individual genes can be teased out. Williams syndrome is one such example.

People with Williams syndrome (MIM 194050) are moderately mentally retarded but speak fluently with a characteristic 'cocktail party' manner. They have a recognizable face, they are growth retarded, as infants they may have life-threatening hypercalcemia, and they often have a heart defect, supravalvular aortic stenosis (SVAS). SVAS also occurs as an isolated mendelian dominant condition, which was mapped to 7q11 and shown to result from disruption of one copy of the elastin gene. In Williams syndrome, a microdeletion at 7q11 removes the elastin gene. Hemizygosity for elastin accounts for the SVAS and probably much of the facial appearance in Williams syndrome, but not the hypercalcemia or mental retardation. These are likely to be caused by lack of expression of other adjacent genes.

Table 15.7: Syndromes often caused by autosomal chromosomal microdeletions

Syndrome	MIM no.	Chromosomal anomaly
Wolf–Hirschhorn	194190	Deletion of 4p16.3
Cri du chat	123450	Deletion of 5p15.2–p15.3
Williams	194050	Deletion of 7q11.23 including the elastin gene
WAGR (Wilms tumor, aniridia, genital anomalies, growth retardation)	194072	Deletion of 11p13 including *WT1* and *PAX6* genes
Prader–Willi	176270	Lack of paternal genes at 15q11–q13
Angelman	105830	Lack of maternal genes at 15q11–q13
Rubinstein–Taybi	180849	Deletion of 16p13.3
Miller–Diecker lissencephaly	247200	Deletion of 17p13.3
Smith–Magenis	182290	Deletion of 17p11.2
Alagille	118450	Deletion at 20p12.1–p11.23
Di George, velocardiofacial, Schprinzen ('Catch 22')	192430	Deletion at 22q11.21–q11.23

In some syndromes, such as Alagille syndrome, only a minority of patients have a deletion, suggesting that loss of function of a single dosage-sensitive gene causes the phenotype. In other cases, such as Williams syndrome, all patients with the full syndrome have a microdeletion, suggesting that more than one locus is involved in generating the phenotype.

The relationship between mutations and syndromes

Locus heterogeneity is common in syndromes that result from failure of a complex pathway

Although the one gene–one enzyme hypothesis of Beadle and Tatum proved the key to understanding biochemical genetics back in the 1940s, it would be a mistake to think of human genetics in terms of one gene–one disease. Diseases and developmental defects represent the failure of a pathway, usually involving many genes. A defect in any of the genes in a pathway may lead to similar clinical results.

Profound congenital hearing loss is often genetic, and when genetic it is usually autosomal recessive – but when two deaf people marry, as they often do, the children usually have normal hearing (*Figure 15.8*). This is an example of **complementation** (page 65). The fact that the children are normal suggests that the parents have defects in different genes. Such **locus heterogeneity** should be distinguished from **allelic heterogeneity** (many different mutations within a given gene produce the disease, the situation illustrated in *Figure 15.1*). It is easy to imagine that there would be many different genes involved in making so exquisite a machine as the cochlear hair cell work. Complementation is not, however, a wholly reliable pointer to nonallelism. Alleles at the same locus can occasionally complement each other (**interallelic complementation**), for example in ataxia telangiectasia (page 473).

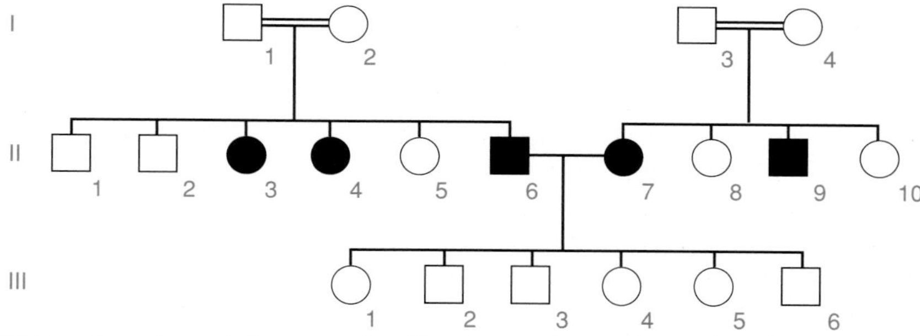

Figure 15.8: Complementation: parents with autosomal recessive profound hearing loss often have children with normal hearing.

II_6 and II_7 are offspring of unaffected but consanguineous parents, and each has affected sibs, making it likely that each has autosomal recessive hearing loss. All their children are unaffected, showing that II_6 and II_7 have nonallelic mutations. This is an example of complementation (see *Box 3.2*, page 66).

Locus heterogeneity is only to be expected in conditions like deafness, blindness or mental retardation, where a rather general pathway has failed; but even with more specific pathologies, multiple loci are very frequent. A striking example is Usher syndrome, an autosomal recessive combination of hearing loss and retinitis pigmentosa. Mutations of at least six unlinked loci can each cause Usher syndrome: *USH1A* (14q31–qter), *USH1B* (11q13–q14) USH1C (11p13–p14), *USH2A* (1q32–q34), *USH2B* (not mapped) and *USH3* (3q).

Mutations often affect only a subset of the tissues in which the gene is expressed

The pattern of tissue-specific expression of a gene is a poor predictor of the clinical effects of mutations. Tissues where the gene is not expressed are unlikely to suffer primary pathology, but the converse is not true. Usually only a subset of expressing tissues are affected. The *HD* gene is widely expressed, but HD affects only limited regions of the brain. The retinoblastoma (*RB1*) gene (page 466) is ubiquitously expressed, but only the retina is affected by inherited mutations. This is also strikingly seen in the lysosomal disorders. Gene expression is required in a single cell type, the macrophage, which is found in many tissues, but not all macrophage-containing tissues are abnormal in affected patients.

This reminds us that simple correlations between a genotype at a locus and the phenotype of a person are the exception, not the rule. This holds both at the level of cells and tissues and for the whole person. The genome contains maybe 70 000 genes (page 154) but only about 6000 mendelian phenotypes are known. Even if half of all genes produce no effect or lethality when mutated, the majority of all potentially pathogenic mutations in the remainder must be involved in complex 'multifactorial' phenotypes. The mendelian phenotypes differ only in that there is at least one tissue or organ which does show simple effects, but again this is likely to be the exception among tissues or organs rather than the rule.

Collagen defects illustrate the complex relationship between tissue-specific syndromes and tissue-specific genes

Over 150 different osteochondrodysplasias (inherited abnormalities of bone development) have been described (International Working Group, 1992). Any or all of these might be caused by collagen abnormalities. *Table 15.8* lists a very small selection, and also a few selected abnormalities of skin and joints, also candidate collagen abnormalities. This is a clinical classification, based on the combined experience of eminent physicians who have examined innumerable patients and their families over many years. *Table 15.9* lists the syndromes which result from a selection of mutations in collagen genes. It is evident that the clinical and molecular classifications are very different.

A first step in understanding the molecular pathology is to consider the structure and anatomical roles of different types of collagen. At least 18 types of collagen

Table 15.8: Clinical classification of connective tissue diseases

Disease	MIM no.	Features	Inheritance
OI type I (divided on dental findings into IA, IB, IC)	166200 166240	Mild–moderate bone fragility; blue sclerae; normal stature; hearing loss (50%)	AD
OI type II	166210 (259400)	Very severe bone fragility; perinatal lethal	AD Rarely AR
OI type III	(166230) (259420)	Moderate–severe bone fragility; progressive deformity; very short stature; often hearing loss	AD Rarely AR
OI type IV (divided on dental findings into IVA, IVB)	166220	Mild–moderate bone fragility; normal sclerae; variable stature	AD
SED	183900	Short stature, short neck, SED	AD
Stickler syndrome	108300	Mild SED, cleft palate, high myopia, hearing loss	AD
Kniest dysplasia	156550	Disproportionate short stature; short neck; SED, etc.	AD
EDS type IV	130050	Fragile bruisable skin, ruptured arteries	AD
EDS type VII	130060	Lax joints and skin	AD
DEB (Hallopeau–Siemens)	226600	Severe mucocutaneous blistering	AR
Alport syndrome	301050 104200 203780	Nephritis with hearing loss	XL AD? AR

This table lists only a small selection of conditions, chosen for purposes of comparison with the genetic classification in *Table 15.9*.
AD, autosomal dominant; AR, autosomal recessive; DEB; dystrophic epidermolysis bullosa; EDS, Ehlers–Danlos syndrome; OI, osteogenesis imperfecta; SED, spondyloepiphyseal dysplasia; XL, X-linked.
For a much fuller list and descriptions see Royce and Steinmann (1993).

Table 15.9: Consequences of mutations in collagen genes

Gene	Location	Mutations	Syndrome
COL1A1	17q22	Null alleles	OI type I
		Partial deletions; C-terminal substitutions	OI type II
		N-terminal substitutions	OI types I, III or IV
		Deletion of exon 6	EDS type VII
COL1A2	7q22.1	Splice mutations; exon deletions	OI type I
		C-terminal mutations	OI type II, IV
		N-terminal substitutions	OI type III
		Deletion of exon 6	EDS type VII
COL2A1	12q13	Point mutations	SED
		Nonsense mutation	Stickler syndrome
		Defect in conversion	Kniest dysplasia
		Missense	Achondrogenesis II, spondylo-meta-epiphyseal dysplasia
COL3A1	2q31	Deletions, point mutations	EDS type IV
COL4A3	2q36	Deletions, nonsense mutations	AR Alport syndrome
COL4A4	2q36	Missense and nonsense mutations	AR Alport syndrome
COL4A5	Xq22	Deletions, substitutions, splicing mutations	XL Alport syndrome
COL5A1	9q34	Splicing mutations	EDS type I/II
COL7A1	3p21.3	Missense mutations	Dominant DEB
		Frameshifts	Recessive DEB
COL10A1	6q21–q22	Intragenic deletion, nonsense mutations	Schmid metaphyseal chondrodysplasia
COL11A2	6p21.3	Splicing mutation	Stickler syndrome

This is a partial list, made for purposes of comparison with the clinical classification in *Table 15.8*. See Byers (1993) for detail of *COL1A1* and *COL1A2* mutations, which have been simplified here.

and 35 collagen genes have been described; here we concentrate on a few major examples.

- **Type I collagen** is the major collagen of skin, bone, tendon and ligaments. Its construction from heterotrimeric triple helices, and the effects of mutations, have been discussed earlier (*Figure 15.7*). All mutations have a dominant-negative effect. Mutations in either the *COL1A1* or *COL1A2* genes cause osteogenesis imperfecta (brittle bone disease) if they disrupt the packing of the procollagen chains into triple helices and then collagen fibrils. Assembly of the procollagen chains into triple helices starts near the C-terminal end, and so mutations located closer to this end of either polypeptide chain are generally more disruptive and tend to produce the more severe types of the disease (although the correlation is far from perfect). Collagen genes comprise numerous small exons, each encoding the amino acids $(Gly–X–Y)_6$, where X and Y are variable. Exon deletions often have relatively mild effects because deletion does not alter the reading frame and generally just reduces the number of repeating units in the polypeptide. Mutations replacing glycines of the (Gly–X–Y) units with other, bulkier, amino acids often severely disrupt packing of the triple helix. Deletion of one N-terminal exon of either *COL1A1* or *COL1A2* leads to normal bone but abnormally lax ligaments, a condition called Ehlers–Danlos syndrome type VII.
- **Type II collagen** forms fibrils in cartilage and other tissues including the vitreous of the eye. Type II collagen is made of homotrimeric triple helices of COL2A1 chains, co-polymerized or cross-linked to type IX and type XI chains. Mutations in these genes result in an overlapping spectrum of skeletal dysplasias including Marshall–Sticker syndrome, spondyloepiphyseal dysplasia and Kniest dysplasia. Which syndrome is produced depends on the overall effect on the final collagen fibrils, and not on which gene is mutated.

- **Type III collagen** is found in skin, tendons, the aorta, etc., and defects produce Ehlers–Danlos syndrome type IV, with fragile and easily bruisable skin.
- **Type IV collagen** forms a 'chicken-wire' meshwork in basement membranes. Patients with Alport syndrome have nephritis because of an abnormality of the glomerular basement membrane, and hearing loss because of an abnormality of the cochlear basement membrane. The genetics of Alport syndrome had long been controversial, with claims of X-linked dominant, autosomal dominant and autosomal recessive inheritance. Type IV collagen includes one chain encoded by an X-linked gene and two encoded on chromosome 2. Mutations in each of these three genes have been found in Alport patients. The autosomal genes usually give recessive phenotypes, which is perhaps unexpected.
- **Type VII collagen** forms fibrils that anchor together the different layers of skin and mucous membranes. Defects are seen in several variants of dystrophic epidermolysis bullosa, in which the skin and mucous membranes blister after very minor trauma. Other forms of epidermolysis bullosa are caused by defects in noncollagen proteins, keratins and laminins, which give skin its mechanical toughness (Compton, 1994).

The collagen diseases provide illuminating insights into the relationship between clinical and molecular classifications. Molecular analysis illuminates rather than supersedes the clinical classification and allows much more accurate counseling. For example, molecular analysis shows that unaffected parents who have more than one affected child are not carriers of a recessive form of osteogenesis imperfecta, but are germinal mosaics (*Figure 3.8* shows an example of mosaicism). For molecular pathology, the collagen diseases show how important it is to think about the biological role of a gene product, and the molecules with which it interacts, rather than expecting clinically defined syndromes to fit into any simplistic one gene–one syndrome scheme.

Allelic series: different mutations in the same gene can cause different diseases

We have already seen how different mutations in either chain of type I collagen can produce four different types of osteogenesis imperfecta or Ehlers–Danlos syndrome. *Table 15.10* lists a selection of other examples of genes responsible for more than one disease. The topic has been reviewed by Romeo and McKusick (1994).

Loss-of-function and gain-of-function mutations in the same gene

Several examples are known where loss-of-function and gain-of-function mutations in the same gene have different pathological effects. The effects of differential dosage of the peripheral myelin protein (*PMP22*) gene have already been discussed (page 412, *Figure 15.6*). Loss-of-function mutations in the *PAX3* gene have also been discussed (page 403, *Figure 15.1*); they cause the developmental abnormality type 1 Waardenburg syndrome. A totally different phenotype is seen in a gain-of-function mutation, when a chromosomal translocation creates a novel chimeric gene by fusing *PAX3* to another transcription factor gene, *FKHR* (*Table 17.3*). The *RET* gene is another example (see van Heyningen, 1994). Loss-of-function mutations are one cause of Hirschsprung's disease, via a dominant-negative effect (see above, page 415), but a set of very specific amino acid substitutions, which must cause some gain of function, are seen in several thyroid and other endocrine cancers (page 463).

Table 15.10: Examples of genes responsible for more than one disease

Gene	Location	Diseases	Symbol	MIM no.
PAX3	2q35	Waardenburg syndrome type 1	WS1	193500
		Alveolar rhabdomyosarcoma	RMS2	268220
CFTR	7p31.2	Cystic fibrosis	CF	219700
		Bilateral absence of vas deferens		
RET	10q11.2	Multiple endocrine neoplasia type 2A	MEN2A	171400
		Multiple endocrine neoplasia type 2B	MEN2B	162300
		Medullary thyroid carcinoma	FMTC	155420
		Hirschsprung disease	HSCR	142623
PMP22	17p11.2	Charcot–Marie–Tooth neuropathy type 1A	CMT1A	118220
		Tomaculous neuropathy	HNPP	162500
SCN4A	17q23.1–q25.3	Paramyotonia congenita	PMC	168300
		Hyperkalemic periodic paralysis	HYPP	170500
		Acetazolamide-responsive myotonia congenita		
PRNP	20p12–pter	Creutzfeldt–Jakob disease	CJD	123400
		Familial fatal insomnia	FFI	176640
GNAS1	20q13.2	Albright hereditary osteodystrophy	AHO	103580
		McCune–Albright syndrome	PFD	174800
AR	Xcen–q22	Testicular feminization syndrome	TFM	313700
		Spinobulbar muscular atrophy	SBMA	313200

The last word in molecular pathology: the hemoglobinopathies

Hemoglobinopathies occupy a special place in clinical genetics for many reasons. They are by far the most common serious mendelian diseases on a world-wide scale. Globins illuminate important aspects of evolution of the genome (*Figures 9.15, 9.19–9.21*) and of diseases in populations. Developmental controls for globin genes are comparatively well understood (*Figures 7.8, 8.5 and 8.6*). More mutations and more diseases are described for hemoglobins than for any other gene family. Their special relevance in the present context is that the relationship between molecular and clinical events is clearer for the hemoglobinopathies than for most other diseases. Molecular understanding of the hemoglobinopathies started earlier and has reached a nearer to perfect state than for any other diseases. Molecular pathology is also simplified by the fact that clinical symptoms follow very directly from malfunction of the protein, which is itself unusually abundant at 15 g per 100 ml of blood.

The thalassemias illustrate a remarkable variety of genetic mechanisms.

- α-Thalassemias are usually caused by gene dosage effects. Unequal crossing over between repetitive sequences flanking the duplicated α-globin genes (*Figure 8.6*) creates haplotypes carrying one or no α-globin genes. The general mechanism is illustrated in *Figure 10.6*. Normal people have four α-globin genes (αα/αα). People with two α-globin genes (whether the phase is α–/α– or αα/– –) suffer mild symptoms; those with only one gene (α–/– –) have severe disease, and lack of all α genes (– –/– –) causes lethal hydrops fetalis.
- Some α-thalassemia is caused by an unstable hemoglobin mRNA or protein. Hb Constant Spring (HBA2 0001), for example, has a UAA →CAA mutation of an α-globin stop codon. Translation continues for an extra 30 codons until another stop codon is encountered. The product is unstable. HBA2 0024 has a mutation AATAAA →AATGAA of the polyadenylation signal.
- A rare variety of α-thalassemia with mental retardation (MIM 301040) is caused by an X-linked mutation, illustrating nonallelic control.
- β-Thalassemia is often caused by mutations producing an unstable product. Nonsense mutations, promoter mutations, abolition of normal splice sites and

activation of cryptic splice sites have all been described. Some examples are listed in *Table 16.4*.

- Lepore hemoglobins are encoded by δ–β *fusion genes* produced by unequal crossover between these closely related genes (see *Figure 10.6* for the general principle). They have the low-activity promoter of the δ gene, leading to β-thalassemia.
- The severity of β-thalassemia is often reduced by hereditary persistence of fetal hemoglobin (HPFH). The causes of HPFH are various and complex, but can include deletions which affect operation of the locus control region (*Figure 8.6*), and occasionally mutations in the γ-globin promoter.

Amino acid substitutions, which alter the properties but not the production of globins, produce phenotypes whose variety has shed much light on hemoglobin function.

- Sickle cell disease is caused by replacing a polar glutamic acid residue with a nonpolar valine on the outer surface of the β-globin molecule. This causes increased intermolecular adhesion, leading to aggregation of deoxyhemoglobin and distortion of the red cell. Sickled red cells have decreased survival time (leading to anemia) and tend to occlude capillaries, leading to ischemia and infarction of organs downstream of the blockage.
- The finely tuned oxygen affinity of hemoglobin can be affected by missense mutations. Considerable changes in tertiary structure accompany the transition between oxygenated and deoxygenated forms, and substitution of any of the amino acids involved can affect the equilibrium. Mutants with too high an oxygen affinity fail to unload oxygen in the target tissues, whereupon a feedback mechanism causes erythrocytosis (usually benign). Low-affinity mutants can cause anemia and cyanosis.
- Many variant hemoglobins are unstable, usually because the heme group is held insufficiently tightly, or sometimes because of weakening of the tertiary or quaternary (tetrameric) structure. The unstable hemoglobin precipitates to form intracellular aggregates (*Heinz bodies*). This causes chronic hemolytic anemia, often mild but seriously exacerbated by drugs or infections.
- M hemoglobins have amino acid changes in the crevice holding the heme group, such that the iron atom is able to assume the nonphysiological ferric (Fe^{3+}) state. This state of methemoglobinemia produces chronic but usually benign cyanosis.

Globin mutations can be found which illustrate virtually every process described in this book. In the space available here, it is impossible to do justice to the breadth and subtlety of globin molecular pathology. Readers are recommended to consult one of the excellent reviews of this topic (Higgs *et al.*, 1989; Cao *et al.*, 1994; see also Phillips and Kazazian, 1990).

Further reading

Further reading

Humphries S, Malcolm S. (1994) *From Genotype to Phenotype*. BIOS Scientific Publishers, Oxford.

Stewart A. (ed.) (1995) *Trends in Genetics Nomenclature Guide*. Elsevier, Cambridge.

References

Ambrose CM, Duyao MP, Barnes G *et al.* (1994) *Somat. Cell Mol. Genet.*, **20**, 27–38.

Ballabio A, Andria G. (1992) *Hum. Mol. Genet.*, **1**, 221–227.

Bardoni B, Zanaria E, Guioli S *et al.* (1994) *Nature Genetics*, **7**, 497–501.

Beaudet A, Tsui LC. (1993) *Hum. Mutat.*, **2**, 645.

Boucher CA, King SK, Carey N *et al.* (1995) *Hum. Mol. Genet.*, **4**, 1919–1925.

Byers PH. (1993) In: *Connective Tissue and its Heritable Disorders: Molecular, Genetic and Medical Aspects* (eds PM Royce, B Steinmann), Wiley-Liss, New York.

Cao A, Galanello R, Rosatelli MC. (1994) *Blood Rev.*, **8**, 1–12.

Compton J. (1994) *Nature Genetics*, **6**, 6–7.

Cooper DN, Krawczak M. (1995) *Human Gene Mutation*, 2nd Edn. BIOS Scientific Publishers, Oxford.

de Lugenbeel KA, Peier AM, Carson NL, Chudley AE, Nelson DL. (1995) *Nature Genetics*, **10**, 483–485.

Dietz HC, Kendzior RJ. (1994) *Nature Genetics*, **8**, 183–188.

Feng Y, Zheng F, Lokey LK, Chastain JL, Lakkis L, Eberhart D, Warren ST. (1995) *Science*, **268**, 731–734.

Fisher E, Scambler P. (1994) *Nature Genetics*, **7**, 5–7.

Francke U, Ochs HD, de Martinville M *et al.* (1985) *Am. J. Hum. Genet.*, **37**, 250–267.

Hemesath TJ, Steingrimsson E, McGill G *et al.* (1994). *Genes Dev.*, **8**, 2770–2780.

Higgs DR, Vickers MA, Wilkie AO, Pretorius IM, Jarman AP, Weatherall DJ. (1989) *Blood*, **73**, 1081–1104

International Working Group on Constitutional Diseases of Bone. (1992) *Am. J. Med. Genet.*, **44**, 223–229.

Jones C, Penny L, Mattina T *et al.* (1995) *Nature*, **376**, 145–149.

Koenig M, Beggs AH, Moyer M *et al.* (1989) *Am. J. Hum. Genet.*, **45**, 498–506.

Krawczak M, Reiss J, Cooper DN. (1992) *Hum. Genet.*, **90**, 41–54.

Lalande M. (1994) *Nature Genetics*, **8**, 5–7.

McAlpine PJ *et al.* (1992) In: *Genome Priority Reports*, pp. 11–142 (eds AJ Cuticchia, PL Pearson, HP Klinger). Karger, Basel.

McClatchey AI, McKenna-Yasek D, Cros D *et al.* (1992) *Nature Genetics*, **2**, 148–152.

Owen MC, Brennan SO, Lewis JH, Carrell RW. (1983) *New Engl. J. Med.*, **309**, 694–698.

Pasini B, Borrello MG, Greco A *et al.* (1995) *Nature Genetics*, **10**, 35–40.

Patel PI, Lupski JR. (1994). *Trends Genet.*, **10**, 128–133.

Phillips JA, Kazazian HH (1990) In: *Principles and Practice of Medical Genetics*, 2nd Edn, pp. 1315–1342 (eds AEH Emery, DL Rimoin). Churchill Livingstone, Edinburgh.

Poulton J. (1995) *Nature Genetics*, **8**, 313–133.

Prezant TR, Agapian JV, Bohlman MC *et al.* (1993) *Nature Genetics*, **4**, 289–293.

Romeo G, McKusick VA. (1994) *Nature Genetics*, **7**, 451–453.

Royce PM, Steinmann B. (eds) (1993) *Connective Tissue and its Heritable Disorders: Molecular, Genetic and Medical Aspects*. Wiley-Liss, New York.

Sege-Peterson K, Hyhan WL, Page T. (1993) In: *The Molecular and Genetic Basis of Neurological Disease* (eds RN Rosenberg, SB Prusiner, S DiMauro, RL Barchi, LM Kunkel). Butterworth-Heinemann, Boston.

Stott K, Blackburn JM, Butler PJG, Perutz M. (1995) *Proc. Natl Acad. Sci. USA,* **92**, 6509–6513.

Tassabehji M, Newton VE, Liu XZ *et al.* (1995) *Hum. Mol. Genet.*, **4**, 2131–2137.

Van Heyningen V. (1994) *Nature*, **367**, 319–320.

Weinstein LS, Shenker A, Gejman PV, Merino MJ, Friedman E, Spiegel AM. (1991) *New Engl. J. Med.*, **325**, 1688–1695.

Wilkie AOM. (1994) *J. Med. Genet.*, **31**, 89–98.

Genetic testing in individuals and populations

Chapter 16

Geneticists have no monopoly on DNA-based diagnosis. For microbiologists and virologists, for example, PCR is a central tool for identifying pathogens (see Chapter 11 of Newton and Graham, 1994), while hematologists, oncologists and other pathologists all use DNA testing as a basis for diagnosis. However, here we will restrict ourselves to DNA-based diagnosis in the context of human molecular genetics.

Clinical diagnosis and laboratory tests

Clinical definitions of diseases are always somewhat arbitrary. Essentially the clinician is saying, more or less convincingly, 'this patient resembles a group of others whom I have seen or read about, and who are labeled in a certain way'. A good physician is highly skilled at making such phenotypic definitions. For many purposes, particularly patient management, this is the most useful method of labeling patients, and in some branches of medicine there is no basis for any more fundamental classification of phenotypes.

Geneticists are fortunate in that, at least for mendelian diseases, there is a natural underlying basic principle. All mendelian diseases can be classified, first by the locus involved and second by the particular mutant allele at that locus. Because diagnostic labels are not simply conventions, they evolve as knowledge of the underlying genetics advances. Diseases are lumped together (Duchenne and Becker muscular dystrophy; see *Figure 15.2*) or split (retinitis pigmentosa is listed in OMIM under seven autosomal dominant, one autosomal recessive and three X-linked headings). Sometimes the original clinical-based classification of a whole spectrum of diseases is completely changed by molecular analysis, as has happened with osteogenesis imperfecta (*Table 15.9*).

Clinical and genetic classifications of diseases are not in competition with each other. Each has its uses, and a satisfactory classification must include both aspects. For clinical management and prognosis, classification of diseases into allelic groups may not be very helpful. As the examples from Chapter 15 show, molecular pathology is a very imperfect science. Problems include:

- mutations at several different loci may produce the same clinical syndrome (for example, congenital deafness).

- Different mutations at the same locus may produce different clinical syndromes, as in hemoglobinopathies (page 423) and *RET* mutations (*Table 15.10*).
- Genetic diseases are often variable even within families, so that knowing the mutation does not necessarily predict which features a patient will show. Neurofibromatosis type 1 (MIM 162200) is one example among many.

Thus DNA-based descriptions of genetic diseases complement rather than supersede traditional clinical descriptions. However, one area where identification of the affected locus and, where possible, the exact mutation is crucially important is in genetic prediction. Genetics is an unusual branch of medicine in that many of the consultations involve phenotypically normal people. They want to know whether their unborn child is affected or possibly whether they will develop a late-onset disease. Diagnosing the family condition requires traditional clinical skills and, for genetic counseling, a statistical risk can be based on standard mendelian principles or empirical data – but only laboratory tests can give a definite answer. Demonstrating that a patient has or has not inherited a mutation which is known to cause a disease in his family is the fundamental requirement for accurate genetic prediction.

Direct and indirect

How much information the DNA tests can give depends on the state of knowledge about the gene(s) involved, but in principle DNA-based diagnosis can be made in two essentially different ways.

- **Direct testing:** DNA from a consultand is tested to see whether or not it carries a given pathogenic mutation.
- **Indirect testing (gene tracking):** linked markers are used in family studies to discover whether or not the consultand inherited the disease-carrying chromosome from a parent.

In general, direct testing is to be preferred but, as shown in *Figure 16.1* and discussed below, it may not always be possible. Even when direct testing is scientifically possible, it may not always be practicable in the context of a routine diagnostic service.

Direct testing

The optimal, though not always practicable, method of DNA-based diagnosis is to test a person's DNA directly to see whether his gene sequence is normal or mutant. We must of course know which gene to examine and we must know the relevant 'normal' (wild-type) sequence (*Figure 16.1*). Assuming the gene has been cloned, the main obstacle to direct mutation testing is allelic heterogeneity, as discussed below.

Direct testing is almost always done by PCR – the few applications of Southern blotting include testing for fragile X and myotonic dystrophy full mutations (see below, *Figure 16.6B*) and for some chromosomal translocations in cancer (*Figure 17.7*). The sensitivity of PCR allows us to use a wide range of samples. These can include:

- **blood samples** – the most widely used source of DNA from adults.

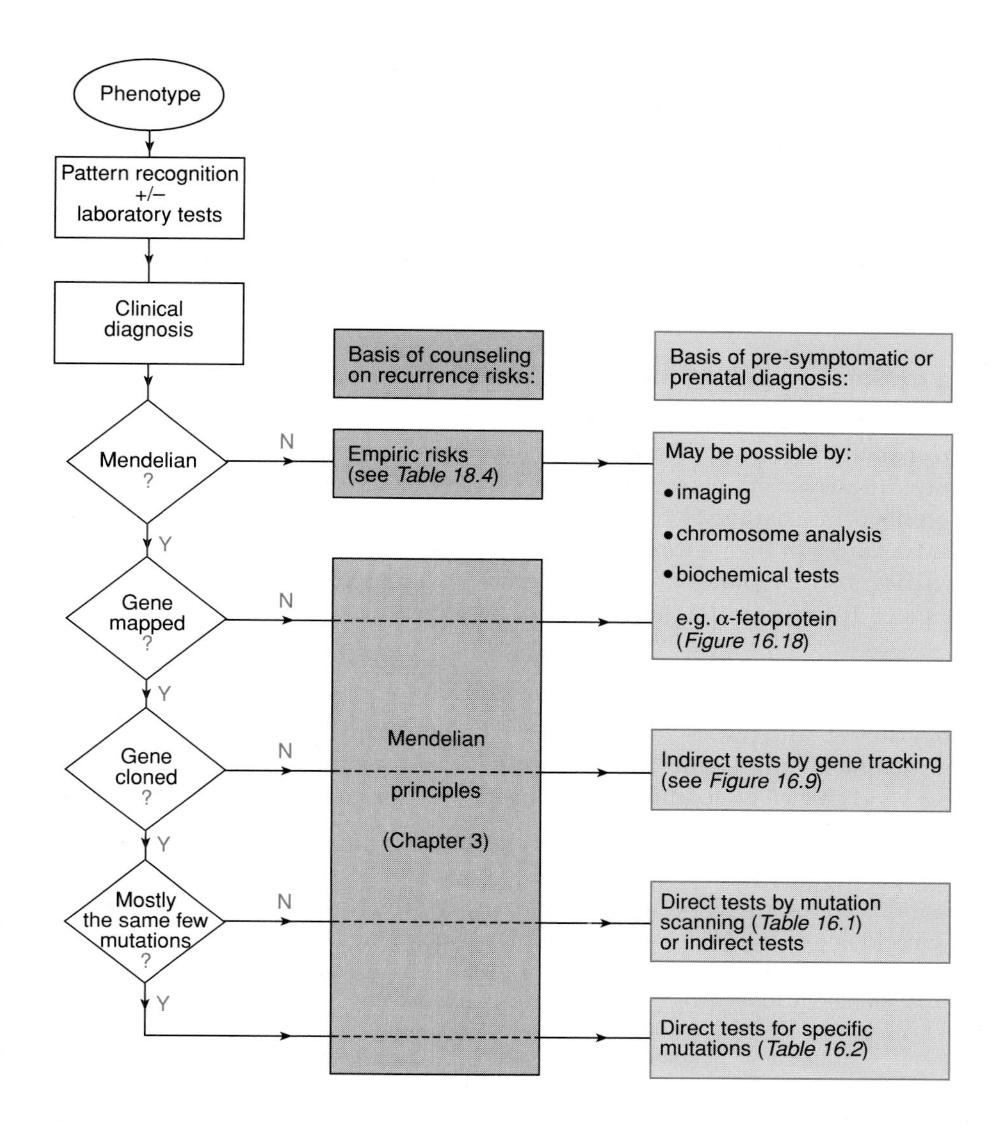

Figure 16.1: Genetic diagnosis, counseling and prediction.

The flow diagram shows how a clinical diagnosis is made starting from the patient's phenotype and then, depending on the mode of inheritance and the state of genetic knowledge, various means are deployed for counseling and predictive tests.

- **Mouthwashes or buccal scrapes** – these yield sufficient DNA for one or two tests; they are especially favored for population screening programs (see page 450).
- **Chorionic villus biopsy samples** – these are the best source of fetal DNA (better than amniocentesis specimens).
- **One or two cells** removed from eight-cell stage embryos – for pre-implantation diagnosis after *in vitro* fertilization.
- **Hair, semen, etc.** for criminal investigations.
- **Archived pathological specimens** can be used where necessary to type dead people when no DNA has been stored. Amplification from fixed or degraded DNA works best for short sequences, say 100 bp or less.
- **Guthrie cards** – these are the cards on which a spot of dried blood is sent to a laboratory for neonatal screening for phenylketonuria (PKU) in the UK and elsewhere; not all the blood spot is used for the screening test. They are a possible source of DNA from a dead child.

Not all mutation tests use DNA. Testing RNA by RT-PCR (*Figure 6.7*) has advantages when screening genes with many exons or seeking splicing mutations. However, RNA is much less convenient to obtain, requiring special sample handling and rapid processing. It is also worth noting that protein assays may stage a come-back after the eclipse of the past 20 years. A protein-based functional assay might classify the products of a highly heterogeneous allelic series into two simple groups, functional and nonfunctional – which is, after all, the essential question in most diagnosis.

Testing for known and unknown mutations in a gene poses different problems and requires different methods

Mutation screening

The many different techniques which can be used to screen a gene for mutations were described in Chapter 14 (page 394ff.). *Table 16.1* summarizes the advantages and disadvantages of the methods which have been widely used, from the point of view of a diagnostic laboratory. Their use in research may produce a rather different balance sheet. For routine diagnostic use, all these methods suffer two limitations:

- they are quite laborious and expensive for use in a diagnostic service, which needs to produce answers quickly and within a modest budget.
- They detect differences between the patient's sequence and the published 'normal' sequence – but they do not generally distinguish between pathogenic and chance nonpathogenic changes.

None of these problems is insuperable. With each passing year, the labor and cost per genotype diminish as new automated methods of mutation screening are developed. Several groups are working on developing **'DNA chips'**, comprising arrays of oligonucleotides attached to a silicon or glass substrate. These could be used to

Table 16.1: Methods for scanning a gene for point mutations

Method	Advantages	Disadvantages
Sequencing	Detects all changes Mutations fully characterized	Laborious Generates excessive information
Heteroduplex gel mobility	Very simple	Short sequences only (<200 bp) Limited sensitivity Does not reveal position of change
Single-strand conformation polymorphism (SSCP) analysis	Simple	Short sequences only (<200 bp) Does not reveal position of change
Denaturing gradient gel electrophoresis (DGGE)	High sensitivity	Choice of primers is critical GC-clamped primers expensive Does not reveal position of change
Mismatch cleavage		
(i) chemical	High sensitivity	Toxic chemicals
(ii) enzymic	Shows position of change	Experimentally difficult
Protein truncation test (PTT)	High sensitivity for chain terminating mutations Shows position of change	Chain terminating mutations only Expensive Experimentally difficult Best with RNA

The table summarizes the advantages and disadvantages of each method for use in a routine diagnostic service. See pages 394–398 for descriptions of the methods and discussion of their use in research. *Note* that heteroduplex gel mobility and SSCP can be performed simultaneously on a single gel (see *Figure 14.14*).

screen a sequence for any mutation in a single reaction. One scheme envisages hybridizing the target to the array, extending by one nucleotide using DNA polymerase and a fluorescent chain terminator, and checking locations where no extension occurred. These should mark mismatches caused by deviation of the target sequence from the standard wild-type sequence (Metspalu *et al.*, 1995). Also, with advancing knowledge of gene function and an increasing database of disease-related mutations, the task of deciding whether a given sequence change is pathogenic is becoming rather easier. Nevertheless, for the forseeable future, testing for unknown mutations in a service laboratory remains a considerable problem.

Mutation detection

Testing a patient sample for the presence or absence of a *known* sequence change is a different and much simpler problem than scanning a gene for the presence of any mutation. *Table 16.2* lists the main methods which are available to do this, and they are described below. Testing for a known change is possible for:

- diseases where there is presumed to be allelic homogeneity, so that all affected people in the population have one particular DNA sequence change.
- Diseases where most affected people in the population have one of a limited number of specific mutations.
- Diagnosis within a family. General mutation detection methods may be needed to define the family mutation but, once it is characterized, other family members normally need be tested only for that particular mutation.
- In research, for testing control samples. A common problem in positional cloning is that a candidate gene has been identified and a change has been seen in this gene in a patient with the relevant disease. The question then arises, is this change the pathogenic mutation (confirming that the candidate gene is the disease gene) or might it be a nonpathogenic polymorphism? (see page 393). One common approach is to screen a panel of 100 or so normal control samples for the presence of the change (which does not, of course, solve the problem of rare neutral variants).

Table 16.2: Methods of testing for a specified mutation

Method	Comments
Restriction digestion of PCR-amplified DNA; check size of products on gel	Only when the mutation creates or abolishes a natural restriction site or one engineered by use of special PCR primers (*Figure 16.2*)
Hybridize PCR-amplified DNA to allele-specific oligonucleotides (ASO) on dot–blot, slot–blot or Southern blot	General method for point mutations (*Figure 16.3*)
PCR using allele-specific primers (ARMS test)	General method for point mutations; primer design critical (*Figure 16.4*)
Oligonucleotide ligation assay (OLA)	General method for point mutations (*Figure 16.5*)
PCR with primers located either side of a suspected deletion or translocation breakpoint	Successful amplification shows presence of the specified rearrangement
Check size of expanded trinucleotide repeat	Trinucleotide repeat diseases (page 266) only; large expansions require Southern blots, smaller ones can be done by PCR

See text for details.

Several methods allow easy testing for a specified DNA sequence change

Presence/absence of a restriction site

As we saw in Chapter 5 (*Figure 5.12*), the single base change responsible for sickle cell disease abolishes a site, CCTNAGG, for the restriction enzyme *Mst*II. This forms a simple direct test for the presence of the sickle cell mutation. Normally the test would be done by PCR (*Figure 6.4*) rather than Southern blotting.

Although hundreds of restriction enzymes are known (page 87), they almost all recognize symmetrical palindromic sites, and most pathogenic point mutations will not happen to affect such sequences. Also, sites for very rare and obscure restriction enzymes are unsuitable for routine diagnostic use because the enzymes are expensive and often of poor quality. Sometimes, however, a diagnostic restriction site can be introduced by a form of PCR mutagenesis (see page 144) using carefully chosen primers. *Figure 16.2* shows an example.

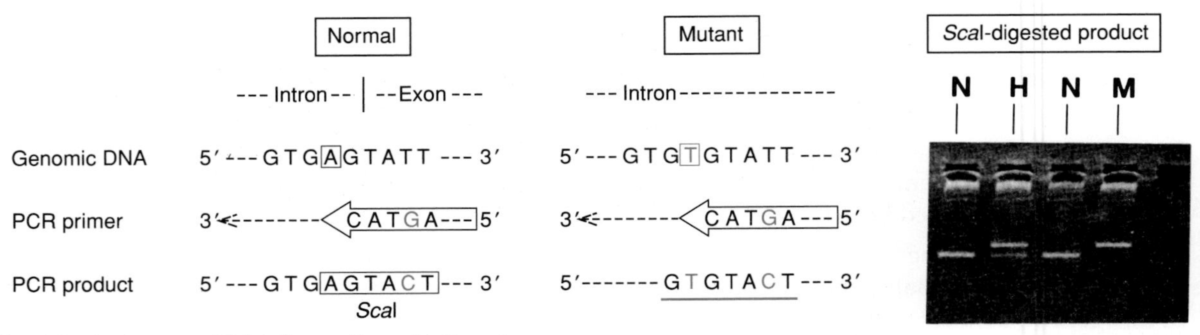

Figure 16.2: Introducing an artificial diagnostic restriction site.

An A →T mutation in the intron 4 splice site of the *FACC* gene does not create or abolish a restriction site. The PCR primer stops short of this altered base, but has a single base mismatch (red G) in a noncritical position which does not prevent it hybridizing to and amplifying both the normal and mutant sequences. The mismatch in the primer introduces an AGTACT restriction site for *Sca*I into the PCR product from the normal sequence. The *Sca*I-digested product from homozygous normal (N), heterozygous (H) and homozygous mutant (M) patients is shown. Courtesy of Dr Rachel Gibson, Guy's Hospital, London.

Use of allele-specific oligonucleotide probes

These short synthetic probes hybridize only to a perfectly matched sequence. Their use has been described in Chapter 5 (page 119, *Figures 5.9* and *5.10*). For diagnostic purposes, a reverse dot–blot procedure is often used (*Figure 16.3*).

ARMS test: PCR amplification using allele-specific oligonucleotide primers

The principle of this method was described in Chapter 6 (*Figure 6.8*). Paired PCR reactions are carried out. One primer (the common primer) is the same in both reactions, the other exists in two slightly different versions, one specific for the normal sequence and the other specific for the mutant sequence. An additional control pair of primers is usually included, which will amplify some unrelated sequence from every sample, as a check that the PCR reaction has worked. The location of the common primer can be chosen to give a product of any desired size, so that it is easy to design **multiplex reaction** systems. With multiplexing, the additional control pair of primers is unnecessary, because normal and mutant-specific primers can be combined so as to ensure that something will amplify in every tube. The ARMS method is well suited to screening large numbers of samples for a given panel of mutations, using multiplexed PCR (*Figure 16.4*). Many variations have been devised by companies working on producing diagnostic kits.

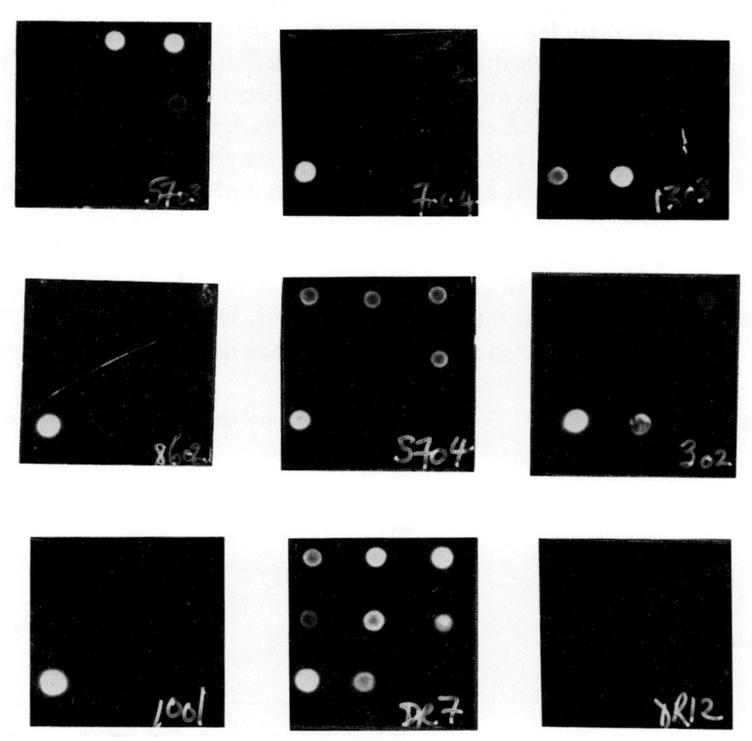

Figure 16.3: Reverse dot–blot for detection of specific mutations.

Nine oligonucleotides are bound to a small filter. Each hybridizes specifically to a sequence found in one or more HLA-DR alleles. A set of such filters has been hybridized with PCR-amplified and labeled DNA of nine patients; the figure shows a negative image of the resulting autoradiographs. These oligonucleotides are used to distinguish sub-types of HLA-DR4; DR12 (bottom right) is negative while DR7 (bottom center) hybridizes to most of the oligos. Commercial versions of reverse dot–blotting produce a colored bar-code when exposed to test DNA. Courtesy of Dr Paul Sinnott, St Mary's Hospital, Manchester.

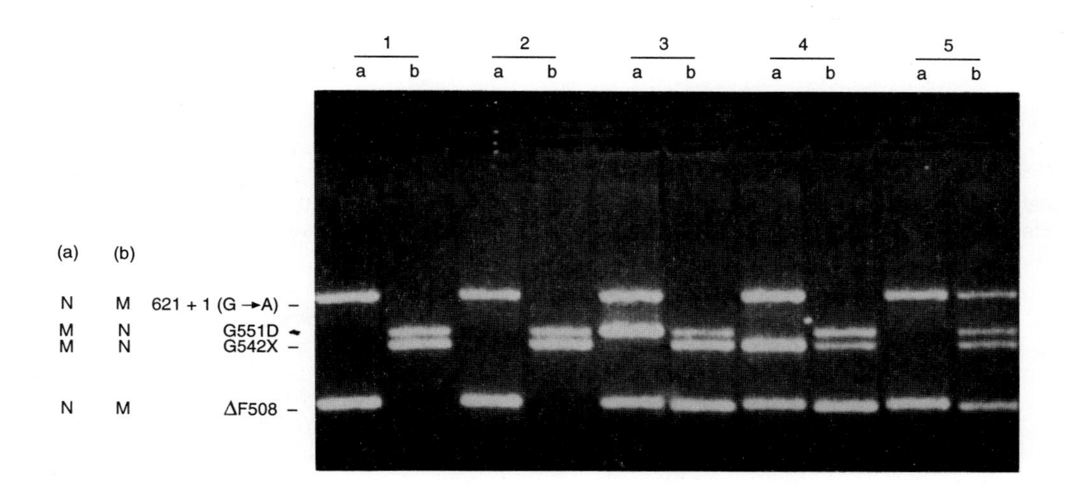

Figure 16.4: Multiplex ARMS test to detect four cystic fibrosis mutations.

The common primers have been designed so that each reaction gives a product of a different size, allowing multiplexing. Each tube contains primers for two normal and two mutant sequences. PCR products, from top to bottom, test for the mutations 621+1 (G →A), G551D, G542X and ΔF508. In the left track of each pair the ARMS primers amplify the normal counterparts of the 621+1 (G →A) and ΔF508 sequences, and the mutants G551D and G542X; in the right hand track the primers amplify the opposite allele in each case. Tracks 1, 2: no mutation detected; track 3: compound heterozygote for ΔF508 and G551D; track 4: compound heterozygote for ΔF508 and G542X; track 5: compound heterozygote for ΔF508 and 621+1 (G →A). Courtesy of Dr Andrew Wallace, St Mary's Hospital, Manchester; data obtained using a kit supplied by Cellmark Diagnostics, Abingdon, UK.

Oligonucleotide ligation assay (OLA)

In the OLA test for a base substitution mutation, two 20-mer oligonucleotides are constructed which hybridize to adjacent sequences in the target, with the join sited at the position of the mutation. DNA ligase will covalently join the two oligonucleotides only if they are perfectly hybridized (Nickerson *et al.*, 1990). The test is done on a PCR-amplified template, and a single ligation reaction is carried out. One oligonucleotide is tagged with biotin and the second carries a reporter molecule such as digoxygenin (see *Figure 5.5*). Reaction products are transferred to streptavidin-coated microtiter plates which bind biotin. If ligation occurs, the reporter will be retained in the well after washing (*Figure 16.5*). The OLA is well suited to large-scale automated testing, since it does not involve electrophoresis or centrifugation.

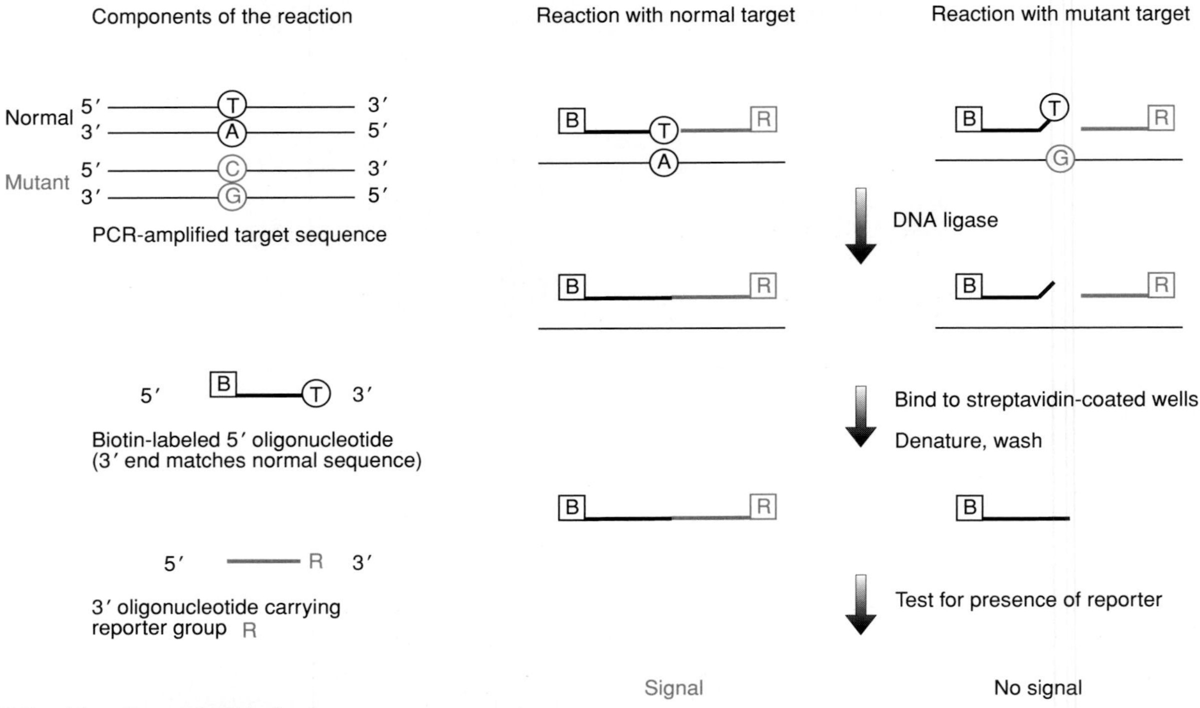

Figure 16.5: The oligonucleotide ligation assay.

This method is suitable for automated testing, using multiwell microtiter plates on a robot workstation. A typical reporter would be digoxygenin, detected by an alkaline phosphatase-conjugated antibody with a chromogenic substrate.

Direct testing for known mutations allows easy diagnosis in relatively homogeneous diseases

For a few diseases, most or all affected people have the same mutation, or one of a limited number of different mutations. *Table 16.3* lists some examples. This desirable situation arises under three sets of circumstances:

- the disease depends on a specific molecular mechanism;
- the nature of the gene is such that one particular mutation occurs repeatedly;
- affected people carry the same ancestral disease mutation, or one of a limited number of ancestral mutations.

Huntington disease and the other diseases caused by expansion of a $(CAG)_n$ trinucleotide repeat (page 266) are good examples of diseases which are uniform because they depend on a specific mechanism (even though the mechanism is currently unknown). Sometimes although many different pathogenic mutations are

Table 16.3: Examples of diseases which show a limited range of mutations

Disease	Cause	Comments
Huntington disease, myotonic dystrophy, fragile X, etc.	Unstable expanded trinucleotide repeat	See pages 266 and 408
Charcot–Marie–Tooth (HMSN1)	Duplication of 1.5 Mb at 17p11.2	Sometimes involves point mutations or other disease loci
α-Thalassemia	Various gene deletions	See page 423
β-Thalassemia	Mostly point mutations	Different common mutations in different populations (see Table 16.4)
Sickle cell disease	E6V in *HBB* gene	See *Figure 5.10*
Achondroplasia	G380R in *FGFR3* gene	Two distinct changes, both causing G380R in *FGFR3* gene; Shiang *et al.* (1994)
Cystic fibrosis	ΔF508 in *CFTR* gene	Major mutation in northern Europeans (see *Table 16.5*)
Hemophilia A	Large inversion in *F8* gene and flanking DNA	50% of mutations (see *Figure 10.18*)
Tay–Sachs disease	*HEXA* gene: 4-bp insertion in exon 11, exon 11 donor splice site G →C	73% in Ashkenazi Jews 15% in Ashkenazi Jews
Leber's optic atrophy	Mitochondrial mutations at nucleotide positions np 3460, 11 778 or 14 484	Found in 80–90% of cases (page 411)
21-Hydroxylase deficiency	About 30% are large deletions; others include only a few examples of point mutations copied from a pseudogene	See *Figure 10.15*

known in a gene, only a given mutation produces the pathology defining a certain condition, as in sickle cell disease. Achondroplasia is another case: virtually all affected people have one of two mutations in the same codon, which both replace glycine with arginine at codon 380 of the fibroblast growth factor receptor gene *FGFR3* (Shiang *et al.*, 1994). Other mutations in the *FGFR3* gene produce other syndromes (and for unknown reasons, the mutation rate for the G380R change is extraordinarily high, so that achondroplasia is one of the commoner genetic abnormalities, despite requiring very specific DNA sequence changes). We would expect all such diseases to be caused by gain-of-function mutations (page 403). Where the phenotype depends on loss of gene function, the same result should be attainable in many different ways.

Fragile X disease (page 408) exemplifies the case where one particular molecular event is a much more common cause of mutations than any other. The syndrome is caused by loss of function of the *FMR1* gene. This can be caused by deletions, point mutations or promoter deletions, but silencing by expansion of the $(CGG)_n$ trinucleotide repeat is by far the commonest cause.

In many other diseases, there are a limited number of common mutations, so that most cases can be picked up by a small panel of tests. Often the total number of different mutations recorded in the literature is large, but in any particular population only a few mutations are common, so the mutation screen uses a panel of tests appropriate for the ethnic origin of the patient. This is a common situation with autosomal recessive diseases. β-Thalassemia is a classic example (see below and *Table 16.4*).

Trinucleotide repeat diseases

Expanded trinucleotide repeat diseases are ideal subjects for direct testing because the mutation is virtually always the same, the only question being how many repeat

Table 16.4: The main β-thalassemia mutations in different countries

Population	Mutation	MIM no.	Frequency (%)	Clinical effect
Sardinia	Codon 39 (C→T)	.0312	95.7	β^0
	Codon 6 (–A)	.0327	2.1	β^0
	Codon 76 (–1)	.0330	0.7	β^0
	Intron –110 (G→A)	.0364	0.5	β^+
	Intron 2 –745 (C→G)	.0367	0.4	β^+
Greece	Intron 1 –110 (G→A)	.0364	43.7	β^+
	Codon 39 (C→T)	.0312	17.4	β^0
	Intron 1 –1 (G→A)	.0346	13.6	β^0
	Intron 1 –6 (T→C)	.0360	7.4	β^+
	Intron 2 –745 (C→G)	.0367	7.1	β^+
China	Codon 41/42 (–TCTT)	.0326	38.6	β^0
	Intron 2 –654 (C→T)	.0368	15.7	β^0
	Codon 71/72 (+A)	.0328	12.4	β^0
	–28 (A→G)	.0381	11.6	β^+
	Codon 17 (A→T)	.0311	10.5	β^0
Pakistan	Codon 8/9 (+G)	.0325	28.9	β^0
	Intron 1 –5 (G→C)	.0357	26.4	β^+
	619-bp deletion	–	23.3	β^+
	Intron 1 –1 (G→T)	.0347	8.2	β^0
	Codon 41/42 (–TCTT)	.0326	7.9	β^0
US black African	–29 (A→G)	.0379	60.3	β^+
	–88 (C→T)	.0372	21.4	β^+
	Codon 24 (T→A)	.0369	7.9	β^+
	Codon 6 (–A)	.0327	0.8	β^0

The nomenclature used is that current for hemoglobinopathies. Mutations of the β-globin (*HBB*) gene are listed and referenced in OMIM under the number shown, e.g. codon 39 (C→T) is 141900.0312. Data courtesy of Dr J. Old, Institute of Molecular Medicine, Oxford.

units are present. *Figure 16.6* shows direct testing for Huntington disease (HD) and fragile X. For HD, normal alleles have between nine and 35 CAG repeats and pathological ones have 37 up to about 100 repeats (*Table 10.6*). A single PCR reaction makes the diagnosis (*Figure 16.6A*). The other $(CAG)_n$ repeat diseases listed in *Table 10.6* can be diagnosed in exactly the same way.

Fragile X is a little more difficult for two reasons. First, unlike HD, FRAXA mutations cause loss of function (of the *FMR1* gene, page 408) and occasional affected patients have deletions or point mutations which would be missed by this test. Second, full mutations have hundreds or thousands of CGG repeats and do not readily amplify by PCR (because of both their length and high GC content). Normal and pre-mutation alleles give clean PCR products, but full mutations have to be detected by Southern blotting (*Figure 16.6B*). This is true also for myotonic dystrophy, where full mutations have extremely large expansions.

β-Thalassemia

Very many different mutations are known which can cause β-thalassemia (page 423). However, β-thalassemia has spread by selection for resistance to *Falciparum* malaria, and in most populations only a few different mutations are common. DNA testing is not needed to diagnose carriers or affected people – orthodox hematology does this perfectly well – but it is the method of choice for prenatal diagnosis. Provided one has DNA samples from the parents and knows their ethnic origin, the parental mutations can often be found using only a small cocktail of direct tests, after which the fetus can be readily checked.

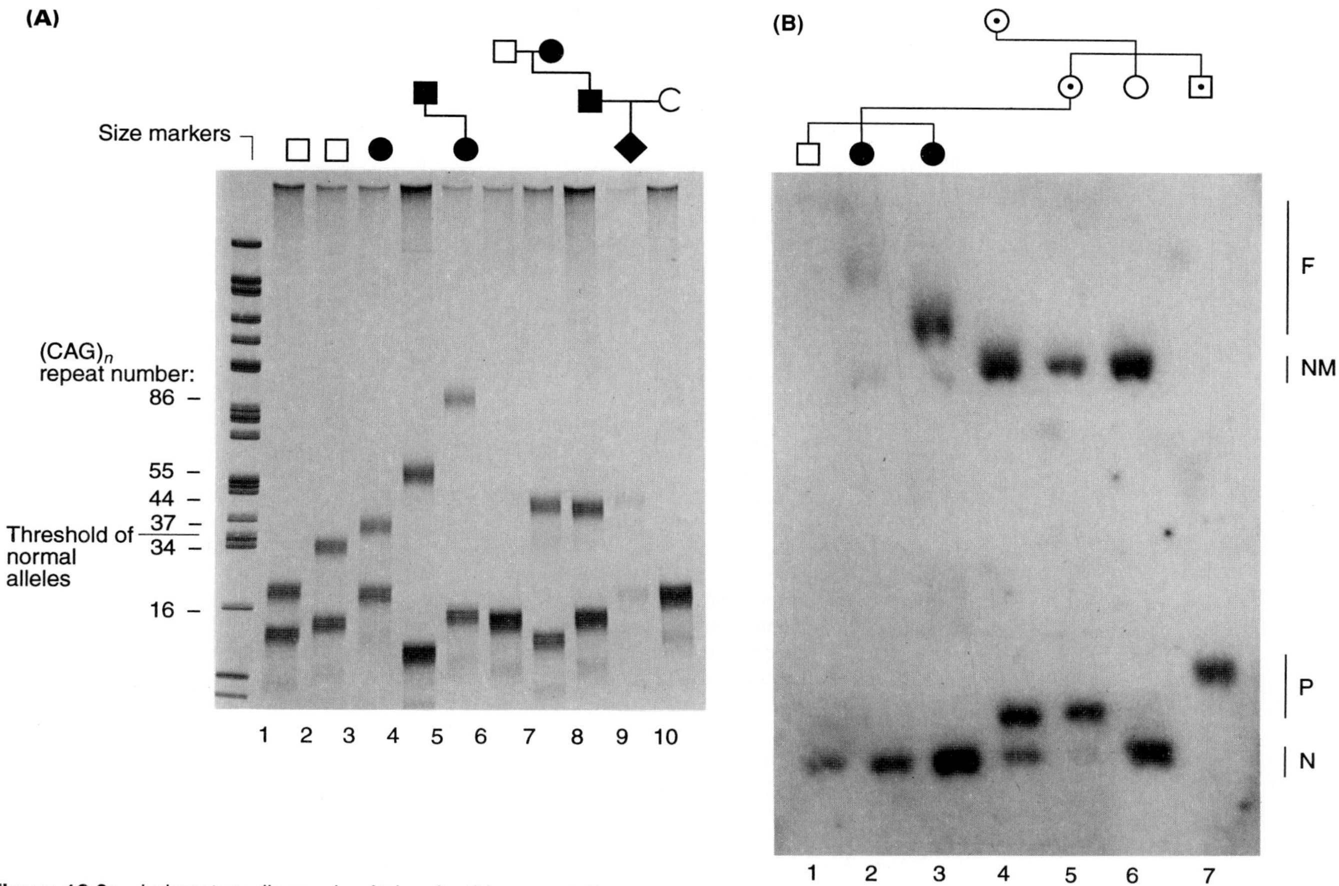

Figure 16.6: Laboratory diagnosis of trinucleotide repeat diseases.

(A) Huntington disease. A fragment of the gene containing the $(CAG)_n$ repeat has been amplified by PCR and run out on a polyacrylamide gel. Bands are revealed by silver staining. The scale shows numbers of repeats. Lanes 1, 2, 6 and 10 are from unaffected people, lanes 3, 4, 5, 7 and 8 are from affected people. Lane 5 is a juvenile onset case; her father (lane 4) had 45 repeats but she has 86. Lane 9 is an affected fetus, diagnosed prenatally. Courtesy of Dr Alan Dodge, St Mary's Hospital, Manchester.
(B) Fragile X. Southern blot of DNA digested simultaneously with *Eco*RI and *Ecl*XI. The *Ecl*XI site at the 5′ end of the gene cuts only when unmethylated. The X in a normal male (track 7) and the active normal X in a female (tracks 2, 3, 4, 6) give a small fragment (N). Unmethylated premutation alleles (P) give a slightly larger band – tracks 4 and 5 are female premutation carriers, track 7 a normal transmitting male. Methylated (inactive) X sequences do not cut with *Ecl*XI, and give larger bands. NM, methylated normal or premutation sequences; F, methylated full mutation sequences. Courtesy of Dr Simon Ramsden, St Mary's Hospital, Manchester.

Cystic fibrosis

Some 500 mutations in the *CFTR* gene have been described. Many are *private mutations*, detected in just one family (and in many cases it is by no means certain that they are in fact pathogenic; all one can say is that they are sequence variants found on a chromosome known to carry a CF disease allele). Different populations vary in the proportions of each mutation (CF Genetic Analysis Consortium, 1994). In most of northern Europe, the ΔF508 mutation accounts for more than half of all mutations. Diagnostic laboratories decide on a limited list of typically four to 10 mutations for routine testing. The number is determined mainly by the limits on multiplexing, and in future one might envisage much larger multiplexes being used. *Figure 16.4* shows a multiplex designed to detect four of the commonest mutations in the UK. As *Table 16.5* shows, there is no obvious natural cut-off in terms of diminishing returns on extra tests. Outside northern Europe, the greater diversity of mutations makes testing more difficult, although in general the disease is less frequent and so presents less of a problem. The impact of this diversity on proposals for population screening is discussed below (page 453).

Table 16.5: Distribution of *CFTR* mutations in 300 CF chromosomes from the north-west of England

Mutation	Exon	MIM no.	Frequency (%)	Cumulative frequency (%)
ΔF508	10	.0001	79.9	79.9
G551D	11	.0013	2.6	82.5
G542X	11	.0009	1.5	84.0
G85E	3	.0038	1.5	85.5
N1303K	21	.0032	1.2	86.7
621+1 (G→T)	4	–	0.9	87.6
1898+1(G→A)	12	.0064	0.9	88.5
W1282X	21	.0022	0.9	89.4
Q493X	10	.0003	0.6	90.0
1154insTC	7	.0030	0.6	90.6
3849+10 kb (C→T)	Intron 19	.0062	0.6	91.2
R553X	10	.0014	0.3	91.5
V520F	10	.0046	0.3	91.8
R117H	4	.0005	0.3	92.1
R1283M	20	.0063	0.3	92.4
R347P	7	.0006	0.3	92.7
E60X	3	–	0.3	93.0
Unknown/private	–	–	7.0	100

CF is more homogeneous in this population than in most others. See *Table 5.1* for nomenclature of mutations. *CFTR* gene mutations are listed and referenced in OMIM under the number shown, e.g. N1303K is 219700.0032. Data courtesy of Dr Andrew Wallace, St Mary's Hospital, Manchester.

Direct mutation detection is less straightforward in diseases with extensive allelic heterogeneity

For many diseases no single mutation is particularly frequent. Autosomal dominant and X-linked diseases are usually subject to strong selection, with a correspondingly high frequency of new mutations (page 72). In these cases, two unrelated affected people usually carry different mutations (but usually in the same gene). Any of the general mutation detection methods listed in *Table 16.1* could be used for direct testing, but this may not be practicable in the setting of a service diagnostic laboratory. Dystrophin illustrates the problems and some possible approaches in a highly heterogeneous disease.

Dystrophin mutations

Although dystrophin mutations are extremely heterogeneous, 60–65% of all mutations are deletions of one or more exons and, as shown in *Figure 15.2*, these preferentially affect certain exons. Two multiplex PCR reactions (the one shown in *Figure 16.7* and another which tests exons in the 5′ part of the gene) reveal 98% of all deletions in affected males. Clearly there is a risk of false positive results, if an exon fails to amplify because of a PCR failure or because of a nonpathogenic change in the intron sequence to which the primer binds. Most deletions remove more than one exon; deletions of noncontiguous exons are suspicious and deletions of just a single exon may need to be confirmed.

The high frequency of deletion mutations in DMD/BMD helps mutation detection in males, but makes it more difficult in females, since the normal X chromosome masks any deletion present on the other X. Deciding whether female members of DMD families are carriers is a major diagnostic problem. The difficulty in measuring gene dosage in these cases illustrates a difference between research and diagnostic service. Quantitative Southern blotting or PCR can easily be done well enough to guide research, but it is not easy to get a result so unambiguous that a mother could prudently base reproductive decisions on it. PCR can be quantitative

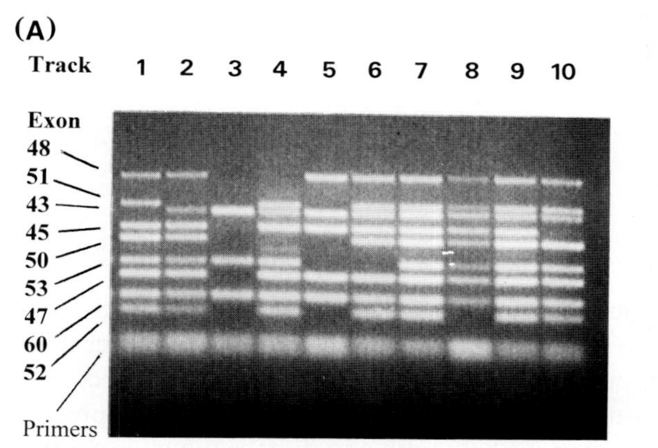

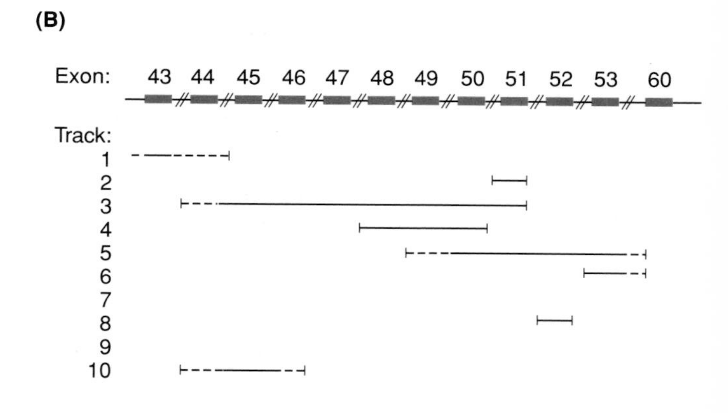

Figure 16.7: Multiplex screen for dystrophin gene deletions.

(A) Products of multiplex PCR amplification of nine exons, using samples from 10 unrelated patients with Duchenne/Becker muscular dystrophy. PCR primers have been designed so that each exon, with some flanking intron sequence, gives a different sized PCR product. Courtesy of Dr R. Mountford, St Mary's Hospital, Manchester.
(B) Interpretation: solid lines show exons definitely deleted, dotted lines show possible extent of deletion running into untested exons. No deletion is seen in samples 7 and 9 – these patients may have point mutations, or deletions of exons not examined in this test. Exon sizes and spacing are not to scale.

provided the number of cycles is restricted, so that the reaction does not reach a plateau, but making quantitative PCR reliable enough for diagnostic work is not easy. Fluorescent sequencing machines (*Figure 11.23*), with their inherently quantitative mode of operation and their ability to measure small quantities of product, probably provide the best method for detection of deletion carriers (and duplications).

Alternatively, if there is a known deletion in the family, typing at-risk females for microsatellites mapping within the deletion may reveal heterozygosity (showing that both X chromosomes are intact) or apparent nonmaternity, where a mother has transmitted no marker allele to her daughter because of the deletion (*Figure 16.8*). A final option is FISH (page 280), using a probe which does not hybridize to the deleted chromosome.

If the multiplex screen fails to show a deletion, then defining the mutation is difficult. With so many exons, screening the DNA exon by exon using SSCP or other standard methods is excessively laborious. Probably the best method is to use the protein truncation test (page 397). Missense mutations rarely or never cause DMD,

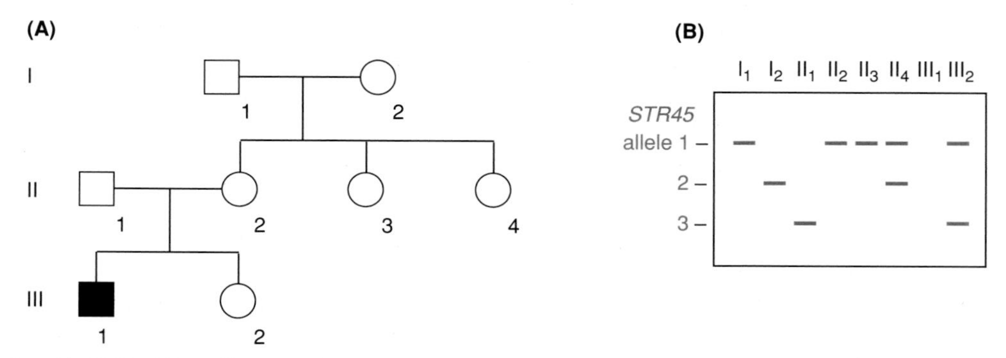

Figure 16.8: A family with DMD. **(A)** Pedigree and **(B)** results of typing with the intragenic marker *STR45*.

The affected boy III_1 has a deletion which includes *STR45* (lane 7 of the gel is blank). His mother II_2 and his aunt II_3 inherited no allele of *STR45* from their mother I_2, showing that the deletion is being transmitted in the family. I_2 is apparently homozygous for this highly polymorphic marker (lane 2), but in fact is *hemizygous*. The other aunt II_4 and the sister III_2 are heterozygous for the marker, and therefore do not carry the deletion.

so most nondeletion mutations are revealed in this way. The dystrophin (*DMD*) gene is primarily expressed in muscle, but in skilled hands low-frequency *ectopic ('illegitimate') transcripts* can be amplified from lymphocytes.

Indirect testing: gene tracking

Gene tracking was historically the first type of DNA diagnostic method to be widely used. Most of the mendelian diseases which form the bread-and-butter work of diagnostic laboratories went through a phase of gene tracking, then moved on to direct tests once the genes were cloned. DMD, HD, CF and myotonic dystrophy are familiar examples. A similar progression is likely with any disease which is studied by the classic 'reverse genetic' approach of linkage analysis followed by positional cloning. However, the progression is not inevitable.

With some diseases, even though the gene has been cloned, mutations are hard to find. In some cases the known mutations are scattered widely over a large gene and, for others, mutation detection, for unknown reasons, has not so far been very successful. Now that we have highly informative microsatellite markers which can be scored by PCR (page 197), gene tracking is not nearly as laborious or frustrating as it was 10 years ago. Thus gene tracking is not just a stop-gap approach for diseases which have been mapped but not yet cloned. Linked markers still have their place in modern molecular diagnosis.

Gene tracking involves three logical steps

In 1982 a RFLP detected by hybridizing *Taq*I-digested DNA to the anonymous genomic clone RC8 was shown to be linked to the locus for DMD (Murray *et al.*, 1982). For the first time it became possible to track segregation of the *DMD* gene in families. *Figure 16.9* shows an early example. Individual II_2 is an obligate carrier since she has an affected brother and an affected son. Her daughter III_2 therefore has a 1:2 risk of being a carrier. The RC8 types, shown in *Figure 16.9*, enable us to refine this risk estimate.

The argument set out in the legend to *Figure 16.9* illustrates the basic logic of gene tracking. Always there is a *consultand*, whose status we wish to know, and there is at least one parent (two for autosomal recessive diseases) who is heterozygous for the disease and so may or may not have passed on the disease gene to the consultand. The investigation always goes through three stages:

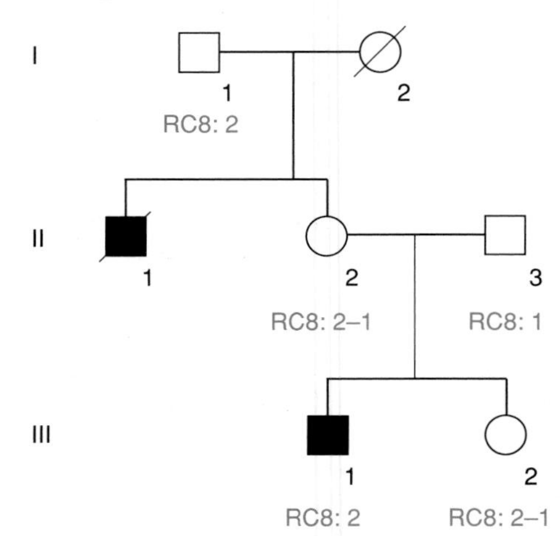

Figure 16.9: Gene tracking: a family with DMD, showing types for the RC8-*Taq*I RFLP.

II_2 inherited an X chromosome which carried DMD from her mother and an X chromosome carrying RC8 allele 2 from her father. The paternal X chromosome carries a normal allele at the *DMD* locus, while the maternal X chromosome must have carried RC8 allele 1. This establishes the phase of II_2. The consultand III_2 is heterozygous for the RC8 polymorphism. She must have inherited allele 1 from her father, therefore she inherited allele 2 from her mother. RC8 allele 2 marks the unaffected X chromosome in II_2, therefore these data suggest that III_2 has not inherited the DMD mutation from her mother (but see text).

(i) *distinguish the two chromosomes of the parent(s) who may have transmitted the disease gene.* Find a polymorphism, linked to the disease locus, for which the parent is heterozygous. In the early days of gene tracking, this was the limiting step. Not many markers were known, and those were mostly two-allele RFLPs, for which at most 50% of people would be heterozygous ($2pq$ has a maximum value of 0.5 at $p = q = 0.5$, see *Table 12.1*). Nowadays, with dense maps of highly informative microsatellite markers, this limitation has largely disappeared.

(ii) *Work out the* ***phase.*** That is, work out which marker allele segregates with the disease allele. This is done by typing suitable relatives. Precisely who is the best relative to type depends on the mode of inheritance and the pedigree structure. For X-linked diseases the father is always best, if available, because he has only one X chromosome which he necessarily passes on to his daughter. Of course, this assumes that he really is the father, which introduces a delicate aspect of gene tracking. Where deductions depend critically on paternity, it seems prudent to check paternity (see below, page 448). Demonstrated nonpaternity usually renders gene tracking impossible. However, it is important that the genetic team should have a clearly thought-out policy on what to do if nonpaternity is discovered. Should the family be told? Practice here may differ from country to country, depending on the legal framework within which the genetics service operates. In the UK, families would normally not be told.

(iii) *Discover which one was passed on to the proband.* Use the marker to identify which chromosome is transmitted to the consultand. This introduces a fundamental limitation of gene tracking. In *Figure 16.9*, II_2 transmitted marker allele 2 to her daughter and we take this as an optimistic finding. However, she also transmitted allele 2 to her affected son III_1. The explanation of course lies in recombination. In the meiosis producing the egg from which III_1 developed, there was a crossover between the *DMD* and RC8 loci; the egg received a recombinant X chromosome carrying RC8 allele 2 and the DMD disease allele (*Figure 16.10*). If the recombination fraction (page 313) is θ, there is also a chance θ that the same thing happened in the meiosis producing III_2. Therefore III_2's carrier risk is not zero but θ.

Recombination sets a fundamental limit on the accuracy of gene tracking

Because the DNA marker sequence used for gene tracking is not the sequence which causes the disease, there is always the possibility of recombination between the disease and the marker. The recombination fraction can be estimated from family studies by standard linkage analysis (page 320). RC8 turned out to show 20% recombination with the *DMD* locus, therefore although it was historically important it was quickly superseded for diagnostic purposes by markers more tightly linked to the disease. If one nucleotide in 300 is polymorphic (page 241), and loci 1 Mb apart show approximately 1% recombination (page 316), there should be a large number of markers less than 1 cM from any given disease locus.

Gene tracking is normally done nowadays with markers showing at most 1% recombination with the disease locus. Ideally one uses an intragenic marker, such as a microsatellite within an intron. If a gene is very large or if it occupies a recombination hotspot, even intragenic markers can recombine at significant frequency with the disease mutation. DMD is a particularly difficult case, with an average of 5% recombination even with intragenic markers.

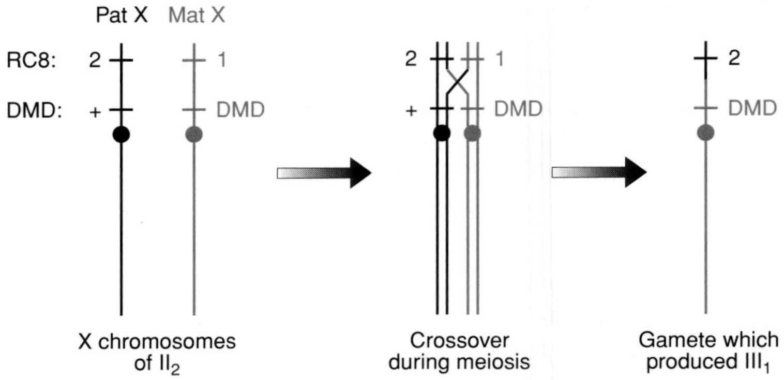

Figure 16.10: Crossover between *DMD* and RC8 loci in the family in *Figure 16.9*.

There may well have been other crossovers between the two X chromosomes of II_2, in addition to the one shown here, but only this one influences the outcome of gene tracking using RC8.

Using flanking markers reduces the error rate due to recombination

Recombination between marker and disease can never be ruled out, even for very tightly linked markers. The error rate can be greatly reduced by using two marker loci, situated on opposite sides of the disease locus. With such flanking or bridging markers, a recombination between either marker and the disease will also produce a marker–marker recombinant, which can be detected (e.g. III_1, *Figure 16.11*). If a marker–marker recombinant is seen in the consultand, then no prediction can be made about inheritance of the disease, but at least a false prediction has been avoided. If no marker–marker recombinant is seen the only residual risk is that of double recombinants. The true probability of a double recombinant is very uncertain because of the conflicting possibilities of interference (page 315) reducing the likelihood below $\theta_A\theta_B$ and various more speculative factors, such as gene conversion (page 283) or locus heterogeneity (*Figure 12.10*), which might increase it. In practice, whatever the precise final risk, it is likely to be small compared with the risk of a wrong prediction due to human error in obtaining and processing the DNA samples.

Gene tracking can be applied to autosomal diseases

Exactly the same principles apply to gene tracking in autosomal diseases as to X-linked diseases. As always the analysis follows the three steps:

(i) distinguish the two chromosomes in the relevant parent(s);

(ii) determine phase;

(iii) work out which chromosome the consultand received.

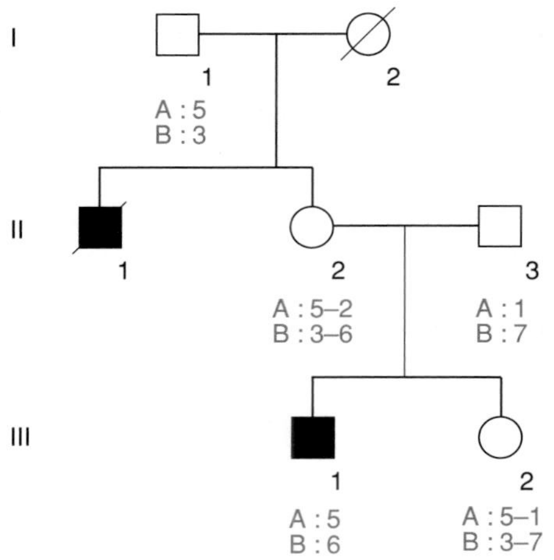

Figure 16.11: Gene tracking using flanking markers.

The family shown in *Figure 16.9* has been typed for two polymorphisms A and B which flank the disease locus. III_2 can have inherited DMD only if she has one recombination between marker A and DMD and another between DMD and marker B. If the recombination fractions are θ_A and θ_B respectively, then the probability of a double recombinant is of the order $\theta_A\theta_B$, which typically will be well under 1%. III_1 can now be seen to be recombinant.

Huntington disease

Figure 16.12 illustrates both gene tracking and direct testing in a family with HD. The same logic applies to any dominant disease, although the late onset of HD causes special problems for counseling. The example has been chosen to illustrate the principle. Now that the *HD* gene has been cloned, **predictive tests** are done by a direct test, measuring the size of the $(CAG)_n$ trinucleotide repeat in a DNA sample from the consultand, II_1. This is more reliable than gene tracking and avoids involving the rest of the family.

Not everybody would wish to know whether they carry the *HD* gene. II_5 has taken a **fetal exclusion test**. This prenatal test enables her to produce only healthy children, without revealing whether or not she herself will develop HD. A single linked marker has been used, so the error rate (per meiosis) is equal to the recombination fraction θ. As before, this could be substantially reduced by using flanking markers. Marker allele 2 in II_5 marks the chromosome that she inherited from her unaffected father. If the fetus III_3 inherits this allele from her, then it is not at risk of HD (barring recombination). The fetal exclusion test carries a price. If the fetus inherited allele 1, then it would be at the same 50% risk as II_5 and the mother would have chosen termination of the pregnancy. For the benefit of avoiding a predictive test on herself, she pays the price of a 50% chance that any terminated fetus would in fact have been normal. We could have tested the fetus directly for the expanded trinucleotide repeat and thus avoided termination of normal fetuses. However, if we did, then a positive result would tell the mother that she herself is also carrying the disease gene.

This family illustrates how genetic testing often involves much more than just the science. Laboratory workers must remember that they are testing people, not DNA. No predictive results must be generated which have not been explicitly requested, after due consideration, by a fully informed adult.

Gene tracking for autosomal recessive diseases

The pedigrees in *Figure 16.13* emphasize the need for both an appropriate pedigree structure (DNA must be available from the affected child) and informative marker types. Even if the affected child is dead, if the Guthrie card (see page 429) can be retrieved, DNA sufficient for PCR typing can usually be extracted from the dried blood spot. Again, using multi-allele microsatellites would make most families informative, provided the pedigree structure was appropriate.

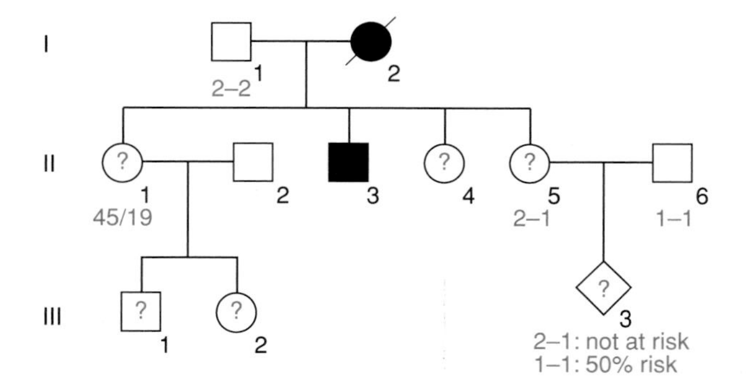

Figure 16.12: Genetic tests in a family with Huntington disease.

II_1 has requested a predictive test. This reveals alleles of 19 and 45 repeats (*Figure 16.6A*), showing that she carries the disease allele and will eventually develop HD. Each of her children is at 50% risk of having the disease allele, but predictive tests would not be carried out on children. II_5 is pregnant and requests a fetal exclusion test. I_1, II_5, II_6 and III_3 are typed for a marker linked to the HD locus. The results on the fetus III_3 will define it as at zero or 50% risk, without revealing whether or not II_5 carries the disease allele. If the fetus is at high risk, II_5 wishes to terminate the pregnancy.

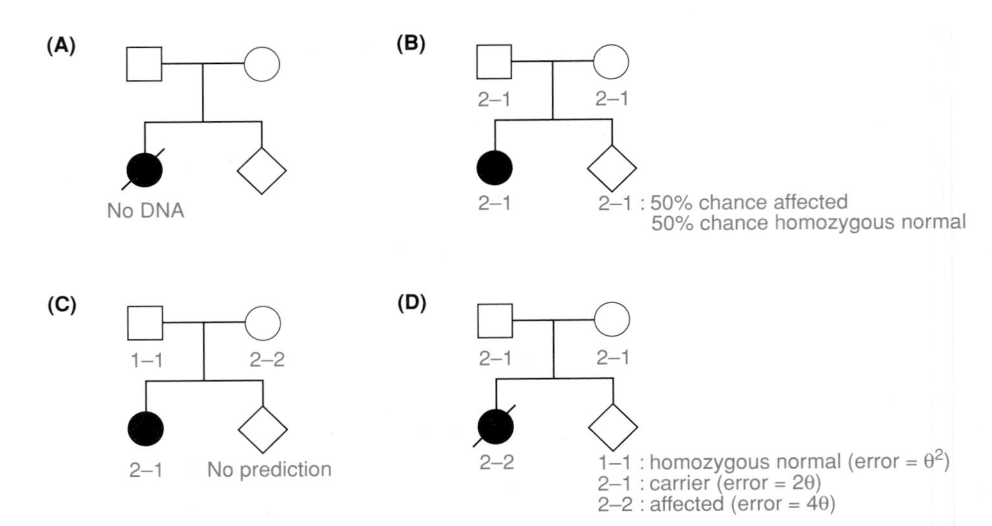

Figure 16.13: Gene tracking for prenatal testing in an autosomal recessive disease.

Four families each have a child affected with a recessive disease.
(A) No diagnosis is possible if there is no sample from the affected child.
(B) If everybody has the same heterozygous genotype for the marker, the result is not clinically useful.
(C) If the parents are homozygous for the marker, no prediction is possible with this marker.
(D) Successful prediction. These examples emphasize the need for both an appropriate pedigree structure (DNA must be available from the affected child) and informative marker types.

The error rates shown are the chance of having a clinically affected child when the prediction was of an unaffected child (carrier or homozygous normal) or vice versa. They are obtained as follows:

- four meioses are involved in producing the two offspring. A prediction that the fetus was affected would be invalidated by a recombination in any one of them, so the error rate is 4θ.
- If the prediction was that the fetus was a carrier, only half the possible 4θ errors would be clinically important (affected child), so the practical error rate is 2θ.
- Two recombinations would be needed to make the fetus affected when it was predicted to be homozygous normal, giving a negligible practical error rate of θ^2.

All these error rates could be substantially reduced by using flanking markers, though working out risks in such a pedigree if a marker–marker recombinant is detected can be difficult.

Calculating risks in gene tracking

Unlike direct testing, gene tracking always involves a calculation. Factors to be taken into account in assessing the final risk include:

- the probability of disease–marker and marker–marker recombination, as in the autosomal recessive example above.
- Uncertainty, due to imperfect pedigree structure or limited informativeness of the markers, about who transmitted what marker allele to whom (see *Figure 12.4* for an example).

- Uncertainty as to whether somebody in the pedigree carries a newly mutant disease allele (see *Figure 3.9* for an example).

Two alternative methods are available for performing the calculation.

Bayesian calculations (see Box 16.1*)*

Bayes' theorem provides a general method for combining probabilities into a final overall probability. The theory and procedure are shown in *Box 16.1*, and a sample calculation is set out in *Figure 16.14*. A very detailed set of calculations covering almost every conceivable situation in DNA diagnostics can be found in the book by Bridge (1994), which the interested reader should consult.

For simple pedigrees, Bayesian calculations give a quick answer, but for more complex pedigrees the calculations can get very elaborate. Few people feel fully confident of their ability to work through a complex pedigree correctly, although the attempt is a valuable mental exercise for teasing out the factors contributing to the final risk. An alternative is to use a linkage analysis program.

Box 16.1: Use of Bayes' theorem for combining probabilities

A formal statement of Bayes' theorem is:

$$P(H_i \mid E) = P(H_i) \cdot P(E \mid H_i) \;/\; \Sigma\, P(H_i) \cdot P(E \mid H_i)$$

$P(H_i)$ means the probability of the *i*th hypothesis, and the vertical line means 'given', so that $P(E \mid H_i)$ means the probability of the evidence (E) given hypothesis H_i.

The steps in performing a Bayesian calculation are:

(i) set up a table with one column for each of the alternative hypotheses. Cover all the alternatives.

(ii) Assign a **prior probability** to each alternative. The prior probabilities of all the hypotheses must sum to 1. It is not important at this stage to worry about exactly what information you should use to decide the prior probability, as long as it is consistent across the columns. You will not be using all the information (otherwise there would be no point in doing the calculation because you would already have the answer) and any information not used in the prior probability can be used later.

(iii) Using one item of information not included in the prior probabilities, calculate a **conditional probability** for each hypothesis. The conditional probability is the probability of the information, given the hypothesis $P(E \mid H_i)$ [*not* the probability of the hypothesis given the information, $P(H_i \mid E)$]. The conditional probabilities for the different hypotheses do not necessarily sum to 1.

(iv) If there are further items of information not yet included, repeat step (iii) as many times as necessary until all information has been used once and once only. The end result is a number of lines of conditional probabilities in each column.

(v) Within each column, multiply together the prior and all the conditional probabilities. This gives a **joint probability**. The joint probabilities do not necessarily sum to 1 across the columns.

(vi) If there are just two columns, the joint probabilities can be used directly as odds. Alternatively the joint probabilities can be scaled to give **final probabilities** which do sum to 1. This is done by dividing each joint probability by the sum of all the joint probabilities.

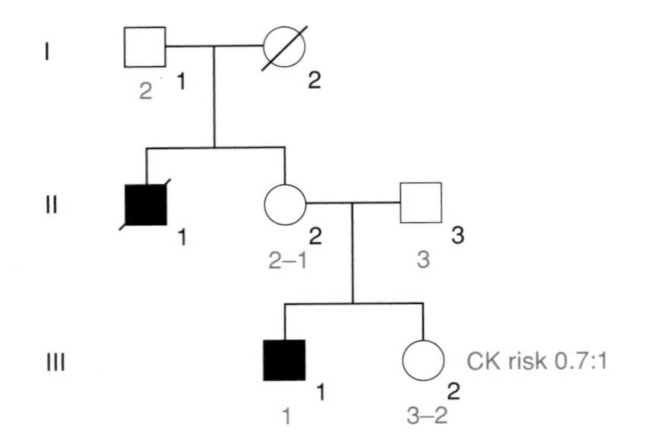

Hypothesis: III_2 is	A carrier	Not a carrier
Prior probability	1/2	1/2
Conditional (1): DNA result	0.05	0.95
Conditional (2): CK data	0.7	1
Joint probability	0.0175	0.475
Final probability	0.0175/0.4925	0.475/0.4925
	= 0.036	= 0.964

Figure 16.14: A Bayesian calculation of genetic risk.

III_2 wishes to know her risk of being a carrier of DMD, which affected her brother III_1 and uncle II_1. Serum creatine kinase testing (an indicator of subclinical muscle damage common in DMD carriers) gave carrier:noncarrier odds of 0.7:1. A DNA marker which shows on average 5% recombination with DMD gave the types shown. The risk calculation, following the guidelines in *Box 16.1*, gives her overall carrier risk as 3.6%.

Using linkage programs for calculating genetic risks

At first sight it may seem surprising that a program designed to calculate lod scores can also calculate genetic risks – but in fact the two are closely related (*Figure 16.15*). Linkage analysis programs are general-purpose engines for calculating the likelihood of a pedigree, given certain data and assumptions. For calculating the likelihood of linkage we calculate the ratio:

$$\frac{(\text{likelihood of data} \mid \text{linkage, recombination fraction } \theta)}{[\text{likelihood of data} \mid \text{no linkage } (\theta = 0.5)]}$$

For estimating the risk that a proband carries a disease gene, we calculate the ratio:

$$\frac{(\text{likelihood of data} \mid \text{proband is a carrier, recombination fraction } \theta)}{(\text{likelihood of data} \mid \text{proband is not a carrier, recombination fraction } \theta)}$$

As in *Box 16.1*, the vertical line | means 'given'.

DNA tests for identity and relationships

DNA profiling can be used for identifying individuals and determining relationships

We use the term **DNA profiling** to refer to the general use of DNA tests to establish identity or relationships. **DNA fingerprinting** is reserved for the technique invented by Jeffreys *et al.* (1985) using multilocus probes. For more detail on this whole topic, the reader is referred to the book by Krawczak and Schmidtke (1994). Three types of genetic markers are widely used, as described in the following sections.

Minisatellite DNA fingerprinting probes

These probes contain the common core sequence of a hypervariable dispersed repetitive sequence (GGGCAGGANG), first discovered by Jeffreys *et al.* (1985) in the myoglobin gene. They give an individual-specific fingerprint of bands when hybridized to Southern blots (*Figure 16.16*). Their chief disadvantage is that it is not possible to tell which pairs of bands in a fingerprint represent alleles. Thus, when

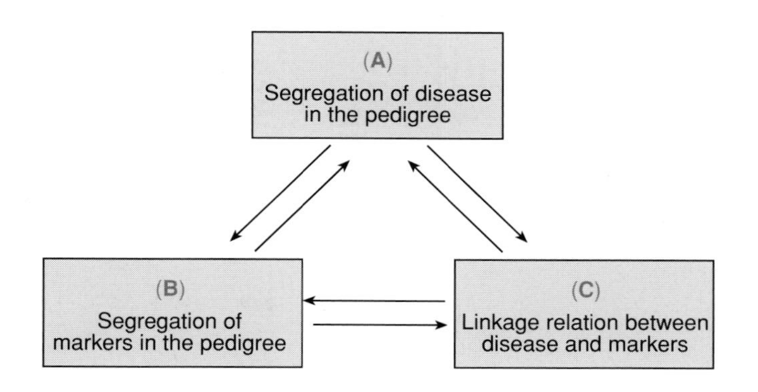

Figure 16.15: Use of linkage analysis programs for calculating genetic risks.

Given information on any two of these subjects, the program can calculate the third. For linkage analysis, the program is given (A) and (B), and calculates (C).
For calculating genetic risks, the program is given (B) and (C), and calculates (A).

matching DNA fingerprints, one compares the position and intensity of each band individually. Other hypervariable repeated sequences have been used in the same way, for example those detected by the synthetic oligonucleotide $(CAC)_5$ (Krawczak *et al.*, 1993).

Minisatellite probes

Minisatellite or VNTR probes (page 196) recognize single-locus hypervariable tandem repeats on Southern blots. Each person's DNA should give two bands, representing the two alleles. Profiling is based on four to 10 different VNTR polymorphisms. These probes allow exact calculations of probabilities (of paternity, of the suspect not being the rapist, etc.), if the gene frequency of each allele in the population is known. For matching alleles between different gel tracks, the continuously variable distance along the gel has to be divided into a number of 'bins'. Bands falling within the same bin are deemed to match. It is imperative that the criteria used for judging matches in each profiling test should be the same binning criteria that were used to calculate the population frequencies of each allele. The binning criteria can be arbitrary within certain limits, but they must be consistent.

Microsatellite markers

Microsatellite polymorphisms (page 197) are based on short tandem repeats, usually di-, tri- or tetranucleotides. They have the advantages over minisatellites that they can be typed by PCR and that discrete alleles can be defined unambiguously by the precise repeat number. This avoids the binning problem and makes it easier to relate the results to population gene frequencies. Minor variations within repeated units of some microsatellites potentially allow an almost infinite variety of alleles to be discriminated, so that the genotype at a single locus might suffice to identify an individual (Jeffreys *et al.*, 1991).

DNA profiling can be used to determine the zygosity of twins

In studying nonmendelian characters (Chapter 18), and sometimes in genetic counseling, it is important to know whether a pair of twins are monozygotic (MZ, identical) or dizygotic (DZ, fraternal). Traditional methods depended on an assessment of phenotypic resemblance or on the condition of the membranes at birth (twins contained within a single chorion are always MZ, though the converse is not true). Errors in zygosity determination systematically inflate the estimate of the role of genetic factors in nonmendelian characters, because very similar DZ twins are wrongly counted as MZ , while very different MZ twins are wrongly scored as DZ.

Genetic markers provide a much more reliable test of zygosity. The extensive literature on using blood groups for this purpose is summarized by Race and Sanger (1975). DNA profiling is nowadays the method of choice. The Jeffreys' fingerprint-

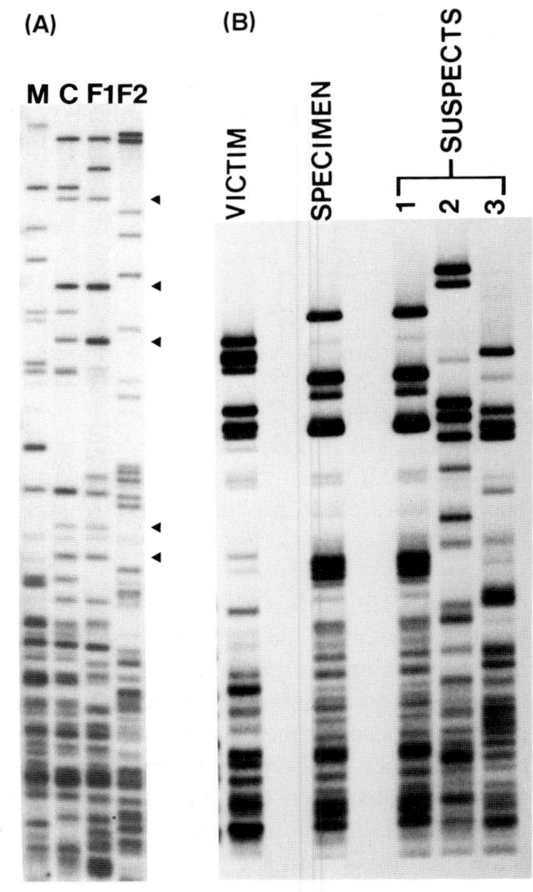

Figure 16.16: Legal and forensic use of DNA fingerprinting.

(A) A paternity test. Fingerprints are shown from the mother (M), child (C) and two possible fathers (F1, F2). The DNA fingerprint of F2 contains all of the paternal bands found in the child, whereas that of F1 contains only one such paternal band. **(B)** A rape case. The fingerprint of suspect A exactly matches that from the semen sample S on a vaginal swab from the victim. As a result of this evidence, Suspect A was charged with rape and found guilty. Photograph courtesy of Cellmark Diagnostics, Abingdon, Oxfordshire.

ing probe allows a very simple test – samples from MZ twins look like the same sample loaded twice, and samples from DZ twins show some differences. An error rate could be calculated from empirical data on band sharing by unrelated people, using some defined binning strategy (see above).

When single-locus markers are used, if twins give the same types, the probability that they are MZ can be calculated. For each locus, the probability that DZ twins would type alike is calculated. If the parents have been typed, this is a simple mendelian ratio; otherwise for each possible parental mating the probability of DZ twins typing the same must be calculated and weighted by the probability of that parental mating calculated from population gene frequencies. The resultant probabilities for each (unlinked) locus are multiplied, to give an overall likelihood P_I that DZ twins would give the same results with all the markers used. The probability that the twins are MZ is then:

$$P_m = m \,/\, [m + (1 - m)P_I]$$

where m is the proportion of twins in the population who are MZ (about 0.4 for like-sex pairs). Sample calculations are given in Appendix 5 of Vogel and Motulsky (1979).

DNA profiling can be used to disprove or establish paternity

Excluding paternity is fairly simple – if the child has a marker allele not present in either the mother or alleged father then, barring new mutations, the alleged father is not the biological father. Proving paternity is in principle impossible – one can never prove that there is not another man in the world who could have given the child that particular set of marker alleles. All one can do is establish a probability of nonpaternity which is low enough to satisfy the courts and, if possible, the putative father.

DNA fingerprinting probes have been widely used for this purpose (*Figure 16.16A*). Bands must be binned according to an arbitrary but consistent scheme, as explained above, to decide whether or not each nonmaternal band in the child fits a band in the alleged father. Then if, say, 10/10 bands fit, the odds favoring paternity are $1{:}p^{10}$, where p is the chance that a random man from the population would have a band matching a given band in the child. Even if p is as high as 0.2, p^{10} is only 10^{-7}. Single-locus probes allow a more explicit calculation of the odds (*Figure 16.17*). A series of four to 10 unlinked single-locus markers can give overwhelming odds favoring paternity if all the bands fit.

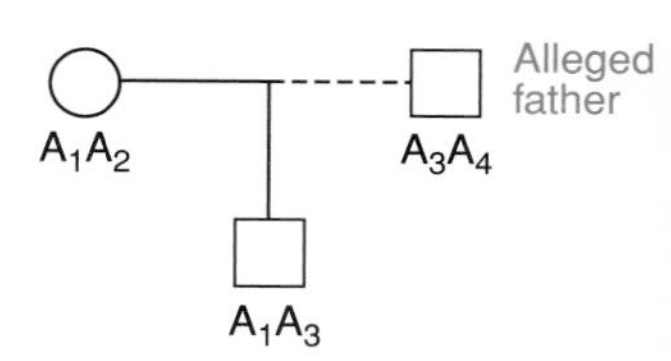

Figure 16.17: Using a single-locus marker for a paternity test.

The odds that the alleged father, rather than a random member of the population, is the true father are $\frac{1}{2}{:}q_3$, where q_3 is the gene frequency of A_3. A series of n markers would be used and, if paternity were not excluded, the odds would be $(\frac{1}{2})^n : q_A.q_B.q_C \ldots q_N$.

Although DNA evidence is universally accepted in paternity cases, it is in principle open to the same objections as have been raised in criminal cases. These are discussed below.

DNA profiling is a powerful tool for forensic investigations

DNA profiling for forensic purposes follows the same principles as paternity testing. Scene-of-crime material (bloodstains, hairs or a vaginal swab from a rape victim) are typed and matched to a DNA sample from the suspect. If the bands all match, the odds that the criminal is the suspect rather than a random member of the population are calculated in the same way as shown above for paternity testing. Objections have been raised to three features of these calculations.

(i) The multiplicative principle, that the overall probabilities can be obtained by multiplying the individual probability for each band or locus, depends on the assumption that bands are independent. If the population were actually stratified into reproductively isolated groups, each of whom tended to have a particular subset of bands or alleles, the calculation would be misleading. This is serious because it is the multiplicative principle which allows such exceedingly definite likelihoods to be given.

(ii) For single-locus markers, the probability depends on the gene frequencies. DNA profiling laboratories maintain databases of gene frequencies – but were these determined in an appropriate ethnic group for the case being considered?

(iii) The odds calculations assume the alternative to the suspect is a random member of the population. If the alternative were the suspect's brother, the odds would look very different.

These issues have been debated at great length, especially in the American courts. The first two objections are true in principle, but the question is whether they make any difference in practice. The US National Research Council (1992) made a well-intentioned attempt to kill the argument by proposing an *ad hoc* 'ceiling principle' (see *Box 16.2*), which they hoped would include enough intentional weakening of the evidence to satisfy any possible objections by defending lawyers, while still giving probabilities high enough to convict. In the event, they unleashed a storm of criticism from statisticians and population geneticists (summarized by Morton, 1995). It is unlikely that the ceiling principle will survive. Meanwhile it would be ironic if courts, seeing opposing expert witnesses giving odds of correct identification differing a million-fold ($10^5{:}1$ versus $10^{11}{:}1$), were to decide that DNA evidence is hopelessly unreliable and turn instead to eye-witness identification (odds of correct identification < 50:50).

Box 16.2: Prudent compromise or unscientific fudge? – the 'ceiling principle' proposed for DNA profiling evidence

(i) Unbiased estimates of gene frequencies should be arbitrarily replaced by the upper 97.5% confidence limit.

(ii) Random samples of 100 persons from each of 15–20 relatively homogeneous populations should be typed. The ceiling gene frequency should be taken as the largest frequency in any of the populations, or 5%, whichever is the greater.

These principles, proposed by the US National Research Council (1992), proved extremely controversial, and were rejected by population geneticists.

Population screening

Population screening follows naturally from the ability to test directly for the presence of a mutation. DNA tests are rather different from many other screening tests because there are no separate screening and diagnostic tests (see below). However, proposals to introduce any population screening test still need to satisfy the same criteria (*Table 16.6*), regardless of the technology used.

Table 16.6: Requirements of a population screening program

Requirement	Example
A positive result must lead to some useful action	Preventive treatment, e.g. special diet for PKU Review and choice of reproductive options in CF carrier screening
The whole program must be socially and ethically acceptable	Subjects must opt in with informed consent Screening without counseling is unacceptable There must be no pressure to terminate affected pregnancies Screening must not be seen as discriminatory, e.g. sickle-cell screening needs full consent
The test must have high sensitivity and specificity	Tests with many false positives, even if these are subsequently filtered out by a definitive diagnostic test, can create unacceptably high levels of anxiety among normal people
The benefits of the program must outweigh its costs	It is unethical to use limited health care budgets in an inefficient way

Screening and diagnostic tests are often different

Traditionally a distinction is drawn between screening and diagnosis. For example, pregnant women are often offered screening for fetal neural tube defects by measuring the level of α-fetoprotein (AFP) in the mother's serum (*Figure 16.18*). Women whose serum AFP level is above an arbitrary cut-off point are checked for wrong dates, multiple pregnancy and fetal death, all of which can produce high serum AFP readings. If these do not account for the raised serum AFP, a second serum AFP test is run and, if the level is still high, the woman is offered amniocentesis or high-resolution ultrasonography. This is a diagnostic test, giving a definitive answer. Only a small proportion (2–5%) of women positive on the screening test are positive on the diagnostic test.

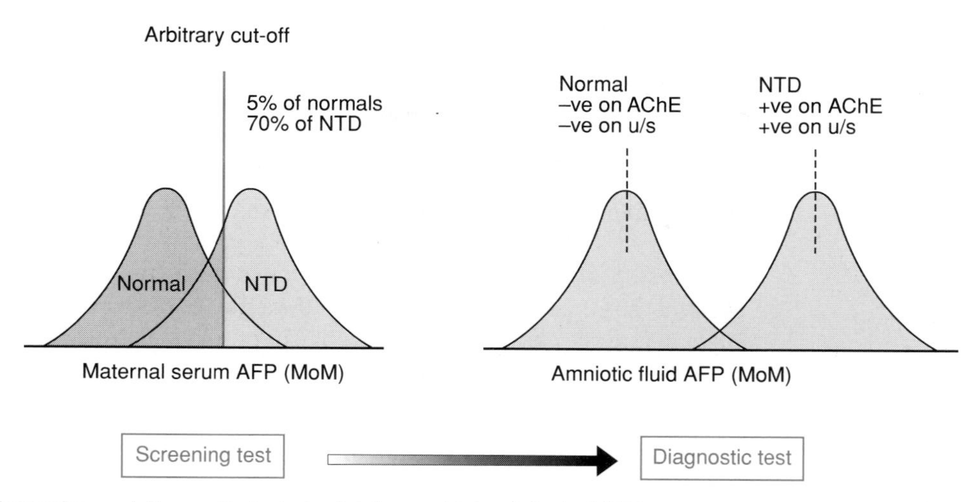

Figure 16.18: Screening and diagnostic tests for fetal neural tube defects (NTD).

Measuring serum AFP is the screening test. An arbitrary cut-off level is defined, typically 2.5 multiples of the median (MoM), which picks up <5% of unaffected pregnancies and 70% of pregnancies where the fetus has an open neural tube defect. Women whose serum AFP level is above the threshold and who fail a series of supplementary checks (see text) are offered amniocentesis. Measuring amniotic AFP is the diagnostic test. It can be supplemented by checking for acetylcholinesterase (AChE) in the fluid and by ultrasound (u/s) scanning. High-resolution ultrasound can alternatively be used directly for screening and diagnosis.

Acceptable screening programs must fit certain criteria

What would screening achieve?

The most important single function of any screening program is to produce some useful outcome. It is quite unacceptable to tell people out of the blue that they are at risk of something unpleasant unless the knowledge enables them to do something about the risk. Proposals to screen for genes conferring susceptibility to breast cancer or heart attacks must be assessed stringently against this criterion. Predictive testing for HD might appear to break this rule – but it is offered only to people who are at high risk of HD and who are suffering such agonies of uncertainty that they request a predictive test, and persist despite counseling in which all the disadvantages are pointed out.

Ideally the useful outcome is treatment, as in neonatal screening for phenylketonuria. Increased medical surveillance is a useful outcome only if it greatly improves the prognosis. A special case is screening for carrier status, where the outcome is the possibility of avoiding the birth of an affected child. People unwilling to accept prenatal diagnosis and termination of affected pregnancies would not see this as a useful outcome, and in general should not be screened.

An ethical framework for screening

Ethical issues in genetic population screening have been discussed recently by a committee of distinguished American geneticists, clinicians, lawyers and theologians, and the reader is referred to their report for a very detailed survey (Andrews *et al.*, 1994). It is in the nature of ethical problems that they have no solutions, but certain principles emerge.

- Any program must be voluntary, with subjects taking the positive decision to opt in.
- Programs must respect the autonomy and privacy of the subject.

- People who score positive on the test must not be pressured into any particular course of action. For example, in countries with insurance-based health care systems, it would be unacceptable for insurance companies to put pressure on carrier couples to accept prenatal diagnosis, or financial pressure or inducements to terminate affected pregnancies.
- Information should be confidential. This may seem obvious, but it can be a difficult issue – we like to think that drivers of heavy trucks or jumbo jets have been tested for all possible risks. Societies with insurance-based health care systems have particular problems about the confidentiality of genetic data, since insurance companies will argue that they are penalizing low-risk people by not loading the premiums of high-risk people.

Specificity and sensitivity measure the technical performance of a screening test

Compared with the ethical problems, the technical questions in population screening are fairly simple. The performance of a test can be measured by its sensitivity and specificity (*Figure 16.19*).

	Affected	Not affected
+ve on test	a	b
–ve on test	c	d

Sensitivity of test = a/(a + c)

Specificity of test = a/(a + b)

Figure 16.19: Sensitivity and specificity of a screening test.

Specificity of a test

Unexpectedly, perhaps, false positive test results can pose a more serious problem than false negatives. If the specificity is low, then a positive test result does not mean much. Even if the false positives can be filtered out subsequently by a diagnostic test, many people will have been worried unnecessarily. *Table 16.7* shows that if a test does have a significant false positive rate, then the specificity is hopelessly low except when testing for very common conditions. DNA tests are potentially valuable for population screening because, compared with biochemical tests, they should generate very few false positives. The most likely causes of false positives in DNA testing are laboratory or clerical errors.

Sensitivity of a test

A test must pick up a reasonable proportion of its intended target (i.e. the sensitivity must be high). While the specificity of DNA tests looks encouraging, the sensitivity usually depends on the degree of allelic heterogeneity. Unless a disease is unusually homogeneous (*Table 16.3*), it is not practicable to test for every conceivable mutation, especially in a large-throughput population screening program. Normally only a subset of mutations will be tested for. *Figure 16.20* shows how the choice of mutations can affect the outcome of a CF carrier screening program.

Table 16.7: A test which performs well in the laboratory may be useless for population screening

Prevalence of condition	True positives in population screened	True positives detected by screening	True negatives in population screened	False positives detected by screening	Specificity of test
1/1000	1000	990	999 000	9990	0.09
1/10 000	100	99	999 900	9999	0.0098
1/100 000	10	10	999 990	10 000	0.001

In a laboratory trial on a panel of 100 affected and 100 control people, this hypothetical test was 99% accurate (it gave a positive result for 99% of true positives, but also for 1% of true negatives). The table shows results of screening 1 million people. The great majority of all people positive on the test are false positives. Such a test is unlikely to be acceptable socially or viable financially for any mendelian disease (these typically affect less than 1 person in 1000).

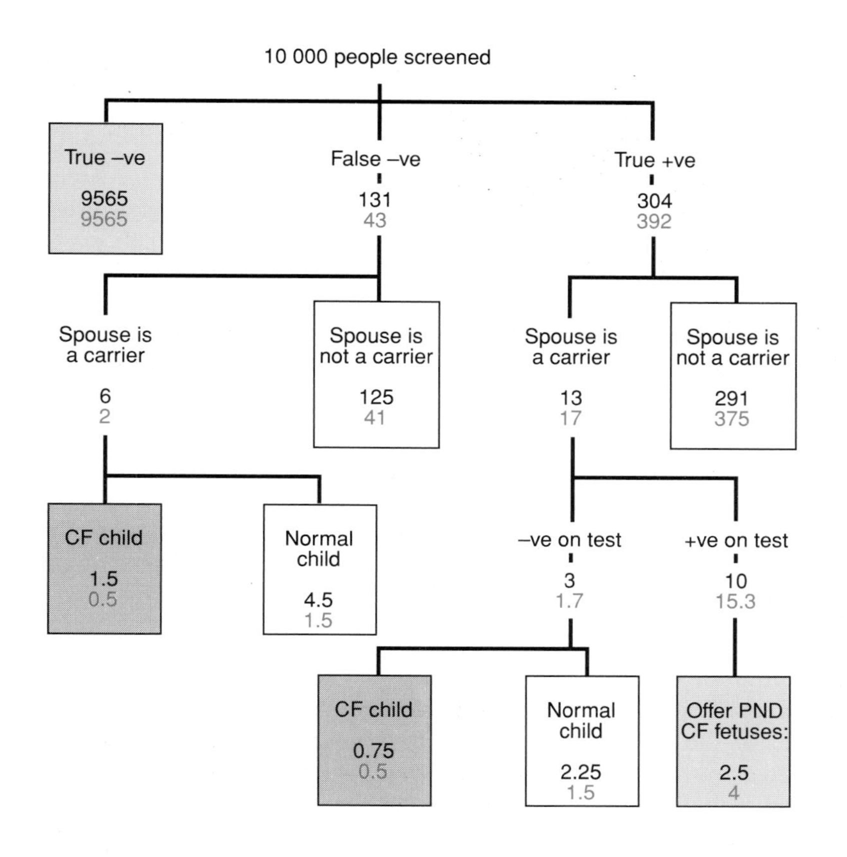

Figure 16.20: Flowchart for CF population screening.

Results of screening 10 000 people, 1:23 of whom is a carrier. If a person tests positive, his/her spouse is then tested. Black figures show results using a test which detects 70% of CF mutations (i.e. testing for ΔF508 only); red figures show results for a test with 90% sensitivity. Pink boxes represent cases which would be seen as successes for the screening program (regardless of what action they then take), gray boxes represent failures. PND, prenatal diagnosis.

It is clear that simply testing for the commonest mutation, ΔF508, would not produce an acceptable program. More affected children would be born to couples negative on the screening program than would be detected by the screening. Whether or not such a program were financially cost-effective, it would surely be socially unacceptable. What constitutes an acceptable program is harder to define. One suggestion (Ten Kate, 1990) focuses on '+/– ' couples, (i.e. couples with one known carrier and the partner negative on all the tests). The partner still might be a carrier of a rare mutation. Ten Kate suggested that an acceptable screening program is one in which the risk for such +/– couples is no higher than the general population risk before screening. That would require a sensitivity of about 95%.

Organization of a genetic screening program

Assuming the proposed program looks ethically acceptable and cost-effective, who should be screened? Three examples highlight some of the options.

Neonatal screening: screening for phenylketonuria

All babies in the UK are tested a few days after birth for PKU. A blood spot from a heel-prick is collected on a card (the Guthrie card) during a home visit and sent to a central laboratory. The phenylalanine level in the blood is measured by chromatography or a bacterial growth test. Babies whose level is above a threshold are called in for further tests. Only a small proportion eventually turn out to have PKU. The lack of informed consent on the part of the infant is justified by the benefit it receives from dietary treatment (Scriver and Clow, 1980).

Prenatal screening: screening for β-thalassemia

Carriers of β-thalassemia can be detected by conventional hematological testing, either before marriage or in the antenatal clinic. Carrier–carrier couples can be offered prenatal diagnosis by DNA analysis. Pre-implantation diagnosis or fetal stem-cell transplants may become alternatives to termination of affected pregnancies. Two ethnic groups in the UK have a high incidence of β-thalassemia: Cypriots and Pakistanis. Screening was quickly accepted by Cypriots in the UK but uptake has been slower among Pakistanis. The comparison illustrates the complex social questions surrounding genetic screening and the relevance of cultural background. Importantly, long-term studies of the Cypriot community show how the success of screening can be measured, not by counts of affected fetuses aborted, but by counts of couples having normal families. Before screening was available, many carrier couples opted to have no children; now they are using screening and having normal families (Modell *et al.*, 1984; Petrou *et al.*, 1990).

Population screening for carriers: proposals for cystic fibrosis

It is now technically feasible and financially worthwhile to screen northern European populations to detect CF carriers. Surveys in the UK suggest that most carrier–carrier couples would opt for prenatal diagnosis and would value the opportunity to ensure that they did not have affected children. This view might change if treatment becomes more effective, for example using gene therapy.

If a screening program is to be introduced, two sets of questions must be considered. How many mutations should the laboratory test for, and who should be offered the test? The problems raised by allelic heterogeneity have been discussed above (see *Figure 16.20*). On the question of who to screen, *Table 16.8* shows some possibilities considered in the UK. Naturally the way health care delivery is organized in each country will determine the range of possibilities. Preliminary results from controlled pilot studies suggest that none of the methods has had the negative effects (increased anxiety) sometimes predicted.

Table 16.8: Possible ways of organizing population screening for carriers of CF

Group tested	Advantages	Disadvantages
Neonates	Easily organized	No consequences for 20 years Many families would forget the result Unethical to test children
School leavers	Easily organized Inform people before they start relationships	Difficult to conduct ethically Risk of stigmatization of carriers
Couples from physicians' lists	Couple is unit of risk Stresses physicians' role in preventitive medicine Allows time for decisions	Difficult to control quality of counseling
Women in antenatal clinic	Easily organized Rapid results	Bombshell effect for carriers Partner may be unavailable Time pressure on laboratory
Adult volunteers ('drop-in CF center')	Few ethical problems	Bad framework for counseling No targeting to suitable users May be inefficient use of resources

Further reading

Andrews LB, Fullarton JE, Holtzman NA, Motulsky AG. (1994) *Assessing Genetic Risks – Implications for Health and Social Policy*. National Academy Press, Washington, DC.

Bridge PJ. (1994) *The Calculation of Genetic Risks – Worked Examples in DNA Diagnostics*. Johns Hopkins University Press, Baltimore, MD.

Krawczak M, Schmidtke J. (1994) *DNA Fingerprinting*. BIOS Scientific Publishers, Oxford.

Newton CR, Graham A. (1994) *PCR*. BIOS Scientific Publishers, Oxford.

References

CF Genetic Analysis Consortium (1994) *Hum. Mutat.*, **4**, 166–177.

Jeffreys AJ, Wilson V, Thein LS. (1985) *Nature*, **314**, 67–73.

Jeffreys AJ, MacLeod A, Tamaki K, Neil DL, Monckton DG. (1991) *Nature*, **354**, 204–209.

Krawczak M, Böhm I, Nürnberg P *et al.* (1993) *Forensic. Sci. Int.*, **59**, 101.

Metspalu A, Shumaker J, Caskey CT. (1995) *Med. Genet.*, **2**, 108.

Modell B, Petrou M, Ward RH, Fairweather DV, Rodeck C, Varnavides LA, White JM. (1984) *Lancet*, **ii**, 1383–1386.

Morton NE. (1995) *Eur. J. Hum. Genet.*, **3**, 139–144.

Murray JM, Davies KE, Harper PS, Meredith L, Mueller CR, Williamson R. (1982) *Nature*, **300**, 69–71.

National Research Council (1992) *DNA Technology in Forensic Science*. National Academy of Sciences, Washington, DC.

Newton CR, Graham A, Heptinstall LE. (1989) *Nucleic Acids Res.*, **17**, 2503–2516.

Nickerson DA, Kaiser R, Lappin S, Stewart J, Hood L. (1990) *Proc. Natl Acad. Sci. USA*, **87**, 8923–8927.

Oostra RJ, Bolhuis PA, Wijburg FA, Zorn-Ende G, Bleeker-Wagemakers EM. (1994) *J. Med. Genet.*, **31**, 280–286.

Petrou M, Modell B, Darr A, Old J, Kin E, Weatherall DJ. (1990) *Ann. N.Y. Acad. Sci.*, **612**, 251–263.

Race RR, Sanger R. (1975) *Blood Groups in Man*, 6th Edn. Blackwell, Oxford.

Scriver CR, Clow CL. (1980) *New Engl. J. Med.*, **303**, 1336–1342, 1394–1400.

Shiang R, Thompson LM, Zhu Y-Z, Church DM, Fielder T, Bocian M, Winokur ST, Wasmuth JJ. (1994) *Cell*, **78**, 335–342.

Ten Kate LP. (1990) *Am. J. Hum. Genet.*, **47**, 359–361.

Vogel F, Motulsky AG. (1979) *Human Genetics*. Springer, Berlin.

Somatic mutations and cancer

Chapter 17

Somatic mutations are not merely frequent, they are inevitable. Human mutation rates are typically 10^{-5} to 10^{-7} per gene per generation, and our bodies contain perhaps 10^{14} cells. It follows that every one of us must be a somatic mosaic for innumerable genetic diseases (see page 242). This should cause no anxiety. If a cell in one's finger mutates to the HD genotype or a cell in one's ear picks up a CF mutation, there are absolutely no consequences for the individual concerned. Only if a somatic mutation results in the emergence of a substantial clone of mutant cells can there be a risk to the whole organism. This can happen in two ways:

- the mutation causes abnormal proliferation of a cell which would normally replicate little or not at all, thus generating a clone of mutant cells.
- The mutation occurs in an early embryo, affecting a cell which is the progenitor of a significant fraction of the whole organism.

We will consider both classes of somatic mutation in this chapter, beginning with those that cause cancer. See *Box 18.1* for a summary of the ways in which cells differ genetically within a single individual (or a pair of identical twins).

Cancer: natural selection among the somatic cells of an organism

Cells of a multicellular organism collaborate for the greater good – but only to a degree. Evolution by natural selection applies to the constituent cells as well as to the whole organism. In the long term, cells can leave progeny only by co-operating to make an organism capable of reproduction, and so they have necessarily evolved a genome which programs them to co-operate. However, in the short term, if a cell has a mutation that confers a reproductive advantage, it will found a successful clone. Its descendants will take over the organism unless prevented by higher controls which have evolved in the interests of the overall organism.

Cancer is the natural result of this selection among somatic cells. Probably cancer is the normal end-state of any multicellular organism which lives long enough – but successful organisms contrive to postpone this fate. During a billion years of evolution, they have evolved many layers of sophisticated controls, so that clinical cancer results only when several independent controls have been lost. Principal among the higher level controls is a mechanism by which potentially cancerous cells commit suicide by programmed cell death (**apoptosis**). Successful cancer cells must have developed ways of inactivating or by-passing this mechanism (page 471).

Several decades ago, studies of the age dependence of cancer suggested that an average of between six and seven successive mutations are needed to convert a normal cell into an invasive carcinoma (Armitage and Doll, 1954). The chance of a single cell undergoing so many independent mutations is negligible; however, two general mechanisms exist which can make the progression more likely (*Box 17.1*).

Box 17.1: Two ways of making a series of successive mutations more likely

Turning a normal cell into a malignant cancer cell requires perhaps six specific mutations in the one cell. With typical mutation rates of 10^{-6} per gene per cell, it is vanishingly unlikely that any one cell should suffer so many mutations (which is why most of us are alive). The probability of this happening to any one of the 10^{-14} cells in a person is $10^{-14} \times 10^{-36}$, or $1:10^{22}$. Cancer nevertheless happens because of a combination of two mechanisms:

- some mutations enhance cell proliferation, creating an expanded target population of cells for the next mutation (*Figure 17.1*).
- Some mutations affect the stability of the entire genome, increasing the overall mutation rate.

Recent studies have identified three groups of genes which are frequently mutated in cancer :

- **oncogenes** (page 459ff.). These are genes whose action positively promotes cell proliferation. The normal nonmutant versions are properly called *proto-oncogenes*. The mutant versions are excessively or inappropriately active. A single mutant allele may affect the phenotype of the cell.
- **Tumor suppressor (TS) genes** (466ff.). TS gene products inhibit cell proliferation. Mutant versions in cancer cells have lost their function. Both alleles of a TS gene must be inactivated to change the behavior of the cell.
- **Mutator genes** (471ff.). These are responsible for maintaining the integrity of the genome and the fidelity of information transfer. Loss of function of both alleles makes a cell error prone. Among the random errors may be mutations in oncogenes and TS genes.

By analogy with a bus, one can picture the oncogenes as the accelerator and the TS genes as the brake. Jamming the accelerator on (a dominant gain-of-function mutation in an oncogene) or having all the brakes fail (a recessive-loss-of function mutation in a TS gene) will make the bus run out of control. Alternatively, a saboteur could simply loosen nuts and bolts at random and wait for a disaster to happen (a mutator gene mutation).

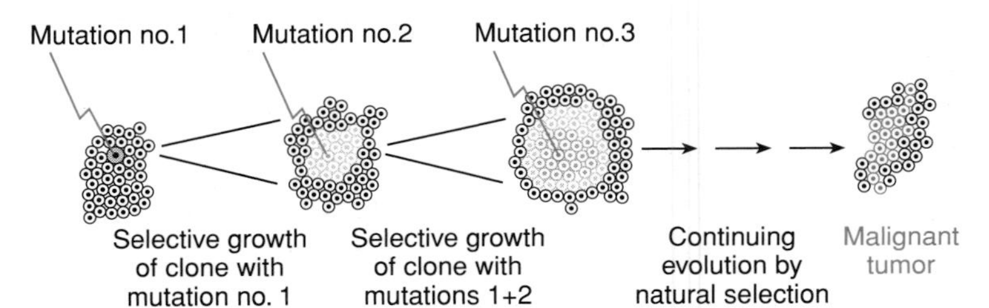

Figure 17.1: Multistage evolution of cancer.

Each successive mutation gives the cell a growth advantage, so that it forms an expanded clone, thus presenting a larger target for the next mutation.

Oncogenes

Animal tumor viruses provided the first evidence of oncogenes

For many years it has been known that some animal leukemias, lymphomas and cancers are caused by viruses. A few human examples are also known (*Table 17.1*). Tumor viruses fall into three broad classes:

- **DNA viruses** normally infect cells lytically. They cause tumors by rare anomalous integrations into the DNA of nonpermissive host cells (cells that do not support lytic infection). One way or another, integration of the viral genome implants the transcriptional activation or replication signals of the virus into the host genome and triggers cell proliferation. Some of the viral genes involved have been identified, such as those for the T antigen of SV40 or E1A and E1B of adenoviruses. Unlike the classic retroviral oncogenes (see below), these genes are virus specific and do not have exact cellular counterparts.
- **Retroviruses** have a genome of RNA. They replicate via a DNA intermediate, which is made using a viral reverse transcriptase (*Figure 17.2*). These viruses do not normally kill the host cell (HIV is an exception) and only rarely transform it. The genome of a typical retrovirus consists of three genes, *gag*, *pol* and *env* (*Figure 17.3A*).
- **Acute transforming retroviruses** are retrovirus particles which, unlike normal retroviruses, transform the host cell rapidly and with high efficiency. Their genomes include an additional gene, the oncogene (*Figure 17.3B*). Usually the oncogene replaces one or more essential viral genes, so that these viruses are replication defective. To propagate them, they are grown in cells which are simultaneously infected with a replication-competent helper virus that supplies the missing functions. Studies of acute transforming retroviruses have revealed more than 50 different oncogenes.

An *in vitro* transfection assay confirmed that cancer cells contain activated oncogenes

An entirely independent way of discovering oncogenes came from a cell transformation assay. The NIH-3T3 mouse cell line readily undergoes transformation *in*

Table 17.1: Human and animal tumor viruses

Species	Disease	Virus	Type	Oncogene
Monkey	Sarcoma	SV40	DNA	T antigen
Mouse	(Transformation *in vitro*)	Adenoviruses	DNA	E1A, E1B
Man	Cervical cancer	Papilloma virus HPV16	DNA	E6, E7
Man	Nasopharyngeal cancer	Epstein–Barr virus	DNA	*BNLF-1* (?)
Man	T-cell leukemia	HTLV-1, HTLV-2	RNA	–
Man	Kaposi sarcoma	HHV8 or KSAAV	RNA	–
Chicken	Sarcoma	Rous sarcoma virus	ATR	*src*
Rat	Sarcoma	Harvey rat sarcoma virus	ATR	H-*ras*
Mouse	Leukemia	Abelson leukemia virus	ATR	*abl*
Monkey	Sarcoma	Simian sarcoma virus	ATR	*sis*
Chicken	Erythroleukemia	Erythroleukemia virus	ATR	*erbB*
Chicken	Sarcoma	Avian sarcoma virus 17	ATR	*jun*
Mouse	Osteosarcoma	FBJ osteosarcoma	ATR	*fos*
Cat	Sarcoma	McDonough feline sarcoma virus	ATR	*fms*
Chicken	Myelocytoma	Avian myelocytomatosis virus	ATR	*myc*

ATR, acute transforming retrovirus, HHV8, human herpesvirus type 8; KSAAV, Kaposi sarcoma-associated herpesvirus.

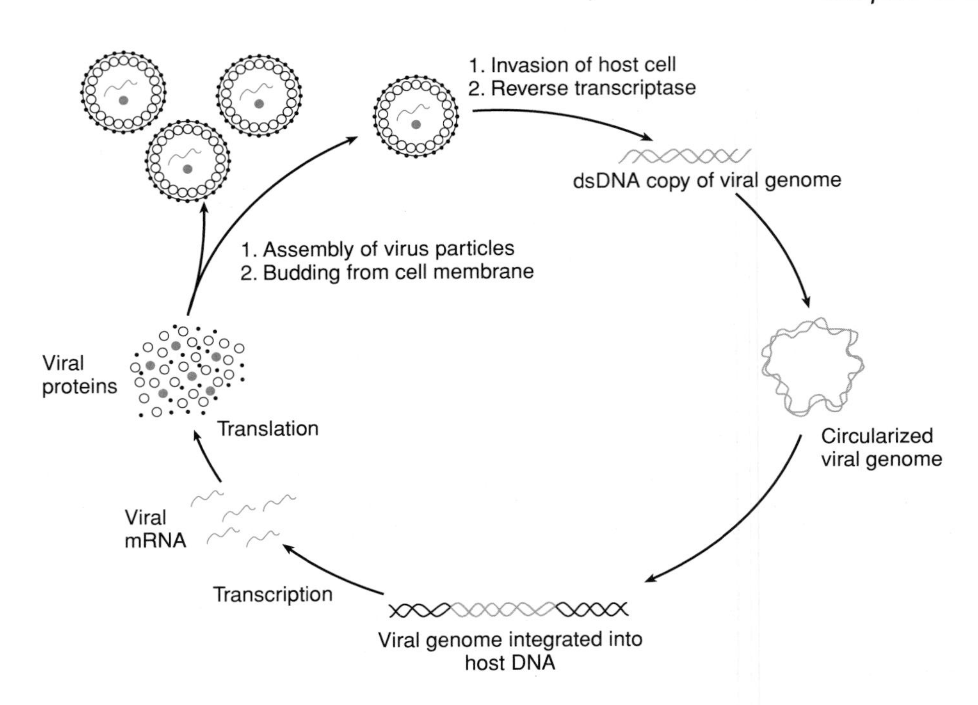

Figure 17.2: Retroviral life cycle.

The virus particle (top) contains the RNA genome and viral reverse transcriptase within an outer lipoprotein envelope and inner protein capsid. A double-stranded DNA copy of the viral genome integrates into the host DNA. Here it directs synthesis of viral RNA and proteins, which self-assemble and bud off from the cell membrane. The host cell is not killed.

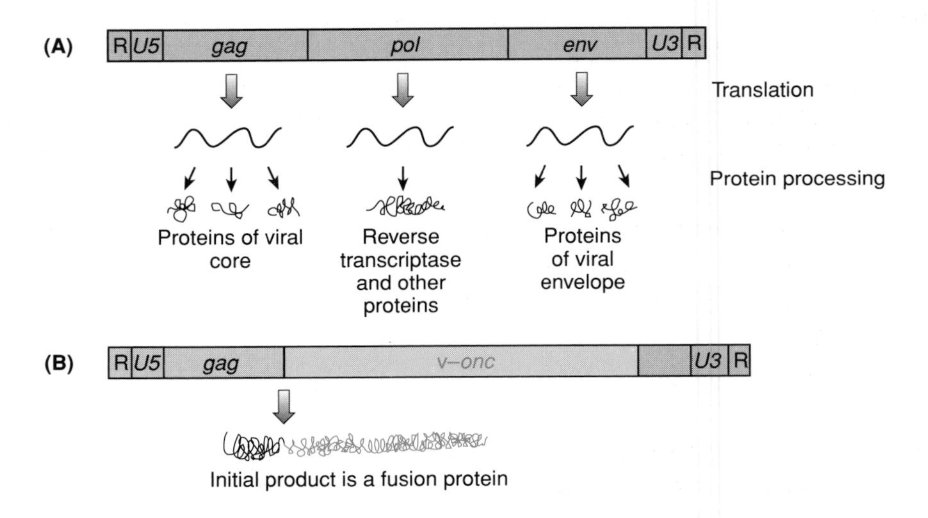

Figure 17.3: A normal and an acute transforming retrovirus.

The RNA genome has terminal repeats (R), subterminal unique sequences (U5, U3) and three genes, *gag*, *pol* and *env*. A complicated scheme of splicing and post-translational processing results in a variety of protein products. In an acute transforming retrovirus (bottom), one or more of the viral genes is replaced by a transduced cellular sequence, the oncogene. Initially it is translated into a fusion protein.

vitro – probably it has already acquired several of the successive genetic changes on the pathway to cancer and one further change suffices to transform it. In the 3T3 test, the cells are transfected with random DNA fragments from human cancer cells. Potential oncogenes can then be identified by selecting transformants and recovering the human DNA present in them (*Figure 17.4*). Transformants are obtained when DNA from tumor cells is used, but not with DNA from nontumor cells. Thus, tumor cells, even from nonviral tumors, contain activated oncogenes. This route led to the identification of essentially the same set of oncogenes as were found in acute transforming retroviruses.

Oncogenes are mutated versions of genes involved in a variety of normal cellular functions

It quickly became apparent that normal cells had counterparts of all the retroviral oncogenes (*Table 17.2*), and in fact that v-*onc* genes were transduced cellular genes.

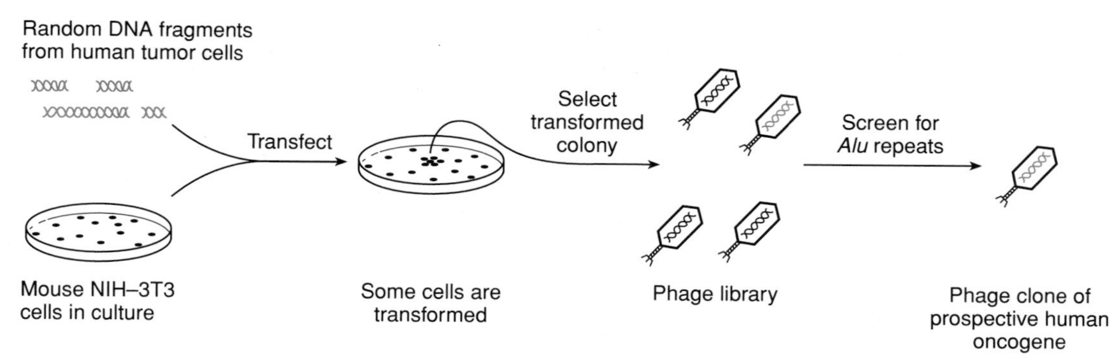

Figure 17.4: The NIH-3T3 assay.

Mouse 3T3 cells are transfected with random fragments of DNA from a human tumor. Any transformed cells (identified by their altered growth) are isolated and a phage library constructed from their DNA. Phage are then screened for the human-specific *Alu* repeat to identify those containing human DNA, which potentially contain oncogenes.

With a few exceptions, the v-*onc* gene products differ from their c-*onc* (proto-oncogene) counterparts by amino acid substitutions or truncations, which serve to activate the proto-oncogene.

Functional understanding of oncogenes began with the discovery that the viral oncogene v-*sis* was derived from the normal cellular platelet-derived growth factor B (*PDGFB*) gene (Doolittle *et al.*, 1983). Uncontrolled overexpression of a growth factor would be an obvious cause of cellular hyperproliferation. The roles of many cellular oncogenes (strictly speaking, proto-oncogenes) have now been elucidated (*Table 17.2*). Gratifyingly, they turn out to control exactly the sort of cellular functions which would be predicted to be disturbed in cancer. Five broad classes can be distinguished:

- secreted growth factors (e.g. *SIS*);
- cell surface receptors (e.g. *ERBB, FMS*);
- components of intracellular signal transduction systems (e.g. the *RAS* family, *ABL*);
- DNA-binding nuclear proteins, including transcription factors (e.g. *MYC, JUN*);
- components of the network of cyclins, cylin-dependent kinases and kinase inhibitors that govern progress through the cell cycle, for example *PRAD1* (Kamb, 1995; Müller, 1995).

Table 17.2: Viral and cellular oncogenes

Viral disease	v-*onc*	c-*onc*	Location	Function
Simian sarcoma	v-*sis*	*PDGFB*	22q13.1	Platelet-derived growth factor B subunit
Chicken erythroleukemia	v-*erbB*	*EGFR*	7p13–q22	Epidermal growth factor receptor
McDonough feline sarcoma	v-*fms*	*CSF1R*	5q33	Macrophage colony-stimulating factor receptor
Harvey rat sarcoma	v-*ras*	*HRAS1*	11p15	p21 component of G-protein signal transduction
Abelson mouse leukemia	v-*abl*	*ABL*	9q34.1	Protein tyrosine kinase
Avian sarcoma 17	v-*jun*	*JUN*	1p32–p31	AP-1 transcription factor
Avian myelocytomatosis	v-*myc*	*MYC*	8q24.1	DNA-binding protein (transcription factor; see Rabbitts, 1994)
Mouse osteosarcoma	v-*fos*	*FOS*	14q24.3–q31	DNA-binding transcription factor

The viral genes are sometimes designated v-*src*, v-*myc*, etc., and their cellular counterparts c-*src*, c-*myc*, etc. The forms of the c-*onc* genes in normal cells are properly termed **proto-oncogenes**. Nowadays it is common to ignore these distinctions and simply use the term **oncogenes** for the normal genes. The abnormal versions can be described as activated oncogenes.

Activation of proto-oncogenes

Some of the best illustrations of molecular pathology in action are furnished by the various ways in which proto-oncogenes can become activated. Activation involves a gain of function. This can be quantitative (an increase in the production of an unaltered product) or qualitative (production of a subtly modified product as a result of a mutation or production of a novel product from a chimeric gene created by a chromosomal rearrangement). These changes are dominant and normally affect only a single allele of the gene.

Activating mutations in oncogenes (unlike mutations in TS genes, see below) are almost invariably somatic events. Constitutional mutations would probably be lethal. Note, however, that this applies only to mutations which activate proto-oncogenes. Other mutations may be inherited constitutionally, and may have effects entirely unrelated to cancer. For example, mutations that inactivate the *kit* oncogene produce piebaldism (MIM 172800), whilst loss-of-function mutations in the *RET* oncogene produce Hirschsprung's disease. Both these conditions are inherited in a mendelian dominant fashion (with low penetrance in the case of *RET* mutations).

Only one case is known where an activated oncogene can be inherited so as to cause familial cancer. That is the *RET* gene, implicated in multiple endocrine neoplasia and familial thyroid cancer (see below). This behavior is so atypical that one might question whether *RET* should really be classified as an oncogene. In fact, *RET* illustrates the limitations of forcing all genes governing cell proliferation into just two categories, oncogenes or TS genes. Another gene that is hard to classify is *TP53* (see below); however *TP53* fits the TS category distinctly better than the oncogene category.

Activation of some oncogenes can occur by amplification

Many cancer cells contain multiple copies of structurally normal oncogenes. Breast cancers often amplify *ERBB2* and sometimes *MYC*; a related gene *NMYC* is usually amplified in late-stage neuroblastomas. Hundreds of extra copies may be present. They can exist as small separate chromosomes (**double minutes**) or as insertions within the normal chromosomes (**homogeneously staining regions, HSRs**). The genetic events producing HSRs may be quite complex because they usually contain sequences derived from several different chromosomes (reviewed by Pinkel, 1994). Similar gene amplifications are often seen in noncancer cells exposed to strong selective regimes (e.g. amplified dihydrofolate reductase genes in cells selected for resistance to methotrexate). In all cases, the result is to greatly increase the level of gene expression.

An alternative technique, **comparative genome hybridization (CGH)** (Kallioniemi *et al.*, 1992), can in principle reveal all regions of amplification in a single experiment, together with any regions of allele loss or aneuploidy, which may point to TS genes (see below). The CGH test (*Figure 17.5*) uses a mixture of DNA from matched normal and tumor cells in competitive FISH. With the aid of image-processing software, chromosomal regions can be picked out where the ratio of FISH signal from normal and tumor DNA deviates from expectation. Depending on the direction of deviation, these mark regions of amplification or of allele loss in the tumor.

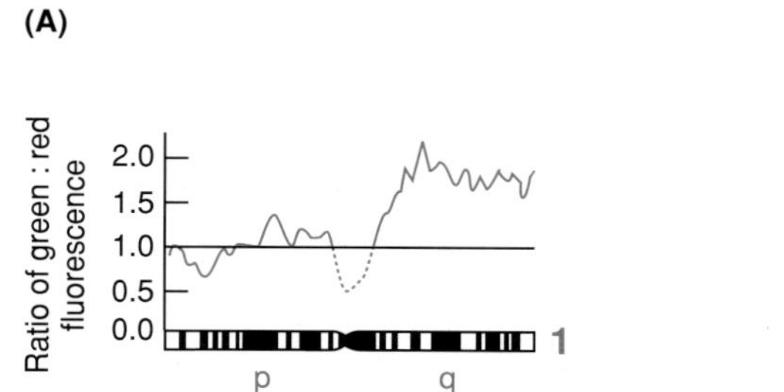

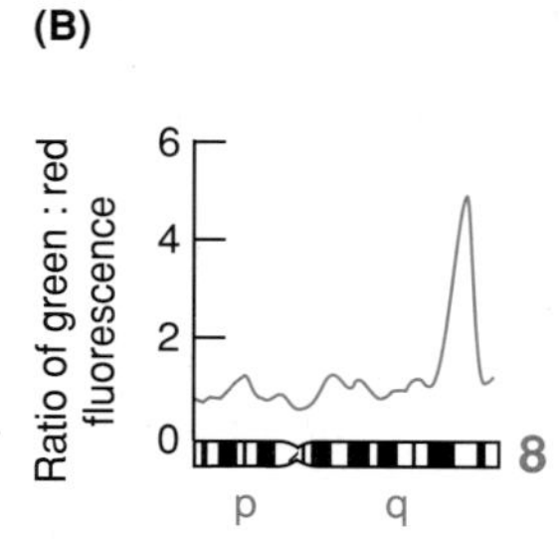

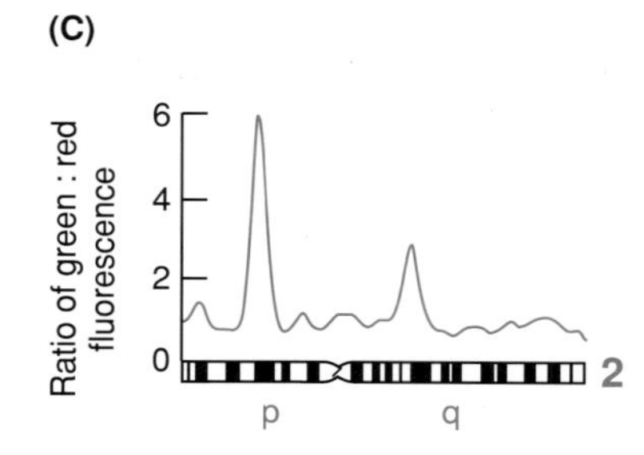

Figure 17.5: Comparative genome hybridization.

Tumor DNA and normal control DNA were labeled with green and red fluorescent labels, respectively, then hybridized *in situ* together in equal quantities to chromosomes of a normal cell. The curves show computer-generated scans of the ratio of green:red fluorescence intensity along three chromosomes.
(A) Chromosome 1: the breast cancer cell line 600PE carries an extra copy of 1q, and a possible interstitial deletion of proximal 1p34–p36.
(B) Chromosome 8: the colon cancer cell line COLO320HSR shows amplification of the *MYC* region at 8q24.
(C) Chromosome 2: the small cell lung cancer cell line NIH-H69 shows three regions of amplification. *NMYC* is amplified at 2p24; the content of the other two amplified regions is not known. Redrawn with permission from Kallioniemi *et al.* (1992). Copyright 1992 American Association for the Advancement of Science.

Some oncogenes are activated by point mutations

The *HRAS* gene (*Table 17.2*) is one of a family of *ras* genes, all of which encode p21 proteins involved in signal transduction from G-protein-coupled receptors. A signal from the receptor triggers binding of GTP to the RAS protein and GTP–RAS transmits the signal onwards in the cell. RAS proteins have GTPase activity, and GTP–RAS is rapidly converted to the inactive GDP–RAS. Specific point mutations in *ras* genes are frequently found in cells from a variety of tumors including colon, lung, breast and bladder cancers. These lead to amino acid substitutions which decrease the GTPase activity of the RAS protein. As a result, the GTP–RAS signal is inactivated more slowly, leading to excessive cellular response to the signal from the receptor.

Another oncogene which is activated by point mutations is *RET*. This gene encodes a cell surface tyrosine kinase that is a receptor for a currently unknown ligand. Mutations leading to amino acid substitutions at certain specific cysteine residues are found in multiple endocrine neoplasia type 2 and in medullary thyroid cancer. The *RET* gene is interesting as an example of a gene where different mutations cause different phenotypes (page 422).

Chromosomal translocations can create novel chimeric genes

Tumor cells typically have grossly abnormal karyotypes (*Figure 17.6*), with multiple extra and missing chromosomes, many translocations, etc. Most of these changes are random and reflect a general genomic instability which is a normal part of carcinogenesis (see below). A huge research effort has been devoted to picking out tumor-specific changes superimposed on the background of random changes. Over 150 different tumor-specific breakpoints have now been recognized (Mitelman *et al.*, 1991), and they reveal an important common mechanism in tumorigenesis.

The best known tumor-specific rearrangement produces the Philadelphia (Ph^1) chromosome, a very small acrocentric chromosome seen in 90% of patients with chronic myeloid leukemia. This chromosome turns out to be produced by a balanced reciprocal 9;22 translocation. The breakpoint on chromosome 9 is within an intron of the *ABL* oncogene. The translocation joins most of the *ABL* genomic sequence on to a gene called *BCR* (breakpoint cluster region) on chromosome 22, creating a novel fusion gene (Chissoe *et al.*, 1995). This chimeric gene is expressed to

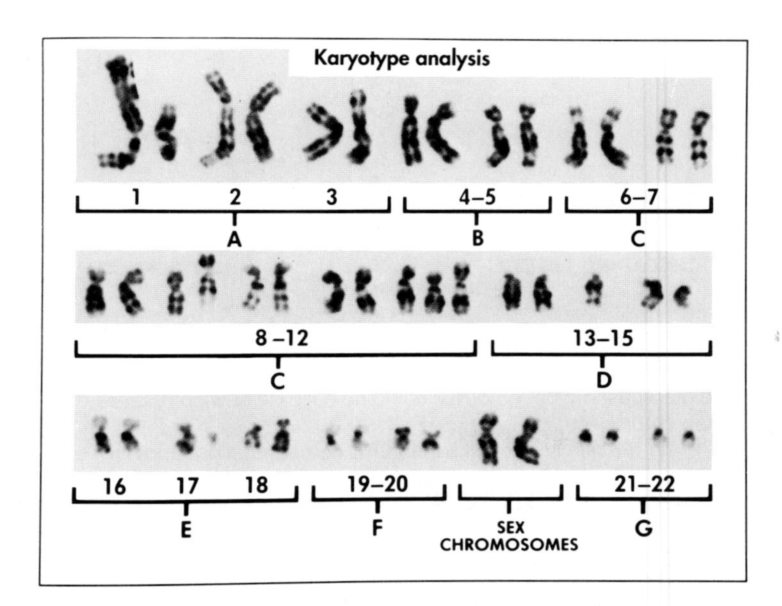

Figure 17.6: Cancer cell karyotype.

Reproduced from *Medical Genetics: an Illustrated Outline* by Andrew P. Read, 1989. Gower Medical Publishers Ltd, London, UK.

produce a tyrosine kinase related to the ABL product but with abnormal transforming properties (*Figure 17.7A*).

Many other rearrangements are known which produce chimeric genes (*Table 17.3*). The products are normally transcription factors (or sometimes tyrosine kinases) which take their target specificity from one component gene, but couple it to an activation or ligand-binding domain from the other. This has been one of the most satisfying stories to emerge from cancer research, with several examples of clinical phenotypes being elegantly explained by a combination of cytogenetic and molecular genetic findings. The whole topic of chromosomal translocations in cancer and the underlying genetic events has been well reviewed by Rabbitts (1994).

Oncogenes can be activated by transposition to an active chromatin domain

Burkitt's lymphoma is a childhood tumor common in malarial regions of central Africa and Papua New Guinea. Mosquitos and Epstein–Barr virus are believed to play some part in the etiology, but activation of the *MYC* oncogene is a central event. A characteristic chromosomal translocation, t(8;14)(q24;q32) is seen in 75–85% of patients (*Figure 17.7B*). The remainder have t(2;8)(p12;q24) or t(8;22)(q24;q11). Each of these translocations puts the *MYC* oncogene close to an Ig locus, *IGH* at 14q32, *IGK* at 2p12 or *IGL* at 22q11. Unlike the tumor-specific translocations shown in *Table 17.3*, the Burkitt's lymphoma translocations do not create novel chimeric genes. Instead, they put the oncogene in an environment of chromatin which is actively transcribed in antibody-producing B cells. Usually exon 1 (which is noncoding) of the *MYC* gene is not included in the translocated material. Deprived of its normal upstream controls, and placed in an active chromatin domain, *MYC* is expressed at an inappropriately high level.

Many other chromosomal rearrangements put one or another oncogene into the neighborhood of either an immunoglobin (*IGG*) or a T-cell receptor (*TCR*) gene (Rabbitts, 1994). Presumably the rearrangements arise by random malfunctioning of the recombinases that rearrange *IGG* or *TCR* genes during maturation of B and T cells (see *Figure 7.18*), and are then selected for their growth advantage. Predictably, these rearrangements are characteristic of leukemias and lymphomas, but not solid tumors.

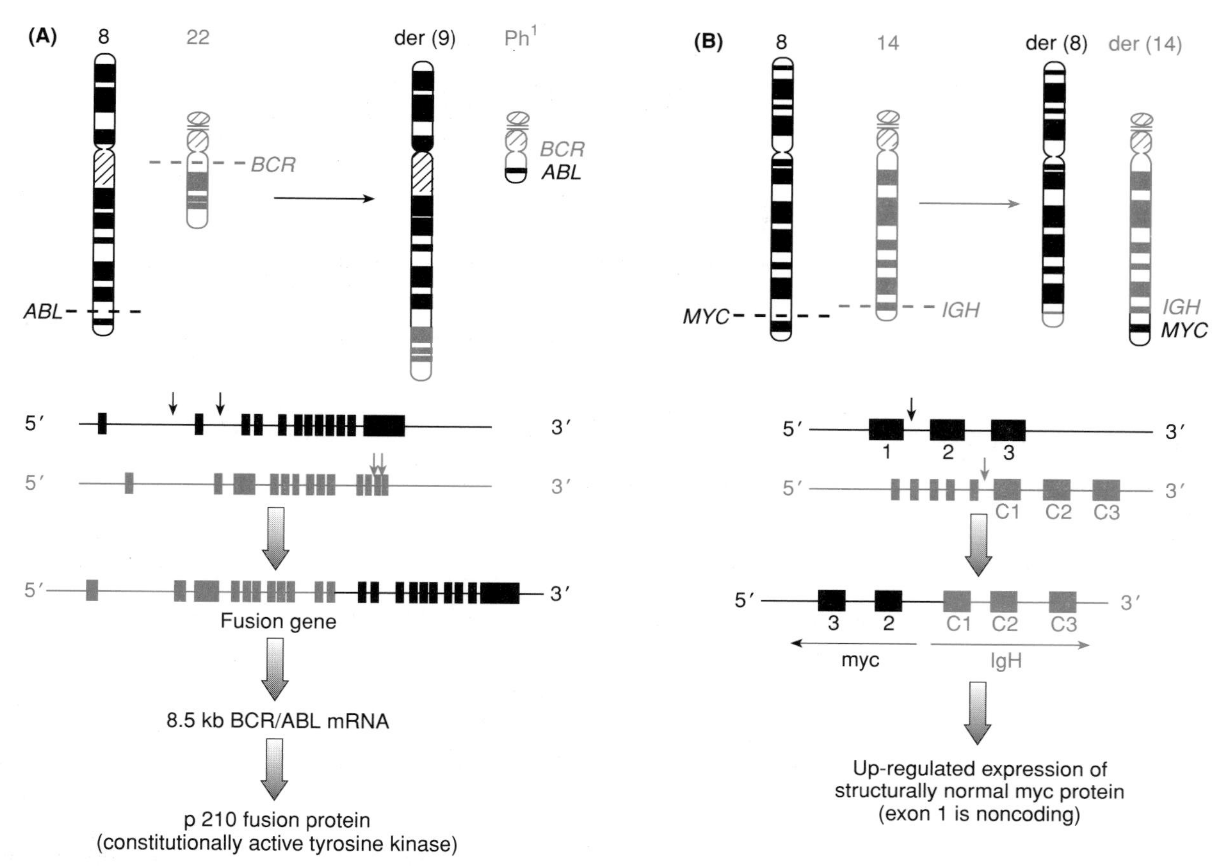

Figure 17.7: Chromosomal translocations which activate oncogenes.

(A) Activation by qualitative change in the t(9;22) in chronic myeloid leukemia. The chimeric *BCR–ABL* fusion gene on the Philadelphia chromosome encodes a tyrosine kinase which does not respond to normal controls [see Chissoe *et al.* (1995) for more detail].
(B) Activation by quantitative change in the t(8;14) in Burkitt's lymphoma. The *MYC* gene from chromosome 8 is translocated into the Ig heavy chain gene. In B cells this region is actively transcribed, leading to overexpression of *MYC*.

Table 17.3: Chimeric genes produced by cancer-specific chromosomal rearrangements

Tumor	Rearrangement	Chimeric gene	Nature of chimeric product
CML	t(9;22)(q34;q11)	*BCR–ABL*	Tyrosine kinase
Ewing sarcoma	t(11;22)(q24;q12)	*EWS–FLI1*	Transcription factor
Ewing sarcoma (variant)	t(21;22)(q22;q12)	*EWS–ERG*	Transcription factor
Malignant melanoma of soft parts	t(12;22)(q13;q12)	*EWS–WTF1*	Transcription factor
Desmoplastic small round cell tumor	t(11;22)(p13;q12)	*EWS–WT1*	Transcription factor
Liposarcoma	t(12;16)(q13;p11)	*FUS–CHOP*	Transcription factor
AML	t(16;21)(p11;q22)	*FUS–ERG*	Transcription factor
Papillary thyroid carcinoma	inv(1)(q21;q31)	*NTRK1–TPM3* (*TRK* oncogene)	Tyrosine kinase
Pre-B cell ALL	t(1;19)(q23;p13.3)	*E2A–PBX1*	Transcription factor
ALL	t(X;11)(q13;q23)	*MLL–AFX1*	Transcription factor
ALL	T(4;11)(q21;q23)	*MLL–AF4*	Transcription factor
ALL	t(9;11)(q21;q23)	*MLL-NEF9*	Transcription factor
ALL	t(11;19)(q23;p13)	*MLL-NL*	Transcription factor
Acute promyelocytic leukemia	t(15;17)(q22;q12)	*PML-NEARA*	Transcription factor + retinoic acid receptor
Alveolar rhabdomyosarcoma	t(2;13)(q35;q14)	*PAX3–FKHR*	Transcription factor

Note how the same gene may be involved in several different rearrangements. For further details see Rabbitts (1994). CML, chronic myeloid leukemia; ALL, acute lymphoblastoid leukemia; AML, acute myelocytic leukemia

Tumor suppressor genes

Cell fusion experiments show that the transformed phenotype can often be corrected *in vitro* by fusion of the transformed cell with a normal cell. This provides evidence that tumorigenesis involves not only dominant activated oncogenes, but also recessive, loss-of-function mutations in other genes. These other genes are the tumor suppressor (TS) genes. Sometimes TS genes are called anti-oncogenes, but that is an unhelpful name because it wrongly implies that they are all specific antagonists or inhibitors of oncogenes. Some may be but, like oncogenes, TS genes can have a variety of functions (see below). The methods for discovering TS genes were developed in classic studies of the rare eye tumor, retinoblastoma.

Retinoblastoma exemplifies Knudson's two-hit hypothesis

Retinoblastoma (MIM 180200) is a rare, aggressive childhood tumor of the retina. Sixty per cent of cases are sporadic and unilateral; the other 40% are inherited as an imperfectly penetrant autosomal dominant trait. In familial retinoblastoma bilateral tumors are common. A prophetic study by Knudson (1971) concluded that two successive mutations ('hits') were required to turn a normal cell into a tumor cell (*Figure 17.8*). Investigations in retinoblastoma families localized the gene to 13q14, and then a brilliant study by Cavenee *et al.* (1983) both proved Knudson's hypothesis and established the paradigm for all subsequent investigations of TS genes.

Cavenee and colleagues typed surgically removed tumor material from patients with sporadic retinoblastoma, using a series of markers from chromosome 13. When they compared the results on blood and tumor samples from the same patients, they noted several cases where the constitutional (blood) DNA was heterozygous for one or more chromosome 13 markers, but the tumor cells were apparently homozygous. Cavenee *et al.* reasoned that what they were seeing was Knudson's second hit: loss of the remaining functional copy of a TS gene. Combining cytogenetic analysis with studies of markers from different regions of 13q, they were able to suggest a number of mechanisms for the second hit (*Figure 17.9*).

Rare familial cancers identify many TS genes

Although activated oncogenes are rarely or never transmitted as constitutional mutations (*RET* is a possible exception, see above), many rare mendelian cancers are believed to involve TS genes via a two-hit mechanism (*Table 17.4*). Mapping the genes in these rare families opens the way to identification and cloning of the TS

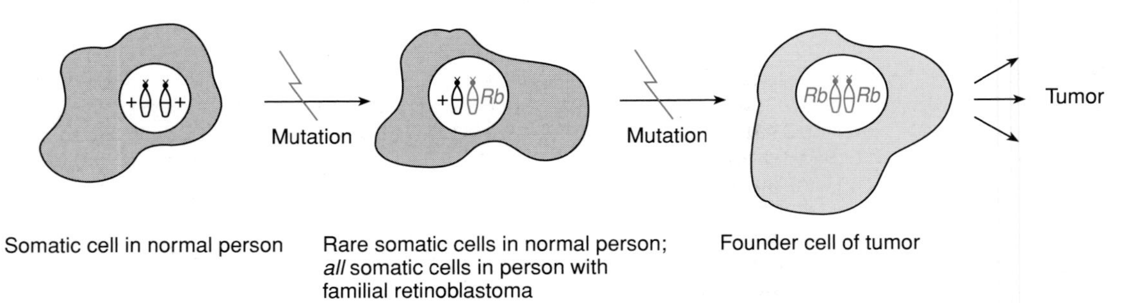

Figure 17.8: Knudson's two-hit hypothesis.

Suppose there are 1 million target cells and the probability of mutation is 10^{-5} per cell. Sporadic retinoblastoma requires two hits and will affect 1 person in 10 000 ($10^6 \times 10^{-5} \times 10^{-5} = 10^{-4}$), while the familial form requires only one hit and will be quite highly penetrant ($10^6 \times 10^{-5} = >1$).

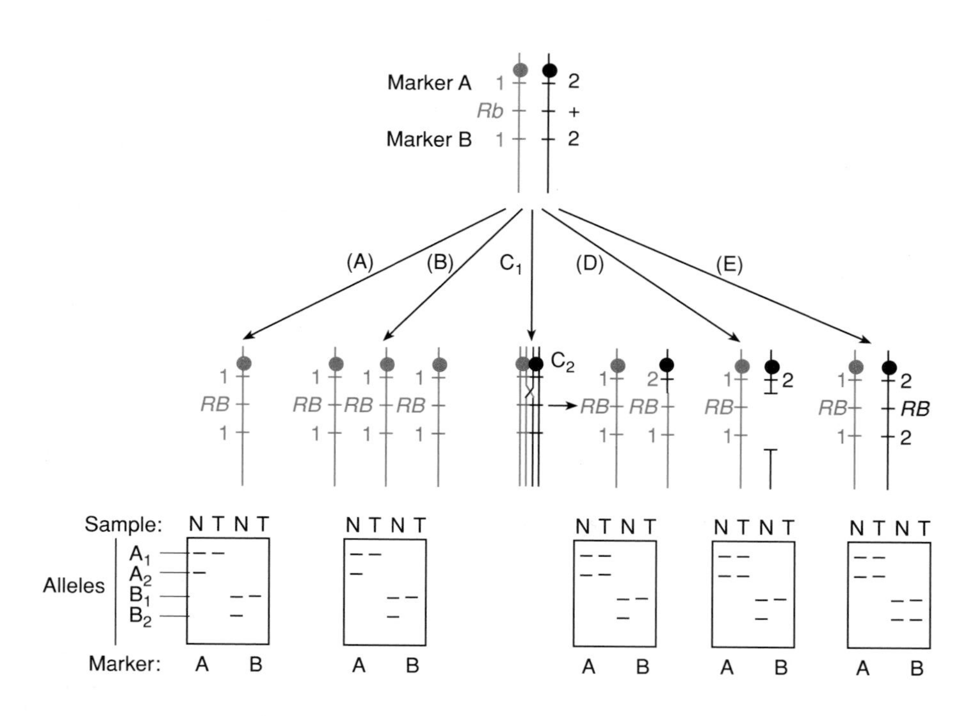

Figure 17.9: Mechanisms of loss of wild-type allele in retinoblastoma.

(A) Loss of a whole chromosome by mitotic nondisjunction.
(B) Loss followed by reduplication to give (in the case studied by Cavenee *et al.*) three copies of the *RB1* chromosome.
(C) Mitotic recombination proximal to the *RB* locus (C_1), followed by segregation of both *RB1*-bearing chromosomes into one daughter cell (C_2); this was the first demonstration of mitotic recombination in humans, or indeed in mammals.
(D) Deletion of the wild-type allele.
(E) Pathogenic point mutation of the wild-type allele (adapted from Cavenee *et al.*, 1983).
The figures underneath show results of typing normal (N) and tumor (T) DNA for the two markers A and B located as shown. *Note* the patterns of loss of heterozygosity.

gene. It should be noted, however, that few of these conditions are as simple genetically as retinoblastoma. The case of the *BRCA1* breast and ovarian cancer gene is described in Chapter 18 (page 494).

Loss of heterozygosity assays identify locations of TS genes

Commonly the first (inherited) mutation is a point mutation or some other small change confined to the TS gene – large deletions would presumably be harmful if carried in every cell of the body. Often, however, the second mutation, whether in a familial or a sporadic case, involves loss of all or part of a chromosome. The mechanism, as in *Figure 17.9*, may be nondisjunction (leading to loss of a whole chromosome), mitotic recombination (leading to loss of those parts of the chromosome distal to the crossover) or a *de novo* interstitial deletion. In each case, one allele is lost of any marker close to the TS gene. Thus, if the patient was heterozygous for a marker, the tumor tissue loses heterozygosity, becoming homozygous or hemizygous. Homozygous deletion of markers (loss of both alleles) is unusual, even in tumor cells.

Loss of heterozygosity (LOH) is a key pointer to the existence of TS genes. By screening paired blood and tumor samples with markers spaced across the genome, we can discover candidate locations for TS genes. Some of these will be tumor-specific (e.g. LOH at 5q21 near the *APC* gene in colon cancer), while others are common to many different tumors (e.g. LOH on 17p near the *TP53* gene). Of course, if the constitutional (blood) DNA is homozygous for a particular marker, that marker gives no information about allele loss in the tumor. Using highly polymorphic microsatellites minimizes the proportion of these uninformative samples.

Classic LOH studies suffer from the need to test large panels of markers and to analyze the data by quantitation of band intensities. Most pathological tumor samples contain a mixture of tumor and nontumor (stromal) tissue, so that what one sees is often a decreased relative intensity, rather than total loss of the band from one allele

Table 17.4: Rare familial cancers caused by TS gene mutations

Disease	MIM no.	Map location	Gene
von Hippel–Lindau disease	193300	3p25–p26	*VHL*
Familial adenomatous polyposis coli	175100	5q21	*APC*
Familial melanoma	600160	9p21	*CDKN2*
Gorlin's basal cell nevus syndrome	109400	9q22–q31	(*BCNS*)
Multiple endocrine neoplasia 2	171400	10q11.2	*RET*
Wilms' tumor	194070	11p13	*WT1*
Multiple endocrine neoplasia 1	131100	11q13	(*MEN1*)
Ataxia telangiectasia	208900	11q22–q23	*ATM*
Breast cancer (early onset)	600185	13q12–q13	(*BRCA2*)
Li–Fraumeni syndrome	151623	17p13	*TP53*
Neurofibromatosis I (von Recklinghausen disease)[a]	162200	17q12–q22	*NF1*
Breast–ovarian cancer	113705	17q21	*BRCA1*
Neurofibromatosis 2 (bilateral acoustic neuromas)	101000	22q12.2	*NF2*

References to the genes and diseases may be found in OMIM under the numbers cited. Genes named in brackets had not been characterized at the time of writing.
[a] Malignant tumors and neurofibromas in NF1 patients show allele loss at 17q.

(*Figure 17.10A*). Comparative genome hybridization (*Figure 17.5*) promises to allow much easier detection of large deletions, but at present lacks the resolution required for picking up small deletions.

There are thus three sources of information on likely locations of TS genes:

(i) linkage analysis in rare mendelian cancers;
(ii) LoH analysis of paired tumor and blood samples;
(iii) cytogenetic or comparative genome hybridization studies to define tumor-specific deletions.

These three complementary approaches have suggested the existence of a surprisingly large number of TS genes, most of which are not yet identified. As an example, *Figure 17.11* shows how they have suggested the existence of TS genes at three separate locations on the short arm of chromosome 3.

The function of TS genes

The biochemical action of TS gene products has proved harder to unravel than the function of oncogene products. Some appear to have simple roles, like *APC* and *DCC* (page 474), which are probably cell adhesion molecules. Others are involved in the complex control of cell cycle progression, as negative regulators. We describe

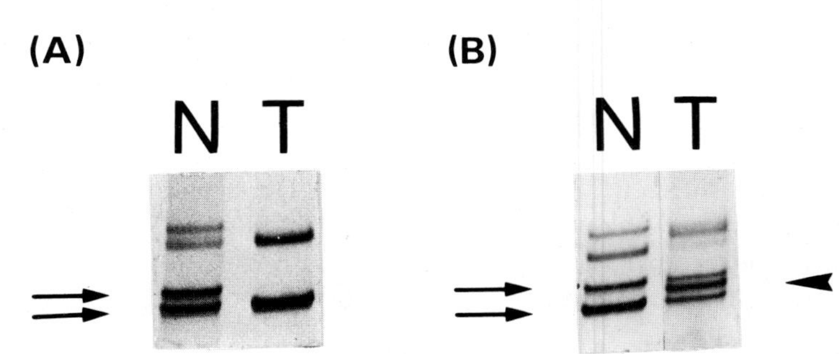

Figure 17.10: Genetic changes in tumors.

(A) Loss of heterozygosity. The normal tissue sample (N) is heterozygous for the marker D8S522 (arrows), whereas the tumor sample (T) has lost the upper allele. The bands higher up the gel are 'conformation bands', subsidiary bands produced by alternately folded sequences of each allele.
(B) Microsatellite instability. The normal tissue sample (N) is heterozygous for the marker D8S552 (arrows), while the tumor sample (T) has gained an extra band (arrowhead). Again, conformation bands are seen higher up the gel. Photographs courtesy of Dr Nalin Thakker, St Mary's Hospital, Manchester.

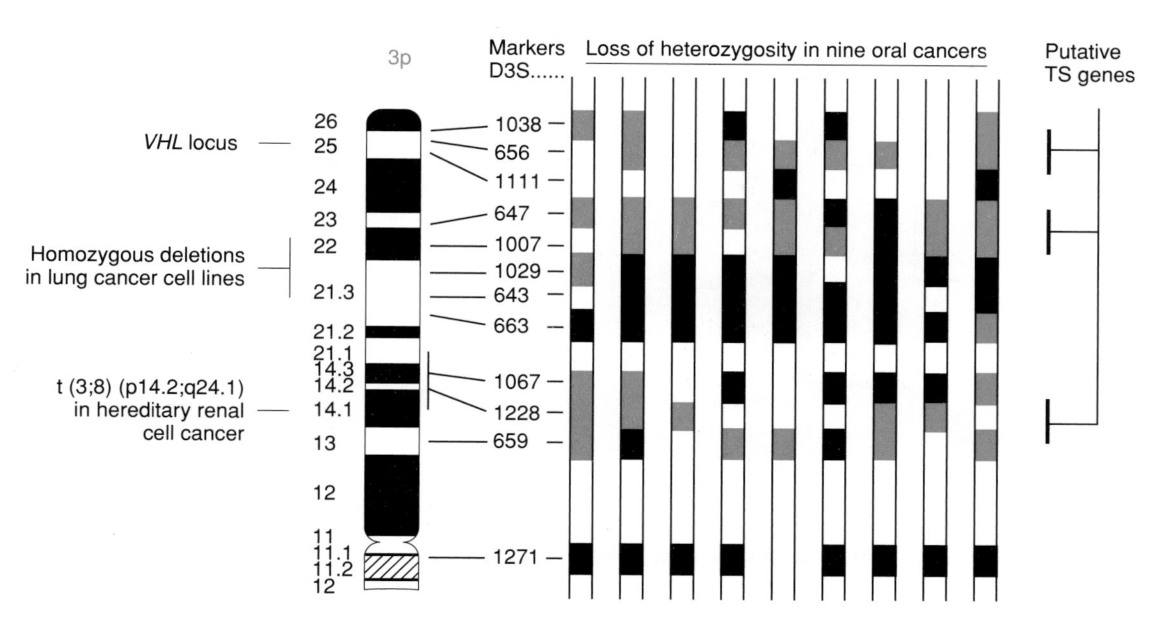

Figure 17.11: Possible TS genes on chromosome 3p.

On the right are the results of typing constitutional and tumor DNA from a series of patients with oral tumors, using various markers (D3S1038, etc.) from 3p (Wu *et al.*, 1994). Red signifies LOH, black signifies retention and clear areas are where markers were uninformative because of constitutional homozygosity. *Note* the complex pattern of LOH in these tumors, which is typical of such studies. Three distinct regions of LOH are seen, and these appear to correspond to the three regions (left side of diagram) defined by a translocation breakpoint in a family with dominant renal cell carcinoma, by homozygous deletions in some lung cancer cell lines and by the map location of the *VHL* tumor suppressor gene discovered by linkage in families with von Hippel–Lindau syndrome (MIM 193300).

here two of the most important, *RB* and *TP53*. The products of these two genes play a pivotal role in the control of cell cycle progression. Simultaneous inactivation of both products is frequently observed in tumor cells, and their functions partially overlap.

Function of the RB1 *gene product*

The *RB1* gene is widely expressed, encoding a 110-kDa nuclear protein (pRb) that is believed to play a key role in controlling cell proliferation. At least part of this role is to bind and inactivate the group of cellular transcription factors called E2F, which are required for cell cycle progression (*Figure 17.12*). In normal cells pRb is inactivated by phosphorylation and activated by dephosphorylation. Two to four hours before a cell enters S phase of the cell cycle (*Figure 2.2*), pRb is phosphorylated. Phosphorylation of pRb releases the inhibition of E2F and allows the cell to proceed to S phase. Phosphorylation is governed by a whole series of cyclins, cyclin-dependent kinases and cyclin kinase inhibitors (Kamb, 1995; Müller, 1995; Weinberg, 1995). This seems to constitute the most crucial single checkpoint in the cell cycle.

The product of the *MDM2* oncogene (which is amplified in many sarcomas) also binds and inhibits pRb, thus favoring cell cycle progression. In cancer cells, pRb may be inactivated in other ways. Several viral oncoproteins (adenovirus E1A, SV40 T antigen, human papillomavirus E7 protein) bind and sequester or degrade pRb, or it may be directly inactivated by loss-of-function mutations in the *RB1* gene. It is not clear why constitutional mutation of a gene so fundamental to cell cycle control should result specifically in retinoblastoma and a small number of other tumors, principally osteosarcomas. However, this is a common theme in molecular pathology: mutation of a gene produces a phenotypic effect in only a subset of the cells or tissues in which the gene is expressed and appears to have a function (page 419).

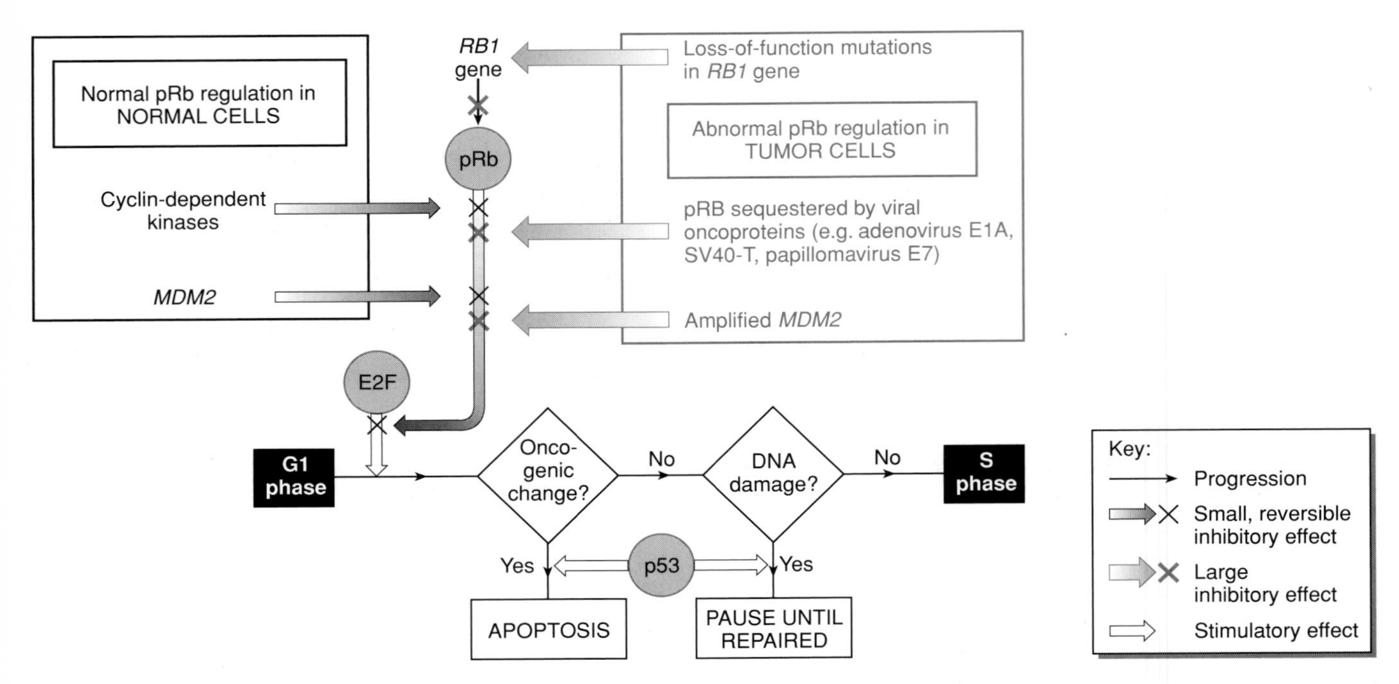

Figure 17.12: Part of the control mechanism for cell cycle progression (highly schematic).

Although we do not fully understand how the cell cycle is regulated in higher eukaryotes, three proteins (E2F, pRb and p53) are known to play key roles. The E2F protein acts as a transcription factor and is required for progression from the G1 phase of the cell cycle to the S phase. This effect is inhibited by pRb which in turn is temporarily inactivated at different stages in normal cells as a result of phosphorylation by cyclin-dependent kinases, or inhibition by the *mdm2* gene product (top left panel). In cancer cells, pRb is inappropriately inactivated by various mechanisms (top right panel), E2F continues to be activated and cells proliferate excessively. p53 prevents cells with potentially oncogenic changes or with DNA damage from progressing to S phase. This scheme shows what is probably only one facet of the activities of pRb and p53 and the G1–S checkpoint.

p53 and apoptosis

p53 was first described in 1979 as a protein found in SV40-transformed cells, where it associated with the T antigen. Later, the *TP53* gene which encodes p53 appeared as a dominant transforming gene in the 3T3 assay (*Figure 17.4*), and so was classed as an oncogene. Subsequently it transpired that while p53 from some tumor cells was oncogenic, p53 from normal cells positively suppressed tumorigenesis. LOH assays confirmed the status of *TP53* as a TS gene. *TP53* maps to 17p12 and this is one of the commonest regions of LOH in a wide range of tumors. Tumors which have not lost *TP53* very often have mutated versions of it. To complete the picture of *TP53* as a TS gene, constitutional mutations in *TP53* are found in families with the dominantly inherited Li–Fraumeni syndrome (Malkin, 1994). Affected family members suffer multiple primary tumors, typically including soft tissue sarcomas, osteosarcomas, tumors of the breast, brain and adrenal cortex, and leukemia (*Figure 17.13*).

Loss or mutation of *TP53* is probably the most common single genetic change in cancer. The only clearly identified biochemical function of p53 is as a transcription factor. Tetramers of p53 bind DNA and can activate transcription of reporter genes placed downstream of a p53 binding site. However, p53 is believed to have a much broader role in the cell, which has been summarized as 'the guardian of the genome'. One of its guardian functions is to stop cells replicating damaged DNA (*Figure 17.12*). p53 is involved in a checkpoint at the G1/S stage of the cell cycle. Normal cells with damaged DNA arrest at this point until the damage is repaired, but cells that lack p53 or contain a mutant form do not arrest at G1. Replication of damaged DNA presumably leads to random genetic changes, some of which are oncogenic, similar to cells with a defective mismatch repair system (see above).

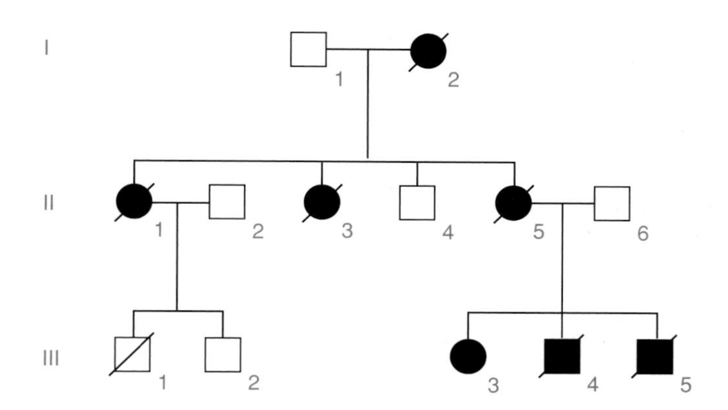

Figure 17.13: A typical pedigree of Li–Fraumeni syndrome.

Malignancies typical of Li–Fraumeni syndrome include bilateral breast cancer diagnosed at age 40 (I_2); a brain tumor at age 35 (II_1); soft tissue sarcoma at age 19 and breast cancer at age 33 (II_3); breast cancer at age 32 (II_5); osteosarcoma at age 8 (III_3); leukemia at age 2 (III_4); soft tissue sarcoma at age 3 (III_5). I_1 had cancer of the colon diagnosed at age 59 – this is assumed to be unrelated to the Li–Fraumeni syndrome. Pedigree from Malkin (1994).

Probably related to this is a crucial role of p53 in cell death. In response to oncogenic stimuli, cells undergo apoptosis (programmed cell death). This is one of the higher level controls which protect the organism against the consequences of natural selection among its constituent cells (page 458). Apoptosis has come to occupy a central place in our understanding of the cancer process (reviewed by Fisher, 1994). A common pathway in carcinogenesis is loss of this control, and cells lacking p53 do not undergo apoptosis. p53 may be knocked out by deletion, by mutation or by the action of an inhibitor such as the *mdm2* gene product (which also binds pRb, see above) or the E6 protein of papillomavirus.

Mutator genes

We saw earlier (*Box 17.1*) that cancer arises only if something happens to counteract the apparent near-impossibility of accumulating half a dozen specific mutations in a single cell. Mutations of oncogenes and TS genes create expanded clones of cells as targets for subsequent mutations (*Figure 17.6*). These genes are directly involved in the cell cycle controls which go wrong in cancer. The third class of genes which are commonly mutated in cancer cells are not part of these pathways. Instead, they have a general role in ensuring the integrity of the genetic information. Mutations in these genes lead to inefficient replication or repair of DNA.

It has long been known that cancer cells show a general genetic instability. Tumor cells typically have bizarrely abnormal karyotypes, with many losses, gains and rearrangements of chromosomes, only a few of which seem to be causally connected with the cancer (*Figure 17.6*). The genes responsible for chromosomal instability have not yet been identified, but two diseases, colon cancer and ataxia telangiectasia, have provided pointers to genes responsible for instability at the DNA level.

Colon cancer

Most colon cancer is sporadic. Familial cases fall into two categories:

(i) **Familial adenomatous polyposis (FAP or APC)** is an autosomal dominant condition in which the colon becomes carpeted with hundreds or thousands of polyps. The polyps (adenomas) are not malignant but, if left in place, one or more of them is virtually certain to evolve into invasive carcinoma. The condition has been mapped to 5q21 and the gene responsible, *APC*, identified.

(ii) **Hereditary nonpolyposis colon cancer (HNPCC)** is also autosomal dominant and highly penetrant but, unlike FAP, there is no preceding phase of polyposis. HNPCC genes were mapped to two locations, 2p15–p22 and 3p21.3.

LOH studies on HNPCC using microsatellite markers showed an entirely unexpected phenomenon in some patients. Rather than lacking alleles present in the constitutional DNA, some tumor specimens appeared to contain extra, novel, alleles. LOH is a property of certain chromosomal regions, but the microsatellite instability (MIN) seemed to be general. Tumors could be classified into MIN^+ and MIN^-. MIN^+ tumors showed gain of alleles for a good proportion of all markers tested, regardless of their chromosomal location. MIN is also seen in other tumors. *Figure 17.10B* shows an example from an oral tumor.

In a brilliant piece of lateral thinking, Fishel *et al.* (1993) related the MIN^+ phenomenon to so-called mutator genes in *E. coli* and yeast. These genes encode an error correction system which checks the DNA for mismatched base pairs (Modrich, 1991; *Figure 17.14*). Because of DNA methylation (the *E. coli* Dam system methylates adenine in GATC sequences, but not until some time after DNA replication), the system can identify the template strand and selectively correct the newly synthesized strand if it has not yet been methylated. Mismatches are excised and replaced.

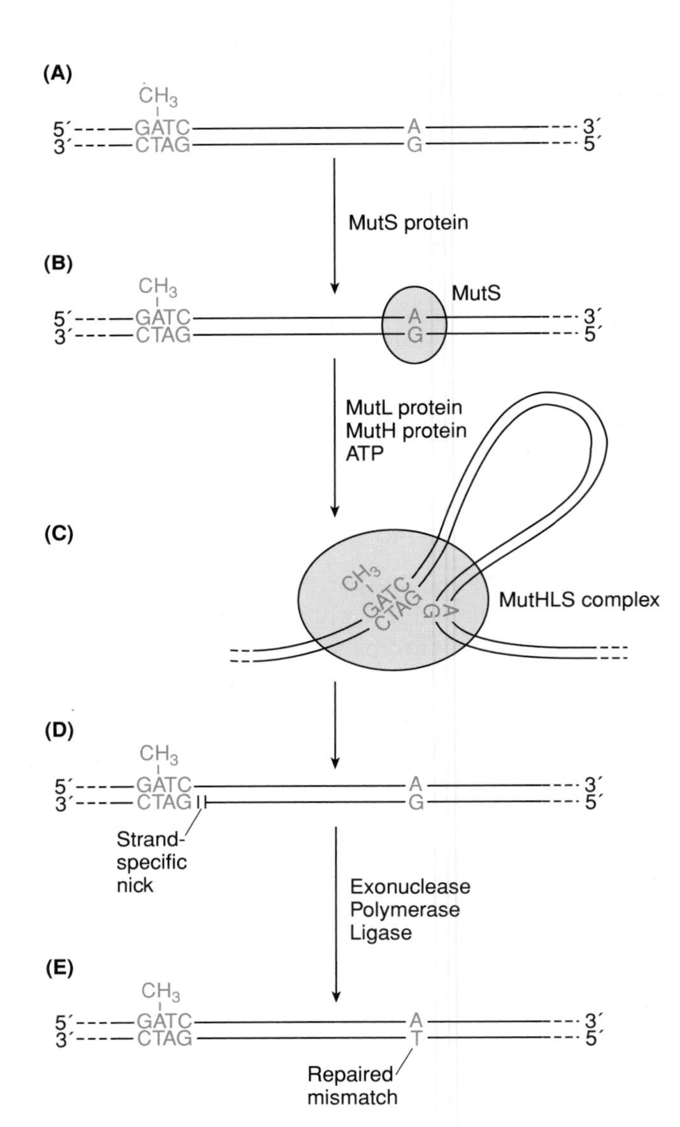

Figure 17.14: The *MutHLS* error correction system in *E. coli.*

The MutS protein binds to mismatched base pairs **(B)**. In an ATP-dependent reaction, a MutS–MutL–MutH complex is formed which probably brings any GATC sequence located within 1 kb either side of the mismatch into a loop **(C)**. MutH makes a single-strand cut 5′ to the GATC sequence **(D)**. The *E. coli* Dam methylation system methylates A in GATC, but in newly synthesized DNA only the template strand is methylated. MutH specifically cuts the unmethylated (newly synthesized) strand **(D)**. Exonucleases, DNA polymerase and DNA ligase then strip back and repair the DNA **(E)**. See Modrich (1991).

Mutations in the genes that encode the *MutHLS* error correction system lead to a 100- to 1000-fold general increase in mutation rates. Fishel and colleagues cloned the human homolog of one of these genes, *Mut*S, and showed that it mapped to the location on 2p of one of the HNPCC genes and was constitutionally mutated in some HNPCC families. Almost simultaneously, the same gene was identified independently through positional cloning. Three other mutator genes were rapidly identified, all of which are related to the *E. coli MutL* gene (*Box 17.2*).

Like the TS gene mutations, mutator gene mutations are recessive and require a two-hit mechanism. Patients with HNPCC are constitutionally heterozygous for a loss-of-function mutation. Their normal cells still have a functioning mismatch repair system and do not show the MIN^{+} phenotype. In a tumor, the second copy is lost by one of the mechanisms shown in *Figure 17.9*.

Box 17.2: Mutator genes in colon cancer

E. coli	Human	Location in man	Per cent of HNPCC
MutS	*MSH2*	2p15–p22	50–60%
MutL	*MLH1*	3p21.3	30–40%
MutL	*PMS1*	2q31–q33	5%?
MutL	*PMS2*	7p22	5%?

Ataxia telangiectasia

Ataxia telangiectasia (AT, MIM 208900) is a rare recessive disorder characterized by neurological signs (notably progressive cerebellar ataxia) and dilation of the blood vessels (telangiectasia) in the conjunctiva and eyeballs. There is also marked immunodeficiency, growth retardation and sexual immaturity. The relevance to the present chapter is that AT patients have a strong predisposition to cancer. Homozygotes usually die of malignant disease before age 25. In addition, there have been suggestions that AT heterozygotes have an increased risk of cancer – for example a 3.9-fold increased risk of breast cancer among women (Easton, 1994). AT affects about one person in 100 000 in the UK and USA, so the Hardy–Weinberg distribution (page 76) suggests that one person in 158 of the population is heterozygous. If their raised risk of cancer is confirmed, this would represent a significant cancer risk at the population level.

In vitro, cells of AT patients show chromosomal instability, with breaks and translocations, often involving the *TCR* or *IGG* genes. The cells are hypersensitive to ionizing radiation or radiomimetic chemicals, even though DNA repair appears to be normal (reviewed by Shiloh, 1995). Although AT has been divided into at least four complementation groups based on cell fusion studies, patients in all complementation groups turn out to have mutations in the same gene, *ATM*, at 11q22–q23 (Savitsky *et al.*, 1995). Thus the complementation is intragenic. ATM shows sequence homology to yeast phosphatidylinositol 3′ kinase, a signal transduction enzyme involved in cell cycle control and meiotic recombination. Its exact function remains to be defined; the AT phenotype suggests a role in some fundamental high-level control of genetic integrity (see Lehman and Carr, 1995).

The multistep evolution of cancer

In the case of colorectal cancer, some details can be filled in to flesh out the generalized scheme we saw in *Figure 17.1*. Malignant carcinomas develop from benign epithelial growths called adenomas, and adenomas can in turn be classified into early (less than 1 cm in size), intermediate (more than 1 cm but without foci of carcinoma) or late (more than 1 cm and with foci of carcinoma). At the genetic level there is not one invariant sequence of mutations in the development of every colorectal carcinoma, but the most likely sequence is one where each successive step confers a growth advantage on the cell. Pointers to the most common sequence include the following observations (Fearon and Vogelstein, 1990):

- constitutional loss of one copy of the *APC* gene on 5q21 is sufficient to carpet the colon with adenomatous polyps. This suggests that loss or mutation of *APC* may be an early event in the development of sporadic cancers.
- About 50% of intermediate and late adenomas, but only about 10% of early adenomas, have mutations in the *KRAS* oncogene (a relative of *HRAS*, *Table 17.2*). Thus *KRAS* mutations may often be involved in the progression from early to intermediate adenomas.
- About 50% of late adenomas and carcinomas show loss of heterozygosity on 18q. This is relatively uncommon in early and intermediate adenomas. The putative TS gene involved has been characterized and named *DCC* (deleted in colon cancer). *DCC* encodes a protein with homologies to cell surface glycoproteins involved in cell adhesion.
- Colorectal cancers, but not adenomas, have a very high frequency of mutations in the *TP53* gene (see above).

These are not the only changes seen in colorectal carcinomas, but they are the ones which can be most readily associated with specific stages, and they lead to the model shown in *Figure 17.15*. It is assumed that the role of the mutator genes is to make each transition more likely by raising the general mutation rate, rather than being directly involved in particular stages of the pathway in *Figure 17.15*.

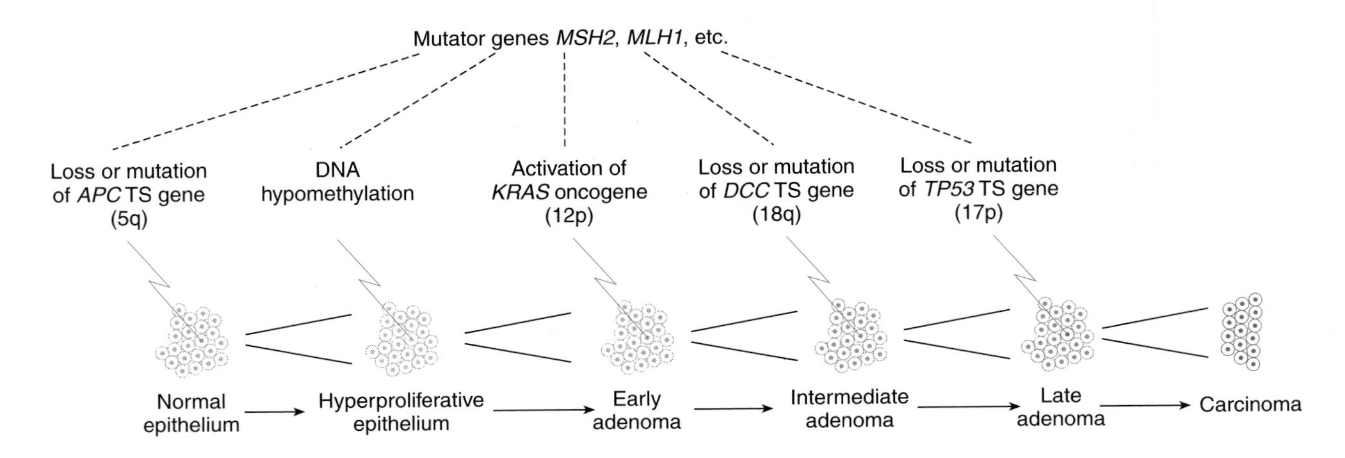

Figure 17.15: Fearon and Vogelstein's model for the development of colorectal cancer.

This is primarily a tool for thinking about how tumors develop, rather than a firm description. Every colorectal carcinoma is likely to have developed through the same progression of histological stages, but the underlying genetic changes are much less predictable. The sequence shown illustrates changes frequently involved in particular stages of progression. Mutations in *MSH2*, *MLH1* and other mutator genes do not play a direct role in the oncogenic pathway, but by increasing the overall mutation rate they make each individual transition more likely.

Somatic mutations in noncancerous diseases

We saw at the start of this chapter that every one of us carries innumerable somatic mutations, but that these are generally of no consequence. The tiny minority which do cause problems usually do so by fostering uncontrolled cell growth. Occasionally, however, a somatic mutation that does not confer any growth advantage will occur early enough in embryonic life and in a small enough stem cell population for it to generate a sizeable clone of mutant cells. These embryos can give rise to clinically abnormal people. Depending whether their germline also contains mutant cells (**gonosomal mosaicism**) they may also be at risk of having affected children.

High-level somatic mosaics may be clinically abnormal

A few aneuploid cells can be found in any perfectly normal person, and their number increases with age. Mosaics for Down syndrome, Klinefelter syndrome, Turner syndrome and so forth are commonly discovered when abnormal patients are referred for cytogenetic investigation. Their phenotype lies somewhere between normal and the phenotype of the constitutional abnormality, but cannot be predicted from observations of the proportion of abnormal cells in blood or the other tissues usually tested. All these chromosomal mosaics are caused by nondisjunction or other events which happen post-zygotically.

Mosaics for single gene mutations tend to be discovered by following a mutation through a family. In serious autosomal dominant and X-linked diseases, where affected people have few or no children, the genes are maintained in the population by new mutations (page 72). Often the new mutation first appears in a mosaic form, usually in a clinically normal person, who then has a constitutionally mutant child. Examples for osteogenesis imperfecta and Duchenne muscular dystrophy are illustrated in *Figures 3.8* and *3.9*.

More exotic abnormalities may be known only in mosaic form – presumably they would be lethal if constitutional. An example is trisomy 8. Mosaic trisomy 8 gives a recognizable phenotype, but constitutional trisomy 8 is not known in liveborns. Children with Pallister–Killian syndrome are mosaic for tetrasomy of 12p, with an extra isochromosome of 12p in affected cells. The phenotype is recognizable, but the chromosome abnormality is seen only in cultured skin fibroblasts, never in lymphocytes. Presumably the abnormal cells are unable to contribute to the blood lineage.

McCune–Albright syndrome (MIM 174800) is an example of a lethal single-gene mutation surviving in mosaics. The condition occurs sporadically and the clinical picture is very variable. Affected people have patchy skin pigmentation, bone problems (polyostotic fibrous dysplasia) and endocrine abnormalities. Affected tissue has an activating mutation in the *GNAS1* gene, encoding a receptor-coupled G-protein (Schwindiger *et al.*, 1992). The tissue responds abnormally to physiological controls and, if enough tissue is affected, the result is a clinical problem. Some pituitary and thyroid tumors have a similar activating mutation in *GNAS1*, extending the parallel with activated oncogenes in cancer.

Low-level somatic mosaicism may be a source of normal variation

The skin is a highly favorable organ for genetic investigation, being more laid out on view than any other organ. Most of us have imperfections in our skin, and some of these may represent somatic mosaicism. Happle *et al.* (1990) have drawn attention to the variety of nevi, spots and patches which normal people often bear. Particularly interesting are twin spots, where two patches of abnormal skin with apparently complementary abnormalities lie side by side. This is precisely what one would expect from segregation after mitotic recombination in a heterozygote (cf. *Figure 17.9*). Patches of daughter cells of opposite homozygous types would arise. The phenomenon is well known in *Drosophila*, where it forms the basis of a test for mutagenicity.

Two-hit mechanisms may explain patchy mendelian phenotypes

A final connection between cancer and noncancer patchy phenotypes is the possibility that a two-hit mechanism similar to the classic retinoblastoma mechanism (*Figure 17.8*) may explain phenotypes which are mendelian in families but patchy in individuals. Why for example does piebaldism produce patches and spots of depigmented skin, rather than general albinism? Why does polycystic kidney disease produce a limited number of grossly dilated tubules in the kidney, and polyposis coli a patchwork of adenomatous polyps growing on apparently normal epithelium? In each of these cases the primary genetic defect is assumed to be present in every cell, but only some show the phenotype. It is tempting to postulate a two-hit mechanism (though this is unproven) and it should be noted that in the cases cited the second hit would need to be a very frequent event. X inactivation provides definite examples of such a mechanism – female carriers of X-linked recessive diseases show a patchy distribution of any nondiffusible manifestation of the condition (see page 73).

Further reading

Lasko D, Cavenee W, Nordenskjöld M. (1991) Loss of constitutional heterozygosity in human cancer. *Annu. Rev. Genet.*, **25**, 281–314.

References

Armitage P, Doll R. (1954) *Br. J. Cancer*, **8**, 1–12.

Cavenee WK, Dryja TP, Phillips RA, Benedict WF, Godbout R, Gallie BL, Murphree AL, Strong LC, White RL. (1983) *Nature*, **305**, 779–784.

Chissoe SL, Bodenteich A, Wang Y-F *et al.* (1995) *Genomics*, **27**, 67–82.

Doolitle RF, Hunkapiller MW, Hood LE, Devare SG, Robbins KC, Aaronson SA, Antoniades HN. (1983) *Science*, **221**, 275–277.

Easton DF. (1994) *Int. J. Radiat. Biol.*, **66**, S177–S182.

Fearon ER, Vogelstein B. (1990) *Cell*, **61**, 759–767.

Fishel R, Lescoe MK, Rao MRS, Copeland NG, Jenkins NA, Garber J, Kane M, Kolodner R. (1993) *Cell*, **75**, 1027–1038.

Fisher DE. (1994) *Cell*, **78**, 539–542.

Happle R, Koopman R, Mier PD. (1990) *Lancet*, **335**, 376–378.

Kallioniemi A, Kallioniemi O-P, Sudar D, Rurovitz D, Gray JW, Waldman F, Pinkel D. (1992) *Science*, **258**, 818–821.

Kamb A. (1995) *Trends Genet.*, **11**, 136–140.

Knudson AG. (1971) *Proc. Natl Acad. Sci. USA*, **68**, 820–823.

Lehman AR, Carr AM. (1995) *Trends Genet.*, **11**, 375–377.

Malkin D. (1994) *Annu. Rev. Genet.*, **28**, 443–465.

Mitelman F, Kaneko Y, Trent J. (1991) *Cytogenet. Cell Genet.*, **58**, 1053–1079.

Modrich P. (1991) *Annu. Rev. Genet.*, **25**, 229–253.

Müller R. (1995) *Trends Genet.*, **11**, 173–178.

Pinkel D. (1994) *Nature Genetics*, **8**, 107–108.

Savitsky K, Bar-Shira A, Gilad S *et al.* (1995) *Science*, **268**, 1749–1753.

Rabbitts TH. (1994) *Nature*, **372**, 143–149.

Read AP. (1989) *Medical Genetics: an Illustrated Outline*. Gower Medical Publishing, London.

Shiloh Y. (1995) *Eur. J. Hum. Genet.*, **3**, 116–138.

Weinberg RA. (1995) *Cell*, **81**, 323–330.

Wu CL, Sloan P, Read AP, Harris R, Thakker N. (1994) *Cancer Res.*, **54**, 6484–6488.

Complex diseases

Chapter 18

Deciding whether a nonmendelian character is genetic: the role of family, twin and adoption studies

Nobody would dispute the involvement of genes in a character which consistently gives mendelian pedigree patterns or which is associated with a chromosomal abnormality. However, with nonmendelian characters, whether continuous (quantitative) or discontinuous (dichotomous), it is necessary to prove claims of genetic determination. The obvious way to approach this is to show that the character runs in families. The degree of family clustering is expressed as λ_R, the ratio of the risk for relatives of patients to the population prevalence. Different λ parameters are calculated for different degrees of relationship (sibs, children, etc). *Table 18.1* shows results from early studies of schizophrenia. Family clustering is evident from the raised λ values and, as expected, λ values drop back towards 1 for more distant relationships – though in this data, not at the rate predicted for a purely genetic character.

Table 18.1: Early studies of familial risk in schizophrenia

Dates	Studies	Relation	Incidence	λ[a]
1928–62	14	Parents	336/7675 = 4.36%	5.45
			(corrected value[b] = 14.12%)	17.65[b]
1928–62	12	Sibs	724/8504 = 8.51%	10.6
1921–62	5	Children	151/1226 = 12.31%	15.4
1930–41	4	Uncles, aunts	68/3376 = 2.01%	2.5
1916–46	3	Half-sibs	10/311 = 3.22%	4.0
1926–38	5	Nephews, nieces	52/2315 = 2.25%	2.8
1928–38	4	Grandchildren	20/713 = 2.81%	3.5
1928–41	4	First cousins	71/2438 = 2.91%	3.6

[a] λ values are calculated assuming a population incidence of 0.8%.
[b] Correction allows for the fact that once schizophrenia has developed, people seldom have children.
Data assembled by Slater and Cowie (1970).

The importance of shared family environment

Geneticists must not forget that humans give their children their environment as well as their genes. Many characters run in families because of the shared family environment – whether one's native language is English or Chinese, for example. One has therefore always to ask whether shared environment might be the explanation for a familial character. This is especially important for behavioral attributes like IQ or schizophrenia, which depend at least partly on upbringing, but it cannot be ignored even for physical characters or birth defects: a family might share an unusual diet or some traditional medicine which could cause developmental defects.

Polygenic theory (page 482ff) allows prediction of the degree to which relatives should resemble each other, and segregation analysis (page 491) can attempt to estimate this under varying genetic or environmental hypotheses, but in practice these analyses rarely lead to the unambiguous and uncontroversial demonstration of genetic factors. Thus, something more than a familial tendency is usually necessary to prove that a nonmendelian character is under genetic control. Unfortunately, these reservations are not always as clearly stated in the medical literature as perhaps they should be.

Twin studies suffer from many limitations

Francis Galton, who laid so much of the foundation of quantitative genetics, pointed out the value of twins for human genetics. Monozygotic (MZ) twins are genetically identical clones and will necessarily be concordant (both the same) for any genetically determined character, with a few exceptions (*Box 18.1*). This is true regardless of the mode of inheritance or number of genes involved; the only exception is for characters dependent on post-zygotic somatic mutations (Chapter 17). Dizygotic (DZ) twins share half their genes on average, the same as any pair of sibs. Genetic characters should therefore show a higher concordance in MZ than DZ twins, and many characters do (*Table 18.2*).

Box 18.1: Genetic differences between identical twins

All individuals, even monozygotic twins, differ in:

- their repertoire of antibodies and T-cell receptors (because of epigenetic rearrangements and somatic cell mutations, see page 77).
- The numbers of mitochondrial DNA molecules (epigenetic partitioning).
- Somatic mutations in general (Chapter 17).
- The pattern of X inactivation, if female.

Higher concordances in MZ compared with DZ twins might also be possible if the character is determined by environmental factors. For a start, half of DZ twins are of unlike sex, whereas all MZ twins are the same sex. Even if the comparison is restricted to same-sex DZ twins (as it is in the studies shown in *Table 18.2*), at least for behavioral traits the argument can be made that MZ twins are more likely to be very similar, to be dressed and treated the same, and thus to share more of their environment than DZ twins.

Table 18.2: Twin studies in schizophrenia

Study	Concordant MZ pairs	Concordant DZ pairs
Kringlen, 1968	14/55 (21/55)	4–10%
Fisher, 1969	5/21 (10/21)	10–19%
Tienari, 1975	3/20 (5/16)	3/42
Farmer, 1987	6/16 (10/20)	1/21 (4/31)
Onstad, 1991	8/24	1/28

The numbers show pairwise concordances, i.e. counts of the number of concordant (+/+) and discordant (+/–) pairs ascertained through an affected proband. Calculating concordances probandwise (counting a pair twice if both were probands) gives higher values for the MZ concordance. Probandwise concordances are thought to be more comparable with other measures of family clustering. Only the studies of Onstad and Farmer use the current standard diagnostic criteria, DSM-III. Figures in brackets are obtained using a wider definition of affected, including borderline phenotypes.
For references, see Onstad *et al.* (1991) and Fischer *et al.* (1969).

MZ twins separated at birth and brought up in entirely separate environments would provide the ideal experiment (Francis Crick once made the tongue-in-cheek suggestion that one of each pair of twins born should be donated to science for this purpose). Such separations happened in the past more often than one might expect because the birth of twins was sometimes the last straw for an overburdened mother. Fascinating television programs can be made about twins reunited after 40 years of separation, who discover they have similar jobs, wear similar clothes and like the same music. As research material, however, separated twins have many drawbacks:

- there are very few of them, so any research is based on small numbers of arguably exceptional people.
- The separation was often not total – often they were separated some time after birth, and brought up by relatives.
- There is a bias of ascertainment – everybody wants to know about strikingly similar separated twins, but separated twins who are very different are not newsworthy.
- Even in principle, research on separated twins cannot distinguish intrauterine environmental causes from genetic causes. This may be important, for example in studies of sexual orientation ('the gay gene'), where some people have suggested that maternal hormones may affect the fetus *in utero* so as to influence its future sexual orientation.

Thus, for all their anecdotal fascination, separated twins have contributed relatively little to human genetic research.

Adoption studies are the most powerful way to disentangle genetic and environmental factors

If separating twins is an impractical way of disentangling heredity from family environment, adoption is much more promising. Two study designs are possible:

- find adopted people who suffer from a particular disease known to run in families, and ask whether it runs in their biological family or their adoptive family.
- Alternatively, start with affected subjects whose children have been adopted away from the family, and ask whether being adopted away from the family saves a child from the risk of the disease.

A celebrated (and controversial) study by Rosenthal and Kety (1968) used the first of these designs to test for genetic factors in schizophrenia. The diagnostic criteria used in this study have been criticized; there have also been claims (disputed) that not all diagnoses were made truly blind. However, an independent re-analysis using DSM-III diagnostic criteria (Kendler *et al.*, 1994) reached substantially the same conclusions. *Table 18.3* shows the results of a later extension of this study (Kety *et al.*, 1994).

Table 18.3: An adoption study in schizophrenia

	Schizophrenia cases among biological relatives	Schizophrenia cases among adoptive relatives
Index cases (chronic schizophrenic adoptees)	44/279 (15.8%)	2/111 (1.8%)
Control adoptees (matched for age, sex, social status of adoptive family and number of years institutionalized)	5/234 (2.1%)	2/117 (1.7%)

The study involved 14 427 adopted persons aged 20–40 in Denmark; 47 of them were diagnosed as chronic schizophrenic. The 47 were matched with 47 nonschizophrenic control subjects from the same set of adoptees.

The main obstacle in adoption studies is the lack of information about the biological family. This is frequently compounded by the undesirability of approaching them with questions. A secondary problem is selective placement, where the adoption agency, in the interests of the child, chooses a family likely to resemble the biological family. Thus organizing a study may be very difficult. Nevertheless, in countries where efficient adoption registers exist, this is unquestionably the most powerful method for checking whether a character is genetically determined. Quantitative characters can be similarly investigated by comparing the correlation between an adoptee and his/her biological or adoptive relatives.

The next section describes the main theory which has been used to explain how nonmendelian characters might be genetically determined.

Polygenic theory of quantitative traits

As we saw in Chapter 3, page 80, the early years of this century were marked by divergence between two traditions of genetics. The followers of Bateson saw genetics as the study of the transmission and segregation of mendelian genes, even if the phenotypic variants involved were rare or trivial, while the school of biometricians founded by Francis Galton saw genetics as the statistical study of evolutionarily important variation, normally in quantitative characters. Theoretical unification was achieved by Fisher in 1918, who demonstrated that the characters favored by the biometricians could be described in mendelian terms if they were **polygenic**, (i.e. governed by the simultaneous action of many gene loci) (Fisher, 1918).

Any variable character which depends on the additive action of a large number of individually small independent causes will be distributed in a Normal (Gaussian) distribution in the population. This can be seen in the highly simplified example shown in *Figure 18.1*. We suppose the character depends on alleles at a single locus, then at two loci, then at three. As more loci are included we see two consequences:

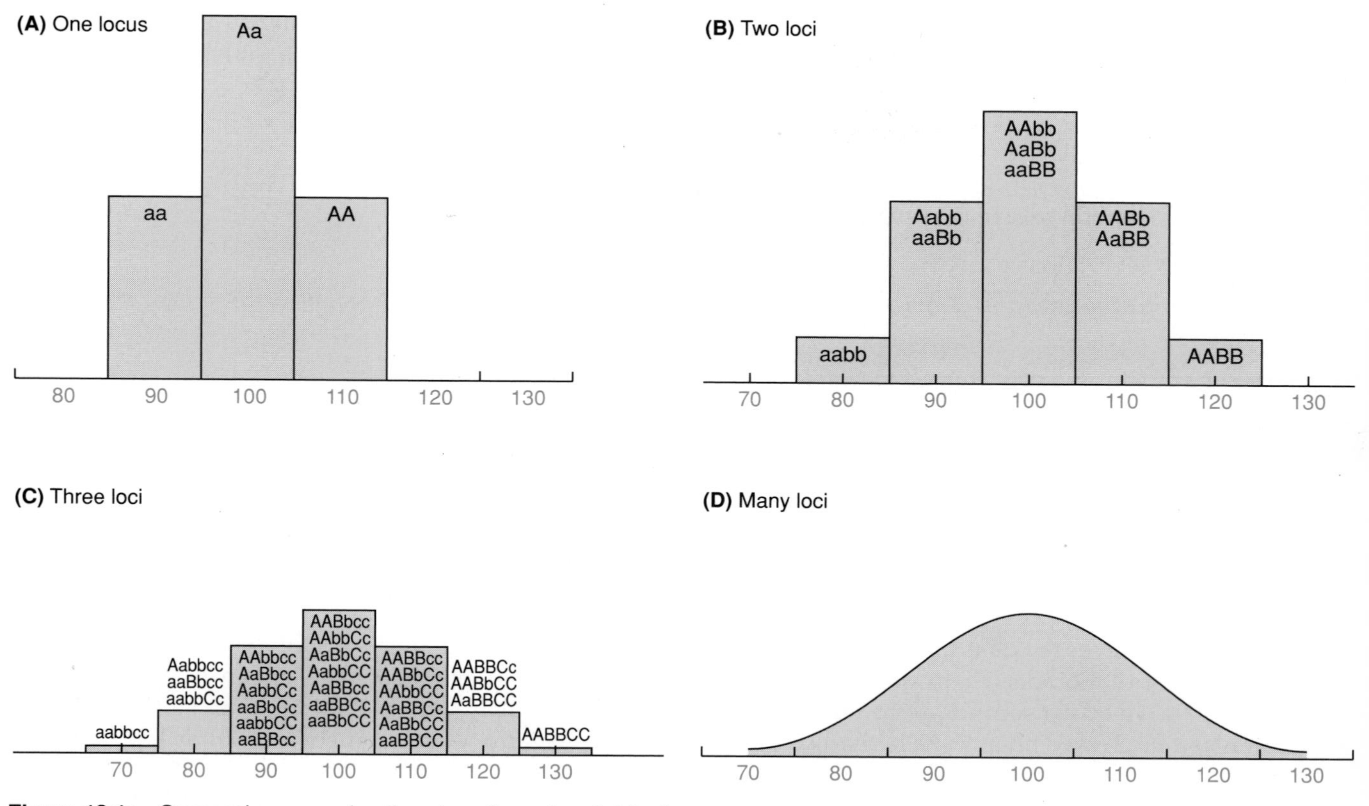

Figure 18.1: Successive approximations to a Gaussian distribution.

The charts show the distribution in the population of a hypothetical character which has a mean value of 100 units. The character is determined by the additive (co-dominant) effects of alleles. Each upper case allele adds 5 units to the value, and each lower case allele subtracts 5 units. All allele frequencies are 0.5. **(A)** The character is determined by a single locus. **(B)** Two loci. **(C)** Three loci: addition of a minor amount of 'random' (environmental or polygenic) variation produces the Gaussian curve **(D)**.

(i) the simple one-to-one relationship between genotype and phenotype disappears: except for the extreme phenotypes, it is not possible to infer the genotype from the phenotype.

(ii) As the number of loci increases, the distribution looks increasingly like a Gaussian curve. Addition of a little environmental variation would smooth out the three-locus distribution into a good Gaussian curve.

A more sophisticated treatment allowing dominance and varying gene frequencies leads to the same conclusions. Since relatives share genes, their phenotypes are correlated, and Fisher's paper predicted the size of the correlation for different relationships.

Phenotypes where relatives share some but not all determinants will show regression to the mean, whether the determinants are genetic or environmental

A much misunderstood feature, both of biometric data and of polygenic theory, is **regression to the mean**. We will use IQ as an example for discussion. Although IQ is of no great intrinsic interest (it measures only a person's ability and willingness to tackle IQ tests, not their 'intelligence' in any general sense, still less their personal worth), arguments about IQ allegedly based on genetic concepts are presented to the general public for political reasons (Herrenstein, 1994). On the simplest genetic assumptions, if one surveys mothers with an IQ of 120, their children would have an average IQ of 110, half way between the mothers' value and the population mean. Mothers with an IQ of 80 would have children with an average IQ of 90. Regression to the mean is often misinterpreted (see *Box 18.2*).

Box 18.2: Two common misconceptions about regression to the mean

- After a few generations everybody will be exactly the same.
- Regression to the mean is a genetic phenomenon. If a character shows regression to the mean, it must be genetic.

Figure 18.2 shows that the first of these beliefs is wrong. In a simple genetic model:

- the overall distribution is the same in each generation.
- Regression works both ways: for each class of children, the average for their mothers is half way between the children's value and the population mean. This may sound paradoxical but it can be confirmed by inspecting, for example, the right-hand column of the bottom histogram in *Figure 18.2* (children of IQ 120). One-quarter of their mothers have IQ 120, half 110 and one-quarter 100, making an average of 110.

Regarding the second of these beliefs, regression to the mean is not a genetic mechanism but a purely statistical phenomenon. Whatever the determinants of IQ are (genetic or environmental or any mix of the two), if we take an exceptional group of mothers (i.e. those with an IQ of 120), then these mothers must have had an exceptional set of determinants. If we take a second group who share half those determinants (their children, their sibs or either of their parents), the average phenotype in this second group will deviate from the population mean by half as much. Genetics provides the figure of one half, but not the principle of regression.

Figure 18.2 shows regression in our simplified two-locus model. For each class of mothers, the average IQ of their children is half way between the mother's value and the population mean. However, this depends on the hidden assumption in this model, that is that there is random mating. For each class of mothers, the average IQ of their husbands is supposed to be 100. Thus the average IQ of the children is the mid-parental IQ, as common sense would suggest. In the real world, highly intelligent women tend to marry men of above average intelligence (assortative mating) and we would not expect regression half way to the population mean, even if IQ were a purely genetic character.

The simplified model of *Figure 18.2* assumes there is no dominance. Thus each person's phenotype is the sum of the contribution of each allele at the relevant loci. If we allow dominance, the effect of some of a parent's genes will be masked by dominant alleles and invisible in their phenotype, but they can still be passed on and can affect the child's phenotype. Given dominance, the expectation for the child is no longer the mid-parental value. Our best guess about the likely phenotypic effect of the masked recessive alleles is obtained by looking at the rest of the population. Therefore, the child's expected phenotype will be displaced from the mid-parental value towards the population mean. How far it will be displaced depends on how important dominance is in determining the phenotype.

The heritability is the proportion of variance due to additive genetic effects

Gaussian curves are specified by only two parameters, the *mean* and the *variance* (or the standard deviation which is the square root of the variance). Variances have the

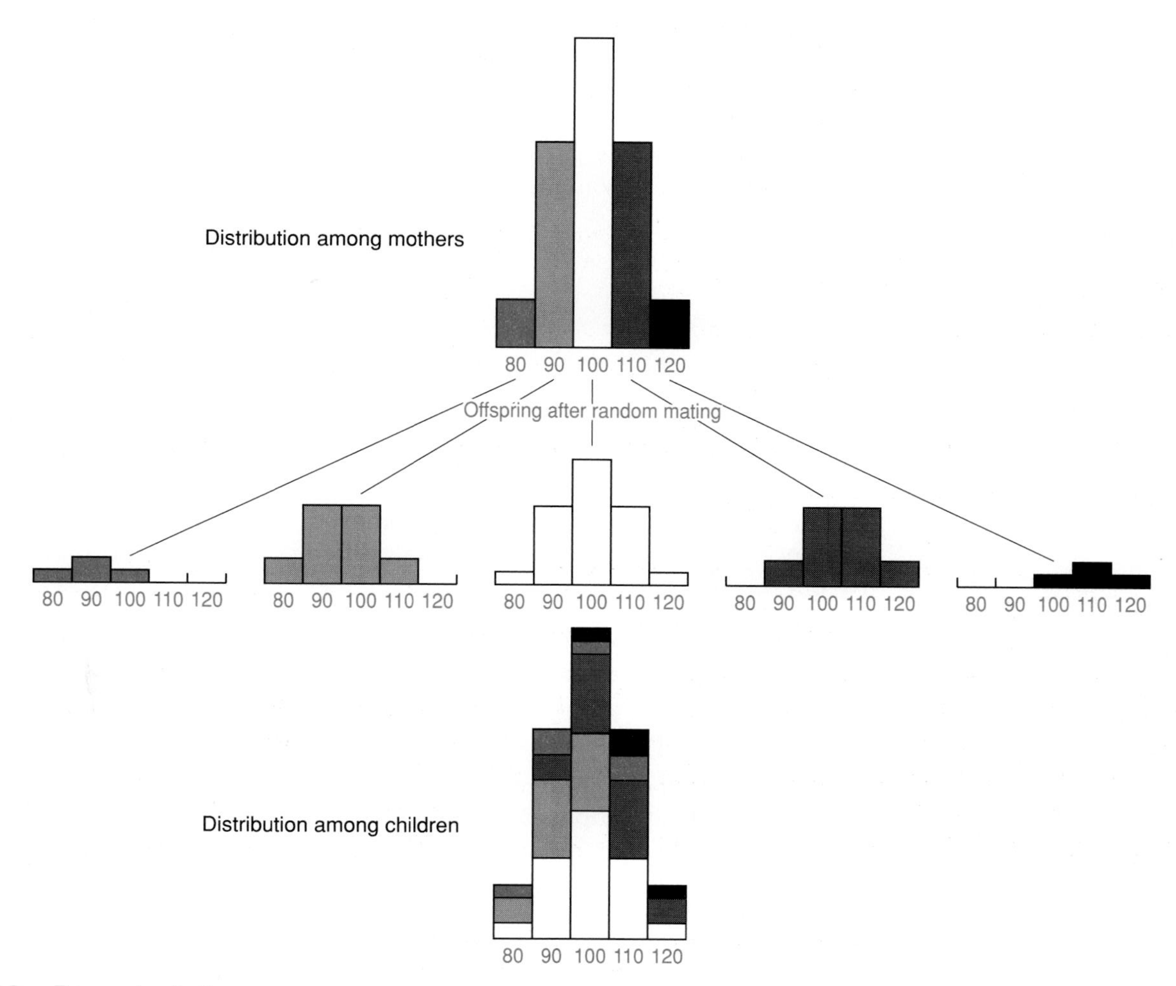

Figure 18.2: Regression to the mean.

The same character as in *Figure 18.1B*: mean 100, determined by co-dominant alleles A, a, B and b at two loci, all gene frequencies = 0.5. Top: distribution in a series of mothers. Middle: distributions in children of each class of mothers, assuming random mating. Bottom: summed distribution in the children. *Note* that: (a) the distribution in the children is the same as the distribution in the mothers; (b) for each class of mothers, the mean for their children is half way between the mothers' value and the population mean (100); and (c) for each class of children (bottom), the mean for their mothers is half way between the children's value and the population mean.

useful property of being additive when they are due to independent causes (*Box 18.3*). Thus the overall variance of the phenotype V_P is the sum of the variances due to the individual causes of variation – the environmental variance V_E and the genetic variance V_G. V_G can in turn be broken down to a variance V_A due to simply additive genetic effects and an extra term V_D due to dominance effects. The **heritability** (h^2) of a trait is the proportion of the total variance which is genetic, that is V_G/V_P. For animal breeders interested in breeding cows with higher milk yields, this is an important measure of how far a breeding program can create a herd in which the average animal resembles today's best. Strictly, V_G/V_P is the broad heritability. Dominance variance cannot be fixed by breeding, so the selection response is determined by the narrow heritability, V_A/V_P. Heritabilities of human traits are often estimated as part of segregation analysis (see page 491 and *Table 18.5*).

The term 'heritability' is often misunderstood. Heritability is quite different from the mode of inheritance. The mode of inheritance (autosomal dominant, polygenic, etc.) is a fixed property of a trait, but heritability is not. 'Heritability of IQ' is shorthand for heritability of variations in IQ. In different social circumstances, the heritability of IQ will differ. In an egalitarian society, we would expect IQ to have a

Box 18.3: Partitioning of variance

Variance of phenotype (V_P) = Genetic variance (V_G) + Environmental variance (V_E)

V_G = Variance due to additive genetic effects (V_A) + Variance due to dominant effects (V_D)

$\rightarrow V_P = V_A + V_D + V_E$

Heritability (broad) = V_G / V_P

Heritability (narrow) = V_A / V_P

higher heritability than in a society where access to education depended on accidents of birth. If everybody has equal opportunities, a number of the environmental differences between people have been removed. Therefore more of the remaining differences in IQ will be due to the genetic differences between people. Contrast the two questions:

- *to what extent is IQ genetic?* This is a meaningless question.
- *How much of the differences in IQ between people are caused by their genetic differences, and how much by their different environments and life histories?* This is a meaningful question, even if difficult to answer.

For many human behavioral traits, the simple partitioning of variance into environmental and genetic components is not applicable. Parents give their children both their genes and their environment, and genetic and environmental factors are often correlated. Genetic disadvantage and social disadvantage tend to go together. V_P does not equal $V_G + V_E$ and researchers are forced to include additional interaction variances in their analyses. A proliferation of variances can rapidly reduce the explanatory power of the models and in general this has been a difficult area in which to work.

Polygenic theory of discontinuous characters

Most of the classical 'polygenic' continuously variable characters like height or weight are of little interest to medical geneticists. Much more interesting are the innumerable diseases and malformations which tend to run in families, but which do not show mendelian pedigree patterns. A major conceptual tool in nonmendelian genetics was provided by Falconer's extension of polygenic theory to **dichotomous (discontinuous) characters** (those you either have or do not have).

Polygenic threshold theory can account for dichotomous nonmendelian characters

Falconer extended polygenic theory to discontinuous nonmendelian characters by postulating an underlying continuously variable **susceptibility** (summarized in Falconer, 1981). You may or may not have a cleft palate, but every embryo has a certain susceptibility to cleft palate. The susceptibility may be low or high; it is polygenic and follows a Gaussian distribution in the population. Together with the polygenic susceptibility, Falconer postulated the existence of a **threshold**. Embryos whose susceptibility exceeds a critical threshold value develop cleft palate; those

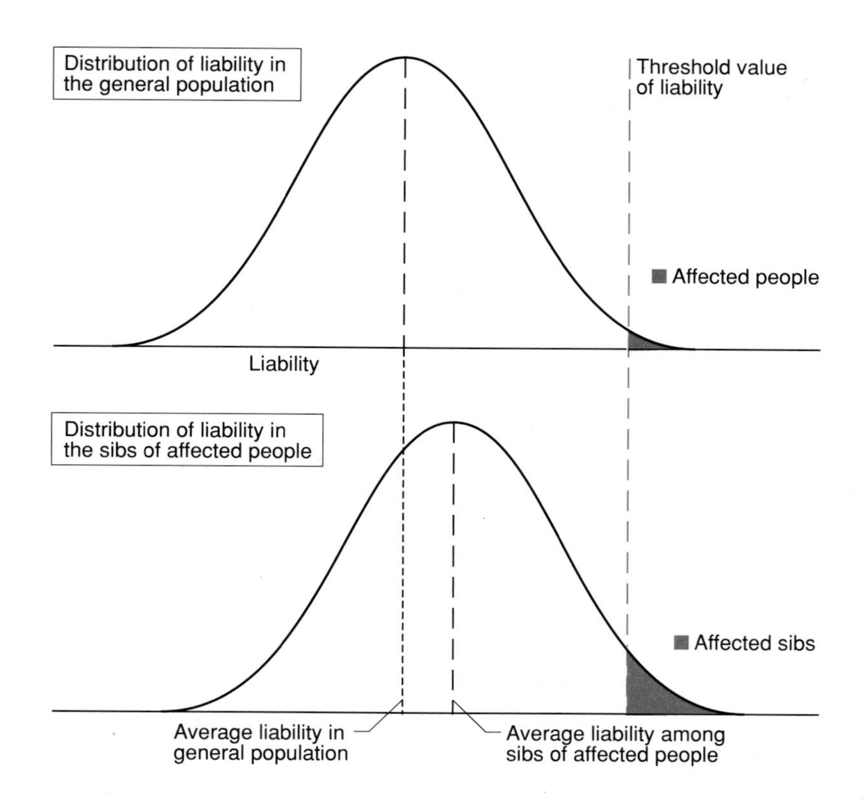

Figure 18.3: Falconer's polygenic threshold model for dichotomous nonmendelian characters.

Liability to the condition is polygenic and normally distributed (upper curve). People whose liability is above a certain threshold value are affected. Their sibs (lower curve) have a higher average liability than the population mean and a greater proportion of them have liability exceeding the threshold. Therefore the condition tends to run in families.

whose susceptibility is below the threshold, even if only just below, avoid cleft palate.

For cleft palate, a polygenic threshold model seems intuitively reasonable (Fraser, 1980). All embryos start with a cleft palate. During early development the palatal shelves must become horizontal and fuse together. They must do this within a specific developmental window of time. Many different genetic and environmental factors influence embryonic development, so it seems reasonable that susceptibility should be polygenic [although more recent evidence suggests at least one major determinant linked to the transforming growth factor-α (*TGFA*) locus; Shiang, 1993]. Whether the palatal shelves meet and fuse with ample time to spare, or whether they only just manage to fuse in time, is unimportant – if they fuse then a normal palate forms, and if they do not fuse then a cleft palate forms. Thus there is a natural threshold superimposed on a variable developmental process.

Affected people have inherited an unfortunate combination of high-susceptibility genes. Their relatives who share genes with them will also, on average, have a raised susceptibility, the divergence from the population mean depending on the proportion of shared genes. Thus polygenic threshold characters tend to run in families (*Figure 18.3*). The mathematical properties of the Gaussian distribution predict the λ parameter (the ratio of incidence in relatives to incidence in the general population).

For first degree relatives, as a rough rule of thumb the expected incidence is the square root of the population incidence. A polygenic threshold condition affecting 0.1% of newborns should affect about 1:30 of their sibs, parents or children. This prediction can then be compared with results of epidemiological surveys and any discrepancy used to calculate the heritability of the condition (*Figure 18.4*). Note, however, that the estimate of heritability depends on the assumption that the condition runs in families because of shared genes and not because of shared environment. As we saw on pages 479–482, proving that assumption is not always easy.

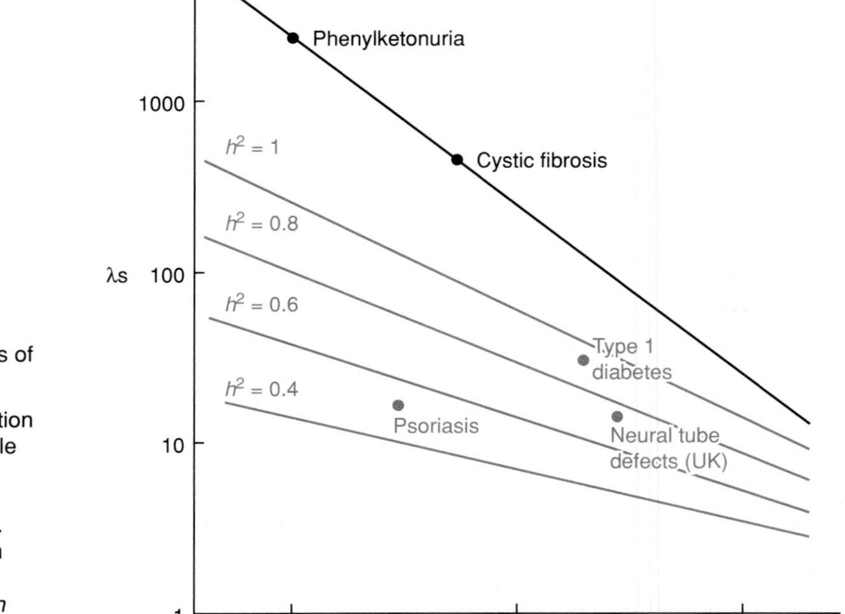

Figure 18.4: Familial clustering: relationship between population incidence and incidence in sibs of affected probands.

λ_s is the ratio of the incidence in sibs to the population incidence. The lines show the relationship for simple mendelian recessive or polygenic threshold inheritance. For a given population incidence, the higher the heritability, the greater is the value of λ_s. The graph can be used to estimate heritability from epidemiological data. Adapted from Cavalli-Sforza and Bodmer (1971). From: *The Genetics of Human Population* by Cavalli-Sforza and Bodmer. Copyright © 1971 by W.H. Freeman and Company. Used with permission.

Falconer's theory admits a number of refinements. Parents who have had several affected children may just have been unlucky, but on average their susceptibility (and hence their children's susceptibility) must have been higher than parents with only one affected child. The threshold is unchanged, but the average susceptibility, and hence the recurrence risk, rises with an increasing number of previous affected children.

Sex-specific thresholds are postulated for conditions where incidence differs between the sexes

Many supposed threshold conditions have different incidences in the two sexes. This implies sex-specific thresholds. Congenital pyloric stenosis, for example, is five times more common in boys than girls. The threshold must be higher for girls than boys, therefore relatives of an affected girl have a higher average susceptibility than relatives of an affected boy (*Figure 18.5*). The recurrence risk is correspondingly higher, although in each case the risk that a baby will be affected is five times higher if it is male (*Table 18.4*).

Alternatives to mendelian and polygenic models

Many characters are neither mendelian nor polygenic

The polygenic threshold theory exemplified in *Figure 18.3* has had a two-edged impact on thinking about nonmendelian disorders. On the one hand, it has provided an elegant quantitative framework within which to consider genetic

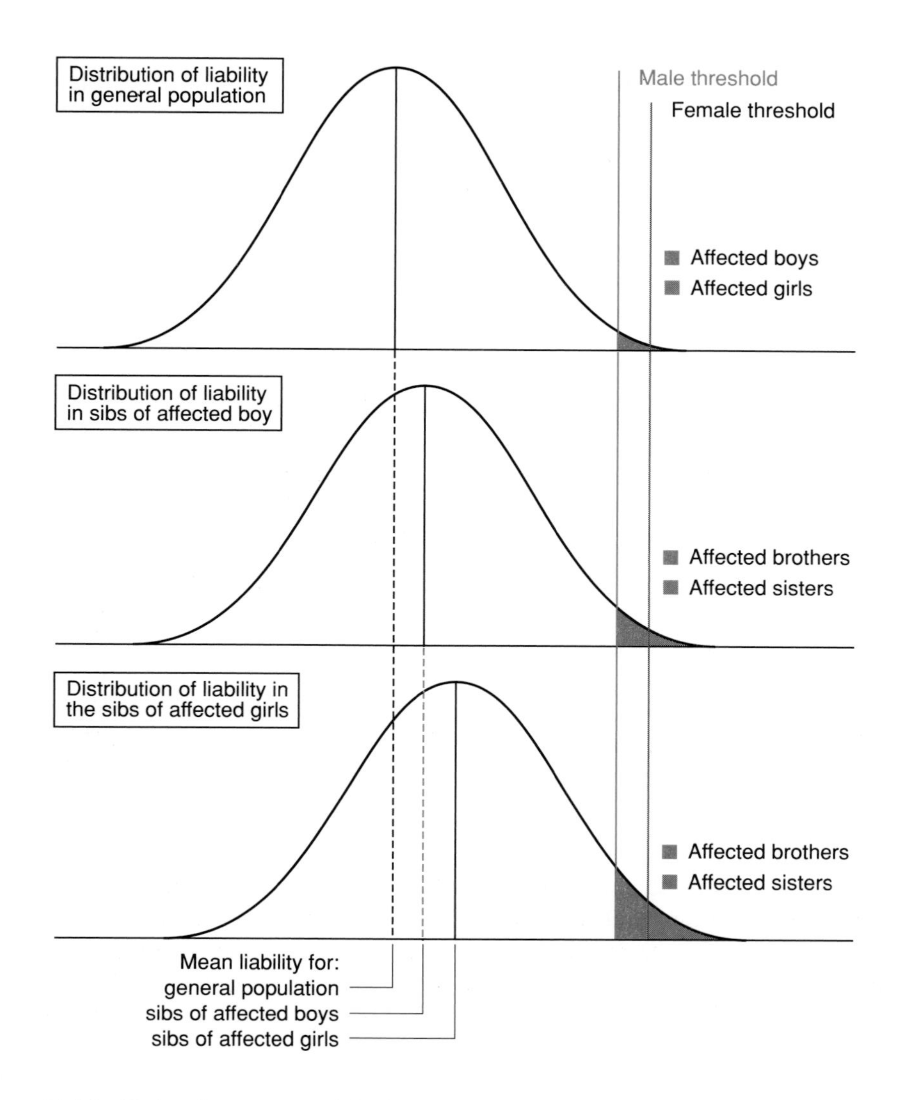

Figure 18.5: A polygenic character with sex-specific thresholds.

If a nonmendelian dichotomous character affects predominantly males, this is accommodated in multifactorial threshold theory by postulating a lower threshold for males than females. It follows that recurrence risks are higher for relatives of affected females, but the majority of these recurrent cases will be male. See *Table 18.4* for an example of data fitting this interpretation.

Table 18.4: Recurrence risks for pyloric stenosis

Relatives of	Sons	Daughters	Brothers	Sisters
Male proband	19/296 (6.42%)	7/274 (2.55%)	5/230 (2.17%)	5/242 (2.07%)
Female proband	14/61 (22.95%)	7/62 (11.48%)	11/101 (10.89%)	9/101 (8.91%)

More boys than girls are affected, but the recurrence risk is higher for relatives of an affected girl. Data fit a polygenic threshold model with sex-specific thresholds (*Figure 18.5*). Data from Fuhrmann and Vogel (1976).

susceptibility and its relation to phenotype. On the other hand, the simplicity and generality of the model encourage its uncritical application to all nonmendelian characters. Its elastic parameters allow predictions to be fitted to most diseases, leading to the impression (sometimes fostered by textbooks) that a polygenic Gaussian susceptibility distribution with a threshold underlies every nonmendelian character.

The assumption that if a human genetic character is not mendelian it must be polygenic is unfortunate. It fosters undue pessimism about the chances of uncovering the genetic factors underlying many human ills. If no single locus makes a major contribution to Falconer's susceptibilty, there is no point in looking for individual susceptibility genes. But in fact there is no *a priori* reason to suppose that there are no major genes underlying common nonmendelian birth defects and adult diseases. In reality, as we saw in Chapter 3 (*Figure 3.10*), between pure mendelian and pure polygenic characters there lies a whole spectrum of traits governed by major susceptibility loci operating over a possibly polygenic genetic background, and sometimes subject to major environmental determinants. Understanding these **oligogenic characters** is probably the most important soluble problem facing human genetics at the present time. Rhesus hemolytic disease presents a successful example.

Rhesus hemolytic disease – interaction of two mendelian loci and obstetric history

Rhesus hemolytic disease of the newborn runs in families but does not follow a mendelian pedigree pattern. The mechanism primarily involves a single locus, *RH*. The *D* allele encoding the D rhesus antigen is dominant over the *d* allele, a gene deletion resulting in absence of D antigen. Hemolytic disease affects rhesus-positive babies of rhesus-negative mothers (*Figure 18.6*).

The matings at risk are dd mother × Dd or DD father. However, hemolytic disease is a risk only if the mother has previously been sensitized to D. Sensitization normally occurs at the time of birth of a previous rhesus-positive baby. Thus rhesus hemolytic disease rarely affects the firstborn child. Sometimes transplacental bleeds, miscarriages or unmatched blood transfusions may sensitize the mother and put her first baby at risk.

Nowadays, sensitization is prevented by giving the mother an injection of anti-D antibody. The antibody clears rhesus-positive cells from her circulation before they can sensitize her. If the mother and baby are ABO-incompatible, the mother's innate ABO antibodies will usually prevent sensitization. Thus susceptibility to rhesus hemolytic disease depends on interactions between the genotypes of mother and baby at two loci, and on the mother's previous obstetric history. If the mechanism had not been unraveled, we would probably be drawing Gaussian curves and thresholds.

Heterogeneity – the example of anemia

Unrecognized heterogeneity is another reason why a disease may appear genetically complex. A character can be 'polygenic' in two different ways. If a condition is sufficiently heterogeneous in a collection of families, many different loci may be

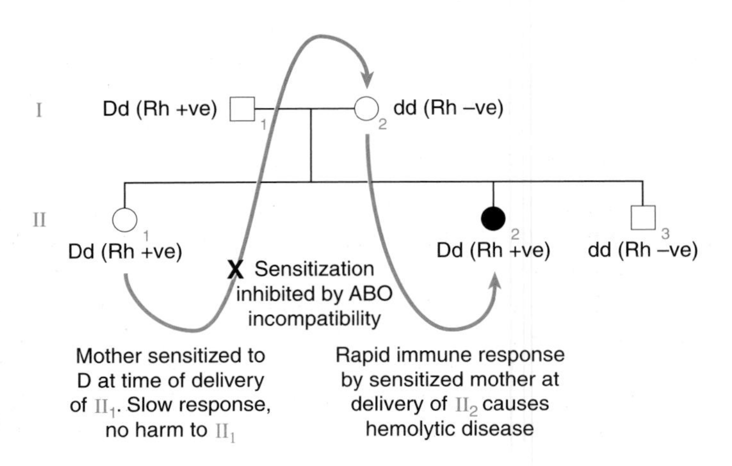

Figure 18.6: Mechanism of rhesus hemolytic disease of the newborn.

Rhesus disease occurs when an Rh-positive fetus bleeds into a previously sensitized Rh-negative mother. Sensitization usually occurs at the time of delivery of the first Rh-positive baby, but is normally avoided if the mother and baby are ABO incompatible and can be prevented by giving the mother an injection of anti-D antibody.

involved, even if in any one family only a single locus is usually involved. This apparently polygenic determination can be difficult to distinguish, except in retrospect, from true polygenic determination where many loci contribute to the character in each individual. Anemia would be an example. There are of course many causes of anemia, ranging from simple iron deficiency through many different mendelian conditions to pernicious anemia, which might be truly polygenic. Because we understand the heterogeneity, we do not describe anemia as a single polygenic condition. Maybe the same would be true of schizophrenia if we understood it.

Major susceptibility loci

Discovering a **major susceptibility locus** can be the key to progress in understanding the causes of a disease. Identifying major susceptibility loci can involve two approaches. In segregation analysis, collections of family data are analyzed for statistical evidence of major loci. This approach can suggest the existence of a major locus and at least partly define its properties, but it cannot identify genes. The loci must be identified through the alternative approach of testing candidate genes or by using markers to localize genes and then positional cloning (Chapter 14) to identify the gene. Both the top-down segregation analysis and bottom-up linkage approaches face difficulties with nonmendelian conditions. Segregation analysis requires large datasets and is very sensitive to subtle biases in the way the data are collected. With marker studies, as we saw in Chapter 12, uncertainty over the precise genetics of the susceptibility usually precludes the use of conventional lod score analysis.

Segregation analysis estimates the most likely mix of genetic factors in pooled family data

Analyzing data on the relatives of a large collection of people affected by a familial but nonmendelian disease is not a simple task. There could be both genetic and environmental factors at work; the genetic factors could be polygenic, mendelian with any mode of inheritance or any mixture of these, while the environmental factors may include both familial and nonfamilial variables. Various methods and programs are available for segregation analysis, depending on the range of possibilities which need to be explored.

Complete ascertainment of a recessive condition: the truncate binomial distribution

A simple type of segregation analysis tests for autosomal recessive inheritance of a character. At first sight this would seem a trivial task. If a condition is autosomal recessive, then one in four of the children of carrier parents should be affected. However, if one collects a set of families where children have a known autosomal recessive condition, the proportion of affected children will not be 1:4 (*Figure 18.7*). The problem is **bias of ascertainment**.

Assuming there is no independent way of recognizing carriers, the families will be identified through an affected child. Thus the families shown in black in *Figure 18.7* will not be ascertained, and the observed segregation ratio in these two-child families is not 1/4 but 8/14. Families with three children, ascertained in the same way, would give a different segregation ratio, 48/111. The ratio for any given family size can be estimated from the truncated binomial distribution, a binomial expansion of $(\frac{1}{4} + \frac{3}{4})^n$ in which the last term (no affected children) is omitted. Experimental data can be corrected for this bias, most simply by the method of Li and Mantel (see *Notebox*).

> ***Notebox:*** correcting the segregation ratio
>
> Complete truncate ascertainment:
> $p = (R-S)/(T-S)$
>
> Single selection:
> $p = (R-N)/(T-N)$
>
> p = true (unbiased) segregation ratio.
> R = number of affected children.
> S = number of affected singletons (children who are the only affected child in the family).
> T = total number of children.
> N = number of sibships.

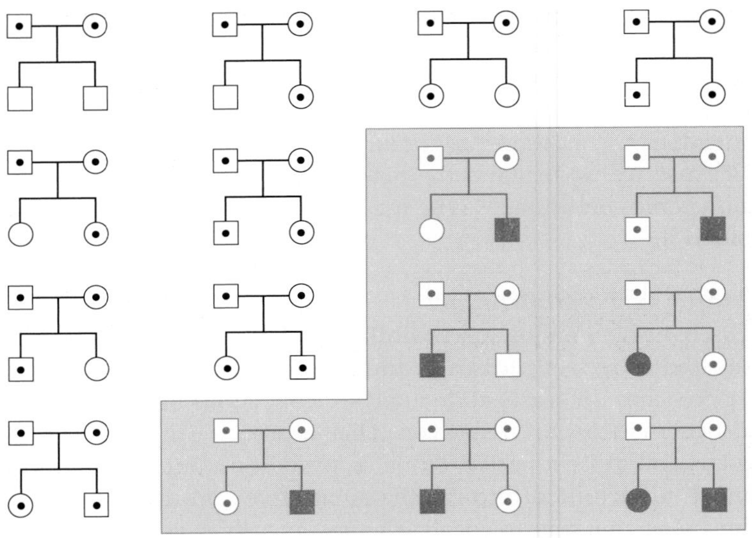

Total children : AA $\frac{8}{32}$ Aa $\frac{16}{32}$ aa $\frac{8}{32}$

Ascertained : AA $\frac{2}{14}$ Aa $\frac{4}{14}$ aa $\frac{8}{14}$

Correction (Li–Mantel): $p = (R - S)/(T - S) = (8 - 6)/(14 - 6) = \frac{2}{8} = 0.25$

Figure 18.7: Bias of ascertainment.

Both parents are carriers of an autosomal recessive condition. Overall one child in four is affected, but if families are ascertained through affected children, only the red families will be picked up and the proportion of affected children is 8/14.

Single selection

Suppose affected children were ascertained by counting the first 100 to be seen in a busy clinic (so that many more could have been ascertained from the same population by carrying on for longer). A family with two affected children is twice as likely to be picked up as one with only a single affected child and one with four affected is four times as likely. This introduces a different bias of ascertainment and requires a different statistical correction (see *Notebox*). The Li–Mantel formula applies to **complete truncate ascertainment**, where every affected case in the population is ascertained; the present case uses **single selection**, where the probability of being ascertained is proportional to the number of affected children in the family.

Complex segregation analysis

In real data collections, the mode of ascertainment is often mixed, falling somewhere between complete truncate and single selection. Moreover, the examples considered so far have included only a single family structure and a single genetic mechanism. In **complex segregation analysis** a whole range of possible mechanisms, gene frequencies, penetrances, etc., are allowed, and the computer performs a maximum likelihood analysis to find the mix of parameter values which gives the greatest overall likelihood for the observed data. *Table 18.5* shows an example. As with lod score analysis (Chapter 12), the question asked is how much more likely the observations are on one hypothesis compared with another.

In the example of *Table 18.5*, the ability of specific models (sporadic, polygenic, dominant, recessive) to explain the data was compared with the likelihood calculated by a general model ('mixed model'), in which the computer could freely optimize the mixture of single-gene, polygenic and random environmental causes. All models were constrained by overall incidences, sex ratios and probabilities of ascertainment estimated from the collected data. A single-locus dominant model is not significantly

Table 18.5: Complex segregation analysis (Badner *et al.*, 1990)

Model	*d*	*t*	*q*	*H*	*z*	*x*	χ^2	*p*
Mixed	1.00	7.51	9.6×10^{-6}		0.01	0.15		
Sporadic							334	$<1 \times 10^{-5}$
Polygenic				1.00	1.00		78	$<1 \times 10^{-5}$
Major recessive locus	0.00	8.22	3.8×10^{-3}				35	$<1 \times 10^{-5}$
Major dominant locus	1.00	7.56	1.2×10^{-5}			0.19	2.8	0.42

Data are for families ascertained through a proband with long-segment Hirschsprung's disease. Parameters which can be varied are *t* (the difference in liability between people homozygous for the low-susceptibility and the high-susceptibility alleles of a major susceptibility gene, measured in units of standard deviation of liability), *d* (the degree of dominance of any major disease allele), *q* (the gene frequency of any major disease allele), *H* (the proportion of total variance in liability which is due to polygenic inheritance, in adults), *z* (the ratio of heritability in children to heritability in adults) and *x* (the proportion of cases due to new mutation). A single major locus encoding dominant susceptibility explains the data as well as a general model in which a mix of all mechanisms is allowed.

worse than the mixed model at explaining the data ($\chi^2 = 2.8$, $p = 0.42$), while models assuming no genetic factors, pure polygenic inheritance or pure recessive inheritance perform very badly. Thus the analysis suggests the existence of a major dominant susceptibility to Hirschsprung's disease. Two such factors have been identified by linkage analysis, the *RET* and *EDNRB* genes (Edery *et al.*, 1994; Puffenberger *et al.*, 1994), though they probably account for only a minority of the susceptibility.

Pitfalls of segregation analysis

However clever the segregation analysis program, it can only maximize the likelihood with the parameters it was given. If a major factor is omitted, the result can be misleading. This was well illustrated by the data of McGuffin and Huckle (*Table 18.6*). They asked their classes of medical students which of their relatives had attended medical school. When they fed the results through a segregation analysis program, it came up with results apparently favoring the existence of a recessive gene for attending medical school. Though amusing, this was not done as a joke or to discredit segregation analysis. The authors did not allow the computer to consider the likely true mechanism, shared family environment. The computer's next best alternative was mathematically valid but biologically unrealistic. The serious point McGuffin and Huckle were making was that there are many pitfalls in segregation analysis of human behavioral traits and incautious analyses can generate spurious genetic effects.

Table 18.6: A recessive gene for attending medical school?

Model	*d*	*t*	*q*	*H*	χ^2	*p*
Mixed	0.087	4.04	0.089	0.008		
Sporadic					163	$<1 \times 10^{-5}$
Polygenic				0.845	14.4	<0.005
Major recessive locus	0.00	7.62	0.88		0.11	N.S.

Data of McGuffin and Huckle (1990) from a survey of medical students and their families. Meaning of symbols is as in *Table 18.5.* 'Affected' is defined as attending medical school. The analysis appears to support recessive inheritance, since this accounts for the data equally well as the unrestricted model. The point of this work was to illustrate how analysis of family data can produce spurious results if shared family environment is ignored (see text).

Using lod score analysis to map susceptibility genes

If a limited number of loci are major determinants of susceptibility to a nonmendelian disease, then it should be possible to map the loci by linkage analysis. The strategy is to seek families where there is such a strong history of the disease that it appears to be a mendelian character, albeit with somewhat reduced penetrance. Arguably these families have an unusually high background level of susceptibility, such that the presence or absence of a single major susceptibility allele is sufficient to tip them over the threshold into disease. In these families it acts as a mendelian determinant, although in other families with a lower background susceptibility, the presence of the same major susceptibility allele on its own would not be enough to cause disease.

If this analysis is correct, the major gene can be located by standard linkage analysis using a battery of markers in these families. Standard lod score analysis is used, with the details of the genetic model (gene frequencies, penetrances, etc.) being supplied by segregation analysis. This general strategy for finding susceptibility loci has been the focus of a major research effort over the past 15 years. It has seen some resounding successes, such as the mapping and later identification of the *BRCA1* breast cancer susceptibility gene on chromosome 17. It has also seen some embarassments, particularly the false mapping of a gene for schizophrenia to chromosome 5.

The *BRCA1* gene shows how susceptibility genes can be identified by segregation analysis followed by linkage analysis and positional cloning

Breast cancer has long been known to show a slight tendency to run in families, especially cases with an unusually early age of onset. Overall the disease is certainly not mendelian, but it seemed plausible that there might be a subset of mendelian families among a majority of sporadic cases. A large-scale segregation analysis of 1500 families supported this view (Newman *et al.*, 1988). The analysis suggested that 4–5% of breast cancer might be attributable to inherited factors and that the proportion would be highest among early-onset cases.

Families with near-mendelian pedigree patterns were therefore collected for linkage analysis (*Figure 18.8*). This work culminated in 1990 in the mapping of a susceptibility gene to 17q21 (Hall *et al.*, 1990). The linkage was seen only in families with onset by age 45. Later-onset families gave negative lod scores. The localization was soon confirmed by other groups, and over the next 2 years analysis of many families defined a candidate region 8 cM long in males and 17 cM long in females, located between markers D17S588 and D17S250. It appeared that the 17q21 gene, called *BRCA1*, might account for 80–90% of families with both breast and ovarian cancer, but only 50% of families with breast cancer alone.

At the same time, this large effort defined another susceptibility locus, *BRCA2* at 13q12. Overlapping with the linkage phase, candidate genes were isolated by positional cloning (Chapter 14) and tested for mutations in affected women from 17q-linked families. Eventually one candidate gene, *BRCA1*, was shown to carry mutations in five independent cases (Miki *et al.*, 1994). Many other *BRCA1* mutations have been characterized since the initial discovery of the gene.

The success of this work gives encouragement that other heterogeneous and apparently nonmendelian conditions can be tackled successfully. One element in the

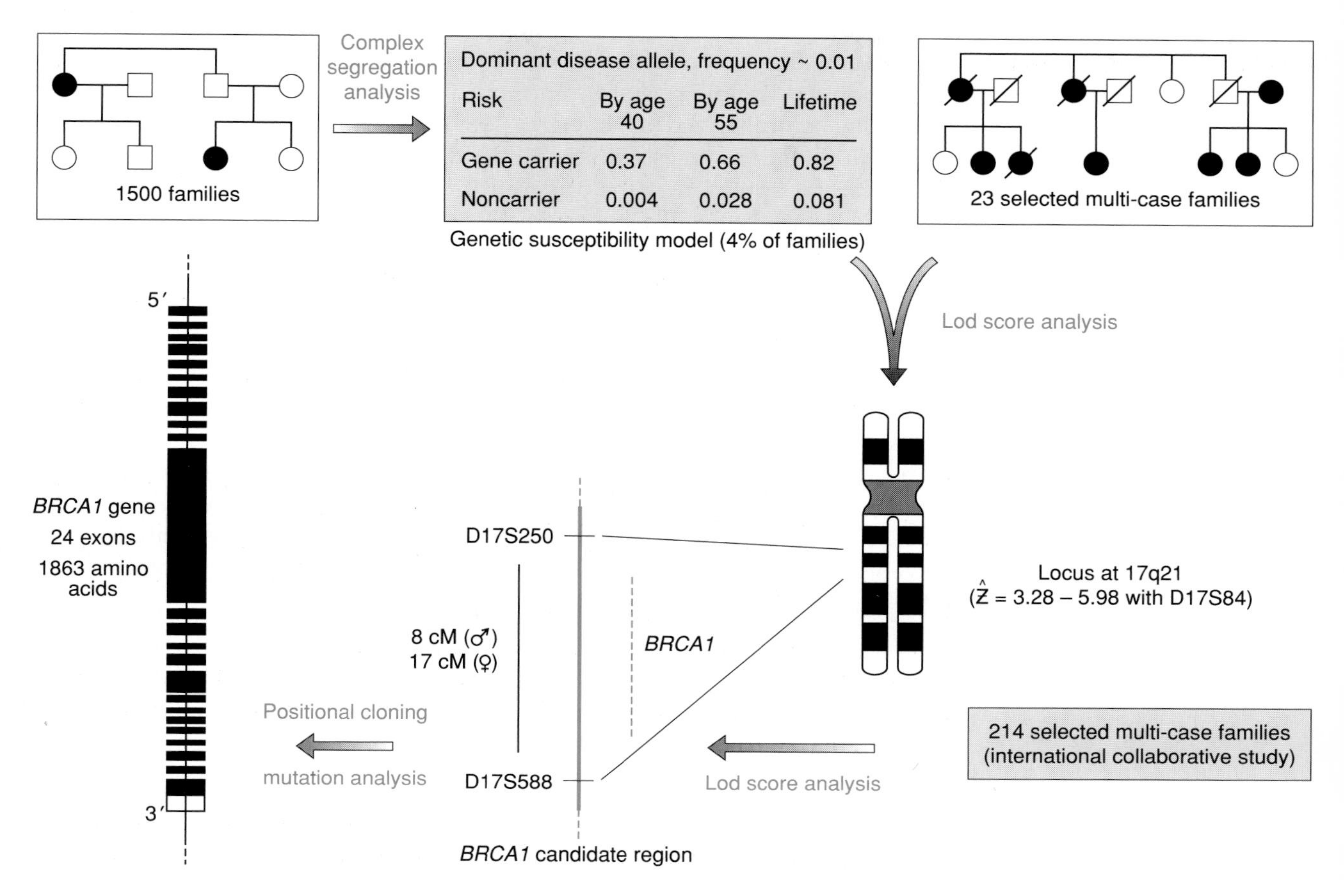

Figure 18.8: How the *BRCA1* gene was found.

This is an example of the successful use of mapping by lod score analysis followed by positional cloning to identify a gene conferring susceptibility to a common disease.

success was the choice of diagnostic grouping, emphasizing age of onset, which turned out to be a key identifier of 'genetic' families. It also helped that, within the genetic families, the penetrance was high. Finally it is probably fair to say that there was a degree of good luck in the original linkage work. Given that even in these near-mendelian families some of the cases of this common disease will be sporadic (so confusing the linkage analysis), and given that it now appears that *BRCA1* mutations occur in only 50% of families with monogenic early-onset breast cancer, the researchers were probably lucky that in their sample 7/7 families with average onset by age 45 gave positive lod scores with the 17q markers.

Lod score analysis gave a false positive result with schizophrenia

Schizophrenia affects around 1% of people in most countries of the world, and it causes enormous distress to patients and their families, not to mention huge costs to society. Not surprisingly, many researchers have devoted themselves to trying to discover the causes of this disorder (Gottesmann and Shields, 1982). A large body of family, twin and adoption studies, parts of which were summarized earlier in this chapter (*Tables 18.1, 18.2* and *18.3*), shows that it tends to run in families and strongly hints at genetic susceptibility. Although schizophrenia, taken as a whole, is non-mendelian, families exist where it appears to be an imperfectly penetrant mendelian dominant character.

Testing seven such families (two English and five Icelandic) with markers from chromosome 5 gave a peak lod score of 6.5 (Sherrington *et al.*, 1987), well above the conventional threshold of significance of 3.0. Subsequent events proved highly disappointing (summarized by Byerley, 1989). Other investigators tested their own pseudo-dominant families with the same markers, with uniformly negative results. It is generally agreed that the apparent linkage was a false positive, a statistical fluke. A similar fate befell a study which had suggested an X-linked locus for manic-depressive psychosis (reviewed by Baron *et al.*, 1994).

Lod score analysis may be inappropriate for nonmendelian characters

The underlying problem lies in the application of lod score analysis to inappropriate material. Lod score analysis requires a precise genetic model to be specified, including the disease gene frequency and the penetrance of each genotype. It also requires people to be classified clearly as affected or unaffected. Diagnostic criteria can be a problem with any disease, but the problems are especially acute in psychiatric genetics. Standardized diagnostic criteria exist, but they are essentially arbitrary conventions. Adhering to them helps make different studies comparable, but does not guarrantee that the right genetic question is being asked.

Thus none of the components of a precise genetic model can be provided for schizophrenia. Instead, the researcher first selects 'suitable' looking families, and then tries a range of models and diagnostic criteria in the analysis, to see which gives the most significant result. This is not unlike the way parameters are maximized in segregation analysis and it is not inherently wrong; however, it does invalidate the conventional tests of significance. Although the original researchers appreciated that this risk existed, they and other workers underestimated the magnitude of the risk.

The schizophrenia debacle has led to some re-evaluation of the way linkage analysis should be used in complex diseases (Lander and Schork, 1994). One precaution is to suspend judgement until the same linkage has been confirmed in an independent set of families – but failure to confirm does not refute the original conclusion, because it is entirely likely that different major genes are segregating in different families. Some researchers have suggested studying other nearby markers (in the same families as showed the original linkage), but this will not help: the lod scores show that a particular chromosome segment is tending to segregate with the disease, and studying more markers tells us nothing new. The general view is that lod score analysis, with its requirement for a precise genetic model, is an inappropriate tool for analyzing nonmendelian characters. Recent work has therefore focused on using batteries of markers with a model-free analysis (page 329) and this approach appears to be bearing fruit.

Using affected sib pair analysis to map susceptibility genes: the example of diabetes

As we saw in Chapter 12 (page 330), the simplest method of model-free linkage analysis is to study haplotype sharing by affected sib pairs. This has recently been applied on an heroic scale to try to map susceptibility genes for diabetes.

The term diabetes embraces two different common diseases, together with one or more other rare conditions

Family and twin studies showed that diabetes tended to run in families, but for many years diabetes remained, as it was famously described, the geneticists' nightmare. The first step towards understanding the genetics was to distinguish the different types of diabetes (*Table 18.7*). Type 1 and type 2 diabetes are totally different diseases, with different causes, different natural histories and different genetics. Maturity-onset diabetes of the young (MODY) is a rare variant having the clinical features of type 2 diabetes but the age of onset of type 1.

Table 18.7: Clinical classification of diabetes

Type 1 diabetes	Type 2 diabetes	MODY
Juvenile onset	Maturity onset (> 40 years)	Juvenile onset
0.4% of UK population	6% of US population	Rare
Requires insulin	Usually controllable by oral hypoglycemics	As type 2 diabetes
No obesity	Strong association with obesity	No obesity
Familial:	Familial:	Familial:
MZ concordance 30%	MZ twin concordance 40–100%	autosomal
sib risk 6–10%	sib risk 30% (maybe subclinical)	dominant?
Associated with HLA-DR3 and DR4	No HLA association	As type 2 diabetes

Type 2 diabetes, the usually undramatic maturity-onset disease, results from a decreased number of pancreatic β cells and/or the development of insulin resistance in the end organs. Age, obesity, physical inactivity and unknown genetic factors all play contributory roles in a complex and doubtless heterogeneous pathogenesis. The incidence rises rapidly with age; possibly if we all lived to 150 years, type 2 diabetes would appear as a late-onset mendelian dominant condition. The rare MODY variant runs strongly in families; in a few of them MODY is caused by mutations in the glucokinase gene.

Type 1 diabetes is quite different. There is strong familial clustering; λ_s, the risk to sibs compared with the general population risk, is about 15. The pathology involves autoimmune destruction of pancreatic islets, and there is a strong association with possession of HLA-DR3 and/or -DR4 antigens. In the UK, about 95% of patients with type 1 diabetes have the DR3 and/or DR4 antigens, compared with 45–54% of the general population (reviewed in Todd *et al.*, 1987). This was the first demonstration of a major genetic factor in diabetes. It turns out that chromosomes carrying HLA haplotypes associated with a low risk of diabetes all carry an allele at the DQ_β locus in which amino acid 57 is aspartic acid, while chromosomes carrying high-risk haplotypes have DQ_β alleles with some other amino acid at position 57 (Todd *et al.*, 1987). This is believed to be the true HLA-linked susceptibility factor, since it is the only feature which consistently distinguishes high-risk from low-risk HLA haplotypes. Exactly how the absence of aspartic acid at position 57 of the DQ_β product predisposes to diabetes is not clear.

The nonHLA susceptibility loci for type 1 diabetes are being mapped by sib pair studies

About 40% of the family clustering in type 1 diabetes is accounted for by the *HLA-*DQ_β susceptibility locus. To map the other susceptibility loci, several groups have systematically searched the whole genome by typing collections of affected sib pairs with a large panel of markers. Progress has been reviewed by Thomson (1994).

Table 18.8: Type 1 diabetes susceptibility loci suggested by sib pair analysis

Locus	Location	Max log likelihood	λ_s	Susceptibility (%)
IDDM1	6p21 (HLA)	7.3	3.1	42
IDDM2	11p15 (INS)	2.1	1.3	10
IDDM3	15q26	?	?	?
IDDM4	11q13	3.4	1.3	?
IDDM5	6q24–q27	2.0	?	?
Potential loci	1q, 2q, 3q, 4q, 6q, 7q, 8p, 8q, 10p, 10cen, 13q, 14q, 16q, 17p, 18q, 19q, X			

The very large number of 'potential loci' illustrate the difficulty of distinguishing signal from noise in these analyses. Data from Thomson (1994) and Davies *et al.* (1994).

We will use the work of Davies *et al.* (1994) to illustrate the scale, power and problems of such studies. Initially these workers typed 96 pairs of affected sibs with 290 markers spread across the whole genome. Among the 55 000 genotypes generated, affected sib pairs showed significant sharing ($p < 0.05$) of markers from 20 different chromosome regions. Ten of the 20 candidate regions showed sharing significant at the $p < 0.005$ level (before correction for the number of markers tested, see below). Two previously established susceptibility loci were noted: the HLA region and the region of the insulin gene on 11p15 (where previous data were suggestive but inconsistent).

Evidence that the other eight regions contained true susceptibility loci and not statistical artifacts was sought by typing in two other independent series of 102 and 84 affected sib pairs. Because of the overwhelmingly strong effect of the HLA-linked *IDDM1* locus, sib pairs were stratified according to whether they shared two, one or no *HLA* alleles and according to HLA type (*DR3*, *DR4*, both or neither). The resulting large number of comparisons gave many potentially interesting departures from the null hypothesis, but interpreting their significance was problematic. The proposed *IDDM4* and *IDDM5* loci (*Table 18.8*) are the two most confident identifications to date. The *IDDM3* locus identified by other workers (see Thomson, 1994) was not detected by Davies *et al.* Many weak associations, in this and other similar studies, may point to further loci or may just be statistical noise (*Table 18.8*).

Identifying susceptibility genes in the regions defined by sib pair analysis may not be easy

Approaches similar to that of Davies *et al.* (1994) are being pursued with most important complex diseases. The next few years will undoubtedly see large numbers of candidate susceptibility regions defined. Sorting out the false positives and identifying the true susceptibility genes presents a considerable challenge. Although one can always collect ever more sib pairs for analysis, the chromosomal regions defined by this approach will seldom be small enough to allow easy positional cloning of the gene responsible. The next step is likely to be to seek associations in the general population between the disease and alleles of candidate genes or markers within the susceptibility region.

Population association studies for mapping susceptibility loci

An alternative to linkage mapping in families is to look for statistical associations in the general population between the disease and some marker genotype. As we saw in Chapter 12 (page 331), linkage and association are different phenomena. The essential difference is that linkage is a relationship between loci, and association is a relationship between alleles. Thus the 21-hydroxylase *locus* (page 268) is *linked* to the *HLA-DR locus*; but rheumatoid arthritis is *associated* with the *HLA-DR4 allele*. Lod score analysis shows no evidence for a rheumatoid arthritis locus linked to the *HLA* locus.

Disease–marker associations at the population level can have various causes

Even very close linkage is not sufficient to cause a population association. Population associations between an allele A and disease D can arise for three reasons:

(i) allele A can directly cause susceptibility to D. Generally, possession of A is neither necessary nor sufficient for somebody to develop disease D, but it increases the likelihood. In this case one would expect to see the same allele A associated with the disease in any population studied (unless the causes of the disease vary from one population to another).

(ii) Very close linkage can produce allelic association at the population level, provided that most disease-bearing chromosomes in the population are descended from one or a few ancestral chromosomes (page 331). If linkage disequilibrium is the cause of the association, a gene should be discovered near to the A locus which has mutations in people with disease D. The particular allele at the A locus which is associated with disease D may be different in different populations.

(iii) People with the disease and people without the disease may be genetically different subsets of the population, who coincidentally also differ in the frequency of allele A (**population stratification**). Lander and Schork (1994) give the example of the association in the San Francisco Bay area between *HLA-A1* and ability to eat with chopsticks. *HLA-A1* is more frequent among Chinese than among Caucasians.

Population-based case–control studies of disease–marker associations became discredited because of statistical problems

Disease–marker associations are found by comparing the frequencies of a particular marker allele in a series of patients and in a series of healthy controls. Such population-based case–control studies fell out of favor some years ago as two sets of problems became apparent:

- problems with selection of controls mean they are very bad at distinguishing linkage disequilibrium from population stratification.
- Statistical analysis was often insufficiently rigorous, with inadequate correction being made for the number of questions asked (*Table 18.10*).

Selecting suitable controls is crucial. Many studies in the past have used published gene frequencies, often without adequate certainty that these frequencies are representative of the population from whom the patients were recruited. Alternatively, students or staff from the investigator's university may be used as a control series. Again, this is undesirable because they may well not be typical of the population from which the patients were drawn. Thus, when an association is found, it may be impossible to know whether it is caused by linkage disequilibrium with a susceptibility locus or by inadequately matched controls. Recently, methods using internal controls have been developed, and these have led to a renaissance of disease–marker association studies.

Association studies with internal controls overcome many of the problems of classical disease–marker association studies

We have seen that the choice of control group in association studies is crucial to avoiding the risk that the study and control groups were drawn from genetically distinct subpopulations. Recently, a clutch of methods have been developed which largely circumvent this problem (see Lander and Schork, 1994). Collectively they can be called association studies with internal controls (ASIC). The two principal methods are the haplotype relative risk (HRR) test and the transmission disequilibrium test (TDT).

Either of these two simple procedures avoid most of the problems of standard association studies. The result is unaffected by population stratification. However, the tests cannot distinguish associations caused by linkage disequilibrium from those where the marker is itself a susceptibility factor. They involve 50% more work than standard case–control studies because three people (proband and parents) are typed in each family – a small price to pay for the gain in reliability. The only serious problem is that parents must be available, which may be difficult for late-onset diseases. Both tests as described here test a proposed association with a single allele M_1 at the marker locus.

The HRR test

The HRR test (Khoury, 1994) takes affected individuals and their two parents. All three are typed for a marker, one allele of which is believed to be associated with the disease. The 'control' is not a real individual, but consists of the two parental alleles which were not transmitted to the affected person. A 2 × 2 table is made of marker frequencies in the probands and controls, and significance tested by a simple χ^2 test (*Table 18.9*). In its simplest form, as shown, no distinction is made between probands homozygous or heterozygous for the marker, nor between affected and unaffected parents.

Table 18.9: Tests to determine whether marker allele M_1 is associated with a disease

HRR test	Controls	
Cases	M present	M absent
M present	*a*	*b*
M absent	*c*	*d*
TDT test	**Nontransmitted allele**	
Transmitted allele	M_1	Not M_1
M_1	*a*	*b*
not M_1	*c*	*d*

For either test, χ^2 (1 degree of freedom) = $(b-c)^2 / (b+c)$

The HRR test: the controls are made from the two marker alleles which the parents of the affected proband did not transmit to the proband. The TDT: families are selected where affected probands have at least one parent who is heterozygous for M_1. The transmitted and nontransmitted parental alleles are compared.

The TDT

The TDT (Spielman *et al.*, 1993) starts with families with one or more affected offspring, where at least one parent is heterozygous at the marker locus for the marker allele (M_1) which is suspected of being associated with the disease. One of the two marker alleles of each heterozygous parent is transmitted to each affected offspring and one is not. The test compares the frequency of M_1 among the transmitted and nontransmitted alleles. Again, the significance of the association is tested by a simple χ^2 test (*Table 18.9*).

Probabilities calculated from association studies must be corrected for the number of questions asked

A mendelian condition must map somewhere so, in linkage analysis, no matter how many markers are used in finding the location, the risk of false positive results remains manageably low (page 323). This is not the case for association studies. There may well be no association to find, and so each test performed carries an independent risk of a false positive result. To avoid errors, a Bonferroni correction (Kidd and Ott, 1984) has to be applied. The threshold of significance is set, not at the conventional $p = 0.05$, but at $p = 0.05/n$, where n is the number of independent potential associations checked (*Table 18.10*). All too few published disease association studies apply the rigorous correction factor, $n(m - 1)$ for the testing of n loci with m alleles each and all too often associations reported in one study cannot be confirmed in a second independent sample of patients.

Table 18.10: *p* values from a hypothetical association study

	D_1	D_2	D_3	D_4	D_5	D_6	D_7	D_8	D_9	D_{10}
M_1	0.29	0.47	0.80	0.47	0.36	0.13	0.93	0.15	0.08	0.08
M_2	0.21	0.26	0.38	0.55	0.96	0.61	0.46	0.28	0.10	0.40
M_3	0.36	0.87	0.61	0.76	0.80	0.51	0.44	0.11	0.76	0.99
M_4	0.12	0.77	0.20	0.68	0.88	0.47	0.39	0.05	0.50	0.53
M_5	0.09	0.56	0.01	0.93	0.24	0.81	0.18	0.28	0.04	0.18
M_6	0.61	0.83	0.27	0.95	0.66	0.03	0.24	0.05	0.03	0.87
M_7	0.63	0.64	0.12	0.33	0.76	0.09	0.54	0.77	0.42	0.09
M_8	0.24	0.12	0.06	0.65	0.98	0.52	0.91	0.63	0.68	0.23
M_9	0.36	0.03	0.15	0.62	0.68	0.88	0.15	0.96	0.94	0.55
M_{10}	0.27	0.94	0.31	0.32	0.54	0.06	0.20	0.63	0.53	0.38

Panels of patients with diseases (D_1–D_{10}) and a panel of controls were typed for markers (M_1–M_{10}). For each possible association the *p* value is tabulated. In reality none of the diseases is associated with any of the markers, but five of the 100 *p* values are significant at the 5% level and one at the 1% level. This is of course exactly what is expected of a series of 100 random numbers. If *n* questions are asked, the appropriate threshold of significance is 0.05/*n*.

Linkage disequilibrium operates over distances less than 1 cM

Allelic associations reflect conserved ancestral chromosome segments

As we saw in Chapter 12 (page 331), allelic associations caused by linkage disequilibrium are seen only when a significant proportion of the apparently independent chromosomes we examine in a population are in fact copies of the same ancestral chromosome. To reiterate, linkage disequilibrium is not an inevitable consequence of close linkage. Alleles at two tightly linked loci will show associations only if they mark shared ancestral chromosomes.

Apparently unrelated people share only very short chromosomal segments

All humans are related, if we go back far enough. In the UK two 'unrelated' people would typically share common ancestors not more than 22 generations ago. If fully outbred, they would have 2^{22} = 4 million ancestors each at that time. Twenty-two generations is about 500 years and in the 15th century the population of Britain was around 4 million. Therefore not more than 44 meioses separate our two unrelated people, assuming that the UK population interbreeds freely.

Loci showing 1% recombination per meiosis would have a better than 50% chance of remaining in the same combination through 44 meioses, since $(0.99)^{44} = 0.64$. If this argument is valid, it suggests that allelic associations reflecting sharing of ancestral chromosomes might begin to be noticeable for loci within 1 cM of each other.

Whole genome scans for linkage disequilibrium are not statistically feasible

Given the total human genetic map length of 2644 cM (page 316), a complete genomic scan for markers in linkage disequilbrium with a disease would require at least 3000 markers. Even if large-scale automated laboratories could take on such a search, they would face a statistical problem. As we saw above (*Table 18.10*), the significance of a p value depends on the number of questions asked. A whole genome scan for linkage disequilibrium would ask untold thousands of questions. Only extraordinarily strong associations would be significant after correction.

Testing for associations will always be restricted to candidate loci

It follows from the statistical argument above that testing for linkage disequilibrium will always be restricted to candidate chromosomal regions or loci, where there is some prior reason to suspect that an association might exist. A common strategy is to use a genome-wide linkage search, maybe using affected sib pairs, to define rather broad candidate regions. These are too big for positional cloning, but can be searched for associations. If associations exist (and they may not), their short range becomes an advantage because they can narrow down the location of the susceptibility gene to a region of DNA short enough to search for genes.

Identifying susceptibility genes

Positional cloning (Chapter 14), even within a narrow region defined by linkage disequilibrium, is not an easy way to identify susceptibility genes. With mendelian diseases, any candidate gene can be screened for mutations in families known by linkage analysis to have a mutation within the candidate region. However, association studies do not allow this selection. In most cases no one susceptibility gene will be mutated in every patient with the disease. Thus, although candidate susceptibility genes may be defined by positional cloning, proving that the candidate is the right one is harder than with mendelian diseases. It usually depends on producing a plausible physiological reason why mutations in the gene in question should predispose to the disease.

Fortunately, large-scale cDNA sequencing is likely to have identified a good proportion of all human genes within the very near future, at least as expressed sequence tags (ESTs; page 302). Databases of partial gene sequences which have been mapped physically (by FISH or using hybrid cells as described in Chapter 11) are rapidly expanding. Thus investigators will be able to pick likely candidate genes from their susceptibility region.

Four sets of advances make us reasonably optimistic of seeing good progress in this difficult area:

- with automated gel analysis and dense genetic maps, marker studies can be carried out on a far larger scale and to a far higher resolution than ever before.
- The newer methods for association studies with internal controls are much more reliable than previous methods. Linkage methods are also being refined, for example by two-locus lod score analysis in oligogenic diseases (Tienari *et al.*, 1994).
- The value of isolated homogeneous populations is being exploited in linkage disequilibrium studies.
- Candidate loci are being defined by studies of animal models.

Table 18.11 lists some examples. The major disappointment has been in psychiatric genetics where, to date, a very large effort on family studies has yielded few solid results. Probably this reflects the special difficulty of defining homogeneous diseases in psychiatry and making unambiguous diagnoses. The paper of Berrettini *et al.* (1994), describing a possible linkage of manic-depressive illness to markers on chromosome 18, gives a good account of the problems and possible solutions. Lander and Schork (1994) give an excellent overview of genetic dissection of complex traits, elaborating many of the points made in this chapter.

Table 18.11: Examples of susceptibility loci in complex diseases

Disease	Risk locus	Risk allele	Relative risk	Method used
Ankylosing spondylitis	*HLA-B*	*B27*	80	P
Type 1 diabetes	*HLA-DR*	*DR3, DR4*	15 (DR3 and/or 4)	S, P
Alzheimer's (late onset)	Apolipoprotein E (*APOE*)	ε4	2.8 (heterozygote) 8.0 (homozygote)	S, P
Myocardial infarction	Angiotensin-converting enzyme (*ACE*)	*D*	2.5 (DD vs. II)	P
Essential hypertension	Angiotensinogen (*AGT*)	*M235T*	1.6	S, P
Osteoporosis	Vitamin D receptor (*VDR*)	*B*	4.4 (BB vs. bb)	P
Breast cancer	*BRCA1*	Various	146 (age 40)	L
Multiple sclerosis	Myelin basic protein (*MBP*), *HLA-DQA1*	*MBP*:1.27 kb *DQA1*0102*	MBP: 3.3 HLA: 3.8	L, P

Method: P, population association; S, allele-sharing within families (usually sib pairs); L, lod score analysis. Only association studies can identify a risk allele; allele-sharing or linkage studies identify a risk locus or chromosomal region. *Note* that relative risk estimates may apply only within multicase families, and often decrease with more studies.
References: ankylosing spondylitis, Schlosstein *et al.* (1973); diabetes, Todd *et al.* (1987); Alzheimer's disease, Corder *et al.* (1993); myocardial infarction, Cambien *et al.* (1992); hypertension, Jeunemaitre *et al.* (1992); osteoporosis, Morrison *et al.* (1994); breast cancer, Easton *et al.* (1993); multiple sclerosis: Tienari *et al.* (1994).

Identifying disease susceptibility loci is a step towards identifying the genetic and environmental causes of a disease

When a disease susceptibility locus is mapped, the first topic in public discussion is often population screening. However, this is very seldom a good idea. The conditions under which population screening is socially useful and ethically acceptable were discussed in detail in Chapter 16. The most essential single point, more important than any technical issues, is that identifying somebody as at risk should lead to some useful action. Usually this means action to avoid the risk. One cannot – yet – change one's genes, so knowing one has a high-risk genotype is not in general helpful.

The true value of discovering susceptibility genes is, paradoxically, to assist investigations of environmental factors. A typical investigation of environmental risk factors would follow a cohort of people and retrospectively try to detect the difference between the environments of those who did and did not develop the disease. However, the people who did not develop the disease may be a mixture of those who avoided the environmental trigger and those who encountered the trigger but did not develop disease because they were not genetically susceptible. The analysis would be far more sensitive if a cohort of susceptible people could be defined. Genetic markers of susceptibility may make this possible.

It is hoped that the identification of genetic factors will also shed light on the pathogenesis of the disease and, maybe, suggest novel treatments. However, in most cases it is likely to be easier to alter the environment, once the relevant factors are identified. At this stage, population screening might conceivably be useful, to identify the minority of people whose environment must be changed, while allowing the genetically resistant majority to lead their lives as before. Neonatal screening for phenylketonuria (page 453) is a familiar example.

Further reading

Kety SS, Rowland LP, Sidman RL, Matthysse SW. (1983) (eds) *Genetics of Neurological and Psychiatric Disorders*. Raven Press, New York.

King RA, Rotter JI, Motulsky AG. (1992) (eds) *The Genetic Basis of Common Disease*. Oxford University Press, Oxford.

References

Badner JA, Sieber WK, Garver KL, Chakravarti A. (1990) *Am. J. Hum. Genet.*, **46**, 568–580.

Baron M, Freimer NF, Risch N *et al.* (1994) *Nature Genetics*, **3**, 49–55.

Berrettini WH, Ferraro TN, Goldin LR, Weeks DE, Detera-Wadleigh S, Nurnberger JI, Gershon ES. (1994) *Proc. Natl Acad. Sci. USA*, **91**, 5918–5921.

Byerley WF. (1989) *Nature*, **340**, 340–341.

Cambien F, Poirier O, Lecerf L *et al.* (1992) *Nature*, **359**, 641–646.

Cavalli-Sforza L, Bodmer W. (1971) *Genetics of Human Populations*. Freeman, San Francisco.

Corder EH, Saunders AM, Strittmatter WJ, Schmechel WD, Gaskell PC, Small GW, Roses AD, Haines JL, Pericak-Vance MA. (1993) *Science*, **261**, 921–923.

Davies JL, Kawaguchi Y, Bennett ST *et al.* (1994) *Nature*, **371**, 130–136.

Easton DF, Ford D, Peto J. (1993) *Cancer Surv.*, **18**, 95–113.

Edery P, Lyonnet S, Mulligan LM *et al.* (1994) *Nature*, **367**, 378–380.

Falconer DS. (1981) *Introduction to Quantitative Genetics*, 2nd Edn. Longman, London.

Fischer M, Harvald B, Hauge M. (1969) *Br. J. Psychiatr.*, **115**, 981–990.

Fisher RA. (1918) *Trans. R. Soc. Edin.*, **52**, 399–433.

Fraser FC. (1980) *Am. J. Hum. Genet.*, **32**, 796–813.

Fuhrmann W, Vogel F. (1976) *Genetic Counselling*. Springer, New York.

Gottesmann II, Shields J. (1982) *Schizophrenia: the Epigenetic Puzzle*. Cambridge University Press, Cambridge.

Hall JM, Lee MK, Newman B, Morrow JE, Anderson LA, Huey B, King MC. (1990) *Science*, **250**, 1684–1689.

Herrenstein RJ, Richard J. (1994) *The Bell Curve: Intelligence and Class Structure in American Life*. Free Press, New York.

Jeunemaitre X, Soubrier F, Kotelevtsev YV *et al.* (1992) *Cell*, **71**, 169–180.

Kendler KS, Gruenberg AM, Kinney DK. (1994) *Arch. Gen. Psychiatr.*, **51**, 456–468.

Kety SS, Wender PH, Jacobsen B, Ingraham LJ, Jansson L, Faber B, Kinney DK. (1994) *Arch. Gen. Psychiatr.*, **51**, 442–455.

Khoury MJ. (1994) *Am. J. Hum. Genet.*, **55**, 414–415.

Kidd KK, Ott J. (1984) *Cytogenet. Cell Genet.*, **37**, S10.

Lander ES, Schork NJ. (1994) *Science*, **265**, 2037–2048.

McGuffin P, Huckle P. (1990) *Am. J. Hum. Genet.*, **46**, 994–999.

Miki Y, Swensen J, Shattuck-Eidens D *et al.* (1994) *Science*, **266**, 66–71.

Morrison NA, Qi JC, Tokita A, Kelly PJ, Crofts L, Nguyen TV, Sambrook PN, Eisman JA. (1994) *Nature*, **367**, 284–287.

Newman B, Austin MA, Lee M, King MC. (1988) *Proc. Natl Acad. Sci. USA*, **85**, 3044–3048.

Onstad S, Skre I, Torgersen S, Kringlen E. (1991) *Acta Psychiatr. Scand.*, **83**, 395–401.

Puffenberger EG, Hosoda K, Washington SS *et al.* (1994) *Cell*, **79**, 1257–1266.

Rosenthal D, Kety SS. (1968) *The Transmission of Schizophrenia*. Pergamon Press, Oxford.

Schlosstein L, Terasaki PI, Bluestone R, Pearson CM. (1973) *New Engl. J. Med.*, **288**, 704–706.

Sherrington R, Brynjolfsson J, Petursson H, Potter M, Dudleston K, Barraclough B, Wasmuth J, Dobbs M, Gurling H. (1988) *Nature*, **336**, 164–167.

Shiang R, Lidral AC, Ardinger HH, Buetow KH, Romitti PA, Munger RG, Murray JC. (1993) *Am. J. Hum. Genet.*, **53**, 836–843.

Slater E, Cowie V. (1970) *Genetics of Mental Disorders*. Oxford University Press, Oxford.

Spielman RS, McGinnis RE, Ewens WJ. (1993) *Am. J. Hum. Genet.*, **52**, 506–516.

Thomson G. (1994) *Nature Genetics*, **8**, 108–110.

Tienari PJ, Terwilliger JD, Ott J, Palo J, Peltonen L. (1994) *Genomics*, **19**, 320–325.

Todd JA, Bell JI, McDevitt HO. (1987) *Nature*, **329**, 599–604.

Studying human gene structure and function and creating animal models of disease

Chapter 19

The structure and function of human genes can be studied by a variety of methods. Well-established methods involve manipulations *in vitro* and the use of cultured cells. More recently, gene expression and function have been studied in whole animals following the introduction of exogenous genes or the modification of endogenous genes by transgenic technology and gene targeting. The latter methods have also been used to create specific animal models of disease.

Obtaining human gene clones

Human gene clones can be isolated by a variety of different routes

Human gene clones can be obtained by cell-based or PCR-based DNA cloning of either genomic DNA or cDNA. Comprehensive studies of gene structure and function normally use cell-based DNA cloning to provide the starting material. Cell-based DNA cloning, unlike PCR, permits very large amounts of a DNA clone to be produced, offers a wide range in the sizes of DNA fragments that can be cloned and permits rapid subcloning into useful vectors which are designed to allow expression of the cloned gene in cells or even whole animals. PCR may provide a rapid way of getting an initial gene clone, but subsequent studies will usually use the PCR products as a way of screening genomic and cDNA libraries to identify and isolate clones of cells containing the gene sequence of interest. A variety of different routes have been used to obtain gene clones, as follows.

cDNA enrichment cloning

The first human gene clones were obtained by this method. Although the genomic DNA of all nucleated human cells consists of basically the same collection of sequences, the mRNA populations and the collection of proteins made by different cells may be very different. As a result, total cDNA from different cell types is enriched for certain gene sequences at the expense of others, and the abundance of specific polypeptides and proteins may vary enormously. For example, the human α- and β-globin gene sequences each comprise less than 0.0001% of the genomic DNA in a human cell, but there is about 15 g of hemoglobin per 100 ml of serum and α- and β-globin sequences are correspondingly abundant in the mRNA populations of red blood cells. A cDNA library from red blood cells therefore affords a good chance of identifying globin cDNA clones, and so it was not surprising that the globin genes were among the first human genes to be cloned. More recently, PCR-based methods have been devised to selectively amplify cDNAs corresponding to subsets of transcriptionally active genes. Such *mRNA differential display* methods permit analysis of gene expression patterns (see page 519) and enable the isolation of genes with particular expression properties found in some cells but not in others (see, for example, Aiello *et al.*, 1994).

Protein-directed cloning

It may be possible to isolate and purify a specific protein from a cell or, more conveniently, from extracellular fluids in the case of secreted proteins. A variety of different protein fractionation techniques can be used, exploiting differences between proteins in size, solubility, charge and binding affinity (see *Table 19.1*). At each stage in the purification, the separated fractions are assayed for a distinctive property of the protein of interest (e.g. enzymatic activity), to enable identification of a fraction that is enriched in the desired protein.

A partially purified protein can then be used in two ways to isolate the corresponding cDNA. The purified protein preparation can be injected into rabbits or mice in order to raise specific antibodies, which can then be used to screen an expression cDNA library (see *Figure 19.1*). Alternatively, the amino acid sequence of a polypeptide is determined as explained on page 370 (see Stryer, 1995). The amino acid sequence is inspected to design gene-specific oligonucleotides which can then be used as hybridization probes for isolating a corresponding cDNA or genomic DNA clone (see *Figure 14.2*).

Location-directed cloning

Physical mapping and positional cloning projects (see pages 372–387) can involve attempts to identify gene clones at specific subchromosomal locations.

Homology-based cloning

Individual human genes may be particularly closely related in sequence to individual genes in other species, (*orthologous genes*) notably mammalian species, and to other members of the same human gene family (*paralogous genes*). Because of evolutionary pressure to conserve gene function, it is the coding sequences which define the close sequence relationship; introns are generally not well conserved unless they are very short. Because of the close homology between certain genes within and between species, an isolated DNA clone can be used to screen for other clones related to it in sequence. This can be done using molecular hybridization or PCR.

Table 19.1: Common protein fractionation methods (see Stryer, 1995)

Method
Gel filtration chromatography. The protein preparation is fractionated according to size after being applied to the top of a column consisting of porous beads made of an insoluble polymer such as dextran or agarose (commercial examples include Sephadex, Sepharose and Bio-gel)
Salting out. The protein preparation is fractionated according to differential solubility at high salt concentrations. For example, 0.8 M ammonium sulfate is sufficient to precipitate fibrinogen out of solution, but 2.4 M is needed to precipitate serum albumin
Ion-exchange chromatography. Separation occurs according to the net charge. Negatively charged proteins can be separated on positively charged diethylaminoethyl (DEAE)–cellulose columns; positively charged proteins can be separated on negatively charged carboxymethyl–cellulose columns
Affinity chromatography. This powerful and versatile method depends on the high binding affinities of many proteins for particular chemical groups. Thus, if the desired protein is known to bind chemical group X with high affinity, then a custom-designed column can be made in which group X or a derivative is covalently attached to the column beads and used to fish out the desired protein from a complex mixture

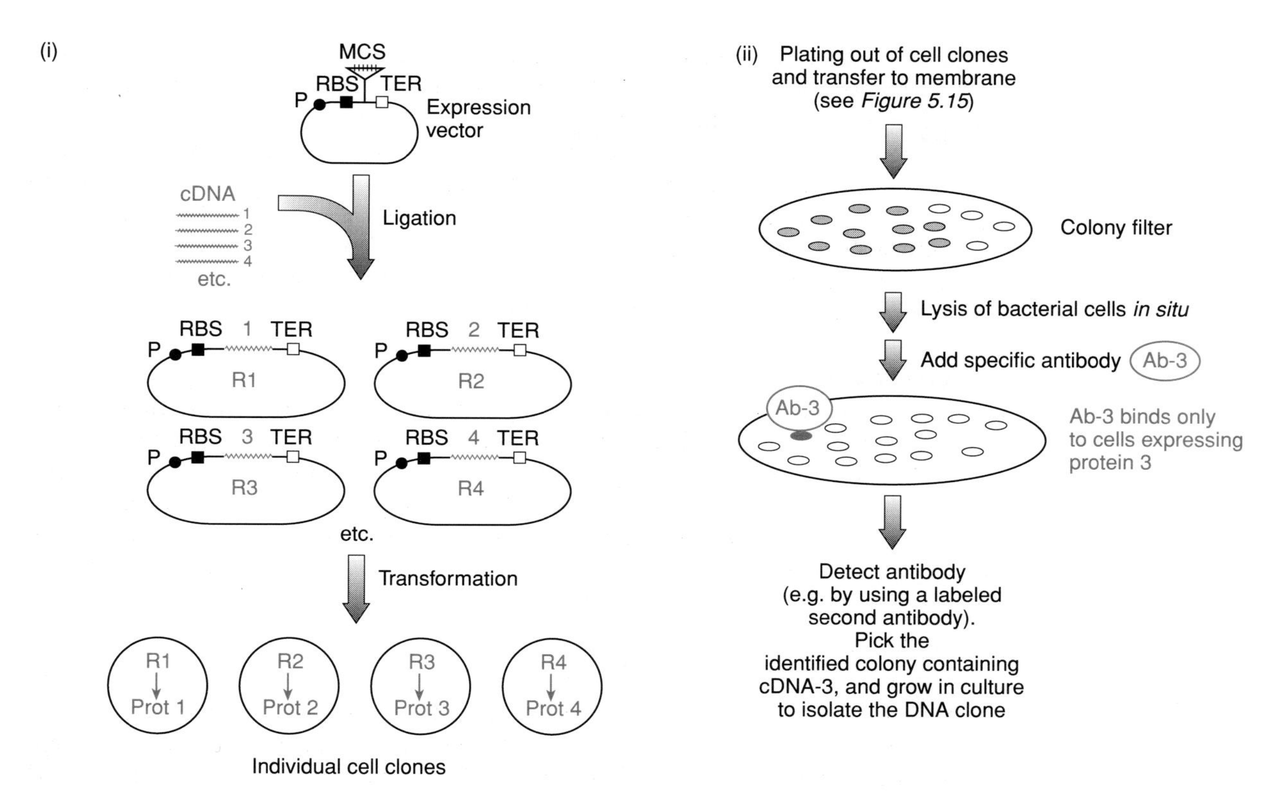

Figure 19.1: An antibody that detects a partially purified gene product can be used to screen an expression cDNA library to identify the corresponding cDNA clone.

General purpose cDNA libraries use cloning vectors which are not designed to express the insert DNA, but just to propagate it. *Expression vectors*, however, are deliberately designed to allow insert cDNAs to be expressed by the host cell to give a foreign polypeptide product. To do this, the vector requires suitable expression sequences flanking the insert. For expression in a bacterial host cell, these will include a strong, usually inducible, bacterial or phage promoter (P) plus a ribosome-binding site (RBS) upstream of the multiple cloning site (MCS). In addition, downstream transcriptional terminator sites (TER) are required to prevent readthrough from disrupting plasmid replication. Cloning of cDNA molecules using such vectors produces an *expression cDNA library* with cells containing recombinant molecules (R1, R2, R3, R4 in figure) able to express the insert to give a protein (prot1, prot2, etc.). Individual bacterial colonies can be separated by plating out and the colonies transferred to colony filters (see *Figure 5.15*). The bacterial cells are lysed *in situ* by treatment with lysozyme and washed to remove cell debris. Exposure to a suitable antibody can result in specific antibody binding to a colony expressing the protein of interest, and detection of bound antibody. The identified colony containing the cDNA of interest is grown in culture to enable isolation of the desired cDNA clone.

Hybridization-based homology cloning. The DNA clone is labeled and used as a hybridization probe to screen a genomic or cDNA library. A labeled genomic DNA clone can identify a cognate cDNA clone and vice versa and, if the clone is a member of a gene family, clones representing other members of the family may be identified. DNA clones from other species can be used as hybridization probes to screen human DNA libraries. Provided the probes are not too short, stable heteroduplex formation can occur between related sequences: sequence matching at even 70% of the nucleotide positions often permits adequate hybridization signals. This is the basis of *zoo-blot hybridization* – see *Figure 11.16*). Many human genes have been isolated as a result of homology with mammalian orthologs. In addition, certain genes may have motifs that are extremely well conserved during evolution and have been cloned by homology ultimately with genes in distantly related species such as *Drosophila melanogaster* (e.g. *HOX* genes, *PAX* genes, *WNT* genes, etc.).

PCR-based homology cloning. In some cases, it has been possible to design degenerate oligonucleotide primers corresponding to short sequences of amino acids that are well conserved between the products of the different members of a gene family. The following two examples illustrate the power of this approach:

(i) Some gene families encode products where there are two or more highly conserved regions separated by regions that are less well conserved. Degenerate PCR primers are chosen from *two conserved motifs*, that is a forward primer corresponds to one motif and the reverse primer corresponds to the second one (see *Figure 19.2*). Starting with a suitable DNA source (e.g. cDNA from a tissue where the gene family members are expected to be expressed), the degenerate primers can be used to amplify the intervening sequence from different transcripts containing the conserved regions. Cloning of the PCR products and sequencing individual products can then be used as a start point for isolating novel members of the gene family (see Gavin and McMahon, 1993).

(ii) A related approach involves a PCR assay in which the two primers are degenerate oligonucleotide primers corresponding to a *single* evolutionarily conserved region (sometimes described as blocks or ACRs – ancient conserved regions; examples include zinc fingers, homeoboxes, etc.). The idea here is to screen genomic clones from a defined chromosomal region, for example a collection of YACs in that region, for the presence of such motifs. This can lead to the identification of novel genes in that region (see d'Esposito *et al.*, 1994).

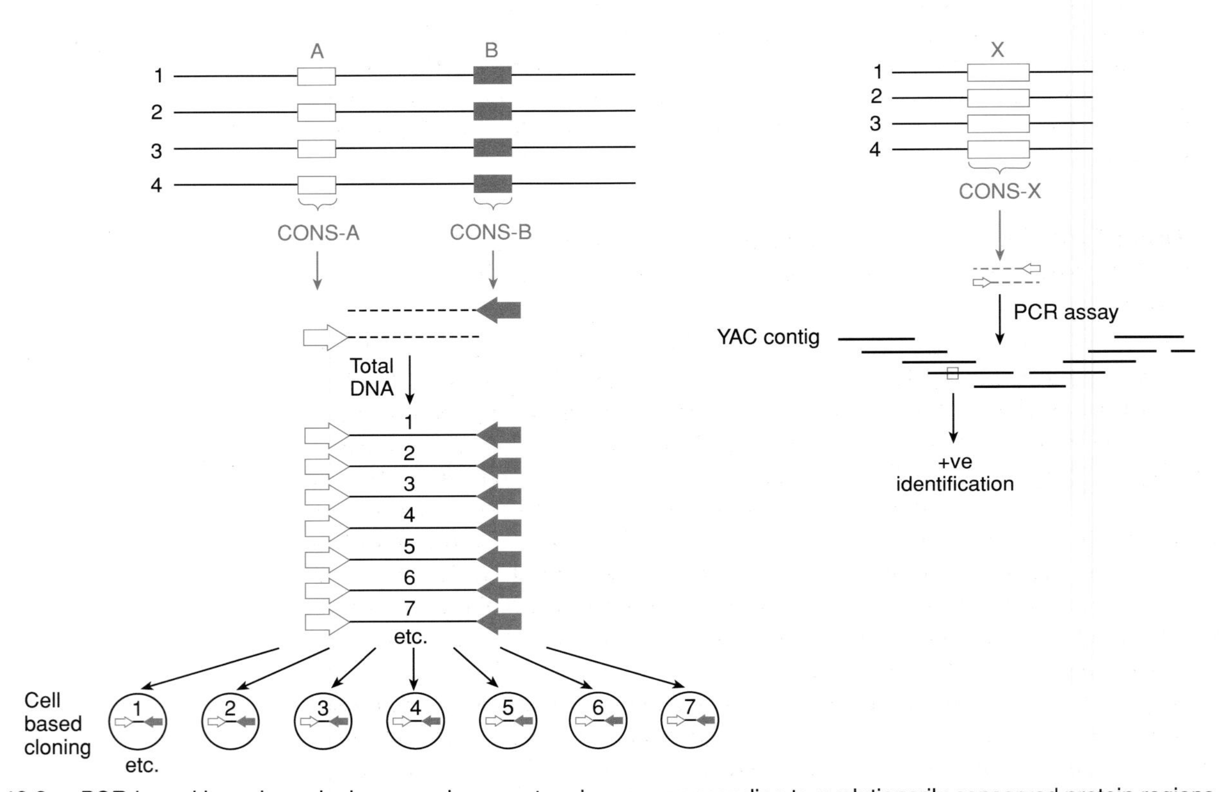

Figure 19.2: PCR-based homology cloning uses degenerate primers corresponding to evolutionarily conserved protein regions.

Left panel. This example shows members of a gene family which encode proteins with two or more evolutionarily well-conserved regions, A and B. From the available information, say from sequences 1–4, consensus sequences are derived for regions A (CONS-A) and B (CONS-B). Degenerate oligonucleotide forward and reverse primers are designed to correspond to the CONS-A and CONS-B sequences, respectively. When used to amplify sequences from total cDNA prepared from a suitable cell type, they may generate novel products, say 5–7, in addition to the known sequences 1–4. Cell-based cloning of the PCR products allows sequencing of individual clones, and the new sequence information is used to isolate large locus-specific clones, for example by RACE-PCR (see *Figure 19.4*). *Right panel.* Members of some gene families encode proteins with significantly large well-conserved domains. From the available information, degenerate forward and reverse oligonucleotide primers are designed to correspond to the conserved sequence (constituting a kind of degenerate STS). The primers can then be used to assay for novel genes containing the conserved motif. For example, the individual clones in a newly characterized YAC contig can be assayed in turn to see if any of them contain a gene that encodes a specific motif, such as a *zinc finger* (see D'Esposito *et al.*, 1994).

Random clone characterization

The above methods have been and continue to be very useful for isolating gene clones. They are now, however, being overtaken by large-scale characterization of random cDNA clones. The human EST (expressed sequence tag) projects (see page 345) involve random sequencing of clones from cDNA libraries, as a prelude to more comprehensive clone characterization. Approximately 80% of all human genes will be obtained by this route (the other genes may be not expressed or poorly expressed in the range of cDNA libraries which are being screened for the EST projects).

Gene structure and transcript mapping studies

After identification of a genomic or cDNA clone containing an uncharacterized gene, the first step is to obtain the corresponding cDNA or genomic DNA clones respectively by screening suitable DNA libraries. An initial priority is to obtain a full-length cDNA sequence and define translational initiation and termination sites, polyadenylation site(s) and exon/intron boundaries. Subsequently, attempts may be made to identify and characterize promoter elements and other regulatory elements in the 5′-flanking sequence and to define the transcriptional initiation site. Especially in the case of small genes, a complete nucleotide sequence is often established; for larger genes, it is usual to focus initially on obtaining just the full-length cDNA sequence.

Completion of full-length cDNA sequences can be achieved by assembling overlapping cDNA clones and by the RACE-PCR procedure

Many of the clones found in cDNA libraries contain partial length sequences. A frequent method of constructing cDNA libraries uses an oligo(dT) primer to bind to the 3′ poly(A) tail of the mRNA and prime first strand synthesis by reverse transcriptase (see *Figure 4.7*). However, very large mRNAs can be a problem because of the difficulty in completing cDNA synthesis, and such libraries typically show a general *3′ end bias*: many of the cDNA clones contain comparatively short sequences containing only the 3′ end of the cDNA. One way of addressing this problem has been to prime first strand synthesis using, as an alternative to an oligo(dT) primer, a degenerate hexanucleotide primer of the type commonly used for labeling DNA *in vitro* (see page 109). Such primers hybridize at essentially random sites along the RNA, and can generate cDNA sequences of a sufficient size to be useful when cloned. Obtaining the sequence of a full-length cDNA can be achieved by sequencing a variety of different cDNA clones and arranging them to form a series of overlapping cDNA clones, as in the case of the cystic fibrosis gene (Riordan *et al.*, 1989; see *Figure 19.3*). This process can therefore be regarded as the cDNA equivalent of *chromosome walking* using genomic DNA clones (see *Figure 11.13*).

Because full-length copies of low abundance mRNAs have been difficult to obtain by conventional cloning, rapid PCR-based methods have been developed. One popular method of obtaining full-length cDNA sequences is the **RACE (rapid amplification of cDNA ends)** technique (also sometimes described as *anchor-PCR*) (Frohman *et al.*, 1988). RACE-PCR is a form of RT-PCR which amplifies sequences between a single previously characterized region in the mRNA and either the 5′ or the 3′ end. A primer is designed from the known internal sequence and the second primer is selected from an *anchor sequence* which is artificially added on to one end of the cDNA (see *Figure 19.4*).

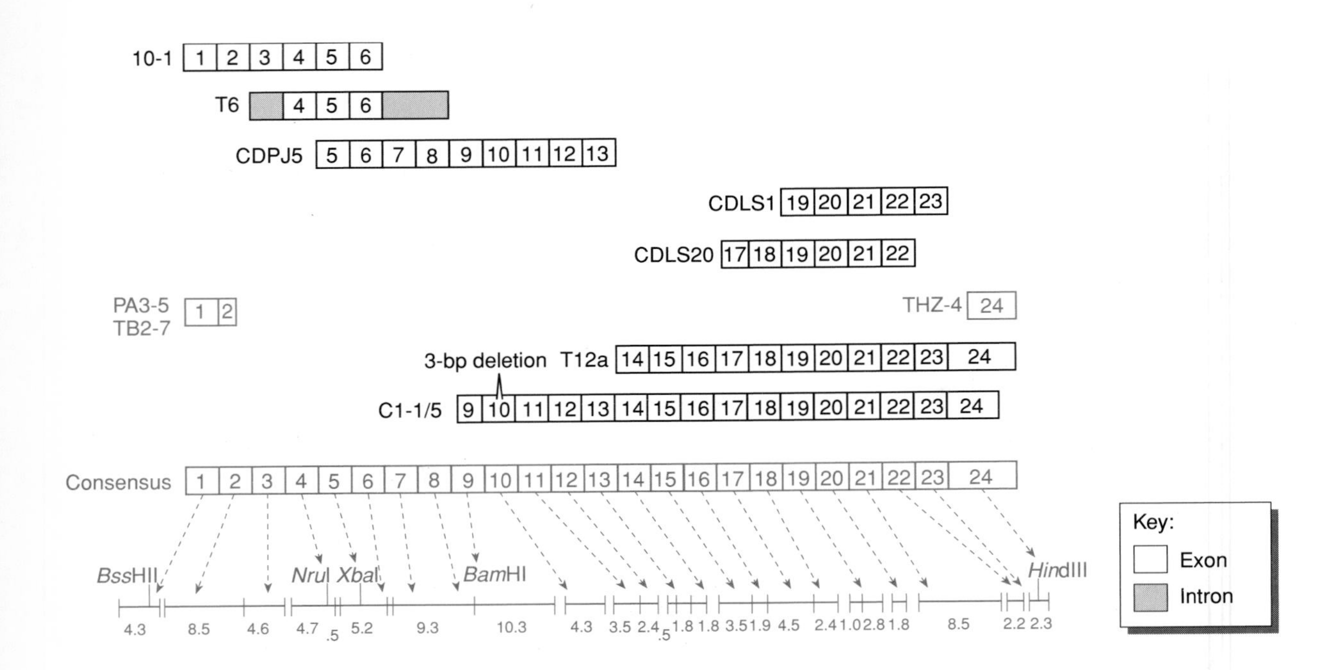

Figure 19.3: A consensus full-length cDNA sequence for the cystic fibrosis gene (*CFTR*) was obtained by assembling a series of overlapping cDNA clones.

Individual cDNA clones are shown as series of boxes with open numbered boxes corresponding to exons. The first *CFTR* cDNA clone to be isolated, 10-1, was obtained from screening a sweat gland cDNA library using conserved sequences identified in the overlapping genomic clones E4.3 and H1.6 (see *Figure 14.12*). This was then used as a hybridization probe to identify overlapping clones in various cDNA libraries, and new clones in turn were used to identify more distal clones. Clones were derived from various sources: T84 (colon carcinoma cell line, clones beginning with the letter T); lung (clones beginning with CDL); pancreas (CDPJ5). In addition, clones in red were generated by RACE-PCR (see *Figure 19.4*). They included clones PA3-5 and TB2-7 (by 5′ RACE-PCR using an exon 2-specific primer) and THZ-4 (by 3′ RACE-PCR using a primer from the 3′ UTS). *Note* the presence of intron sequences in some clones (this is possible because of reverse transcription of precursor RNA sequences, or occasional alternatively spliced sequences) and extraneous sequences in others. Hybridizing clones or subclones back to genomic DNA restriction fragments (bottom) helped to orientate the exons in the genomic DNA (see *Figure 14.12*). For the sake of clarity many of the clones presented in the original figure have been omitted. Derived with permission from Riordan *et al.* (1989). Copyright 1989, American Association for the Advancement of Science.

Transcription start sites can be mapped by nuclease S1 protection and by primer extension.

Important regulatory sequences are often located close to transcription start sites and can often be identified after mapping the latter. Although 5′ RACE-PCR can permit rescue of sequences corresponding to the 5′ end of a mRNA (and therefore, the transcriptional start site), two major methods are preferentially used to define the transcriptional start site.

Nuclease S1 protection

The endonuclease S1 is an enzyme from the mold *Aspergillus oryzae* which cleaves single-stranded RNA and DNA but not double-stranded molecules. In order to map the transcription start site for a gene, a genomic DNA clone suspected of containing the start site is required. The DNA clone is then digested with a suitable restriction endonuclease to generate a fragment that is expected to contain the transcription start site. As shown in *Figure 19.5A*, hybridization to the cognate mRNA and S1 nuclease digestion defines the distance of the transcription start site from the unlabeled end of the restriction fragment. If more precise localization is required, the labeled DNA fragment in the heteroduplex can be sequenced by a chemical method of DNA sequencing (see Maxam and Gilbert, 1980). Note that, in much the same

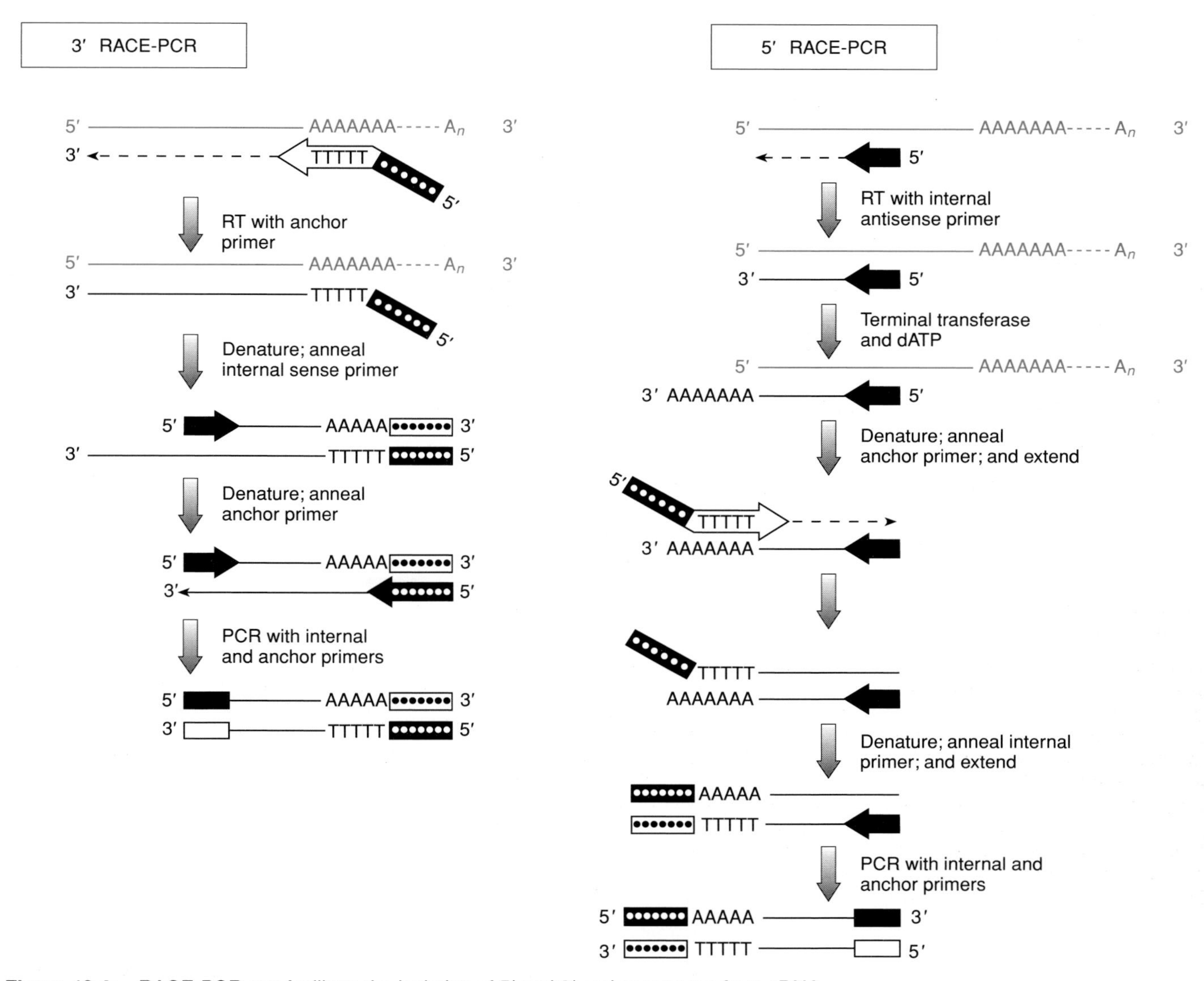

Figure 19.4: RACE-PCR can facilitate the isolation of 5′ and 3′ end sequences from cDNA.

A preliminary step in RACE (*rapid amplification of cDNA ends*)-PCR involves the introduction of a specific sequence at either the 3′ or the 5′ end by what is effectively a form of *5′-add-on mutagenesis* (see pages 144–145).
(A) 3′ RACE-PCR uses a starting antisense primer with a specific 5′ extension sequence (*anchor sequence*, often >15 nucleotides long) which becomes incorporated into the cDNA transcript at the reverse transcriptase step. An internal sense primer is then used to generate a short second strand ending in a sequence complementary to the original anchor sequence. Thereafter, PCR is initiated using the internal sense primer and an anchor sequence primer.
(B) 5′ RACE-PCR. Here an internal antisense primer is used to prime synthesis from a mRNA template (red) of a partial first cDNA strand (black). A poly(dA) is added to the 3′ end of the cDNA using terminal transferase. Second strand synthesis is primed using a sense primer with a specific extension (anchor) sequence. This strand is used as a template for a further synthesis step using the internal primer in order to produce a complementary copy of the anchor sequence. PCR can then be accomplished using internal and anchor sequence primers.

way, nuclease S1 mapping can also be used to map other boundaries between coding and noncoding DNA such as exon/intron boundaries (see below) and the 3′ end of a transcript.

Primer extension

The method is very similar to the nuclease S1 protection method. In this case, the chosen restriction fragment must be shorter than the mRNA and the overhang is filled in using reverse transcriptase (*Figure 19.5B*). Again a more accurate location is

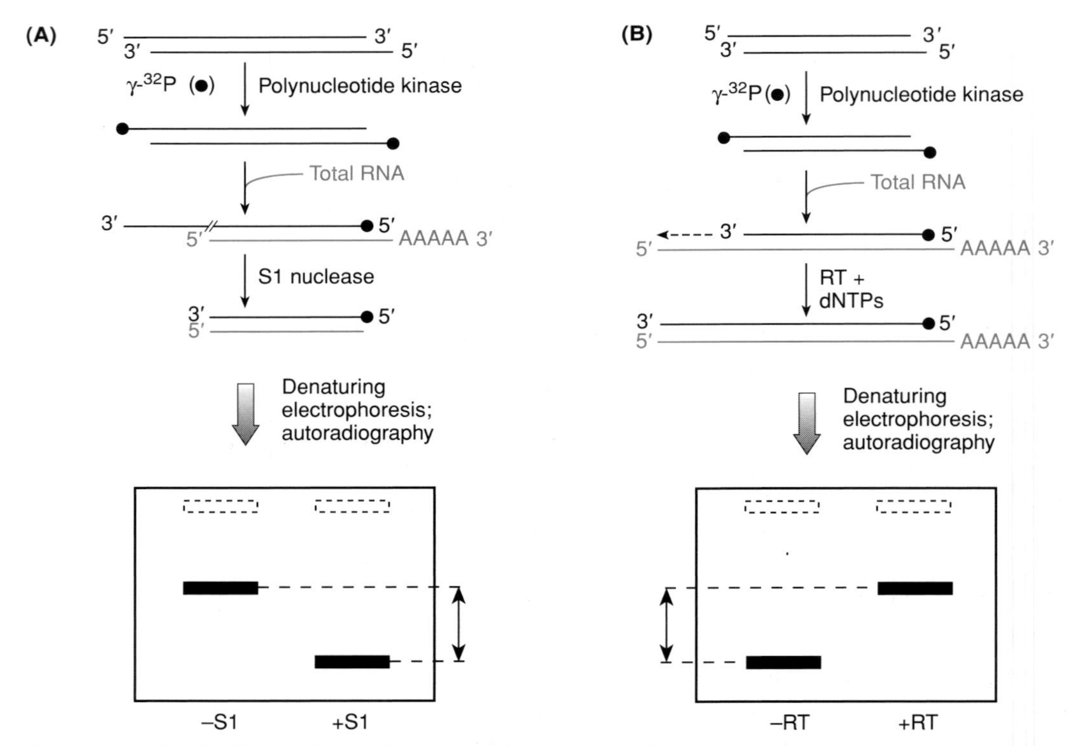

Figure 19.5: The transcriptional initiation site can be mapped by nuclease S1 protection or primer extension assays.

(A) Nuclease S1 protection assay. A restriction fragment from the 5′ end of a cloned gene is suspected of containing the transcription initiation site. It is end-labeled at the 5′ ends, then denatured and mixed with total RNA from cells in which the relevant gene is thought to be expressed. The cognate mRNA can hybridize to the antisense DNA strand to form an RNA–DNA heteroduplex. Subsequent treatment with nuclease S1 results in progressive cleavage of the overhanging 3′ DNA sequence until the point at which the DNA is hybridized to the 5′ end of the mRNA. Size-fractionation on a denaturing electrophoresis gel can identify the size difference between the original DNA and the DNA after nuclease S1 treatment.
(B) Primer extension assay. In this case, the restriction fragment suspected of containing the transcriptional initiation site is deliberately chosen to be small. Hybridization with a cognate mRNA will then leave the mRNA with an overhanging 5′ end. The DNA can serve as a primer for reverse transcriptase (RT) to extend its 3′ end until the 5′ end of the mRNA is reached. The size increase after reverse transcriptase treatment (+RT) compared with before treatment (–RT) maps the transcription initiation site. More precise mapping is possible in both methods by sequencing the DNA following treatment with S1 or RT.

possible by using the chemical sequencing method of Maxam and Gilbert (1980) to sequence the labeled DNA strand.

Exon/intron boundaries can be mapped by a variety of different methods.

An early requirement in investigating gene structure is to isolate suitable genomic DNA clones covering the length of the gene. The availability of both genomic and cDNA clones then provides the opportunity of mapping exon/intron boundaries. Not all human genes have introns (see *Table 7.6*). Where introns are present, however, they are often comparatively large. The presence of introns can usually be inferred from comparative mapping of cognate genomic and cDNA clones. Once the full cDNA sequence has been established, sequencing primers can be designed from various segments of the cDNA and used in *cycle DNA sequencing* with denatured cosmid clones as the DNA sequencing templates (see page 309). The sequences obtained should cross an exon/intron boundary, unless the exon is very large, in which case additional sequencing primers may be required.

Other methods of mapping exon/intron boundaries include nuclease S1 protection (see page 512) and the PCR-based *genomic walking* method. The latter includes techniques such as *bubble-linker PCR* or *inverse-PCR* (see *Figures 6.11* and *6.12*), which can be used with an exon-specific primer to amplify a short stretch of neighboring intronic sequence from a YAC clone template (or even from total genomic DNA) and then sequence across the exon/intron boundaries. This was the way, for example, in which comprehensive exon/intron boundaries were established for the dystrophin gene (Roberts *et al.*, 1993).

Computer-based sequence analysis programs and database searches can illuminate aspects of gene structure, function and evolution

Once the cDNA sequence has been obtained, visual inspection almost always enables identification of a single long ORF in one of the three possible translational reading frames. In some cases, such as the *H19* and *XIST* genes, no such ORF may be found, indicating that the gene may not encode a polypeptide. An alternative possible explanation, in some cases, may be that the gene specifies a very short polypeptide product, the translational reading frame for which may not be easily distinguished from the other two frames. Computer-based sequence analyses may clarify the likeliest possibility (see below).

Once the correct reading frame has been identified, the translational termination site can be identified easily by the presence of one of the recognized termination codons. Thereafter, possible polyadenylation sites can usually be distinguished by the presence of the well-conserved AAUAAA sequence (often located ~400–800 bp downstream of the termination codon in mammalian cDNAs), which defines the end of the 3′-UTS. However, the identification of the translational start point and the extent of the 5′-UTS may be less straightforward. In the former case, AUG (which specifies methionine) is found in the vast majority of cases, *but not all* – rare alternatives which have been found include ACG (Thr), CUG (Leu) and GUG (Val). Even if the initiation codon is AUG, it should also be noted that the N-terminal amino acid of the mature polypeptide may not be methionine [e.g. see the example of β-globin where the N-terminal methionine is cleaved to generate the mature polypeptide (see *Figure 1.19*)].

Inspection of the nucleotide sequence at the 5′ end of a cDNA or in the corresponding sequence of a genomic DNA clone may reveal a stop codon in the same reading frame as the presumed coding sequence. Clearly, the initiating codon must lie downstream from this and, unless there are several possible codons for methionine in this area, the identification of a methionine codon downstream of this and forming part of a large ORF is usually straightforward. If, however, there are several possible AUG codons, the initiator codon may have to be identified by characterization of the polypeptide product in a suitable expression cloning system. Other important sequences are most easily identified by computer-based sequence analysis.

Sequence analysis programs

As mentioned previously, the correct reading frame of the coding DNA can be established by looking for large ORFs, a procedure which is often accomplished by a simple computer program that translates all three possible reading frames of the cDNA. In addition, simple programs exist for constructing detailed restriction maps, and for searching for regions of internal repetition. Tandemly repeated small segments, such as triplet repeats, are usually evident by cursory inspection of the sequence. However, large repeats, distantly spaced repeats and repeats that have undergone substantial divergence may not be readily apparent until analyzed by computer. The output of such searches is often in the form of a **dot-matrix analysis**.

A dot is marked on the chart wherever a string of nucleotides or amino acids at one position in the gene or protein matches a string at another position. Matching can be defined as identity, or as matching above a pre-set value. Internal repeats generate diagonal lines of contiguous dots superimposed on the random background. Dot-matrix analysis can be used to compare one sequence with another (see *Figure 19.6*).

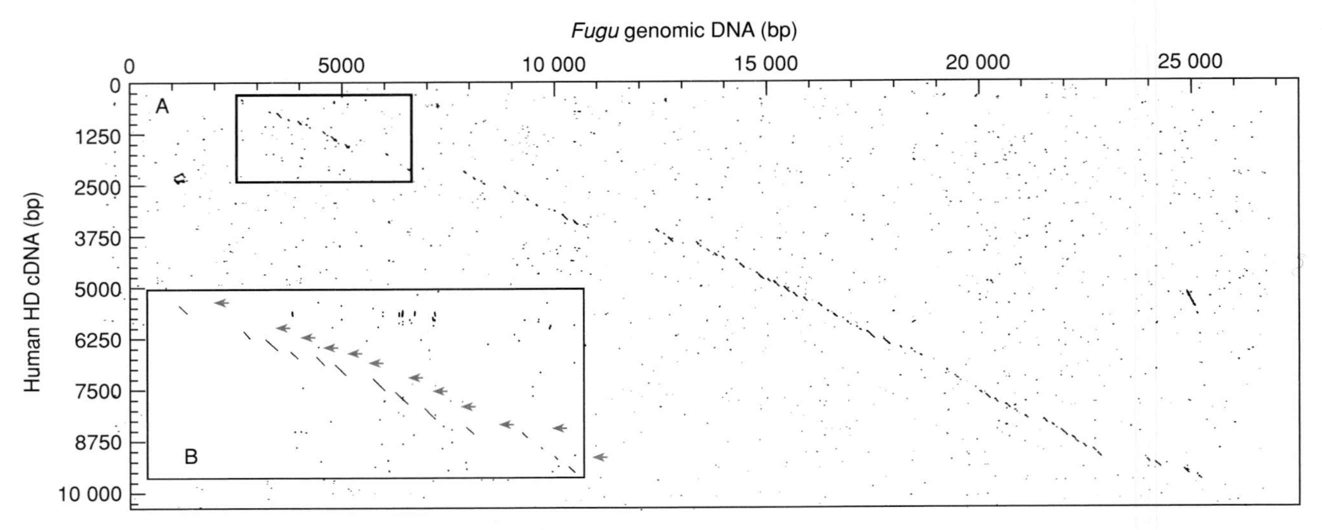

Figure 19.6: Dot-matrix analysis is a useful graphical method for comparing sequences.

The example shows comparison of the human Huntington's disease cDNA sequence with the genomic DNA sequence for the homologous gene in the pufferfish. The lines along the diagonal correspond to the regions of homology permitted within the parameters of the program, and define exons in this case. The insert (B) shows a tenfold increase in resolution of the first 12 exons (boxed in A) of the gene (indicated by arrows). *Note* that although the *Fugu* HD gene spans only 23 kb of genomic DNA compared to 170 kb for the human gene, all 67 exons are conserved. Reproduced from Baxendale *et al.* (1995) with permission from Nature America Inc.

Database homology searches

Some computer programs have been devised to permit searching of nucleic acid and protein sequence databases (see *Table 13.4*) for significant *sequence homology* (i.e. a significantly high degree of sequence matching) between a sequence under investigation and any sequence in the database. Popularly used programs are the **FASTA** program and the different **BLAST** programs (see Ginsburg, 1994 and *Table 19.2*). Depending on the length of the sequences under investigation a variety of different levels of sequence homology searching can be conducted.

Table 19.2: The BLAST and FASTA programs for sequence comparisons

Program	Compares
FASTA	A nucleotide sequence against a nucleotide sequence database, or an amino acid sequence against a protein sequence database
TFASTA	An amino acid sequence against nucleotide sequence databases
BLASTN	A nucleotide sequence against a nucleotide sequence database
BLASTP	An amino acid sequence against a protein sequence database
BLASTX	A nucleotide sequence translated in all reading frames against a protein sequence database
TBLASTN	A protein sequence against a nucleotide sequence database translated in all reading frames

Note that because the design of comparable programs such as FASTA and BLASTN is different, they may give different results (see Ginsburg, 1994). All of the above programs are accessible through the Internet from various centers, such as the European Bioinformatics Institute (world-wide web address is http://www.ebi.ac.uk/) and the US National Center for Biotechnology Information (world-wide web address is http://www.ncbi.nih.gov/). Recently, an enhanced BLAST-based search tool, known as **BEAUTY**, has been developed and greatly improves the identification of weak, but functionally significant matches in BLAST database searches (Worley *et al.*, 1995).

Homolog and domain searching. A full-length or partial DNA sequence (or the interpreted amino acid sequence) for a human gene can be entered into a program that compares it against all other sequences stored in electronic sequence databases. Such comparisons will detect sequences that are significantly related to a substantial component (domain) or even all of the sequence of interest (homolog), either in the same species or in different species. Searching across large databases often provides the first indication of the function of a sequence. One recent example concerns the normal function of the neurofibromatosis type 2 gene (*NF2*). Homology searching of databases using the newly established cDNA sequence identified related sequences, notably those of moesin, ezrin and radixin (see, for example, Rouleau *et al.*, 1993). These proteins were known to act as structural links between cell membrane proteins and intermediate filament proteins, thereby providing some initial clues to the function of the *NF2* gene product.

Motif searching. This means a homology search for a short nucleic acid or amino acid sequence (motif) that is indicative of a specific function. For example, searching at the nucleotide level could include scanning sequences flanking the coding DNA for ubiquitous promoter elements (such as CCAAT, TATA and GC boxes – see page 14), tissue-specific regulatory elements (see page 162, the AATAAA polyadenylation signal, etc. At the amino acid level, searches could, for example, check for the presence of specific protein destination signals such as sequences which indicate export to the nucleus (see *Table 1.6*) or glycosylation signals, etc. (see *Table 1.5*), or the search could scan for extended runs of hydrophobic amino acids, membrane-spanning sequences, etc. Because there are 20 different amino acids, even very short sequences would not occur very frequently just by chance, and so may be of considerable significance if known to be associated with a particular function or family of genes with related functions. For example, the DEAD sequence (one letter code for Asp–Glu–Ala–Asp – see page 185) would occur by chance in only about one in 170 human genes (on the basis of known amino acid frequences and an average size of 500 amino acids in a protein). This motif is found, along with a variety of other motifs, in a family of genes that encode RNA helicases (see *Figure 8.1*).

Studying gene expression and gene regulation at the nucleic acid level using cultured cells or cell extracts

The isolation of gene clones allows a variety of studies of gene expression and gene function. Increasingly, gene expression and gene function are modeled in whole organisms using transgenic animals (see Chapter 19). The present section describes studies which are carried out in cultured cells or using cell extracts. Once a gene clone becomes available, it is customary to examine its expression patterns in different tissues. Often this involves preparing a suitably locus-specific DNA probe, or an antibody may be raised that can specifically detect the protein product. Other studies may focus on identifying regulatory elements and on mapping functional elements within the gene (e.g. by developing a functional assay and using *in vitro* mutagenesis to introduce predetermined changes). Expression cloning systems may be used to prepare large quantities of gene product, and assays can be developed to identify other proteins which bind the gene product under study or to identify specific nucleic acid sequences which can be bound by the gene product.

Gene-specific nucleic acid or oligonucleotide probes are often used to study the expression patterns of a gene

Once a genomic or cDNA clone becomes available, it can be labeled and used as a probe to track the expression of that gene in tissues and cells. If the gene is a single-copy gene, then a cDNA clone is usually used as a probe, or an *antisense riboprobe* is used (see below). Sometimes, however, the gene in question is a member of a gene family and other closely related gene sequences are expressed. If so, some effort may be required to prepare a suitably *locus-specific* probe. This involves comparing the sequences, where available, of other members of the gene family and designing a probe to represent regions that are not well conserved between the family members. In some difficult cases it may be necessary to use oligonucleotide probes. Two major methods use a labeled nucleic acid probe to track the expression of a gene:

- **Northern blot hybridization**. Total RNA, or poly(A)$^+$ mRNA [purified by binding to a column of oligo(dT)–cellulose or poly(U)–Sepharose] is prepared from extracts of different tissues and cell types. The RNA is fractionated according to size on a denaturing gel (e.g. a formaldehyde–agarose gel; see *Box 11.4*). The separated RNA molecules are transferred to a membrane and then probed with a labeled gene probe. Initial studies may focus on a range of adult tissues (see *Figure 19.7*). Subsequently, the studies may be extended to include a survey of different developmental stages (embryonic, fetal and adult) and more comprehensive surveys of cell types (e.g. different types of brain tissue).
- **Tissue *in situ* hybridization.** Thin tissue sections are cut with a microtome, fixed, mounted on microscope slides and then exposed to a solution containing a labeled gene probe. After washing off excess probe, the dried slide is submitted to autoradiography (in the case of a radiolabeled probe; see *Figure 5.16* for an example) or fluorography. Usually, the labeled probe is an **antisense riboprobe**, a labeled antisense RNA, or an oligonucleotide probe. In the latter case, however, very long exposure times may be required to obtain an adequately strong signal, unless the expression of the gene being studied is particularly strong.

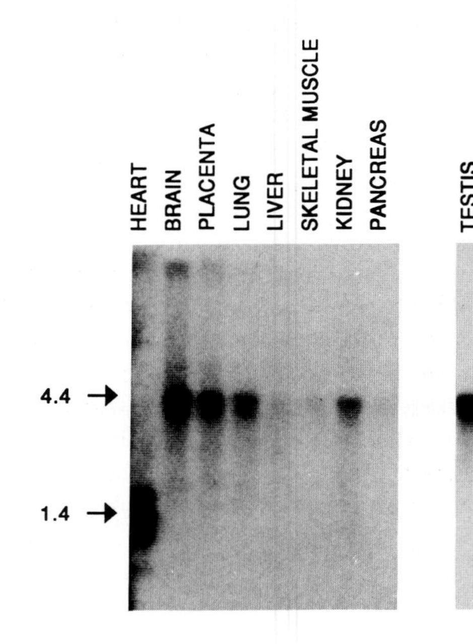

Figure 19.7: Northern blot hybridization is used to evaluate the gross expression patterns of a gene.

Northern blotting involves size-fractionation of samples of total RNA [or purified poly(A)$^+$ mRNA], transfer to a membrane and hybridization with a suitable labeled nucleic acid probe. The example shows the the use of a labeled cDNA probe from the *FMR1* (fragile-X mental retardation syndrome) locus. Highest levels are detected in the brain and testis, with decreasing expression in the placenta, lung and kidney respectively. Multiple smaller transcripts are present in the heart. Reproduced from Hinds *et al.* (1993) with permission from Nature America Inc.

RT-PCR and mRNA differential display are PCR-based methods which allow the study of gene expression in cells

As described in Chapter 6, the great advantages of PCR are its speed, sensitivity and simplicity. Although it is not appropriate for providing spatial patterns of expression (in the way that tissue *in situ* hybridization, for example, does), it can provide rapid gross patterns of expression which may be valuable.

Studying gene expression by conventional RT-PCR

As described on pages 137 and 393, RT-PCR has been extremely useful in permitting amplification of cDNA sequences corresponding to both abundant and rare transcripts, thereby providing an important source of DNA for mutation screening. It can also be used for providing rough quantitation of expression of a particular gene, which can be very useful in the case of cell types or tissues that are not easy to access in great quantity (such as early stage pre-implantation human embryos, e.g. see Daniels *et al.*, 1995). In addition, RT-PCR can be useful for identifying and studying different *isoforms* of an RNA transcript. For example, different mRNA isoforms may be produced by alternative splicing and can be identified when exon-specific primers identify extra amplification products in addition to the expected products (for an example, see Pykett *et al.*, 1994).

Studying gene expression by mRNA differential display

NonPCR-based methods of distinguishing mRNA in comparative studies rely largely on *subtractive hybridization* or *differential hybridization*, technically difficult procedures which have, however, been successful in isolating some important genes (see Lee *et al.*, 1991). **mRNA differential display** is a form of RT-PCR in which reverse transcriptase catalyzes cDNA synthesis by using a modified oligo(dT) primer which binds to the poly(A) tail of a subset of mRNAs (Liang *et al.*, 1993). For example, if the oligonucleotide TTTTTTTTTTTCA is used as a primer, it will preferentially prime cDNA synthesis from the poly(A) tails of mRNAs whose 3′ sequence ends in TG. The second primer which is used is usually an arbitrary short sequence (often 10 nucleotides long but, because of mismatching, especially at the 5′ end, it can bind to many more sites than expected for a decamer). The resulting amplification patterns are deliberately designed to produce a complex ladder of bands when size-fractionated in a long polyacrylamide gel.

The most interesting application of mRNA differential display is the comparison of gene expression in cells at different physiological or developmental stages. This allows identification of a small subset of genes whose expression patterns are different between the cell types. By isolating specific bands from the gel and submitting the DNA to futher PCR cycles, gene clones can be obtained (for an example, see Aiello *et al.*, 1994).

The use of reporter genes and deletion analysis provides a first approach to identifying regulatory gene sequences

In addition to studying gene expression *per se*, it is important to be able to study the control of gene expression. This can be done in different ways. A simple way of (tentatively) identifying control sequences is to use computer programs to scan the DNA sequence of a gene and its neighboring regions for the presence of sequence elements known to have a role in gene expression. A more general approach, however, is to use an artificial gene expression system and to study how deleting different segments of DNA upstream of the gene, or occasionally in the first intron, affects gene expression. Clearly, human cells are the most appropriate system for studying the expression of human genes, but this then leaves the problem of how to follow expression of the transfected human gene in the presence of an endogenous homolog which may be expressed in the same cells. As a result, therefore, it is usual to clone the presumptive regulatory sequences into a vector which contains a **reporter gene** downstream of the cloning site. The reporter gene is deliberately designed to be one which is not found in human cells and which can be assayed simply. Commonly used reporter genes include the bacterial *CAT* (chloramphenicol acetyl transferase) gene, derived from the Tn9 transposon of *E. coli,* the β-galactosidase gene and the firefly luciferase gene. Luciferase has the advantage of providing a very sensitive assay: it catalyzes the oxidation of luciferin with the emission of yellow–green light which can be detected easily and at low levels.

In such artificial expression systems, the vector is designed so that expression of the reporter gene is controlled almost entirely by the introduced upstream human sequences. If all the necessary elements are present for expression of the human gene, high level expression of the reporter gene will result when the construct is transfected into an appropriate type of cultured human cell (see *Figure 19.8*). Often the test construct is simply transfected into HeLa cells, a well-established cell line derived from a human cervical carcinoma. However, if the gene is known to be expressed predominantly in a certain type of cell, say hepatocytes, it is usual to use a more appropriate recipient cell line, a hepatoma cell line in this case. A series of progressive deletion constructs can then be made using the enzyme *exonuclease III* from *E. coli* which is inactive on single-stranded DNA but progressively cleaves from the 3′ end of double-stranded DNA. This allows mapping of sequences which control gene expression. Alternatively, a series of PCR amplification products can be designed covering the regions of interest and then cloned into a suitable expression vector (see *Figure 19.9*).

Gel retardation, DNase footprinting and methylation interference assays can identify protein-binding sites on a DNA molecule

The deletion analysis described above usually provides only a rough indication of the location of a sequence that regulates gene expression. Because such control sequences can specifically bind regulatory proteins, another way of identifying them involves screening for sequences that can specifically bind proteins. Often this involves using synthetic oligonucleotides corresponding to sequences from a region thought to contain a regulatory sequence. Different methods can be used as follows.

DNase I footprinting

When a protein binds specifically to a DNA sequence, only a few nucleotides of the DNA are involved in DNA–protein contacts. The bound protein, however, renders the underlying DNA segment relatively resistant to cleavage by pancreatic *deoxyribonuclease I* (DNase I), when compared with naked DNA. Short cloned DNA fragments (usually a few hundred bases long) are used as targets and are end-labeled at one end only. The cloned fragments are individually incubated in the presence or absence of a protein extract (often a nuclear extract from human HeLa cells) and subsequently exposed briefly to low concentrations of DNase I. Such *partial digestion* conditions ensure that, for any one DNA fragment, each DNA molecule is cut only rarely and at a random position if no protein is bound. The digestion products are then size-fractionated on long denaturing polyacrylamide gels, prior to autoradiography. Control samples will show a series of bands corresponding to DNA fragments of every possible length. The corresponding test lanes will, however, reveal gaps where no fragments are seen (*footprints*), as the DNase I was not able to cut at these positions because of steric inhibition by the bound protein (see *Figure 19.10*).

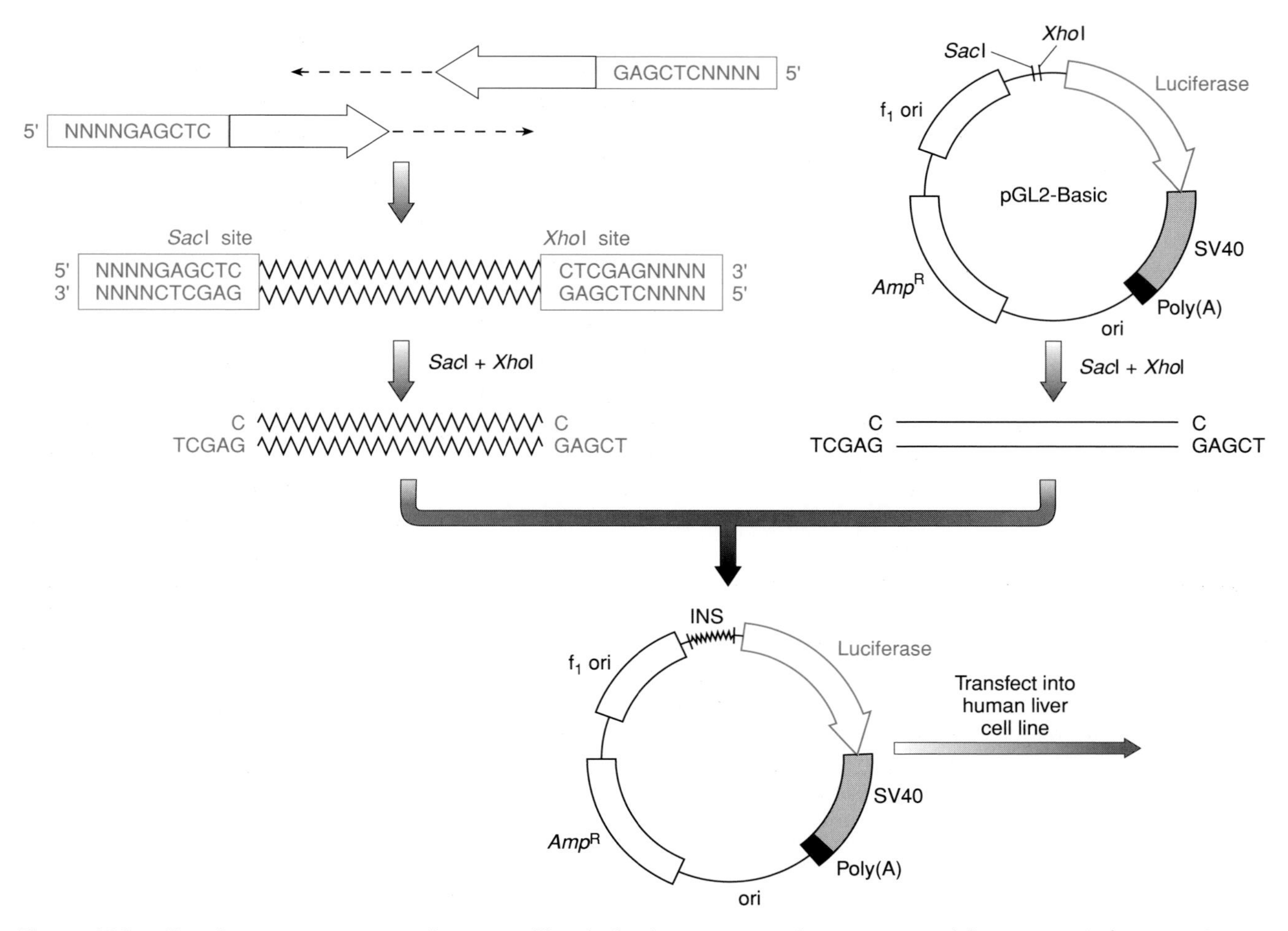

Figure 19.8: Regulatory sequences can be mapped by cloning into an expression vector containing a promotorless reporter gene.

The expression vector pGL2-Basic is a eukaryotic expression vector designed to assay for promoter sequences. It contains a promoterless luciferase reporter gene, a downstream SV40 polyadenylation signal [poly(A)], a conventional origin of replication (ori), an f1 replication origin for producing single-stranded DNA if required, and an ampicillin resistance gene. The figure shows the way in which PCR amplification products corresponding to sequences upstream of the human factor VIII gene (*F8*) were cloned into this vector by Figueiredo and Brownlee (1995). The PCR primers were modified at their 5′ ends by a 10-nucleotide extension (in red) which is not related to the target sequences, but is simply designed to include a recognition sequence for *Sac*I (GAGCTC) or for *Xho*I (CTCGAG). Amplification with these primers constitutes an example of *add-on mutagenesis* (see pages 144–145): the 'foreign' decanucleotide sequence becomes incorporated into the amplification product. The PCR products were cloned by double digestion with *Sac*I and *Xho*I and ligation to similarly cut vector DNA. The recombinants were transfected into human liver-derived cell lines and the inserts were assayed for *cis*-acting regulatory sequences which, in the presence of complementary *trans*-acting factors provided by the cell, could drive expression of the luciferase gene (see *Figure 19.9*).

Gel retardation assay

This technique is alternatively known as the *electrophoretic mobility shift assay*, and its rationale is that binding of a protein to a DNA fragment reduces its mobility during gel electrophoresis. A DNA fragment from a genomic clone or a corresponding synthetic oligonucleotide, which is suspected of containing regulatory sequences, is end-labeled and then mixed with a protein extract. The resulting preparation is size-fractionated by PAGE in parallel with a control sample in which the DNA was not mixed with the protein. DNA fragments which have bound protein will be identifiable as low mobility bands. Since DNA is a highly charged molecule, there is the opportunity for nonspecific binding, and controls are required whereby a nonspecific competitor DNA is added in increasing concentrations (see *Figure 19.11*).

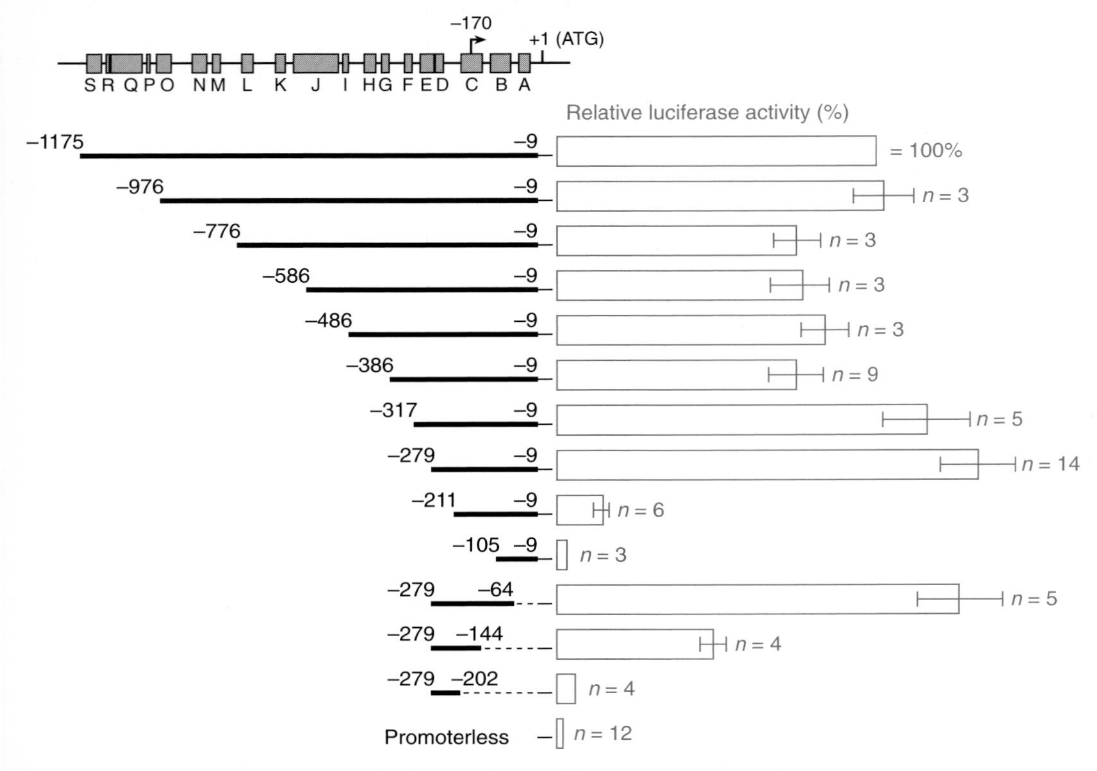

Figure 19.9: Deletion analysis of the human factor VIII gene promoter region.

Solid bars to the left indicate variously sized sequences upstream of the human factor VIII gene (*F8*) which had been cloned into the expression vector pGL2-Basic (see *Figure 19.8*). Co-ordinates –1175 to –9 are referenced against the initiator methionine, arbitrarily labeled as +1. Boxes on top indicate sequences upstream which were suggested to be protein-binding sites on the basis of *DNase I footprinting assays* (see *Figure 19.10*). Boxes on the right indicate the level of luciferase activity relative to the intact sequence, based on *n* replicate experiments. On the basis of observed luciferase activity for the different expression clones, the deletion mapping shows that all the necessary elements for maximal promoter activity are located in the region from –279 to –64, including protein-binding sites B, C and D. Reproduced from Figueiredo and Brownlee (1995) with permission from The American Society for Biochemistry and Molecular Biology.

Methylation interference assay

This technique complements the two outlined above by defining which guanines are implicated in the protein-binding site (see Piccolo *et al.*, 1995, for a recent application).

Once a DNA region has been identified as capable of binding protein, comparison of its sequence to other known protein-binding sequences may indicate the identity of the binding protein. A more general screening method for identifying a protein which binds to a given stretch of DNA has been described as **Southwestern screening**. The method involves cloning cDNA into a suitable expression vector such as λgtII. The resulting expression library is plated out on to nutrient agar in large culture dishes, resulting in well-separated phage plaques in a lawn of bacteria. A nitrocellulose membrane is placed on top of the agar surface for a brief period, then peeled off to give an imprint of phage material originally present in the plaques (this so-called ***plaque-lift*** procedure is the phage equivalent of the *colony blot* procedure used to transfer separated bacterial colonies to a nitrocellulose or nylon membrane; see *Figure 5.15*). Because the foreign cDNA sequences in the phage are expressed to give protein, expressed fusion proteins are absorbed on to the nitrocellulose membrane. The membrane is incubated with a radiolabled *duplex* DNA oligonucleotide representing the known protein-binding sequence. Under suitably stringent conditions, the DNA probe will bind selectively to protein expressed by an individual plaque, leading to identification of a specific binding protein (see Old and Primrose, 1994).

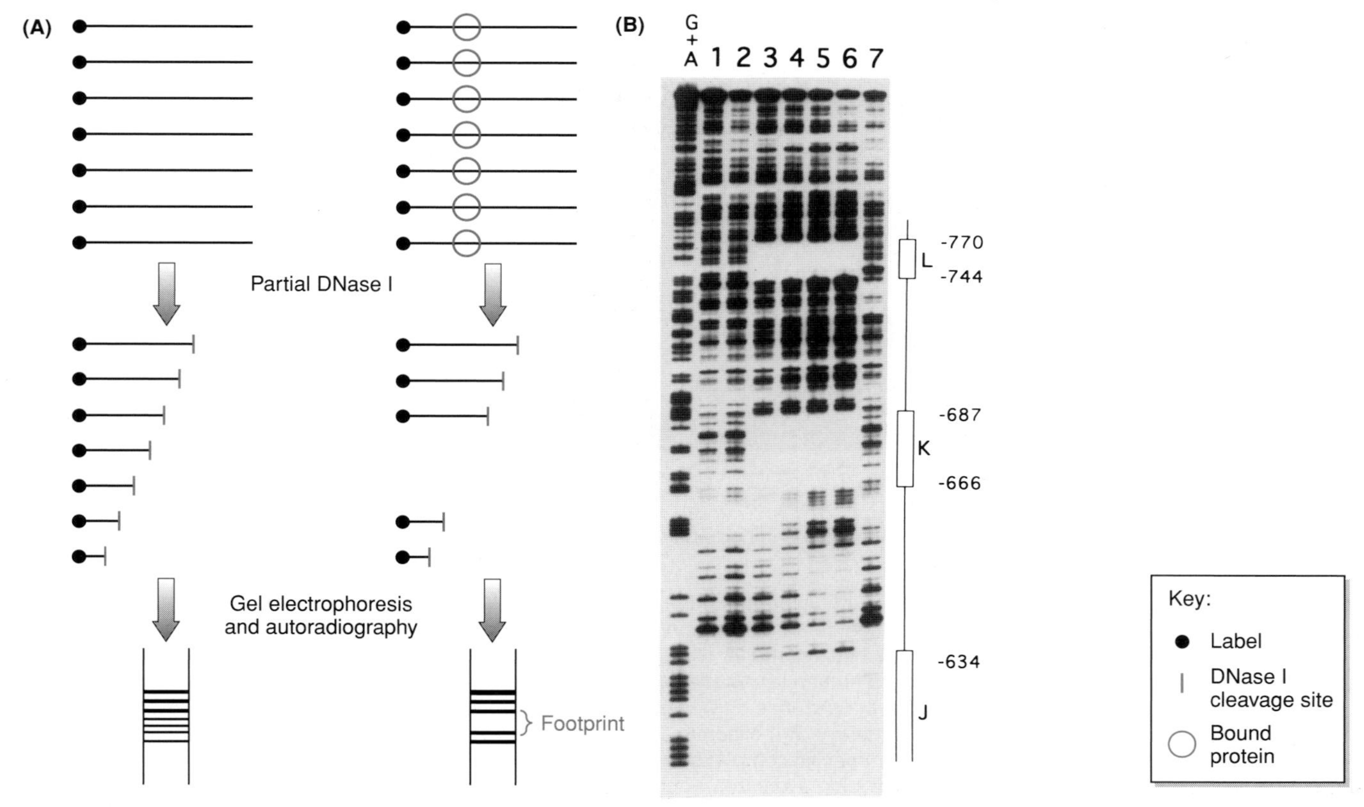

Figure 19.10: DNase I footprinting identifies protein-binding regions in a DNA molecule by their ability to confer resistance to cleavage by DNase I.

(A) Basis of method. DNA molecules are labeled at a single end. If the DNA is not complexed with protein (left), *partial digestion* with pancreatic DNase I can result in a ladder of DNA fragments of different sizes because the location of a cleavage site (red vertical bars) on an *individual* DNA molecule is essentially random. If, however, they contain a protein-binding site, then incubation with a suitable protein extract may result in binding of protein at a specific region (right). Subsequent partial digestion with pancreatic DNase I does not produce random cleavage: the bound protein protects the underlying DNA sequence from cleavage. As a result, the spectrum of fragment sizes is skewed against certain fragment sizes, leaving a gap when size-fractionated on a gel, a DNase I 'footprint'.
(B) Practical example. The example shows DNase I footprinting analysis of the human factor VIII promoter using liver cell nuclear extracts. Three protein-binding sites were identified in the region spanning –800 to –600: J, K and L (see also *Figure 19.9*). Lanes were: *G + A*, Maxam and Gilbert chemical sequencing reaction; *1* and *2*, control reactions, lacking nuclear extract; *3–6*, test reactions with increasing amounts of liver nuclear extract; *7*, control reaction in which the nuclear extract had been heated for 10 min at 75°C in order to denature the protein. Reproduced from Figueiredo and Brownlee (1995) with permission from The American Society for Biochemistry and Molecular Biology.

Studying gene expression and gene function at the protein level using cultured cells or cell extracts

Human gene products can be detected using antibodies obtained by immunization of animals or by genetic engineering methods

Because of their exquisite diversity and sensitivity in detecting proteins, antibodies have numerous applications in research, and their therapeutic potential is considerable (see page 553). Essentially, the applications fall into three classes:

- *tracking gene expression by detecting a gene product.* This application will be discussed in full in this section.
- *Binding to a protein in such a way that the protein can be purified.* This ability is exploited with great success in *affinity chromatography* where the antibody is coupled to the matrix of a chromatography column and used as a hook to pull

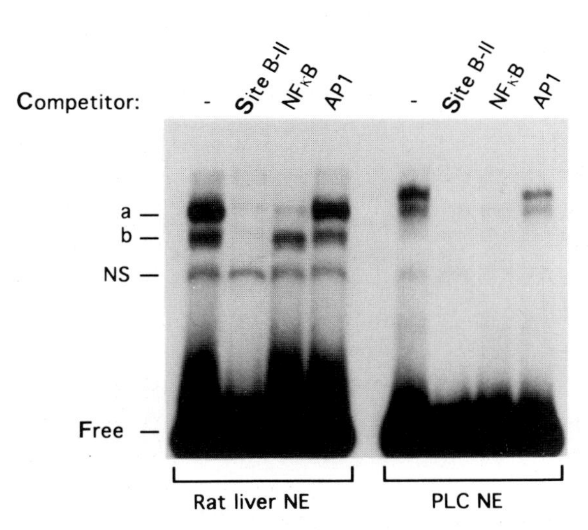

Figure 19.11: Protein-binding regions in a DNA molecule can be identified by gel retardation assays and the identity of the protein can be verified by competitive binding studies.

The figure shows investigation of a protein-binding site upstream of the factor VIII gene (site B; see *Figure 19.9*). This site had been identified by DNase I footprinting analysis and the gel retardation assay (electrophoretic mobility shift assay; EMSA) confirmed its existence. A sequence in the B site resembles the binding site for the ubiquitous transcription factor NF-κB. The EMSA assay used a nuclear protein extract (NE) from rat liver or from a human liver-derived cell line (PLC) and a labeled oligonucleotide from the factor VIII upstream site B (B-II). Assays were conducted in the absence or presence of an excess of competitor oligonucleotide. Most of the label runs near the bottom of the gel at the position of the free B-II oligonucleotide. Upper bands indicate the presence of complexes of protein bound to the labeled B-II oligonucleotide: a and b, specific complexes with rat nuclear extract; c and d, specific complexes with the human liver cell extract; NS, nonspecific protein binding. *Note* that when using the human liver cell nuclear extract, complex formation is strongly inhibited by both the competitor B-II oligonucleotide (control) and also by an oligonucleotide representing the consensus NF-κB binding sequence (PLC NE). Further experiments using purified NF-κB instead of the nuclear extracts verified the result. Reproduced from Figueiredo and Brownlee (1995) with permission from The American Society for Biochemistry and Molecular Biology.

out a desired protein, leading to its purification. It can also be applied to identifying an unknown protein which interacts with a protein of interest.

(iii) *Catalyzing chemical reactions*. This is a relatively recent application but the potential of **catalytic antibodies** (also known as **abzymes**) is great (see Benkovic, 1992).

Traditionally, antibodies have been isolated by immunizing animals, but increasingly genetically engineered antibodies are being used (see *Box 19.1*). Antibodies have been used to detect proteins by different methods.

- **Western blotting** (or ***immunoblotting***). In this method proteins from cell extracts are fractionated according to size. Usually this is achieved by by one-dimensional *SDS–PAGE*, a form of polyacrylamide gel electrophoresis in which the mixture of proteins is first dissolved in a solution of sodium dodecyl sulfate (SDS), an anioinic detergent that disrupts nearly all noncovalent interactions in native proteins. Mercaptoethanol or dithiothreitol is also added to reduce disulfide bonds. Following electrophoresis, the fractionated proteins can be visualized by staining with a suitable dye (e.g. Coomassie blue) or a silver stain. Two-dimensional gels may also be used: the first dimension involves *isoelectric focusing*, that is separation according to charge in a pH gradient, and the second dimension, at right angles to the first, involves size-fractionation by SDS–PAGE (see Stryer, 1995). In this case the fractionated proteins are transferred ('blotted') to a sheet of nitrocellulose and then exposed to a specific antibody (see *Figure 19.12*).

Box 19.1: Obtaining antibodies

Traditional methods

Traditionally, antibodies to human gene products have been obtained by repeatedly injecting suitable animals (e.g. rodents, rabbits, goats, etc.) with a suitable *immunogen*. Two types of immunogen are commonly used:

- *synthetic peptides*. The amino acid sequence (as inferred from the known cDNA sequence) is inspected and a synthetic peptide (often 20–50 amino acids long) is designed. The idea is that, when conjugated to a suitable molecule (e.g. keyhole limpet hemocyanin), the peptide will adopt a conformation that resembles that of the corresponding segment of the native polypeptide. This approach is relatively simple, but success in generating suitably specific antibodies is far from assured and difficult to predict.
- *Fusion proteins*. An alternative approach is to clone a suitable cDNA sequence into a bacterial gene contained within an appropriate expression cloning vector. The rationale is that a hybrid mRNA will be produced which will be translated to give a *fusion protein* with an N-terminal region derived from the bacterial gene and the remainder derived from the inserted gene. The N-terminal bacterial sequence is often designed to be quite short, but may nevertheless confer some advantages. For example it can provide a signal sequence to ensure secretion of the fusion protein into the extracellular medium, thereby simplifying its purification, and it may protect the foreign protein from being degraded within the bacterium. Because the fusion protein contains most or all of the desired polypeptide sequence, the probability of raising specific antibodies may be reasonably high.

If the animal's immune system has responded, specific antibodies should be secreted into the serum. The antibody-rich serum (*antiserum*) which is collected contains a heterogeneous mixture of antibodies, each produced by a different B lymphocyte [*because immunoglobulin gene rearrangements are cell-specific as well as cell type (B-lymphocyte)-specific*, see page 177]. The different antibodies recognize different parts (*epitopes*) of the immunogen (**polyclonal antisera**). A homogeneous preparation of antibodies can be prepared, however, by propagating a clone of cells (originally derived from a single B lymphocyte). Because B cells have a limited life-span in culture, it is preferable to establish an immortal cell line: antibody-producing cells are fused with cells derived from an immortal B-cell tumor. From the resulting heterogeneous mixture of hybrid cells, those hybrids that have both the ability to make a particular antibody and the ability to multiply indefinitely in culture are selected. Such **hybridomas** are propagated as individual clones, each of which can provide a permanent and stable source of a single type of **monoclonal antibody**.

Genetic engineering methods

Antibodies generated by the above classical approaches originate from animals. Once the various immunoglobulin genes had been cloned, however, DNA cutting and ligation technology could be used to generate new antibodies including both *partially humanized antibodies* and *fully human antibodies* (see page 554). In addition, novel approaches can bypass the need for hybridoma technology, and, even immunization altogether. The powerful *phage display technology* permits the construction of a virtually limitless repertoire of human antibodies with specificities against both foreign and self-antigens (see Winter *et al*., 1994). The essence of this method is that the gene segments encoding antibody heavy and light chain variable sequences are cloned and expressed on the surface of a filamentous bacteriophage, and rare phage are selected from a complex population by binding to an antigen of interest (see page 528 for a fuller explanation).

- **Immunocytochemistry**. This technique is concerned with studying the overall expression pattern of a gene within a tissue or other multicellular structure. It can therefore be regarded as the equivalent of tissue *in situ hybridization* (see page 126) but uses an antibody instead of a nucleic acid as the probe.
- **Immunofluorescence microscopy**. This method can give a rough indication of subcellular location for a protein of interest. A suitable fluorescent dye, such as *fluorescein* or *rhodamine* is coupled to the desired antibody, enabling the relevant protein to be localized within the cell by light microscopy. Light of

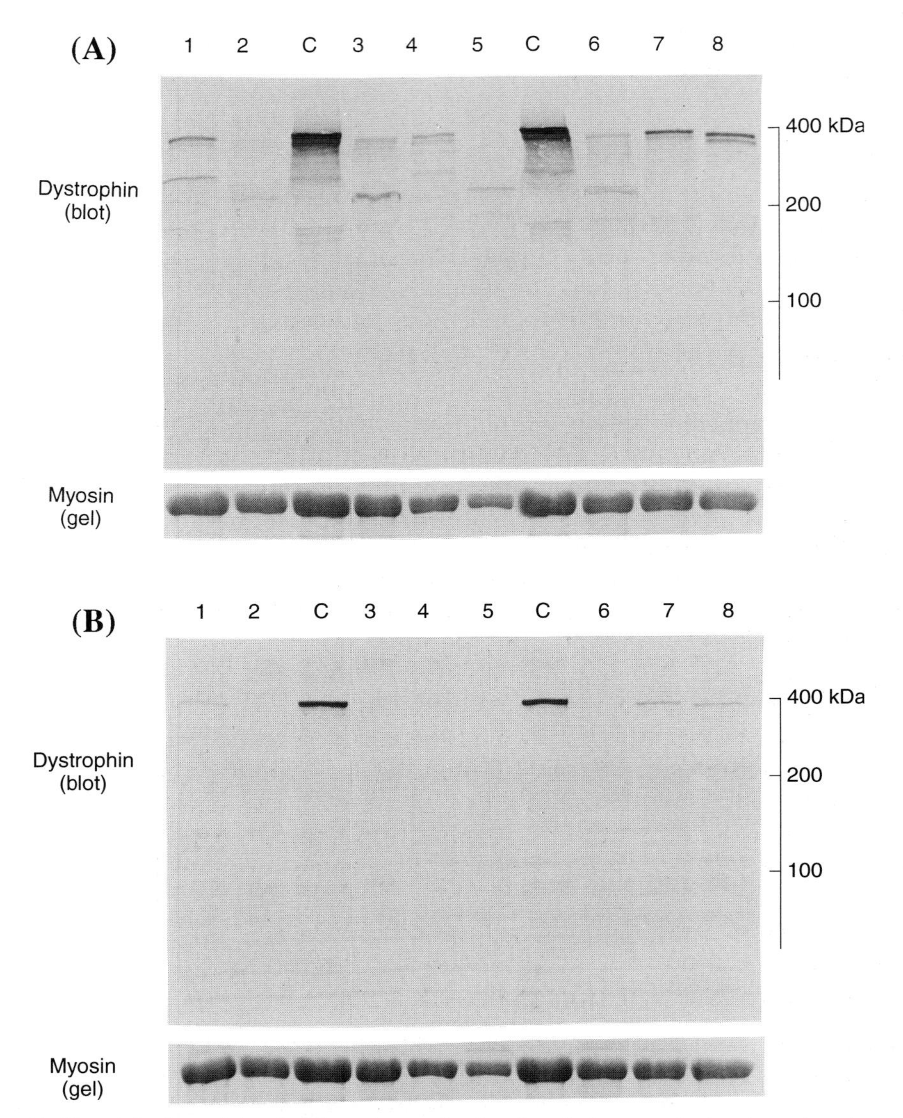

Figure 19.12: Western blotting detects proteins that have been size-fractionated on an electrophoresis gel.

Western blotting (*immunoblotting*) involves detection of polypeptides after size-fractionation in a polyacrylamide gel and transfer ('blotting') to a membrane. This example illustrates its application in detecting dystrophin using two antibodies. The Dy4/6D3 antibody is specific for the rod domain and was generated by using a fusion protein immunogen (see *Box 19.1*). The Dy6/C5 antibody is specific for the C-terminal region and was generated by using a synthetic peptide immunogen. Reproduced from Nicholson *et al.* (1993) with permission from the BMJ Publishing Group. The photograph was kindly provided by Louise Anderson (formerly Nicholson), University of Newcastle upon Tyne.

suitable wavelength causes the dye to fluoresce, emitting light at a specific and longer wavelength which is recorded.

- **Ultrastructural studies**. Higher resolution still of the intracellular localization of a gene product or other molecule is possible using electron microscopy. The antibody is typically labeled with an electron-dense particle, such as colloidal gold spheres.

In vitro mutagenesis is a powerful method for studying gene function and for producing novel proteins

If a gene has been cloned and a functional assay of the product is available, molecular genetic techniques permit exquisite dissection and mapping of the functional elements of the product. **Site-directed mutagenesis** techniques can specifically alter any nucleotide in a DNA clone in a predetermined way, either by cell-based cloning or PCR techniques. Either a single nucleotide or a limited run of nucleotides can be altered. If the mutation is engineered into a coding sequence, a form of protein engineering is possible (see Fersht and Winter, 1992). Expression cloning of the mutated gene provides a mutant product that can be assayed for function, using the normal gene product as a control. An alternative to producing a single mutation is to generate simultaneously a large series of related mutant sequences. These can be used to synthesize a library of different polypeptides, in the hope of producing a novel protein superior in some way to any naturally occurring equivalent.

Producing a predetermined sequence change

Basic PCR-based techniques for site-specific mutagenesis have been described in Chapter 6, page 144ff. Here we describe alternative methods, including site-specific mutagenesis using cell-based cloning systems. **Oligonucleotide mismatch mutagenesis** relies on using a synthetic mutagenic oligonucleotide which is designed to be almost perfectly complementary to a short sequence in the gene of interest. The gene is cloned using a vector such as the bacteriophage M13 or a *phagemid* (see page 307), which permits recovery of single-stranded recombinant DNA. The mutagenic oligonucleotide is allowed to anneal to its complementary sequence in the single-stranded recombinant, and then to prime new DNA synthesis to create a complementary full-length sequence containing the desired mutation. The newly formed heteroduplex is used to transform cells, and the desired mutant genes can be identified by screening for the mutation (see *Figure 19.13*). Using this method, a single nucleotide substitution can be introduced at a predetermined location or other small-scale mutations can be introduced, such as deletion of a codon.

Oligonucleotide mismatch mutagenesis has been very powerful as a tool for altering a predetermined amino acid in order to test its contribution to the function of a protein. Genes encoding enzymes have been favorite targets, because the results can be tested by a simple enzyme assay *in vitro*. Another useful application is for testing the effect of a sequence variant found while investigating a patient with a genetic disease. Unlike nonsense or frameshifting mutations, the effect of missense mutations on gene function may be hard to predict, and it may be unclear whether a sequence variant is pathogenic or just a neutral variant. Site-directed mutagenesis can be used to create a replica of the variant for testing *in vitro*.

Producing libraries of random variants in a predetermined part of the gene

Several methods have been developed for generating a mixed pool of random variants. Individual clones can then be assayed for function.

- ***Oligonucleotide cassette mutagenesis***. A chosen segment of the gene of interest is replaced by an artificially designed segment consisting of a random mixture of oligonucleotides, synthesized using redundancy at individual positions. The total set of oligonucleotides can be ligated, albeit inefficiently, into suitable double-stranded DNA molecules (see Stemmer, 1994, for references).

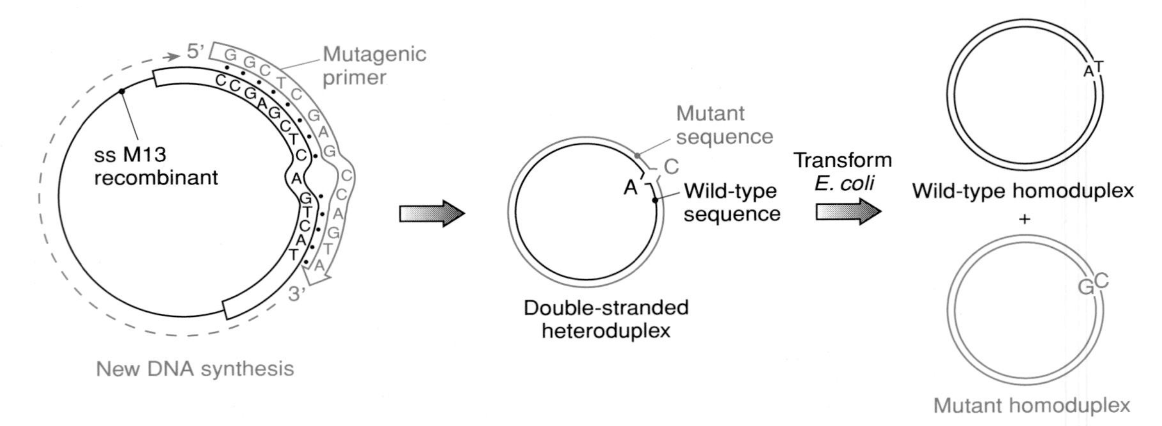

Figure 19.13: Oligonucleotide mismatch mutagenesis can create a desired point mutation at a unique predetermined site within a cloned DNA molecule.

The figure illustrates only one of many different methods of cell-based oligonucleotide mismatch mutagenesis (for alternative PCR-based site-directed mutagenesis, see page 144). The example illustrates the use of a mutagenic oligonucleotide to direct a single nucleotide substitution in a gene. The gene is cloned into M13 in order to generate a single-stranded recombinant DNA. An oligonucleotide primer is designed to be complementary in sequence to a portion of the gene sequence encompassing the nucleotide to be mutated (A) and containing the desired noncomplementary base at that position (C, not T). Despite the internal mismatch, annealing of the mutagenic primer is possible, and second strand synthesis can be extended by DNA polymerase and the gap sealed by DNA ligase. The resulting heteroduplex can be transformed into *E. coli*, whereupon two populations of recombinants can be recovered: wild-type and mutant homoduplexes. The latter can be identified by molecular hybridization (by using the mutagenic primer as an allele-specific oligonucleotide probe; see *Figure 5.10*) or by PCR-based allele-specific amplification methods (see *Figure 6.8*).

- ***Error-prone PCR mutagenesis.*** Many thermostable DNA polymerases, such as *Taq* DNA polymerase, are prone to errors in the fidelity of replication. Methods have been developed to exploit this tendency in order to create a set of amplification products differing in sequence (see Stemmer, 1994, for references).
- ***Coupled* in vitro *mutagenesis and recombination.*** A method of *in vitro* homologous recombination of pools of selected mutant genes by random fragmentation and PCR-mediated reassembly has been described recently. In a test case to design novel variants of the enzyme β-lactamase, this method has been spectacularly successful (Stemmer, 1994): it resulted in some mutant variants whose catalytic activity was much greater than that of the natural enzyme, and much greater than variants produced by oligonucleotide cassette mutagenesis or by error-prone PCR.

Phage display technology is a powerful aid to protein engineering

Phage display is a form of expression cloning of foreign genes using phage (see Clackson and Wells, 1994). Genetic engineering techniques are used to insert foreign DNA fragments into a suitable phage coat protein gene. The modified gene can then be expressed as a fusion protein which is incorporated into the virion and displayed on the surface of the phage which, however, retains infectivity (**fusion phage**). If an antibody is available for a specific protein, phage displaying that protein can be selected by preferential binding to the antibody: affinity purification of virions bearing a target determinant can be achieved from a 10^8-fold excess of phage not bearing the determinant, using even minute quantities of the relevant antibody. Initially, phage display involved the use of filamentous phages such as fd, f_1, M13, etc., where the foreign gene was incorporated into a gene specifying a minor coat protein such as the gene III protein (see *Figure 19.14*). More recently, the method has

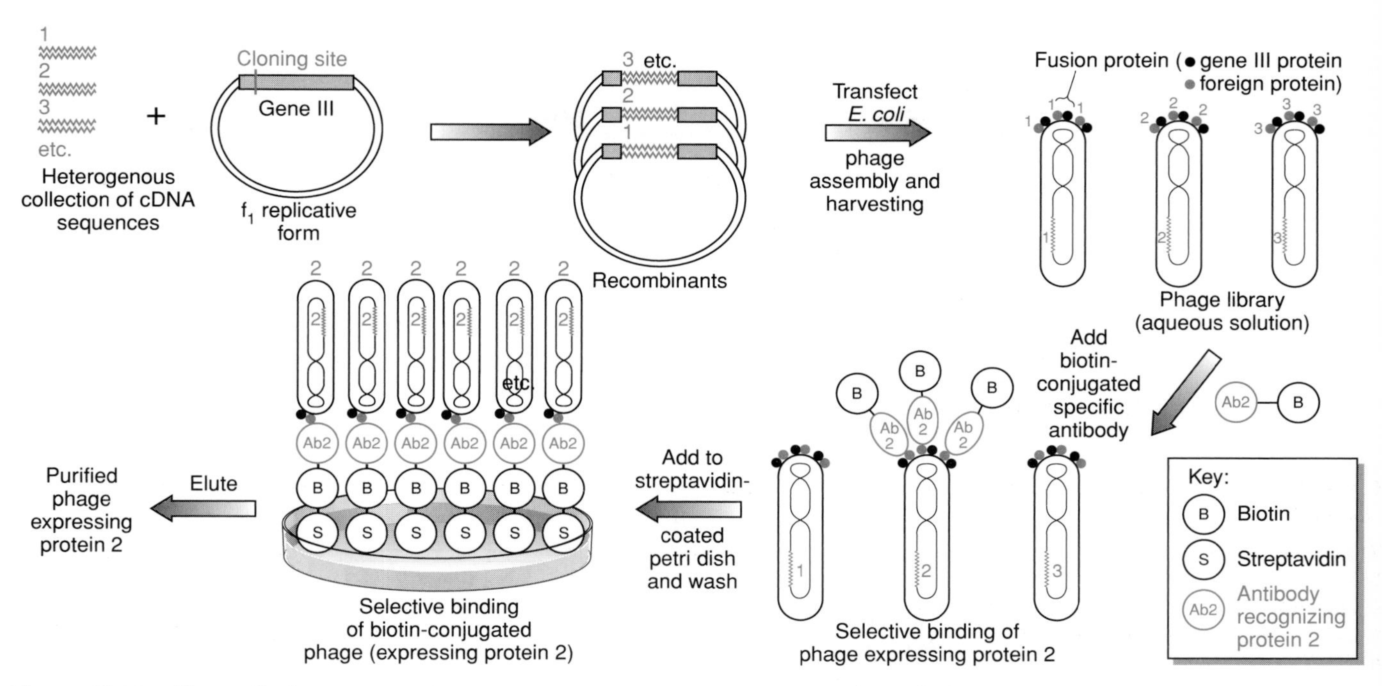

Figure 19.14: Phage display

Phage display is a form of expression cloning which involves cloning cDNA into phage vectors and expressing foreign proteins on the phage surface. The figure illustrates one popular approach whereby the DNA is cloned into gene III of the filamentous phage f1 (or M13, etc.), a gene which encodes a minor phage coat protein. The cloning site is usually designed by site-specific mutagenesis to occur at a position corresponding to the extreme N-terminal sequence of the gene III protein. Following transfection of *E. coli*, phage assembly, extrusion from the cells and phage harvesting (see *Figure 11.25* for the general scheme of cloning using filamentous phage vectors), a phage library is produced. Recombinants with inserts which do not produce a frameshift in the reading frame may often be expressed to give a *fusion protein* in which the N-terminal component consists of a foreign protein sequence. An antibody which specifically recognizes one of the foreign protein sequences can then be used to bind specifically to the phage which displays the sequence, leading to its purification. Such *affinity purification* permits identification of cDNA sequences encoding an uncharacterized protein of interest (see Parmsley and Smith, 1988).

been extended to other phages, notably phage λ. Among many useful applications are:

- **Antibody engineering**. Phage display is proving a powerful alternative source of constructing antibodies, including humanized antibodies, bypassing immunization and even hybridoma technology (see Winter *et al.*, 1994).
- **General protein engineering**. Phage display is a powerful adjunct to random mutagenesis programs as a way of selecting for desired variants from a library of mutants.
- **Studying protein–protein interactions** (see next section).

Several methods are available for identifying and studying protein–protein interactions

The function of a gene may not be obvious even after comprehensive study of its structure and expression patterns. In such cases, it may be advantageous to attempt to identify those proteins with which it may interact specifically, especially if these turn out to be ones that have previously been well studied and whose functions are known. Different approaches can be taken, including physical methods and library-based methods (see Phizicky and Fields, 1995).

Co-immunoprecipitation

The basis of this classical physical method of detecting protein–protein interactions is simple: cell lysates are generated, antibody is added, the antigen is precipitated and washed, and bound proteins are eluted and analyzed. If a *polyclonal antibody* is used, it is important to show that the co-precipitated proteins have been precipitated by the expected antibody, rather than a contaminating one. It also needs to be established that the antibody itself does not recognize the co-precipitated protein.

Phage display technology

This is a library-based method which can be used to identify proteins that interact with a given protein. In the same way that antibodies can be used in affinity screening, a desired protein or any other molecule to which a protein can bind can be used as the selective agent. The protein can select fusion phage which display any other proteins that significantly bind to it.

The yeast two-hybrid system

Also called the *interaction trap system,* this is a widely used library-based method (see Fields and Sternglanz, 1994). It can be used to identify proteins that bind to a protein under study, or to delineate domains or residues crucial for interaction. Proteins that physically bind to one another are detected by their ability, when bound, to activate transcription of a reporter gene. The key to the method is the observation that transcriptional activation factors require two separate domains, a DNA binding domain and an activation domain. In natural transciption factors these are usually part of the same molecule, but an active transcription factor can equally be constituted out of two proteins that associate together, one of which carries the DNA binding domain and the other carries the activation domain. To use the two-hybrid system, a recombinant gene is constructed, using standard methods, that encodes the protein under study coupled to one of the necessary domains. Cells are co-transfected with this gene and with random genes coupled to the other necessary domain, in the presence of a reporter construct, expression of which signals the presence of an active transcription factor (*Figure 19.15*). The transfected DNA in cells in which the reporter is expressed should encode proteins that physically associate in the cell.

Using transgenic technology and gene targeting to study gene expression and function

Exogenous DNA can be transfected into cultured cells using a variety of approaches and, in some cases, the DNA can *integrate* into the chromosomes of the host cell and be transmitted to daughter cells during cell division (see *Figure 20.3*). During the 1980s, techniques were refined for artificially inserting foreign DNA into the chromosomes of certain **totipotent** animal cells. These are cells which have the capacity to differentiate into the different cells of an adult animal [such as fertilized oocytes, the cells of early stage embryos and **embryonic stem (ES) cells**]. By arranging for such genetically modified cells to contribute to the development of a whole animal, it was possible to generate so-called **transgenic animals**. Initially, partially transgenic animals are often produced. These are mosaics, with the exogenous gene in only a proportion of their cells. Crossing these mosaic animals generates fully transgenic offspring.

Fully transgenic animals are able to transmit genetically modified chromosomes in a mendelian fashion to subsequent generations. If the inserted material is a foreign gene, a **transgene**, its expression can be studied in a variety of different cellular

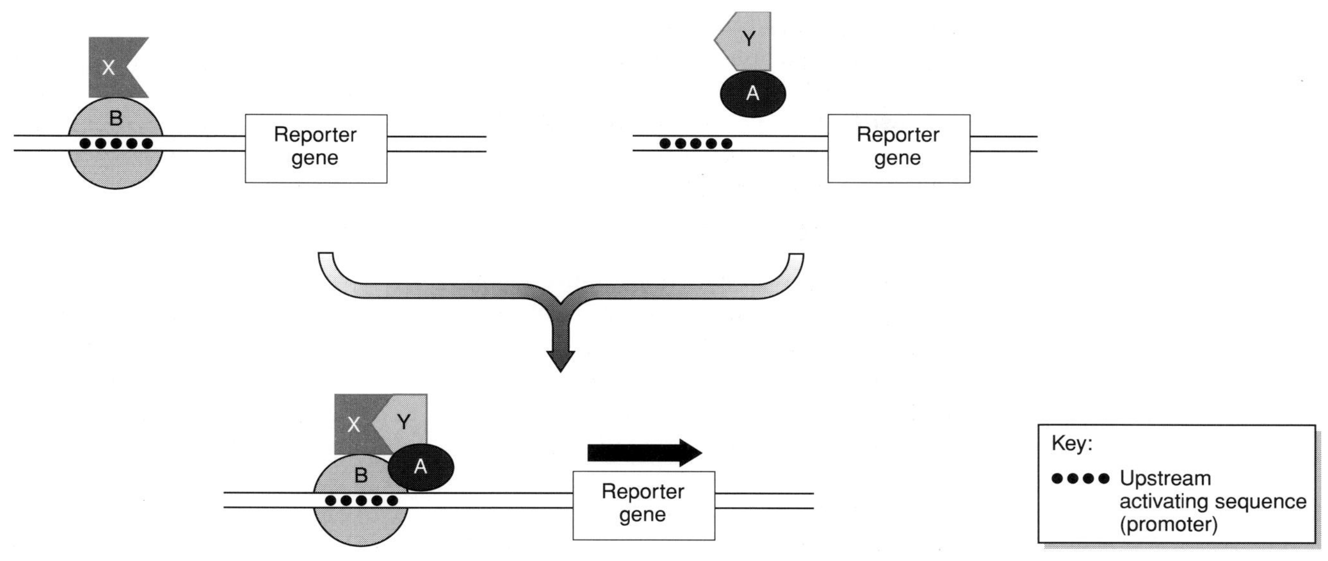

Figure 19.15: The yeast two-hybrid system is designed to identify proteins that interact with the product of a gene of interest.

Transcription factors (TFs) have two complementary domains: a DNA-binding domain and an activation domain, both of which are required for normal function. The individual domains can be bound separately to proteins to form two types of hybrid, neither of which has the original TF activity. If two proteins with complementary TF domains interact, the two TF domains interact to restore the original TF function.
Top left. A hybrid protein is generated consisting of a protein (X) and a DNA-**b**inding domain (B). The hybrid can bind to a promoter sequence upstream of a reporter gene, but cannot activate transcription in the absence of the activation domain.
Top right. A second hybrid protein is formed, consisting of a protein Y and a transcription **a**ctivation domain (A). This hybrid protein cannot activate transcription of the reporter gene in the absence of a DNA-binding domain.
Bottom. Mixing hybrids with complementary domains may result in specific association of one protein (X) with an attached DNA-binding domain (B), and a second protein (Y) with an attached activation domain, resulting in activation of the downstream reporter gene.

environments in a whole animal. Transgenic animal technology also offers the possibility of a whole animal expression–cloning system (see below).

Although transgenes often integrate into the host chromosomes without affecting the expression of any endogenous genes, occasionally the integration event alters endogenous gene expression (**insertional mutation**), producing a recognizable phenotype. This constitutes a form of *in vivo* mutagenesis, albeit at an unselected target gene. **Gene targeting** was developed as a method of *in vivo* mutagenesis in which the mutation is introduced into a *pre-selected* endogenous gene. This can be achieved in somatic cells, but gene targeting in ES cells is particularly powerful because it can lead to the construction of an animal in which all nucleated cells contain a mutation at the desired locus (see below). Because the introduced DNA is normally a mouse gene, and because the resulting mouse may sometimes not contain any nonmouse sequences, the term *transgenic animal* may not always be applicable.

Transgenic mice can be produced following transfer of cloned DNA into fertilized oocytes, embryos or embryonic stem cells

Because of its amenability to genetic experimentation (see *Boxes 14.4* and *19.2*), the laboratory mouse has been the preferred animal for transgenic studies. In order to construct a fully transgenic mouse (one which contains the inserted DNA in all nucleated cells), two broad approaches can be taken. One way involves random integration of the exogenous DNA into the chromosomal DNA of a fertilized oocyte. The alternative involves integration of the DNA at the post-zygotic level, either by using early stage embryos or ES cells. The latter procedure results in a partially transgenic animal but, as long as a reasonable proportion of germline cells contain the introduced DNA, the animals can be bred to produce fully transgenic offspring. The three methods which have been used to make transgenic mice are given below.

Pronuclear microinjection

This method has been used to prepare a variety of different transgenic animals as well as transgenic mice. For example, it is possible to generate *transgenic sheep* and other livestock animals which can serve as **bioreactors,** whole-animal expression cloning systems in which introduced genes are expressed to give large amounts of therapeutic or commercially valuable gene products (see page 552). In addition, it has been used to model disease phenotypes resulting from a *gain-of-function* mutation by simply adding the mutant gene (see page 544).

To obtain transgenic mice by this route, females are superovulated, mated to fertile males and sacrificed the next day. Fertilized oocytes are recovered from excised oviducts. The DNA of interest is then microinjected using a micromanipulator into the male pronucleus of individual oocytes. Surviving oocytes are re-implanted into the oviducts of foster females and allowed to develop into mature animals (see *Figure 19.16* and Gordon, 1992). During this procedure, the microinjected DNA (**transgene**) randomly integrates into chromosomal DNA, almost always at a single site, although rarely two sites of integration are found in a single animal. Individual insertion sites contain multiple copies of the transgenes integrated into chromosomal DNA as head-to-tail concatemers (it is not unusual to find 50 or more copies at a single insertion site). As a result of chromosomal integration, the transgenes can be passed on to subsequent generations in mendelian fashion: if the foreign DNA has integrated at the one-cell stage, it should be transmitted to 50% of the offspring.

Retroviral transfer into pre- or post-implantation embryos

Retroviruses are RNA viruses which can undergo an intermediate DNA form that allows them to integrate into cellular genomes (see *Figure 17.2*). Infection of pre-implantation mouse embryos with a retrovirus such as Moloney murine leukemia virus or injection of the retrovirus into early post-implantation mouse embryos results in mosaic offspring. Retroviruses should integrate rarely and at random into accessible cells, and the use of replication-defective retroviruses provides heritable markers for clonal descendants of the target cell (unlike wild-type viruses which spread from cell to cell). This approach has been used, therefore, for studying cell lineage using reporter genes.

Injection of genetically modified embryonic stem cells into host blastocysts

The microinjection of foreign DNA into fertilized oocytes is technically difficult and not suited to large-scale production of transgenic animals or to sophisticated genetic manipulation. A popular alternative, which has so far been restricted to the construction of transgenic mice, involves transferring the foreign DNA initially into ES cells. Mouse ES cells are derived from 3.5-day post-coitum embryos and arise from the *inner cell mass* of the blastocyst (see *Figure 19.17*). The ES cells can be cultured *in vitro* and retain the potential to contribute extensively to all of the tissues of a mouse, *including the germline*, when injected back into a host blastocyst and re-implanted in a pseudopregnant mouse. The developing embryo is a **chimera**: it contains two populations of cells derived from different zygotes, those of the blastocyst and the implanted ES cells. If the two strains of cells are derived from mice with different coat colors, chimeric offspring can easily be identified (see *Figure 19.17*).

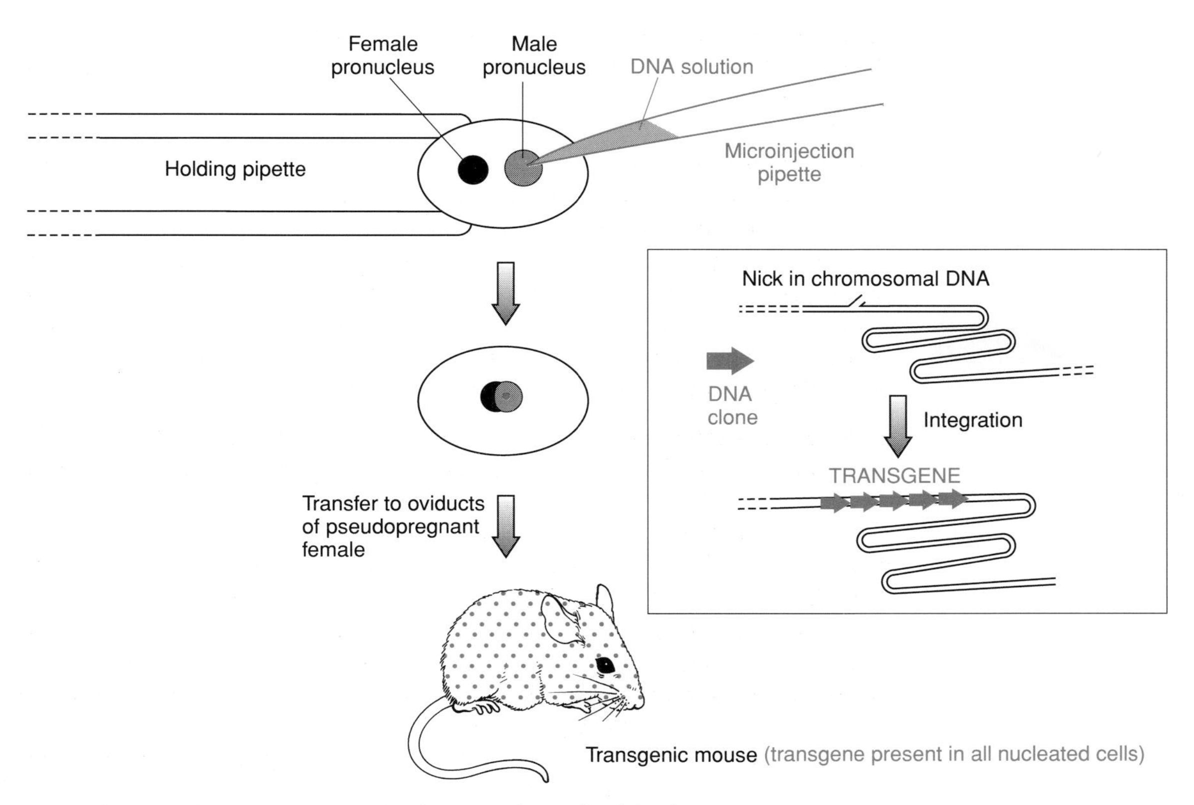

Figure 19.16: Construction of transgenic mice by pronuclear microinjection.

Very fine glass pipettes are constructed using specialized equipment: one, a holding pipette, has a bore which can accommodate part of a fertilized oocyte, and thereby hold it in place, while the microinjection pipette has a very fine point which is used to pierce the oocyte and thence the *male* pronucleus (because it is bigger). An aqueous solution of the desired DNA is then pipetted directly into the pronucleus. The introduced DNA clones can integrate into chromosomal DNA at nicks, forming transgenes, usually containing multiple head-to-tail copies. Following withdrawal of the micropipette, surviving oocytes are re-implanted into the oviducts of *pseudopregnant* foster females (which have been mated with a vasectomized male; the mating act can initiate physiological changes in the female which stimulate the development of the implanted embryos). Newborn mice resulting from development of the implanted embryos are checked by PCR for the presence of the desired DNA sequence. See Gordon (1992).

Use of genetically modified ES cells results in a partially transgenic mouse. Because the injected ES cells can form all or part of the functional germ cells of the chimera, it is possible to derive fully transgenic mice. This is usually accomplished by screening the offspring of matings between chimeras (usually males) and mice with a coat colour recessive to that of the strain from which the ES cells were derived (see *Figure 19.17*).

The ES cell approach to constructing transgenic mice was made possible by the successful establishment in the early 1980s of cell lines from isolated mouse ES cells. As ES cells can be grown readily in culture, a variety of manipulations can be conducted to ensure that transformed cells are selected. For example, the desired gene can be ligated to a marker gene, such as the *neo* gene, enabling a positive selection for transformants (see *Box 11.1*). The presence of the desired gene can also be verified quickly by a PCR-based assay. ES cells also offer the possibility of *gene targeting* by homologous recombination, a method which can permit the creation of a desired animal model of disease by selective inactivation of a specific predetermined gene (see below).

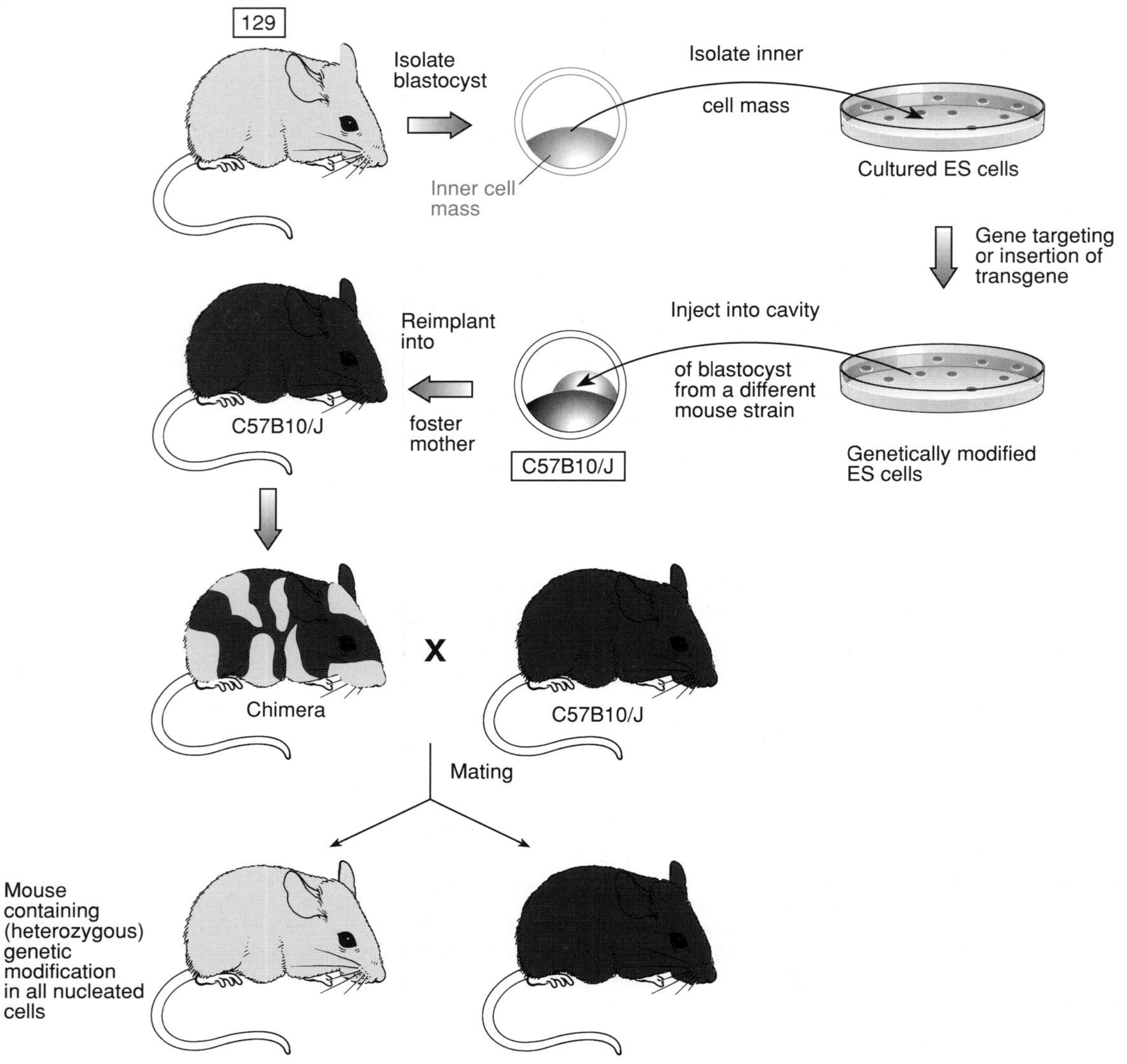

Figure 19.17: Genetically modified ES cells as a route for transferring foreign DNA or specific mutations into the mouse germline.

Cells from the *inner cell mass* were cultured following excision of oviducts and isolation of blastocysts from a suitable mouse strain (129). Such embryonic stem (ES) cells retain the capacity to differentiate into, ultimately, the different types of tissue in the adult mouse. ES cells can be genetically modified while in culture by insertion of foreign DNA or by introducing a subtle mutation. The modified ES cells can then be injected into isolated blastocysts of another mouse strain (e.g. C57B10/J which has a black coat color that is recessive to the agouti color of the 129 strain) and then implanted into a pseudopregnant foster mother of the same strain as the blastocyst. Subsequent development of the introduced blastocyst results in a *chimera* containing two populations of cells (including germline cells) which ultimately derive from different zygotes (normally evident by the presence of differently colored coat patches). Backcrossing of chimeras and subsequent interbreeding can produce mice that are heterozygous or homozygous for the genetic modification.

The use of inducible and tissue-specific promoters and of YAC transgenes has greatly extended the scope for studying gene expression and gene function in transgenic animals

Transgenic animals have been important for analyzing human genes (see Camper, 1995; Theuring, 1995), and have helped greatly in our understanding of a variety of fundamental biological processes, notably in immunology, neurobiology, cancer and developmental studies. They have been helpful in clarifying species differences in regulatory sequences and in dissecting the roles of *cis*-acting elements, both in and around genes, and in mediating the tissue specificity of expression (the equivalent expression systems in cultured cells have always been regarded as likely to be considerably different from the situation *in vivo*). They have also been extremely important in studying human gene function, by providing models of:

- *gain of function*. This includes studying ectopic expression (the gene is linked to a tissue-specific promoter which results in it being expressed in a cell type in which it is not normally expressed); overexpression (the gene is attached to a powerful promoter); oncogene expression and dominantly acting mutant genes (see page 544).
- *Loss of function*. This includes insertional inactivation of a gene by gene targeting – see below; antisense constructs and mutated transgenes. In addition, **genetic ablation** is a procedure where a sequence encoding a toxin, for example diphtheria toxin subunit A, ricin, etc., is linked to a tissue-specific promoter. When the promoter becomes active at the appropriate stage of tissue differentiation, the toxin is produced and kills the cells. Thus, certain cell types can be eliminated. Proponents claim that this approach can be useful in evaluating the functions of cells that express specific genes.

Inducible promoters

For many applications it is desirable to have a transgene expressed under the control of a tissue-specific promoter or one that is inducible. In the former case, genetic engineering can be used to splice known tissue-specific promoters from cloned genes to the gene of interest (see *Table 7.10* for examples). In some cases, coupled regulatory elements can confer both position-independent and tissue-specific expression as with sequence elements from the β-globin locus control region (Grosveld *et al.*, 1987).

Various attempts have also been made to create inducible transgenic mice (e.g. by using heavy metal ions to induce expression of an integrated gene which has a coupled metallothionein promoter, etc.). Generally, the use of **inducible promoters** has been hampered by 'leakiness' in gene expression and by relatively low levels of induction, and they have often been applicable to a limited range of tissues. Recently, however, more promising systems have been developed. For example, novel methods employing tetracycline-regulated inducible expression have permitted construction of both highly inducible transfected cells (with much greater efficiency than the constitutive system) and transgenic mice (Shockett *et al.*, 1995). In the latter case the expression of a reporter gene, such as the luciferase gene, can be controlled by altering the concentration of tetracycline in the drinking water of the animals.

YAC transgenics

Early studies of gene expression and regulation in transgenic animals involved transfer of small genes. However, expression of small transgenes often fails to follow the normal temporal and spatial patterns of expression or match the expression level of the endogenous homolog. Increasingly there was the recognition that human genes are often very large (see *Figure 7.6*). Even in the case of small genes, important regulatory elements that are required for correct expression may be located many kilobases upstream of the coding sequence (see *Figure 8.6*). In order to be able to study the expression and regulation of a human gene under the control of its own *cis*-acting regulatory elements, it was therefore necessary to establish transfection conditions which would allow the transfer of large DNA clones.

A major breakthrough in transgenic studies was the development of so-called **YAC transgenics** (see Lamb and Gearhart, 1995). The first report to be published described transfer of a 670-kb YAC containing the human *HPRT* (hypoxanthine phosphoribosyltransferase) gene into mouse ES cells (Jakobovits *et al.*, 1993). This was accomplished by *spheroplast fusion* (i.e. fusion of ES cells with YAC-containing yeast cells that have been stripped of the hard cell wall; see page 103). Fragments from the yeast genome can integrate at the same time, however, and so alternative methods have sought to purify an individual YAC by size-fractionation on a preparative gel using pulsed-field gel electrophoresis (assuming that the YAC migrates at a position in the gel that is different from any yeast chromosome). The purified YAC can be inserted into a fertilized oocyte by pronuclear microinjection (see above). This method is, however, limited to small YACs: the DNA of large YACs is more likely to fragment following microinjection with very fine micropipettes. Alternatively, purified YACs have been transferred into ES cells by using *liposomes*, artificial lipid vesicles that are used to transport molecules into a cell following fusion of the lipid coat with the plasma membrane of the recipient cell (see *Figure 20.6*).

YAC transgenics have permitted study of large genes such as the 400-kb human *APP* gene (amyloid protein precursor, one of the genes known to contribute to Alzheimer's disease) which showed tissue- and cell type-specific expression patterns closely mirroring that of the endogenous mouse gene (for this and other examples, see Lamb and Gearhart, 1995). Long-range gene regulation mechanisms (locus control regions, imprinting and other chromatin domain effects; see Chapter 7) can be modeled. An interesting application has been in the production of fully human antibodies in the mouse by transfer of human YACs containing large segments of the human heavy and kappa light chain immunoglobulin loci into mouse ES cells, and thence the creation of transgenic mice able to produce human antibodies (see page 554). Finally, YAC transgenics may also find a role in modeling disease caused by large-scale gene dosage imbalance (see page 545).

Gene targeting by homologous recombination in ES cells can be used to produce mice with a mutation in a predetermined gene

Gene targeting involves engineering a mutation in a *pre-selected* gene within an intact cell. It can therefore be viewed as a form of artificial *site-directed* in vivo *mutagenesis* (by comparison with the various methods of site-directed *in vitro* mutagenesis described in pages 526–528). The mutation may result in inactivation of gene expression (a **'knock-out' mutation**), or altered gene expression, and so can be useful for studying gene function (see below). In addition, the same method can be used to *'correct'* a pathogenic mutation by restoring the normal phenotype, and so has therapeutic potential (see page 571). Gene targeting typically involves introducing a mutation by **homologous recombination** a cloned gene (or gene segment) which is closely related in sequence to an endogenous gene is introduced into the cells, and cells are selected in which homologous recombination has occurred between the introduced gene and its corresponding chromosomal homolog. Gene targeting by homologous recombination has been achieved in some somatic mammalian cells, such as myoblasts (see Arbones *et al.*, 1994). However, the most important application involves mouse ES cells: once a mutation has been engineered into a specific mouse gene within the ES cells, the modified ES cells can then be injected into the blastocyst of a foster mother and eventually a mouse can be produced with the mutation in the desired gene in all nucleated cells (see Capecchi, 1989; Melton, 1994).

Homologous recombination in mammalian cells is a very rare occurrence (unlike in yeast cells, for example, where it is much more frequent, enabling sophisticated genetic manipulation). The frequency of homologous recombination is increased, however, when the degree of sequence homology between the introduced DNA and the target gene is very high. As a result, the introduced DNA clone is a *mouse gene* (or gene segment) usually derived from a strain that is *isogenic* with the strain used for gene targeting. Even then, the frequency of genuine homologous recombination events is very low and may be difficult to identify against a sizeable background of random integration events. To assist identification of the desired homologous recombination events, therefore, the *targeting vector* (often a plasmid vector) contains a marker gene, such as the *neo* gene (see *Box 11.1*), which permits selection for cells that have taken up the introduced DNA, and PCR assays screen for evidence of a homologous recombination event (by using a marker-specific primer plus a primer derived from a sequence within the chromosomal gene of interest, but not in the introduced gene segment). The targeting construct is transferred into cultured mouse ES cells by **electroporation**, a method in which pulses of high voltage are delivered to cells, causing temporary relaxation of the selective permeability properties of the plasma membranes. Two basic approaches have been used:

(i) **Insertion vectors** target the locus of interest by a single reciprocal recombination, causing insertion of the entire introduced DNA including the vector sequence. This is the most secure and favorite way of causing a knock-out mutation (*Figure 19.18A*).

(ii) **replacement vectors** are designed to enable replacement of some of the sequences originally present in the chromosomal gene by a homologous sequence from the introduced DNA, either as a result of a double reciprocal recombination or by gene conversion. This method can be used to inactivate a gene by ensuring that the introduced sequence contains one or more premature termination codons or lacks critical coding sequences. It can also be used to correct a pathogenic mutation (*Figure 19.18B*).

The replacement vector approach, as well as the insertion vector method, can result in leaving residual foreign sequences in the target locus. In some cases, however, a more subtle mutation is required. For example, it may be desirable to investigate the effect of changing a single codon. Various two-step recombination techniques can be used to accomplish this method (see *Figure 19.19* and Melton, 1994).

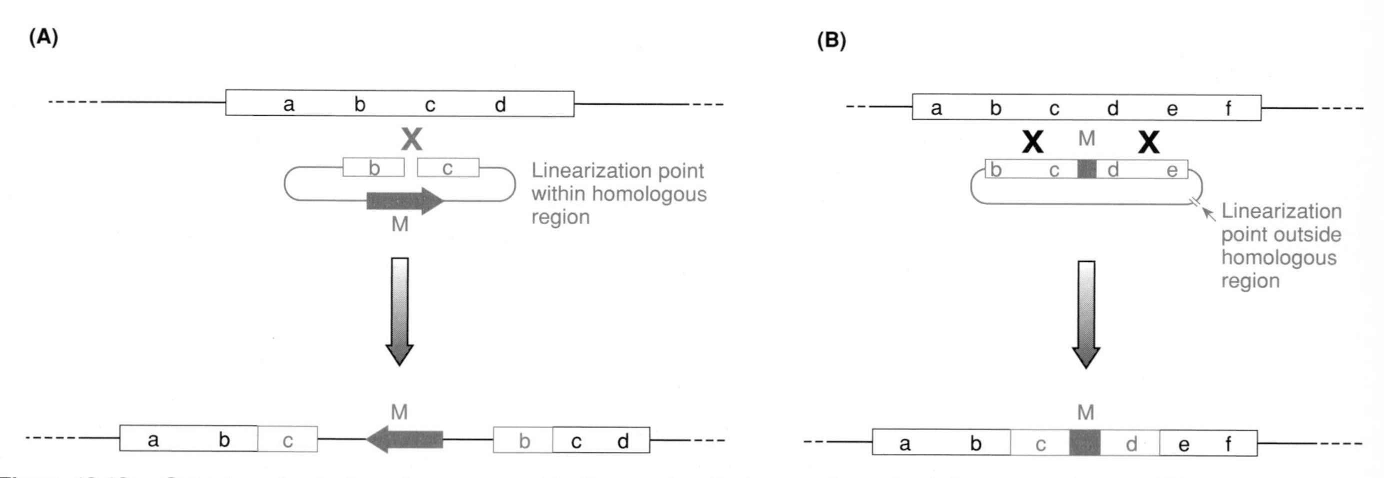

Figure 19.18: Gene targeting by homologous recombination can inactivate a predetermined chromosomal gene within an intact cell.

(A) Integration vector method. The introduced vector DNA (red) is cut at a unique site within a sequence which is identical or closely related to part of a chromosomal gene (black). Homologous recombination (X) can occur, leading to integration of the entire vector sequence including the marker gene (M). *Note* that the letters do not represent exons but are simply meant to indicate linear order within the gene.

(B) Replacement vector method. In this case, the marker gene is contained within the sequence homologous to the endogenous gene, and the vector is cut at a unique location outside the homologous sequence. A double recombination or gene conversion event (X X) can result in replacement of internal sequences within the chromosomal gene by homologous sequences from the vector, including the marker gene.

Gene targeting in mice is popularly used for producing artificial mouse models of human disease (see page 543). In addition, it provides a powerful general method of studying gene function: the gene in question is selectively inactivated to produce a **'knock-out' mouse**, and the effect of the mutation on the development of the mouse is monitored carefully. Sometimes there is little or no phenotypic consequence after inactivating a gene that would be expected to be crucially important, such as the gene for a transcription factor known to be expressed in early embryonic development. The lack of effect is thought to be due mostly to **genetic redundancy**: another gene is able to carry out the function of the gene that has been knocked out. For example, the mouse *Engrailed* genes, *En-1* and *En-2*, are developmental control genes which encode homeodomain-containing proteins that are considered to play crucial roles in brain formation. *En-1* knock-outs have serious abnormalities but *En-2* knock-outs have only minor problems. Expression of the *En-1* gene is switched on 8–10 h before that of the *En-2* product, suggesting that perhaps the *En-1* product can compensate for the lack of *En-2* product in *En-2* knock-outs. To test for this, the newly developed **'knock-in'** technique was used: an *En-2* gene clone was spliced into the DNA used to create a knock-out of the *En-1* gene. The result was that the endogenous *En-1* gene was inactivated but at the same time the *En-2* gene introduced into the middle of it was activated by the *En-1* regulatory sequences and expressed before the endogenous *En-2* gene was switched on. The resulting mouse was normal, confirming that the two genes are functionally equivalent (Hanks *et al.*, 1995).

Site-specific recombination systems, notably the Cre–*lox*P system, extend the power of gene targeting

Several site-specific recombination systems from bacteriophages and yeasts have been characterized and are promising tools for genome engineering (Kilby *et al.*, 1993). Thus far, the Cre–*lox*P recombination system from bacteriophage P1 has been the most widely used. The natural function of the Cre (causes recombination) recombinase is to mediate recombination between two *lox*P sequences that are in the same orientation, leading to excision of the intervening sequence between the two *lox*P sites (see *Figure 4.14*). Using gene targeting, *lox*P sequences can be stitched into a desired gene or chromosomal location, and the subsequent provision of a gene

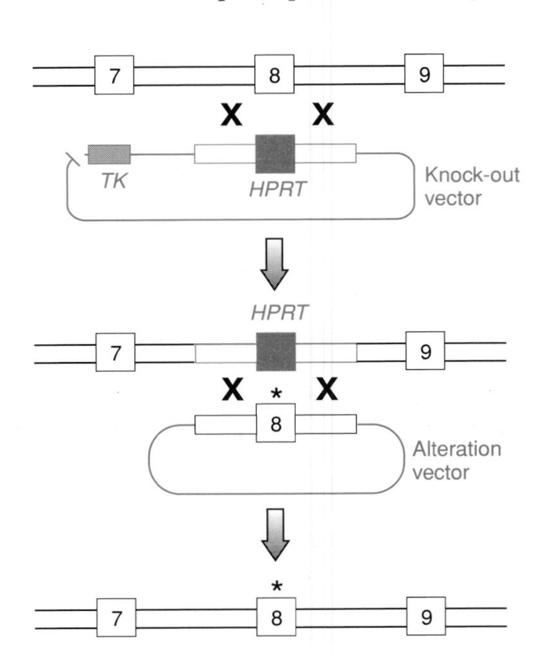

Figure 19.19: Double replacement gene targeting can be used to introduce subtle mutations.

Both the methods in *Figure 19.18* result in introduction of a substantial amount of exogenous sequence within the endogenous gene. To introduce a subtle mutation without leaving residual exogenous sequence, a double replacement method with positive and negative selection can be used (see Melton, 1994). Exons in the endogenous gene are represented as numbered large boxes, and introns as long thin boxes. In order to introduce a subtle mutation, such as a single nucleotide substitution in exon 8, a replacement knock-out vector is used with a marker gene (e.g. the *HPRT* gene) flanked by homologous sequences from introns 7 and 8, and a second marker, such as the herpes simplex thymidine kinase (*TK*) gene outside the homologous region. Gene conversion, or double crossover within the flanking intron sequences, can lead to replacement of exon 8 by the *HPRT* gene, and can be selected for if a mutant *HPRT*⁻ ES cell is used. A positive–negative selection system can be used: selection in the first step is for $HPRT^{+}$ TK^{-} cells. Cells containing random vector integrations will contain the *TK* gene and can be killed with the thymidine analog *gancyclovir* (see *Figure 20.15*). The second replacement involves introducing an altered exon 8 with a point mutation (*) to replace the *HPRT* gene and can be screened by identifying $HPRT^{-}$ cells. *Note* that mice engineered in this way cannot be described as transgenic because of the lack of foreign sequences in the germline.

encoding the Cre product can result in an artificially generated site-directed recombination event (see Chambers, 1994). Several applications can be envisaged.

Tissue- and cell type-specific knock-outs

Some genes are vital to early development and simple knock-out experiments are generally not helpful: death ensues at the early embryonic stage. To overcome this problem, methods have been developed to inactivate expression of the target gene in only selected, predetermined cells of the animal. The animal can therefore survive and the effect of the knock-out can be studied in a tissue or cell type of interest. An initial example has been the programmed knock-out of the DNA polymerase β gene in only the T cells of transgenic mice (Gu *et al.*, 1994). In this case, the strategy taken was to replace part of the endogenous gene by an introduced homologous segment flanked by *lox*P sequences. Mice carrying this targeted mutation were then mated with a strain of transgenic mice which carried a transgene, containing the *Cre* gene under the control of a promoter which is expressed specifically in T cells. Offspring with the *lox*P-flanked pol β sequences plus the Cre^{lck} transgene were identified and survived to adulthood: the Cre product was expressed only in T cells, leading to excision of the sequences of the DNA polymerase β gene between the two *lox*P sequences (see *Figure 19.20* for the general method).

Tissue- and cell type-specific gene activation

This approach is the opposite to that described above: it involves selective *activation* of a gene in certain cells of the animal to switch on a foreign gene only in predetermined cells of the animal (see Barinaga, 1994).

Site-directed chromosome rearrangements

Another important recent development is a strategy for chromosome engineering in ES cells which relies on sequential gene targeting and Cre–*lox*P recombination. Gene targeting is used to integrate *lox*P sites at the desired chromosomal locations and, subsequently, transient expression of Cre recombinase is used to mediate a selected chromosomal rearrangement (Smith *et al.*, 1995). This strategy offers the exciting possibility of creating novel mouse lines with specific chromosomal abnormalities for genetic studies.

Creating animal models of disease using transgenic technology and gene targeting

Animal models of human disease are important to medical research because they allow detailed examination of the physiological basis of disease and offer a front-line testing system for studying the efficacy of novel treatments (see Chapter 20) before conducting clinical trials on human subjects. Although many individual human disorders do not have a good animal model, animal models exist for some representatives of all the major human disease classes: genetically determined diseases, disease due to infectious agents, sporadic cancers and autoimmune disorders (see Leiter *et al.*, 1987; Darling and Abbot, 1992; Clarke, 1994). Some animal models of human disease originated spontaneously; others have been generated artificially by a variety of different routes (*Table 19.3*).

Until recently, the great majority of available animal disease models were ones which arose spontaneously. In addition, some disease models were obtained as a result of artificially induced exposure to mutagenic sources, including certain chemicals and X-ray irradiation. However, there are major difficulties with both reliance on spontaneous mutants and those induced by exposure to chemicals or X-rays.

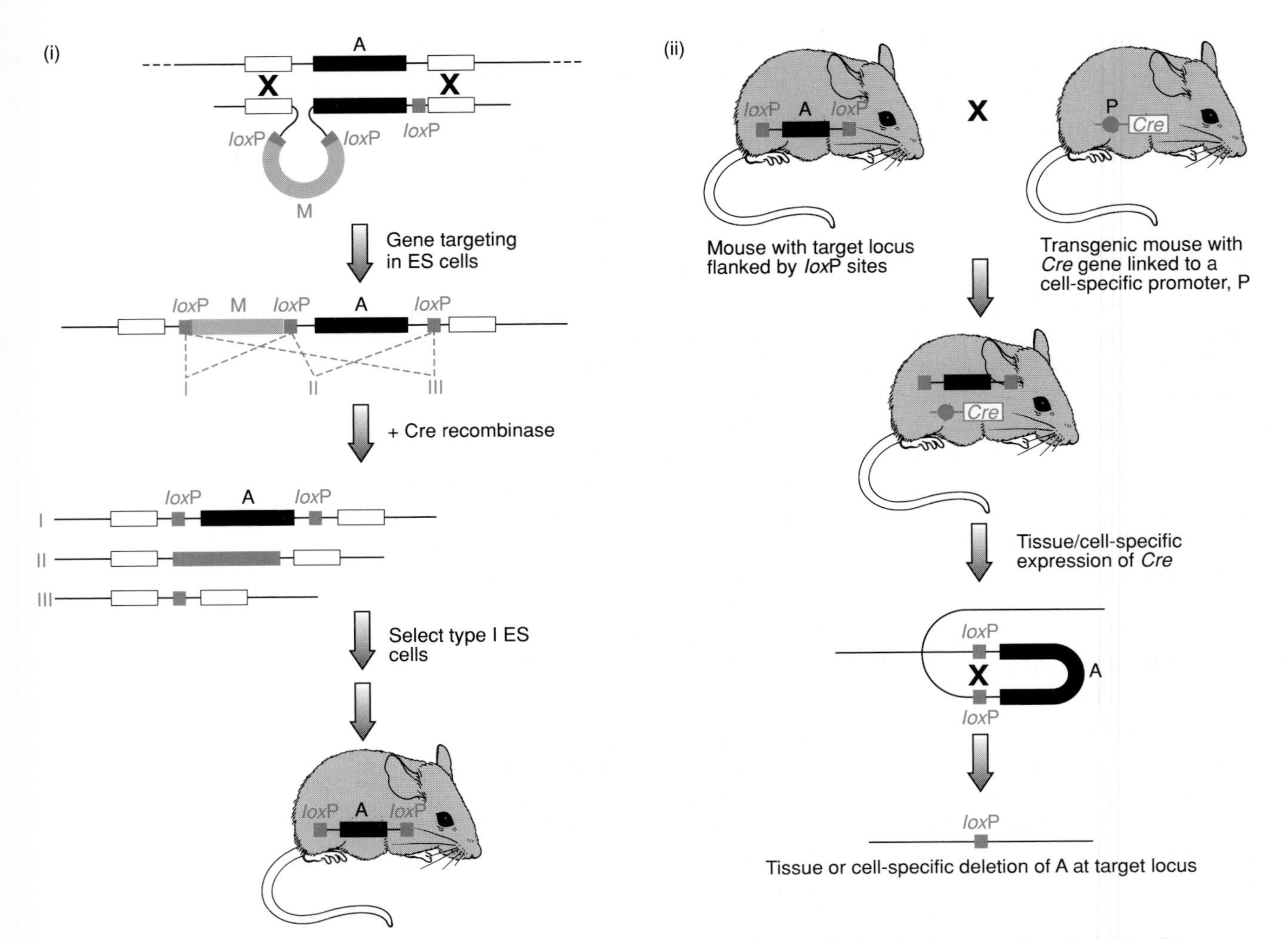

Figure 19.20: Gene targeting using the cre-*lox*P recombination system can be used to inactivate a gene in a desired cell type.

Illustration of a standard homologous recombination method using mouse ES cells, in which three *lox*P sites are introduced along with a marker M at a target locus A (typically a small gene or an internal exon which if deleted would cause a frameshift mutation). Subsequent transfection of a *Cre* recombinase gene and transient expression of this gene results in recombination between the introduced *lox*P sites to give different products. Type I recombinants are used to generate mice in which the target locus is flanked by *lox*P sites. Such mice can be mated with previously constructed transgenic mice which carry an integrated construct consisting of the *Cre* recombinase gene linked to a tissue-specific promoter. Offspring which contain both the *lox*P-flanked target locus plus the *Cre* gene will express the *Cre* gene in the desired tissue type, and the resulting recombination between the *lox*P sites in these cells results in tissue-specific inactivation of the target locus A.

Table 19.3: Classes of animal model of disease

Origin	Type	Comments
Spontaneous	Germline mutation → inherited disorder Somatic mutation → cancer	See *Table 19.4.*
Artificial intervention or artificially generated	Selective breeding to obtain strains that are genetically susceptible to disease	
	Infect animal strain with relevant microbial pathogen Manipulate environment to induce disease without causing mutation	
	In vivo mutagenesis using a strong mutagen such as X-rays or powerful chemical mutagens such as ethyl nitrosurea (ENU)	See page 541
	Genetic modification of ES cells or fertilized egg cells and subsequent animal breeding (*transgenic* and gene targeting technologies)	See above

Recently gene targeting and transgenic technologies have provided directed ways of obtaining animal models of disease, and targeted mutations in the mouse have been particularly valuable. Interestingly, it has become increasingly clear that disease phenotypes due to mutations in human and mouse gene homologs often show considerable differences.

Animal models of disease occurring spontaneously or as a result of exposure to chemical mutagens or radiation may be difficult to identify

Spontaneous animal disease models

Mutant human phenotypes, especially those associated with obvious disease symptoms, are subject to intense scrutiny: many individuals who suffer from a disorder seek medical advice. If they present with a previously undescribed phenotype their case may well be referred to experts who often will document the phenotype in the medical literature. Given the motivation of both affected individuals and their families, physicians and interested medical researchers, and the large population size for screening (current total global population is $\sim 6 \times 10^9$ individuals), there is a remarkably effective screening for mutant human phenotypes. In contrast, many animal disease phenotypes will go unrecorded. Only a small percentage of the animal population is in captivity, and the documentation of mutant animal phenotypes is largely dependent on identification of mutants within animal colonies bred for research purposes and, to a lesser extent, within livestock and pet populations. Only mutants with obvious external anomalies are likely to be noticed.

Despite the difficulty in identifying spontaneous animal mutants, a number of animal phenotypes have been described as likely models of human diseases (see *Table 19.4* for some examples). In some cases, the animal mutant phenotype closely parallels the corresponding clinical phenotype, but in others there is considerable divergence because of species differences in biochemical and developmental pathways (see Erickson, 1989). Additionally, phenotypic differences may result because of different classes of mutation at orthologous loci.

Chemical and irradiation mutagenesis

Classical methods of producing animal mutants have involved controlled exposure to mutagenic chemicals, notably ethyl nitrosurea and ethyl methylsulfonate (EMS) or to high doses of X-rays. Large numbers of *Drosophila* and mouse mutants have been obtained by this method and, very recently, efficient large-scale mutagenesis and saturation screening have also been conducted on zebrafish. A major problem with chemically induced and irradiation-induced mutations, however, is that they are generated essentially randomly. In order to identify a mutant phenotype of interest, a laborious screen for mutants needs to be conducted by close examination of the phenotypes following mutagenesis. The mutant phenotypes which have been described in these studies, as well as for spontaneous mutants, show a clear bias towards phenotypes with obvious external abnormalities, simply because of the ease of identifying them. Nevertheless, several important mouse models of human disease have been created using such methods. Zebrafish offer the advantage that the embryos are transparent and large-scale screens can identify many mutants with abnormalities of internal organs such as the heart (Mullins *et al.*, 1994).

Rodents, especially mice, have been used widely as animal models of human disease

Spontaneous and artificially produced disease phenotypes have been described in a wide range of animal species with differing potentials for modeling human disease. In some cases, the species may be too evolutionarily remote from humans to provide useful disease models. For example, numerous *Drosophila* mutants have been generated and studies of some developmental mutants have enabled the identification of human genes that are important in development, but they cannot serve as useful models of human disease. Other species, such as the zebrafish, offer some advantages, even although they are evolutionarily remote from us. Mammals would be expected to provide better disease models but, for a variety of reasons, our closest relatives, the great apes, have not been very useful in providing disease models. Instead, other mammals, notably mice, have been used widely to model human disease (see *Box 19.2*).

Box 19.2: The potential of animals for modeling human disease

Primates *should* provide the best animal models of human disease because they are so closely related to us (see pages 234–237). Humans and great apes show extensive developmental, anatomical, biochemical and physiological similarities. Primates are expensive to breed, however, and the population sizes in captivity are very small. Public sensitivity also plays a part: while some are opposed, in principle, to all animal experimentation, those that feel it is justified in the interests of medical research are generally more comfortable with experimentation on small laboratory animals such as mice and rats. Most importantly, primates are not well suited to experimentation: they are comparatively long lived and less fecund than rodents, and so breeding experiments are more difficult to organize. Given that novel therapeutic approaches are rapidly being developed for a range of human disorders (see Chapter 20), the long delay required to perform test experiments on primates has prompted the study of alternative models.

Mice have been the most widely used animal models of human disease. They are small and can be maintained in breeding colonies comparatively cheaply. They have a short life-span (~2–3 years), a short generation time (~3 months) and are prolific (an average female will produce four to eight litters with an average litter size of about six to eight pups). Because they can be bred easily, complex breeding programs can be arranged to produce *recombinant inbred strains* and *congenic* strains (see Taylor, 1989 and *Box 14.4*) and their short generation time and life-span means that the effects of transmitting a pathogenic mutation through several generations can be monitored relatively easily. As a result, the genetics of the laboratory mouse have been studied extensively for decades, and the phenotypes of many mutants have been recorded (Lyon and Searle, 1989). Most such mutants have originated spontaneously within breeding colonies. A few have also been produced artificially, initially by X-ray/chemical mutagenesis, but increasingly by gene targeting. Mapping of the mouse mutants is facilitated by *interspecific backcross mapping* (see Avner *et al.*, 1988, and *Box 14.4*) and by the availability of numerous polymorphic markers (>5000 dinucleotide repeat markers have been mapped). Because regions which show conservation of synteny between mouse and humans have been well documented (see *Figure 14.13*), this information is useful in identifying genuinely homologous single gene disorders in mouse and man.

Rats are comparatively large and have been more amenable to physiological, pharmacological and behavioral studies, especially in cardiovascular and neuropsychiatric studies. They have a longer generation time (11 weeks), and breeding colonies are more expensive. Some classes of human disorders (e.g. hypertension, behavioral disorders, etc.) have no good mouse models and instead have relied on rat models. A dense map of genetic markers is rapidly being constructed.

Zebrafish genetics has developed only extremely recently: a genetic linkage map and efficient mutagenesis and saturation screening procedures were first published in 1994. However, mutant screening is facilitated in zebrafish because the embryos are transparent, enabling identification of abnormalities in internal organs during embryonic development. Inevitably, because of the considerable evolutionary divergence between humans and zebrafish, the relevance of zebrafish mutants to human disease will be expected to be confined to certain developmental disorders.

Transgenic technology and gene targeting have permitted the artificial creation of specific predetermined animal models of human disease

In the case of animal disease models which are artificially induced by exposure to mutagenic chemicals or radiation, or which originate spontaneously, there is little or no artificial control over the resulting phenotype and frequently, the identification of an animal disease model is serendipitous.

The great advantage of transgenic/gene-targeted mouse models of disease is that *specific disease models can be constructed to order*: provided that the relevant gene clones are available, including mutant genes in some cases, mice can be generated with a desired alteration in a chosen target gene. All the major classes of disease, inherited disorders, cancers, infectious diseases and autoimmune disorders can be modeled in this way (see Smithies, 1993; Clarke, 1994). In most cases, the transgenic/gene targeting approaches have been used to model single gene disorders but, increasingly, attempts are being made to produce mouse models of complex genetic diseases, such as Alzheimer's disease, atherosclerosis and essential hypertension effects.

Table 19.4: Examples of spontaneous animal mutants

Animal mutant	Phenotypic features and molecular pathogenesis
NOD mouse	Diabetic, but without being obese. Mimics human insulin-dependent diabetes mellitus
mdx mouse	X-linked muscular dystrophy due to mutations in mouse dystrophin gene. The original *mdx* mutant has a nonsense mutation but phenotype is much milder than Duchenne muscular dystrophy (DMD)
Hemophiliac dog	Missense mutation in canine factor IX gene causes complete loss of function. Human homolog is hemophilia B
Watanabe heritable hyperlipidemic (WHHL) rabbit	Hyperlipidemic as a result of a deletion of four codons of the low density lipoprotein receptor gene (*LDLR*); human homolog is familial hypercholesterolemia
Atherosclerotic pigs	Marked hypercholesterolemia. Normal LDL receptor activity, but variant apolipoproteins, including apolipoprotein-B
Splotch mouse	Abnormal pigmentation; phenotypic overlap with Waardenburg syndrome suggested that it was an animal model for this disease, and confirmed by identification of mutations in homologous human and murine *PAX* genes
NF damselfish	Extensive neurofibromas suggest that it could be a homolog of human neurofibromatosis type I

Single gene disorders can be modeled conveniently by gene targeting in mice (loss-of-function mutations) or by integration of mutant genes (gain-of-function mutations)

Modeling loss-of-function mutations by gene targeting in mice

Many disease phenotypes, including those of essentially all recessively inherited disorders and many dominantly inherited disorders, are thought to result from loss of gene function. The simplest way of modeling the disease for single gene disorders of this type is to make a knock-out mouse. The first step is to introduce an inactivating mutation into the appropriate endogenous gene in mouse ES cells using gene targeting. Following injection of the genetically modified ES cells into the blastocyst of a foster mother, and continued development, *founder mice* are obtained with the targeted mutation in a sizeable proportion of their germ cells. These mice can be interbred and the offspring can be screened for the presence of the desired mutation, and for the presence of the wild-type allele using PCR assays of cells collected from tail bleeds.

The gene targeting event is intended to create a *null allele*, one in which there is complete absence of gene expression, but sometimes the result may be a **'leaky mutation'**, so that the mutant allele retains some gene expression. This may explain why gene targeting has produced mouse models of a disease with divergent phenotypes.

Table 19.5: Examples of transgenic or gene-targeted mouse models of human disease

Human disease or abnormal phenotype	Gene	Method of constructing model
Cystic fibrosis	*CFTR*	Insertional inactivation by gene targeting
β-Thalassemia	*HBB* (β-globin)	Insertional inactivation by gene targeting
Hypercholesterolemia and atherosclerosis	Apolipoprotein genes, e.g. *APOE*	Insertional inactivation by gene targeting
Gaucher's disease		Insertional inactivation by gene targeting
Fragile-X syndrome	*FMR1*	Insertional inactivation by gene targeting
Gerstmann–Sträussler–Scheinker (GSS) syndrome	Prion protein gene (*PRNP*)	Integration of mutant mouse prion protein gene with missense mutation
Spinocerebellar ataxia type 1 (SCA1)	*SCA1* (ataxin)	Integration of mutant human ataxin gene with expanded triplet repeat
Alzheimer's disease	*APP* (β-amyloid precursor protein)	Integration of mutant full-length *APP* cDNA under control of a platelet-derived growth factor promoter.

For example, some mouse models of cystic fibrosis obtained by gene targeting show considerably different phenotypes, such as the mild phenotype reported by Dorin *et al.* (1992) and the severe phenotype described by Snouwaert *et al.* (1992). The difference is known to be due to a leaky mutation in the mouse with mild phenotype. Differences in phenotype may also occur because of *modifier genes* using different mouse strains (see page 547).

Modeling gain-of-function mutations by insertion of a mutant gene

This general experimental design has been used frequently in conjunction with the pronuclear microinjection technique of gene transfer (*Figure 19.16*). The disease to be modeled must be one where the presence of an introduced DNA is itself sufficient to induce pathogenesis, and can include inherited gain-of-function mutations (page 412), oncogenes, etc. To model such disorders, it is necessary to isolate a mutant gene or, if necessary, design one by *in vitro* mutagenesis (see page 526), and transfer a construct containing this into fertilized oocytes. Because there is no requirement for the introduced mutant gene to integrate at a specific location, human mutant genes will suffice although, in some cases, mouse mutant genes have been used. Two examples illustrate this approach:

(i) an early example was intended to assess the significance of a mutation in the human prion protein gene: a leucine substitution at codon 102 was found in patients with Gerstmann–Sträussler–Scheinker (GSS) syndrome, a rare dominantly inherited neurodegenerative disease, but was apparently absent in normal individuals. To test the hypothesis that the mutation caused GSS, the mouse prion protein gene was modified so that the equivalent codon had a leucine substitution. The mutant mouse prion protein gene was then injected into fertilized oocytes to produce transgenic mice which went on to develop spontaneous neurodegeneration, confirming the original hypothesis (Hsiao *et al.*, 1990).

(ii) The transgenic approach has been used recently to test the gain-of-function hypothesis for neurodegenerative disorders arising from expanded triplet repeats, as in the case of spinocerebellar ataxia type 1 (SCA1). SCA1 is a dominantly inherited disorder characterized by degeneration of cerebellar Purkinje cells, spinocerebellar tracts and some brainstem neurons, and results from expansion of a CAG triplet repeat in the ataxin gene. Transgenic mice were produced with the normal human ataxin gene (*SCA1*), or with a mutant ataxin gene containing an expanded CAG repeat. Both types of transgene were stable in parent to offspring transmissions, but only those with the expanded allele developed ataxia and Purkinje cell degeneration, confirming the gain-of-function hypothesis (Burright *et al.*, 1995).

Considerable effort is being devoted currently to constructing mouse models of cancers and other complex genetic disorders

Modeling human cancers

- *Gain of function.* Disease due to inappropriate inactivation of a proto-oncogene can be modeled by constructing a transgenic mouse: the appropriate oncogene is introduced into the mouse genome by standard pronuclear microinjection techniques.
- *Loss of function.* Disease due to inactivation of tumor suppressor genes can be modeled by constructing knock-out mice through gene targeting. For example, several models have been generated by inactivating the mouse homologs of the *TP53* and *RB1* genes but the phenotypes show only broad similarity to the homologous human phenotypes, respectively Li–Fraumeni syndrome and retinoblastoma.

Modeling chromosomal disorders

Existing mouse models for human chromosomal disorders are sparse. In some cases this is due to insuffcent conservation of synteny between the two species. Taking the example of Down syndrome (human trisomy 21, human chromosome 21 shares a large region of genetic homology with mouse chromosome 16 (see *Figure 14.13*), but trisomy 16 (Ts16) mice are not good models because they die *in utero*. The Ts16 mouse could never be expected to be a good model of human trisomy 21: it is not trisomic for all human chromosome 21 genes (the genes in the distal 2–3 Mb of human chromosome 21 have orthologs on mouse chromosomes 17 and 10) and is trisomic for some genes which have human orthologs on chromosomes other than chromosome 21.

In order to produce a better mouse Down syndrome model, attention has focused on the Down syndrome *critical region* at 21q21.3–q22.12 (deduced from observing the phenotypes of rare Down syndrome patients with partial trisomy 21). Recently, a segmental trisomy 16 mouse has been produced by Reeves *et al.* (1995) which shows learning and behavior deficits. This was possible by standard methods of irradiating mice and screening for chromosomal abnormalities involving chromosome 16, notably translocations. However, the recent advances in transgenic and gene targeting technology offer potentially powerful approaches for the future.

- *Gene targeting using Cre–*lox*P.* As described above (page 359), this system offers tremendous potential for genome engineering and can be used to engineer chromosome translocations at defined positions on pre-selected chromosomes.
- *YAC transgenics.* YACs which span the sequences on mouse chromosome 16 corresponding to the critical Down syndrome region can be introduced into the germline of mice to construct transgenic animals with partial trisomy 16. Other human disorders involving dosage imbalance over large regions can be modeled in the same way. Possible examples include human trisomies 13 and 18 (assuming that the pathogenesis is particularly confined to comparatively small chromosome regions) and Charcot–Marie–Tooth type 1A, which is due to overexpression as a result of a 1.5-Mb duplication in the *PNP22* gene region (see *Figure 15.6*).

Modeling complex diseases

Increasingly, the focus in human genetics is moving towards understanding the pathogenesis of complex genetic diseases such as atherosclerosis, essential hypertension, diabetes, etc. Such disorders have a complex etiology with multiple genetic and environmental components. However, gene targeting approaches are expected in the future to provide badly needed models (see Smithies and Maeda, 1995). As long as suitably promising genes can be identified as being involved in the pathogenesis, then breeding experiments can be used to bring different combinations of disease genes together, and the effect of different genetic backgrounds in different strains of mice and of different environmental factors can be assessed. This approach may not be so daunting as it sounds because, increasingly, many complex disease phenotypes are considered to be due mostly to the combination of only a very few major susceptibility genes. Recently, for example, a digenic model of spina bifida occulta was generated serendipitously in offspring obtained by crossing a mouse heterozygous for the *Patch* mutation (*pdgfrb*; platelet-derived growth factor receptor) with a mouse homozyogus for the *undulated* mutation (*pax-1*) (see Helwig *et al.*, 1995).

Mouse models of human disease may be difficult to construct because of a variety of human/mouse differences

It is not uncommon for spontaneous or artificially generated mouse models of disease to show phenotypes that are considerably different from the homologous human disorders. For example, gene targeting to inactivate several mouse tumor suppressor genes has often produced disappointing mouse models, as in the case of *TP53* and *RB1* (retinoblastoma) knock-outs. There are several areas where differences between mice and humans could be expected to result in divergent disease phenotypes for mutations in orthologous genes (see Erickson, 1989).

- *Biochemical pathways.* Although biochemical pathways in mammals are generally well conserved, some differences are known between the pathways of humans and mice. The human retina appears to depend heavily on the accurate function of the *Rb* gene product, but other vertebrate retinas do not. As a result, spontaneous retinoblastoma mouse mutants have not been described, and retinoblastoma is not a feature of *rb-1* knock-out mice. Another useful example concerns mouse models of Lesch–Nyhan syndrome. This X-linked recessive disorder is due to an absence of HPRT, an enzyme involved in purine salvage, and is characterized by mental retardation and self-mutilation (*Table 15.5*). *HPRT*$^-$ mice produced by gene targeting did not, however, show evidence of Lesch–Nyhan characteristics. Mice are more tolerant of HPRT deficiency because they are more reliant than humans on an alternative purine salvage pathway, involving adenosine phosphoribosyltransferase (APRT). When an APRT inhibitor is administered to *HPRT*$^-$ mice, a more effective mouse model of Lesch–Nyhan is produced (Wu and Melton, 1993).
- *Developmental pathways.* The differences in human and mouse developmental pathways are not well understood but are expected to be significant.
- *Absolute time.* Because of the huge difference in the average lifespans of mice and humans, certain human disorders in which the disease is of late onset may possibly be difficult to model in mice.
- *Genetic background: the importance of modifier genes.* Most human populations are outbred. Laboratory strains of mice, however, are very inbred. Often a particular phenotype can vary considerably in different strains of mice because of differences in their alleles at other loci (**modifier genes**), which can interact with the locus of interest.

A useful example of the importance of genetic background is the *Min* (*multiple intestinal neoplasia*) mouse which was generated by ENU mutagenesis (see page 541) and results from mutations in the mouse *Apc* gene. Mutations in the orthologous human gene, *APC*, cause adenomatous polyposis coli and related colon cancers and the *Min* mouse has been regarded as a good model for such disorders. The phenotype of the *Min* mouse is, however, dramatically modified by the genetic background. For example the number of colonic polyps in mice carrying APC^{Min} is strikingly dependent on the strain of mouse. Similar phenotypic variability is found in human families where different members of the same family may have strikingly different tumor phenotypes although they possess identical mutations in the *APC* gene. Some of the variabilility could be due to environmental factors, but the involvement of modifier genes had been strongly suspected. The *Min* mouse provides a well-defined genetic system for mapping and identifying modifier genes (Dietrich *et al.*, 1993; MacPhee *et al.*, 1995).

Further reading

Brandon EP, Idzerda RL, McKnight GS. (1995) Targeting the mouse genome: a compendium of knock-outs (parts 1–3). *Curr. Biol.* **5**, 625–634; 758–765; and 873–881.

Farzaneh F, Cooper DN. (1995) (eds) *Functional Analysis of the Human Genome*. BIOS Scientific Publishers, Oxford.

First NL, Haseltine FP. (1991) (eds) *Transgenic Animals*. Butterworth-Heinemann, Boston, MA.

Grosveld F, Kollias G. (1992) (eds) *Transgenic Animals*. Academic Press, London.

Joyner AL. (1993) (ed.) *Gene Targeting: a Practical Approach*. IRL Press, Oxford.

TBASE: an electronic transgenic/targeted mutation database. World-Wide Web address. http://www.gdb.org/Dan/tbase. html

References

Aiello LP, Robinson GS, Lin Y-W, Nishio Y, King GL. (1994) *Proc. Natl Acad. Sci. USA*, **91**, 62316235.

Arbones ML, Austin HA, Capon DJ, Greenburg G. (1994) *Nature Genetics*, **6**, 90–97.

Askew GR, Doetschman T, Lingrel JB. (1993) *Mol. Cell Biol.*, **13**, 4115–4124.

Avner P, Amar L, Dandolo L, Guenet JL. (1988) *Trends Genet.*, **4**, 18–23.

Barinaga M. (1994) *Science*, **265**, 26–28.

Baxendale S, Abdulla S, Elgar G *et al.* (1995) *Nature Genetics*, **10**, 67–75.

Benkovic SJ. (1992) *Annu. Rev. Biochem.*, **61**, 29–54.

Burright EN, Clark HB, Servadio A *et al.* (1995) *Cell*, **82**, 937–948.

Camper SA, Saunders TL, Kendall SK *et al.* (1995) *Biol. Reprod.* **52**, 246–257.

Capecchi M. (1989) *Trends Genet.*, **5**, 7076.

Chambers CA. (1994) *BioEssays*, **16**, 865–868.

Clackson T, Wells JA. (1994) *Trends Biotechnol.*, **12**, 173–184.

Clarke AR. (1994) *Curr. Opin. Genet. Dev.*, **4**, 453–460.

Daniels R, Kinis T, Serhal P, Monk M. (1995) *Hum. Molec. Genet.*, **4**, 389–393.

Darling SM, Abbott CM. (1992) *BioEssays* **14**, 359–366.

Dietrich WF, Lander ES, Smith JS *et al.* (1993) *Cell*, **75**, 631–639.

Dorin JR, Dickinson P, Alton EW *et al.* (1992) *Nature*, **359**, 211–215.

d'Esposito M, Pilia G, Schlessinger D. (1994) *Hum. Mol. Genet.*, **3**, 735–740.

Erickson RP. (1989) *Trends Genet.*, **5**, 13.

Fersht A, Winter G. (1992 *Trends Biochem.*, **17**, 292–294.

Fields S, Sternglanz R. (1994) *Trends Genet.*, **10**, 286–292.

Figueirdo MS, Brownlee GG. (1995 *J. Biol. Chem.*, **270**, 11828–11838.

Frohman MA, Dush MK, Martin GR. (1988) *Proc. Natl Acad. Sci. USA*, **85**, 8998–9002.

Gavin BJ, McMahon AP. (1993) *Meth. Enzymol.*, **225**, 653–663.

Ginsburg M. (1994) In: *Guide to Human Genome Computing*. (ed. MJ Bishop), pp. 215–248 Academic Press, New York.

Gordon JW. (1992) *Methods. Enzymol.*, **225**, 747–771.

Grosveld F, van Assendelft GB, Greaves DR, Kolias G. (1987) *Cell*, **51**, 975–985.

Gu H, Marth JD, Orban PC, Mossmann H, Rajewsky K. (1994) *Science*, **265**, 103–107.

Hanks M, Wurst W, Anson-Cartwright L, Auerbach AB, Joyner AL. (1995) *Science*, **269**, 679–682.

Hermes JD, Blacklow SC, Knowles JR. (1990) *Proc. Natl Acad. Sci. USA*, **87**, 696–700

Helwig U, Imai K, Schmahl W, Thomas BE, Varnum DS, Nadeau JH, Balling R. (1995) *Nature Genetics*, **11**, 60–63

Hinds HL, Ashley CT, Sutcliffe JS *et al.* (1993) *Nature Genetics*, **3**, 36–43.

Hsiao KK, Scott M, Foster D, Groth DF, Dearmond SJ, Prusiner SB. (1990) *Science*, **250**, 1587–1590.

Jakobovits A, Moore AL, Green LL *et al.* (1993) *Nature*, **362**, 255–258.

Kilby NJ, Snaith MR, Murray JAH. (1993) *Trends Genet.*, **9**, 413–421.

Lamb BT, Gearhart JD. (1995) *Curr. Opin. Genet. Dev.*, **5**, 342–348.

Lee SW, Tomasetto C, Sager R. (1991) *Proc. Natl Acad. Sci. USA*, **88**, 2825–2829.

Leiter EH, Beamer WG, Shultz LD, Barker JE, Lane PW. (1987) *Birth Defects*, **23**, 221–257.

Liang P, Averboukh L, Pardee AB. (1993) *Nucleic Acids Res.*, **21**, 3269–3275.

Lyon MF, Searle AG. (1989) *Genetic Variants and Strains of the Laboratory Mouse*, 2nd Edn. Oxford University Press, Oxford.

MacPhee M, Chepenik KP, Liddell RA, Nelson KK, Siracusa LD, Buchberg AM (1995) *Cell*, **81**, 957–966.

Maxam AM, Gilbert W. (1980) *Methods Enzymol.*, **65**, 499–560.

Melton DW. (1994) *BioEssays* **16**, 633–638.

Mullins MC, Hammerschmidt M, Haffter P, Nusslein-Volhard C. (1994) *Curr. Biol.*, **4**, 189–202.

Nicholson LVB, Johnson MA, Bushby KMD *et al.* (1993) *J. Med. Genet.*, **30**, 737–744.

Old RW, Primrose SB. (1994) *Principles of Gene Manipulation. An Introduction to Genetic Engineering*, 5th Edn. Blackwell Scientific Publications, Oxford.

Parmsley SF, Smith GP. (1988) *Gene*, **73**, 305–318.

Phizicky E, Fields S. (1995) *Microbiol. Rev.*, **59**, 94–123.

Piccolo S, Bonaldo P, Vitale P, Volpin D, Bressan GM. (1995) *J. Biol. Chem.*, **270**, 19583–19590.

Pykett MJ, Murphy M, Harnish PR, George DL. (1994) *Hum. Mol. Genet.*, **3**, 559–564.

Ramirez-Solis R, Davis C, Bradley A. (1992) *Methods Enzymol.*, **225**, 855–877.

Reeves RH, Irving NG, Moran TH *et al.* (1995) *Nature Genetics*, **11**, 177–183.

Riordan JR, Rommens JM, Kerem B *et al.* (1989) *Science*, **245**, 1066–1073.

Roberts RG, Coffey AJ, Bobrow M, Bentley DR. (1993) *Genomics*, **16**, 536–538.

Rouleau G, Merel P, Lutchman M *et al.* (1993) *Nature*, **363**, 515–521.

Shockett P, Difillppantonio M, Hellman N, Schatz DG. (1995) *Proc. Natl Acad. Sci.* USA, **92**, 6522–6526.

Smith AJH, De Sousa MA, Kwabi-Addo B, Heppell-Parton A, Impey H, Rabbitts P. (1995) *Nature Genetics*, **9**, 376–385.

Smithies O. (1993) *Trends Genet.*, **9**, 112–116.

Smithies O, Maeda N. (1995) *Proc. Natl Acad. Sci. USA*, **92**, 5266–5272.

Snouwaert JN, Brigman KK, Latour AM *et al.* (1992) *Science*, **257**, 1083–1088.

Stemmer WPC. (1994) *Nature*, **370**, 389–391.

Stryer L. (1995) *Biochemistry*, 4th Edn. W.H. Freeman & Co., New York.

Taylor BA. (1989) In: *Genetic Variants and Strains of the Laboratory Mouse*, 2nd Edn (eds MF Lyon, AG Searle), pp. 773–796. Oxford University Press, Oxford.

Theuring F. (1995) In: *Functional Analysis of the Human Genome* (eds F Farzaneh, DN Cooper), pp. 185–205. BIOS Scientific Publishers, Oxford.

Winter G, Griffiths AD, Hawkins RE, Hoogenboom HR. (1994) *Annu. Rev. Immunol.*, **12**, 433–455.

Wu C-L, Melton DW. (1993) *Nature Genetics*, **3**, 235–240.

Worley KC, Wiese BA, Smith RF. (1995) *Genome Res.*, **5**, 173–184.

Gene therapy and other molecular genetic-based therapeutic approaches

Chapter 20

Principles of molecular genetic-based approaches to the treatment of disease

Once a human disease gene has been characterized, molecular genetic tools can be used to dissect gene function and explore the biological processes involved in the normal and pathogenic states. The resulting information can be used to design novel therapies using conventional drug-based approaches. In addition, molecular genetic technologies have recently provided a variety of novel therapeutic approaches that depend on: the ability to clone individual types of gene, transfer them into recipient cells and express them; the ability to re-design proteins; and the ability to inhibit specifically the expression of a predetermined (characterized) gene *in vivo*. Novel molecular genetic-based therapeutic approaches can be categorized into two broad groups, depending on whether the therapeutic agent is a gene product/vaccine or genetic material:

Recombinant pharmaceuticals and vaccines can be provided by gentic engineering. Possible approaches include:

expression cloning of normal gene products – cloning of (usually) human genes and expression in suitable expression systems to make large amounts of a medically valuable gene product; see pages 104–105 and 509.

Genetically engineered antibodies – antibody genes can be manipulated to form novel antibodies, including partially or fully *humanized antibodies*, for use as therapeutic agents; see pages 553–555.

Genetically engineered vaccines – novel cancer vaccines and vaccines against infectious agents.

Gene therapy is the genetic modification of the cells of a patient in order to combat disease. This very broad definition includes many different possible approaches, and can involve transfer of cloned human genes, double-stranded human gene segments, genes from other genomes, oligonucleotides and various artificial genes, such as antisense genes. Such gene transfer can result in genetic modification of the cells of the patient. In most cases, the gene therapy is designed to lead to genetic modification of disease cells, but some approaches are d*eliberately designed to target nondisease cells,* notably immune system cells, and constitute a form of vaccination.

Cloned human genes can be used as a source of medically important products

Once a human gene has been cloned, large amounts of the purified product can be obtained by using a suitable system for *expression cloning* (see page 104). This often involves expressing the desired gene in bacterial cells, which have the advantage that they can be cultured easily in large volumes, and large amounts of product can be obtained. Even if long-standing alternative treatments are available, the gene cloning–expression approach may be preferred because it minimizes safety risks. For example, diabetes sufferers traditionally have been treated with insulin prepared from cows or pigs. Because of species differences in the amino acid sequence of the product, however, animal products are potentially immunogenic and may produce unwanted side effects in highly immunoreactive individuals.

The administration of biochemically purified human products may also be hazardous. Recently, many hemophiliacs have developed acquired immune deficiency syndrome (AIDS) as a result of treatment with factor VIII which was purified from the serum of unscreened human donors. Growth hormone deficiency was treated in the past by purified human growth hormone. However, some patients developed Creutzfeldt–Jakob disease (a rare neurological disorder which is a human counterpart of scrapie in sheep and of 'mad cow disease') because the hormone was extracted from large numbers of pooled cadaver pituitaries.

Cloning of a medically important human gene and expression in suitable expression–cloning systems is therefore attractive. If biochemical purification of the product from a human or animal source is difficult or impossible, it provides the only product-based therapeutic route. Even if biochemical purification of the desired product from animals or humans is possible, large quantities of the human product can be produced by expression–cloning without the attendant hazards of the type described above. Recombinant human insulin was first marketed in 1982 and, subsequently, a number of other cloned human gene products of medical interest have been produced commercially (see *Table 20.1*). Treatment with the products of cloned genes may also pose risks, however. For example, patients who completely lack a normal product may mount a vigorous immune response to the administered pharmaceutical product as in the case of some patients with severe hemophilia A who have been treated with recombinant factor VIII.

Expression cloning often involves the use of microorganisms, but this approach may not always be suitable. For example, expression of a human gene in a bacterial cell can give a product that shows differences from the normal human product: the

Table 20.1: Examples of pharmaceutical products obtained by expression cloning

Product	For treatment of
Blood clotting factor VIII	Hemophilia A
Blood clotting factor IX	Hemophilia B
Erythropoietin	Anemia
Insulin	Diabetes
Growth hormone	Growth hormone deficiency
Tissue plasminogen activator	Thrombotic disorders
Hepatitis B vaccine	Hepatitis B
α-Interferon	Hairy cell leukemia; chronic hepatitis
β-Interferon	Multiple sclerosis
γ-Interferon	Infections in patients with chronic granulomatous disease
Interleukin-2	Renal cell carcinoma
Granulocyte colony-stimulating factor (G-CSF)	Neutropenia following chemotherapy
DNase (deoxyribonuclease)	Cystic fibrosis

polypeptide may have the same sequence of amino acids but patterns of glycosylation, etc., produced by post-translational processing may be different. This may mean that the gene product is not particularly stable in a human environment, or it may provoke an immune response, or its biological function may be less effective than desired. Alternative expression systems have been utilized, and increasing attention has been paid to constructing **transgenic livestock** whose post-translational processing systems are more similar to analogous human systems. For example, a cloned human gene can be fused to a sheep gene specifying a milk protein then inserted into the genome of the sheep germline. The resultant transgenic sheep can secrete large quantities of the fusion protein in its milk. The design of the fusion gene normally permits the ability to cleave the secreted fusion protein using a specific protease in order to generate the human protein residue which can be purified easily (see Moffat, 1991).

Antibody engineering has permitted the construction of humanized antibodies and fully human antibodies

Antibodies are natural therapeutic agents which are produced by B lymphocytes. In each B-cell precursor, a cell-specific rearrangement of antibody gene components occurs (see page 177ff.). Additional diversity is provided by other mechanisms, including frequent somatic mutation events. As a result, each individual has a population of B cells which *collectively* ensures a huge repertoire of different antibodies as a defense system against a diverse array of foreign antigens. The antibody may be thought of as an *adaptor molecule*: it contains binding sites for foreign antigen at the variable (V) end, and for effector molecules at the constant (C) end. Binding of an antibody may by itself be sufficient to neutralize some toxins and viruses, but it is more common for the antibody to trigger the complement system and cell-mediated killing.

Artificially produced **therapeutic antibodies** are designed to be *monospecific* (i.e. they will recognize a single type of antigenic site) and can be employed specifically to recognize particular disease-associated antigens, leading to killing of the disease cells. Notable targets for such therapy are cancers, especially lymphomas and leukemias, infectious disease (using antibodies raised against antigens of the relevant pathogen) and autoimmune disorders (antibody recognition of inappropriately expressed host cell antigens). A favorite way of producing immortal monospecific antibodies is to fuse individual antibody-producing B lymphocytes from an immunized mouse or rat with cells derived from an immortal mouse B-lymphocyte tumor. From the fusion products, a heterogenous mixture of hybrid cells, those hybrids that have both the ability to make a particular antibody and the ability to multiply indefinitely in culture, are selected. Such **hybridomas** are propagated as individual clones, each of which can provide a permanent and stable source of a single type of **monoclonal antibody**.

Until recently, the therapeutic antibody approach was not straightforward. It has proved very difficult to make human monoclonal antibodies (mAbs) using hybridoma technology. Although rodent mAbs can be created against human pathogens and cells, they have limited clinical utility: the rodent mAbs have a short half-life in human serum; only some of the different classes can trigger human effector functions; and they can elicit an unwanted immune response in patients (human anti-mouse antibodies). Once immunoglobulin genes had been cloned, however, the possibility of designing artificial combinations of immunoglobulin gene segments arose (**antibody engineering**). The process was assisted by the use of different exons to encode different domains of an antibody molecule: domain swapping could be done easily at the DNA level by artificial exon shuffling between different antibody genes. The resulting recombinant antibody genes could then be expressed to give *chimeric antibodies* (see Winter and Harris, 1993).

Humanized antibodies

One immediate goal of antibody engineering was the production of **humanized antibodies**, that is rodent–human recombinant antibodies (see Winter and Harris, 1993). Humanizing of rodent antibodies allows, in principle, access to a large pool of well-characterized rodent mAbs for therapy, including those with specificities against human antigens that are difficult to elicit from a human immune response. Early versions contained the variable domains of a rodent antibody attached to the constant domains of a human antibody: the immunogenicity of the rodent mAb is reduced, while allowing the effector functions to be selected for the therapeutic application. A further stage of humanizing antibodies is possible. The essential antigen-binding site is a subset of the variable region characterized by hypervariable sequences, the complementarity-determining regions (CDRs). Accordingly, second generation humanized antibodies were CDR-grafted antibodies: the hypervariable antigen-binding loops of the rodent antibody were built into a human antibody. Chimeric V/C antibodies and CDR-engrafted antibodies have been constructed against a wide range of microbial pathogens and against human cell surface markers, including tumor cell antigens and, in some cases, their use has already been demonstrated in the clinic (see *Table 20.2*).

Fully human antibodies

Two approaches have been taken towards the construction of fully human antibodies. One approach, *phage display technology* bypasses hybridoma technology, and even immunization. Instead, antibodies are made *in vitro* by mimicking the selection strategies of the immune system (see page 528), a procedure which should facilitate the construction of human antibodies of therapeutic value, as well as research potential. A second recent approach has involved using *transgenic mice* (Green *et al.*, 1994). This strategy involves transferring yeast artificial chromosomes containing large segments of the human heavy and kappa light chain immunoglobulin loci into mouse embryonic stem cells, and subsequent production of transgenic mice (see page 531 for the principle of constructing transgenic mice). The resulting mice produce a diverse repertoire of human heavy and light chain immunoglobulins and, upon immunization with tetanus toxin, can be used to derive antigen-specific fully human monoclonal antibodies. By breeding the mice to mice that are engineered by

Table 20.2: Examples of the clinical potential of humanized antibodies

Target	Clinical potential
CDw52	Lymphomas, systemic vasculitis, rheumatoid arthritis
CD3	Organ transplantation
CD4	Organ transplants, rheumatoid arthritis, Crohn's disease
IL-2 receptor	Leukemias and lymphomas, organ transplants, graft-versus-host disease
TNF-α	Septic shock
HIV	AIDS
RSV	Respiratory syncytial virus infection
HSV	Neonatal, ocular and genital herpes infection
Lewis-Y	Cancer
p185^{HER2}	Cancer
PLAP	Cancer
CEA	Cancer

TNF, tumor necrosis factor; HIV, human immunodeficiency virus; RSV, Rous sarcoma virus; HSV, herpes simplex virus; p185^{HER2}, human epidermal growth factor receptor 2; PLAP, placental alkaline phosphatase; CEA, carcinoembryonic antigen. Derived from Winter and Harris (1993).

gene targeting (see page 536) to be deficient in mouse immunoglobulin production, a mouse strain was obtained in which high levels of antibodies were produced, mostly with both human heavy and light chains. Such strains should permit the development of fully human monoclonal antibodies with therapeutic potential.

Genetically engineered vaccines have great therapeutic promise

Recombinant DNA technology is also being applied to the construction of novel vaccines. Several different strategies are being used:

- *direct DNA injection*. Direct injection of a piece of influenza virus DNA (which was conserved between several different strains) into muscle cells in mice has resulted in a potent antibody response against influenza. If this approach works as effectively in humans, the therapeutic potential is very considerable.
- *Genetic modification of antigen*. This can be achieved, for example, by fusion with a cytokine gene to increase antigenicity.
- *Genetic modification of microorganisms*. This can involve two approaches:
  - (i) genetically disabling an organism (e.g. by removing genes required for pathogenesis or survival). This is a genetic method of *attenuation* so that a live vaccine can be used without undue risk.
  - (ii) Inserting an exogenous gene that will be expressed in bacteria or parasites.

One promising application is the use of genetically modified *bacille Calmette-Guérin* (BCG) as a vehicle for immunization. BCG is a live attenuated tubercle bacillus which is the most widely used vaccine in the world: it is inherently immunostimulatory, and has a very low incidence of serious complications. Recombinant BCG strains have been developed using expression vectors containing BCG regulatory sequences coupled to genes encoding foreign antigens. Such strains can elicit long-lasting antibody and cell-mediated immune responses to foreign antigens in mice, and hold the promise of allowing simultaneous expression of multiple protective antigens of different pathogens (Stover *et al.*, 1991). Recombinant organisms of this type could be used to carry genes not only for infectious agents, but also for tumors, providing *cancer vaccines*, and, theoretically, autoantigens. Note that some gene therapy approaches, such as *adoptive immunotherapy*, are effectively forms of genetically engineered vaccination (see page 580).

The different strategies for gene therapy

The term *gene therapy* is a broad one: it encompasses many different strategies, all of which are designed to overcome or alleviate disease by a procedure in which genes, gene segments or oligonucleotides are introduced into the cells of an affected individual. The genetic material may be transferred directly into cells within a patient (***in vivo* gene therapy**), or cells may be removed from the patient and the genetic material inserted into them *in vitro*, prior to replacing the cells in the patient (***ex vivo* gene therapy**). Because the molecular basis of diseases can vary widely some gene therapy strategies are particularly suited to certain types of disorder, and some to others. Major disease classes include:

- **infectious diseases** (as a result of infection by a virus or bacterial pathogen);
- **cancers** (inappropriate continuation of cell division and cell proliferation as a result of activation of an oncogene or inactivation of a tumor suppressor gene or an apoptosis gene – see Chapter 17);

- **inherited disorders** (genetic deficiency of an individual gene product or genetically determined inappropriate expression of a gene);
- **immune system disorders** (includes allergies, inflammations and autoimmune diseases – the inappropriate destruction of body cells by immune system cells).

A major rationale for the development of gene therapy has been the need to treat diseases for which there is no effective treatment. Gene therapy has the potential to treat all of the above classes of disorder. Depending on the basis of pathogenesis, different gene therapy strategies can be considered (*Box 20.1*). Current gene therapy is exclusively somatic gene therapy, the introduction of genes into somatic cells of an affected individual. The prospect of human germline gene therapy raises a number of ethical concerns, and is currently not sanctioned (see page 585).

Box 20.1: General gene therapy strategies (see also *Figure 20.1*)

Gene augmentation therapy (GAT)

For diseases caused by loss of function of a gene, introducing extra copies of the normal gene may increase the amount of normal gene product to a level where the normal phenotype is restored (see *Figure 20.1*). As a result GAT is targeted at clinical disorders where the *pathogenesis is reversible*. It also helps to have no precise requirement for expression levels of the introduced gene and a clinical response at low expression levels. GAT has been particularly applied to autosomal recessive disorders where even modest expression levels of an introduced gene may make a substantial difference. Dominantly inherited disorders are much less amenable to treatment: gain-of-function mutations are not treatable by this approach and, even if there is a loss-of-function mutation, high expression efficiency of the introduced gene is required: individuals with 50% of normal gene product are normally affected, and so the challenge is to increase the amount of gene product towards normal levels.

Targeted killing of specific cells

This general approach is popular in cancer gene therapies. Genes are directed to the target cells and then expressed so as to cause cell killing.

Direct cell killing is possible if the inserted genes are expressed to produce a lethal toxin (**suicide genes**), or a gene encoding a **prodrug** is inserted, conferring susceptibility to killing by a subsequently administered drug.

Indirect cell killing uses immunostimulatory genes to provoke or enhance an immune response against the target cell (see page 580ff.).

Targeted mutation correction

If an inherited mutation produces a *dominant-negative effect* (see page 403), gene augmentation is unlikely to help. Instead the resident mutation must be corrected. Because of practical difficulties, this approach has yet to be applied but, in principle, it can be done at different levels: at the gene level (e.g. by *gene targeting* methods based on homologous recombination – see page 536); or at the RNA transcript level (e.g. by using particular types of therapeutic *ribozymes* – or therapeutic RNA editing, see pages 572–573).

Targeted inhibition of gene expression

If disease cells display a novel gene product or inappropriate expression of a gene (as in the case of many cancers, infectious diseases, etc.), a variety of different systems can be used specifically to block the expression of a single gene at the DNA, RNA or protein levels (see page 565ff.). Allele-specific inhibition of expression may be possible in some cases, permitting therapies for some disorders resulting from dominant negative effects.

The technology of classical gene therapy

We make a distinction between gene therapy approaches which simply rely on gene transfer and expression (which we somewhat arbitrarily describe as *classical gene therapy*; see Mulligan, 1993) and other approaches (targeted inhibition of gene expression *in vivo* and targeted gene mutation correction *in vivo*) which are considered separately beginning on page 565.

Genes can be inserted into the cells of patients by direct and indirect routes, and the inserted genes can integrate into the chromosomes or remain extrachromosomal

An essential component of classical gene therapy is that cloned genes have to be introduced and expressed in the cells of a patient in order to overcome the disease. Practically, this usually involves targeting the cells of diseased tissues. However, in some cases, unaffected tissues are deliberately targeted. For example, it is sometimes advantageous to target genes to healthy immune system cells in order to enhance immune responses to certain cancer cells and infectious agents. Additionally, genes may be targeted initially to one type of tissue while the gene products may be delivered to a remote location. For example, the myonuclei in muscle fibers have the advantage of being very long lived. Genetically engineered myoblasts therefore have the potential to ameliorate some nonmuscle diseases through long-term expression of exogenous genes which encode a product secreted into the blood stream (see for example, Jiao *et al.*, 1993).

Two major general approaches are used in the transfer of genes for gene therapy:

(i) ***ex vivo*** **gene transfer**. This initially involves transfer of cloned genes into cells grown in culture. Those cells which have been transformed successfully are selected, expanded by cell culture *in vitro*, then introduced into the patient. To avoid immune system rejection of the introduced cells, *autologous cells* are normally used: the cells are collected initially from the patient to be treated and grown in culture before being re-introduced into the same individual (see *Figure 20.2*). Clearly, this approach is only applicable to tissues that can be removed from the body, altered genetically and returned to the patient where they will engraft and survive for a long period of time (e.g. cells of the hematopoietic system, skin cells, etc.).

(ii) ***In vivo*** **gene transfer**. Here the cloned genes are transferred directly into the tissues of the patient. This may be the only possible option in tissues where individual cells cannot be cultured *in vitro* in sufficient numbers (e.g. brain cells) and/or where cultured cells cannot be re-implanted efficiently in patients. Liposomes and certain viral vectors are increasingly being employed for this purpose. In the latter case, it is often convenient to implant **vector-producing cells (VPCs)**, cultured cells which have been infected by the recombinant retrovirus *in vitro:* in this case the VPCs transfer the gene to surrounding disease cells. As there is no way of selecting and amplifying cells that have taken up and expressed the foreign gene, the success of this approach is crucially dependent on the general efficiency of gene transfer and expression.

Principles of gene transfer

Classical gene therapies normally require efficient transfer of cloned genes into disease cells so that the introduced genes are expressed at suitably high levels. In principle, there are numerous different physico-chemical and biological methods that can be used to transfer exogenous genes into human cells. The size of DNA

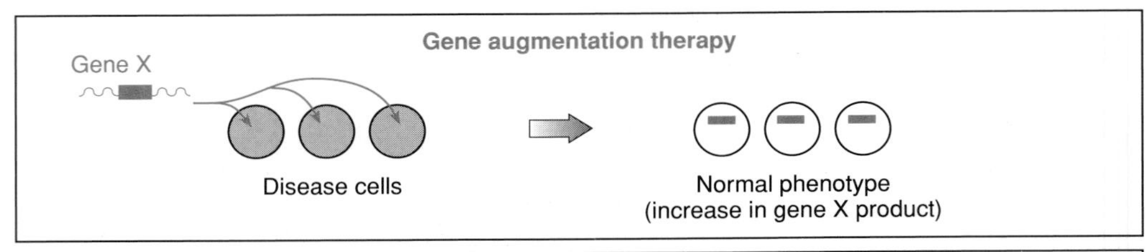
Gene augmentation therapy
Gene X
Disease cells
Normal phenotype
(increase in gene X product)

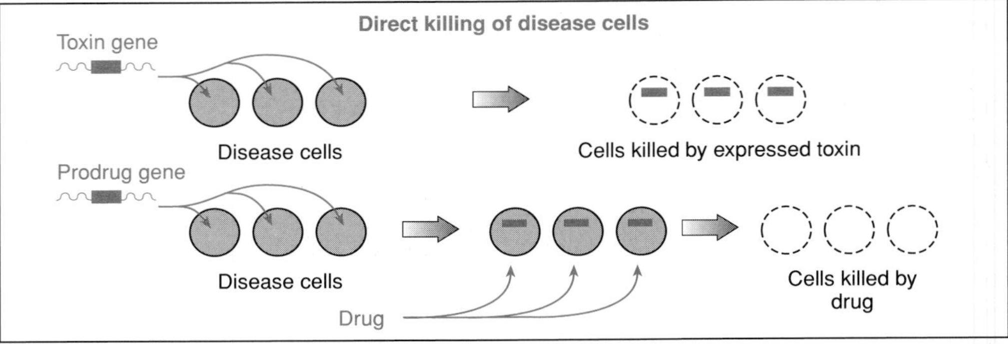
Direct killing of disease cells
Toxin gene
Disease cells
Cells killed by expressed toxin
Prodrug gene
Disease cells
Drug
Cells killed by
drug

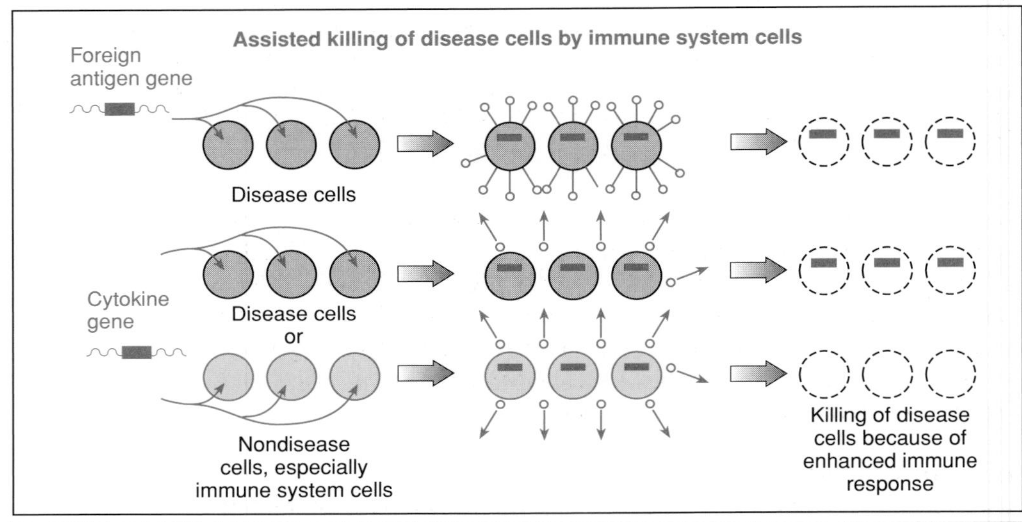
Assisted killing of disease cells by immune system cells
Foreign
antigen gene
Disease cells
Cytokine
gene
Disease cells
or
Nondisease
cells, especially
immune system cells
Killing of disease
cells because of
enhanced immune
response

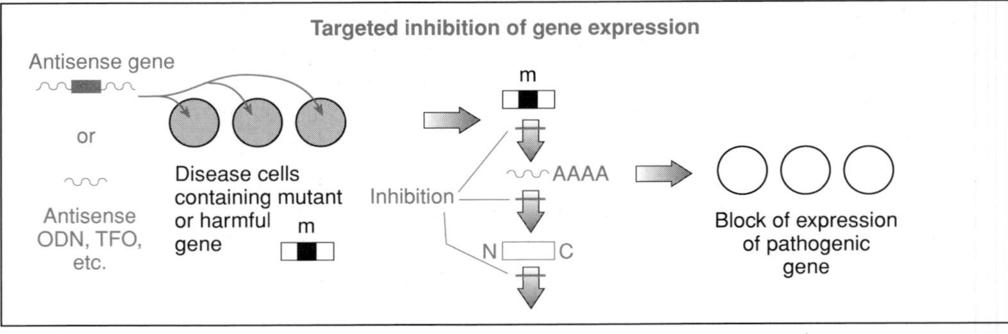
Targeted inhibition of gene expression
Antisense gene
or
Antisense
ODN, TFO,
etc.
Disease cells
containing mutant
or harmful
gene
m
Inhibition
AAAA
N
C
Block of expression
of pathogenic
gene

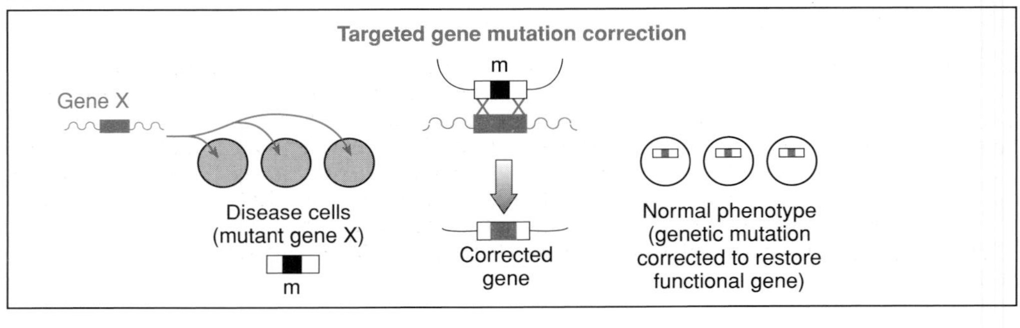
Targeted gene mutation correction
Gene X
Disease cells
(mutant gene X)
m
Corrected
gene
Normal phenotype
(genetic mutation
corrected to restore
functional gene)

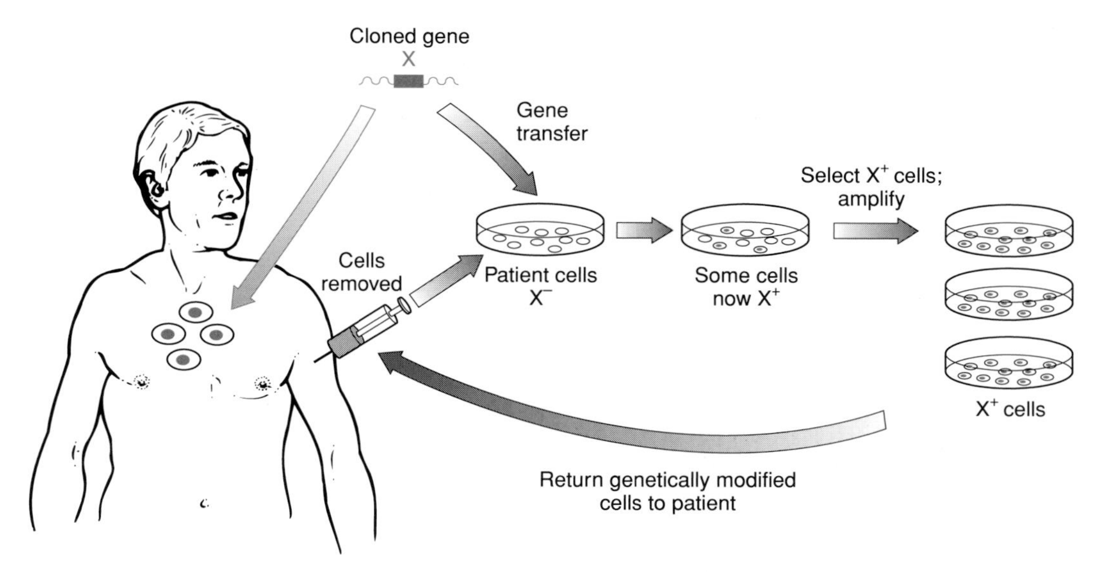

Figure 20.2: *In vivo* and *ex vivo* gene therapy.

In vivo gene therapy (red arrow) entails the genetic modification of the cells of a patient *in situ*. *Ex vivo* gene therapy (black arrows) means that cells are modified outside the body before being implanted into the patient. The figure shows the usual situation where *autologous cells* are used, i.e. cells are removed from the patient, cultured *in vitro*, before being returned to the patient. Occasionally, however, the cells that are implanted do not belong to the patient but are *allogeneic* (from another human source) in which case HLA matching is routinely required to avoid immune rejection.

fragments that can be transferred is in most cases comparatively very limited, and so often the transferred gene is not a conventional gene. Instead, an artificially designed **minigene** may be used: a cDNA sequence containing the complete coding DNA sequence is engineered to be flanked by appropriate regulatory sequences for ensuring high level expression, such as a powerful viral promoter. Following gene transfer, the inserted genes may integrate into the chromosomes of the cell, or remain as extrachromosomal genetic elements (*episomes*).

Genes integrated into chromosomes

The advantage of integrating into a chromosome is that the gene can be perpetuated by chromosomal replication following cell division (*Figure 20.3*). As progeny cells also contain the introduced genes, long-term stable expression may be obtained. As a result, gene therapy using this approach may provide the possibility of a cure for some disorders. For example, in tissues composed of actively dividing cells, the key is to target the **stem cells** (a minority population of undifferentiated precursor cells which gives rise to the mature differentiated cells of the tissue). This is so because stem cells not only give rise to the mature tissue cells, but during this procedure they also renew themselves. As a result, they are an immortal population of cells from which all other cells of the tissue are derived. High efficiency gene transfer into

Figure 20.1: Five approaches to gene therapy.

Of the five illustrated approaches, four have been used in clinical trials. Gene augmentation therapy by simple addition of functional alleles has been used to treat several inherited disorders caused by genetic deficiency of a gene product. Artificial cell killing and immune system-assisted cell killing have been popular in the treatment of cancers. The former has involved transfer to cells of genes encoding toxic compounds (*suicide genes*), or *prodrugs* (reagents which confer sensitivity to subsequent treatment with a drug). Targeted inhibition of gene expression is particularly suitable for treating infectious diseases and some cancers. Targeted gene mutation correction, the repair of a genetic defect to restore a functional allele, is the exception: technical difficulties have meant that it is not sufficiently reliable to warrant clinical trials. The example shows correction of a mutation in a mutant gene by *homologous recombination*, but mutation correction may also be possible at the RNA level (see *Figure 20.11*). ODN, oligodeoxynucleotide; TFO, triplex-forming oligonucleotide.

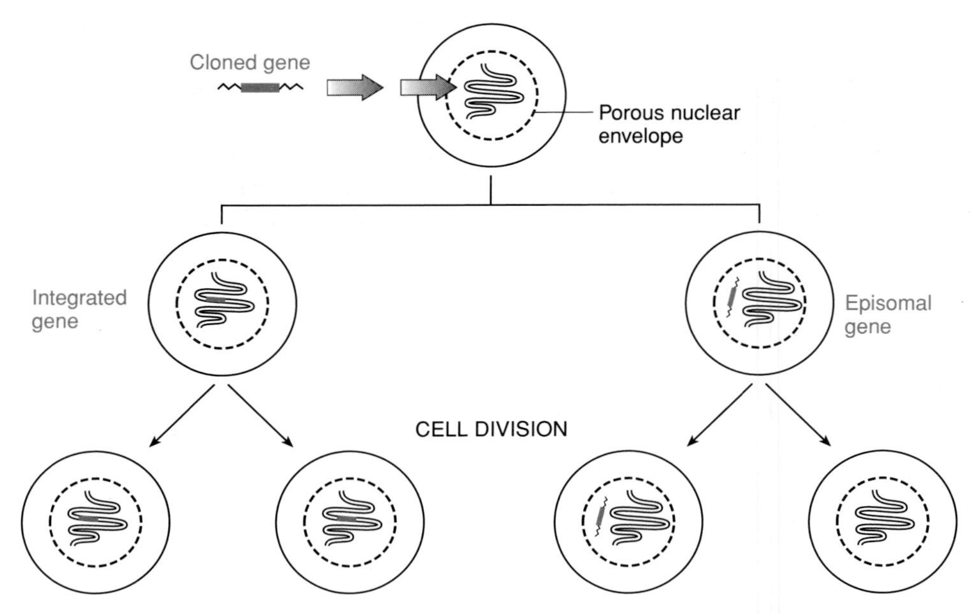

Figure 20.3: Exogenous genes that integrate into chromosomes can be stably transmitted to all daughter cells, unlike episomal (extrachromosomal) genes.

The figure illustrates two possible fates of genes that have been transferred into nucleated cells. If the cells are actively dividing, any genes which integrate stably into chromosomal DNA can be replicated under the control of the parent chromosome (during the S phase of the cell cycle). Following each cell division, an integrated gene will be stably inherited by both daughter cells. As a result, all cells that descend from a single cell in which stable integration took place, will contain the integrated gene. Gene therapy involving chromosomal integration of exogenous genes offers the possibility of continued stable expression of the inserted gene and a permanent cure, but carries certain risks, notably the possibility that one of the integration events may result in cancer (see text). By contrast, episomal genes which do not integrate but replicate extrachromosomally (under the control of a vector origin of replication) may not segregate to all daughter cells during subsequent mitoses. As a result, this type of approach has been particularly applied in gene therapies where the target tissue consists of nondividing cells (see text).

stem cells, and subsequent stable high level expression of a suitable introduced gene, can therefore provide the possibility of curing a genetic disorder.

Chromosomal integration has its disadvantages, however, because the insertion often occurs almost randomly: the location of the inserted genes can vary enormously from cell to cell. In many cases, inserted genes may not be expressed because of integration into a highly condensed heterochromatic region. In some cases, the integration event can result in death of the host cell (for example, by insertion into a crucially important gene, thereby inactivating it). Such an event has consequences only for the single cell in which the integration occurred. A greater concern is the possibility of cancer: an integration event in one of the many cells that are targeted could disturb the normal expression patterns of genes that control cell division or cell proliferation, for example by activating an oncogene or inactivating a tumor suppressor gene or a gene invoved in apoptosis (programmed cell death). *Ex vivo* gene therapy at least offers the opportunity for selecting cells where integration has been successful, amplifying them in cell culture and then checking the phenotypes for any obvious evidence of neoplastic transformation, prior to transferring the cells back into the patient.

Nonintegrated genes

Some gene transfer systems are designed to insert genes into cells where they remain as extrachromosomal elements and may be expressed at high levels (see

Table 20.3). If the cells are actively dividing, the introduced gene may not segregate equally to daughter cells and so long-term expression may be a problem. As a result, the possibility of a cure for a genetic disorder may be remote: repeated treatments involving gene transfer will be necessary. In some cases, however, there may be no need for stable long-term expression. For example, cancer gene therapies often involve transfer and expression of genes into cancer cells with a view to killing the cells. Once the malignancy has been eliminated, the therapeutic gene may no longer be needed.

Table 20.3: Properties of major methods of gene transfer used in gene therapy and their applications

Method of gene transfer	Efficiency of gene transfer[a]	Efficiency of integration[a]	Applications in gene therapy[b]	
			Ex vivo	*In vivo*
Retrovirus vectors	High	High	+	?
Adenovirus vectors	High	Low	±	+
Herpes simplex vectors	Low	Low	±	+
Adeno-associated virus vectors	High*	High*	+	?
Receptor-mediated endocytosis	High	Low	−	+
Liposomes	Low	Low	±	+

[a] As determined from gene transfer into cultured cells
[b] Classifications are: (+) major application; (±) some application; (−) little or no application.
* At least for wild-type.

Most gene therapy protocols have used mammalian viral vectors because of their high efficiency of gene transfer

The method chosen for gene transfer depends on the nature of the target tissue and whether transfer is to cultured cells *ex vivo* or to the cells of the patient *in vivo*. No one gene transfer system is ideal; each has its limitations and advantages (see Hodgson, 1995). However, mammalian viral systems have been particularly attractive because of their high efficiency of gene transfer into human cells. The major classes of recombinant virus vectors are based on certain retroviruses which are integrating but only infect actively dividing cells, and adenoviruses which infect a wide range of cell types but do not integrate efficiently. In addition, increasing use is made of other systems such as adeno-associated viruses, while some viruses are particularly suitable for infecting specific tissues, as in the case of herpes simplex virus (HSV; see below).

Retrovirus vectors

Retroviruses are RNA viruses which possess a reverse transcriptase function, enabling them to synthesize a complementary DNA form that can integrate into chromosomal DNA (see Boris-Lawrie and Temin, 1994, and *Figure 17.2*). They are very efficient at transferring DNA into cells, and integration of viral DNA occurs usually at a single chromosomal site. The integrated DNA can be stably propagated, offering the possibility of a permanent cure for a disease. Simple injection of retroviral vectors is usually inappropriate for *in vivo* gene therapy because they can generally be killed by human complement.

Retroviruses can only be produced at relatively low titers and only infect actively dividing cells, thereby excluding their use in treating tissues composed of non-dividing cells (e.g. neurons, etc.). This same property is, however, beneficial to gene therapy for cancers of tissues that normally have nondividing cells: the actively dividing cancer cells can be selectively infected and killed without major risk to the nondividing cells of the normal tissue (see page 582). Widely used murine retrovirus vectors can accommodate inserts up to 8 kb.

Adenovirus vectors

Adenoviruses are DNA viruses that produce infections of the upper respiratory tract and have a natural tropism for respiratory epithelium, the cornea and the gastro-intestinal tract (see Brody and Crystal, 1994). Unlike retroviruses, which can only infect actively dividing cells, adenoviruses can infect a very wide variety of cell types. Entry into cells occurs by receptor-mediated endocytosis (*Figure 20.4;* see also below) and is efficient, but the inserted DNA does not appear to integrate and so expression of inserted genes can only be sustained over short periods. Adenovirus vectors can be produced at very high titers, typically accept insert sizes up to 7–8 kb and, because of their ability to infect many different types of cell, have found widespread applications, notably in *in vivo* gene therapy strategies. Because they can infect virtually all human cells, cancer gene therapies involving cell killing without causing toxicity to normal surrounding cells could be a problem. Another problem is that adenovirus vectors can induce significant inflammatory responses as happened when used to treat cystic fibrosis (see page 577).

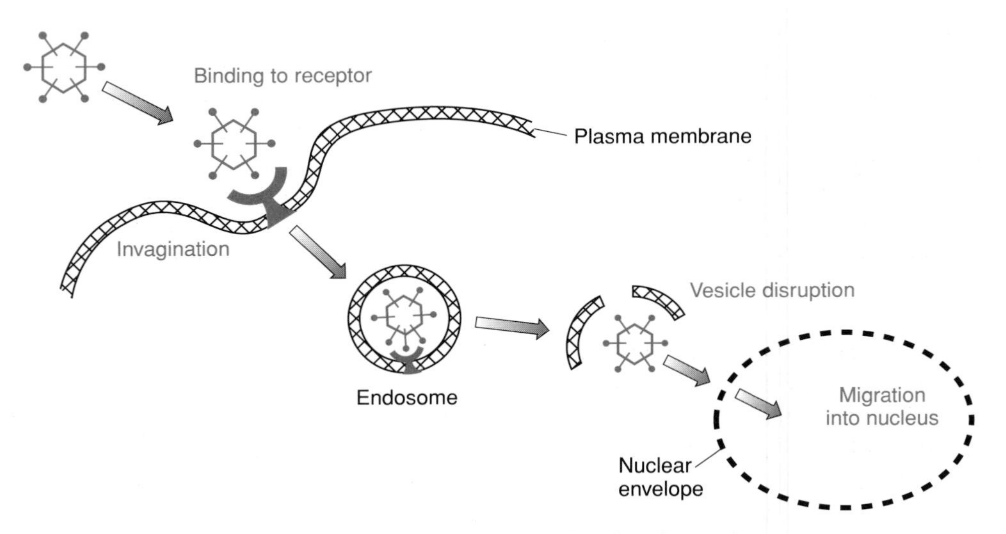

Figure 20.4: Adenoviruses enter cells by receptor-mediated endocytosis.

Binding of viral coat protein to a specific receptor on the plasma membrane of cells is followed by *endocytosis*, a process in which the plasma membrane invaginates and then pinches off to form an intracellular vesicle (*endosome*). Subsequent vesicle disruption by adenovirus proteins allows virions to escape and migrate towards the nucleus where viral DNA enters through pores in the nuclear envelope. Adapted from Curiel (1994) with permission from the New York Academy of Sciences.

Herpes simplex virus vectors

HSV vectors are tropic for the central nervous system (CNS) and can establish lifelong latent infections in neurons. They are nonintegrating and so long-term expression of transferred genes is not possible. Their major applications are expected to be in delivering genes into neurons for the treatment of neurological diseases, such as Parkinson's disease, and for treating CNS tumors. They have a comparatively large insert size capacity (>20 kb).

Adeno-associated virus vectors

Adeno-associated viruses (AAVs) are a group of small, single-stranded DNA viruses which cannot usually undergo productive infection without co-infection by a *helper virus*, such as an adenovirus or HSV. In the absence of co-infection by a helper virus, unmodified human AAV integrates into chromosomal DNA, usually at a specific site on 19q13.3–qter. Subsequent superinfection with an adenovirus can activate the integrated virus DNA, resulting in progeny virions. AAV vectors can only accommodate inserts up to 4.5 kb, but they have the advantage of providing the possibility of long-term gene expression with a high degree of safety: they integrate into chromosomal DNA but, because 96% of the parental AAV genome has been deleted, the AAV vectors lack any viral genes and recombinant AAV vectors only contain the gene of interest (see Kaplitt *et al.*, 1994).

> ***Notebox***: a **helper virus** is a virus that provides certain viral functions *in trans* (e.g. enzymes involved in viral DNA replication, etc.) which are essential for *productive infection* (including viral DNA replication, viral assembly and infection of new cells) by certain natural viruses, such as AAV, or artificially disabled viruses.

Concerns over the safety of recombinant viruses have prompted increasing interest in nonviral vector systems for gene therapy

Increasingly concern has been expressed regarding the safety of viral vector systems. The recombinant viruses which are used for *ex vivo* gene therapy are designed to be disabled: typically some viral genes required for viral replication are deleted, and the therapeutic genes that are to be transferred are inserted in their place. The resulting *replication-incompetent* viruses are then intended to infect individual cells. In the case of retrovirus vectors, chromosomal integration is still possible but they, like other replication-incompetent virus vectors, should not be able to undergo a productive infection in which they replicate, assemble new virions and infect new cells. However, there is the remote possibility that the introduced viruses can recombine with endogenous retroviruses, resulting in recombinant progeny that can undergo productive infection. Additionally, adenoviruses are generally nonintegrating and the repeated injections that may be required may provoke severe inflammatory responses to the recombinant adenoviruses as has happened recently in a gene therapy trial for cystic fibrosis. Increasingly, therefore, attention has been focused towards studying alternative methods of gene transfer.

Direct injection/particle bombardment

In some cases, DNA can be injected directly with a syringe and needle into a specific tissue, such as muscle. This approach has been considered, for example, in the case of DMD, where early studies investigated intramuscular injection of a dystrophin minigene into a mouse model, *mdx* (Acsadi *et al.*, 1991). An alternative direct injection approach uses particle bombardment techniques: DNA is coated on to metal pellets and fired from a special gun into cells. Successful gene transfer into a number of different tissues has been obtained using this approach. Such direct injection techniques are simple and comparatively safe. However, there is poor efficiency of gene transfer, and a low level of stable integration of the injected DNA. The latter property is particularly disadvantageous in the case of proliferating cells, and would necessitate repeated injections. It may be less of a problem in tissues such as muscle which do not regularly proliferate, and in which the injected DNA may continue to be expressed for several months.

Receptor-mediated endocytosis

The DNA is coupled to a targeting molecule that can bind to a specific cell surface receptor, inducing endocytosis and transfer of the DNA into cells. Coupling is normally achieved by covalently linking polylysine to the receptor molecule and then arranging for (reversible) binding of the negatively charged DNA to the positively charged polylysine component. For example, hepatocytes are distinguished by the presence on the cell surface of asialoglycoprotein receptors which clear asialoglycoproteins from the serum. Coupling of DNA to an asialoglycoprotein via a polycation such as polylysine can target the transfer of exogenous DNA into liver cells. The complexes can be infused into the liver either via the biliary tract or vascular bed, whereupon they are taken up by hepatocytes. A more general approach utilizes the transferrin receptor which is expressed in many cell types, but is relatively enriched in proliferating cells and hemopoietic cells (see *Figure 20.5*). Gene transfer efficiency may be high but the method is not designed to allow integration of the transferred genes. A further problem has been that the protein–DNA complexes are not particularly stable in serum. Additionally, the DNA conjugates may be entrapped in endosomes and degraded in lysosomes, unless previously co-transferred with, or physically linked to, an adenovirus molecule (see *Figure 20.5* and Curiel, 1994).

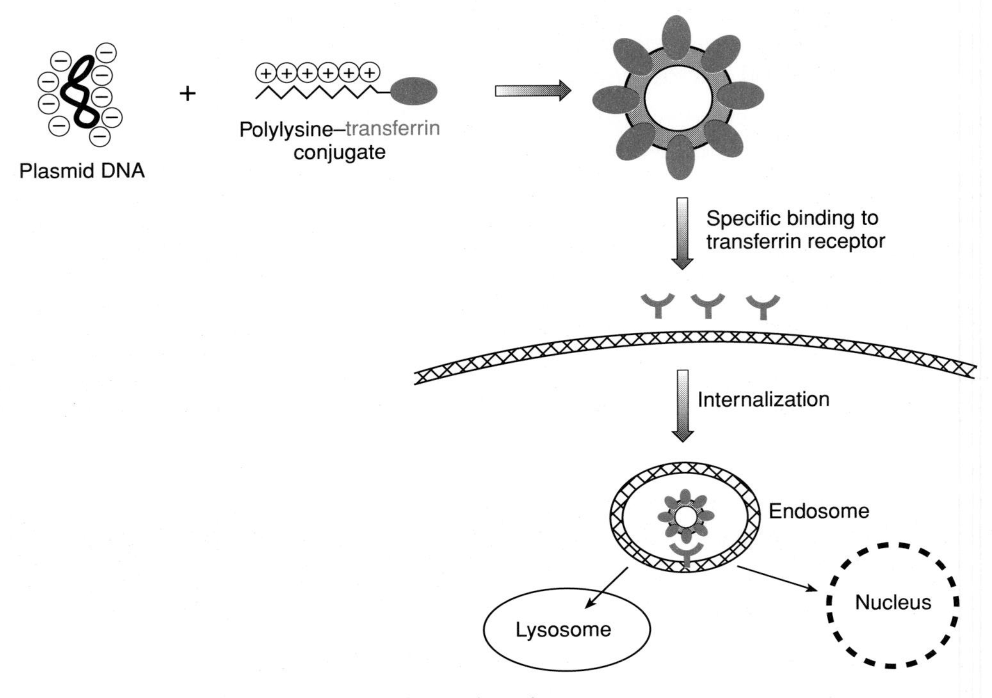

Figure 20.5: Gene transfer via the receptor-mediated endocytosis pathway.

The negatively charged plasmid DNA can bind reversibly to the positively charged polylysine attached to the transferrin molecule. During this process, the DNA is condensed into a compact circular toroid with the transferrin molecules located externally and free to bind to cell surface transferrin receptors. Following initial endosome formation, a portion of the endocytosed conjugates can migrate to the nucleus, although a very significant fraction is alternatively transferred to lysosomes where the DNA is degraded. The efficiency of transfer can be increased by the further refinement of coupling an inactivated adenovirus to the DNA–transferrin complex: following endocytosis and transport to lysosomes, the added adenovirus causes vesicle disruption (see *Figure 20.4*), allowing the DNA to avoid degradation and to survive in the cytoplasm. Adapted from Curiel (1994) with permission from The New York Academy of Sciences.

Liposomes

Liposomes are spherical vesicles composed of synthetic lipid bilayers which mimic the structure of biological membranes. The DNA to be transferred is packaged *in vitro* within the liposomes and used directly for transferring the DNA to a suitable target tissue *in vivo* (*Figure 20.6*). The lipid coating allows the DNA to survive *in vivo*, bind to cells and be endocytosed into the cells. Liposomes have become popular vehicles for gene transfer in *in vivo* gene therapy because of the safety concerns when using recombinant viruses. However, the efficiency of gene transfer is low, and the introduced DNA is not designed to integrate into chromosomal DNA, so expression of the inserted genes is transient.

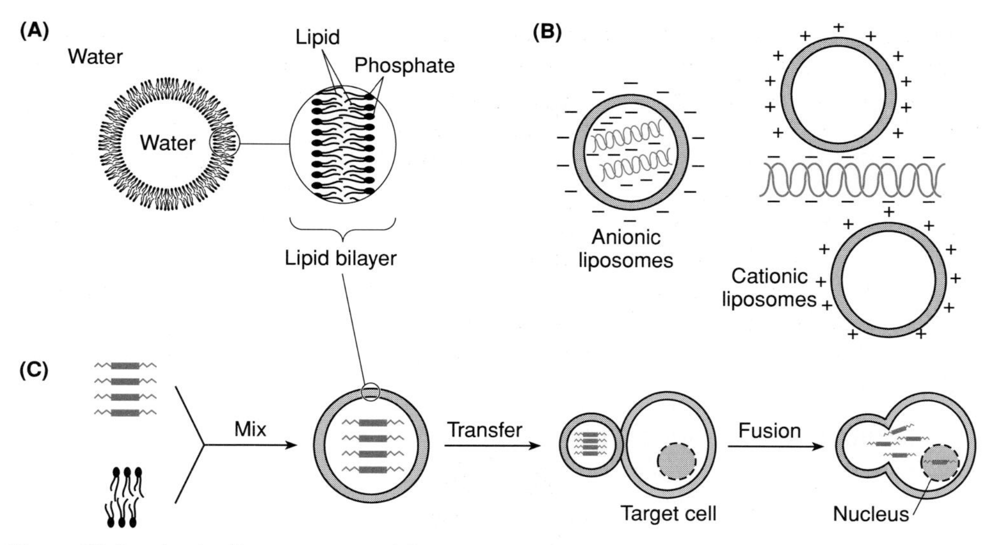

Figure 20.6: *In vivo* liposome gene delivery.

(A) and **(B)** Structure of liposomes. Liposomes are synthetic vesicles which can form spontaneously in aqueous solution following artificial mixing of lipid molecules. In some cases, a phospholipid bilayer is formed, with hydrophilic phosphate groups located on the external surfaces and hydrophobic lipids located internally (left). In other cases there is a multilamellar lipid envelope. Anionic liposomes have a negative surface charge and when the lipid constitutents are mixed with negatively charged DNA molecules (see panel C below), the DNA is internalized. Cationic liposomes have a surface positive charge and DNA molecules bind to the surface of liposomes.
(C) Use of liposomes to transfer genes into cells. This figure illustrates the use of anionic liposomes to transfer internally located DNA into cells. The plasma membranes of cells are fluid structures whose principal components are phospholipids, and so mixing of cells and liposomes can result in occasional fusion between the lipid bilayer of the liposome and the plasma membrane. When this happens the cloned genes can be transferred into the cytoplasm of a cell, and can thence migrate to the nucleus by passive diffusion through the pores of the nuclear envelope. Note that, in practice, cationic liposomes have been more widely used for transferring DNA into cells.

Therapeutics based on targeted inhibition of gene expression and mutation correction *in vivo*

Principles and applications of therapy based on targeted inhibition of gene expression *in vivo*

One way of treating certain human disorders is to selectively inhibit the expression of a predetermined gene *in vivo*. In principle, this general approach is particularly suited to treating cancers and infectious diseases, and some immunological disorders. In such cases, the basis of the therapy is to knock out the expression of a specific gene that allows the cancerous cells, infection, allergy, inflammation, etc., to flourish, without interfering with normal cell function. For example, attention could be focused on selectively inhibiting the expression of a particular viral gene that is necessary for viral replication, or an inappropriately activated oncogene, etc.

In addition to the above, targeted inhibition of gene expression may offer the possibility of treating certain dominantly inherited disorders. If a dominantly inherited disorder is the result of a *loss-of-function* mutation, treatment may be possible using conventional gene augmentation therapy. However, since heterozygotes with 50% of normal gene product can be severely affected, successful gene therapy for heterozygotes requires efficient expression of the introduced genes. However, dominantly inherited disorders which arise because of a *gain-of-function* mutation (see pages 402 and 412ff.) may not be amenable to simple addition of normal genes. Instead, it may be possible, in some cases, to specifically inhibit the expression of the mutant gene, but the expression of the normal allele must be maintained. Such *allele-specific* inhibition of gene expression is facilitated if the pathogenic mutation results in a significant sequence difference between the alleles.

The expression of a selected gene might be inhibited by a variety of different strategies. One possible type of approach involves specific *in vivo mutagenesis* of that gene, altering it to a form that is no longer functional. Gene targeting by homologous recombination offers the possibility of *site-specific* mutagenesis to inactivate a gene (see page 536). However, this technique has only very recently become feasible with normal diploid somatic cells and is still very inefficient. Instead, methods of blocking the expression of a gene without mutating it are currently preferred. In principle, this can be accomplished at different levels: at the DNA level (by blocking transcription); at the RNA level (by blocking post-transcriptional processing, mRNA transport or engagement of the mRNA with the ribosomes); or at the protein level (by blocking post-translational processing, protein export or other steps that are crucial to the function of the protein). Therapy by selective inhibition of gene expression is technically possible at all three expression levels (see *Figure 20.7*):

(i) *triple helix therapeutics* (involves binding of gene-specific oligonucleotides to double-stranded DNA in order to inhibit transcription of a gene).

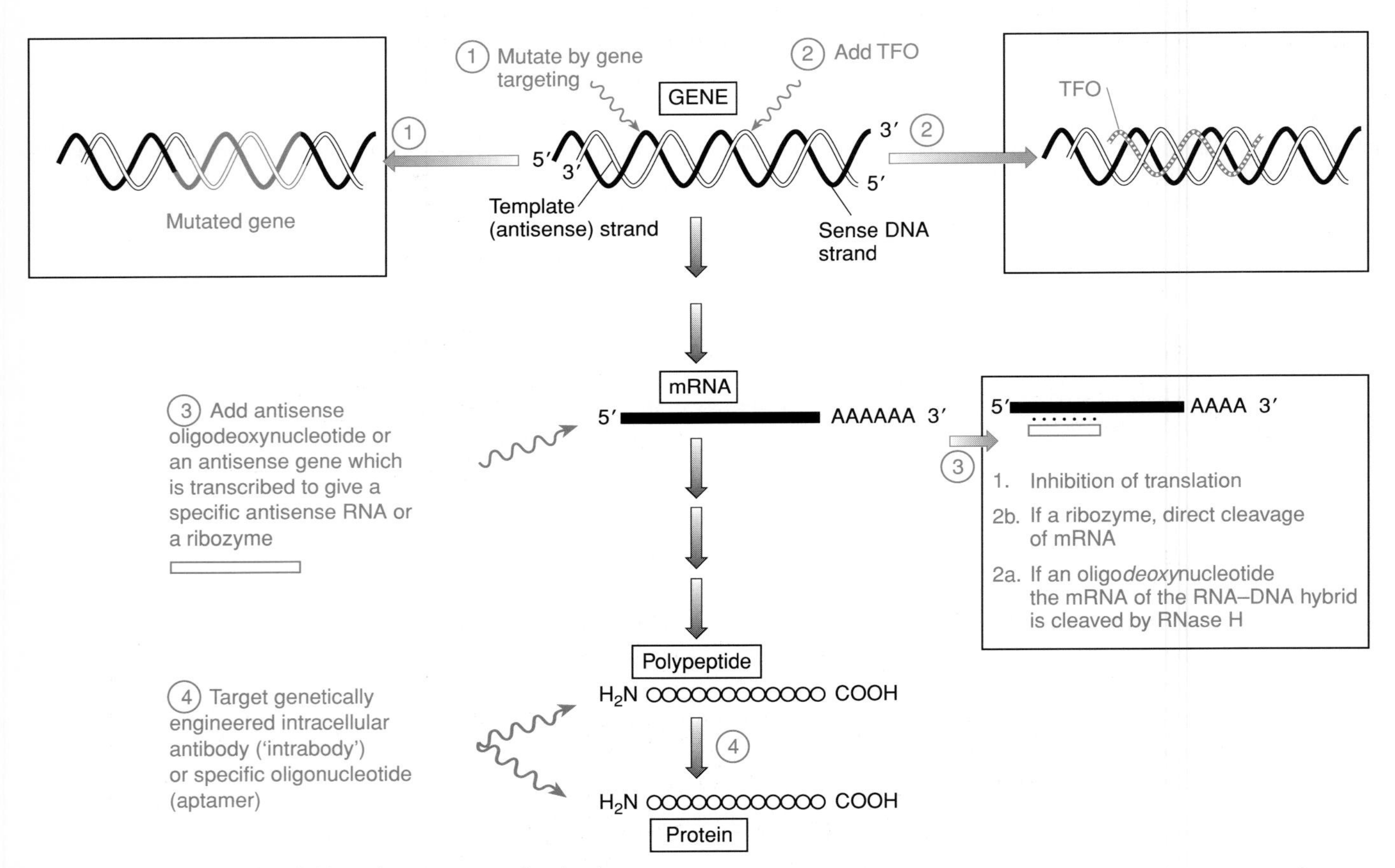

Figure 20.7: Targeted inhibition of gene expression *in vivo.*

Gene therapy based on selective inhibition of a predetermined gene *in vivo* can be achieved at several levels. In principle, it is possible to mutate the gene via homologous recombination-mediated gene targeting to a nonfunctional form (1). In practice, however, it is more convenient to block expression at the level of transcription as a result of binding of a gene-specific triplex-forming oligonucleotide (TFO; see *Figure 20.8*) to the promoter region (2), or of a gene-specific antisense oligonucleotide or RNA to the mRNA (3). In the latter case, an antisense gene is normally provided (see *Figure 20.9*) which can encode a simple antisense RNA or a ribozyme (*Figure 20.10*). In each case, the binding interferes with the ability of the mRNA to direct polypeptide synthesis, and may ensure its destruction: a bound oligodeoxynucleotide makes the mRNA susceptible to cleavage by RNase H, while a bound ribozyme cleaves the RNA directly. The technology for specific inhibition at the polypeptide/protein level is less well developed but is possible using genes which encode intracellular antibodies or oligonucleotide aptamers which specifically bind to the polypeptide and inhibit its function.

(ii) *Antisense therapeutics* (involves binding of gene-specific oligonucleotides or polynucleotides to the RNA; in some cases, the binding agent may be a specifically engineered **ribozyme**, a catalytic RNA molecule that can cleave the RNA transcript).

(iii) Use of **intracellular antibodies** (intrabodies) and **oligonucleotide aptamers** (involves the construction of antibodies that can be directed to specific locations within cells in order to bind a specific protein, or oligonucleotide *aptamers*, which can bind specifically to a selected polypeptide).

Triple helix therapeutics relies on binding of gene-specific oligonucleotides to the major groove of the double helix

Synthetic short oligonucleotides (15–27 nucleotides long) are capable of specifically binding to a sequence of double-stranded DNA, forming a *triple helix*. The oligonucleotide binds by **Hoogsteen hydrogen bonds** to the double-stranded DNA, without disrupting the original Watson–Crick hydrogen bonding. The most stable Hoogsteen-bonded structures are G bound to a GC base pair and a T bound to an AT base pair (see *Figure 20.8*). Although such structures can inhibit DNA replication *in vitro*, helicases can unwind triple strand structures *in vivo*. However, triplex formation has been shown to block binding of transcription factors *in vitro*, and also, at

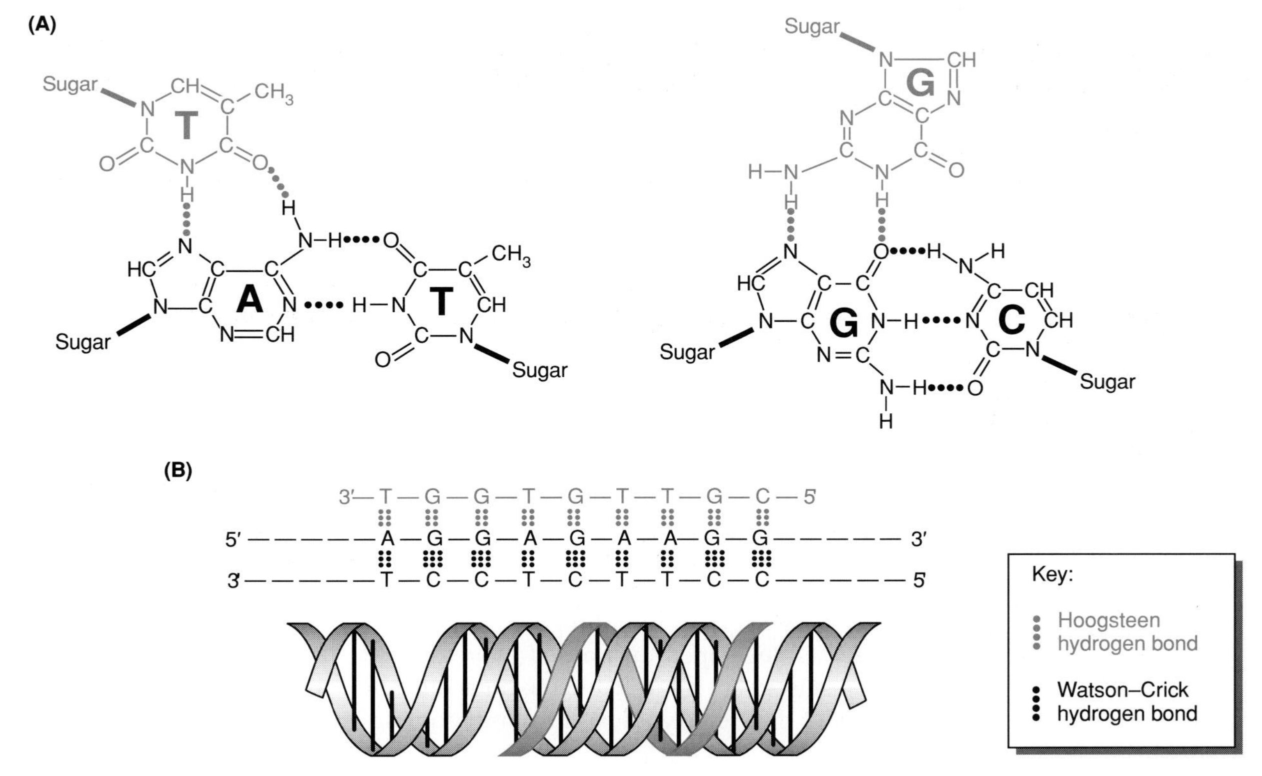

Figure 20.8: Triple helix formation at physiological pH.

(A) Hoogsteen hydrogen bonding. At physiological pH, a thymine can bind to an A–T base pair to give T–A–T, and a guanine can bind to a G–C base pair to give G–G–C. In each case, the Watson–Crick hydrogen bonds present in the original base pairs are maintained, and the third base is bound by two additional hydrogen bonds (*Hoogsteen hydrogen bonds*). Reverse Hoogsteen hydrogen bonding is also possible and involves rotation of the third base through 180° about a vertical axis. For example, in a reverse TAT structure, it is the carbonyl bond at carbon number 2 of the thymine which is involved in Hoogsteen bonding not the one at carbon number 4 as shown. **(B)** Stable triplex formation. The example shown should be very stable, because Hoogsteen hydrogen bonding is possible at all base positions. *Note* that this means that the recognition sequence in the double-stranded DNA target must be unusual: the strand to which the TFO (triplex-forming oligonucleotide) binds contains only purines. The TFO binds to the *major groove* (see *Figure 1.5*) of the double helix.

least in some cases, evidence has been obtained for gene-specific inhibition of transcription in intact cells (see Chubb and Hogan, 1992).

Oligonucleotides are large polyanionic hydrophilic structures and so are not ideally suited to diffusing across the highly hydrophobic plasma membrane. Direct delivery into the cytoplasm using cell permeabilization techniques provides the most efficient approach to enable subsequent transfer into the nucleus, and delivery using liposomes (see page 563) is a popularly used method. Thereafter, the oligonucleotides can migrate rapidly to the nucleus (by passive diffusion through the pores of the nuclear envelope). Inside the cell, the oligonucleotides are exposed to nuclease attack, notably from exonucleases, and the half-life of conventional oligonucleotides with phosphodiester bonds is typically about 20 min. Accordingly, it is usual for the 3′ and 5′ ends of the oligonucleotides to be chemically modified to protect against nuclease attack. Often, chemical modification involves incorporation of sulfur-containing *phosphorothioate bonds* to generate so-called **S-oligonucleotides**.

Although the technology is improving rapidly, some general difficulties need to be overcome. Inhibition of gene expression requires comparatively large amounts of oligonucleotide. More worrying is the limitation imposed by Hoogsteen hydrogen bonding: the target sequences need to carry virtually all their purine bases on one DNA strand. Preliminary attempts to solve this problem include replacement of the phosphate groups by different chemical groupings that allow triplex-forming oligonucleotides to 'hop' from one strand of the bound DNA duplex to the other.

Antisense oligonucleotides or polynucleotides can bind to a specific mRNA, inhibiting its translation and, in some cases, ensuring its destruction

During transcription, only one of the two DNA strands in a DNA duplex, the *template strand*, serves as a template for making a complementary RNA molecule. As a result, the base sequence of the single-stranded RNA transcript is essentially identical (except that U replaces T) to the other DNA strand, commonly called the *sense strand* (see page 11). Any oligonucleotide or polynucleotide which is complementary in sequence to an mRNA sequence, including the template strand of the gene, can therefore be considered to be an *antisense* sequence.

Binding of an antisense sequence to the corresponding mRNA sequence would be expected to interfere with translation, and thereby inhibit polypeptide synthesis. Indeed, naturally occurring antisense RNA is known to provide a way of regulating the expression of genes in some plant and animal cells, as well as in some microbes. Synthetic oligonucleotides can be designed to be complementary in sequence to a specific mRNA and, when transferred into cells, show evidence of inhibition of expression of the corresponding gene. As a result, the concept of **antisense therapeutics** was developed: unwanted expression of a specific gene in disease tissues could be selectively inhibited using an artificially gene-specific antisense sequence. A variety of different types of antisense sequence can be used.

Antisense oligodeoxynucleotides

The use of artificial antisense oligonucleotides is often favored, simply because they can be synthesized so simply. They can be transferred efficiently into the cytoplasm of cells using liposomes, and their intracellular stability is improved by using chemically modified oligonucleotides, notably S-oligonucleotides (see above; note that even although antisense oligonucleotides migrate to the nucleus, they do not bind the double-stranded DNA because they are not designed to participate in Hoogsteen hydrogen bonding). Antisense oligo*deoxy*nucleotides (ODNs) are preferred to oligoribonucleotides because they are generally less vulnerable to nuclease attack, and importantly because they have the additional advantage of inducing the destruction of an mRNA to which they bind. This is so because an ODN–mRNA hybrid, like all DNA–RNA hybrids, is vulnerable to attack and selective cleavage of the RNA strand by a specific class of intracellular ribonuclease, RNase H. Despite teething problems in early studies, refinement of the technology has meant that antisense ODNs are now considered to have great therapeutic potential, and clinical trials are now in progress for several human diseases (see Wagner, 1994 and *Tables 20.5* and *20.7*).

Antisense genes

Antisense oligonucleotides, even when chemically modified, are not stable indefinitely. One way of ensuring a continuous supply of antisense sequence is a form of expression cloning in which a specially designed **antisense gene** is transferred into the relevant cells. Such a gene can be engineered simply by constructing a minigene in which an inverted coding sequence is placed downstream of a powerful promoter. The DNA strand that normally serves as the sense strand is now transcribed to give an antisense RNA which can be synthesized repeatedly (*Figure 20.9*). If the antisense gene is provided using an integrative vector, long-term production of antisense RNA may be obtained.

Ribozymes

Increasingly, it is becoming clear that RNA molecules are functionally different from DNA molecules: collectively they can serve diverse functions, rather than simply being involved in transfer of genetic information. Some RNA molecules are able to lower the activation energy for specific biochemical reactions, and so effectively function as enzymes (*ribozymes*). For example, the transcripts of group I introns (see page 225) are *autocatalytic* and self-splicing. Other ribozymes which cleave RNA,

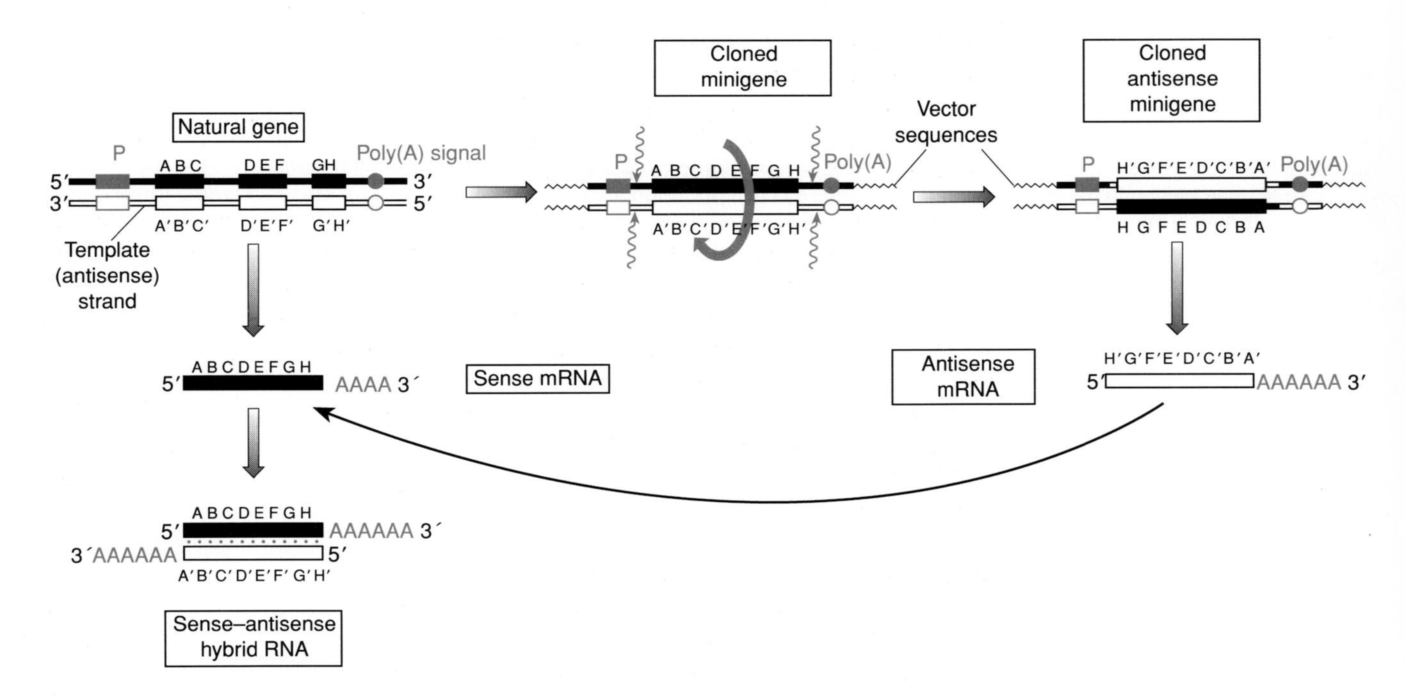

Figure 20.9: Artificially designed antisense genes can be transcribed to give antisense mRNA.

An antisense minigene can be obtained by cloning a coding DNA sequence from a cDNA clone into a vector which contains an insertion site flanked by a promoter sequence (P) on one side and a polyadenylation sequence on the other. In addition to normal sense clones, some clones will contain the insert in the inverse orientation, i.e. the downstream sequences will be adjacent to the promoter. Transcription from such a gene produces an antisense RNA.

are *trans-acting*, that is they cleave an RNA sequence on a different molecule. They contain two essential components: target recognition sequences (which base-pair with complementary sequences on target RNA molecules), and a catalytic component, much like the active site of an enzyme (which cleaves the target RNA molecule while the base-pairing holds it in place). The cleavage leads to inactivation of the RNA, presumably because of subsequent recognition by intracellular nucleases of the two unnatural ends. Examples include human ribonuclease P and various ribozymes obtained from plant *viroids* (virus-like particles).

Genetic engineering can be employed to custom design the recognition sequence so that it contains antisense sequences that can base-pair to a specific mRNA molecule, while retaining the catalytic site (see *Figure 20.10*). Engineered genes which can be transcribed to produce the desired ribozyme can then be transfected into suitable cells. One application has been the design of ribozymes against specific oncoproteins. Pilot studies have shown that transfection of an anti-Ras ribozyme gene into human bladder carcinoma cells with the *ras* mutation resulted in blocking of Ras production and reversal of the metastatic, invasive and tumorigenic properties of the cells. Early problems in the efficiency of targeting to their targets inside the cell are currently being addressed (see Barinaga, 1993), and clinical trials have already been initiated in some cases, such as in gene therapy for AIDS.

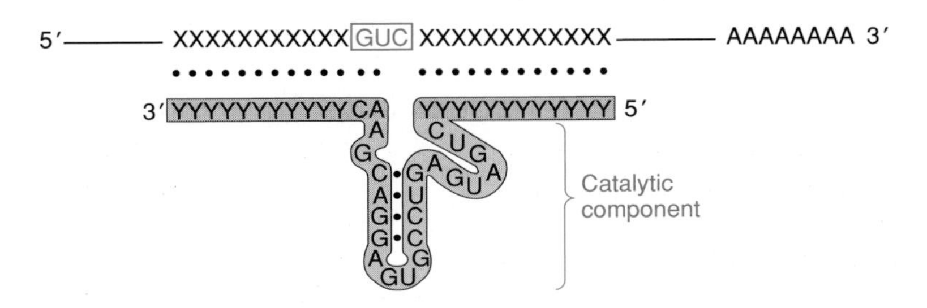

Figure 20.10: Genetically engineered hammerhead ribozymes.

The hammerhead ribozyme is a constituent of some plant viroids (virus-like particles) and is so called because of the shape of its catalytic component. It is *trans*-acting and cleaves specific target RNA molecules, whose recognition sequence contains a centrally located triplet: GUC (boxed) or a close variant. Recognition is achieved by two sequences which flank the catalytic component and which permit base pairing to the target sequence. In order to custom design an artificial ribozyme that will cleave a predetermined mRNA sequence, it is simply necessary to choose a suitable GUC (or variant triplet) in the target and then design the ribozyme to contain the usual catalytic component flanked by sequences (YYYYY...) which are complementary to the sequences flanking the chosen triplet in the target sequence (XXXXX...). Because of the comparative lability of RNA, gene therapy experiments using ribozymes to inhibit gene expression may involve synthesis and transfer of genes encoding the desired ribozyme.

Artificially designed intracellular antibodies (intrabodies), oligonucleotides (aptamers) and mutant proteins can inhibit the function of a specific polypeptide

Intracellular antibodies (intrabodies)

Antibody function is normally conducted extracellularly: upon synthesis, antibodies are normally secreted into the extracellular fluid or remain membrane bound on the B-cell surface as antigen receptors. Recently, however, antibody engineering (see page 529) has been extended to the design of genes encoding intracellular antibodies, or *intrabodies*. This achievement raises the possibility of using antibodies within cells to block the construction of viruses or harmful proteins, such as oncoproteins. The first example of this approach involved engineering the antibody F105 which binds to gp120, a crucial human immunodeficiency virus (HIV) envelope protein that the AIDS virus uses to attach to and infect its target cells (Marasco *et al.*,

1993). This envelope protein is derived from a larger precursor gp160 which is synthesized in the endoplasmic reticulum. Marasco and co-workers designed a novel F105 gene which encoded an antibody that was stably expressed and retained in the endoplasmic reticulum without being toxic to the cells. The engineered antibody binds to the HIV envelope protein within the cell and inhibits processing of the gp160 precursor, thereby substantially reducing the infectivity of the HIV-1 particles produced by the cell.

Oligonucleotide aptamers

Fully degenerate oligonucleotides can be synthesized by delivering 25% each of the four bases A, C, G and T at each base position during oligonucleotide synthesis. As a result, the number of sequence permutations which can be generated (4^n where n is the chosen length of oligonucleotide) can be enormous. The resulting mixture of oligonucleotides can be used to screen for the ability to bind to a selected target protein (**protein epitope targeting**). In practice, the use of *partially degenerate* oligonucleotides is preferred so that the concentration of individual oligonucleotides is not too low. In effect, this means simultaneous screening of many thousands of oligonucleotides, and so the chance of at least one epitope of the target protein being specifically bound by an oligonucleotide can be high. The bound oligonucleotide (sometimes known as an *adaptamer* or ***aptamer***) can be eluted from the protein and sequenced to identify the specific recognition sequence. Transfer of large amounts of a chemically stabilized aptamer into cells can result in specific binding to a predermined polypeptide, thereby blocking its function.

An initial success was the identification of oligonucleotides that could bind to and inhibit the protease thrombin, which is part of the blood coagulation cascade (Bock *et al.*, 1992). Thrombin functions in serum, and extracellular applications of this type are no different, in principle, from standard drug therapy. However, the future use of oligonucleotide aptamers to inhibit specific intracellular protein targets will inevitably involve genetic modification of cells, and can therefore be considered as a form of gene therapy.

Mutant proteins

Naturally occurring gain-of-function mutations can involve the production of a mutant polypeptide that binds to the wild-type protein, inhibiting its function. In many such cases, the wild-type polypeptides naturally associate to form multimers, and incorporation of a mutant protein inhibits this process (see page 412ff.). In some cases, gene therapy may be possible by designing genes to encode a mutant protein that can specifically bind to and inhibit a predetermined protein, such as a protein essential for the life-cycle of a pathogen. For example, one form of gene therapy for AIDS involves artificial production of a mutant HIV-1 protein in an attempt to inhibit multimerization of the viral core proteins (see page 585).

Artificial correction of a pathogenic mutation *in vivo* is possible, in principle, but is very inefficient and not readily amenable to clinical applications

Certain disorders are not easy targets for gene therapy. For example, dominantly inherited disorders where a simple mutation results in a pathogenic gain of function cannot be treated by gene augmentation therapy, and targeted inhibition of gene expression may be difficult to achieve. Target inhibition is best suited to inhibiting novel or inappropriate gene expression in human cells, for example expression of viral genes, oncogenes, etc. Expression of a gain-of-function mutant allele may need to be inhibited while retaining expression of a very similar wild-type allele. If the mutant allele carries a significant change in sequence at the site of the mutation, it may be possible to achieve selective inhibition but if the change is a simple mutation, say a single nucleotide substitution, other approaches may be needed. One possible approach is **targeted mutation correction** by inserting some reagents into cells in order to change the mutant sequence back to a form that is compatible with normal function.

In principle, there are several different ways in which a specific mutation can be corrected selectively *in vivo*, mostly at the DNA level. One way is to use *gene targeting* techniques based on *homologous recombination* (see page 536). Because this approach offers the ability to make *site-specific* modifications of endogenous genes, it represents a potentially powerful method for gene therapy: both acquired and inherited mutations could be corrected, and novel alterations could be engineered into the genome. Thus far, this technique has been limited largely to pluripotent mouse embryonic stem cells, although recently it has been applied to normal diploid somatic cells (Arbones *et al.*, 1994). However, the enormous inefficiency of this procedure (even when using the ideal target of cells cultured *in vitro*) and the need to correct the defect in many different cells *in vivo* has meant that clinical applications are a long way off.

An alternative approach is to repair the genetic defect *at the RNA level.* One possibility is to use a therapeutic ribozyme. One method envisages using a class of ribozyme known as group I introns, which are distinguished by their ability to fold into a very specific shape, capable of both cutting *and splicing* RNA (Cech, 1995). If a transcript has, for example, a nonsense or a missense mutation, it may be possible to design specific ribozymes that can cut the RNA upstream of the mutation and then splice in a corrected transcript, a form of *trans-splicing* (see *Figure 20.11*). Thus far, this technology is in its infancy, and catalytic efficiency needs to be improved. Another possibility is *therapeutic RNA editing.* This involves using a complementary

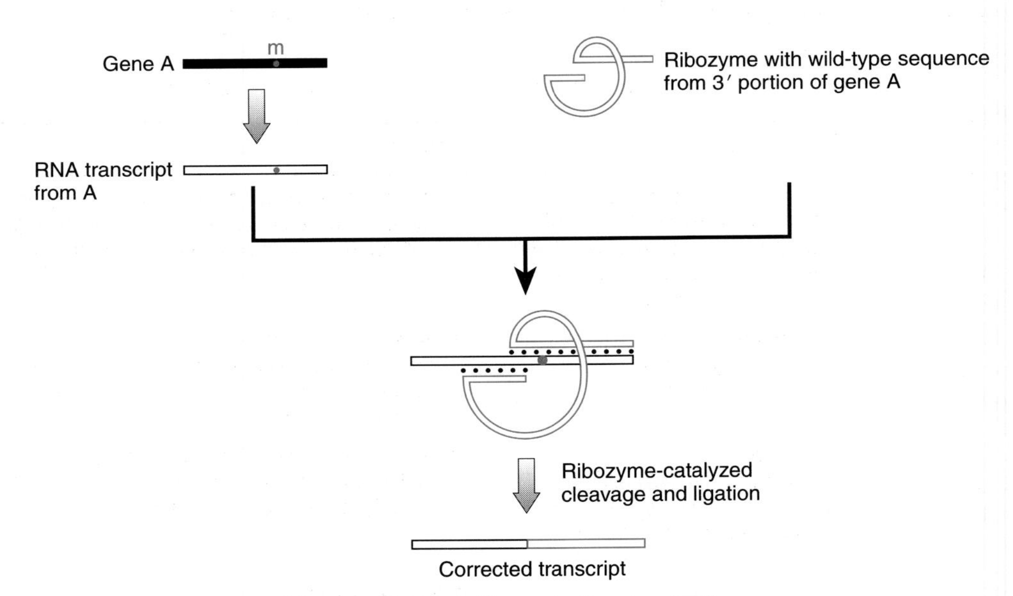

Figure 20.11: Some ribozymes also have the potential of repairing mutations in mRNA.

Group I introns are a class of self-splicing intron. The RNA transcript acts as a ribozyme by catalyzing the cleavage of the RNA and subsequent splicing (see *Box 9.2*). They possibly could be used as therapeutic agents capable of repairing certain mutations at the mRNA level (Cech, 1995). In this example, gene A has a missense or a nonsense mutation (m) and is transcribed to give a mutant RNA. The therapeutic ribozyme is designed so that its flanking recognition sequences are complementary to the 3′ end of the wild-type mRNA sequence for gene A, encompassing the location of the mutation. The ribozyme is designed to cleave the mutant mRNA at a position upstream of the mutation site. Subsequent ligation of the 5′ end of the mutant mRNA to the 3′ end sequence carried by the ribozyme can result in repair of the mutation at the mRNA level.

RNA oligonucleotide to bind specifically to a mutant transcript at the sequence containing the pathogenic point mutation, and an RNA editing enzyme, such as double-stranded RNA adenosine deaminase, to direct the desired base modification (Woolf *et al.*, 1995). Again this technology is in its infancy and formidable technical difficulties need to be overcome before clinical applications can be envisaged.

Gene therapy for inherited diseases

Different genetic disorders are amenable to varying degrees to treatment by gene therapy. Common nonmendelian genetic diseases may involve a complex interplay between different genetic loci and/or environmental factors, and so possible gene therapy approaches may not be straightforward. Single gene disorders, where individuals are severely affected and where there is no effective treatment, are more obvious candidates for gene therapy. Increasingly, genes underlying a variety of such diseases are being isolated and characterized as a result of positional cloning and candidate gene approaches (Chapter 14). However, differing pathogeneses means that certain single gene disorders will be more amenable to gene therapy approaches than others (*Table 20.4*).

Recessively inherited disorders are conceptually the easiest inherited disorders to treat by gene therapy

Those disorders where the disease results from a simple deficiency of a specific gene product are generally the most amenable to treatment: high level expression of an introduced normal allele should be sufficient to overcome the genetic deficiency. Recessively inherited disorders have been of particular interest as candidates for gene therapy because the mutations are almost always simple loss-of-function

Table 20.4: Factors governing the amenability of single gene disorders to gene therapy approaches

Factor	Most amenable	Least amenable
Mode of inheritance	Recessive: affecteds usually have no or extremely little gene product, so that even low level expression of introduced genes can have an effect	Dominant: even where the mutation is a loss-of-function mutation most affected people are heterozygotes, with at least 50% of the normal gene product already present
Nature of mutation	Loss of function: can be treated simply by gene augmentation therapy (see *Box 20.1)*	Gain of function – novel mutant protein or toxic product, etc. may not easily be treated by simply adding normal genes. Instead, may need specifically to block expression of mutant gene or repair genetic defect
Accessibility of target cells and amenability to cell culture	Readily accessible tissues, e.g. blood, skin, etc. Cells that can be cultured readily and re-inserted in the patient permit *ex vivo* gene transfer	Tissues that are difficult to access (e.g. brain), or to derive cell cultures which can be re-implanted (thereby excluding *ex vivo* gene therapy)
Size of coding DNA	Small coding DNA size means easy to insert into vector e.g. β-globin = ~0.5 kb	Large coding DNA; may be difficult to insert into suitable vector
Control of gene expression	Loose control of gene expression with wide variation in normal expression levels, e.g. ADA expression (page 575)	Tight control of gene expression, e.g. in the case of β-globin (page 574)

mutations. Affected individuals have deficient expression from both alleles and so the disease phenotype is due to complete or almost complete absence of normal gene expression. Heterozygotes, however, have about 50% of the normal gene product and are normally asymptomatic. Additionally, there is, in at least some cases, wide variation in the normal levels of gene expression, so that a comparatively small percentage of the average normal amount of gene product may be sufficient to restore the normal phenotype (see opposite). It is also often observed that the severity of the phenotype of recessive disorders is inversely related to the amount of product that is expressed (see *Table 15.5*). As a result, even if the efficiency of gene transfer is low, modest expression levels for an introduced gene may make a substantial difference. This is quite unlike dominantly inherited disorders where heterozygotes with loss-of-function mutations have 50% of the normal gene product and may yet be severely affected.

Although recessively inherited disorders are, in principle, amenable to gene augmentation therapy, certain disorders are less amenable than others. In addition to the question of accessibility of the disease tissue, some disorders may be difficult to treat for other reasons. A good example is provided by β-thalassemia which results from mutations in the β-globin gene, *HBB*. This is a severe disorder affecting hundreds of thousands of people world-wide, and superficially would appear to be an excellent candidate for gene therapy: the gene is very small and has been characterized extensively, the disorder is recessively inherited and affects blood cells. An initial attempt at gene therapy for this disorder in 1980 failed, largely because of inefficient gene transfer and poor expression of the introduced β-globin genes. Even though we now know much about how this gene is expressed, there have been no subsequent gene therapy attempts. This is due to the problem of the very tight control of gene expression required following insertion of a normal β-globin gene into the desired cells: the amount of β-globin product made must be equal to the amount of α-globin. If too much β-globin were to be made, the imbalance between β-globin and α-globin chains would result in an α-thalassemia phenotype.

The first apparently successful gene therapy was initiated in 1990 for adenosine deaminase deficiency

For years we have been accustomed to the applications of molecular genetics in the diagnosis of disease (see Chapter 16), and, more recently, to the isolation and characterization of novel disease genes (Chapter 14). Now we are living in a decade where molecular genetics is poised, at last, to deliver novel treatments for human disorders. Exciting though this prospect is, the limitations of the current technologies are apparent (see Marshall, 1995). Even now, gene therapy has not *cured* any patient. Instead, current gene therapy trials are providing forms of *treatment* for some disorders: there may be amelioration of the disease, but the effects are temporary, and treatments have to be repeated at regular intervals.

The first apparently successful gene therapy was initiated on 14 September 1990. The patient, Ashanthi DeSilva, was just 4 years old and was suffering from a very rare recessively inherited disorder, adenosine deaminase (ADA) deficiency (see *Figure 20.12*). ADA is involved in the purine salvage pathway of nucleic acid degradation, and is a housekeeping enzyme which is synthesized in many different types of cell. An inherited deficiency of this enzyme has, however, particularly severe consequences in the case of T lymphocytes, one of the major classes of immune system cells. As a result, ADA^- patients suffer from severe combined immunodeficiency. This severe disorder was particularly amenable to gene therapy for a variety

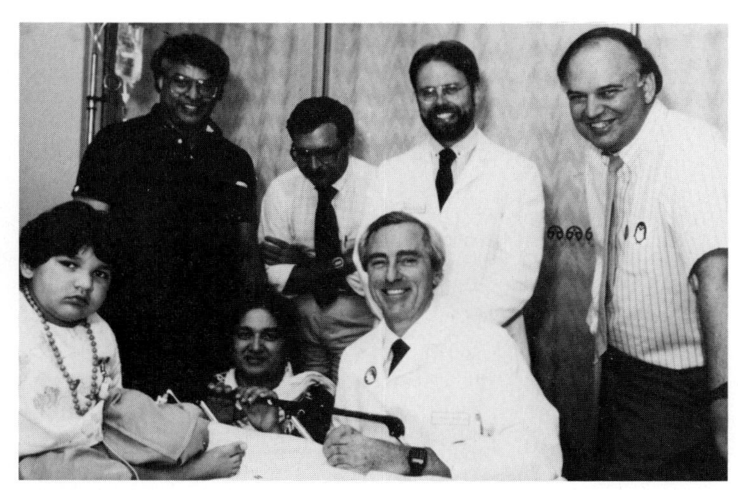

Figure 20.12: The first apparently successful human gene therapy involved treatment of adenosine deaminase deficiency.

Ashanthi DeSilva, 4 years old, was the first recipient of apparently successful human gene therapy in a clinical trial commencing on 14 September, 1990. Others present in the photograph are her parents Raj and Van (left), and doctors Melvin Berger (back, second from left), Kenneth Culver (back, second from right), R. Michael Blaese (back, right) and W. French Anderson (front). The basis of the gene therapy is illustrated in *Figure 20.13.* The efficacy of the gene therapy has been difficult to assess because of concurrent treatment with PEG-ADA (see text). Reproduced with permission from Mary Ann Liebert Inc.

of reasons: the *ADA* gene is small, and had previously been cloned and extensively studied; the target cells are T cells which are easily accessible and easy to culture, enabling *ex vivo* gene therapy; the disorder is recessively inherited and, importantly, gene expression is not tightly controlled (normal individuals show a huge range in enzyme levels, from 10 to 5000% of the average levels). The observation that allogeneic bone marrow transplantation can cure the disorder suggested that engraftment of T cells alone may be sufficient, and transfer of normal *ADA* genes into ADA^- T cells was noted to result in restoration of the normal phenotype.

Alternative treatments for ADA deficiency do exist. Indeed, the treatment of choice is bone marrow transplantation from a perfectly HLA-matched sibling donor, which provides a cure in about 80% of cases. For children where this is not an option, an alternative is *enzyme replacement therapy*, consisting of weekly intramuscular injections of ADA conjugated to polyethylene glycol (PEG). PEG stabilizes the ADA enzyme, allowing it to survive and function in the body for days. Inevitably, however, enzyme replacement therapy does not provide full immune reconstitution and so life expectancy is still likely to be shortened (T cells are required for mounting effective immune responses against invading microorganisms, and in preventing cancer).

The ADA gene therapy approach involved essentially four steps:

(i) cloning a normal ADA gene into a retroviral vector;

(ii) transfecting the ADA recombinant into cultured ADA^- T lymphocytes from the patient;

(iii) identifying the resulting ADA^+ T cells and expanding them in culture;

(iv) re-implanting these cells in the patient (see *Figure 20.13*).

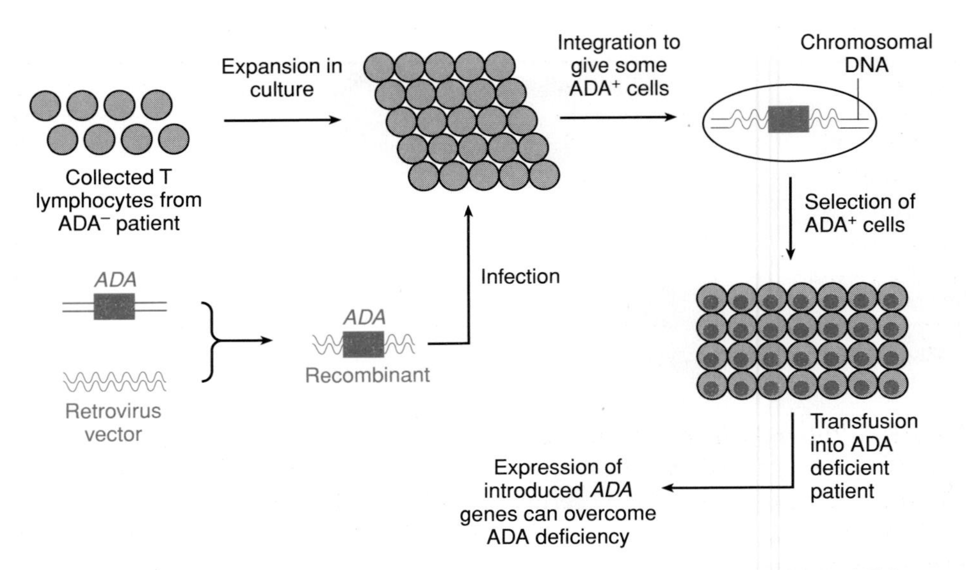

Figure 20.13: *Ex vivo* gene augmentation therapy for adenosine deaminase (ADA) deficiency.

Note that identification of suitably transformed cells is helped by having an appropriate selectable marker in the retrovirus vector, such as a *neo*^{R} gene which confers resistance to the neomycin analog G418 (see *Box 11.1*). Following infection, the target cells can be cultured in a medium containing G418 to select for the presence of retroviral sequences, and then assayed by PCR for the presence of the inserted ADA gene. Suitable ADA$^+$ cells can then be expanded in culture before being re-introduced into the patient.

This approach is necessarily a treatment, not a cure – which would instead require successful transfer into bone marrow stem cells. The trouble here is that human bone marrow stem cells are very difficult to isolate, although enrichment is possible using the monoclonal antibody CD34 (which selectively binds a population of cells that includes totipotent stem cells). There is also the problem that the insertion of retroviral vectors into bone marrow stem cells is very inefficient. The compromise of targeting the differentiated T lymphocytes has meant that stable expression of the introduced ADA genes can be maintained over several weeks. As a result, however, repeated injections were given, initially every 1–2 months, and subsequently once every 3–6 months. Evidence that the treatment was having the desired effect in kick-starting the patient's immune system was obtained from various measures of antibody and T-cell function. In parallel, there has been evidence of clinical improvement: the frequency of infections has dramatically decreased when compared with the incidence before treatment. The efficacy of *ADA* gene therapy is still difficult to assess because Ashanthi and other patients were simultaneously treated with PEG-ADA (see Marshall, 1995).

Since the pioneering work on ADA deficiency, gene therapy trials have been initiated for a few inherited disorders

For the reasons given above, ADA deficiency presented a favorable target for gene therapy. Since this pioneering work, gene therapy has been initiated for a few additional inherited disorders (*Table 20.5*), while progress for others has been frustratingly slow. Different recessively inherited disorders have been targets for *in vivo* or *ex vivo* gene augmentation therapy and, in the one case where a dominantly inherited disorder has been treated, the patient was a homozygote. The following examples are simply illustrative of current progress and difficulties.

Table 20.5: Examples of current gene therapy trials for inherited disorders

Disorder	Cells altered	Gene therapy strategy
ADA deficiency	T cells and hemopoietic stem cells	*Ex vivo* GAT using recombinant retroviruses containing an *ADA* gene
Cystic fibrosis	Respiratory epithelium	*In vivo* GAT using recombinant adenoviruses or liposomes to deliver the *CFTR* gene
Familial hypercholesterolemia	Liver cells	*Ex vivo* GAT using retrovirus to deliver the LDL receptor gene (*LDLR*)
Gaucher's disease	Hemopoietic stem cells	*Ex vivo* GAT using retroviruses to deliver the glucocerebrosidase gene (*GBA*)

GAT, gene augmentation therapy.

Familial hypercholesterolemia (FH)

This disorder is caused by a dominantly inherited deficiency of low density lipoprotein (LDL) receptors, which are normally synthesized in the liver, and is characterized by premature coronary artery disease. About 50% of heterozygous affected males die by 60 years of age, unless treated. Because FH is such a common single gene disorder, homozygotes are occasionally seen. They suffer precocious onset of disease and increased severity, with death from myocardial infarction commonly occurring in late childhood. Gene therapy for FH was first applied with some apparent success to a 28-year-old woman who is homozygous for a pathogenic missense mutation in the *LDLR* gene. She suffered a myocardial infarction at the age of 16 and required coronary artery bypass surgery at the age of 26.

The liver, being a solid internal organ, may not seem to be an ideal choice for targeting gene therapy, and its major cell population, the differentiated hepatocyte, is refractory to infection with retroviruses, the most widely used vector system. However, hepatocytes can be cultured *in vitro* and, under such conditions, are susceptible to retroviral infection. *Ex vivo* gene therapy became a possibility when animal experiments showed that cultured hepatocytes could be injected via the portal venous system – the veins which drain from the intestine directly into the liver – after which they appear to seed in the liver. The gene therapy involved surgical removal of a sizeable portion of the left lobe of the patient's liver, disaggregation of the liver cells and plating in cell culture prior to infection with retroviruses containing a normal human *LDLR* gene (Grossman *et al.*, 1994). The genetically modified cells were infused back into the patient through a catheter implanted into a branch of the portal venous system. The patient's LDL/high density lipoprotein (HDL) ratio subsequently declined from 10–13 before gene therapy to 5–8, and such improvement was maintained over a long period.

Cystic fibrosis

Cystic fibrosis is an autosomal recessive disorder that results in defective transport of chloride ions through epithelial cells, and results from mutations in a gene, *CFTR*, which encodes a cAMP-regulated chloride channel. The primary expression of the defect is in the lungs: a sticky mucus secretion accumulates which is prone to chronic infections. Because there are no methods to culture lung cells routinely in the laboratory, *in vivo* gene therapy approaches have been adopted. As respiratory epithelial cells are differentiated, retroviral vectors cannot be used. Instead, gene therapy trials have used adenovirus vectors or liposomes to transfer a suitably sized *CFTR* minigene, either through a bronchoscope or through the nasal cavity. The first adenovirus-based protocol began in 1993 and, although preliminary data have confirmed gene transfer into respiratory epithelium *in vivo*, there have been major concerns regarding the safety of the procedure. The first patient to be treated with a high dose of recombinant adenovirus experienced transient pulmonary infiltrates and alterations in vital signs, before recovering uneventfully. This experience prompted recognition of the need to confirm the maximum tolerated adenovirus dose. The liposome-based gene therapy trials are regarded as safer procedures, but the efficiency of gene transfer is expected to be much lower.

Duchenne muscular dystrophy

DMD is a severe X-linked recessive disorder: affected males suffer progressive muscle deterioration, are confined to a wheelchair in their teens and die usually by the third decade. The target tissue is skeletal muscle, and initial interest in treatment for this disorder focused on *cell therapy* because of the unique cell biology of muscle (see Miller and Boyce, 1995). As well as muscle fibers (or myofibers – very long, post-mitotic, multinucleate cells), skeletal muscle contains mononucleate myoblasts which are normally quiescent but can divide and subsequently fuse with myofibers to repair muscle damage. Although implanting normal or genetically modified myoblasts into diseased muscles appeared attractive, difficulties have been evident with this approach in humans, despite promising pilot studies with myoblast transfer in mice. Suitable gene therapy approaches have also been difficult to conceive, largely because of the lack of a suitable gene transfer system. Retroviral vectors cannot be used because adult skeletal muscle fibers are post-mitotic and hence not susceptible to retroviral infection. Adenovirus vectors have been used to deliver genes to muscle fibers *in vivo* and, although the post-mitotic state of muscle nuclei allows the expression to persist, the need for expression to continue over the course of a lifetime (which would be required for successful therapy) remains doubtful. For this, and many other disorders, one would like to see a vector system that combined the stable expression conferred by integrative vectors such as retroviruses with the wide target cell range of vectors such as adenoviruses. A final problem is the sheer size of the dystrophin coding sequence (~14 kb), although a very large central segment appears not to be crucially important (see England *et al.*, 1990).

Gene therapy for neoplastic disorders and infectious disease

General principles of gene therapy for neoplastic disorders and infectious disease

Cancer gene therapies

Many different approaches can be used for cancer gene therapy (see Culver and Blaese, 1994 and *Table 20.6*) and, in marked contrast to the few gene therapy trials for inherited disorders, numerous cancer gene therapy trials are currently being conducted (*Table 20.7*). This reflects partly the severity of the disorders that are being treated and the considerable funding for cancer research, and partly reflects the comparative ease in applying treatments based on targeted killing of disease cells, artificially or by enhancing an immune response. In a few cases, the gene therapy approach has focused on targeting single genes, such as *TP53* gene augmentation therapy and delivery of antisense *KRAS* genes in the case of some forms of non-small cell lung cancer. In most cases, however, targeted killing of cancer cells has been conducted without knowing the molecular etiology of the cancer.

In principle, two types of cancer gene therapy strategy can be envisaged:

- **tumor reduction strategies**. Many current cancer gene therapy strategies are not expected to result in 100% success at targeting the tumor cells. For

Table 20.6: Potential applications of gene therapy for the treatment of cancer

General approaches

Artificial killing of cancer cells

Insert a gene encoding a toxin (e.g. diphtheria A chain) or a gene conferring sensitivity to a drug (e.g. herpes simplex thymidine kinase) into tumor cells

Stimulate natural killing of cancer cells

Enhance the immunogenicity of the tumor by, for example, inserting genes encoding foreign antigens or cytokines

Increase anti-tumor activity of immune system cells by, for example, inserting genes that encode cytokines

Induce normal tissues to produce anti-tumor substances (e.g. interleukin-2, interferon)

Production of recombinant vaccines for the prevention and treatment of malignancy (e.g. BCG-expressing tumor antigens)

Protect surrounding normal tissues from effects of chemotherapy/radiotherapy

Protect tissues from the systemic toxicities of chemotherapy (e.g. multiple drug resistance type 1 gene)

Tumors resulting from oncogene activation

Selectively inhibit the expression of the oncogene

Deliver gene-specific antisense oligonucleotide or ribozyme to bind/cleave oncogene mRNA

Inhibit transcription by triple helix formation following delivery of a gene-specific oligonucleotide

Use of intracellular antibodies or oligonucleotide aptamers to specifically bind to and inactivate the oncoprotein

Tumors arising from inactivation of tumor suppressor

Gene augmentation therapy

Insert wild-type tumor suppressor gene

Table 20.7: Examples of current cancer gene therapy trials

Disorder	Cells altered	Gene therapy strategy
Brain tumors	Tumor cells *in vivo* Tumor cells *ex vivo* Hematopoietic stem cells *ex vivo*	Implanting of murine fibroblasts containing recombinant retroviruses to infect brain cells and ultimately deliver HSV-tk gene DNA transfection to deliver antisense *IGF1*
Breast cancer	Fibroblasts *ex vivo* Hematopoietic stem cells *ex vivo*	Retroviruses to deliver *MDR1* gene Retroviruses to deliver *IL4* gene
Colorectal cancer	Tumor cells *in vivo* Tumor cells *ex vivo*	Retroviruses to deliver *MDR1* gene Liposomes to deliver genes encoding HLA-B7 and β_2-microglobulin
Malignant melanoma	Fibroblasts *ex vivo* Tumor cells *in vivo* Tumor cells *ex vivo* Fibroblasts *ex vivo*	Retroviruses to deliver *IL2* or *TNF* gene Retroviruses to deliver *IL2* or *IL4* genes Liposomes to deliver genes encoding HLA-B7 and β_2-microglobulin Retroviruses to deliver *IL2* gene
Myelogenous leukemia	T cells/tumor cells *ex vivo*	Retroviruses to deliver *IL4* gene
Neuroblastoma	Tumor cells	Retroviruses to deliver *TNFA* gene
Nonsmall cell lung cancer	Tumor cells Tumor cells *in vivo*	Retroviruses to deliver HSV-tk gene Retroviruses to deliver antisense *KRAS*
Ovarian cancer	Tumor cells *in vivo* Tumor cells *ex vivo*	Retroviruses to deliver wild-type TP53 gene Retroviruses to deliver HSV-tk gene
Renal cell carcinoma	Hematopoietic stem cells *ex vivo* Tumor cells *ex vivo*	Retroviruses to deliver *MDR1* gene Retroviruses to deliver *IL2* or TNF genes
Small cell lung cancer	Fibroblasts *ex vivo*	Retroviruses to deliver *IL4* gene
Solid tumors	Tumor cells *ex vivo* Tumor cells *in vivo*	DNA transfection to deliver *IL2* gene Liposomes to deliver genes encoding HLA-B7 and β_2-microglobulin

example, gene therapy based on simple gene transfer into tumor cells (to ensure direct killing of the cancerous cells) is, like any other current form of gene transfer, comparatively inefficient and so some tumor cells may not be targeted. Accordingly, such types of treatment may be viewed as a refinement of conventional radiotherapy and surgical treatments.

- **Tumor elimination strategies**. Such approaches are intended to kill 100% of cancer cells. If, for example, immune system cells can be stimulated into a specific immune response against the tumor cells, complete remission may be possible. No matter which method is used, however, the aim of complete elimination of the cancerous cells may not be easy to attain because of the rapid evolution of cancer cells and strong selection for resistant cells (see page 473).

Gene therapy for infectious disorders

The gene therapy approaches for treating infectious disorders are slightly different. In common with cancer gene therapy are the principles of provoking a specific immune response or specific killing of infected cells. In addition, and increasingly popular, are strategies that are intended to affect the life-cycle of the infectious agent, reducing its ability to undergo productive infection. Some infectious agents are genetically comparatively stable. Others, however, may be undergoing rapid evolution, and, much as in the case of cancer cells, present problems for any general therapy. The classic example is AIDS, where the infectious agent, HIV-1, appears to mutate rapidly (see page 583).

Ex vivo cancer gene therapies frequently involve attempts to recruit immune system cells to destroy the tumor cells

Gene transfer into tumor-infiltrating lymphocytes

One of the earliest gene therapy protocols used a population of immune system cells for specifically targeting a foreign protein to a tumor. The therapy could be considered to be a form of adoptive immunotherapy (see below) because a gene encoding a cytokine, tumor necrosis factor-α (TNF-α), was transferred into tumor-infiltrating lymphocytes (TILs) in an effort to increase their anti-tumor efficacy. The TIL population is a natural population of T lymphocytes which can seek out and infiltrate tumor deposits, such as metastatic melanomas. TNF-α is a protein naturally produced by T lymphocytes which, if infused in sufficient amounts in mice, can destroy tumors. However, it is a toxic substance and intravenous infusion of TNF has significant adverse side effects in humans. An attractive alternative was to use TILs as *cellular vectors* for transferring the toxic protein directly to tumors. The gene therapy approach that was used, therefore, involved retroviral-mediated transfer of a TNF gene to a TIL population which had initially been obtained from an excised tumor and then grown in culture. Subsequent transfusion of the genetically modified TILs into a patient with metastatic melanoma was expected to result in the TILs 'homing in' on the melanomas, expression of the introduced TNF gene and tumor regression (*Figure 20.14*). However, the trial has been marked by comparatively poor efficiency of gene transfer into human TILs and a down-regulation of cytokine expression by the TILs.

Adoptive immunotherapy by genetic modification of tumor cells

Animal studies in which murine tumor cells were genetically modified by the insertion of genes encoding various cytokines [several different interleukins (ILs), TNF-α, interferon (IFN)-γ, granulocyte–macrophage colony-stimulating factor (GM-CSF)] and then re-implanted in mice gave cause for encouragement. In each case, the genetically altered tumor cells either never grew, or grew and then regressed.

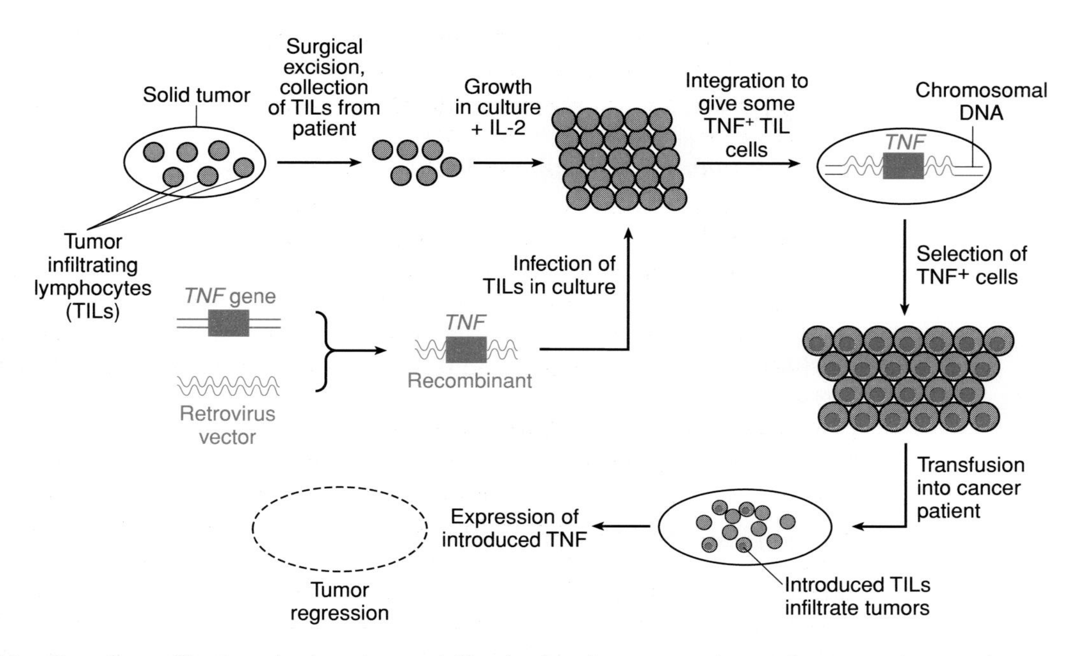

Figure 20.14: Genetic modification of cultured tumor-infiltrating lymphocytes can be used to target therapeutic genes to a solid tumor.

This approach has been used in an attempt at *ex vivo* gene therapy for metastatic melanoma. The tumor-infiltrating lymphocytes (TIL) appear to be able to 'home in' to tumor deposits. In this example, they act as cellular vectors for transporting to the melanomas a retrovirus recombinant which contains a gene specifying the anti-tumor cytokine TNF-α (tumor necrosis factor-α). Problems with the efficiency of gene transfer into the TILs and down-regulation of cytokines limited the success of this approach.

In addition, most of the treated mice were then systemically immune to re-implantation of nonmodified tumors. However, the results were much less satisfactory when animals with established sizeable tumors were treated. Nevertheless, the idea of modifying a patient's own tumor cells for use as a vaccine (**adoptive immunotherapy**) caught on, and human gene therapy trials have been approved for the insertion of cytokine genes using retrovirus vectors for treating a wide variety of cancers (see *Table 20.7*).

In each case, the idea is to immunize the patients specifically against their own tumors by genetically modifying the tumor with one of a variety of genes that are expected to increase the host immune reactivity to the tumor. In addition to cytokine genes, other genes such as foreign HLA antigen genes have been transferred to tumors for the same general reason. Insertion of genes encoding HLA-B7 into tumors of patients lacking HLA-B7 is intended to provoke an immune response to the tumors as a consequence of the presence on the tumor cell surface of the effectively foreign HLA-B7 antigen (see *Table 20.7* for some examples). Such a response is hoped to provide subsequent immunity against the same type of tumor even in the absence of the HLA-B7 antigen.

Adoptive immunotherapy by genetic modification of fibroblasts

One problem with *ex vivo* therapy for tumors is the difficulty in growing tumor cells *in vitro:* less than 50% of tumor cell lines grow in long-term culture. As an alternative, fibroblasts, which are much easier to adapt to long-term tissue culture, have been targeted in some cases. For example, transfer of genes encoding the cytokines IL-2 and IL-4 into skin fibroblasts grown in culture provides the basis of some clinical trials for treatment of breast cancer, colorectal cancer, melanoma and renal cell carcinoma. The IL-2- and IL-4-secreting fibroblasts are then mixed with irradiated autologous tumor cells and injected subcutaneously. In such cases, the hope is that the local production and secretion of cytokines by the transferred fibroblasts will induce a vigorous immune response to the nearby irradiated tumor cells and thereby result in a systemic anti-cancer immune response.

Other immunological approaches

Two other *ex vivo* gene therapy strategies are using immunological approaches to tumor destruction. One involves transferring an antisense insulin-like growth factor-1 (*IGF1*) gene into tumor cells in order to block production of IGF-1. Animal studies have shown that when tumor cells modified in this way are re-implanted *in vivo,* they provoke an immune response which can lead to destruction of nonmodified tumors, but the basis of immunological destruction is not known. A second approach involves the insertion of a co-stimulatory molecule such as B7-1 or B7-2, molecules which are normally present on lymphocytes, being required for full T lymphocyte activation.

In vivo gene therapy may be the only feasible approach for some cancers

Currently, a variety of different gene therapy approaches are being used involving genetic modification of tumor cells *in vivo.* In some cases, adoptive immunotherapy approaches are being employed, as in the case of increasing the immunogenicity of melanoma, colorectal tumors and a variety of solid tumors by the direct injection of liposomes containing a gene which encodes HLA-B7. The tumor cells take up the liposomes by phagocytosis and express the foreign HLA-B7 antigen transiently on their cells. More recent modifications include the additional insertion of a gene encoding the conserved light chain of HLA antigens, β_2-microglobulin.

A second approach has been the use of retrovirus-mediated transfer of a gene encoding a '**prodrug**', a reagent that confers sensitivity to cell killing following subsequent administration of a suitable drug. In one recent example, the target cells were brain tumor cells, notably recurrent glioblastoma multiforme, and the retroviruses were provided in the form of murine fibroblasts that are producing retroviral vectors (retroviral vector-producing cells or VPCs). The cells were directly implanted into multiple areas within growing tumors using stereotactic injections guided by magnetic resonance imaging (*Figure 20.15*). Once injected, the VPCs continuously produce retroviral particles within the tumor mass, transferring genes into surrounding tumor cells. Although retroviruses are not normally used for *in vivo* gene therapy because of their sensitivity to serum complement, they are comparatively stable in this special environment and have the advantage that, since they only infect actively dividing cells, the tumor cells are a target, but not nearby brain cells (which are usually terminally differentiated).

The prodrug gene that was transferred is a HSV gene which encodes thymidine kinase (HSV-tk). HSV-tk confers sensitivity to the drug gancyclovir by phosphorylating it within the cell to form gancyclovir monophosphate which is subsequently converted by cellular kinases to gancyclovir triphosphate. This compound inhibits DNA polymerase and causes cell death (see *Figure 20.15*). Such therapy appears to benefit from a phenomenon known as the **by-stander effect**: adjacent tumor cells that have not taken up the HSV-tk gene may still be destroyed. This is thought to be due to diffusion of the gancyclovir triphosphate from cells which have taken up the HSV-tk gene, perhaps via gap junctions.

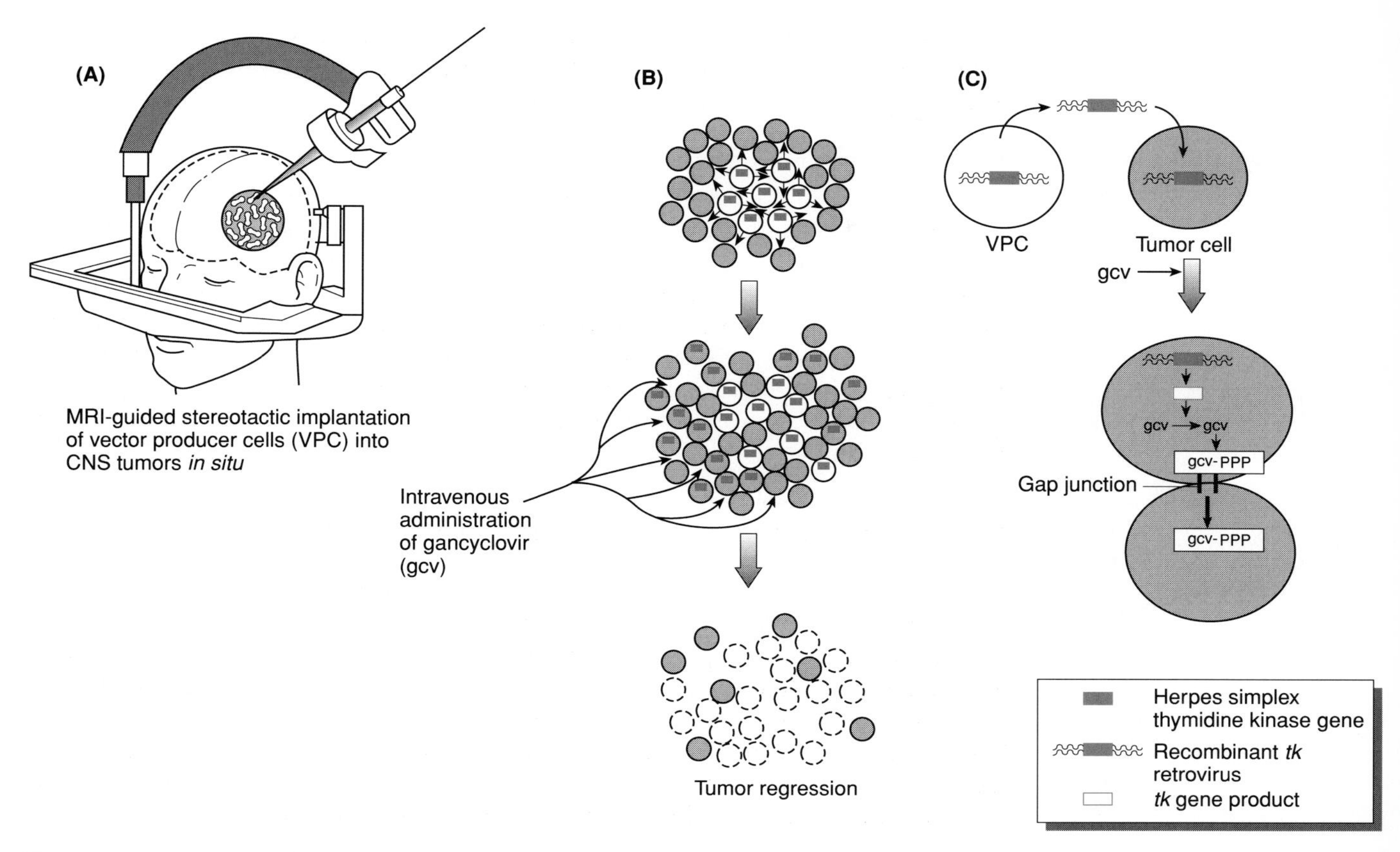

Figure 20.15: *In vivo* gene therapy for brain tumors.

This example shows a strategy for treating glioblastoma multiforme *in situ* using a delivery method based on magnetic resonance imaging-guided stereotactic implantation of retrovirus vector-producing cells (VPCs). The retroviral vectors produced by the cells were used to transfer a gene encoding a 'prodrug', herpes simplex thymidine kinase (HSV-tk), into tumor cells. This reagent confers sensitivity to the drug gancyclovir: HSV-tk phosphorylates gancyclovir (gcv) to a monophosphorylated form gcv-P and, thereafter, cellular kinases convert this to gancyclovir triphosphate, gcv-PPP, a potent inhibitor of DNA polymerase which causes cell death. Because retroviruses infect only dividing cells, they infect the tumor cells, but not normal differentiated brain cells. The implanted VPCs transferred the HSV-tk gene to neighboring tumor cells, rendering them susceptible to killing following subsequent intravenous administration of gancyclovir. In addition, it was found that uninfected cells were also killed by a *by-stander effect*: the gancyclovir triphosphate appeared to diffuse from infected cells to neighboring uninfected cells, possibly via gap junctions. Reproduced in part from Culver *et al.* (1994) with permission from Mary Ann Liebert Inc.

Gene therapy for infectious disorders is often aimed at selectively interfering with the life-cycle of the infectious agent

Current gene therapy trials for infectious disorders are conspicuously targeted at treating AIDS patients. The infectious agent for this usually fatal disorder is a class of retrovirus known as *HIV-1* which can infect helper T lymphocytes, a crucially important subset of immune system cells (see *Figure 20.16*). Two features of HIV-1 make it especially deadly: it eventually kills the helper T cells (thereby rendering patients susceptible to other infections), and the provirus tends to persist in a latent state before being suddenly activated (the lack of virus production during the latent state complicates anti-viral drug treatment). A major problem is that the HIV genome is mutating at a very high rate (Hahn *et al.*, 1986).

In principle, a variety of gene therapy strategies can be envisaged for treating AIDS. As in the case of cancer gene therapy, infected cells can be killed directly (by insertion of a gene encoding a toxin or a prodrug; see above) or indirectly, by enhancing an immune response against them. For example, this can involve transferring a gene that encodes an HIV-1 antigen, such as the envelope protein gp120, and expressing

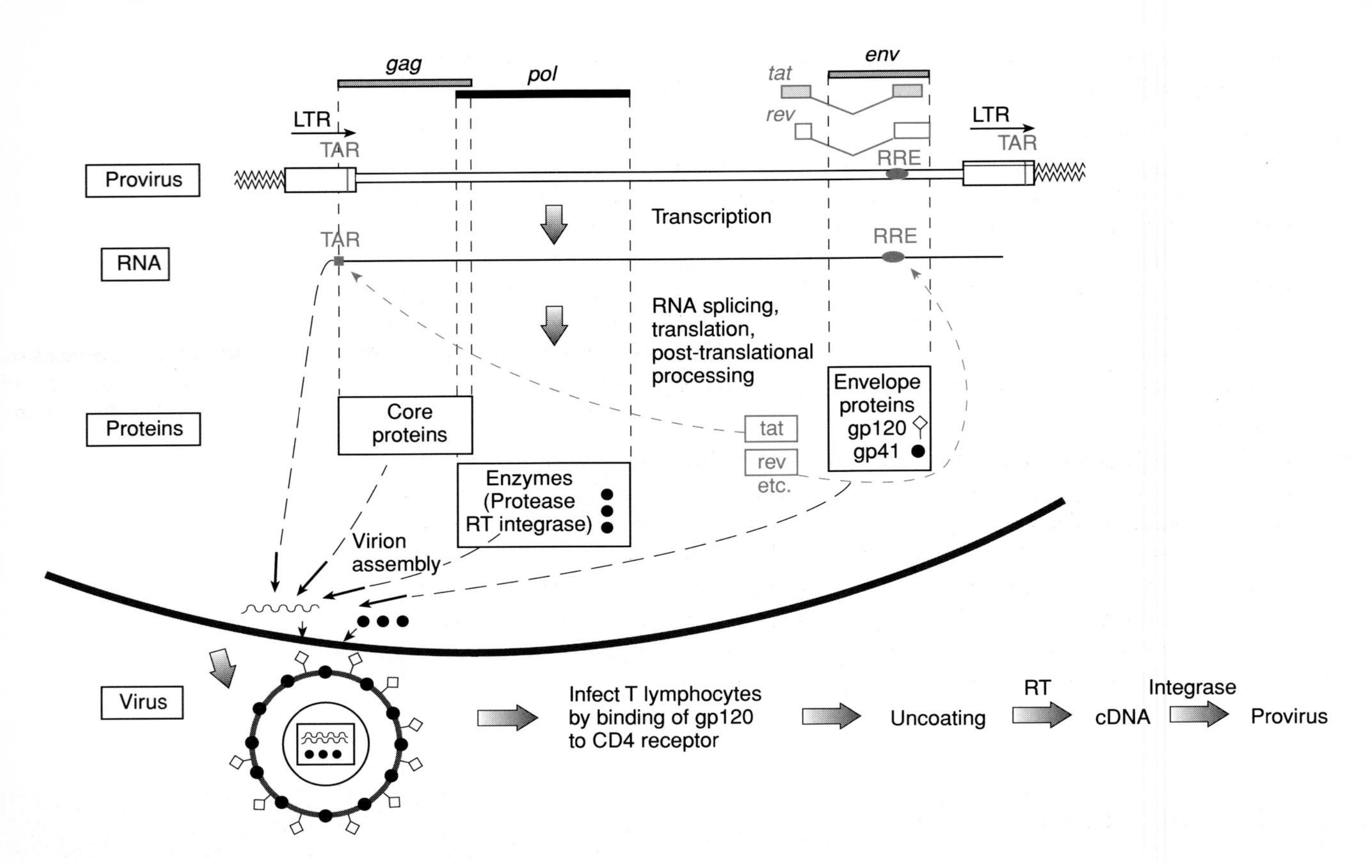

Figure 20.16: The HIV-1 virus life-cycle.

The HIV-1 virus is a retrovirus which contains two identical single-stranded viral RNA molecules and various viral proteins within a viral protein core, which itself is contained within an outer envelope. The latter contains lipids derived from host cell plasma membrane during budding from the cell, plus viral coat proteins gp120 and gp41. Penetration of HIV-1 into a T lymphocyte is effected by specific binding of the gp120 envelope protein to the CD4 receptor molecules present in the plasma membrane. After entering the cell, the viral protein coat is shed, and the viral RNA genome is converted into cDNA by viral reverse transcriptase (RT). Thereafter a viral *integrase* ensures integration of the viral cDNA into a host chromosome. The resulting **provirus** (see top) contains two long terminal repeats (LTRs), with transcription being initiated from within the upstream LTR. For the sake of clarity, the figure only shows some of the proteins encoded by the HIV-1 genome. In common with other retroviruses, are the gag (core proteins), pol (enzymes) and env (envelope proteins). Tat and rev are regulatory proteins which are encoded in each case by two exons, necessitating RNA splicing.
The tat protein functions by binding to a short RNA sequence at the extreme 5′ end of the RNA transcript, known as TAR (***t**rans*-**a**cting **r**esponse element); the rev protein binds to an RNA sequence, RRE (**r**ev **r**esponse **e**lement), which is encoded by sequence transcribed from the *env* gene.

it in the patient in order to provoke an immune response against the HIV-1 virus, or the patient's immune system can be boosted by transfer and expression of a gene encoding a cytokine, such as an interferon. Another general approach, which is applicable to all disorders caused by infectious agents, is to find a means of interfering with the life-cycle of the infectious agent.

Gene therapy strategies designed to interfere with the HIV-1 life-cycle

A wide variety of such strategies are available (see Gilboa and Smith, 1994). Basically inhibition has been envisaged at three major levels:

- *Blocking HIV-1 infection.* HIV-1 normally infects T lymphocytes by binding of the viral gp120 envelope protein to the CD4 receptor on the cell membrane. Transfer of a gene encoding a soluble form of the CD4 antigen (sCD4) into T lymphocytes or hemopoietic cells and subsequent expression will result in circulating sCD4. If the levels of circulating sCD4 are sufficiently high, binding of sCD4 to the gp120 protein of HIV-1 viruses could be imagined to inhibit infection of T lymphocytes without compromising T lymphocyte function.

- *Inhibition at the RNA level*. The production of HIV-1 RNA can be selectively inhibited by standard antisense/ribozyme approaches (see page 568), and also by the use of **RNA decoys**. The latter strategy exploits unique regulatory circuits which operate during HIV replication. Two key HIV regulatory gene products are tat and rev which bind to specific regions of the nascent viral RNA, known as TAR and RRE respectively (*Figure 20.16*). Artificial expression of short RNA sequences corresponding to TAR or RRE will generate a source of decoy sequences which can compete for binding of tat and rev, and possibly thereby inhibit binding of these proteins to their physiological target sequences.
- *Inhibition at the protein level*. There are numerous different strategies. One strategy involves designing *intracellular antibodies* (see page 570), against HIV-1 proteins, such as the envelope proteins. Another involves introducing genes that encode dominant-negative mutant HIV proteins which can bind to and inactivate HIV proteins (**transdominant proteins**). For example, transdominant mutant forms of the gag proteins have been shown to be effective in limiting HIV-1 replication, possibly by interfering with multimerization and assembly of the viral core (see Gilboa and Smith, 1994).

The ethics of human gene therapy

All current gene therapy trials involve treatment for somatic tissues (**somatic gene therapy**). Somatic gene therapy, in principle, has not raised many ethical concerns other than its possible application in *enhancement genetic engineering* (any treatment involving genetic modification of an individual's cells in order to enhance some trait, such as height, without attempting to treat disease; see Anderson, 1985). Clearly, every effort must be made to ensure the safety of the patients, especially since the technologies being used for somatic gene therapy are far from perfect. However, confining the treatment to somatic cells means that the consequences of the treatment are restricted to the individual patient who has consented to this procedure. Many, therefore, view the ethics of somatic gene therapy to be at least as acceptable as, say, organ transplantation, and feel that ethical approval is appropriate for carefully assessed proposals. Patients who are selected for such treatments have severely debilitating, and often life-threatening, disease for which no effective conventional therapy is available. As a result, despite the obvious imperfections of the technology, it may even be considered to be unethical to refuse such treatment.

Germline gene therapy, involving the genetic modification of germline cells (e.g. in the early zygote), is considered to be entirely different. It has been successfully practised on animals (e.g. to correct β-thalassemia in mice – see Constantini *et al.*, 1986). However, thus far, it has not been sanctioned for the treatment of human disorders, and approval is unlikely to be given in the near future, if ever (see next section).

Human germline gene therapy has not been practised because of ethical concerns and limitations of the technology for germline manipulation

The lack of enthusiasm for the practice of germline gene therapy can be ascribed to three major reasons.

The imperfect technology for genetic modification of the germline

Germline gene therapy requires modification of the genetic material of chromosomes (most easily by chromosomal integration of an introduced gene). However, vector systems for accomplishing this do not allow accurate control over the integration site or event. In somatic gene therapy, the only major concern about lack of control over the fate of the transferred genes is the prospect that one or more cells undergoes neoplastic transformation. However, in germline gene therapy, genetic modification has implications not just for a single cell. In addition to cancer, accidental insertion of a recombinant retrovirus within an important gene could result in a novel inherited pathogenic mutation, for example.

The questionable ethics of germline modification

Genetic modification of human germline cells may have consequences not just for the individual whose cells were originally altered, but also for all individuals who inherit the genetic modification in subsequent generations. Germline gene therapy would inevitably mean denial of the rights of these individuals to any choice about whether their genetic constitution should have been modified in the first place (Wivel and Walters, 1993). Some, however, have considered that the technology of germline modification will inevitably improve in the future to an acceptably high level and, provided there are adequate regulations and safeguards, there should then be no ethical objections (see, for example, Zimmerman, 1991). Others perceive that, in addition to the question of the rights of individuals in the future, this technology will inevitably lead to a slippery slope towards genetic enhancement. This would entail a program of *positive eugenics,* whereby planned genetic modification of the germline could involve artificial selection for genes that are thought to confer advantageous traits. Inevitably, even if this were judged to be acceptable in principle, the question is raised of who decides what traits are advantageous. The horrifying nature of past *negative eugenics* programs (most recently in Nazi Germany, and in many states of the USA where compulsory sterilization of individuals adjudged to be feeble-minded was practised well into the present century) serves as a reminder to many of the potential Pandora's box of ills that could be released if ever human germline gene therapy were to be attempted.

The questionable need for germline gene therapy

Germline genetic modification may be considered as a possible way of avoiding what would otherwise be the certain inheritance of a known harmful mutation. However how often does this situation arise and how easy would it be to intervene? A 100% chance of inheriting a harmful mutation could most likely occur in two ways. One is when an affected woman is *homoplasmic* for a harmful mutation in the mitochondrial genome (see page 66) and wishes to have a child. The trouble here is that, because of the multiple mitochondrial DNA molecules involved, we still have very little to offer in the way of providing gene therapy for such disorders. A second situation concerns inheritance of mutations in the nuclear genome. To have a 100% risk of inheriting a harmful mutation would require mating between a man and a woman both of whom have the same recessively inherited disease, an extremely rare occurrence. Instead, the vast majority of mutations in the nuclear genome are inherited with at most a 50% risk (for dominantly inherited disorders) or a 25% risk (for recessively inherited disorders). *In vitro* fertilization provides the most accessible way of modifying the germline. However, if the chance that any one zygote is normal is as high as 50 or 75%, gene transfer into an unscreened fertilized egg which may well be normal would be unacceptable: the procedure would inevitably carry some risk, even if the safety of the techniques for germline gene transfer improves markedly in the future. Thus, screening using sensitive PCR-based techniques would be required to identify a fertilized egg with the harmful mutation. Inevitably, the same procedure can be used to identify fertilized eggs that lack the harmful mutation. Since *in vitro* fertilization generally involves the production of several fertilized eggs, it would be much simpler to screen for normal eggs and select these for implantation, rather than to attempt genetic modification of fertilized eggs identified as carrying the harmful mutation.

Further reading

Crooke ST, Lebleu B. (1993) *Antisense Research and Applications.* CRC Press, Boca Raton, FL.

Culver KW. (1994) *Gene Therapy. A Handbook for Physicians.* Mary Ann Liebert, Inc., New York.

Friedmann T. (1994) Gene therapy for neurological disorders. *Trends Genet.*, **10**, 210–214.

Huber BE, Lazo JS. (1994) Gene therapy for neoplastic disorders. *Ann. NY Acad. Sci.*, **716**.

Kay MA, Woo SLC. (1994) Gene therapy for metabolic disorders. *Trends Genet.*, **10**, 253–257.

Wilkinson GWG, Darley RL, Lowenstein P. (1994) Viral vectors for gene therapy. In: *From Genetics to Gene Therapy* (ed. DS Latchman), pp. 157–188. BIOS Scientific Publishers, Oxford.

Wolff JA. (1994) *Gene Therapeutics.* Birkhauser Boston, Boston.

References

Acsadi G, Dickson G, Love DR *et al.* (1991) *Nature*, **352**, 815–818.

Anderson WF. (1985) *J. Med. Philosoph.*, **10**, 275–191.

Arbones ML, Austin HL, Capon DJ, Greenburg G. (1994) *Nature Genetics*, **6**, 90–97.

Barinaga M. (1993) *Science*, **262**, 1512–1514.

Bock LC, Griffin LC, Latham JA, Vermaas EH, Toole JJ. (1992) *Nature*, **355**, 564–566.

Boris-Lawrie K, Temin HM. (1994) *Ann. NY Acad. Sci.*, **716**, 59–71.

Brody SL, Crystal R. (1994) *Ann. NY Acad. Sci.*, **716**, 90–103.

Cech T. (1995) *Biotechnology*, **13**, 323–326.

Chubb JM, Hogan ME. (1992) *Trends Biotechnol.*, **10**, 132–136.

Culver KW, Blaese RM. (1994) *Trends Genet.*, **10**, 174–178.

Curiel DT. (1994) *Ann. NY Acad. Sci.*, **716**, 36–58.

England SB, Nicholson LV, Johnson MA *et al.* (1990) *Nature*, **343**, 180–182.

Gilboa E, Smith C. (1994) *Trends Genet.*, **10**, 139–144.

Green LL, Hardy MC, Maynard-Currie CE *et al.* (1994) *Nature Genetics*, **7**, 13–21.

Grossman M, Raper RE, Kozarsky K *et al.* (1994) *Nature Genetics*, **6**, 335–341.

Hahn BH, Shaw GM, Taylor ME *et al.* (1986) *Science*, **232**, 1548–1553.

Hodgson CP. (1995) *Biotechnology*, **13**, 222–225.

Jiao S, Gurevich V, Wolff JA. (1993) *Nature*, **362**, 450–453.

Kaplitt MG, Leone P, Samulski RJ, Xiao X, Pfaff DW, O'Malley KL, During MJ. (1994) *Nature Genetics*, **8**, 148–153.

Marasco WA, Haseltine WA, Chen S. (1993) *Proc. Natl Acad. Sci. USA*, **90**, 7889–7893.

Marshall E. (1995) *Science*, **269**, 1050–1055.

Miller JB, Boyce FM. (1995) *Trends Genet.*, **11**, 163–165.

Moffat AS. (1991) *Science*, **254**, 35–36.

Mulligan RC. (1993) *Science*, **260**, 926–932.

Wagner RW. (1994) *Nature*, **372**, 333–335.

Winter G, Harris W. (1993) *Immunol. Today*, **14**, 243–246.

Woolf TD, Chase JM, Stinchcomb DT. (1995) *Proc. Natl Acad. Sci.* USA, **92**, 8298–8302.

Wivel NA, Walters L. (1993) *Science*, **262**, 533–538.

Zimmerman B. (1991) *J. Med. Philos.*, **16**, 593–612.

Glossary

Acrocentric (of chromosome): having the centromere close to one end.

Add-on mutagenesis: a form of PCR mutagenesis in which the 5′ end of a primer is deliberately designed to introduce into the amplification product a DNA sequence not present in the target DNA. See *Figure 6.14.*

Allele: one of several alternative forms of a gene or DNA sequence at a specific chromosomal location (*locus*). At each autosomal locus an individual possesses two alleles, one inherited from the father and one from the mother.

Allele-specific oligonucleotide (ASO): a synthetic oligonucleotide, often about 20 bases long, which hybridizes to a specific target sequence and whose hybridization can be disrupted by a single base pair mismatch under carefully controlled conditions. ASOs are often labeled and used as allele-specific hybridization probes (see *Figure 5.10*). They can also be designed to act as allele-specific primers in certain PCR applications. See *Amplification refractory mutation system.*

Allelic association: any significant association between specific alleles at two or more neighbouring loci.

Allelic exclusion: the mechanism in B lymphocytes (or T lymphocytes) whereby an immunoglobulin chain (or T-cell receptor) is synthesized by either a paternal homolog or the maternal homolog, but not by both.

Alphoid DNA (also called **α-satellite DNA**): a class of satellite DNA with an average repeat length of about 170 bp; found at centromeres. See *Figure 8.8.*

Alternative splicing: the natural usage of different sets of splice junction sequences, to produce more than one product from a single gene. See *Figure 7.12.*

***Alu* repeat** (or **sequence**): one of a family of about 750 000 interspersed sequences in the human genome which are thought to have originated from the 7SL RNA gene. See *Figures 8.12* and *9.24*

***Alu*-PCR**: a PCR reaction in which an oligonucleotide primer with a sequence derived from the *Alu* repeat is used to amplify sequences between pairs of neighboring *Alu* sequences which are in converging opposite orientations. See *Figure 6.13* for the general principle.

***Amber* mutation**: a mutation which creates a UAG nonsense mutation within a coding DNA sequence.

Amplification refractory mutation system (ARMS): an allele-specific PCR amplification reaction. See *Figure 6.8.*

Amplimer: an oligonucleotide used as a primer of DNA synthesis in the polymerase chain reaction. See *Figure 6.1.*

Annealing: the association of complementary DNA (or RNA) strands to form a double-stranded structure.

Anonymous DNA: DNA not known to have a coding function.

Anticipation: a phenomenon in which the age of onset of a disorder is reduced and/or the severity of the phenotype is increased in successive generations.

Anticodon: a sequence of three consecutive bases in a tRNA molecule which specifically binds to a complementary codon sequence in mRNA. See *Figures 1.6* and *1.20.*

Antisense strand: the DNA strand of a gene which, during transcription, is used as a template by RNA polymerase to synthesize a complementary RNA strand. See *Figure 1.11.*

Antisense oligonucleotide (or **RNA**): a synthetic oligonucleotide or an RNA molecule which has a sequence that is complementary to a naturally occurring mRNA molecule. See *Figure 20.7.*

Apoptosis: programmed cell death.

ARMS: see *Amplification refractory mutation system.*

Autosome: any chromosome other than the sex chromosomes, X and Y.

Autozygosity: in an inbred person, homozygosity for alleles identical by descent.

Autozygosity mapping: a form of genetic mapping for autosomal recessive disorders in which affected individuals are expected to have two identical disease alleles by descent.

Bacterial artificial chromosome (BAC): a recombinant plasmid which permits propagation of very large inserts (up to 300 kb) in bacterial cells. The plasmid vector is selected to have a low copy number origin of replication.

Bacteriophage: a virus which infects bacterial cells. Often used in the abbreviated form *phage*. See *Figure 4.10* for an example.

Bubble-linker PCR (also called **vectorette-PCR**): a method of PCR amplifying DNA adjacent to a known sequence. See *Figure 6.12.*

Candidate gene: any gene which by virtue of a known property (function, expression pattern, chromosomal location, structural motif, etc.) is considered as a possible locus for a given disease.

Cap: a specialized chemical group that is naturally added to the 5′ end of mRNA. See *Figure 1.17.*

cDNA: DNA which is synthesized by the enzyme reverse transcriptase using mRNA as a template. The initial product is a single-stranded DNA which is complementary in sequence to the mRNA. See *Figure 4.7.*

cDNA selection (also called **direct selection**): a hybridization-based method for retrieving genomic clones containing an expressed sequence by heteroduplex formation with a suitable cDNA source. See *Figure 11.20.*

CentiMorgan (cM): a unit of genetic distance equivalent to a 1% probability of recombination during meiosis (*see* Morgan). One centiMorgan is equivalent, on average, to a physical distance of approximately 1 megabase in the human genome.

CentiRay (cR): a mapping unit when using radiation hybrids, which is dependent on the intensity of the irradiation. A distance of 1 cR_{8000} represents 1% frequency of breakage between two markers after exposure to a dose of 8000 rad.

Centromere: the primary constriction of a chromosome, separating the short arm from the long arm. Its major function is to ensure correct segregation of homologous chromosomes during meiosis and mitosis. See *Figure 2.3*.

CEPH (Centre d'Études du Polymorphisme Humain): an institution in Paris, France, which has played a leading role in genetic and physical mapping of the human genome.

CEPH families: a panel of 60 families, each comprising at least six children, their parents and their four grandparents from which samples have been made available by CEPH for genetic mapping by the scientific community.

Chimera: an organism derived from more than one zygote.

Chimeric DNA clone: a DNA clone in which insert sequences derive from more than one locus, either as a result of *co-ligation* of noncontiguous DNA fragments or as a result of recombination between different recombinant DNA molecules in a single cell.

Chromatid/chromosome: from the anaphase stage of mitosis through G1 to the early part of the S stage of the cell cycle, a chromosome takes the form of a single, linear double-stranded DNA molecule complexed with chromosomal proteins (a chromatid). Following DNA duplication at the S stage of the cell cycle, a chromosome consists of two identical sister chromatids joined at the centromere. At the anaphase stage of mitosis the centromere divides to give two chromosomes (chromatids). See *Figure 2.2*.

Chromosome jumping: a technique whereby a genomic DNA clone is used to identify other clones containing sequences originally located tens or hundreds of kilobases along the chromosome. See *Figure 14.11*.

Chromosome painting: fluorescent labeling of whole chromosomes by a *FISH* procedure in which labeled probes each consist of a complex mixture of different DNA sequences from a single chromosome. See *Figure 11.6*.

Chromosome walking: a method of assembling clone contigs by using individual genomic DNA clones as hybridization probes for screening a genomic DNA library. See *Figure 11.13*.

***Cis*-acting**: pertaining to a nucleic acid sequence which regulates the expression or function of a gene present on the same molecule, often as a result of binding a *trans*-acting protein.

Coding DNA: the segment of a gene whose sequence is decoded during gene expression to give a polypeptide or mature RNA product.

Codon: a nucleotide triplet which specifies an amino acid or a signal for terminating the synthesis of a polypeptide. See *Figure 1.22*.

Cognate sequences (or **DNA** or **clones)**: two or more sequences which ultimately derive from the same locus and share a common sequence of genetic information (e.g. a β-globin cDNA clone and a YAC containing the β-globin gene, *HBB*).

Colony hybridization: a form of *molecular hybridization* in which the target DNA is present within individual bacterial colonies that have been transferred to a nitrocellulose or nylon membrane. See *Figure 5.15*.

Comparative genome hybridization (CGH): use of competitive fluorescence *in situ* hybridization to detect chromosomal regions that are amplified or deleted, especially in tumors. See *Figure 17.5*.

Complementary DNA (or RNA) strands: strands which form a stable double-stranded structure. See *Figure 1.5*.

Concerted evolution: the process whereby individual members of a DNA family within on species are more closely related to each other than to members of the same type of DNA family in other species. See *Figure 9.14*.

Conservative substitution: a mutation causing a codon to be replaced by another codon that specifies a different amino acid, but one which is related in chemical properties to the original amino acid. A *nonconservative substitution* results in replacement of a codon by another which specifies an amino acid with different chemical properties. See *Box 10.2*.

Constitutional mutation: a mutation which is inherited and therefore present in all cells containing the relevant nucleic acid.

Constitutive expression: a state whereby a gene is permanently active. Mutations which result in inappropriate constitutive expression may often be pathogenic.

Contig: continuous region of genomic DNA which has been cloned as a series of identifiable overlapping DNA clones. Contigs can be assembled by a directional process (*chromosome walking*) or by random *clone fingerprinting* of all clones in a genomic DNA library. See *Figures 11.12* and *11.15*.

Copy number: the number of different copies of a particular DNA sequence in a genome. *Reassociation kinetics* permit fractionation of complex eukaryotic genomes into broad fractions with different copy number, such as single copy, intermediate repetitive and highly repetitive.

Cos sequence: cohesive termini at the extremities of linear lambda DNA molecules.

COS cells: a cell line established from monkey kidney cells, commonly used as in eukaryotic expression cloning.

Cosmid: a vector constructed by inserting the *cos* sequences of bacteriophage lambda into a plasmid. Permits cloning of foreign DNA in the size range of about 30–46 kb. See *Figure 4.13*.

Cot: the product of concentration and time in experiments which measure the rate of reassociation of denatured complementary DNA strands.

Cot-1 DNA: a fraction of DNA which contains a high proportion of highly repetitive sequences. Obtained from total genomic DNA by selecting for rapidly reassociating DNA sequences during renaturation of DNA. See *Box 5.3*.

CpG dinucleotide (or **doublet**): a dinucleotide with a cytosine at the 5′ end connected by a phosphodiester bond to a guanine at the 3′ end. The CpG dinucleotide is relatively rare in mammalian DNA and is a mutational hotspot because of the tendency for the cytosine to be methylated and subsequently deaminated to thymine.

CpG island: short stretch of DNA, often <1 kb, containing CpG dinucleotides which are unmethylated and present at the expected frequency. CpG islands often occur at transcriptionally active DNA. See *Figure 11.17.*

Cre: a bacteriophage P1 gene whose product facilitates recombination between specific target sequences, known as *lox*P sequences. So called because it creates recombination. See *Figures 4.14* and *19.20.*

Crossover analysis: a common form of genetic mapping in which a disease gene is mapped to a small genetic interval by identifying recombinants with distal and proximal flanking markers. See *Figure 14.5.*

Cryptic splice site: a sequence which resembles an authentic splice junction site and which can, under certain circumstances, participate in an RNA splicing reaction. See *Figure 10.12.*

Degenerate oligonucleotide: a panel of synthetic oligonucleotides designed so that collectively they correspond to various codon permutations for a given sequence of amino acids. See also *DOP-PCR.*

Denaturation: dissociation of complementary strands to give single-stranded DNA and/or RNA.

Differential display: a form of RT-PCR in which reverse transcriptase catalyzes cDNA synthesis by using a modified oligo (dT) primer which binds to the poly(A) tail of a subset of mRNAs. Enables study of mRNA subsets.

Distal (with reference to a chromosomal location): close to, or in the direction of, a telomere.

DNA fingerprinting: use of a hypervariable minisatellite DNA probe (usually those developed by Jeffreys) on a Southern blot to produce an individual-specific series of bands for identification of individuals or relationships (see *Figure 16.16*). See also *Clone fingerprinting.*

DNA footprinting: a method of identifying and localizing sequences within DNA molecules that can specifically bind protein molecules, as a result of the protection that bound protein affords against digestion by nucleases. See *Figure 19.10.*

DNA library: a collection of cell clones containing different recombinant DNA clones which are collectively representative of a complex source of starting DNA.

Dominant: (in human genetics) describes any trait which is expressed in a heterozygote. See also *Semi-dominant.*

Dominant negative mutation: a mutation which results in a mutant gene product which can inhibit the function of the wild-type gene product in heterozygotes. See, for example, *Figure 15.7.*

Dosage effect: see *Gene dosage.*

DOP-PCR (degenerate oligonucleotide primer PCR): a form of PCR in which an individual primer is designed to be a mixture of closely related oligonucleotides corresponding to different codon permutations for an amino acid sequence specified by the target DNA. See *Figure 6.9.*

Dot-blot: a *molecular hybridization* method in which the target DNA is spotted on to a nitrocellulose or nylon membrane.

Ectopic transcription: see *Illegitimate transcription.*

Electroporation: a method of transferring nucleic acid molecules into cells following delivery of a high voltage pulse.

Embryonic stem cells (ES cells): a cell line derived from undifferentiated, pluripotent cells from the embryo. Mouse ES cell lines are commonly used as a vehicle for transferring foreign DNA into the germline in order to generate transgenic mice.

Enhancer: a combination of short sequence elements which stimulate the transcription of a gene and whose function is not critically dependent on their precise position or orientation.

Episome: any DNA sequence which can exist in an autonomous extra-chromosomal form or can be integrated into the chromosomal DNA of the cell. Often used to describe self-replicating and extra-chromosomal forms of DNA.

EST (expressed sequence tag): a short sequence of a cDNA clone for which a PCR assay is available.

Euchromatin: the fraction of the nuclear genome which contains transcriptionally active DNA and which, unlike *heterochromatin*, adopts a relatively extended conformation.

Expression cloning: a form of DNA cloning in which the vector (*expression vector*) contains regulatory sequences designed to ensure expression of a coding DNA sequence insert, resulting in synthesis of foreign polypeptide. The resulting *expression libraries* can be screened using a specific antibody to bind to and identify a bacterial colony which synthesizes the desired polypeptide. See, for an example, *Figure 19.1.*

Ex vivo: pertaining to an experimental design that involves manipulating living cells in culture prior to introducing the manipulated cells into a whole animal or individual. See *Figure 20.2.*

Exon: segment of a gene which is decoded to give an mRNA product or a mature RNA product. Individual exons may contain coding DNA and/or noncoding DNA (untranslated sequences). See *Figure 1.14.*

Exon skipping: a form of alternative splicing in which splice junction sites that are normally involved in RNA splicing are not used by the splicing apparatus, resulting in the loss of whole exon sequences from the spliced RNA. See *Figure 10.11.*

Exon trapping: an artificial RNA splicing assay for detecting sequences within a cloned DNA that are capable of splicing to exons within a specialized vector. See *Figure 11.19.*

FiberFISH: a form of fluorescence *in situ* hybridization in which the target DNA is artificially stretched DNA fibers.

FISH: see *Fluorescence* in situ *hybridization.*

Flow cytometry: the fractionation of chromosomes according to size and base composition in a fluorescence-activated chromosome (or cell) sorter. See *Figure 11.7.*

Fluorescence *in situ* hybridization (FISH): a form of chromosome *in situ* hybridization in which the nucleic acid probe is labeled by incorporation of a *fluorophore*, a chemical group which fluoresces when exposed to UV irradiation.

Frameshift mutation: a mutation which alters the normal translational *reading frame* of a DNA sequence.

Gain-of-function mutation: a mutation which produces a different phenotype to that observed in the case of loss of function at the same locus.

Gel retardation assay: a gel electrophoresis-based assay for identifying sequences within a cloned DNA that can bind protein. See *Figure 19.11*.

Gene: a segment of DNA which normally specifies a functional polypeptide or RNA product.

Gene conversion: a naturally occurring nonreciprocal genetic exchange in which a sequence of one DNA strand (*acceptor sequence*) is altered so as to become identical to the sequence of another DNA strand (*donor sequence*). See *Figure 10.10*.

Gene dosage: the number of copies of a gene. Variation in the normal copy number can result in aberrant gene expression.

Gene targeting: a form of *in vivo* mutagenesis whereby the sequence of a pre-determined gene is selectively modified within an intact cell. See *Figure 19.18*.

Gene therapy: an attempt to treat disease by genetic modification of the cells of a patient. See *Box 20.1*.

Genetic imprinting: in its widest sense, any mechanism by which individual cells express the paternal or the maternal allele of a biallelic gene, but not both. Commonly used to describe autosomal genes, but *X chromosome inactivation* in mammals is a form of imprinting. See *Figure 7.15*.

Genome walking: any method of using a cloned DNA sequence to identify sequences adjacent to it on a chromosome. This may be hybridization-based (chromosome walking) or PCR-based (e.g. the use of *bubble-linker PCR*).

Genotype: (i) the genetic constitution of an individual; (ii) the types of alleles found at a locus in an individual

Germline: the gametes (egg and sperm cells) and precursor cells from which the gametes derive by cell division.

Germline mosaic (also **germinal mosaic, gonadal mosaic, gonosomal mosaic**): an individual who has a subset of germline cells carrying a mutation which is not found in other germline cells.

Haploid: describing a cell (typically a gamete) which has only a single copy of each chromosome (i.e. 23 in man). *Note* that the mammalian Y chromosome is haploid outside the pseudoautosomal region in otherwise diploid cells in males.

Haploinsufficiency: a locus shows haploinsufficiency if producing a normal phenotype requires more gene product than the amount produced by a single copy.

Haplotype: a series of alleles found at linked loci on a single (paternal or maternal) chromosome.

Hardy–Weinberg law: the simple relationship between gene frequencies and genotype frequencies that is found in a population under certain conditions. See *Box 3.3*.

Hemizygous: having only one copy of a gene or DNA sequence in diploid cells. Males are usually hemizygous for sex-linked genes. Deletions occurring on one homolog produces hemizygosity in males and in females.

Heritability: the proportion of the causation of a character that is due to genetic causes. See *Box 18.3* and *Figure 18.4* for examples of how heritability is calculated.

Heterochromatin: a region of the genome which remains highly condensed throughout the cell cycle and shows little or no evidence of active gene expression.

Heteroduplex: double-stranded DNA in which the two DNA strands do not show perfect base complementarity.

Heteroplasmy: the co-existence of wild-type and mutant mitochondrial DNA molecules within a cell or tissue.

Heterozygous: an individual is heterozygous at a locus if (s)he exhibits two different alleles at that locus.

Homeobox: a highly conserved sequence of about 180 bp which encodes a *homeodomain*, a DNA-binding protein domain which can bind to target sequences in other genes and regulate their expression during development.

Homologous chromosomes (homologs): two copies of the same type of chromosome found in a diploid cell, one having being inherited from the father and the other from the mother.

Homologous genes (homologs): two or more genes whose sequences are significantly related because of a close evolutionary relationship, either between species (*orthologs*) or within a species (*paralogs*).

Homozygous: an individual is homozygous at a locus if (s)he exhibits two identical alleles at that locus.

Hotspot: a mutational hotspot is any sequence which is associated with an abnormally high frequency of recombination or mutation.

Housekeeping gene: a gene, whose expression is essential for the function of most or all types of cell.

***HOX* gene**: a homeobox gene found within one of four major homeobox gene clusters in mammals. See *Figure 9.5*.

Hybridization assay: involves mixing single DNA (or RNA) strands from a labeled probe with those of a target DNA (or RNA) sample, then allowing complementary strands to anneal. See *Figure 5.7*.

Identity by descent (IBD): alleles in an individual or in two people that are identical because they have both been inherited from the same common ancestor, as opposed to *identity by state* (IBS; coincidental possession of identical alleles).

Illegitimate transcription (also called **ectopic transcription**): low-level transcription in many cell types of genes which are predominantly expressed in certain types of cell.

***In situ* hybridization**: hybridization of a labeled nucleic acid to a target nucleic acid which is typically immobilized on a microscopic slide, such as the DNA of denatured metaphase chromosomes (*chromosome* in situ *hybridization)* or the RNA in a section of tissue (*tissue* in situ *hybridization*).

Intron: noncoding DNA which separates neighboring exons in a gene. During gene expression introns, like exons, are transcribed into RNA but the transcribed intron sequences are subsequently removed by RNA splicing and are not present in mRNA (see *Figure 1.14*). See *Box 9.2* for different classes of introns.

Isoforms/isozymes: alternative forms of a protein/enzyme.

Ligation: formation of a 3′–5′ phosphodiester bond between nucleotides at the ends of two different molecules (*intermolecular ligation*) or of the same molecule (*intramolecular ligation = cyclization*).

LINE (long interspersed nuclear element): a class of moderately repetitive DNA sequences consisting of different sequence families, each with a fairly long consensus sequence [e.g. the LINE-1 (*Kpn* repeat) family]. See *Figure 8.12.*

Linkage: the tendency of genes or other DNA sequences at specific loci to be inherited together as a consequence of their physical proximity on a single chromosome.

Linkage disequilibrium (also called **allelic association**): nonrandom association of alleles at linked loci.

Linker (or **adaptor oligonucleotide**): a double-stranded oligonucleotide which can be ligated to a cloned DNA of interest in order, for example, to facilitate its ability to be cloned. See, for example, *Figure 6.10.*

Liposome: a synthetic lipid membrane designed to transport a molecule of interest into a cell. See *Figure 20.6.*

Locus: a unique chromosomal location defining the position of an individual gene or DNA sequence.

Locus control region (LCR): a stretch of DNA containing regulatory elements which control the expression of genes in a gene cluster that may be located tens of kilobases away. See, for example, *Figure 8.6.*

Lod score: a measure of the likelihood of genetic linkage between loci. A lod score greater than +3 is often taken as evidence of linkage; one that is less than –2 is often taken as evidence against linkage. See *Figures 12.5* and *12.6.*

Loss of heterozygosity (LOH): loss of alleles on one chromosome detected by assaying for markers for which an individual is constitutionally heterozygous. See *Figure 17.10.*

***lox*P**: a pair of sequences found within the bacteriophage P1 genome which recombine when in the presence of the cre protein. See *Figures 4.14* and *19.20.*

Lyonization: the process of X chromosome inactivation in mammals. See *Figure 7.16.*

Lysogen: a bacterial cell in which a phage genome has integrated into the chromosomal DNA.

M13: a bacteriophage with a single-stranded DNA genome which, as part of the normal life cycle, is converted to a double-stranded form (*replicative* or *RF* form). Vectors based on M13 are particularly valuable for DNA sequencing. See *Figure 11.21.*

Marker: a genetic marker is a polymorphic DNA or protein sequence deriving from a single chromosomal location, which is used in genetic mapping. See *Box 12.1.*

Meiosis: reductive cell division occurring exclusively in testis and ovary and resulting in the production of haploid cells, including sperm cells and egg cells. See *Figures 2.11* and *2.12.*

Mendelian segregation: the process whereby individuals inherit and transmit to their offspring one out of the two alleles present in homologous chromosomes, producing the pedigree patterns shown in *Figure 3.2.*

Metacentric (of chromosome)**:** having the centromere in the middle.

Minisatellite DNA: an intermediate size array (often 0.1–20 kb long) of short tandemly repeated DNA sequences. *Hypervariable minisatellite DNA* is the basis of DNA fingerprinting and many VNTR markers.

Microsatellite DNA: small array (often less than 0.1 kb) of tandem repeats of a very simple sequence, often between 1 and 4 bp.

MIM number: a catalog number for an inherited disorder or phenotypic trait as listed in Victor McKusick's Mendelian Inheritance in Man, available as a book, and electronically. See *OMIM.*

Minigene: an artificially designed recombinant DNA in which the coding sequence of a gene of interest is coupled with regulatory elements in the vector, enabling expression of the insert to give the desired polypeptide.

Missense mutation: a nucleotide substitution which results in an amino acid change. See *Box 10.2.*

Mitosis: cell division in somatic cells. See *Figure 2.6.*

Modifier gene: a gene whose expression can influence a phenotype resulting from mutation at another locus.

Molecular hybridization: any process in which a probe DNA and a target DNA are denatured and allowed to reanneal under conditions which encourage formation of heteroduplexes. See *Figure 5.7.*

Morgan: a unit of genetic distance corresponding to a length of DNA which, on average, undergoes one crossover per *individual* chromatid strand. On average, there are about 52 chiasmata in human male meiosis and so the total male genetic map distance is 26 Morgans or 2600 cM.

Mosaic: a genetic mosaic is an individual who has two or more genetically different cell lines derived from a single zygote. See *Figure 3.7.*

Multifactorial character: a character determined by several factors, normally assumed to include both genetic and environmental factors.

Multigene family: a set of evolutionarily related loci within a genome, at least one of which can encode a functional product.

Mutation: a heritable alteration in the DNA sequence.

Mutator gene: a common class of gene which, when mutated, results in tumor formation.

Natural selection: the process whereby some of the inherited genetic variation within a population will affect the ability of individuals to survive and to reproduce (*fitness*). See *Box 10.1.*

Nested primers: a pair of PCR primers which are used to amplify an internal sequence from within a larger PCR amplification product. The use of internal nested primers is designed to increase the specificity of amplification.

Nonsense mutation: a mutation which occurs within a codon and changes it to a stop codon. See *Box 10.2.*

Northern blot hybridization: a form of molecular hybridization in which the target consists of RNA molecules that have been size fractionated by gel electrophoresis and subsequently transferred to a membrane. See *Figure 19.7.*

Nucleosome: a structural unit of chromatin. See *Figure 2.3.*

Nullizygous: lacking any copy of a gene or DNA sequence normally found in chromosomal DNA. Usually this requires a homozygous deletion, but it can result from a single deletion in the sex chromosomes of males.

Oncogene: a gene involved in control of cell proliferation which, when overactive can help to transform a normal cell into a tumor cell. See *Table 17.2.*

OMIM: On-line Mendelian Inheritance in Man, an electronic catalog of inherited human disorders and phenotypic traits. See also *MIM number.*

Open reading frame (ORF): a significantly long sequence of DNA in which there are no termination codons. Six reading frames are possible for a DNA duplex because each strand can have three reading frames.

Ortholog: one of a set of homologous genes in different species (e.g. *SRY* in humans and *Sry* in mice). See *Box 9.1.*

P1 cloning: a cloning system using bacteriophage P1 vectors to propagate large inserts. See *Figure 4.14.*

PAC (P1 artificial chromosome) cloning: a cell-based DNA cloning system which uses elements of bacteriophage P1 and the F (fertility factor) plasmid and requires electroporation for transferring DNA into cells.

Paralog: one of a set of homologous genes within a single species. See *Box 9.1.*

Penetrance: the frequency with which a gene or gene combination manifests itself in the phenotype of the carriers.

PFGE: see *Pulsed field gel electrophoresis.*

Phage display: an expression cloning method in which foreign genes are inserted into a phage vector and are expressed to give polypeptides that are displayed on the surface (protein coat) of the phage. See *Figure 19.14.*

Phagemid: a plasmid vector designed to contain an origin of replication from a filamentous phage, such as M13 or fd. Phagemids go through a single-stranded form in cells which have been *superinfected* with the relevant phage.

Phase (of an intron): a term used to classify introns found in polypeptide-encoding DNA, depending on the position at which the intron is inserted into the coding DNA.

Phase (of linked markers): is the relation (coupling or repulsion) between alleles at two linked loci. If allele A1 is on the same physical chromosome as allele B1, they are *in coupling*; if they are on different parental homologs they are *in repulsion.*

Phenotype: the physical characteristics of a cell or organism, as determined by the genetic constitution.

PIC (polymorphism information content): a measure of the informativeness of a genetic marker.

Point mutation: a mutation causing a small alteration in the DNA sequence at a locus, often a single nucleotide change.

Polygenic character: a character determined by the combined action of a number of genetic loci. Mathematical polygenic theory (see *Figure 18.1*) assumes there are very many loci, each with a small effect.

Polymerase chain reaction (PCR): an *in vitro* method of DNA cloning involving the use of specific oligonucleotides to prime DNA synthesis from a specified target DNA sequence. See *Figure 6.1.*

Polymorphism: the existence of two or more alleles at significant frequences in the population.

Polyploid: having multiple chromosome sets as a result of a genetic event that is abnormal (e.g. constitutional or mosaic triploidy, tetraploidy, etc.), or programmed (e.g. certain human body cells are naturally polyploid).

Positional cloning: cloning of a gene which is dependent only on knowledge of its subchromosomal location.

Primer: a short nucleic acid sequence, often a synthetic oligonucleotide, which binds specifically to a single strand of a target nucleic acid sequence and initiates synthesis, using a suitable polymerase, of a complementary strand.

Primer extension: a method of identifying the transcription initiation site. See *Figure 19.5.*

Probe: a DNA or RNA fragment which has been labeled in some way, and used in a *molecular hybridization* assay to identify DNA or RNA sequences which are closely related to it in sequence.

Processed pseudogene (also called *retropseudogene*): a pseudogene which lacks intronic sequences and flanking sequences of the related functional gene. Originates by reverse transcription of an RNA transcript from the functional gene and subsequent integration into a different chromosomal location. See *Figure 8.7* and *Box 8.1.*

Proofreading: an enzyme mechanism by which DNA replication errors are identified and repaired.

Promoter: a combination of short sequence elements to which RNA polymerase binds in order to initiate transcription of a gene.

Protein truncation test: a method of screening for chain-terminating mutations by artificially expressing a mutant allele in a coupled transcription–translation system. See *Figure 14.16.*

Proto-oncogene: a cellular gene which when mutated is inappropriately expressed and becomes an *oncogene.*

Provirus: a phage or viral genome which has integrated into the chromosomal DNA of a cell. See *Figures 4.10* and *20.16.*

Proximal (with reference to a chromosomal location): close to, or in the direction of, the centromere.

Pseudoautosomal region: a region on the tips of mammalian sex chromosomes which is involved in recombination during male meiosis. See *Figure 9.7.*

Pseudogene: a DNA sequence which shows a high degree of sequence homology to a nonallelic functional gene but which is itself nonfunctional. Different classes of pseudogene are described in *Box 8.1.*

Pulsed field gel electrophoresis (PFGE): a form of gel electrophoresis which permits size fractionation of large DNA molecules. See *Figures 11.10 and 11.11.*

RACE (rapid amplification of cDNA ends)-PCR: a form of PCR used to retrieve and identify the uncharacterized end of a cDNA molecule. See *Figure 19.4.*

Radiation hybrid: a type of somatic cell hybrid in which fragments of chromosomes of one cell type are generated by exposure to X-rays, and are subsequently allowed to integrate into the chromosomes of a second cell type. See *Figure 11.3.*

Rare cutter: a restriction nuclease which cuts DNA infrequently because the sequence it recognizes is large and/or contains one or more CpGs.

Reading frame: the *translational reading frame* describes the mechanism which moves a ribosome three nucleotides at a time during translation.

Reassociation kinetics: The rates at which complementary DNA strands reassociate. Highly repetitive sequences reassociate rapidly; single copy sequences reassociate slowly.

Renaturation (also known as **reannealing** or **reassociation**): the process whereby individual strands of double-stranded nucleic acids (produced by denaturation) are mixed and allowed to form double-stranded nucleic acids again.

Repetitive DNA: a set of nonallelic DNA sequences which show considerable *sequence homology*.

Reporter gene: a promoterless gene used to test the ability of an upstream sequence joined on to it to act as a promotor. See *Figure 19.8*.

Restriction fragment length polymorphism (RFLP): a polymorphism due to differences in size of allelic restriction fragments as a result of restriction site polymorphism.

Restriction map: a graphical representation of the locations of restriction sites within a DNA sequence.

Restriction site: a short DNA sequence, often 4–8 bp long, which is recognized by a restriction nuclease. See *Table 4.1*.

Restriction site polymorphism (RSP): a polymorphism involving two alleles which differ in possessing or lacking a specific restriction site. See *Figures 5.13* and *6.4*.

Retrotransposon: a transposable DNA element which transposes by means of an RNA intermediate: reverse transcriptase acts on an RNA transcript to make a cDNA copy which then integrates into chromosomal DNA at a different location. See *Figures 8.10* and *8.11*.

Retrovirus: an RNA virus with a reverse transcriptase function, enabling the RNA genome to be copied into cDNA prior to integration into the chromosomes of a host cell. See *Figures 17.2* and *17.3*.

Riboprobe: a labeled RNA probe prepared by *in vitro* transcription from a cloned DNA sequence. See *Figure 5.4*.

Ribozyme: a catalytic RNA molecule. See *Figure 20.10*.

RNA editing: a natural process in which the base sequence of an RNA molecule is altered enzymatically. Known to occur in the transcripts of only a very few human genes. See *Figure 7.13*.

RT-PCR (reverse transcriptase-PCR): a PCR reaction in which the target DNA is a cDNA copied by reverse transcriptase from an mRNA source. Usually uses RNA from blood cells and often relies on the presence of *illegitimate transcription*.

S1 nuclease: an enzyme which cleaves single-stranded DNA or RNA.

S1 nuclease protection assay: a method of assessing the degree of overlap of two differently sized complementary nucleic acid strands. Can be used to identify transcription initiation sites, exon–intron junctions, etc. See *Figure 19.5*.

Satellite DNA: a large array (often 100 kb to several megabases) of tandemly repeated DNA sequences. Satellite DNA is not transcribed and accounts for the bulk of the heterochromatin. See *Figure 8.8*.

Segregation analysis: a statistical technique for investigating the mode of inheritance of a character. See *Table 18.5*.

Selection pressure: *conservative* selection pressure refers to the pressure exerted by *natural selection* to conserve the sequence of a functionally important molecule. See *Box 10.3*.

Semi-dominant: describes a mutation which, in the heterozygote, produces a mutant phenotype that is recognizably different from wild-type but not so severe as the phenotype in homozygotes. Especially used in mouse genetics, but not generally used in human genetics (partly because homozygotes for dominantly inherited disorders are so very rare).

Sense strand: the DNA strand of a gene which is complementary in sequence to the antisense strand, and identical to the transcribed RNA sequence, except that in the latter the base thymine has consistently been replaced by uracil. See *Figure 1.11*.

Sequence homology: a measure of the similarity in the sequences of two nucleic acids or two polypeptides.

Sequence tagged site (STS): any piece of DNA whose sequence is known and for which a specific PCR assay has been designed. See *Box 11.2*.

Shuttle vector: a vector which can be propagated in different types of cell (e.g. eukaryotic and bacterial cells). See *Figures 11.19* and *19.8*.

Sib-pair analysis: a form of genetic analysis in which markers are tested for linkage to a disease or phenotypic trait by measuring high haplotype sharing in a panel of affected sib pairs. See *Figure 12.12*.

Silencer: combination of short DNA sequence elements which suppress the transcription of a gene.

SINE (short interspersed repetitive element): a class of moderate to highly repetitive DNA sequences consisting of different sequence families, each with a fairly short consensus sequence (e.g. the *Alu* repeat family).

Sister chromatids: two chromatids present within a single chromosome and joined by a centromere. Chromatids present on different but homologous chromosomes are described as *nonsister chromatids*.

Slipped strand mispairing (also called **slippage replication** or **replication slippage**): a process in which the complementary strands of a single double helix pair out of register, often at short tandemly repeated sequences. The resulting daughter DNA strands will have corresponding deletions or insertions. See *Figure 10.4*.

SnRNA: small nuclear RNA species, such as the U1, U2, U3 series, which are often involved in the RNA splicing mechanism.

Somatic cell: any cell in the body except the gametes.

Somatic cell hybrid: an artificially constructed cell in which chromosomes have been stably introduced from cells of different species. See *Figure 11.1*.

Southern blot hybridization: a form of molecular hybridization in which the target nucleic acid consists of DNA molecules that have been size fractionated by gel electrophoresis and subsequently transferred to a nitrocellulose or nylon membrane. See *Figure 5.11*.

Splice acceptor site: the junction between the end of an intron terminating in the dinucleotide AG, and the start of the next exon. See *Figure 1.15.*

Splice donor site: the junction between the end of an exon and the start of the downstream intron, commencing with the dinucleotide GT. See *Figure 1.15.*

Spliceosome: a ribonucleoprotein complex used in RNA splicing.

Splicing: conventionally refers to *RNA splicing*, in which RNA sequences transcribed from introns are excised and discarded while those transcribed from exons are spliced together in the same linear order as the exons (see *Figure 1.16*). *DNA splicing* is a much rarer event but naturally occurs in B and T lymphocytes (see *Figure 7.18*).

SSCP (single stranded conformation polymorphism): a commonly used method for point mutation screening. See *Figure 14.14.*

STS: see *Sequence tagged site.*

Subcloning: any process which involves molecular cloning of a fragment of DNA obtained from a starting DNA clone.

Substitution: a mutation which results in replacement of one nucleotide by another. See also *Transition, Transversion, Conservative substitution* and *Synonymous substitution.*

Subtraction cloning: a form of DNA cloning which involves the mixing of two complex DNA populations that have broadly similar collections of DNA sequences but which differ in that one lacks some sequence found in the other. See *Figure 14.9.*

Suppressor tRNA: any transfer RNA molecule which, due to a mutation in the tRNA gene, shows an altered coding specificity and is able to translate a nonsense (or missense) codon. See *Box 4.2.*

Synonymous substitution: a substitution which replaces one codon by another without changing the amino acid that is specified.

Synteny: the property of occurring on the same chromosome. *Conservation of synteny* means that a group of genes which is on a single chromosome in one species is observed to be similarly linked in a second species. See *Figures 9.22* and *14.13.*

Target DNA: (i) template DNA used in a PCR reaction; (ii) DNA to which a probe DNA is designed to hybridize in a molecular hybridization reaction.

TATA box: a common promoter element. See *Table 1.3.*

Telomere: a specialized structure occurring at the tips of linear chromosomes. It consists of an array of short tandem repeats (the hexanucleotide TTTAGG in humans) which are added using the enzyme telomerase. See *Figure 2.14.*

Tissue *in situ* hybridization: a form of molecular hybridization in which the target nucleic acid is RNA within cells of tissue sections immobilized on a microscope slide. See *Figure 5.16.*

***Trans*-acting**: pertaining to the regulation of gene expression by a product that is encoded by a gene at a remote location, usually as a result of binding to *cis*-acting sequences in the vicinity of the gene.

Transcription: the synthesis of RNA from DNA using RNA polymerase.

Transformation (of a cell): a process by which an introduced DNA molecule is taken up by a suitably *competent* cell.

Transgenic animal: an animal in which artificially introduced foreign DNA becomes stably incorporated into the germline. This can be accomplished by different methods, most commonly by pronuclear microinjection or, in the case of transgenic mice, by the use of embryonic stem cells. See *Figures 19.16* and *19.17.*

Transient expression: time-limited expression of a transfected gene as a result of its failure to integrate into chromosomal DNA.

Transition: G⇌A (purine for purine) or C⇌T (pyrimidine for pyrimidine) nucleotide substitution. See *Figure 10.1.*

Translocation: transfer of chromosomal regions between nonhomologous chromosomes. See *Figures 2.20* and *2.21.*

Transversion: a nucleotide substition of purine for pyrimidine or vice versa. See *Figure 10.1.*

Two-hit hypothesis: Knudson's theory that hereditary cancers require two successive mutations to affect a single cell. See *Figure 17.8.*

Two-hybrid system: a yeast-based system for identifying and purifying proteins that bind to a protein of interest.

Untranslated sequences: noncoding sequences found at the 5′ and 3′ termini of mRNA.

Unequal crossover (UEC): recombination between nonallelic sequences on nonsister chromatids of homologous chromosomes.

Unequal sister chromatid exchange (UESCE): recombination between nonallelic sequences on sister chromatids of a single chromosome.

Vectorette PCR: see *Bubble-linker PCR.*

VNTR (variable number of tandem repeats) polymorphism: in its broadest sense this describes polymorphism occurring because of differences in the copy number of any type of tandem repeat, but conventionally used in human genetics to describe highly polymorphic *minisatellite DNA* sequences.

Western blotting: a process in which proteins are size-fractionated in a polyacrylamide gel prior to transfer to a nitrocellulose membrane for probing with an antibody. See *Figure 19.12.*

X chromosome activation (also called *lyonization*): the process in which one of the two X chromosomes in the cells of female mammals is inactivated by a specialized form of *genetic imprinting.* See *Figure 7.16.*

Yeast artificial chromosome (YAC): an artificial chromosome produced by combining large fragments of foreign DNA with small sequence elements necessary for chromosome function in yeast cells. See *Figure 4.15.*

Zinc finger: a polypeptide motif which is stabilized by binding a zinc atom and confers on proteins an ability to bind specifically to DNA sequences. Commonly found in transcription factors. See *Figure 7.9.*

Zoo blot: a Southern blot containing DNA samples from different species. See *Figure 11.16.*

Index